2004 NSTI Nanotechnology Conference and Trade Show
NSTI Nanotech 2004
Volume 1

*An Interdisciplinary Integrative Forum on
Nanotechnology, Biotechnology and Microtechnology*

March 7-11, 2004
Boston Sheraton Hotel
Boston, Massachusetts, USA
www.nsti.org

Nano Science and Technology Institute
Boston • Geneva • San Francisco

TransportXVII (cover):

Two-dimensional electron flow in a semiconductor heterostructure. Electrons were launched from the upper center into a weakly random potential, the randomness caused by positively charged donor atoms in the "delta layer" above the 2 dimensional electron gas. Trajectories were launched evenly over 180 degrees, using strict overwrite of successive trajectories. This gives a three dimensional hidden surface effect to the caustics (cusps) which has much the same topology as an erosion landscape. Color is assigned according to the direction of the trajectory.

2004 NSTI Nanotechnology Conference and Trade Show

NSTI Nanotech 2004

Volume 1

An Interdisciplinary Integrative Forum on
Nanotechnology, Biotechnology and Microtechnology

March 7-11, 2004
Boston, Massachusetts, USA
http://www.nsti.org

NSTI Nanotech 2004 Joint Meeting

The 2004 NSTI Nanotechnology Conference and Trade Show includes:
7th International Conference on Modeling and Simulation of Microsystems, MSM 2004
4th International Conference on Computational Nanoscience and Technology, ICCN 2004
2004 Workshop on Compact Modeling, WCM 2004

NSTI Nanotech 2004 Proceedings Editors:

Matthew Laudon
mlaudon@nsti.org

Bart Romanowicz
bfr@nsti.org

Nano Science and Technology Institute
Boston • Geneva • San Francisco

Nano Science and Technology Institute
One Kendall Square, PMB 308
Cambridge, MA 02139
U.S.A.

The papers in this book comprise the proceedings of the 2004 NSTI Nanotechnology Conference and Trade Show, Nanotech 2004, Boston, Massachusetts, March 7-11 2004. They reflect the authors' opinions and, in the interests of timely dissemination, are published as presented and without change. Their inclusion in this publication does not necessarily constitute endorsement by the editors, Nano Science and Technology Institute, or the sponsors. Dedicated to our family and friends that have supported and suffered all of our late nights. And to Avinash and Nina, study nicely – Ed.

Nano Science and Technology Institute Order Number PCP04020391
ISBN 0-9728422-7-6
ISBN 0-9728422-5-X (www)
ISBN 0-9728422-6-8 (CD-ROM, Vol. 1-3)

Additional copies may be ordered from:

Nano Science and Technology Institute
Publishing Office
One Kendall Sq., PMB 308
Cambridge, MA 02139, USA
http://www.nsti.org

Printed in the United States of America

Nano Science and Technology Institute
Boston • Geneva • San Francisco

Sponsors

Nano Science and Technology Institute

Intel Corporation

Motorola

Texas Instruments

Ciphergen Biosystems, Inc.

iMediasoft Group

Frontier Carbon Corporation

Veeco Instruments

Hitachi High Technologies America, Inc.

Racepoint Group, Inc.

International SEMATECH

FEI Company

Keithley

Saint-Gobain High Performance Materials

Zyvex Corporation

Accelrys, Inc.

ANSYS, Inc.

Atomistix

PolyInsight

Georgia Department of Industry & Trade

Swiss Business Hub USA

Australian Government - Invest Australia

State of Bavaria, Germany - The United States Office For Economic Development

m+w zander

Engis Corporation

COMSOL

Engelhard Corporation

Nanonex, Inc.

nanoTITAN

Tegal Corporation

Umech Technologies

MEMSCAP

Swiss House for Advanced Research and Education SHARE

Basel Area Business Development

Greater Zürich Area

Development Economic Western Switzerland

Location:Switzerland

Nanoworld AG

Nanosensors

Nanofair

Burns, Doane, Swecker & Mathis, L.L.P.

Mintz Levin Cohn Ferris Glovsky and Popeo, PC

Jackson Walker L.L.P.

Nano Science and Technology Institute
Boston • Geneva • San Francisco

Supporting Organizations

American Physical Society

IEEE Electron Devices Society

IEEE Boston Section

The American Ceramic Society

ASME Nanotechnology Institute

NANOPOLIS

The Global Emerging Technology Institute, Ltd.

Nanotechnology Researchers Network Center of Japan

TIMA-CMP Laboratory, France

Squire, Sanders & Dempsey

Boston University

Massachusetts Technology Collaborative

CIMIT (Center for Integration of Medicine & Innovative Technology)

Draper Laboratory

International Scientific Communications, Inc

DARPA (Defense Advanced Research Projects Agency)

Swiss Federal Institute of Technology

Ibero-American Science and Technology Education Consortium

Applied Computational Research Society

Media Sponsors

Technology Review

iMediasoft Group

Taylor & Francis

NanoApex

Institute of Physics

nanotechweb.org

Photonics Spectra

R&D Magazine and Micro/Nano

Earth and Sky Radio Series

Nano Nordic Network

Table of Contents

Bio Molecular Motors

Ion Channels

Bio Nano Computational Methods and Applications

Bio Micro Sensors and Systems

Micro Fluidics and Nanoscale Transport

MEMS Design and Application

Smart MEMS and Sensor Systems

Micro and Nano Structuring and Assembly

Wafer and MEMS Processing

NSTI Nanotech 2004 Program Committee

TECHNICAL PROGRAM CO-CHAIRS
Matthew Laudon	*Nano Science and Technology Institute, USA*
Bart Romanowicz	*Nano Science and Technology Institute, USA*

TOPICAL AND REGIONAL SCIENTIFIC ADVISORS AND CHAIRS

Nanotechnology
Matthew Laudon	*Nano Science and Technology Institute, USA*
Philippe Renaud	*Swiss Federal Institute of Technology, Switzerland*
Mihail Roco	*National Science Foundation, USA*
Wolfgang Windl	*Ohio State University, USA*

Biotechnology
Robert S. Eisenberg	*Rush Medical Center, Chicago, USA*
Srinivas Iyer	*Los Alamos National Laboratory, USA*

Pharmaceutical
Kurt Krause	*University of Houston, USA*

Microtechnology
Narayan R. Aluru	*University of Illinois Urbana-Champaign, USA*
Bernard Courtois	*TIMA-CMP, France*
Anantha Krishnan	*Defense Advanced Research Projects Agency, USA*
Bart Romanowicz	*Nano Science and Technology Institute, USA*

Semiconductors
David K. Ferry	*Arizona State University, USA*
Andreas Wild	*Motorola, Germany*

NANOTECHNOLOGY CONFERENCE COMMITTEE
M.P. Anantram	*NASA Ames Research Center, USA*
Phaedon Avouris	*IBM, USA*
Xavier J.R. Avula	*Washington University, USA*
Wolfgang S. Bacsa	*Université Paul Sabatier, France*
Roberto Car	*Princeton University, USA*
Franco Cerrina	*University of Wisconsin - Madison, USA*
Murray S. Daw	*Clemson University, USA*
Alex Demkov	*Motorola, USA*
Toshio Fukuda	*Nagoya University, Japan*
David K. Ferry	*Arizona State University, USA*
Sharon Glotzer	*University of Michigan, USA*
William Goddard	*CalTech, USA*
Gerhard Goldbeck-Wood	*Accelrys, Inc., UK*
Niels Gronbech-Jensen	*UC Davis and Berkeley Laboratory, USA*
Karl Hess	*University of Illinois at Urbana-Champaign, USA*
Charles H. Hsu	*Maximem Limited, Taiwan*
Hannes Jonsson	*University of Washington, USA*
Anantha Krishnan	*Defense Advanced Research Projects Agency, USA*
Alex Liddle	*Lawrence Berkeley National Laboratory, USA*
Chris Menzel	*Nano Science and Technology Institute, USA*
Stephen Paddison	*Los Alamos National Laboratory, USA*
Sokrates Pantelides	*Vanderbilt University, USA*
Philip Pincus	*University of California at Santa Barbara, USA*
Joachim Piprek	*University of California at Santa Barbara, USA*
Serge Prudhomme	*University of Texas at Austin, USA*
Nick Quirke	*Imperial College, London, UK*
PVM Rao	*Indian Institute of Technology, Delhi, India*
Philippe Renaud	*Swiss Federal Institute of Technology of Lausanne, Switzerland*
Robert Rudd	*Lawrence Livermore National Laboratory, USA*
Douglas Smith	*University of San Diego, USA*
Clayton Teague	*National Nanotechnology Coordination Office, USA*
Dragica Vasilesca	*Arizona State University, USA*
Arthur Voter	*Los Alamos National Laboratory, USA*
Phillip R. Westmoreland	*University of Massachusetts Amherst, USA*
Wolfgang Windl	*Ohio State University, USA*

| Gloria Yueh | *Midwestern University, USA* |
| Xiaoguang Zhang | *Oakridge National Laboratory, USA* |

BIOTECHNOLOGY CONFERENCE COMMITTEE

Dirk Bussiere	*Chiron Corporation, USA*
Amos Bairoch	*Swiss Institute of Bioinformatics, Switzerland*
Stephen H. Bryant	*National Institute of Health, USA*
Fred Cohen	*University of California, San Francisco, USA*
Daniel Davison	*Bristol Myers Squibb, USA*
Andreas Hieke	*Ciphergen Biosystems, Inc., USA*
Leroy Hood	*Institute for Systems Biology, USA*
Sorin Istrail	*Celera Genomics, USA*
Srinivas Iyer	*Los Alamos National Laboratory, USA*
Brian Korgel	*University of Texas-Austin, USA*
Kurt Krause	*University of Houston, USA*
Daniel Lacks	*Tulane University, USA*
Mike Masquelier	*Motorola, Los Alamos NM, USA*
Atul Parikh	*University of California, Davis, USA*
Andrzej Przekwas	*CFD Research Corporation, USA*
George Robillard	*BioMade Corporation, Netherlands*
Jonathan Rosen	*Center for Integration of Medicine and Innovative Technology, USA*
Tom Terwilliger	*Los Alamos National Laboratory, USA*
Michael S. Waterman	*University of Southern California, USA*

MICROSYSTEMS CONFERENCE COMMITTEE

Narayan R. Aluru	*University of Illinois Urbana-Champaign, USA*
Xavier J. R. Avula	*University of Missouri-Rolla, USA*
Stephen F. Bart	*Nano Science and Technology Institute, USA*
Bum-Kyoo Choi	*Sogang University, Korea*
Bernard Courtois	*TIMA-CMP, France*
Robert W. Dutton	*Stanford University, USA*
Gary K. Fedder	*Carnegie Mellon University, USA*
Elena Gaura	*Coventry University, USA*
Steffen Hardt	*Institute of Microtechnology Mainz, Germany*
Lee W. Ho	*Corning Intellisense, USA*
Eberhard P. Hofer	*University of Ulm, Germany*
Michael Judy	*Analog Devices, USA*
Yozo Kanda	*Toyo University, Japan*
Jan G. Korvink	*University of Freiburg, Germany*
Anantha Krishnan	*Defense Advanced Research Projects Agency, USA*
Mark E. Law	*University of Florida, USA*
Mary-Ann Maher	*MemsCap, France*
Kazunori Matsuda	*Naruto University of Education, Japan*
Tamal Mukherjee	*Carnegie Mellon University, USA*
Andrzej Napieralski	*Technical University of Lodz, Poland*
Ruth Pachter	*Air Force Research Laboratory, USA*
Michael G. Pecht	*University of Maryland, USA*
Marcel D. Profirescu	*Technical University of Bucharest, Romania*
Marta Rencz	*Technical University of Budapest, Hungary*
Siegried Selberherr	*Technical University of Vienna, Austria*
Sudhama Shastri	*ON Semiconductor, USA*
Armin Sulzmann	*Daimler-Chrysler, Germany*
Mathew Varghese	*The Charles Stark Draper Laboratory, Inc., USA*
Dragica Vasilesca	*Arizona State University, USA*
Gerhard Wachutka	*Technical University of Münich, Germany*
Jacob White	*Massachusetts Institute of Technology, USA*
Thomas Wiegele	*BF Goodrich Aerospace, USA*
Wenjing Ye	*Georgia Institute of Technology, USA*
Sung-Kie Youn	*Korea Advanced Institute of Science and Technology, Korea*
Xing Zhou	*Nanyang Technological University, Singapore*

CONFERENCE OPERATIONS MANAGER

| Sarah Wenning | *Nano Science and Technology Institute, USA* |

NSTI Nanotech 2004 Proceedings Topics

2004 NSTI Nanotechnology Conference and Trade Show

NSTI Nanotech 2004, Vol. 1, ISBN 0-9728422-7-6

- Bio Nano Systems and Chemistry
- Bio Nano Analysis and Characterization
- Bio Molecular Motors
- Ion Channels
- Bio Nano Computational Methods and Applications
- Bio Micro Sensors and Systems
- Micro Fluidics and Nanoscale Transport
- MEMS Design and Application
- Smart MEMS and Sensor Systems
- Micro and Nano Structuring and Assembly
- Wafer and MEMS Processing

NSTI Nanotech 2004, Vol. 2, ISBN 0-9728422-8-4

- Advanced Semiconductors
- Nano Scale Device Modeling
- Compact Modeling
- Circuits
- System Level Modeling
- MEMS Modeling
- Modeling Fundamental Phenomena in MEMS
- Computational Methods and Numerics

NSTI Nanotech 2004, Vol. 3, ISBN 0-9728422-9-2

- Nano Photonics, Optoelectronics and Imaging
- Nanoscale Electronics and Quantum Devices
- Atomic and Mesoscale Modelling of Nanoscale Phenomena
- Nano Devices and Systems
- Carbon Nano Structures and Devices
- Nano Composites
- Surfaces and Films
- Nano Particles and Molecules
- Characterization and Parameter Extraction
- Micro and Nano Structuring and Assembly
- Commercial Tools, Processes and Materials

Message from the Program Committee

The Nano Science and Technology Institute is proud to present the 2004 Nanotechnology Conference and Trade Show (Nanotech 2004). The charter of the Nanotech conference, and its numerous sub-conferences, remains the same since its original conception in 1997. The Nanotech provides for a single interdisciplinary integrative community, allowing for core scientific advancements to disseminate into a multitude of industrial sectors and across the breadth of traditional science and technology domains converging under Nanotechnology, Biotechnology and Microtechnology.

The Nanotech Program Committee makes every effort to provide a scientifically outstanding environment, through its review and ranking process. The Nanotech received a total of 656 abstracts for the 2004 event. All abstracts submitted into the Nanotech are reviewed and scored by a minimum of three (3) expert reviewers. Of the 656 submitted abstracts, 25% were accepted as oral presentations, 40% were accepted as poster presentations resulting in a 35% rejection rate. In addition to the regular program there are a number of invited sessions, panels, overview presentations and tutorials provided for completeness. The Nanotech program committee thanks the hard work of all of this year's reviewers (116 active reviewers in total). This grassroots self-review process by the nano, bio and micro technology communities is a source of pride for the Nanotech conference. This process is in place to provide for a yearly evolution and technical validation of the Nanotech conference content. We thank the authors for submitting their latest work, making a meeting of this caliber possible.

We hope the reader will find the papers assembled in these proceedings rewarding to read, and that the conference continues to foster further advances in this fascinating and multi-disciplinary field. Although the Nanotech conference makes every effort to be as comprehensive as possible, due to the rapid advancements in science and industry, there will inevitably be under-represented sections of this event. We look to you, as the true science and business leaders of small-technologies, to assist us in identifying the needs of our participants so that we can continue to grow the content of this event to best serve this community.

We would like to take this opportunity to thank the many individuals who have worked so hard to make this meeting happen, and to welcome the new members of the program committee. Conferences of this scope are possible only because of the continuing interest and support of the community, expressed both by their submission of papers of high quality and by their attendance. The Nanotech 2004 program committee is grateful to all keynote speakers, authors and session chairs for contributing to the success of the event. We are also indebted to the foundations, agencies and companies whose financial contributions made this meeting possible. We encourage your feedback and participation. Additionally, if you have an interest in conference or session organizational assistance, please contact the conference manager and we will attempt to accommodate. Information concerning next year's conference is posted online at URL: http://www.nsti.org. We look forward to seeing you again next year, and thank you for your continuing support and participation.

Technical Program Co-chairs, Nanotech 2004

Matthew Laudon, Nano Science and Technology Institute, USA

Bart Romanowicz, Nano Science and Technology Institute, USA

NSTI
Nano Science and Technology Institute

Forisomes: Mechano-Proteins that Exert Force in Contraction and Expansion

Michael Knoblauch*, Gundula A. Noll**, Dirk Prüfer*, Hannah Jaag*, Maria E. Fontanellaz*,
Ingrid Schneider-Hüther**, Aart J. E. van Bel**, Winfried S. Peters**

*Fraunhofer Institut für Molekularbiologie und Angewandte Ökologie, Auf dem Aberg 1,
D-57392 Schmallenberg-Grafschaft, Germany, michael.knoblauch@bot1.bio.uni-giessen.de

**Institut für Allgemeine Botanik, Justus Liebig-Universität, Senckenbergstr. 17-21,
D-35390 Gießen, Germany

ABSTRACT

Forisomes are elongate protein bodies of up to 30 μm length from cells of the sieve tube network of higher plants. In this natural microfluidics system, they act as reversible stopcocks by undergoing rapid conformational changes which involve more than three-fold increases in volume. The conformational switch is controlled by Ca^{2+} with a threshhold concentration in the nM range. We here report recent progress in our attempts to define the technological potential of this novel biological actuator.

Keywords: biomimetic actuator, contractile protein, forisome, intracellular Ca^{2+}, microfluidics

1 INTRODUCTION

Sieve elements are highly specialized cells in the phloem of vascular plants. They form a continuous microtube system enabling pressure driven long distance transport of photo-assimilates [1]. Unique elongate protein bodies of up to 30 μm length occur in the sieve elements of legumes (Fabaceae). We showed that these Forisomes act as molecular stopcocks [2]. Under normal physiological conditions, they exist in a condensed state, characterized by a highly ordered fibrillar ultrastructure. At increased intracellular Ca^{2+}, the protein bodies enter a dispersed state, forming plugs in the sieve tube. The reaction is fully reversible.

2 FORISOME CONTRACTION *IN VITRO*

We scrutinized the Ca^{2+} response *in vitro*, using forisomes isolated from broad bean (*Vicia faba*) sieve elements [3]. Surprisingly, forisomes did not only swell radially in response to Ca^{2+}, but also contracted longitudinally. In contraction, forisomes shortened by 29.4 ± 6.7 % and increased in diameter (measured across their center) by 119.5 ± 47.9 % (means ± s.d., n = 15). The response to Ca^{2+} was fully reversible by transfer to Ca^{2+}-free solutions. Contraction-expansion cycles could be repeated at least 50 times in the same forisome without decrease in respon-

siveness. Contraction seemed to follow an all-or-none law: either it occurred to the full extent or not at all. The threshhold concentration for Ca^{2+}-induced contraction was below 100 nM.

3 GENERATION OF FORCE

In forisomes that were attached to one glas pipette tip at each end, contraction to the normal extent could still be induced reversibly by increased Ca^{2+}. As a result, the pipettes became bent (Fig. 1). Thus, contracting forisomes develop significant mechanical forces.

4 pH EFFECTS

In Ca^{2+}-free buffers, forisomes did not react to changes of pH between 5.0 and 9.5. When the pH was increased beyond 9.6, the protein bodies started to swell radially and contracted gradually and reversibly in a pH-dependent fashion. At pH 10.6, contraction reached the degree normally evoked by Ca^{2+} at the physiological pH 7.3.

5 FORISOME CONTRACTION CONTROLLED BY ELECTROTITRATION

Considering technical applicability, it would appear desirable if forisome action could be controlled electrically. We used diffusional electrotitration [3] to drive pH-dependent contraction and expansion in isolated forisomes. Provided an appropriate pH range was covered by the electrically induced proton gradients, forisomes showed pH responses as discussed above (Fig. 2). Voltage pulsed at varying frequency led to corresponding oscillations of the state of the forisome, demonstrating in principle that forisome action can be controlled electrically. At least 4,000 electrically induced contraction–expansion cycles could be observed in single forisomes.

6 OUTLOOK

Present research in our labs and those of our collaborators centers on the mechanism of forisome contraction [4], on the characterisation of forisome molecular structure, on the development of integrated leak-free microfluidic systems in which forisomes serve as reversible switches, and on the construction of prototypic micro-devices in which forisomes act as actuators.

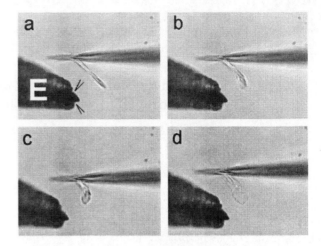

Figure 2: Forisome swelling and curvature induced by electrotitration.

The images show a forisome of 23 μm length in the expanded state (**a**) stuck to a glass pipette, and a shielded platin microelectrode (E; the unshielded, conducting tip is the pointed protrusion between the two arrowheads) driven as the cathode to induce zones of increased pH in its vicinity. The anode was placed 50 μm away and is not visible here. The pH of the weakly buffered (0.4 mM Tris/HCl) medium was 7.3. **a**, No voltage applied. **b–d**, Reactions to 950 μs pulses of 6.2 V applied at increasing frequencies. **b**, 5 Hz; slight curvature of the weakly swollen forisome. **c**, 8 Hz; strong curvature (roughly 60°) exhibited by the significantly contracted forisome. **d**, 20 Hz; the forisome contracts longitudinally and expands radially in response to the zone of increased pH that develops around the electrode.

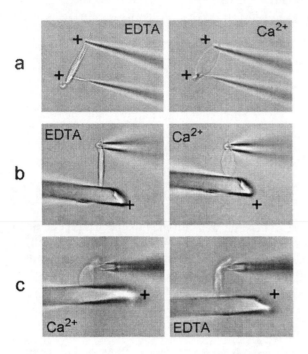

Figure 1: Demonstration of mechanical force exerted by forisomes

a, Forisome fixed between two glass pipette tips in the expanded state (left). Application of Ca^{2+} caused longitudinal contraction by some 30%, resulting in bending of the pipettes (right; black crosses are at identical positions in both images, to indicate the relative movement of the pipette tips). The response was fully reversible. The length of the expanded forisome (left) was 21μm. Repetitions of the experiment in five different forisomes yielded similar results. **b**, Determination of pulling force in a forisome fixed in the expanded state between the tip of a rigid glass pipette (which enters the visual field from the right) and a flexible glass fibre (9μm diameter; entering from the left). On addition of Ca^{2+}, the forisome pulls the fibre towards the pipette (right image). **c**, Analogous experiment to **b**; this time, a forisome is fixed in the contracted state (left). Addition of EDTA chelator causes it to expand, pushing the flexible fibre away from the rigid pipette (right). Black crosses in the pairs of images in **b** and **c** are at identical positions relative to the tip of the rigid glass pipette, respectively. Note that the amount of displacement of the fibre tip is similar in **b** and **c**, indicating that similar forces (about 0.1μN, in this case) are exerted by pulling and pushing forisomes.

REFERENCES

[1] AJE van Bel, K Ehlers and M Knoblauch, "Sieve elements caught in the act", *Trends in Plant Science* **7**: 126-132, 2002.

[2] M Knoblauch, WS Peters, K Ehlers and AJE van Bel, "Reversible calcium-regulated stopcocks in legume sieve tubes", *Plant Cell* **13**: 1221-1230, 2001.

[3] M Knoblauch, GA Noll, T Müller, D Prüfer, I Schneider-Hüther, D Scharner, AJE van Bel and WS Peters, "ATP-independent contractile proteins from plants", *Nature Materials* **2**: 600-603, 2003.

[4] M Knoblauch and WS Peters, "Forisomes, a novel type of Ca^{2+}-dependent contractile protein motor", *Cell Motility and the Cytoskeleton* in press.

Development of a Self-Assembled Muscle-Powered Piezoelectric Microgenerator

Jianzhong Xi [*], Eric Dy[**], Ming-Tsung Hung[**] and Carlo Montemagno[**]

[*] Biomedical Engineering Department, 7523 Boelter Hall UCLA, Los Angeles, CA, USA jzxi@ucla.edu
[**] Biomedical Engineering Department, 7523 Boelter Hall UCLA, Los Angeles, CA, USA

ABSTRACT

As technology advances into a new era of N/MEMS, powering these tiny devices becomes a significant issue. To address this issue, a self-assembled living/nonliving microgenerator has been developed through seamless integration of PVDF-TrFE and neonatal cardiac muscle cell. This microgenerator system can covert glucose into electrical energy, which is of particular interest in engineering applications. Since glucose is ubiquitous inside our body, the microgenerator is especially suitable for powering implantable devices. The calculated peak voltage is approximately 0.45V for each hybrid device with the dimensions 1 mm length, 50μm width, and 10μm thickness. Through monolithic integration with integrated circuits (IC), an integral system of novel BioMEMS can be built that carries potential applications in both health monitoring and biosensing.

Keywords: piezoelectric microgenerator, self-assemble, muscle, PVDF-TrFE

1 INTRODUCTION

The increasingly smaller dimensions of MEMS challenges scientists to develop novel techniques of powering these devices. Current methods of powering these devices become problematic once they become free standing systems and are no longer tethered to a silicon wafer. This is not just an issue of energy, but also one of integration and miniaturization, all of which must be addressed on the basis of compatibility of the components. This is especially challenging when dealing with MEMS devices to be used in biological systems. On the other aspect, when dealing with the medical applications of MEMS devices, it would be ideal to develop a system capable of utilizing glucose as power source. However, this proposal is complicated by the fact that most current inorganic microdevices are driven by electrical energy.

Piezoelectric materials have been extensively utilized because of their unique capability of converting mechanical energy into electrical energy. As components in engineered systems, biological muscles have unique advantages over other inorganic actuators, such as high power density, utilization of biochemical fuels, and self-assembly from single cells. Seamless miniaturized integration of these two components would ultimately solve the above-mentioned problem and promise us to create novel implantable power generators.

2 BACKGROUND

Piezoelectric materials have been utilized to build a power generation system using, as energy harvest becomes more and more important. Polyvinylidene fluoride (PVDF) and its copolymer PVDF (Polyvinylidene fluoride)-TRFE (trifluoroethylene) are of great interest because of their strong piezoelectricity. PVDF is a semicrystalline homopolymer, which was shown to have a strong piezoelectric coefficient in 1969 by Kawai [1]. Its copolymer PVDF-TrFE crystallizes to a much greater extent than PVDF, yielding a higher remanent polarization, lower coercive field and much sharper hysteresis. After applying a high electric field, called poling, the Coulomb interactions between injected, trapped charges and oriented dipoles in crystals make the polarization stable [2].

When the PVDF-TrFE film is stretched in transverse direction, the dipole crystalline structures realign due to the applied stress and an internal field is created, causing charges to build up on the electrodes. This produces a voltage potential between two electrodes. The generated charge Q and voltage V be written as

$$Q = \sigma A d_{31} \tag{1}$$

$$V = Q/C \tag{2}$$

Where σ is the stress, A is the area of cross section, d_{31} is the piezoelectric coefficient, and C is the capacitance of the piezoelectric film.

Mechanical force, vibration, thermal gradient, and fluid flow have been used as energy sources to activate piezoelectric films to generate electrical energy [3]. However, these devices are not suitable for implantable applications due to their demanding operation conditions. In our research, we provide an alternative and compatible way to activate the piezoelectric film - the utilization of muscle power.

As microcomponents in engineered systems, biological muscles have unique advantages such as high efficiency, utilization of biochemical fuel, and self-assembly from single cells, over other inorganic actuators. Successful integration of muscles with inorganic MEMS promises the capability of creating self-assembled autonomous implantable structures, powered by ubiquitous glucose in the human body.

However, the use of mature muscle tissues in these devices is impractical and inefficient, as the tissues must be dissected and incorporated into each device by hand thus creating crude interfaces between the biological tissues and inorganic materials [4-8]. Integration of muscle with fabricated structures would be optimally achieved through patterned growth *in situ*.

A novel system for fabricating self-assembled muscle-powered microdevices has been recently developed in our laboratory [9]. This system has enabled us to selectively direct myocyte growth and differentiation into a single muscle bundle *in situ*, attach these functional bundles to MEMS devices and controllably release hybrid devices. Mechanical studies of cardiac myocytes and fabrication of self-assembled hybrid microrobots have been performed. Here we extend this system to develop self-assembled muscle-powered piezoelectric microgenerators.

3 METHODS AND RESULTS

3.1 Cell Isolation and Culture

The neonatal ventricular myocytes of 1-3-day-old Sprague-Dawley rats (NRVMs), which were prepared and isolated as previously described[10], were kindly provided by Dr. Robert S. Ross at UCLA. The cell cultural medium (NRVMs) and conditions also were followed to the previous description except moderate modification [11]. Briefly, prior to the isolated myocytes being plated, the fabricated devices already glued to ordinary culture dishes were warmed to 37°C. The plated myocytes density is 4.6-6.1 million/cm2 and this culture would be kept at a 37°C incubator supplied with 5% CO2 for 2-3 days.

3.2 Device Fabrication

A novel muscle-powered piezoelectric microgenerator has been fabricated. The creation of the device consists of three main steps: microfabrication of the bendable PVDF-TrFE unimorph, selectively poling of the PVDF-TrFE layer, and self-assembly of muscle cells. The bendable PVDF-TrFE unimorph is a multiple-layer beam structure, which rested on a supporting anchor (Figure 1.). For the microfabrication of process, we determine the shape and thickness of each layer using surface micromachining techniques.

After the fabrication of a Si anchor, a thin layer of poly-N-isopropylacrylamide (PNIPAAm) was coated and dried on the wafer as a sacrificial layer. After patterning PNIPAAm through a shadow mask, a 0.7μm-thick parylene layer was deposited as an electrical insulator. A 200 nm thick Au layer was deposited through a shadow mask to form the first electrode. Following the electrode layer, the 10μm PVDF-TrFE was then deposited. A 200nm-thick Au layer was deposited as the second electrode followed by the deposition of a second 1.5μm parylene. A final 300 nm Au layer was deposited, which was utilized as a mask for the subsequent plasma etching to pattern the unimorph, which is the last step of the microfabrication process.

Unlike the conventional procedure for poling PVDF-TrFE, a 10μm PVDF-TrFE layer were poled at 40 V/μm electric field, at 65 °C for 2 hours because of the low glass temperature of PNIPAAm. The poled devices were fixed on a petri dish and followed by the addition of the cell culture medium mixed with the neonatal cardiac myocytes. After 2~3 days culture, the devices were moved to room temperature. Once PNIPAAm was dissolved, the unimorph powered by muscle began to bend, and the generated voltage was measured.

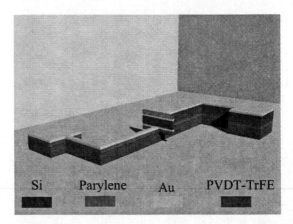

Si   Parylene   Au   PVDT-TrFE

Figure 1. Schematic diagram of a piezoelectric generator. Note: The third Au layer is not shown; and both electrodes are insulated with PDMS prior to the addition of the culture medium.

PNIPAAm, a thermally responsive polymer, has been previously considered as an intelligent substrate to harvest intact cells. A solid at temperatures greater than 32°C, PNIPAAm undergoes a solid-liquid phase transition as it is cooled to lower temperatures and can dissolve in a surrounding liquid medium. Therefore, after the removal of PNIPAAm, our device was a double-ended cantilever beam, which is enclosed by parylene and unpoled PVDF-TrFE (Figure 2).

Si   Parylene   PNI   PVDF-TrFE   Au   Poled PVDF-TrFE

Figure 2.Cross-section of a piezoelectric generator. Muscle bundle self-assembled on the top of Au film is not shown.

Parylene was chosen as an insulation layer because of its unique capabilities of water-resistance, chemical and electrical inertness, large yielding strain, and low Young's modulus [12]. Since our device is the unimorph and requires an elastic layer to shift the stress distribution, the second parylene layer was deposited thicker as part of the elastic layer. When the self-assembled muscle bundle contracts, it introduces a bending moment and cause the unimorph to bend up. The upper layer of the unimorph is subjected to a compressive stress while the bottom layer is tensional stressed. Due to the stress change, the piezoelectric layer will parasically converts the mechanical energy into electrical energy.

3.3 Results

When applying an external moment M from the muscle, the unimorph will generate the charge and voltage calculated by the following equations [13-16]:

$$Q = \frac{6d_{31}L}{t_p^2} \frac{AB(B+1)}{1+4AB+6AB^2+4AB^3+A^2B^4} M \quad (3)$$

$$V = \frac{6d_{31}}{\varepsilon_{33}^X wt_p} \frac{AB(B+1)}{1+4AB+6AB^2+4AB^3+A^2B^4-k_{31}^2AB(1+AB^3)} M \quad (4)$$

where

$$A = \frac{E_m}{E_p}, \quad B = \frac{t_m}{t_p}$$

E_m and E_p are Elastic modulus of the elastic layer and piezoelectric layer; t_m and t_p are thickness of the elastic layer and piezoelectric layer; w and L are the width and length of the unimorph; ε_{33}^X is the dielectric constant of the piezoelectric material under a free condition, k_{31} is the transverse piezoelectric coupling coefficient.

Based on the assumption that the electrode layers and parylene layers are very thin and can be negligible, we can calculate the charge and voltage generated, around 8.3E-15C and 0.48V respectively, by our device. The device itself is 1000µm long, 50um width and 20um thick. A group of such hybrid devices with various dimensions have been fabricated and the preliminary experimental results matched the theoretical calculation very well.

4 DISCUSSION AND FUTURE WORKS

A novel system for the self-assembly of muscle-powered MEMS devices have been previously reported. Through the manipulation of material interfaces and phase, we are able to effect the self-assembly and growth of muscle fibers from individual muscle cells on micro-size lithographically fabricated structures. Using this system, a number of muscle-powered MEMS devices have been created. In this work, we extended this system to develop a self-assembled muscle-powered piezoelectric generator. Like the other muscle-MEMS devices, this novel self-assembled generator system has the following advantages:

non-tethered, simple fabrication, and biologically friendly operating conditions.

Using surface micromachining techniques, we can easily alter the pattern of the gold film, which subsequently determine the dimensions and orientation of the self-assembled muscle bundle. Therefore, the geometries and configuration of the devices can be tailored in series or parallel to produce higher voltage and energy. Finally, since all integration steps including MEMS fabrication, Parylene and PVDF-TrFE deposition, PNIPAAm and Au patterning, and cell self-assembly are relatively independent, a variety of MEMS devices, electronic circuits, and cells can be integrated and packaged, thus showing high versatility and flexibility of our devices. For example, with monolithic integration with IC, an integral system of BioMEMS can be built up, which carries potential applications in health monitoring, and biosensing.

The potential limitation of this piezoelectric generator is its low energy density produced. There are various ways to solve this limitation. The first is to build a piezoelectric bimorph structure, which will reduce the mechanical losses and increase the output power. In addition, the higher crystalline the annealed piezoelectric film has, the higher polarization it can achieve, thus the higher output [15]. Therefore, we can use the poled PVDF-TrFE films to solve the limitation of the low glass temperature of PNIPAAm.

REFERENCES

[1] Kawai, H., The piezoelectricity of Poly (vinylidene fluoride). J. Appl. Phys., 1969. 8: p. 975.

[2] Harrison, J.S. and Z. Ounaies, Piezoelectric polymers. 2001, Institute for Computer Applications in Science and Engineering (ICASE). p. 1-26.

[3] Glynne-Jones, P., S.P. Beeby, and N.M. White, Towards a piezoelectric vibration-powered microgenerator. IEE Proceedings, 2001. 148(2): p. 68-72.

[4] Tashe, E., E. Meyhofer, and B. Brenner, A force transducer for measuring mechanical properties of single cardiac myocytes. Am J Physiol, 1999, 277: p. H 2400-8.

[5] Lin, G., et al., Miniature heart cell force transducer system implemented in MEMS technology. IEEE Transactions on Biomedical Engineering, 2001. 48(9): p. 996-1006.

[6] Lin, G., K.S.J. Pister, and K.P. Roos, Surface micromachined polysilicon heart cell force transducer. Journal of microelectromechanical system, 2000. 9(1): p. 9-17.

[7] van der Velden, J., et al., Force production in mechanically isolated cardiac myocyte from human ventricular muscle tissue. Cardiovasc. Res, 1998. 38(2): p. 414-423.

[8] McMahon, D., et al., C2C12 cells: biophysical, biochemical, and immunocytochemical properties.

the American Physiological Society, 1994. 35: p. C1795-C1802.

[9] Xi, J.Z., J. Schmidt, and C. Montemagno, A Novel System for the Self-Assembly of Muscle on MEMS Devices. (In preparation).

[10] Ross, R.S., et al., ß1 integrins participate in the hypertrophic response of rat ventricular myocytes. Circulation Research, 1998. 82: p. 1160-1172.

[11] Sen, A., et al., Terminally differentiated neonatal rat myocardial cells proliferate and maintain specific differentiated functions following expression of SV40 large T antigen. Journal of Biological Chemistry, 1988. 263(35): p. 19132-19136.

[12] Zou, Q., et al. Implantable bimorph piezoelectric accelerometer for feedback control of functional neuromuscular stimulation. in Transducer '03, the 12th international conference on solid state sensors, actuators and microsystem. 2003. Boston.

[13] Smits, J.G., S.I. Dalke, and T.K. Cooney, The constituent equations of piezoelectric bimorphs. Sensors and Actuators, 1991. A28: p. 41-46.

[14] Smits, J.G. and W.S. Choi, The constituent equations of piezoelectric heterogeneous bimorphs. IEEE Transaction on ultrasonics, ferroelectrics, and frequency control, 1991. 38: p. 256.

[15] Ploss, B. and B. Ploss, Influence of poling and annealing on the nonlinear dielectric permittivities of PVDF-TrFE copolymers. IEEE Transaction on Dielectrics and Electrical Insulation, 1998. 5: p. 91-95.

[16] Wang, Q-M, et al., Theoretical Analysis of the Sensor Effect of Cantilever Piezoelectric Benders. J. Appl. Physics, 1999. 85(3): p. 1702-1712.

Nanosecond Pulsed Electric Fields Trigger Intracellular Signals in Human Lymphocytes

P. T. Vernier[1], Y. Sun[2], L. Marcu[3], C. M. Craft[4], and M. A. Gundersen[5]

[1]MOSIS, Information Sciences Institute, University of Southern California (USC),
Marina del Rey, CA, USA, vernier@mosis.org
[2]Department of Materials Science, USC, Los Angeles, CA, USA, yinghuas@usc.edu
[3]Cedars-Sinai Medical Center, Los Angeles, CA, USA, marculaura@cshs.org
[4]Department of Cell and Neurobiology, Keck School of Medicine, USC, Los Angeles, CA, USA,
ccraft@usc.edu
[5]Department of Electrical Engineering-Electrophysics, USC, Los Angeles, CA, USA, mag@usc.edu

ABSTRACT

We present evidence for a new technological approach to signaling and triggering of *intracellular* events in biological cells. Remotely applied nanosecond, megavolt-per-meter, pulsed electric fields perturb the intracellular environment, causing calcium release, phosphatidylserine externalization, and apoptosis. This technology does not require direct electrical contacts to biological material. Ultra-short pulses with durations less than the charging time constant of the plasma membrane generate voltages across structures in the cell interior, enabling intracellular manipulations with stimuli that are remotely generated and controlled. The work reported here suggests possibilities within the reach of state-of-the-art bioelectrical engineering and nanotechnology for utilizing nanopulse-triggered intracellular electroperturbations in remote bio-systems deployed as reporters and effectors, and for other applications such as wound healing.

Keywords: nanosecond, megavolt-per-meter, calcium burst, electroperturbation, phosphatidylserine translocation

1 INTRODUCTION

This paper summarizes ongoing studies of a new approach for inducing electric field effects within biological cells, without contacts or probes directly attached to the cells. We describe basic observations demonstrating that properly tailored external electrical pulses of short (ns) duration produce clearly demarcated internal effects, such as programmed cell death (apoptosis).

Regulated changes in electrical potential, particularly the transmembrane voltages at the interfaces between intracellular compartments, between the cytoplasm and the external environment, and between neighboring cells, comprise an integral part of the network of mechanisms through which living systems respond to their internal and external environments. The development of practical methods for controlling these cell signaling pathways requires minimally invasive means for accessing them *inside* cells. DC and low frequency electric fields affect cell activities through non-specific effects mediated through receptors in the external membrane.

We demonstrate here that pulsed power physics and electronics offer a convenient and versatile tool for reaching *into* the cell — ultra-short (ns), high-field (MV/m) electrical pulses. Fast-rising pulses of very brief duration bypass the plasma membrane dielectric and at the same time eliminate the complications of heating even with very large pulse amplitudes (high power but *low* total energy).

Critical to this approach is addressing the biology demonstrating the *intracellular* effects of ultra-short external pulses, which bypass the cytoplasmic membrane. The charging time constant for the plasma membrane for many eukaryotic cells is on the order of 100 ns [1]. For these cells a nanosecond pulsed electric field is an *intracellular* stimulus [2], in contrast to the situation with the longer, slower-rising (>µs) pulses used for conventional electroporation [3].

Intracellular effects of nanoelectropulse exposure reported previously include permeation of eosinophil granules [4], calcium bursts in T cells [5], and various indicators of apoptosis in several cell lines [2], [6], [7]. Engineering biologically effective devices and systems in this new ultra-short, high-field regime will require deepening and extending this knowledge.

In this report we concentrate on two immediate effects of remote nanoelectropulse perturbation of the intracellular environment in human lymphocytes (Jurkat T cells): increases in cytoplasmic calcium concentration [5] and translocation of phosphatidylserine (PS) to the external face of the cell membrane [2]. Because calcium ions serve as regulatory messengers in a wide variety of processes across the physiological landscape of the cell, understanding how to manipulate calcium ion release with remotely delivered electrical signals is of great interest. We monitor intracellular calcium concentration changes in live cells during pulse exposure with the calcium-sensitive fluorochromes Calcium Green-1 and rhod-2.

PS externalization modifies the thermodynamic and mechanical equilibria associated with the membrane lipid bilayer and marks the cell physiologically for removal by phagocytic agents. We track this event in real time with FM1-43, a cationic styryl fluorescent dye that localizes in the outer leaflet of the plasma membrane lipid bilayer, and which exhibits increased binding when the PS fraction increases on the external face of the membrane [8].

Intracellular calcium modulation and maintenance of membrane phospholipid asymmetry represent important intra- and extracellular signaling pathways. The ability to control these and other physiological signals remotely, through externally generated electric pulses, opens new avenues for bioengineering in the nanosecond, nanometer regime.

2 MATERIALS AND METHODS

Cell lines and culture conditions. Human Jurkat T lymphocytes (ATCC TIB-152) were maintained as described previously [2].

Pulse generator and pulse exposures. A MOSFET-based, inductive-adding pulse generator with a balanced, coaxial-cable pulse-forming network and spark-gap switch designed and assembled at the University of Southern California, provided trapezoidal electrical pulses to cell suspensions (2 x 107 cells /mL) at room temperature in growth medium in rectangular electroporation cuvettes with a 1-millimeter electrode separation (Bio-Rad).

For microscopic observations, cells were placed in a rectangular channel 100 μm wide, 30 μm deep, and 12 mm long, with gold-plated electrode walls, microfabricated with photolithographic methods on a glass microscope slide. A fast MOSFET MicroPulser [9] was mounted on the microscope stage for delivery of pulses directly to the microchamber electrodes in ambient atmosphere at room temperature.

Fluorescence microscopy. Observations of live cells stained with Calcium Green (Molecular Probes) were made with an epifluorescence microscopy system as described previously [5]. To reduce nonspecific cytoplasmic staining with the calcium-sensitive cationic fluorochrome rhod-2, cells were loaded at 2 μM in growth medium for 5 minutes at 25 °C, 25 minutes at 4 °C, 30 minutes at 37 °C, resuspended in growth medium without rhod-2, and incubated for an additional 60 minutes at 37 °C.

For Mn^{2+} quenching experiments, cells were incubated in 250 μM $MnCl_2$ in growth medium for 15 minutes at 37 °C. To stain mitochondria, cells were loaded with JC-1 (Molecular Probes; 750 nM) or MitoTracker Green (Molecular Probes; 50 nM). FM1-43 (Molecular Probes; λ_{ex} = 480 nm, λ_{em} = 580 nm) was added to the medium at 5 μM 20 minutes before observations of PS externalization.

EGTA and ionomycin were obtained from Calbiochem. $LaCl_3$, $GdCl_3$, cyclosporin A, verapamil, and ruthenium red were from Sigma-Aldrich, and BAPTA-AM and thapsigargin were from Molecular Probes.

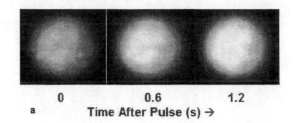

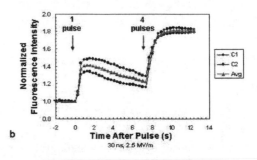

Figure 1: (a) Calcium Green fluorescence images of calcium burst in Jurkat T cells after single pulse, 30 ns, 2.5 MV/m. (b) Normalized calcium burst intensity (Calcium Green fluorescence) after 1 and 4 pulses, 30 ns, 2.5 MV/m.

3 RESULTS

Calcium bursts. Observations with Calcium Green show a uniform release of calcium throughout the cell within milliseconds after the leading edge of a nanosecond electric pulse (Fig. 1). Pulse-induced calcium bursts are not affected by EGTA in the medium or by the calcium channel blockers La^{3+}, Gd^{3+}, or verapamil, or by the mitochondrial permeability transition inhibitor cyclosporin A, or by the mitochondrial calcium transport blocker ruthenium red, but they are inhibited by thapsigargin and cytochalasin D [5].

Cells loaded with rhod-2, a calcium-sensitive fluorochrome with a larger calcium dissociation constant than Calcium Green-1 (570 nM versus 190 nM), present a somewhat different picture (Fig. 2). The nanoelectropulse-induced rhod-2 fluorescence increase is more gradual, spanning several seconds, and in many cells the fluorescence intensification occurs at multiple distinct locations in the cytoplasm. These localized rhod-2 bright spots are not stained by JC-1 or MitoTracker Green, indicating that they are not mitochondria.

Cells loaded with the calcium chelate BAPTA or the calcium competitor and fluorescence quencher Mn^{2+} do not show pulse-induced Calcium Green or rhod-2 fluorescence intensification.

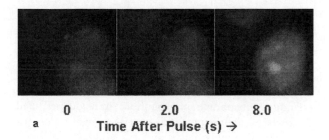

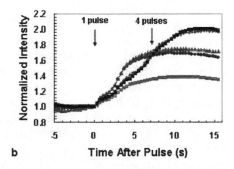

a

b

Figure 2: (a) Rhod-2 fluorescence images of calcium burst in Jurkat T cells after single pulse, 30 ns, 2.5 MV/m. (b) Normalized calcium burst intensity (rhod-2 fluorescence) after 1 and 4 pulses, 30 ns, 2.5 MV/m.

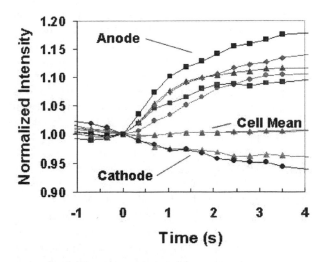

Figure 3: FM-1-43 fluorescence intensity at selected points in a cell after pulsed electric field exposure, 4 pulses, 30 ns, 2.5 MV/m, normalized to the integrated intensity of the whole cell. Points at the anode pole of the cell brighten relative to the integrated intensity of the whole cell, while the response of points at the cathode pole is less than the cell mean.

Longer, microsecond pulses porate the external membrane and permit entry of propidium iodide, Ca^{2+}, Na^+, and Mn^{2+} into the cell. No influx of these species is detected after nanosecond pulses [2], [5].

Membrane phospholipid rearrangement. Nanoelectropulse-induced changes in the intensity and distribution of the membrane-staining dye FM1-43 begin less than one second after pulse exposure. An initial brightening at the anodic pole of the cell appears to dissipate within seconds, but over the next several minutes the fluorescence intensity of FM1-43 in the membrane increases all around the cell (Fig. 3).

These observations would be expected if the interaction of the pulsed field with the cell causes an immediate rearrangement of membrane phospholipids at the pole of the cell nearest the anode, including a translocation of some PS molecules from the cytoplasmic side of the membrane, where they normally are exclusively found, to the outer surface of the cell. This PS externalization would lead to the binding of additional FM1-43, but the initial local concentration disturbance would be quickly dissipated by lateral diffusion in the bilayer. Over the next few minutes, levels of FM1-43 in the membrane would rise, adjusting to the new equilibrium concentration of PS in the membrane

4 DISCUSSION

A conservative interpretation of the evidence reported and summarized here leads to the conclusion that nanoelectropulses cause the release of calcium from intracellular compartments and the translocation of PS in the plasma membrane through directly physical and immediate mechanisms — without any measurable poration of the cell membrane.

Nanoelectropulses could trigger intracellular calcium release in a number of ways. Our data suggests the involvement of the calcium compartments of the endoplasmic reticulum (which cannot be re-supplied in the presence of thapsigargin) and some components of the cytoskeleton (which cannot be dynamically maintained in the presence of cytochalasin D).

Two mechanisms for pulse-induced PS externalization may be considered: nanopore-facilitated "lateral" diffusion and direct electrostatic transport across the dielectric barrier of the lipid bilayer (Fig. 4).

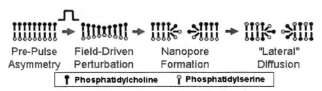

Figure 4: Formation of an aqueous pore with nanometer dimensions permits "lateral" diffusion of phosphatidylserine from the inner leaflet to the external face of the cell. Transmembrane voltages greater than 1 V may provide sufficient energy (50 kT) to drive the charged phospholipid head group directly across the hydrophobic membrane dielectric; no pore formation would be required.

A simple dielectric shell model for nanoporation (nanometer-diameter, nanosecond-duration electropores) predicts PS externalization at both the anode and cathode poles of a nanoelectropulsed cell [5]. FM1-43 fluorescence intensification after pulse exposure always occurs at the anode pole and has never been observed at the cathode pole, consistent with direct, electrostatically driven migration of the negatively charged PS head group across the membrane lipid bilayer.

Ultra-short, high-field electric pulses are not immediately physically damaging to cells and not necessarily lethal in the medium or long term (although they induce apoptosis under the right conditions). At low pulse counts they produce no gross morphological changes. Nanoelectropulsed cells remain in place, actively metabolizing, potentially serving as dynamic environmental sensors and transducers. Appropriately prepared cells, or bio-inspired nanomachines differentially sensitive to a defined set of pulse regimens (Fig. 5), could execute on remote command a variety of biochemical tasks, as standalone "biobots" or as components of integrated biomicroelectromechanical systems or other "lab on a chip" implementations.

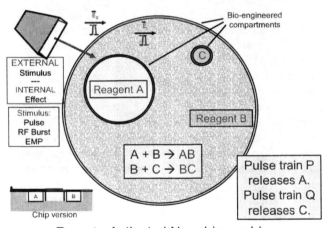

Remote-Activated Nanobiomachine

Figure 5: Nested dielectric shells with biochemically active components — a bionanoelectromechanical system.

5 ACKNOWLEDGEMENTS

This work was supported by grants from the Air Force Office of Scientific Research and the Army Research Office. We gratefully recognize Matthew Behrend for pulse generator design and construction, Sarah Salemi and Mya Thu for assistance with fluorescence microscopy and cell manipulations, and Jingjing Wang and Yushun Zhang for cell culture.

REFERENCES

[1] K. H. Schoenbach, S. J. Beebe, and E. S. Buescher, "Intracellular effect of ultrashort electrical pulses," Bioelectromagnetics, vol. 22, pp. 440-448, 2001.

[2] P. T. Vernier, A. Li, L. Marcu, C. M. Craft, and M. A. Gundersen, "Ultrashort pulsed electric fields induce membrane phospholipid translocation and caspase activation: differential sensitivities of Jurkat T lymphoblasts and rat glioma C6 cells," IEEE Transactions on Dielectrics and Electrical Insulation, vol. 10, pp. 795-809, 2003.

[3] U. Zimmermann, "The effect of high intensity electric field pulses on eukaryotic cell membranes: fundamentals and applications," in Electromanipulation of Cells, U. Zimmermann and G. A. Neil, Eds. Boca Raton, FL: CRC, 1996.

[4] E. S. Buescher and K. H. Schoenbach, "Effects of submicrosecond, high intensity pulsed electric fields on living cells-intracellular electromanipulation," IEEE Transactions on Dielectrics and Electrical Insulation, vol. 10, pp. 788-794, 2003.

[5] P. T. Vernier, Y. Sun, L. Marcu, S. Salemi, C. M. Craft, and M. A. Gundersen, "Calcium bursts induced by nanosecond electric pulses," Biochemical and Biophysical Research Communications, vol. 310, pp. 286-295, 2003.

[6] S. J. Beebe, P. M. Fox, L. J. Rec, K. Somers, R. H. Stark, and K. H. Schoenbach, "Nanosecond pulsed electric field (nsPEF) effects on cells and tissues: Apoptosis induction and tumor growth inhibition," IEEE Transactions on Plasma Science, vol. 30, pp. 286-292, 2002.

[7] S. J. Beebe, P. M. Fox, L. J. Rec, E. L. Willis, and K. H. Schoenbach, "Nanosecond, high-intensity pulsed electric fields induce apoptosis in human cells," FASEB J, vol. 17, pp. 1493-5, 2003.

[8] A. Zweifach, "FM1-43 reports plasma membrane phospholipid scrambling in T-lymphocytes," Biochem J, vol. 349, pp. 255-60, 2000.

[9] M. Behrend, A. Kuthi, X. Gu, P. T. Vernier, L. Marcu, C. M. Craft, and M. A. Gundersen, "Pulse generators for pulsed electric field exposure of biological cells and tissues," IEEE Transactions on Dielectrics and Electrical Insulation, vol. 10, pp. 820-825, 2003.

Viruses as Optical Probes

E-S Kwak*, C. Chen*, C. Cheng Kao**, W. L. Schaich*, and B. Dragnea*

*University of Indiana, US, dragnea@indiana.edu

*University of Texas A&M, TX, US

Abstract

We report here on two new methods, which have the potential to use the advantages of optical spectroscopy (physiological environment, chemical specificity, non-intrusiveness, time resolution) to accede, in-vitro, to the formation of virus capsids, phase transitions and cellular transit, one particle at a time. The first approach is the encapsulation of nanoparticles of different sizes and surface chemistry inside virus capsids. The nanoparticle cores act as spectroscopic enhancers and templates with tunable surface properties for the virus self-assembly from protein subunits. The second approach is to use the optical near-field to trap single particles in a physiological fluid inside nanochannels lithographically patterned on a surface. These two approaches, combined, will open the way to real-time monitoring of virus formation and intracellular imaging using viruses as optical probes.

Encapsulation of optically active nanoparticles in brome mosaic virus capsids

At present, a general goal in molecular materials is to preserve the chemical function of their molecular constituents upon self-assembly (e.g. DNA sensors, enzymatic assays). In the future, however, research inspired by complex molecular systems such as viruses, will take advantage of changes in the molecular characteristics occurring upon self-assembly. These changes may lead to a plethora of functional properties due to self-organization. As an example, a virus whose coat is made out of a single type of molecular building block, has a life-cycle composed of a sequence of stages, each stage being defined by a certain viral function (protection, recognition, etc.). These stages are orchestrated by symmetry changes in the self-assembled protein coat. Addressing the structural transformations, the assembly pathways and the cellular transit characteristics in a way that preserves the physiological environment, has enough chemical specificity and time-resolution to detect intermediates, and enough spatial resolution to investigate one virus at a time, remained an elusive goal until today. Our long-term strategy in overcoming these challenges is to associate the virus capsids with nanoparticles acting as spectroscopic enhancers and then use the advantages of optical spectroscopy (physiological environment, chemical specificity, non-intrusiveness, time resolution) to accede, in-vitro, to the formation, evolution and cellular transit of single virus capsids.

We have recently reported on the first example of Rayleigh resonance spectroscopy on single brome mosaic virus (BMV) capsids (28 nm) with Au particles (2.5-4.5 nm) inside, Fig. 1, [1][2].

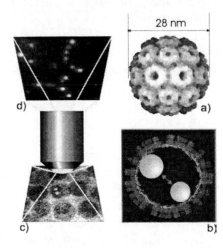

Figure 1: Collage showing pairs of Au particles incorporated during the self-assembly in virus capsids from protein subunits, and their optical signatures. (a) Reconstructed BMV capsid (adapted from [4]). (b) Schematic of two Au nanoparticles enclosed in a capsid. The distance between the particles is an important parameter for the optical spectral signature of the complex. (c) TEM picture of a pair of Au nanoparticles encapsulated in a capsid surrounded by reassembled capsids with RNA inside. (d) Dark-field scanning confocal microscopy of

individual virus capsids containing gold nanoparticles in aqueous buffer.

In this work, inspired by the natural virus particle formation, we developed a procedure to partially replace the negatively charged RNA inside the virions by gold particles. The protocol is adapted from the *in vitro* assembly of cowpea chlorotic mottle virus [3]. The association of negatively charged, citrate-covered, Au particles, 2.5-4.5 nm diameter, with the internal compartment of BMV capsids is done in three steps: disassembly of the BMV virions, reassembly in the presence of Au particles and purification of BMV particles.

If the viral particles are not completely disassembled, but only swollen by the pH change, the association of the Au with the capsid does not occur.

Negatively charged Au sols are obtained by using the method of citrate and tannic acid, which allows for synthesis of particles of uniform and controlled diameter (min. 3 nm ± 12%, max. 17 nm ± 7%) by varying the ratio between the rapid and slow reductants, the tannic acid and the sodium citrate, respectively.

Our procedure, based on self-assembly of protein subunits and metal cores, is novel. As a consequence of the self-assembly principles, this method should be quite general, allowing for incorporation of different types of extant nanoparticles such as quantum dots, and superparamagnetic beads.

Here we ask the question: are the protein-protein interactions still the main driving force for the capsid self-assembly, or the Au-protein interaction is now dominant? To address this issue, we have varied the diameter of the nanoparticle to be incorporated and measured by transmission electron microscopy (TEM) the outer diameter of the nanoparticle/capsid complex. If the Au-protein interaction were the dominant one, one would expect an increase in the capsid diameter proportional to the size of the incorporated particle. We have found that this is not the case. Instead, the diameter of the capsid/nanoparticle complex is constant. Moreover, there is a cutoff size of 12.5 nm for the encapsidated Au particle, which further supports a dominant protein-protein interaction dictating the capsid formation. However, the Au particle encapsidation yield does depend on the Au particle diameter. We have obtained an optimum diameter of 9±1 nm for citrate-covered Au particles incorporation. The existence of an optimum diameter may point to the existence of a large protein precursor from which capsids form.

A 5-fold improvement in the yield of nanoparticle incorporation has been obtained by replacing the citrate with biotinylated DNA containing the high affinity capsid-binding site. The same DNA and biotin/avidin linker has been used to encapsidate CdSe quantum dots whose luminescent properties make them interesting candidates for tracking of particle-modified viruses in the cell. Fig. 2 shows TEM pictures of DNA-coated Au particles and quantum dot particles incorporated in brome

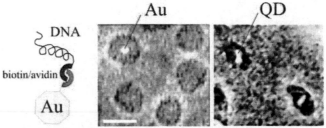

mosaic viruses.

Figure 2: Core/capsid interaction can be adjusted to resemble more those in the wild-type virions by functionalizing the nanoparticle surface with DNA containing the high-affinity capsid-binding site. The DNA-modified nanoparticles can be then encapsidated as shown here for Au cores and CdSe quantum dot cores.

More research on the nature of nanoparticle core virus capsid interactions will unveil ways to make this interaction more similar to that occurring between the RNA molecules and the capsid, thus preserving the functional attributes of the capsid.

An important step forward would be to have access to the capsid formation at the scale of a single virus in real-time and in a physiological environment. Single-virus studies avoid the ensemble averaging of other methods and have direct access to intermediates. In the following, we introduce a novel method that is not limited by diffraction in the size of trapped particles, is non-intrusive and allows for parallel manipulation.

Near-field optical trapping of small particles inside nanochannels

When a small particle with index of refraction greater than that of its environment passes through the waist of a focused Gaussian beam of light, forces resulting from scattering of photons on different regions of the particle tend to immobilize the particle in the most intense part of the focal zone, i.e. the center. This force can exceed the Brownian forces at the room temperature if the field gradient across the diameter of the particle is significant [5]. This method of three-dimensional position control of a particle in a fluid, known as optical tweezers, found an enormous number of applications ranging from fundamental physics of ultracold atoms to subcellular dynamics. The wide acceptance gained in the biology community by the optical tweezers technique, greater than any other proximity probe technique, is due to its non-intrusiveness, quasi three-dimensional character, and possibility of measuring very small forces 10^{-13}-10^{-10} N, relevant for biological matter. However, the requirement that the light intensity gradient across the whole diameter of the particle is non-negligible, implies that particles smaller than the diffraction limit (~250 nm) are difficult to trap. Because of this limitation, only relatively large viruses, such as the tobacco mosaic virus (~500 nm in length), could be trapped with mild-enough light beams to preserve their integrity. To alleviate the particle size problem, several researchers proposed to use near-field optical tweezers relying on the strongly enhanced electric fields present close to metal nanostructures [6][7]. The near-field close to such a metal structure consists of evanescent components, which decay on much shorter distances than the wavelength of the light.

This idea, although very promising, has not been realized in practice yet, with either aperture-based or apertureless near-field scanning optical microscopes, probably because it would be difficult to find in water a rapidly diffusing nanoparticle with a single probe tip.

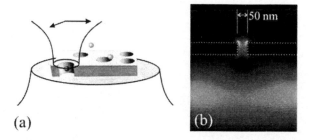

(a)   (b)

Figure 3: (a) Schematic picture of the nanochannel optical trap method: the incident beam illuminates from the bottom a few nanoholes at a time in an opaque metal screen. The nanoholes, with a diameter from 50 to 200 nm, are generated by nanosphere lithography and spaced widely enough to allow single nanohole selection by high numerical aperture collection optics from the top. (b) Vertical cross-section, with calculated density map of the electric field intensity, inside a one-dimensional channel through a 50 nm thick gold film. The incident field polarization is perpendicular to the channel walls; the laser wavelength is 800 nm.

Instead of using a near-field microscope tip to catch a nanoparticle, we have used a large ensemble of nanoholes, or channels, fabricated in a 50-200 nm thick metal film. For fabrication, we are currently employing the nanosphere lithography method, Fig. 3 (a). The particles for study are dispersed in solution in contact with the metal film. Some of them diffuse into the nanochannels. When a laser beam is shone onto the metal film, the particles located inside the nanochannels are trapped due to strong near-field intensity gradients present in the subwavelength cavity, Fig. 3 (b). Although a large ensemble of nanochannels are simultaneously illuminated, they can be "read" one at a time, with a confocal scanning microscope from the other side of the film. We can thus choose a trap that was incidentally populated at the time when the laser was turned on.

Initial experiments on the near-field trapping efficiency have been done on commercially available 200 nm fluorescent latex beads and 500 nm diameter nanochannels in a 250 nm thick Au film deposited on silica. The realization of this initial step in itself is a very promising result since despite a significant number of theoretical papers predicting near-field optical trapping effects, there are no experimental reports to confirm its achievement, with the exception of Ng et al. who observed coherent motion of colloidal particles close to the walls of a waveguide [8]. From measurements of the fluorescence emission from an occupied trap we have found that a) single beads are "sucked" inside the hole and remain there for the time the laser is on, b) the amplitude of the position fluctuations of a bead inside a nanochannel depends on the trapping laser intensity, c) stable trapping is obtained at a fraction of laser power required for the same bead diameter in the classical optical tweezers case, d) a 10-fold fluorescence enhancement is observed as a result of the interaction between the fluorescent bead and the near-field of the metal walls.

In the future, we will couple the nanoparticle/virus approach with the near-field optical scheme to monitor the self-assembly process between a nanoparticle core and protein subunits to make a viral particle with a nanoparticle core

References

[1] Dragnea, B., Chen, C., Kwak, E. S., Stein, B., Kao, C. C., "Gold nanoparticles as spectroscopic enhancers for in vitro studies on single viruses", *J. Am. Chem. Soc.*, *125*, 6374-6375, 2003.

[2] Editorial news, "Virus particles take the gold", *Analytical Chemistry*, *75*, 286A, 2003.

[3] Kao, C. C., Sivakumaran, K. *Molec. Plant. Path.* **2000**, *1*, 91-97.

[4] Reddy, V., Natarajan, P., Okerberg, B., Li, K., Damodaran, K., Morton, R., III, C. B., Johnson, J., "VIrus Particle ExploreR (VIPER), a Website for Virus Capsid Structures and Their Computational Analyses", *J. Virol.*, *75*, 11943-11947, 2001.

[5] Ashkin, A. "History of Optical Trapping and Manipulation of Small Neutral Particles, Atoms and Molecules". In *Single Molecule Spectroscopy - Nobel conference lectures*; Rigler, R., Orrit, M., Basché, T., Ed.; Springer: Berlin, 2002.

[6] Novotny, L., Bian, R. X., Xie, X. S."Theory of nanometric optical tweezers", *Phys. Rev. Lett.* **1997**, *79*, 645-648.

[7] Chaumet, P. C., Rahmani, A., Nieto-Vesperinas, M."Selective nanomanipulation using optical forces", *Phys. Rev. B* **2002**, *66*, art. no.-195405.

[8] Ng, L. N., Zervas, M. N., Wilkinson, J. S., Luff, B. J., "Manipulation of colloidal gold nanoparticles in the evanescent field of a channel waveguide", *Appl. Phys. Lett.* **2000**, *76*, 1993-1995.

pH Dependent Change in the Optical Properties of Surface Modified Gold Nanoparticles Using Bovine Serum Albumin

Yang Xu [*], K.A. Linares [**], K. Meehan [*], B. J. Love [***], and N.G. Love [**]

[*]The Bradley Department of Electrical and Computer Engineering, yaxu5@vt.edu and kameehan@vt.edu
[**]Civil and Environmental Engineering Department, klinares@vt.edu and nlove@vt.edu
[***]Material Science and Engineering Department, blove@vt.edu
Virginia Polytechnic Institute and State University, Blacksburg, VA, USA

ABSTRACT

The effect of coatings of Bovine SerumAlbumin (BSA) on the optical properties and reactivity of surface-modified gold nanoparticles in different pH buffer solutions is presented. Aggregation of uncoated gold nanoparticles is seen in all pH buffer solutions. The presence of these aggregations is observed in the absorption spectra by the appearance of a second, red-shifting peak in the spectra and is confirmed in transmission electron micrographs. Coating of individual gold nanoparticles with BSA prevents the aggregation of nanoparticles in solutions with a pH greater than 5. It is concluded, based upon analysis of visible absorption spectra of the coated and uncoated gold nanoparticles, that the BSA coating on the gold nanoparticles also causes a decrease in the scattering of light in lower pH buffer solutions. Analysis of Mie's theory on the optical extinction of light by metal particles supports the interpretation of the spectra.

Keywords: gold, nanoparticle, aggregation, surface plasmon resonance, BSA

1 INTRODUCTION

Gold nanoparticles have been studied with great interest because of its unusual optical properties [1]. According to Mie, who was the first to theoretically explained those properties in the metal colloid solution [2], spherical metal colloids absorb certain wavelengths via a phenomenon known as surface plasmon resonance (SPR). The excitation wavelength of a surface plasmon wave is a function of the size of the particles and the dielectric constants of both the metal and the media surrounding the metallic particles. By monitoring the intensity and position of the SPR peaks in the absorption spectra of gold nanoparticles attached with certain functional molecules, useful information about the interaction between the molecules and other substance in the solution can be acquired. This phenomenon can be used in many advanced bio/chemical sensors [3-5]

Aggregation of gold nanoparticles is a common concern in biosensor applications [6-8] because the SPR signal can not reflect the interaction that happens on the surface of the particles correctly [9]. If used in live cells, large clusters of gold nanoparticles may not be uptaken as easily as individual particles and may selectively collect near the cell membrane. Surface modification of the gold nanoparticles has been used to eliminate aggregation [10-13]. In *in-vivo* biosensing applications, the coating must be biocompatible and robust over the range of intracellular pH, yet sufficiently thin to enable the intracellular fluid to influence the SPR signal of the gold nanoparticles. In this paper, studies are concentrated on optical properties of Bovine Serum Albumin (BSA)-coated gold nanoparticles in different pH environments. Investigation shows that there are significant differences in the absorption spectra of BSA-coated and uncoated gold nanoparticles in solution. Transmission electron microscopy (TEM) micrographs verify the conclusions drawn from these spectra. Because of the biocompatibility of BSA, the surface modification of the gold nanoparticles with BSA may be useful in the design of a live cell biosensor, which can monitor, for example, fluctuations in the intracellular pH in response to biochemical stimuli.

2 EXPERIMENTAL PROCEDURE

Gold nanoparticles with extremely uniform diameter of 20 nm (Figure 1), synthesized according to the method developed by Beesley [14], are purchased from ICN Biomedicals. 0.1 wt % of gold in solution translates to 10^{12} nanoparticles per milliliter. BSA stock is purchased from Sigma and then diluted with nanopure water to a 0.1mmol/L solution. Several different pH buffer solutions are made of phosphate buffer. The solutions are pH 4, 5, 6, 7, and 8, which covers the pH range expected in a live cell biosensor under development by our group. U V-Vis spectra are acquired with Beckman DU640 spectrophotometer. All spectra presented are normalized using Origin (7th edition). No attempt is made to calibrate the intensities of the spectra.

Carbon thin film coated TEM grids are prepared by drying a small droplet of the solution to be examined in open air. The TEM imaging is performed with Philips EM420T transmission electron microscope. Micro-pipettes are utilized through the experiment to ensure that all of the solutions are accurately measured to within a few il ranges.

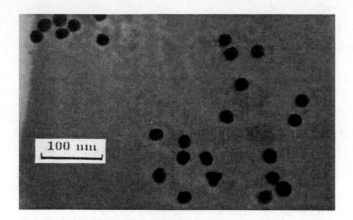

Figure 1: Transmission electron micrograph of uncoated gold nanoparticles on carbon n film coated copper grid.

3 RESULTS AND DISUSSION

2mL of each pH (4, 5, 6, 7, and 8) buffer solution is mixed with 0.25mL of the uncoated gold nanoparticles solution in a disposable cuvette at room temperature. The buffer-nanoparticle solution is then immediately placed in the UV-Vis spectrophotometer and spectra are collected between 450 nm and 750 nm every 10 minutes (each scan takes approximately one minute to collect). The first maximum (around 528 nm) in absorption spectra (Figure 2) is the SPR signal, characteristic of 20 nm gold nanoparticles [15]. The second red-shifting peak is indicative of aggregation of uncoated gold nanoparticles [13, 16]. It is clear that the aggregation of the gold particles happens more slowly with increasing solution pH.

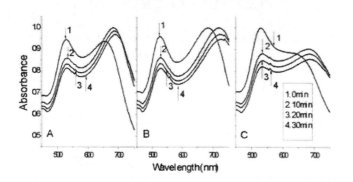

Figure 2: Optical absorption spectra as a function of time of 0.25 mL of gold nanoparticles solution following the addition of 2 mL of pH buffer. A. pH4 B. pH6 C. pH8

In original solution, gold particles are electrostatic stabilized with an electrical double layer on individual particles [17]. Upon mixing the nanoparticles solution with the different pH buffer solutions, the double layer is destroyed as the buffer alters the charge balance in the solution and, thus, aggregation forms.

Figure 3 presents the optical absorption spectra of both uncoated and BSA-coated gold nanoparticles as a function of time. The coating process happens immediately upon mixing the BSA stock and gold colloid solution (1:1 volume ratio). There is a minor change in the spectra of the BSA-coated gold nanoparticles in the first five minutes, after which the absorption spectrum of the mixture is quite stable. The significant increase and the slight red shift of the SPR peak of the BSA-coated nanoparticles as compared to the uncoated particles can be interpreted by Mie's theory.

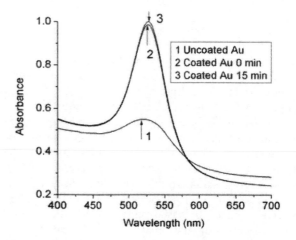

Figure 3: (1) Absorption spectra of uncoated gold particles (2) Same solution upon mixed with BSA stock (3) 15 minutes afterwards

The measured attenuation of light of intensity, I_0, passing through a colloid solution with N particles per unit volume over a light path, d cm, is defined as:

$$A = \log_{10} I_0/I_d = NC_{ext} \, d/2.303 \qquad (1)$$

where C_{ext} is the extinction cross section of a single particle. For spherical particles embedded in a media of dielectric function ε_m, C_{ext} is equal to

$$C_{ext} = 2\pi/k^2 \sum (2n+1) \operatorname{Re}(a_n + b_n) \qquad (2)$$

where $k = 2\pi \sqrt{\varepsilon_m}/\lambda$ and coefficients a_n and b_n are spherical Bessel functions of complex dielectric constant $(\varepsilon = \varepsilon' + i\varepsilon'')$ of colloid metal, which describe the contribution of nth electric and magnetic partial oscillation, respectively.

For particles whose size is far less than the wavelength of light, only the first term, the electric dipole, plays a significant role. Then equation (2) can be simplified to

$$C_{ext} = \frac{24\pi^2 R^3 \varepsilon_m^{3/2}}{\lambda} \frac{\varepsilon''}{\left(\varepsilon' + 2\varepsilon_m\right)^2 + \varepsilon''^2} \qquad (3)$$

where R is the radius of the particle. As the dielectric constant of water (1.33) is little less than that of BSA (1.37), the red-shifted and more intense peak of absorption spectra of BSA-coated gold particles is expected compared to uncoated gold particles.

0.5mL (0.25mL of colloid gold and 0.25mL of BSA) of the coated gold nanoparticles solution is added to 2mL of each of the pH buffer solution. The optical absorption spectra are presented in Figure 4. A red shift of the surface plasmon resonance along with a reduction in the intensity of the resonance is observed in lower pH (pH 4 and 5) buffers. The larger the red shift and the larger the magnitude of the decrease in the maximum of the spectra indicate a higher degree of aggregation of gold particles occurs in the solution [13, 16] (as seen in Figure 5A and 5B). The color of these mixture solutions gradually changes from pink to clear in the first few minutes after mixing, again indicative of less absorption in the visible spectrum. At higher pH value (pH 6, 7 and 8) solutions, the peak of the resonance remains constant with respect to time, up to several days, as does the color of the solution.

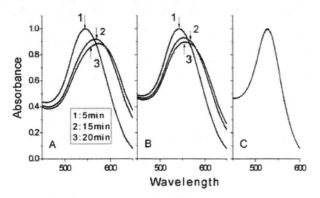

Figure 4: Optical absorption spectra of 0.5 mL of BSA-coated gold nanoparticles solution mixed with 2 mL of pH buffer as a function of time. Spectra of pH 6, 7 and 8 are identical A. pH4 B. pH5 C. pH6, 7 and 8

The isoelectric point of BSA is between 4.5-4.9 [18-20]. Therefore, in the buffer solution with pH value greater than 5, the BSA molecule has a significant number of negative ions attached on the surface, which forms the attractive columbic interaction between the protein and the gold nanoparticles; yet repels similarly BSA-coated gold nanoparticles. Thus, the BSA molecules prevent the gold nanoparticles from interacting with one another to form aggregates.

It should be noted that, although the aggregation of gold still occurs in pH 4 and 5 buffer solutions, the spectra of the BSA-coated gold nanoparticles is not similar to that of the uncoated gold nanoparticles. The second maximum in absorption spectra of uncoated gold particles solution (Figure 2) does not appear in the spectra of the coated particles. We believe this is, in part, due to surface plasmon enhancement of the hyper-Rayleigh scattering (HRS). HRS has been observed to increase by at least a factor of 10 when the physical dimensions of the scatterer is altered from a symmetric to an asymmetric scattering center while there is minimal change in the Rayleigh scattering [21, 22]. As the BSA proteins are likely to attach randomly on the gold nanoparticles and as aggregation of the gold nanoparticles occurs, the symmetry of isolated gold nanoparticles is lost and HRS at the surface plasmon resonance frequency increases. Given the normalization technique used in this study, this will also lead to a decrease in the magnitude of the optical absorption calculated at other wavelengths. An additional effect may be due to a "lens effect" introduced by the BSA coating. Since it has slightly higher index of refraction than water, the thin layer of BSA on the gold nanoparticles will bend the incoming light towards the particles. This increases the optical cross-section of an individual gold nanoparticle, which would increase the number of photons interacting with the individual Au nanoparticle, resulting in a larger surface plasmon wave being excited at 528nm.

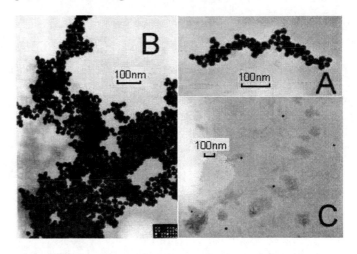

Figure 5: A. BSA-coated gold particle aggregation is observed immediately after mixing with pH 5 buffer; B. Much larger clusters appear in the same pH 5 solution 20 minutes later; C. No aggregation of BSA-coated gold nanoparticles is observed in pH 8 buffer

In order to determine the effect of the concentration of protein on the aggregation of gold particles, two additional mixtures are made. 0.025mL and 0.125mL of 0.1mmol/L BSA stock are mixed with 0.25mL of the gold colloid solution, respectively, prior to the addition of 2mL of pH 4 buffer to the solutions. The peaks of optical absorption spectra (Figure 6) of both solutions are compared to that observed for a solution containing 0.25mL of the gold colloid solution mixed with 0.25mL of the 0.1mmol/L BSA stock and 2mL of pH 4 buffer (Figure 4A). Since the aggregation happens as soon as the coated gold particles solution is introduced to pH 4 buffers, the peak wavelength of each curve is different from each other even at time equals to zero. Based upon the magnitude of the shift in the wavelength peak of the spectra as a function of time, it is observed that the few molecules of BSA introduced into the solution, the faster the aggregation of the gold nanoparticles takes place.

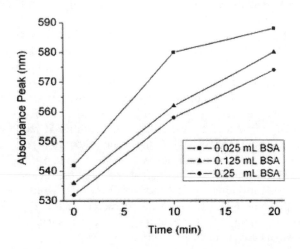

Figure 6: The peak of optical absorption spectra of 0.25mL of the gold particles coated with different amount BSA in 2mL of pH 4 buffers

In conclusion, the pH dependent optical properties of gold nanoparticles coated with BSA are quite different from the uncoated particles. The BSA coating prevents the gold nanoparticles from aggregating in solutions of pH greater than 5 and, thus, eliminating the time-varying red-shifting peak observed in the spectra of the uncoated gold nanoparticles. The absorption peak of the aggregated BSA-coated gold nanoparticles associated with the excitation of the surface plasmons is enhanced and the optical scattering at longer wavelengths by the aggregates is not observed at lower pH value solutions. The present study demonstrates that the BSA-coated gold nanoparticles may be used in biosensors to reduce the aggregation of the gold nanoparticles, which can prevent the uptake of these nanoparticles in live cells and reduce their applicability as bioprobes. More research is needed to identify the role of BSA coating at lower pH level in the aggregation of the nanoparticles. Additional studies are needed to determine an optimal thickness for the BSA coating and to identify alternatives to BSA to prevent gold nanoparticle aggregation at pH lower than 5.

4 ACKNOWEDLEDGEMENT

We kindly acknowledge Dr. Steven McCartney for his assistance on the operation of the TEM and Ms. Jody Smiley for her assistance in the cell culture lab. Financial assistance for this project has been received from the Bradley Department of Electrical and Computer Engineering and the Paul L. Busch Award.

REFERENCES

[1] M.A. Hayat, Colloidal Gold: Principles, Methods, and Applications, Vol. 1 and 2, Academic Press: San Diego, 1989
[2] G. Mie, Ann. Phys, 25, 377, 1908
[3] B. L. Shigekawa and Y. A. Hsieh, Ligand Gold Bonding US Patent# 05294369, 1994
[4] J. M.C. Martin and M. Paqeus, J. Immunoassay 11, 31, 1990
[5] R. Van Erp and T. C. Gribnau, J. Immunoassay 12, 425, 1991
[6] D. A. Weitz, M. Y. Lin and C. J. Sandroff, Surf. Sci, 158, 147, 1985
[7] J. P. Wilconxon, J. E. Martinm and D. W. Schaefer, Phys. Rev. A, 39, 2675, 1989
[8] D. Asnaghi, M. Carpineti and M. Giglio, MRS Bull., 19, 14, 1994
[9] S. Hayashi, R. Koga, M. Ohtuji, K. Yamamoto and M. Fujii, Solid State Commun., 76, 1076, 1990
[10] K. Aslan and V. H. Perez-Luna, Langmuir, 18, 6059, 2002
[11] A. Kumar, P. Mukherjee, A. Guha, S. D. Adyantaya, A. B. Mandale, R. Kumar and M. Sastry, Langmuir, 16, 9775, 2000
[12] W. D. Geoghegan, S. Ambegaonkar and N. J. Calvanico, J. Immunological Methods, 34, 11, 1980
[13] K. S. Mayya and M. Sastry, Langmuir, 13, 3944, 1997
[14] J. E. Beesley, Proc. Royal Micro. Soc., 20, 187, 1985
[15] G. C. Papavassiliou, Prog. Solid St. Chem., 12, 185, 1979
[16] C. S. Weisbecker and M. V. Merritt, Langmuir, 12, 3763, 1996
[17] G. Schmid, Cluster and Colloids: From Theory to Application (Weinheim: VCH), 18, 1994
[18] C. Chaiyasut and T. Tsuda, Chromotography, 22, 91, 2001
[19] J. H. Song, W-S. Kim, Y. H. Park, E. K. Yu and D. W. Lee, Bull. Korean Chem. Soc., 20, 1159, 1990
[20] C. Combes, C. Rey and M. Freche, J. Mater. Sci: Mater. in Med., 10, 153, 1999
[21] F. W. Vance, B. I. Lemon and J. T. Hupp, J. Phys. Chem. B, 102, 10091, 1998
[22] P. Galletto, P. F. Brevet, H. H. Girault, R. Antoine and M. Broyer, Chem. Comm., 581, 1999

Internalization of Nanoparticles in the Middle Ear Epithelium in Response to an External Magnetic Field: Generating a Force.

K. Dormer[*], C. Seeney[**], A. Mamedov[***] and F. Mondalek[****]

[*]University of Oklahoma Health Sciences Center, Department of Physiology Room 634, 940 Stanton L. Young Blvd., Oklahoma City, OK, USA, kenneth-dormer@ouhsc.edu
[**]NanoBioMagnetics INC., 124 N. Bryant Ave. Edmond, OK, nanobio@att.net
[***]Nomadics, Inc., 1024 S. Innovation Way, Stillwater, OK, amamedov@nomadics.com
[***]University of Oklahoma, Department of Chemical Engineering and Materials Science
100 E. Boyd, Room T-335 Norman, OK, fadee@ou.edu

ABSTRACT

In this study we report the first generation of force by long-term intracellular magnetic nanoparticles (MNP) using a guinea pig middle ear model. Synthesis of magnetite nanoparticles was done using a modified precipitation technique. We used different methods to characterize MNP. The mechanism for internalizing 16 ± 2.3 nm diameter, silica coated, superparamagnetic, MNP was likely magnetically enhanced endocytosis. During the surgery a rare earth, permanent magnet (0.35 Tesla) placed under the animal was used to pull the MNP into the tissue. After 8 days, following euthanasia, tissues were harvested and confocal scanning laser interferometry used to verify intracellular MNP. Displacements of the osscicular chain in response to an external sinusoidal electromagnetic field were measured. Laser Doppler interferometry, measured, for the first time, a normal biomechanical function enabled by intracellular MNP.

Keywords: magnetic nanoparticles, ear, hearing, superparamagnetic, intracellular delivery.

MATERIALS AND METHODS

Spherical particles of magnetite (Fe_3O_4) included an external shell of silica (SiO_2) and were 10-20 nm in diameter were synthesized in collaborative effort of NanoBioMagnetics Inc. (NBMI, Edmond, OK) and Nomadics, Inc. (Stillwater, OK). Obtained core-shell MNP then has been treated to form surface amine groups, which provide attachment sites to other molecules. MNP were conjugated with fluorochrome fluoroscein isothiocyanate (FITC) using the amine linkers, following standard conjugation protocol (Molecular Probes, 2003). After the conjugation, several rinses were made using sterile 0.9% saline and the particles were stored in saline suspension at $20\,^{\circ}$C until used.

1.1 MNP Synthesis.
Magnetite nanoparticles were synthesized using a modified procedure of Massart (Massart 1981; Philipse 1994). A 2 M iron (II) sulfate heptahydrate solution was prepared in 2 M HCl and combined with 1 M iron (III) chloride hexahydrate aqueous solution. The solutions were mixed and added to 0.7 M ammonium hydroxide with rapid stirring. The resulting gelatinous precipitate was stirred for 30 minutes then the precipitate was collected using a small magnet with removal of the supernatant. After several such washes, the precipitate was resuspended in 0.7 M ammonium hydroxide, and peptized through the addition of 1 M tetramethylammonium hydroxide aliquots. The total volume of this suspension was then taken to 250 ml.

1.2 Silica Coating.
The magnetite particles were then coated with silica according procedure described elsewhere (Correa-Duarte 1998). Briefly, a suspension of magnetite nanoparticles was vigorously stirred and a 4 ml aliquot taken up to a volume of 100 ml with distilled water. A solution of 0.54% sodium silicate was prepared at pH 10.5, and 4 ml was added to the previously prepared magnetite nanoparticles suspension. The suspension pH was adjusted to 10.0 and stirred for 2 hours and then allowed to stand for 4 days, after which the excess silica was removed by washing several times with distilled water by the aid of a magnet.

1.3 Amine Linkers.
To make the silica coated core-shell MNP biocompatible and suitable for our biological application, amine groups were attached to the surface. Particles were treated with 3-aminopropyl trimethoxy silane then a 1 ml aliquot of

the suspension was brought to a volume of 5 ml with distilled water and sufficient 3-aminopropyl trimethoxy silane added to give a final concentration of 5% (Wang 2001). The reaction system was occasionally stirred at room temperature for an hour. After the incubation period, the particles were washed with distilled water with the aid of a magnet. The Kaiser assay was next used to confirm attachment of functional group onto the surface of the silica coated nanoparticles (Kaiser 1970).

1.4 MNP Characterization

The MNP crystal structure, morphology, size distribution, elemental and chemical composition were investigated using powder x-ray diffractometry (XRD) (Scintag X'TRA), transmission electron microscopy (TEM), and energy dispersive spectroscopy (EDS). Crystal structure (phase) and average MNP size found in a macroscopic sample can be determined using XRD techniques. For this work, crystallinity, phase and crystal size are important properties because they affect the magnetic susceptibility of the particles. TEM analysis allows individual particles to be directly imaged (Figure 1). These images are used to determine the specific morphology, chemical composition, as well as size and size distribution of MNP. High-resolution TEM (HRTEM) and electron diffraction can show whether an individual MNP is a single crystal or poly-crystalline. Of importance here, different materials

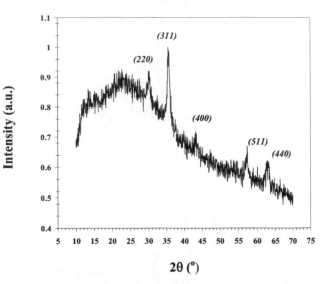

Figure 2 XRD pattern of uncoated iron oxide MNPs, indexing corresponds to Fe3O4 magnetite.

such SiO_2 (silica coating) iron oxide (magnetic core) can be differentiated even on single particles, indirectly through electron diffraction and directly through Z-contrast imaging. Thus under the appropriate imaging conditions silica-coated iron-oxide NPs can be differentiated from non-coated NPs. The elemental constituents can also be unambiguously determined using EDS (typically with resolution larger than the electron beam size itself). In EDS analysis the electron beam interacts with the sample and excites atomic core-shell (e.g. K-shell) x-rays. The peaks in the collected x-ray spectra are characteristic of the elemental constituents of the nanoparticles. The X-ray diffraction spectra (Figure 2) were taken from 10° to 70° (2θ values) using Cu Kα radiation. TEM experiments, including selected area electron diffraction (SAED), HRTEM and EDS, were all performed on a JEOL 2000FX operated at 200 kV equipped with a KEVEX EDS system. Initial assessment of the nature and extent of silica surface treatments was tested. Particle samples that ranged in silica content, based on reaction feed

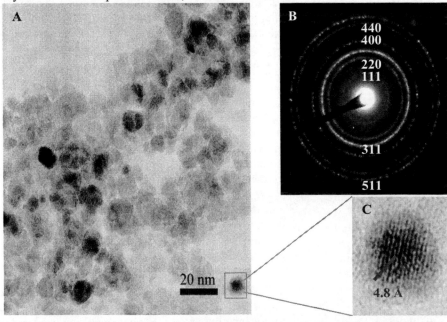

Figure 1 (A) TEM image of uncoated magnetite MNP. (B) SAED pattern from large area of NPs. (C) Zoom-in showing HREM image of an individual MNP as marked in (A).

ratio and time conditions, were dispersed in 0.9% saline and the settling rates were observed as a function of time. Particles with higher ratios in the reaction feed will have the longest settling rates. Hermeticity of the silica coating and nascent resistance to corrosion of uncoated magnetite nanoparticles was tested by soaking both uncoated and silica-coated nanoparticles in 10% NaCl solution for 2 days at room temperature and then 2 days at 40^0C (total of 4 days). The presence of Fe ions was assayed periodically during the 48 hrs. using a standard sodium thiocyanate (NaSCN) test. Long-term tests for corrosion, up to 6 months, were also performed on the stock solution of the MNP, an unprecipitated ferrofluid in tetramethyl ammonium hydroxide. Additionally, precipitated MNP that were placed in aqueous solution, double distilled water at pH = 13, were tested after 5 months of soaking using the sodium thiocyanate test for the Prussian Blue Reaction.

1.5 Cellular Uptake. The MNP, suspended in sterile physiological saline at pH = 7.4 to a flowing paste consistency, were sonicated (Sonicor, Copaigue, NY) for 2-3 minutes before placement pigs. Sonication was used to resuspend particles before placement onto tissues. Next, a 25 µl drop of the MNP suspension was delivered to the epithelial surface whereupon the drop flowed onto the target tissue. This was repeated for an approximate application volume of 50-75 µl, essentially bathing the incus or tympanic membrane with the suspension. During recovery from anesthesia each animal was placed onto the pole face of a 4 x 4 x 4 in. permanent magnet (NdFeBo, 50 MGO) while maintaining the surgical position of experimental ear facing upward. Thus, the upward surface of the incus or tympanic membrane with MNP held by surface tension, then exposed to a magnetic field pulling the MNP downward and into the epithelia. The experimental incus and tympanic membrane at distances of 1 in. from the pole face of the magnet were exposed to a magnetic field of approximately 0.35 Tesla. Each animal was exposed to this external magnetic field for 20-30 minutes, during recovery from anesthesia. They were subsequently returned to their cages for 1-15 days of monitored survival.

1.6 Histology. The experimental incus from the middle ear of each guinea pig was surgically removed and placed in 80% ethanol, 10% normal saline for a minimum of 24 hours. Next the incii were decalcified (Decalcification Solution™, Richard Allen Scientific, Kalamazoo, MI) for 3 days and subsequently automatically processed for paraffin

sectioning (Tissue-Tek VIP™, Sakura Finetek, Torrance, CA) with the exclusion of aldehydes to prevent autofluorescence. Transverse 5 µm microtome sections of the incii or tympanic membrane in paraffin blocks were mounted and either stained using hematoxylin and eosin for histopathology or unstained for confocal laser microscopy. Unstained transverse sections were examined using an Argon laser scanning Spectral Confocal and Multiphoton Microscope (Leica Model TCS SP2, Leica Microsystems, Mannheim, Germany). Stacks of 10-20 scans were made through the epithelium lining the incii, or the tympanic membrane epithelium while looking for intracellular fluorescence, indicating the presence of MNP (Figure 3).

Figure 3   100X confocal micrograph of guinea pig tympanic membrane implanted for 8 days with FITC-labeled MNP

1.7   Laser   Doppler   interferometry. Biomechanical confirmation of tissue implantation was made using laser interferometry. Immediately following euthanasia, an animal was placed in a lateral recumbent position with the implanted ear up. The pinna was removed so as to provide access to the bony ear canal and operating microscope visualization of the tympanic membrane. Next, a 1 x 1 mm piece of reflective tape (3-M, Minneapolis, MN) was placed near the umbo of the tympanic membrane and displacements measured using a single point, helium neon laser Doppler interferometer (LDI, Model OFV 501 and Model 3000 Controller, Polytec PI, Tustin, CA) caused by an external magnetic field interacting with the implanted MNP.   The LDI provided contactless velocity information with a frequency range of 0-150 Hz, with a fringe counter converting velocity into displacement (Gan 1997). The electromagnetic coil from the *Integrated Sound Processor* of a semi-implantable hearing device (SOUNDTEC, *direct system*^{TM}, Oklahoma City, OK) was positioned 1-2 mm from the surface of the tympanic membrane implanted with MNP (Hough 2001). When the coil was activated with a 1000 Hz sine wave at 5-8 volts, peak to peak (Model 80 Function Generator, Wavtek, San Diego, CA and Model 2706 Precision Amplifier, Bruel & Kjaer, Denmark) the interferometer recorded acceleration and calcuated displacements. In two of the animals, interferometry was used to measure displacements of implanted incii. The surgical wound was exposed, accessing the middle ear cavity, and the tip of the electromagnetic coil placed 1-2 mm from the surface of the implanted incus. The same electromagnetic signal was presented and displacements recorded from reflective tape on the lateral surface of the incus. The ossicular chain and tympanic membrane were in tact in all three of these animals; therefore the displacements recorded were movements of the whole middle ear, ossicular chain.

CONCLUSION

We have shown that silica encapsulated, magnetite nanoparticles can be used in conjunction with an external magnetic field to produce a biomechanical force in tissues. MNP, customized for ease of internalization and long-term biocompatibility, can be implanted in epithelia without producing necrosis or apoptosis and remain viable for at least 8 days. Auditory frequencies (subthreshold) have been generated in the in tact ossicular chain of the guinea pig using an electromagnetic coil of the type used in implantable hearing devices.

AKNOWLEDGEMENTS

We gratefully recognize Kevin Lewelling, Ph.D., Guoda Lian, Ph.D., and Matthew Johnson, Ph.D. Department of Physics & Astronomy, Oklahoma University, for their characterization of the nanoparticles using electron microscopy and X-ray diffractometry.

REFERENCES

[1] Aliev, F. G., Correa-Duarte, M.A., Mamedov, A., Ostrander, J.W., Giersig, M., Liz-Martin, L.M., Kotov, N.A. (1999). "Layer-by-layer assembly of core-shell magnetite nanoparticles; Effect of silica coating on interparticle interactions and magnetic properties." Adv. Materials **11**(12): 1006-1010.

[2] Austin, D. F. (1994). "Acoustic mechanisms in middle ear sound transfer." Otolaryngologic Clin N America **27**(4): 641-654.

[3] Correa-Duarte, M. A., Giersig, M., Kotov, N. A., and Liz-Marzan, L. M. Langmuir **14**: 6430-6435, 1998.

[4] Gan, R. Z., Wood, M.W., Ball, G. R., Dietz, T.G., Dormer, K.J. "Implantable hearing device performance measured by laser Doppler interferometry." Ear, Nose & Throat J **75**(5): 297-309, 1997.

[5] Hough, J. V. D., Dyer Jr., R.K., Matthews, P., Wood, M.W. "Early clinical results: SOUNDTEC implantable hearing device phase II study." Laryngoscope **111**: 1-8, 2001.

[6] Kaiser, E. T., Colescott, R. L., Bossinger, C. D., and Cook, P. I. Analytical Biochem **34**: 595-598, 1970.

[7] Mamedov, A., Kotov. N.A. (2000). "Free-standing layer by layer assembled films of magnetite nanoparticles." Langmuir **16**: 5530-5533.

[8] Massart, R. "Preparation of aqueous magnetic liquids in alkaline and acidic media." Transactions on Magnetics **Mag-17**(2): 1247-1249, 1981.

[9] Philipse, A. P., Van Bruggen, M. P. B., and Pathmamanoharan, C. "Magnetic silica dispersions: preparation and stability of surface modified silica particles with a magnetic core." Langmuir **10**(1): 92-99, 1994.

[10] Wang, N., Naruse, K., Stamenovic, D., Fredberg, J.J., Mijailovich S.M., Tolic-Norrelykke, I.M., Polte, T., Mannix, R., Ingber, D.E. "Mechanical behavior in living cells consistent with the tensegrity model." Proc Natl Acad Sci **98**: 7765-7770, 2001.

High Throughput *in-silico* Screening against Flexible Protein Receptors

H. Merlitz and W.Wenzel

Forschungszentrum Karlsruhe, Institute for Nanotechnology
P.O. Box 2640, D-76021 Karlsruhe, Germany
email: wenzel@int.fzk.de

ABSTRACT

We report results for the in-silico screening of a database of 10000 flexible compounds against various crystal structures of the thymidine kinase receptor complexed with 10 known inhibitors. The ligands were docked with the stochastic tunneling method using a first-principle based scoring function. Using a rigid receptor we find large deviations in the rank of the known inhibitors depending on the choice of receptor conformation. We then performed a screen in which critical receptor sidechains were permitted to change conformation and found improved scores for those inhibitors that did not dock well in any of the previous screens. These data demonstrate that the failure to dock did not originate from deficiencies in the scoring function or the docking algorithm, but from the neglect of receptor degrees of freedom.

Keywords: receptor ligand docking, in-silico screening, drug development, stochastic optimization, receptor flexibility

1 INTRODUCTION

Virtual screening of chemical databases to targets of known three-dimensional structure is developing into an increasingly reliable method for finding new lead candidates in drug development [1], [2]. Both better scoring functions [3] and novel docking strategies [4] contribute to this trend, although no completely satisfying approach has been established yet [5]. This is not surprising since the approximations which are needed to achieve a reasonable screening rates impose significant restrictions on the virtual representation of the physical system. Relaxation of these restrictions, such as permitting ligand or receptor flexibility, potentially increase the reliability of the scoring process, but come at a high computational cost.

While ligand flexibility is now routinely considered in many atomistic in-silico screening methods, accounting for receptor flexibility still poses significant challenges [6]–[8]. Using the thymidine kinase inhibitors as a prototypical example we document the shortcoming of rigid receptor screens in a realistic system. We then present screens with the stochastic tunneling method against a subset of up to 10000 ligands of the NCI-Open database considering increasing receptors sidechain flexibility and demonstrate an increased reliability of the screening results.

2 METHOD

We docked the ligands using the stochastic tunneling method (STUN) [9] with flexible ligands (free rotatable bonds). This method was shown to be superior to other competing stochastic optimization methods[10] and had performed adequately in a screening of 10000 ligands to the active site of dihydrofolate reductase (pdb code 4dfr [11]), where the known inhibitor (methotrexate) emerged as the top scoring ligand [12].

The stochastic tunneling technique was proposed as a generic global optimization method for complex rugged potential energy surfaces(PES). In STUN the dynamical process explores not the original, but a transformed PES, which dynamically adapts and simplifies during the simulation. For the simulations reported here we replace the original transformation [9] with:

$$E_{STUN} = \ln\left(x + \sqrt{x^2 + 1}\right) \qquad (1)$$

where $x = \gamma(E - E_0)$. E is the energy of the present conformation and E_0 the best energy found so far. The problem-dependent transformation parameter γ controls the steepness of the transformation [9]. The general idea of this approach is to flatten the potential energy surface in all regions that lie significantly above the best estimate for the minimal energy (E_0). Even at low temperatures the dynamics of the system becomes diffusive at energies $E \gg E_0$ independent of the relative energy differences of the high-energy conformations involved. The dynamics of the conformation on the untransformed PES then appears to "tunnel" through energy barriers of arbitrary height, while low metastable conformations are still well resolved. Applied to receptor-ligand docking this mechanism ensures that the ligand can reorient through sterically forbidden regions in the receptor pocket.

We employed the following simple, first-principle scoring function:

$$S = \sum_{Protein}\sum_{Ligand}\left(\frac{R_{ij}}{r_{ij}^{12}} - \frac{A_{ij}}{r_{ij}^6} + \frac{q_i q_j}{r_{ij}}\right) +$$

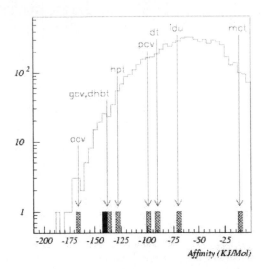

Figure 1: Affinities of the 5353 docked ligands (see text) to the receptor conformation complexed with deoxythymidine (pdb-code: 1ki2).

$$\sum_{h-bonds} \cos \Theta_{ij} \left(\frac{\tilde{R}_{ij}}{r_{ij}^{12}} - \frac{\tilde{A}_{ij}}{r_{ij}^{10}} \right), \qquad (2)$$

which contains the empirical Pauli repulsion (first term), the Van de Waals attraction (second term), the electrostatic potential (third term) and an angular dependent hydrogen bond potential (term four and five). The Lennard-Jones parameters R_{ij} and A_{ij} were taken from OPLSAA [13], the partial charges q_i were computed with InsightII and esff force field, and the hydrogen bond parameters $\tilde{R}_{ij}$, $\tilde{A}_{ij}$ were taken from AutoDock [14]. This force field lacks solvation terms to model entropic or hydrophobic contributions. The omission of such terms has been argued to be appropriate for constricted receptor pockets in which all ligands with high affinity displace essentially all water molecules. Each screen was repeated 6 times to reduce the influence of statistical fluctuations and the best affinity was used for ranking the ligands.

3 RESULTS

3.1 Rigid Receptor Screens

We investigated the degree of database enrichment of 10000 compounds, randomly chosen from the nciopen3D database [15], and 10 known inhibitors when docked to the X-ray TK receptor structure, which was experimentally determined in complex with one of the inhibitors, dt (deoxythymidine, pdb entry 1ki2 [16]). In this screen 5353 ligands attained a stable conformation with negative affinity within the receptor pocket.

Table 1: Ranking of the TK inhibitors in a screen of 10000 randomly chosen ligands of the nciopen3D database. The top row designates the crystal structure of the receptor that was used in the screen. (nd = not docked)

Inhibitor	1kim	1ki2	1ki3	1e2h	flex
acv	719	9	22	2048	199
ahiu	nd[c]	nd	nd	nd	2673
dhbt	4	104	118	38	13
dt	5	1310	2576	2779	681
gcv	3351	78	15	4516	57
hmtt	nd	nd	nd	nd	656
hpt	6	152	266	36	148
idu	515	2436	3272	2913	1365
mct	nd	6074	nd	nd	247
pcv	4845	952	4	4739	1656
Score:	3751	3705	4575	1926	4999

Figure1 shows the number of ligands as a function of affinity and highlights the rank of the known TK inhibitors in the screen. Three structurally similar inhibitors, including the ligand associated with the receptor conformation, are ranked with very high affinity. This result demonstrates that docking method and scoring function are adequate to approximate the affinity of these ligands to the receptor. Four further ligands (idu, acv,gvc, pcv, for a detailed description of TK and its inhibitors we refer to [5]) docked badly, three further ligands did not dock at all according to the criteria above. Repeating the docking simulations for these ligands did not substantially improve their rank in the database, eliminating inaccuracies of the docking algorithm as the source for this failure.

The resulting ranks of this screen are summarized in Table 1 (second column), which displays the rankings of the 10 inhibitors. Three were ranked within the first 1%, 6 were ranked among the first 10% of the database, respectively. This enrichment rate is comparable to the results of other scoring functions that were previously investigated for this system, but the overall performance is disappointing [5].

Inspection of the crystal structures of the different receptor-ligand complexes reveal differences in the conformation of some side groups inside the receptor pocket, depending on the docked inhibitor. This is a well known fact, but it is often assumed that the impact of these conformational variations on the ranking accuracy is moderate. We therefore repeated the screening with the X-ray structure of TK in complex with the inhibitor gcv (ganciclovir, pdb entry: 1ki2 [16]), which had scored particularly bad in the original screen. The results are shown in Table 1 (third column). Now, gcv was ranked

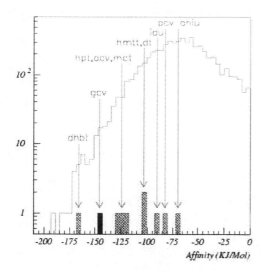

Figure 2: Histogram of the affinities of the docked ligands in the flexible receptor screen.

within the leading 1% of the database, but dt, formerly ranked on position 5, dropped to 1310.

For comparison purposes, we also performed a screen of the ligand free X-ray structure of TK (pdb entry: 1e2h [17]), which would most likely be used in a screen if no inhibitor was known. In this screen the receptor is unbiased to any of the inhibitors, which results in a dramatic loss of screening performance. As shown in column 7 only two ligands scored reasonably well (within the upper 10% of the database), all others would be discarded by any rational criterion as possible lead candidates.

3.2 Flexible Receptor Screens

Next we performed a flexible receptor screen against the same database. We identified the critical amino acid side chains and introduced 23 receptor degrees of freedom into the structure 1ki2, i.e. the dihedral rotations of the amino acids His13(2), Gln76(3), Arg173(4), Glu176(4), Tyr52(3), Tyr123(3) and Glu34(4). The numbers in brackets indicate the degrees of freedom for each sidechain. Each step in the stochastic search now consisted of an additional random rotation for *each* receptor degree of freedom. The results of this screen are summarized in Fig. 2, the scores of the individual inhibitors listed in the column labeled 'free' in the table. The figure demonstrates that in contrast to all rigid receptor screens now all inhibitors dock to the receptor. As expected, the number of false positives also increases, because a flexible conformation of the receptor reduces the bias of the screen against the known inhibitors. It must be noted that the accuracy of the flexible receptor

screen is lower than that of the rigid receptor screens (with the same number of function evaluations) because the number of degrees of freedom has increased. The increased fluctuations in the flexible screen can be best seen for acv, where the optimization method failed to locate the global optimum of the affinity (as independently obtained in a longer scree for just this ligand). We are presently developing algorithms to only selectively move the sidechains to reduce the computational effort in the flexible receptor screen. Even though a larger computational cost will be incurred in flexible receptor screens.

3.3 Comparison

To quantitatively compare different screens against the same ligand database, which used different receptor geometries, scoring functions or docking methods, it is sensible to assign an overall score to each screen which rates its performance [18]. We computed such a "score" for the entire screen from the ranks of the docked known inhibitors among the $N = 1000$ best ligands. This score is computed as the sum of $N - P$ where P is the rank of the known inhibitor and shown in the bottom row of table 1. An inhibitor ranking in the top of the screen contributes a score of 1000 to the sum, an badly ranked inhibitor comparatively little. Because only the best N inhibitors are evaluated, screens which dock many known inhibitors with moderate rank may have comparable scores with screens which perform perfectly for one inhibitor, but fail for all others. For the rigid receptor screens performed here the scores for the entire screen ranged from between 1926 for the screen against 1e2h, the ligand free X-ray structure of TK, to 4575 (1ki3, X-ray structure of TK in complex with pcv), which was arguably the best performing screen of all receptor conformations. Despite the increase of the number false positives the overall score of the flexible receptor screen (4999) was better than that of any rigid receptor screens.

4 CONCLUSIONS

Our results offer a good demonstration that the ranking of known inhibitors can strongly depend on the particular receptor structure used for the screen. The differences in affinity and rank which we find for a given ligand in different receptor conformations are of the same order of magnitude as the affinity of the best ligands. As a result, any given high affinity ligand can rank either at the very top of the database or somewhere in its tail. Our data demonstrates that this variability in rank is not, in general, a shortcoming of either scoring function nor docking methodology. If the three-dimensional structure of the receptor is suitable for a single ligand only but inaccurate for others, a overall scoring of the entire database remains inaccurate, regardless of the quality of the scoring function. As a consequence, differences in the enrichment ratio for different scoring functions may depend more on the suitability of the receptor conformation and environment than on the quality of the scoring function.

Regarding the evaluation of scoring methodologies and the validation scoring functions the results of the screen of the unbiased ligand-free receptor structure are particularly disappointing. The poor ranking of the known ligands in this screen indicates that high enrichment rates for rigid receptor screens against a receptor conformation complexed with a known ligand can be fortuitous. The high ranking obtained for known inhibitors (such as in column 2 of Table 1) is a result of the restriction of the search space which is particularly favorable for the known inhibitors. In the absence of such a restriction (column 6 of the table), the enrichment rate for the same scoring functions drops dramatically. As a consequence, a good enrichment rate in a rigid receptor screen does not necessarily validate a scoring function even for the system under consideration.

These findings suggest the importance of a flexible binding pocket to obtain a better unbiased scoring of high-affinity ligands. The results of the flexible receptor screen reported here suggest that much better accuracy of the scoring process can be achieved when receptor flexibility is considered.

5 SUMMARY

The investigation of the TK inhibitor family furnished a clear example for the impact of receptor conformation in rigid receptor screens. Docking databases including a known ligand into the receptor conformations of this ligand induces a bias into the screening procedure which tends to overestimate the accuracy of the screening methodology. As evidenced in the study here, the use of an unbiased receptor conformation significantly reduces the overall screening performance of the score. Ultimately only the routine use of accurate scoring techniques for flexible receptors will ameliorate this problem. The results presented here that such screens will become feasible with present day computational resources in the near future.

ACKNOWLEDGMENTS

We thank the Fond der Chemischen Industrie, the BMBF, the Deutsche Forschungsgemeinschaft (grant WE 1863/11-1) and the Kurt Eberhard Bode Stiftung for financial support.

REFERENCES

[1] W. Walters, M. Stahl, and M. Murcko, Drug Discovery Today **3**, 160 (1998).

[2] J. Drews, Science **287**, 1960 (2000).

[3] G. Klebe and H. Gohlke, Angew. Chemie (Intl. Ed.) **41**, 2644 (2002).

[4] G. Schneider and H. Boehm, Drug Discovery Today **7**, 64 (2003).

[5] C. Bissantz, G. Folkerts, and D. Rognan, J. Med. Chem. **43**, 4759 (2000).

[6] V. Schnecke and L. Kuhn, Persp. Drug. Des. Discovery **20**, 171 (2000).

[7] H. Claußen, C. Buning, M. Rarey, and T. Lengbauer, J. Mol. Biol. **308**, 377 (2001).

[8] F. Osterberg, Proteins **46**, 34 (2002).

[9] W. Wenzel and K. Hamacher, Phys. Rev. Lett. **82**, 3003 (1999).

[10] H. Merlitz and W. Wenzel, Chem. Phys. Lett. **362**, 271 (2002).

[11] J. Bolin, D. Filman, A. Matthews, R. Hamlin, and J. Kraut, J. Biol. Chem. **257**, 13650 (1982).

[12] H. Merlitz, B. Burghardt, and W. Wenzel, Chem. Phys. Lett. **370**, 68 (2003).

[13] W. Jorgensen and N. McDonald, J. Mol. Struct. **424**, 145 (1997).

[14] G. Morris, D. Goodsell, R. H. andR. Huey, W. Hart, R. Belew, and A. Olson, J. Comput. Chem. **19**, 1639 (1998).

[15] G. Milne, M. Nicklaus, J. Driscoll, and a. D. Z. S. Wang, J. Chem. Inf. Comput. Sci. **34**, 1219 (1994).

[16] J. Champness, M. Bennett, F. Wien, R. Visse, W. Summers, P. Herdewijn, E. de Clerq, T. Ostrowski, and R. J. M. Sanderson, Proteins **32**, 350 (1998).

[17] J. Vogt, R. Perozzo, A. Pautsch, A. Prota, P. Schelling, B. Pilger, G. Folkerts, L. Scapozza, and G. Schulz, Proteins **42** (2000).

[18] R. Knegtel and M. Wagnet, Proteins **37** (1999).

Overlapping Bio and Nano Technologies

Srinivas Iyer

Bioscience Division, Los Alamos National Laboratory, Los Alamos, NM 87545
siyer@lanl.gov

Abstract

"What is the connection between nanoscience and biology?" "What are biologists doing at Nanotech 2004 and why do you have so many "bio" related sessions?" "So biologists are also jumping on the nanoscience bandwagon." These are real questions/comments that are brought up even today. In spite of well established, classic hybrid fields such as biochemistry or biophysics, biology has long been perceived as an isolated science. This was clearly evident during the formative stages of Nanotech (erstwhile MSM/ICCN) conference series which brings together cross-disciplinary scientists to provide a forum for a free-flowing exchange of ideas. A lot has changed since then and rapid strides have been made towards merging bio- and nanotechnologies. The nano-bio association is a two-way street. Modern biology is heavily dependent on technological innovations that have improved analytical processes and, in turn, biological molecules and processes have inspired the synthesis of nanomaterials with novel properties and functionality. The purpose of this brief overview is to highlight a few hybrid topics that could reinforce the belief that nano-biotechnology as a field is here to stay and that its benefits will be clearly visible to the public much sooner than was predicted.

Keywords: biotechnology, nanoscience, protein, materials, interfaces

Nano to Bio?

Modern biology is rapidly moving towards systems level studies. Current challenges include understanding structure-function relationships, generating networks of signals that underlie the response to stimuli and developing crosscuts to other disciplines. The sequencing of a large number of complete genomes has led to an explosion in proteomic projects and new insights are being gained into disease processes. Several major advances that we have seen in the past couple of decades are largely due to increasingly sophisticated analytical methods to probe complex systems. A key component of these methods is the scale at which they operate. Most biological molecules and processes fall into the nano-regime in terms of size and time. Being able to view or manipulate these processes at this scale is a tremendous advantage and nanotechnology offers several attractive options. While many definitions for nanotechnology exist, the National Nanotechnology Initiative [1] calls it "nanotechnology" only if it involves all of the following: a) Research and technology development at the atomic, molecular or macromolecular levels, on the scale of approximately 1 - 100 nm range, b) Creating and using structures, devices and systems that have novel properties and functions because of their small and/or intermediate size, and c) Ability to control or manipulate on the atomic scale.

The National Institutes of Health (NIH) has identified "Imaging biological processes and the effects of disease" as one of the leading opportunities for nanoscience in biomedical research. This is not surprising considering the variety of applications that have benefited from novel imaging tools. Multiple color fluorescence imaging spectroscopy using dyes with overlapping spectra has been used to follow pH-sensitive liposomes within cells [2]. Chen *et al.* [3, 4] used two-photon excitation fluorescence resonance energy transfer (FRET) microscopy to image protein interactions in cells and tissues *in vivo*. The dynamics of protein association and distance measurements can be made by examining donor fluorophore lifetime [4] in the presence of acceptor. On the structural front, protein folding dynamics is critical to understanding its function and ameliorating debilitating conditions such as Alzheimer's disease that are caused by protein misfolding. Laser induced temperature-jump relaxation techniques and time-resolved infrared and fluorescence spectroscopies have been used to probe these dynamics in peptides [5] and small proteins. To do such studies however, it is important to be able to provide biologically relevant experimental conditions. 100 nm sized lipid vesicles have been tethered [6] using biotin-streptavidin chemistry to a supported lipid bilayer and proteins were incorporated in the vesicles to obtain an average 1:1 ratio. Of all biological molecules, DNA is perhaps the most widely studied using imaging tools at the single molecule level. Kim *et al.* [7] have demonstrated the use of scanning near-field optical microscopy to detect damaged bases in DNA at a scale of tens of nanometers. In an elegant experiment, single molecules of fluorescently labeled nucleotides were detected following exonuclease digestion [8]. DNA sequencing has also been achieved using electrochemical detection by passing single-strands through a nanopore and a solid-state nanopore microscope has been described [9] that can probe individual molecules. The application of nanopores in sensors for biomolecular characterization has led to enhanced techniques for fabrication of robust pores with nanometer precision [10].

The sequencing of several complete genomes, significantly the human genome, has revolutionized biology. The emphasis now is on understanding how the various components of a cell interact. Two major approaches stand out in their contributions to enable such studies. The first is mass spectrometry based proteomics, the large scale study of the proteins related to a genome. Proteins are the terminal effectors of the biological functions of genes and can be modified post-translationally, making their study a complex task. While advances in mass spectrometry methods and protein identification software have given protein scientists a great tool to work with, the power of this approach has truly been realized due to efficient, high-throughput upstream separation systems such as capillary electrophoresis (CE) and HPLC with nanoliter flow rates. Multi-dimensional protein identification technology or MudPIT [11] combines strong cation exchange with reversed phase chromatography of tryptic peptides in a nanoflow HPLC format that is coupled either with online nanospray mass spectrometry or with a MALDI-TOF plate spotting robot. CE on the other hand is being used more and more in a microfluidic chip format [12, 13] coupled with mass spectrometry directly. Laser-induced fluorescence is another detection method that is often coupled to CE for protein analysis. The liquid core optical waveguide properties of Teflon™ coated capillaries have been used to perform protein complex separations in fixed two-color laser format using zone electrophoresis [14] or for isoelectric focusing using a rastered laser [15] to image the whole separation. Separations of biomolecules are not typically thought of as a nanoscale science. This is surprising considering that sample volumes can be as low as picoliters, the molecules being analyzed are nanometers in size and

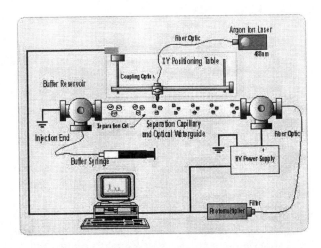

Fig 1: Schematic of a capillary electrophoresis laser-induced fluorescence imaging system. (courtesy of Dr. J.A. Olivares)

that technologies such as interferometric lithography are used to generate these devices. Researchers at the University of New Mexico in Albuquerque have fabricated integrated fluidic systems that allow facile manipulation of fluids and sample injection into nanochannel arrays [16]. The fabrication methods employed are very reproducible and have the potential to yield a new generation of economically viable devices for high throughput biomolecular separation. Most of the advances in microchip separations have their origins in techniques developed in the semiconductor industry. This is now a rapidly growing area of research where a significant amount of sample manipulation is conducted in micro fabricated devices [17] for improved sensitivity, throughput and reproducibility.

Bio to nano?

The transcription of biological designs and functions into robust synthetic systems is one of the key elements of the nanotechnology initiative and a significant research thrust of several agencies including the NIH, DOE etc. Biological systems have the ability to adapt to their environment by their inherent functional flexibility. The design of novel programmable or adaptive materials at the interface of biology and materials science is an exciting area of research with tremendous potential for growth.

Materials with controlled fluid and ionic transport properties are needed in diverse areas such as sensing, drug delivery, fuel cells, artificial photosynthesis and molecular electronics. Although the control of fluidic transport at the nanometer length scales is rarely achieved in synthetic material systems, many biological processes exhibit a remarkable control of both ionic and molecular transport using nanometer-sized pore-channels, coupled exquisitely to specific biomolecular functions. Every cell is enclosed by a plasma membrane that provides a barrier between the environment and the intracellular components. Transmembrane proteins span the lipid bilayer and hence have hydrophobic regions that penetrate the membrane and hydrophilic regions that are exposed to aqueous medium on either side. A significant class of transmembrane proteins form pores in the membrane to allow selective transport of charged species such as ions and macromolecules through the otherwise largely impermeable bilayer. Pore forming proteins display intricate control of transport through them, which is used in several biological systems for a variety of functions. Perhaps the most commonly used feature of such proteins is selective ion transport. This process serves to regulate the electrical potential across the membrane medium by facilitating an active, directional and preferential transport of ions. Channel proteins form water-filled pores across membranes and act as gatekeepers, allowing the flow of ions across the otherwise impermeable membrane. The transport process is not coupled to an energy source and hence is always along the electrochemical concentration gradient. An important feature of ion channels is their selectivity, permitting some ions to pass through but not others. In addition, ion channels are "gated", and their momentary opening and closing is closely regulated in response to specific stimuli. Such gated ion channels have been studied extensively and the synthesis of biomimetic pores in robust materials [18] forms a major area of current research.

While the applications of bio-inspired materials are many, there is a strong thrust by the NIH to develop biomedical applications for such materials. "Biomaterials and tissue engineering" is another of the key opportunities outlined by this agency. These materials may have applications in regenerative medicine due to easier acceptance by the body and maintenance of functionality in that environment. Lipid bilayer vesicles, alluded to earlier, are being developed as drug delivery vehicles [19] in a pH-sensitive format so as to gain control over release of their contents upon cellular internalization. The folate-derivatized liposomes were used to deliver cytosine-b-d-arabinofuranoside to oral cancer cells with elevated folate receptor expression Control over release or activation of specific agents will allow targeted drug delivery and reduce deleterious effects. Sensor development is another area with significant overlaps and of strong interest to several agencies, including the NIH. The integration of molecular recognition elements into portable widgets finds applications in the biothreat reduction world as well as for medical diagnostics. Atomic force microscopy in combination with a recognition based sensor has been reported for the detection of virus particles [20]. This is just one example of several projects that are involved in the development of biosensors (see [21] for an interesting article and a list of companies involved in nanobiotechnology).

Concluding remarks

Of the nine Grand Challenges posed by the NNI focusing on nanotechnology applications with highest potential impact, at least three are directly related to the biosciences. Perhaps the strongest support to this area of research is endorsement from the NIH, without doubt the leading authority in biomedical research. We have barely scratched the surface in terms of the huge potential that this field has to dramatically improve quality of life with advances in areas such as health and novel forms of energy production/conservation. The US Department of Energy is preparing for the nanotechnology revolution by establishing five Nanoscale Research Centers across the country. At least one of these user centers, CINT (Center for Integrated Nanotechnologies, www.cint.lanl.gov) has "Nano-Bio-Micro Interfaces" as one of its key thrust areas. The chief focus of this thrust area is to exploit the conceptual interface between biology and nanoscale materials science. Efforts such as these can only be fruitful with meaningful collaborations across disciplines.

Acknowledgements

In an overview this brief, omissions rather than inclusions are the norm. Hence, apologies are due to groups whose research in this area was not covered. The author acknowledges research support from LANL LDRD and DOE NSET.

References

1.   www.nano.gov, *National Nanotechnology Initiative*.

2.   Huth, U., et al., *Fourier transformed spectral bio-imaging for studying the intracellular.* Cytometry, 2004. **57A**(1): p. 10-21.

3.   Chen, Y., J. Mills, and A. Periasamy, *Protein localization in living cells and tissues using FRET and FLIM.* Differentiation, 2003. **71**(9-10): p. 528-541.

4.   Chen, Y. and A. Periasamy, *Characterization of two-photon excitation fluorescence lifetime imaging.* Microscopy research and technique, 2004. **63**(1): p. 72-80.

5.   Maness, S., et al., *Nanosecond temperature jump relaxation dynamics of cyclic beta-hairpin.* Biophysical journal, 2003. **84**(6): p. 3874-82.

6.   Boukobza, E., A. Sonnenfeld, and G. Haran, *Immobilization in surface-tethered lipid vesicles as a new tool for single biomolecule spectroscopy.* The journal of physical chemistry. B, Materials, surfaces, interfaces, 2001. **105**(48): p. 12165-12170.

7.   Kim, J., et al., *Near-field optical imaging of abasic sites on a single DNA molecule.* FEBS letters, 2003. **555**(3): p. 611-5.

8.   Werner, J., et al., *Progress towards single-molecule DNA sequencing: a one color.* Journal of biotechnology, 2003. **102**(1): p. 1-14.

9.   Li, J., et al., *DNA molecules and configurations in a solid-state nanopore microscope.* Nature materials, 2003. **2**(9): p. 611-5.

10.   Storm, A., et al., *Fabrication of solid-state nanopores with single-nanometre precision.* Nature materials, 2003. **2**(8): p. 537-40.

11.   Washburn, M.P., D. Wolters, and J.R. Yates, 3rd, *Large-scale analysis of the yeast proteome by multidimensional protein identification technology.* Nat Biotechnol, 2001. **19**(3): p. 242-7.

12.   Figeys, D., et al., *An integrated microfluidics-tandem mass spectrometry system for automated protein analysis.* Anal Chem, 1998. **70**(18): p. 3728-34.

13.   Figeys, D., et al., *Microfabricated device coupled with an electrospray ionization quadrupole time-of-flight mass spectrometer: protein identifications based on enhanced-resolution mass spectrometry and tandem mass spectrometry data.* Rapid Commun Mass Spectrom, 1998. **12**(20): p. 1435-44.

14.   Iyer, S., Stark, P., and Olivares, J., *Two-color Capillary electrophoresis with on-column excitation and wave-guide based fluorescent detection.* JALA, 2003. **8**(4): p. 41-45.

15.   Olivares, J.A., P.C. Stark, and P. Jackson, *Liquid core waveguide for full imaging of electrophoretic separations.* Anal Chem, 2002. **74**(9): p. 2008-13.

16.   O'Brien, M., et al., *Fabrication of an integrated nanofluidic chip using interferometric lithography.* Journal of vacuum science, 2003. **21**(6): p. 2941-5.

17.   Li, J., et al., *Integrated system for high-throughput protein identification using a microfabricated device coupled to capillary electrophoresis/nanoelectrospray mass spectrometry.* Proteomics, 2001. **1**(8): p. 975-86.

18.   Steinle, E., et al., *Ion channel mimetic micropore and nanotube membrane sensors.* Analytical chemistry, 2002. **74**(10): p. 2416-22.

19.   Sudimack, J.J., et al., *A novel pH-sensitive liposome formulation containing oleyl alcohol.* Biochim Biophys Acta, 2002. **1564**(1): p. 31-7.

20.   Nettikadan, S.R., *ViriChip: a solid phase assay for detection and identification of viruses by atomic force microscopy.* Nanotechnology, 2004. **15**(3): p. 383.

21.   Stikeman, A., *Nanobiotech makes the diagnosis.* Technology review, 2002. **105**(4): p. 60.

Detection of DNA hybridyzation with impedance amplifying labels

Kuo-Sheng Ma*, Hong Zhou*, Alessandra Fischetti**, Jim Zoval*, Marc Madou*

*Department of Mechanical and Aerospace Engineering
Henry Samueli School of Engineering
University of California, Irvine
**ST Microelectronics, Milan, Italy

ABSTRACT

Traditionally used nucleic acid analysis detection platforms rely on fluorescent or radioactive labels. Optical techniques require large and expensive equipment such as automated colorimetric and fluorimetric instruments that greatly increase the cost of the whole set-up. Working with radioactive labels is an even less attractive option. Electrical measurements of DNA hybridization are expected to involve less complicated and less expensive instrumentation and feature detection limits similar to the traditional optical methods. The diagnostic technique we introduce here will enable low-cost molecular diagnostics methods to become a reality and broaden the extent of the nucleic acid analysis applications.

Keywords: DNA, streptavidin, alkaline phosphatase, electrochemical impedance spectroscopy, impedance

1. INTRODUCTION

Today, the most commonly used method for detection and quantification in nucleic acid analysis involves luminescent measurements using a variety of labels (i.e., fluorescent and/or chemiluminescent probes, photoproteins, etc.). Traditionally used nucleic acid detection techniques rely on fluorescent or radioactive labels [1]. Spectroscopic methods afford the necessary sensitivity but require large and expensive equipment such as automated colorimetric and fluorimetric instruments greatly increasing the cost of the whole detection system. Working with radioactive labels is an even less attractive option since the protocols require well-trained technicians. Electrical detection of oligonucleotide binding is expected to involve less complicated and less expensive instrumentation and have similar detection limits compared to the traditional optical methods.

The success of biosensors based on label-free or label-less impedance sensing has been the topic of debate for years [2][3][4]. The debate is on-going since the promise of a simple impedance detection scheme of this nature remains very attractive. Label-less impedance sensing of affinity binding on functionalized insulators, semiconductors, heterostructures (semiconductor/ dielectric/ electrolyte) and metals have been attempted [5]. While the feasibility of such electrochemical impedance measurements on insulators, functionalized heterostructures and semiconductors was demonstrated [6][7][8], more convincing data were obtained using a metal as support for the DNA hybridization [9]. When metal electrodes such as Pt and Au are used as the substrate for DNA hybridization they exhibited stronger signals (e.g., detection limit of about 10 ng/ml for Si/SiO_2 hetero structures vs. around 10 pg/ml for a gold electrode). A typical example of work in this area by Bergren [9], demonstrates the feasibility of a biosensor for direct detection of DNA hybridization on gold. It must be noted though that even here the specificity was not very high and the reproducibility of the sensors was poor. Martelet et al. [10] presented so far the most convincing antibody/antigen binding detection on gold functionalized electrodes using a differential impedance measurement. This group demonstrated the ability to detect an antibody/antigen bio-recognition event with a functionalized Pt electrode (0.3 mm in diameter) and a functionalized Au electrode (1.0 mm in diameter).

The aim of this work is to provide a more sensitive and more reproducible method for the detection of DNA hybridization by measuring electrical impedance changes upon DNA hybridization.

2. MATERIALS AND METHODS

2.1 Materials

All reagents were purchased from Aldrich Chemical and used as received, unless otherwise noted. All solutions were made up using deionized water (18 MΩcm resistivity). DNA oligonucleotides were purchased from BioSource Inc. (Camarillo, CA). The two different probes used in this work were T1 and T2, DNA sequences (25mers) with the 5' end thio-modified and with a 6-mercaptohexyl spacer. The single-stranded target (CT1) is a 25mer complementary to probe T1 and CT1' has the same sequence as CT1 but with a biotin label on the 3' end. The sequence and modification of each oligo-nucleotide is listed in Table 1.

Name	Labeled	Sequence
T1	5'-SH-(CH$_2$)$_6$-	5'-CACGACGTTGTAAAACGACGACCAG-3'
T2	5'-SH-(CH$_2$)$_6$-	5'-GATGCCCGGGTCCGGCTAGATGATC-3'
CT1	none	5'-CTGGTCGTCGTTTTACAACGTCGTG-3'
CT1'	3'-biotin	5'-CTGGTCGTCGTTTTACAACGTCGTG-3'

Table 1 Oligonucleotides Sequences and Modification

2.2 Electrode preparation

Single-crystal Si (001) covered with a 400 nm thick thermal silicon dioxide was used as the substrate material. Before metal deposition the substrates were cleaned using an RCA clean (5 parts deionized water, 1 part NH_4OH, 1 part H_2O_2). In order to promote adhesion of the gold to the silicon dioxide, a thin Cr adhesion layer (50 nm) was deposited by e-beam evaporation, followed by 200 nm of gold.

The measuring electrodes were patterned using photolithography. The lithography process is schematically illustrated in Figure 1. A thin layer of positive photoresist (Shipley 1827) was spun onto the gold coated substrate at 4000 rpm for 40 seconds, followed by a soft bake at 90 °C on a hot plate for 2 minutes. The photoresist was then UV exposed through a patterned iron oxide mask in a Karl-Suss aligner and subsequently developed in MF 319 developer. Au etchant and then Cr etchant were used to remove the unwanted metal areas. Finally, the remaining photoresist was removed by acetone. The dimensions of the working electrode and counter electrode are 200µm x 200µm.

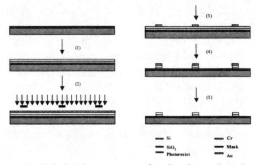

Figure 1. Fabrication process for the detection device

2.3 Self-assembled monolayer films preparation

Prior to the formation of thiolated-ssDNA self-assembled monolayers on the Au electrodes, the Au electrodes were cleaned by dipping into a 3% aqueous HF solution for 10 second followed by extensive rinsing with deionized water (18 MΩcm). Next the electrodes were blown dry using pure nitrogen gas and immediately transferred into a solution of 1µM of oligonucleotide probes in a potassium phosphate buffer (0.5M, pH=7). For the thiolated-ssDNA self-assembled monolayers to form we allowed for 15 hours or more. Finally, the surfaces were rinsed with deionized water for 5 s and dried under a stream of pure nitrogen.

2.4 Procedures

All electrochemical characterizations were performed with an Electrochemical Impedance Spectroscopy Potentiostat from Gamry Inc. (Warminster PA). A probe station (Micromanipulator Inc. Carson City, NV) was used to contact the contact pads of the counter and working electrodes. The solution we used for all the reported impedance measurements is a 100mM NaCl in 1X TE buffer (10mM Tris, pH 7.4 and 1mM EDTA). The bare gold impedance values were measured in this solution after patterning and cleaning. The ssDNA self-assembled monolayer impedance values were measured in the same solution after immobilization of ssDNA onto the electrode surface and after cleaning. Hybridization was performed by spotting the CT1 or CT1' solutions (1µM DNA in 1X TE buffer plus 1M NaCl) on the electrode surfaces. After 4 hours, the electrodes were rinsed with impedance measurement buffer for 10 seconds and dried under a stream of pure nitrogen prior to characterization.

As a method for impedance change amplification upon DNA hybridiziation, we used the commercially available Enzyme-Labeled Fluorescence (ELF 97) signal detection technique (Molecular Probes, Eugene OR). This well-known technology involves the alkaline phosphatase enzyme converting the ELF substrate (ELF-97), a water soluble weak blue fluorescent molecule, to an insoluble, intense green fluorescent molecule. The insoluble product precipitates onto the surface of the solid phase used in the assay. We speculated that this type of a precipitation on an underlying conductor electrode would enhance/amplify the impedance change upon hybridization significantly. The

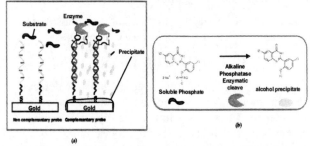

experimental procedure is schematically illustrated in Figure 2(a) and the chemical reaction involved is shown in Figure 2(b).

Figure 2 (a) Schematic of signal amplification using alkaline phosphatase as an enzyme label. (b) Chemical reaction involved in the amplification process.

3. RESULTS AND DISCUSSIONS

3.1 Theory of DNA hybridization detection with electrochemical impedance spectroscopy (EIS) measurements

When a conductor is placed in an electrolyte solution a potential is generated due to an unequal distribution of charges across the interface (Figure 3(a)). Two oppositely charged layers, one on the electrode surface and one inside the electrolyte form a "double layer," which can be modeled as a parallel plate capacitor. When an additional layer is present at the electrode, such as an oxide or a self-assembled monolayer (SAM), an additional capacitance and/or resistance associated with that layer is added to the circuit. In principle, the introduction of molecules, such as DNA, to that interfacial region, will affect the measured complex impedance through changes of the local geometry, the dielectric constant, and the amount of charge at the electrode/electrolyte interface.

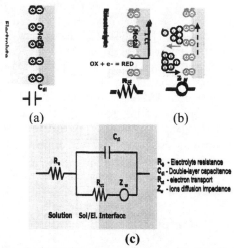

(a) (b)

(c)

Figure 3. Interface phenomena and Randles equivalent circuit. (a) Electrical double layer capacitance C_{dl}. (b) Charge leaking across the double layer represented by "charge transfer" resistance R_{ct} and diffusion of ions to the interface represented by Warburg impedance Z_w. (c) Randles' equivalent circuit.

A perfect biomolecule recognition layer, for example a self-assembled monolayer of single stranded DNA (a SAM of ssDNA), would cover the electrode completely and the selective binding of complementary biomolecules to that layer (i.e., hybridization in the current example) would be the only contributing impedance element. Figure 3(b) reveals possible competing processes, these include (1) a 'charge transfer' resistance R_{ct}, corresponding to charge leakage across the double layer (this could involve electrons and/or ions), given by

$$R_{ct} = RT / nFi_0 ,$$ [1]

Where R is the universal gas constant, T is the absolute temperature, n is the number of charges involved in the electrode reaction, F is Faraday's constant and i_0 is the exchange current density (electrons and/or ions); and (2) the Warburg impedance Z_w, corresponding to the diffusion of ions to the interface from the bulk of the electrolyte. The latter is given by

$$Z_\omega = (1 - j)\sigma\omega^{-1/2} ,$$ [2]

Where ω is the angular frequency (s^{-1}) and σ is the Warburg coefficient ($\Omega\,s^{-1/2}$). The equivalent circuit of such an interface is represented in Figure 3(c), and is known as the Randles equivalent circuit model, featuring all the above elements as well as a series resistance corresponding to the solution resistance (R_e). To maximize the device's selectivity to a bio-recognition event at the electrode

surface, the impedance variation must be dominated by the capacitance change and not by the electrolyte resistance, charge transfer or diffusion effects. To optimize the sensors' sensitivity one consequently works in a frequency regime where the capacitive contributions associated with the binding of biomolecules dominates. For example, in the case of an Au electrode covered with a SAM of ssDNA, at high frequency, the impedance is dominated by the electrolyte resistance R_e. At low frequency the impedance of the sensor exhibits phase and quadrature values that are proportional to $\omega^{-1/2}$, indicative of Warburg impedance control. Only measurements conducted at medium frequencies, with our electrode arrangement we found that to be around 100 Hz, are dominated by the double layer capacitance changes occurring as binding phenomena take place.

3.2 Optimized impedance measurements for DNA hybridization sensing

Based on the theory described in 3.1, the double layer capacitance will change upon surface modifications induced by binding of molecules into that double layer. We will use this property first to measure the double layer capacitance changes upon label-less DNA hybridization onto surface immobilized capture probes. Figure 4 shows the results of impedance measurements made on bare gold electrodes, on gold electrodes with specific and non-specific ssDNA capture probes, and on the same electrodes after hybridization. In the case of specific DNA hybridization, we see an impedance change of 10% at an optimum measuring frequency of 100 Hz . In the case of non-specific DNA hybridization no impedance change is registered. Thus, in principle, label-less DNA hybridization can be monitored using a single frequency impedance measurement.

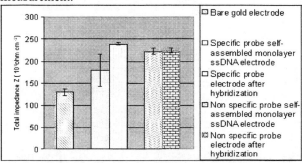

Fig 4. Total impedance measurements at 100Hz made on bare gold electrodes, specific and non-specific ssDNA capture probes on gold electrodes, and the same capture probe electrodes after complementary target strands were hybridized. The dimension of the electrode are 50X 50 um^2. (n=3, error bars = 1 standard deviation)

As pointed out in Section 3.1, in order to selectively monitor a bio-recognition process through its accompanying impedance change, the electrode must be perfectly polarizable. The impedance changes are expected to be larger and more reproducible if the bio-recognition layer is attached to the electrode surface in a site-specific manner and allows for an optimized interaction with the target DNA. The use of alkanethiol self-assembly methods to fabricate DNA probe-modified gold surfaces is known to form a good surface coverage that exhibits high hybridization activity. But even with this type of surface preparation, thio-labeled ssDNA adsorbs onto the gold surface, both via thiol-gold linkage and non-specific interactions, thus introducing undesirable random impedance variations. The impedance amplifying techniques introduced here, should, in principle, decrease the need for an ideally polarizable interface as the signal to noise ratio (S/N) is improved dramatically by the amplification of the impedance signal.

Impedance measurement results, comparing DNA hybridization with and without enzyme amplification are shown in Figure 5 (we worked again at a fixed frequency of 100 Hz). The impedance data for specific DNA hybridization without the enzyme label is used as a reference point here (normalized to 1). With amplification label, specific hybridization leads to an impedance change of a factor 1.6 (and factor of 4 when compared with bare gold !) . Clearly the insoluble precipitate from the enzyme reaction does amplify the impedance change after hybridization and enzymatic reaction. Similar experimental results have been presented by Willner [11], who also reported a 4 times increase in the double layer capacitance with enzyme amplification technique for the sensing of DNA hybridization.

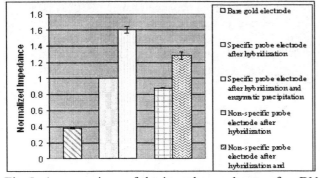

Fig 5. A comparison of the impedance change after DNA hybridization with and without an enzyme labeled. (n=3 error bars = 1 standard deviation)

However, for the non-specific DNA hybridization, the impedance also increased. The reason for that maybe because the streptavidin-alkaline phosphatase conjugate physically adsorbs on the capture probe which also generates some insoluble precipitate onto the electrode.

4. Conclusions

In this work, we have shown that label-less DNA hybridization can be detected using a single frequency impedance measurement. However, the difficulty reproducing such results makes this technique undesirable.

The irreproducibility is due to the difficulty in reproducing high quality DNA capture probe layers.

In the case of enzymatic impedance labels, the signal to noise ratio is improved and this could potentially overcome irreproducibility in SAM manufacture but results are still inconsistent. This may be a consequence of the precipitating film being deposited unevenly over the Au/thio-dsDNA electrode. The solution we have come up with and are currently exploring is to use an "Adjacent Impedance Probing" (AIP) technique. In this case, the DNA hybridization site is optimized for the biorecognition event (this site does not need an underlying conductor) and a bare adjacent electrode is optimized for generating the largest possible impedance change through precipitation or passivation. We believe the AIP technique will dramatically increase the sensitivity and reproducibility of the assay. The technique could lead to a more consistent and inexpensive method for impedance measurement of nucleic acid hybridization.

Reference

1. J. Dicesare, B. Grossman, E. Katz, E. Picozza, R. Ragusa and T. Woudenberg, "A High-Sensitivity Electrochemiluminescence-Based Detection System for Automated Pcr Product Quantitation". *Biotechniques*. 15(1). 152 (1993).
2. J. Janata, "Twenty Years of Ion Selective Field-Effect Transisitors. *Analyst*. 119(11). 2275 (1994).
3. J. Janata and G. F. Blackburn, "Immunochemical Potentiometric Sensors". Ann. NY Acad. Sci. 428 (Jun.). 286 (1984).
4. S. Collins and J. Janata, "A Critical Evaluation of the Mechanism of Potential Response of Antigen Polymer Membrances to the Corresponding Antiserum. *Anal. Chem. Acta*. 136 (Apr.). 93 (1982).
5. C. Berggren, B. Bjarnason and G. Johansson, "Capacitive Biosensors". *Electroanalysis*. 13 (3). 173 (2001).
6. E. Souteyrand, J. P. Cloarec, J. R. Martin, C. Wilson, I. Lawrence, S. Mikkelsen and M. F. Lawrence, "Direct Detection of the Hybridization of Synthetic Homo-Oligomer DNA Sequences by Field Effect". *J. Phys. Chem. B* 101. 2980 (1997).
7. J. P. Cloarec, J. R. Martin, C. Polychronakos, I. Lawrence, M. F. Lawrence and E. Souteyrand, "Functionalization of Si/SiO$_2$ substrates with homooligonucleotides for a DNA biosensor". *Sensors and Actuators B*. 58. 394 (1999).
8. E. Souteyrand, C. Chen, J. P. Cloarec, X. Nesme, P. Simonet, I. Navarro and J. R. Martin, "Comparison Between Electrochemical and Optoelectrochemical Impedance Measurements for Detection of DNA Hybridization". *Appl. Biochem. And Biotech.* 89. 195 (2000).
9. C. Berggren, P. Stalhandske, J. Brundell and G. Johansson, "A Feasibility Study of a Capacitive Biosensor for Direct Detection of DNA Hybridization". *Electroanal.* 11(3). 156 (1999).
10. Bockris J.O'M. Reddy A.K.N. Modern Electrochemistry Plenum Press, New York (1973)
11. F. Patolsky, E. Katz, A. Bardea and I. Willner, "Enzyme-linked amplified electrochemical sensing of oligonucleotide-DNA interactions by means of the precipitation of an insoluble product and using impedance spectroscopy", *Langmuir* **15**, 3703-3706 (1999).

Functional Nanoarrays for Protein Biomarker Profiling

M. Lynch[*], C. Mosher[*], J. Huff[*], S. Nettikadan[*], J. Xu[*], and E. Henderson[*‡]

[*]BioForce Nanosciences, Inc., Ames, IA, USA, mlynch@bioforcenano.com
[‡]Iowa State University, Dept. of Genetics, Developmental and Cellular Biology, Ames, IA, USA

ABSTRACT

The success of microarrays for analyzing nucleic acid and, more recently, protein-protein interactions illustrates the advantages of bioassay miniaturization. However, the current level of miniaturization of useful bioanalytical tools still creates limits in terms of breadth of application. For example, analyses of picoliter sample volumes containing protein analytes that cannot be amplified or replaced (e.g., cancer biopsy material) require unprecedented levels of miniaturization and novel analytical approaches. Novel platform technologies comprised of instrumentation, methods and software must be developed to achieve the goals of ultraminiaturized biodiagnostics. We will describe one such system and its real-world application to biomarker profiling.

Keywords: nanoarray, microarray, protein array, cytokine, AFM

1 INTRODUCTION

One of the greatest benefits that protein array [1] technologies can offer to the public is the potential for biomarker profiling of disease states [2,3]. This information may lead to earlier diagnosis and treatment for ailments that exhibit distinct patterns of protein expression. Monitoring the molecular basis of disease progression will ultimately lead to new therapeutic targets. The predominant protein profiling technology is based on printing microarrays of antibodies generated against specific known proteins. Lysates from the cell population of interest are typically labeled by conjugation with a fluorophore and incubated with the antibody array to reveal relative expression levels. One of the main drawbacks of this protocol is the differential labeling efficiency of each protein. Alternatively, the lysates can be incubated on the array without any chemical labeling step and then detected with a cocktail of antibodies, which adds an additional degree of specificity to each interaction.

Either antibody microarray based protein profiling method will suffer from a requirement for relatively large sample volumes. Ten years ago 50 µl would not have been considered a large volume, however advances in high-throughput screening and microfluidics have redefined our perceptions. Our solution to the problem is the miniaturization of the entire assay. A new contact-printing instrument, the NanoArrayer™ [4] (Figure 1) was devised

to print micron and sub-micron spots of each antibody. These nanoarrays can be created in virtually any pattern, as shown in Figure 2. The 1 µm diameter spots are resolved in the optical image because their spacing (pitch) is greater than 2 µm, the approximate resolving limit of far field optical systems. In this example a protein nanoarray was patterned to encode the nature of the array content, anti-interferon-gamma (IFN-γ). One of the applications we are developing with this technology is a multiplexed, nanoscale cytokine assay. Figure 5 illustrates the sensitivity and dynamic range of an early version. The array is resolvable by optical methods or by more novel approaches such as atomic force microscopy (AFM).

The applicability of nanoarrays to optical detection methods makes them immediately available to a large number of researchers. However, AFM analysis of biological nanoarrays offers unique advantages including the ability to operate in fluids, label-free detection, single molecule detection capability, and nanometer spatial resolution.

Our focus has been on the practical application of the emerging tools and concepts of nanotechnology to the development of ultraminiaturized bioanalytical methods and devices. A key question that remains is "how small is small enough?". Numerous reports have shown that with relatively simple readout methods such as AFM topography and/or force measurements it is possible to explore molecular interactions at the single molecule level.

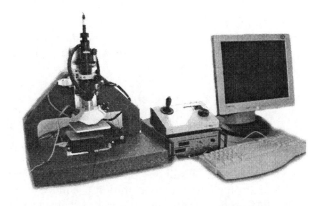

Figure 1. NanoArrayer III incorporates several novel features that create the most optimal NanoArraying platform available. Environmental chamber not shown.

However, the issue of data integrity must be addressed in these cases. As the assay size approaches the single-molecule level, the influence of molecular dynamics and statistics becomes more pronounced. Thus, in our current view, there is a "sweet spot" for ultraminiaturized bioanalysis that exploits the power of numbers (e.g., thousands of molecules per assay) while still reaping the benefits of ultraminiaturization.

Figure 2. Brightfield image of 1 μm diameter mouse anti-human IFN-γ antibody spots arrayed on a gold surface.

2 MATERIALS AND METHODS

2.1 Instrumentation

The results reported here were all obtained using the third generation of an instrument we have constructed and termed the NanoArrayer™. The underlying principle of the NanoArrayer™ is mechanically mediated direct transfer of materials on surfaces with high spatial precision, surface sensing capabilities and environmental control. Initial development of the first hand crafted NanoArrayer (NanoArrayer™ I) was promising and led to development of a second generation NanoArrayer (NanoArrayer™ II), which incorporated closed loop motion control, environmental controls, and surface sensing. A third version (NanoArrayer™ III) has since been developed and the full details of the NanoArrayer™ instrumentation development will be reported elsewhere [5]. Briefly, a microfabricated deposition tool is loaded by immersing the distal end in a drop of the protein solution. The surface to be patterned is positioned on a piezoelectric inchworm-driven XY stage with 20 nm resolution over 25 mm of travel. Manipulating the local humidity at the interface between the deposition tool and surface with gentle bursts of wet or dry air allows precise control of the molecular transfer rate. All stage movement and patterning parameters are controlled from within a custom software package called NanoWare™.

In the proof-of-principle experiments shown here the deposition tools used to physically place proteins on the substrates were 0.58 N/m DNP AFM probes from Digital Instruments. AFM probes were pretreated with UV and ozone for 30 min in a UV-TipCleaner (BioForce Nanosciences) to remove silicone oils and organic debris. While adequate for these demonstrations, commercially available AFM probes have proven to be less than ideal for macromolecular deposition. Therefore, non-AFM probe deposition tools for the NanoArrayer™ are currently under development [6]. These custom microfabricated silicon

dioxide surface patterning tools (SPTs) provide a significant increase in reproducibility and writing duration.

2.2 Surfaces

The nanoarrays presented here were created on 4 mm squares of diced silicon wafer (Montco Silicon Technologies, Inc.) that were first coated with 5 nm of chromium and 10 nm of gold by ion beam sputtering through an alphanumeric indexed electron microscopy specimen grid (Electron Microscopy Sciences). Gold-coated silicon wafers or "chips" were removed from the sputterer and used for arraying (Fig. 2, Fig. 3) or immediately immersed in 0.5 mM solutions of dithiobis-succinimidyl undecanoate (DSU) [7] in 1,4-dioxane. Following successive washes in dioxane and dry 100% ethanol, the chips were blown dry with argon and used for deposition (Fig. 4). The nanoarrays from which the plot in Figure 5 is derived were constructed on glass slides with a 3-D polymer matrix coating (Full Moon Biosystems).

2.3 Proteins

The antibody deposited in the nanoarrays in Figures 2-5 was a monoclonal mouse anti-human interferon-γ at 0.9 mg/ml in 10 mM Tris-HCl pH 8.0, 10 mM NaCl (Pierce Endogen). The Cy3-goat anti-mouse F(ab')₂ used in Figure 3 was obtained from Jackson ImmunoResearch Laboratories. The cytokine sandwich assays in Figure 4 and Figure 5 were performed using matched mouse monoclonal antibody pairs and recombinant proteins generously provided by Pierce Endogen. Our ELISA data from these pairs of antibodies (not shown) indicates high specificity and minimal cross-reactivity with other cytokine matched pairs.

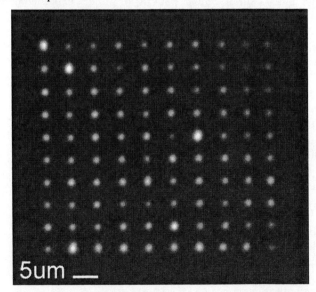

Figure 3. Image of mouse anti-human IFN-γ antibody on gold and labeled with Cy3-goat anti-mouse F(ab')₂.

2.4 Cytokine Profiling Nanoarrays

Overnight hydration of the mouse antibody arrayed chips at 25°C and 60% RH facilitated complete binding of the mouse IgG to the surface. The remaining reactive sites were blocked by incubation in ViriBlock™ (BioForce Nanosciences) for 30 min at 25°C, followed by a 5 min wash in PBST. Chips were incubated with recombinant human interferon-γ (Pierce Endogen) at the indicated concentrations in PBS + 4% IgG-free, protease-free BSA (Jackson ImmunoResearch Laboratories) with agitation for one hour at 25°C. Biotinylated mouse anti-IFN-γ detection antibody (Pierce Endogen) was added to the reaction to 250 ng/ml and incubated with agitation for another hour at 25°C. The chips were then transferred into 500 μl of Cy3-streptavidin (Jackson ImmunoResearch Laboratories) at 1.8 μg/ml in PBST and incubated with agitation for 30 min at 25°C. Three successive washes were carried out for 1 min each in 500 μl of 50 mM Tris pH 8.0 + 0.2% Tween-20 prior to image acquisition.

2.5 Detection

Nanoarrays that utilized fluorescent reporting molecules were visualized on a Nikon TE-2000U inverted microscope equipped with a HYQ Cy3 filter #41007a from Chroma Technology. Images were captured using a Cohu cooled CCD camera. Immediately prior to visualization, chips were briefly rinsed with a few drops of ddH₂O and inverted onto a drop of n-propyl gallate/glycerol anti-fade solution (5% n-propyl gallate, 70% glycerol, 25% 0.5M Tris-HCl pH 9.0, then adjusted to a final pH of 7.4). Fluorescent images were analyzed for spot size, intensity, and coefficient of variance with the Array Pro Analyzer software package (Media Cybernetics).

3 RESULTS

Antibody spots with diameters of 1 μm and estimated volumes of 30 attoliters were deposited on a gold surface with a high degree of precision using NanoArrayer™ III, as illustrated by the brightfield image in Figure 2. The NanoWare™ software interface allowed for creative array design in the form of text. The presence of mouse antibody in the transferred material was confirmed by labeling a 10 x 10 x 5 μm pitch array with Cy3-goat anti-mouse F(ab')₂ as shown in Figure 3 (negative controls not shown here). Incubation of a 5 x 5 x 5 μm antibody array with a biologically relevant concentration of 5 ng/ml IFN-γ in a sandwich style assay demonstrated the functionality of the antibodies (Figure 4). Specificity controls confirmed the ELISA cross reactivity data for the matched antibody pairs (not shown). An eight point standard curve was processed for IFN-γ nanoarrays to determine the sensitivity and dynamic range. The results plotted in Figure 5 suggest that current labeling strategies have sub-ng/ml detection limits and approximately 2.5 logs of dynamic range. The "hook

effect" displayed at higher concentrations of IFN-γ was caused by adding the detection antibody to the chip without first removing the excess IFN-γ.

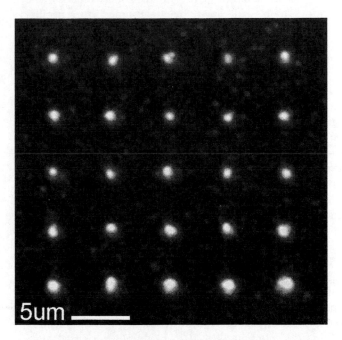

Figure 4. Fluorescent image of an antibody array exposed to 5 ng/ml IFN-γ. Detection provided by a biotinylated detection antibody and Cy3-streptavidin.

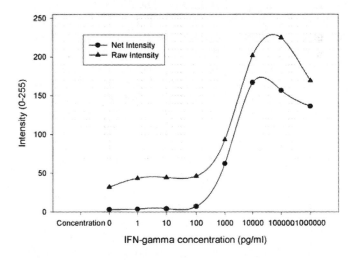

Figure 5. A preliminary concentration series indicates at least 2.5 logs of dynamic range and sub-ng/ml sensitivity.

4 DISCUSSION

In this report we have demonstrated the construction of protein nanoarrays containing 1 μm spots of an antibody directed against the human cytokine IFN-γ. These protein spots can be arranged on the surface as text, in an array, or any desired pattern. A preliminary concentration series for IFN-γ revealed at least 2.5 logs of dynamic range through a biologically relevant concentration range (Fig. 5). The next step in the progression of this technology is multiplexing the assay to allow evaluation of complex biomarker profiles from ≤1 μl of sample. Custom microfabricated surface patterning tools [6] have been designed for parallel deposition of large numbers of individual molecular species. These SPTs are currently being optimized for throughput and reproducibility.

The concept of a profiling panel for screening 100 or more protein biomarkers in the same area occupied by a single microarray spot offers nearly unlimited applications. It is anticipated that this technology will be first introduced for screening biomarkers from tiny protein samples harvested by Laser Capture Microdissection (LCM) of a subpopulation of cancerous cells [8]. Other potential applications include cytokine profiling for small animal models, as well as pre-natal biomarker testing.

The trend toward miniaturization in biotechnology has led to nothing short of a revolution in the way research is conducted. The practical limits of this trend must be carefully considered. Spot size is directly related to the number of available capture molecules and total binding capacity. Based on our best estimates, a 100 μm microarray spot could contain 10^7-10^8 randomly oriented IgG molecules. A similarly packed 1 μm nanoarray spot would contain only 10^3-10^4 IgG molecules. The number of antibody molecules present in a 100 nm spot such as those constructed via dip pen nanolithography (DPN) [9] is easily less than 100. It is reasonable to assume that a certain percentage of capture molecules in any size spot will be inactive or oriented improperly. While "single-molecule analysis" may sound seductive, it lacks statistical validity and value as a quantitative assay. Therefore, we have focused our efforts on the creation of protein nanoarrays with 1 μm spots which will contain enough antibodies to provide an adequate dynamic range. The preliminary standard curve for IFN-γ illustrated in Figure 5 demonstrates that a protein nanoarray on this scale can still offer at least 2.5 logs of dynamic range.

Another factor determining the optimal size of capture domains is the detection scheme. One benefit of protein nanoarrays is their ability to be analyzed by common optical methods [10]. The size scale of nanoarrays makes them more accessible than microarrays since the entire nanoarray field can be read in a single optical microscope image. Moreover, a typical 100 μm diameter microarray spot scanned in an expensive microarray scanner has a pixel density on the order of ~300 pixels/spot at 5 μm resolution or ~80 pixels/spot at 10 μm resolution. In comparison, an optical microscope with a 1.3 megapixel cooled CCD camera yields a pixel density for a 1 μm nanoarray spot of ~50 pixels/spot when viewed with a 60x objective or ~110 pixels when viewed with a 90x objective. This sampling density should be sufficient for statistically significant sampling of each nanoarray spot by optical microscope. Our work to date has utilized standard fluorescent labeling techniques, although greater sensitivity may be achieved through the use of quantum dots, dendrimers, resonance light scattering particles, or alternative methods such as total internal reflectance microscopy (TIRFM).

Atomic force microscopy (AFM) may also emerge as an attractive alternative to optical detection of nanoarrays. Microarrays are much too large to efficiently employ AFM, however an entire nanoarray can easily fit within the limits of a typical 100 μm XY piezo scanner. The principle driving force behind AFM-based detection is the potential for single molecule sensitivity. There are also additional benefits that cannon be realized with optical technologies such as the simultaneous acquisition of friction, adhesion, and viscoelastic data.

Advanced approaches to sample delivery must be integrated to take full advantage of the kinetics [11] and diminutive size scale of nanoarrays. We are presently exploring simple microfluidic technologies that will enable us to constrain small volumes of analyte to the active capture regions. More sophisticated components such as microvalves for sample circulation or oscillators for on-chip agitation have been demonstrated by others and will be incorporated as our design matures.

This work was supported in part by NIH/NIBIB SBIR Phase II grant R44-EB00316, NIH/NIDDK SBIR Phase I grant R43-DK064463, and Army BCRP grant DAMD17-00-1-0362.

REFERENCES

[1] B. Haab, M. Dunham, and P. Brown, Genome Biology, 2, research0004.1-0004.13, 2001.

[2] V. Knezevic, C. Leethanakul, V. Bichsel, J. Worth, et al., Proteomics, 1, 1271-1278, 2001.

[3] T. Urbanowska, S. Mangialaio, C. Hartmann, and F. Legay, Cell Biology and Toxicology, 19, 189-202, 2003.

[4] C. Mosher, M. Lynch, S. Nettikadan, W. Henderson, et al., JALA, 5, 75-78, 2000.

[5] C. Mosher, M. Lynch, S. Nettikadan, J. Xu, et al., in preparation.

[6] J. Xu, M. Lynch, J. Huff, C. Mosher, et al., submitted for publication.

[7] K. Nakano, H. Taira, M. Maeda, M. Takagi, Anal. Sciences, 9, 133-136, 1993.

[8] C. Pawletz, L. Charboneau, V. Bichsel, N. Simone, et al., Oncogene, 20, 1981-1989, 2001.

[9] J.-H. Lim, D. Ginger, K.-B. Lee, J. Heo, et al., Angew. Chem. Int. Ed., 42, 2309-2312, 2003.

[10] M. Lynch, C. Mosher, J. Huff, S. Nettikadan, et al., Proteomics, in press.

[11] R. Ekins, F. Chu, E. Biggart, Clin. Chem. Acta, 1, 91-114, 1990.

An Chip-based Instant Protein Micro Array Formation and Detection System

*Fan-Gang Tseng, *C.E. Ho, M.H. Chen, *K.Y. Hung, *C. J. Su, *Y.F. Chen, **H.M. Huang, and *Ching-Chang Chieng
*Engineering and System Science Department, **Life Science Department
National Tsing Hua University
101, Sec. 2, Kuang Fu Rd., Taiwan, R.O.C , fangang@ess.nthu.edu.tw

ABSTRACT

This paper proposes a novel protein micro arrayer employing 3-in-1 chip system, instantly delivering hundreds of biosamples in parallel for disease diagnosis. The 3-in-1chip system included: micro filling chip, micro stamper array and micro bio-reaction chip, applicable to high throughput disease diagnosis and drug screening. Capillary force is applied to drive liquids flowing through all the chip channel system. The filling chip contains hand-fill-in reservoir and sealed with PDMS thin film for protein storage and loading to arrayer. The uniformity of the printed spots by the stamper array can be within 10% standard deviation. The micro photo sensor array is designed for the fluorescent detection. Two human cancer (HURP and E6) markers have been successfully detected by this micro stamper system. Reproducibility and Surface wettability for micro/nano droplet formation around stamping process are investigated by simulation and experiment in this paper.

1 INTRODUCTION

The development and application of protein arrays have become an important trend in modern biomedical diagnoses because of micro-array's ability of massive parallel process [1]. Protein micro arrays have become powerful tools in biochemistry and molecular biology. Array-based technology involves two principal processes: transferring hundreds of biosamples onto substrates, and immobilizing the biosamples on the substrates. "Micro contact printing" [1-4] is one of the microarray technology to simultaneously and rapidly imprint a vast number of bio-reagents in a short time. The conventional way to produce micro arrays utilized a needle array and computerized robot system to select and spot numerous bio reagents [3]. This arraying system has the drawbacks of high cost, large size variation, and serial and long spotting process, which may malfunction proteins from dry-out. As a result, "Micro contact printing" [1-4] has been becoming an emerging technology to simultaneously and rapidly imprint a vast number of bio-reagents in a short time. Among various micro printing schemes, a new type microarrayer has been proposed by us recently [5] by employing micro stamping method to simultaneously immobilize hundreds of proteins in parallel. However, because of the short distance among stamping heads, it becomes necessary to employ a filler chip[6] for protein

fill-in in batch instead of fill in individually. The combination of the batch fill-in and batch arraying process enable a rapid and high throughput formation of protein array to prevent proteins from de-naturization during the array process.

Besides the batch fill-in, parallel detection of protein array is also important for rapid and high throughput diagnosis/screening. Among many detection methods, optical detection is one of the most applied detection means because of its high sensitivity and reliability. As a result, the proposed system employ micro optical system to directly couple weak fluorescent light emitted from protein arrays through micro optical collectors onto the micro sized photodiode, thus greatly improves the efficiency of light detection by reducing the light path from meter range to sub-millimeter range.

By employing the three-in-one chip system as mentioned above, parallel protein array formation and detection process can be realized, as shown in Fig. 1. Proteins can be loaded and preserved as a library in each fill-in reservoir in the micro filling chip before real application. When protein diagnosis/screen desired, the micro array stamper can be filled in proteins in parallel from the filling chip by piercing the needle array on the top of the stamper into the membrane of the filling chip at once. Protien array can then be formed simultaneously by the gentle contact between the stamper and bio reaction chip. The formed protein array is now ready for parallel bio reaction. Micro optical detection system is setup on the bottom of the bioreaction chip for parallel protein detection. Thus high throughput protein diagnosis/screen process can be finished in hours including bioreaction

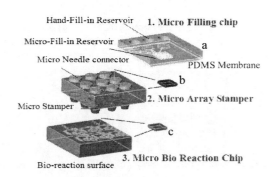

Fig1 integration of chip system

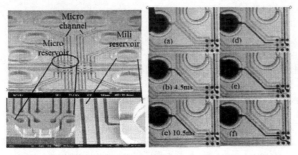

Fig2. micro filling chip and filling test

process for hundreds to thousands of proteins at one time. The detail operation and design of the 3-in-1 chip system is elaborated in the following sections.

2 BIOCHIP SYSTEM DESIGN
2.1 Micro Flling Chip

The operational concept of protein filler chip to perform batch fill-in process for micro arrayer is shown in Fig. 1a. After we dispensing different proteins in the filler chip, parallel protein fill-in process can be carried out by directly contact the filler chip with the arrayer chip, thus tens to hundreds of proteins can be transferred by capillary force from the filler into the arrayer chip in seconds. This process can be repeated to fill another arrayer chip until the empty of the filler chip. The Filler chip is not only employed for protein filling purpose, but also applied to preserve the stored protein library permanently by sealing the top and bottom surface with PDMS thin film until the need for arrayer feeding.

The filler chip contains mili-reservoirs on the chip top for protein sample dispensing, micro reservoirs on the bottom for holding protein solutions, and micro channels connecting each mili-reservoir to the corresponding micro reservoir. Protein samples can be filled-in by capillary force from individual mili-reservoir to the corresponding micro-reservoir automatically without external power. To fill in protein solutions vertically and allow proteins arranged on a two-dimensional array, three-dimensional fluidic structures are desired. The fabricated filling chip and the loading process is shown in Fig. 2.

2.2 Micro Stamper Chip

As shown in Fig. 1b, the stamper chip can be used to spot different kinds of biosamples simultaneously

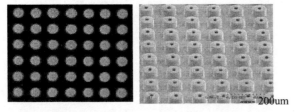

Fig3 Fluoresent stamp result and uniform PDMS micro stamper array

and expanded to stamp hundreds of different biosamples. The micro stamper comprises micro-needles, micro-stamp array, and a microchannel array connecting the reservoirs and micro-stamp array. Different proteins could be dispensed into different channels and driven by capillary force to the tips of the micro stamps through the micro channels. The protein-filled micro stamper was then brought into contact with the bio-assay chip to generate sample arrays for further bio processing. Elastomeric polydimethylsiloxane (PDMS) is used to fabricate the micro stamps, so conformal contact can be achieved on rough surfaces. The uniform fluoresent spot of stamping result and the picture of the PDMS stamper was shown in the Fig3.

2.3 Micro Bio-Reaction Chip

After the proteins were delivered to the bio-assay chip, the binding strength between the bio-assay chip and proteins is important for subsequent bio processing. Protein immobilization means are categorized into three major types: carrier-binding, lattice-type, and micro-capsule-type immobilization. Among those three, carrier-binding is more favorable for protein chip application for its properties of thin molecule film and ease of optical or electrochemical detection. Covalent bond is the strongest bond of carrier-binding. As a result, covalent bond of proteins by SAMs to carriers (glass slide in this study) is chosen for the application of our micro stamping system.

Under the immobilized protein array, micro optical system with APD (avalanche photo diode) sensor arrays are employed for fluorescent detection. APD sensor made by the conventional CMOS process was easy to be integrated with micro array. The avalanche effect under a suitable reverse voltage produces an internal-gain to amplify the weak fluorescent signal. As a result, an integrated micro optical system to detect weak fluorescent signals is designed underneath the micro reaction chip. The conceptual design of the sensor array and the fabricated sensors is shown in Fig. 4.

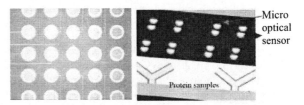

Fig4 (a) Optical microscope picture of APD photo diode sensor (b) the illustration of the photo sensor detection

3 MATERIAL

The surface properties of bioreaction chip greatly affect the stamped droplet size. Thus, surfaces with different wetting properties were prepared by glass slide, APTS (aminopropyltrimethoxysilane) and BS[3] (bis-sulfosuccinimidyl suberate) on silde, APTS and DSC (N, N'-disuccinimidyl carbonate) on slide, and Su8 photo resist (MicroChem) coating on slide. The surface peak roughness is ranged form 73.8nm to 148nm. The deformation of the elastic stamper, which is made of PDMS, can make a conformal contacting with the substrate. The effect of surface roughness to the stamped feature size can be ignored. The hydrophilic properties of those surfaces have been characterized and illustrated in Figure 5, ranging from 18° to 78°, respectively on different surface coating. PBS buffer with 30% glycerol and 1% Cy3 dye is used as the stamping fluid in experiment. Both fluorescent and optical microscope images are used to measure the stamping feature size.

(a)18°   (b)51°   (c)58°   (d)78°

Figure 5. The contact angle of the PBS solution with 30% glycerol on different substrate surface (a) Glass (b) APTS+BS^3 coating (c) APTS+DSC coating (d) Su8 photo resist coating

4 SIMULATION AND EXPERIMENT

To fully understand the flow behaviors of the stamping system, numerical methods are employed to simulate the stamping process for effective design through solving first principle equation. Governing equations as conservation laws of mass and momentum describing both liquid and gas phases are included. Most of all, surface tension force is the major driving force and the total surface force depends on the orientation and surface area of the liquid-gas interface, precise determination of the moving location and shape of the interface is the key issue of accurate computation. Numerical models [7] consist of Volume-of-Fluid (VOF) method for two-phase homogenous flow model and the interface tracking technique in cooperation with CSF Continuum Surface tension Force model. In order to discover the dynamic stamping process and the drop breakdown process, 3D simulation was demonstrated as shown on the Figure6.

5 RESULTS

The high-speed camera under the transparent glass substrate sequences the droplet formation and the separation of the micro-stamper. The dynamic spot size varying with the time can be analyzed and compared with the 3-D dynamic simulation. Meanwhile, the stamped spot size varies with the different substrate wettability as shown in the Figure7. On the Su8 coating substrate, the printed spot size is 47μm, which is much smaller than that on the glass substrate of 123μm.

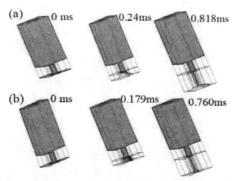

Figure 6. Dynamic simulation of stamping process (a) on higher hydrophilic surface (b) on lower hydrophilic surface

For the minimum spot size of the conventional quill pen microarrayer is 75μm, the physical property of the printing liquid and substrate has to be precisely controlled. Features containing fluorescent dye are printed as shown on the Figure7. The spot size increases with substrate wettability, which is observed in both simulation results and experiments as illustrated in Figure8. The PBS solution plus 30% glycerol (viscosity 3.20cp) makes the feature size smaller than that of the BSA (bovine serum albumin) protein solution (viscosity 1.02cp), and the simulation result for the former solution agrees well with the experiment onc. It demonstrated that the wettability of the substrate surfaces and the viscosity of stamping solution could control the stamping spot size of the micro stamping process.

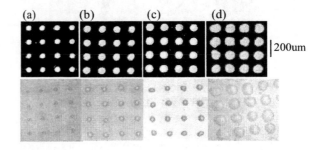

Figure 7. Drop features on different substrate observed from optical microscope and fluorescent image (a)Su8 photo resist (b) APTS+DSC (c) APTS+BS^3 (d) Glass

Fig9 shows the stamping results, and demonstrates a repeatable stamping result within a deviation 10% in size and 25% in intensity. In the repeatability testing of the micro stamping process, the micro stamper arrayer was filled with 2 kinds of fluorescent solution. One is cy3 0.001 μg/ μl with PBS solution and glycerol 30%and the other is cy5 0.1 μg/ μl with PBS solution and glycerol 30%. It is 12 times that the micro stamper was pressed continuously on the substrate without cross contamination as shown on the Fig10. The glass substrate is coated by APTS+BS^3.

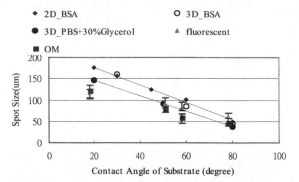

Fig 8. Spot size varied with wettability of substrate surface observed in simulation results and experiments.

In the testing of utility of protein stamper for cancer marker detection, four kinds of solution of the tumor marker antigen (HURP: liver cancer, E6: Papillomavirus antigen) are filled into the micro stamper and printed to transfer the protein on the substrate. Through the ELISA testing, the antigens with different concentration have been recognized as shown on the Fig11 in a concentration of E6 76 ng/μl. It demonstrates that the micro stamper can successfully transfer proteins into array simultaneously for disease detection.

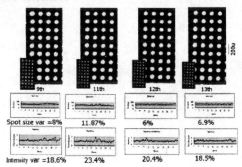

Fig9 . Protein spots uniformity with each stamp result

6 CONCLUSIONS

This paper introduces a 3-in-1 chip system for protein array rapid immobilization and detection. The operation and testing of the three chips, including micro filling chip, micro stamper chip, and bioreaction chip has been successfully carried out. The testing result shows the spot size of soft printing decreasing with the lower hydrophilic substrate surface, demonstrated by both simulation and experiment. The test of stamping uniformity and repeatability has been performed. It demonstrates a repeatable stamping result within a

deviation 10% in size and 25% in intensity. Two human cancer (HURP and E6) markers have been successfully detected by this micro stamper system. The micro stampr array successfully prints the protein to the substrate. It can be applied on the disease detection, diagnosis and drug discovery.

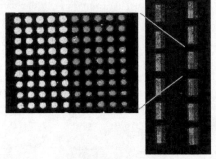

Fig 10   Stamp two types solution   cy3 cy5 simultaneously in 12 times.

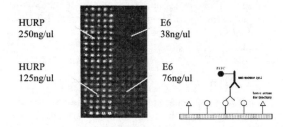

Fig11. ELISA of tumor marker detection. HURP : liver cancer antigen, E6: Papillomavirus antigen

REFERENCES

[1] S.C. Lin. et al.,"Microsized 2D protein arrays immobilized by micro-stamps and micro-wells for disease diagnosis and drug screening" Analytical and Bioanalytical Chemistry, Vol. 371, pp. 202-208, 2001.

[2] F.G. Tseng, et al.,"Protein micro arrays immobilized by μ-stamps and protein wells on PhastGel pad" *Sensors and Actuators B*, Vol. 83, p. 22-29, 2002.

[3] Kane, R.S. et al., "Patterning proteins and cells using soft lithography"*Biomaterials*, Vol. 20,p.2363-2376, 1999.

[4] F.G. Tseng, et al, "Fluid filling into micro-fabricated reservoirs" Sensors and Actuators A, Vol. 97-98, p131-138, 2002

[5] S.C. Lin, et al.,"A novel protein micro stamper with back-filling reservoir for simultaneous immobilization of large protein arrays" *IEEE MEMS*, Kyoto, Japan. p299-302, Jan. 2003

[6] F.G. Tseng,et al., "Micro protein filler chip for protein library preservation and arrayer filling in batch" μTAS, Squaw Valley CA, USA, p639-643, Oct. 2003

[7] Brackbill, J.U. et al.,"A continuum method for modeling surface tension" Journal of computational physics Vol 100, p335-354, 1992

High Resolution Electrochemical *In Situ* Printing

Ryan D. Egeland*, Edwin M. Southern

Oxamer & Oxford Gene Technology,
Begbroke Science Park, Oxford, OX5 1PF, England, ryan.egeland@oxamer.com

ABSTRACT

This work demonstrates a new method for making oligonucleotide microarrays by synthesis in situ. The method uses conventional DNA synthesis chemistry with an electrochemical deblocking step. Acid is delivered to specific regions on a glass slide, thus allowing nucleotide addition only at chosen sites. Deblocking is complete in a few seconds, when competing side-product reactions are minimal.

Features generated in this study are 40 microns wide, with sharply defined edges. The acid used for synthesis is produced by electrochemical oxidation of a reversible redox couple at microelectrodes. Under suitable conditions, the active species is confined to micron-sized features and diffusion does not obscure the surface pattern produced. Experimental results, theoretical analysis, and digital simulation identify a chemical annihilation process critical to ensuring high feature resolution.

The microelectrodes used for synthesis are designed to sustain high currents, chemical attack, and mechanical wear so that many hundreds of syntheses can be performed with the same microelectrode set. Software, custom electronics, and a sealed piston and cylinder fluidics chamber allow synthesis automation. The electrochemical and mechanical system is robust and may be adapted to a range of other biomolecule syntheses.

Keywords: oligonucleotide, microarray, DNA, electrochemical, fabrication

1 INTRODUCTION

Literature accepted for publication at the time this manuscript was submitted describes results of a new method for in situ microarray fabrication and will be presented in the NSTI conference. This manuscript summarises the results and discusses the method in the context of existing microarray fabrication techniques. The method uses microelectrodes to generate acid, which then directs the selective attachment of nucleotide bases to a chip surface.

In summary, a hydroquinone and benzoquinone electrolyte controls diffusion through a reversible redox annihilation process such that acid is limited to a defined and sharp region (see Figure 1). Probe feature sizes are about 2-3 μm. In addition to the synthetic and array chemistry system, fabrication of the microelectrodes (specific for high current synthetic applications), design of a fluidics and positioning system, and accompanying electronics and software will be presented at the conference.

2 RESULTS

The electrochemical system enabled highly-parallel experiments exploring effects of electrolyte composition and electronic parameters such as voltage, current, and pulse timing on patterns produced. Exploration and optimisation of this surface patterning method in the context of oligonucleotide synthesis enabled successful and reproducible electrochemical construction of six-base sequences in high yield. Seventeen base sequences were synthesised as a final demonstration of the entire system, with discrimination of single-base mismatches detected in the human haemoglobin gene sequence.

As the electrochemical systems and techniques introduced in this work are new, evaluation of the results can be considered in the context of other existing micropatterning methods, with a focus on synthetic oligonucleotide DNA microarrays in particular. The immense number of distinct chemical steps involved in production of a single DNA microarray device brings unique mechanical, fluidics handling, synthetic chemistry, and quality control challenges; the apparatus and systems presented in this work go far in solving these. That no single microarray fabrication system has yet proven best for all circumstances, despite a decade of intensive research, demonstrates the challenges and continuing evolution in this fast-moving scientific field.

Ultimately, the practical utility of a device made with the electrochemical methods presented in this work will be determined by its suitability in particular biological applications. Although different applications exploit specific device capabilities to various degrees, certain characteristics are universally desirable, and the electrochemical method presented here may be usefully evaluated in the context of these characteristics.

2.1 Uniformity, Consistency, and Reproducibility

Spotting Arrays. Array analytical sensitivity is important, given the biological significance of subtle variations in messenger RNA (mRNA) levels in gene expression analyses, or of single-nucleotide mismatches in genomic analyses. The method used by many laboratories,

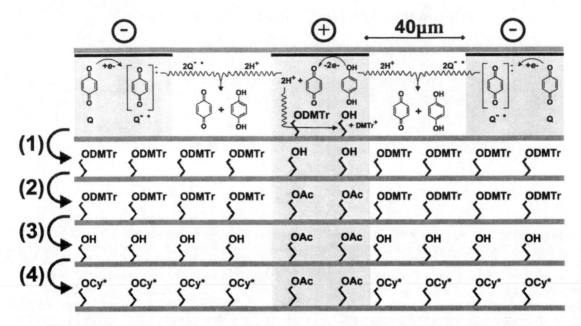

Figure 1 The electrochemical oxidation of hydroquinone at the anodes (+) on an array of microelectrodes delivers acid to regions on a surface (the grey shading symbolises anode regions). The acid removes a dimethoxytrityl (DMTr) group (1) to expose a primary hydroxyl (OH), which is then exposed to an acetylating reagent (2) to attach an acetyl group (Ac). Subsequent treatment of the entire surface with an acid solution (3) followed by coupling of a fluorescent (Cy*) dye (4) allows imaging by confocal microscopy so that regions of acetylation are revealed by diminished fluorescence. The counterelectrode process at the adjacent cathodes (−) is the reduction of benzoquinone, which yields a radical anion reactive with protons, thus depleting acid in the cathode region and regenerating hydroquinone (from *Anal Chem*).

spotting pre-synthesised oligonucleotides or expressed copy DNA (cDNA)(1,2) on glass, is prone to a number of problems. Non-uniform spots, scratches, dye separation, and clumps commonly introduce significant error(3,4), decreasing the power of the analysis. These problems are compounded when the probes are cDNAs, as the process of making the probes, by growing clones or polymerase chain reaction (PCR) amplification, often leads to replacement of the desired probe by a contaminant(5-7). Error rates as high as 40% have been reported when multiple error-prone methods are combined(8,9). Furthermore, as the state of bound DNA is ill-defined for cDNA probes(1), hybridisation to the target may be unpredictable and difficult to control. The ink-jet and photolithographic arrays, given that each probe is designed synthetically, are much more uniform than the spotting arrays.

Electrochemical Arrays. The electrochemical method, in contrast, allows specific direction of probe synthesis *in situ*. The electrochemically directed deprotection reaction is driven by solution chemistry and electronicswitching of microelectrodes, rather than mechanical means of moving and depositing liquids. Because any inconsistency in the electrolyte solution or electronic circuitry will be recognised through significant current fluctuations, the current measured at each microelectrode gives a useful check of chemical yields. Notably, this quality-control check occurs in real-time; any deviations in the array fabrication process

are noticed as the array is made, rather than after hybridisation experiments reveal flawed probes.

2.2 Fidelity of Synthesis and Length of Probes

Relevance of Probe Length. Some applications, such as the analysis of sincle nucleotide polymorphisms (SNPs), exploit minor variations in probe sequence to detect target sequence variation. For these applications, maximal sensitivity is achieved with short probes, as they show relatively large changes in hybridisation yields with only single base variations. In other applications, such as expression analysis, the probes are used primarily to quantify total levels of relatively dissimilar targets. Although short probes may be suitable for some of these applications(10), longer probes minimise effects of site-specific interactions and melting temperature, and ensure high complementarity to target, thus maximising detection sensitivity(11).

Photolithographic Arrays. Existing photolithographic *in situ* fabrication methods rely on a photo-deblocking step that is inefficient relative to conventional solution-phase 5′-dimethoxytrityl deblocking in acid. The resulting low stepwise yields, reported to range from 81-94% depending on the nucleoside and its substituents(12), restrict the maximum oligonucleotide length on these photolithographic arrays to less than 25 nucleotides. More efficient photolabile blocking groups(13), or other photolithographic processes, including the use of photochemically-generated acids to deprotect conventional dimethoxytrityl (DMT) groups(14), may increase yields. Nonetheless, at present, it seems doubtful that any of the photo-deblocking methods will reach the nearly quantitative yields(15) now routine with conventional solution-phase oligonucleotide synthesis. Indeed, the persistence of "descumming" steps (required to remove residual photoresist after exposure and development steps) in conventional microelectronics photolithography, where there are no complex DNA synthesis challenges, indicates the inherent difficulty of achieving quantitative yields with even simple photochemical reaction systems.

Ink-Jet Arrays. In contrast to the photolithographic arrays, ink-jet based arrays rely upon conventional oligonucleotide synthesis chemistry, resulting in correspondingly higher yields. High-quality short probes may be produced for analysis of variation, or probes in excess of 50 nucleotides may be used for expression analysis(16,17). The ink-jet methods require precise mechanical alignment of the print head, process monitoring, and atmospheric control systems to ensure high-fidelity synthesis.

Electrochemical Arrays. This work demonstrates the electrochemical synthesis of 17-mers in high yield with a hand-made prototype device. Measured coupling yields are high (likely greater than 98%). Only the deblocking step, which is performed with an electrochemically generated acid in acetonitrile, differs from conventional oligonucleotide synthesis. The concentration of the acid generated was easily controllable by manipulating the gap distance, applied voltage, or application time; a series of optimisation experiments exploited this flexibility to reach high deblocking efficiencies quickly. As the acid is generated in solution-phase, the deblocking step may be driven to completion by maintaining high acid concentrations. Using excess reactant in this fashion to increase product yields is a commonly used technique in organic chemistry, but has no direct equivalent in the case of photochemical reactions.

2.3 Density of Probes

Fundamental physical dimensions (the diameter of double-stranded DNA is 2.4 nm and 0.34 nm from base to base) place an lower bound on the smallest array feature size. Existing array fabrication techniques, however, do not approach this limit, although it is possible to perform manipulation and hybridisation of DNA at a molecular level in the limited scale of individually interrogated probes(18,19). Thus, there is much room for improvement in array fabrication technology before molecular dimensions are reached; until that time, the primary impediments to increased density are more those of practical engineering concerns than fundamental physics.

Spotting and Ink-Jet Arrays. As cDNA-based arrays typically use a limited number of probes and are fabricated by hand or robotically driven pins, most spotting arrays are fabricated on a macroscopic scale. However, higher-density cDNA arrays can be made by more precise piezoelectric driven "pens," or by ink-jet based methods. The ink-jet based methods typically achieve the smallest features (whether used for spotting cDNA or synthesising oligonucleotide probes in-situ), but imprecision of the ink jet printing head and spread of drops when they hit the surface limits ink-jet array densities to about 10-100 probes per mm^2 (feature sizes greater than 100 μm) (17).

Photolithographic Arrays. Photolithographic arrays have a density of 1000-2000 probes per mm^2 (feature sizes around 20 μm). However, because the photolithographic arrays contain short probes, between 30 and 40 probes must be used per target in a redundant fashion(10,11,20), offsetting the advantage of high probe density. Thus, the overall number of analysed targets per surface area is about the same for ink-jet and photolithographic arrays. Whereas mechanical precision limits the density of ink-jet arrays, diffraction, scatter, and internal reflection in the glass substrate blur the edges of the probes in photolithographic arrays; nonetheless, the sub-micron feature sizes realised through advanced techniques in the microelectronics industry demonstrate the potential for further miniaturisation.

Electrochemical Arrays. The smallest single-step feature demonstrated through the electrochemical method presented in this work is 3 μm (representing a theoretical probe density greater than 100,000 per mm^2), but the lower limit has not been aggressively pursued. The mechanical alignment of the substrate during reagent flushing, not diffusion or size of the microelectrodes, limited the smallest synthesised oligonucleotide probe to 40 μm, with 120 μm centre-to-centre distance. Assuming a large library of probes deposited at this feature size, the equivalent probe density (70-600 probes per mm^2) compares favourably with ink-jet and photolithographic arrays. Nonetheless, both the ability to make very small electrodes and the diffusion-driven nature of the electrochemical process will limit the maximum densities possible through this electrochemical method.

Engineering Considerations. Multiple experiments presented in this work show that feature sizes on the substrate correspond to the dimensions of the microelectrodes. As the microelectrodes used in these experiments were made through conventional clean-room techniques, there is no inherent reason high-tolerance equipment already used routinely in semiconductor manufacture could not be adapted to the process. Fabrication of 5 μm microelectrodes was accomplished in this work with dated processes and equipment. Although clean-room capabilities limited device complexity to a relatively modest design of 96 linear, passively switched electrodes, further integration of switching and measurement circuitry on the electrode device itself could readily enable hundreds or thousands of switchable microelectrode elements(21,22). Thus, the challenges of electrode design in this work did not, and in principle should not, impede further attempts to increase array density.

The high currents necessary to drive the deblocking reaction to completion required special microelectrode materials. Iridium microelectrodes were most successful in these experiments, but as the electrolyte composition, applied current, electrode materials, and electrode geometry all influence the rate of acid reaching the surface, it may be possible to alter the most accessible of these parameters in future device designs. For example, exploring various electrolyte formulations could be continued further than in this work; adding chemical substituents to the benzoquinone, varying the concentration of background electrolyte, or even using a different solvent could improve compatibility with more conventional electrode materials.

Fundamental Theoretical Limits. The 40 μm microelectrode size used in this work was chosen only for convenience of manufacture in the clean room. Simple, single-step electrochemical patterns, with sizes below 5 μm, were created using this technique. Although experimentally difficult to explore synthesis at very small dimensions, a comprehensive theoretical analysis aided by a detailed computer model evaluates the question of density and resolution for this technique. The analysis suggests feature sizes below 2 μm are not unreasonable using the same chemical and electrochemical parameters as routine in this work. Computer simulations strongly suggest that sub-micron features may be possible with higher current, or a slightly modified electrolyte. Although the practical obstacles to manufacturing and controlling a device at that scale may prove severe, theory and simulation suggest that only engineering skill, and not fundamental physics will limit the technique's ultimate potential.

2.4 Ease of Fluidics and System Integration

Challenges in Fluid Handling. Oligonucleotide synthesis, whether performed through photolithographic or conventional deblocking chemistry, requires the delivery of reactive solutions and solvents including acetonitrile, dichloroacetic acid in dichloromethane, iodine, tetrahydrofuran (THF) and pyridine, acetic anhydride, N-methyl imidazole, and lutidine to a solid support (usually glass). Glass materials do not degrade on exposure to these reagents, but other materials used in constructing the fluidics chambers, delivery lines, valves, and seals of a fluidics delivery system must withstand these harsh chemical conditions. Such design challenges limit the utility of microfluidics techniques developed for non-reactive, aqueous liquid handling. Furthermore, the coupling step during synthesis must be performed under anhydrous conditions; it is therefore necessary that fluid handling systems be impervious to outside air, or conducted under inert atmosphere.

Ink-Jet and Photolithographic Arrays. Only relevant to the *in situ* fabrication methods, various systems have been devised to deliver synthesis reagents for array construction. The ink-jet method, as mentioned earlier, relies upon the movement of a mechanical printing head to direct synthesis precursors on a substrate with localised hydro-

phobic and hydrophilic regions(23) to minimise spread of droplets when they hit the surface; the substrate is then taken to a liquid handling station for treatment with the oxidising and deprotecting reagents. One photolithographic system(24) requires the assembly and disassembly of chrome masks at each base addition step, representing the need for 100 chrome masks for arrays of 25-mers (although optimisation techniques can reduce the number required). Other photolithographic systems(25,26) use solid-state micromirror devices(27) in place of chrome masks, thereby alleviating the expense of mask design and complexity of alignment. In these systems, no moving parts are required for photolithographic array fabrication, as synthesis reagents can be flushed through a chamber sealed to one side of a glass slide, with light directed from the opposite side, through the slide.

Fluid handling is also necessary during hybridisation for delivery of reagents, generally with the array mounted in a flow system. In the micromirror-based system(26), the same flow system is used for both hybridisation and synthesis. This can be a disadvantage, as the machine is monopolised during hybridisation and cannot be used to make another array. For both the mask-based photolithographic and ink-jet methods, the arrays are diced into chips which are assembled into a custom flow device for hybridisation and processing.

Electrochemical Arrays. One advantage of the electrochemical method is that the work piece need not be separated from the tool at any stage. The flow cell described in this work allowed positioning of the substrate adjacent to the microelectrode array with an accuracy of about one micron. Although it was necessary to move the substrate further away from the microelectrodes while flushing synthesis reagents through the chamber in this design, there is no inherent reason the fluids could not be delivered through alternative means directly into the gap.

The sealed piston and cylinder used in the work presented here eliminated the need to carefully re-register the lateral position of the substrate each time it was moved back and forth for reagent flushing. Nonetheless, it became apparent that fluctuations in lateral alignment over the 34 piston movements during 17-mer synthesis confounded the ability to direct electrochemical deprotection at precisely the same location on the substrate each time. As the piston and cylinder apparatus was constructed with PTFE ("Teflon"), a relatively deformable polymer, it is not surprising that there was some horizontal play in the piston; measurement of this misalignment by confocal image analysis of defective probe attachments indicated a possible 40 μm maximum deviation with 34 piston motions, with 5-10 μm generally more typical.

Although this horizontal piston deviation proved irritating, it did not prevent successful synthesis of long probes with a careful reproduction and control of the piston motion from step to step. Furthermore, it may not be necessary to allow such continuous control of the substrate in future designs; recall that the system was designed to explore the effects of substrate-microelectrode gap distances on the deprotection reaction. Subsequent flow-cell designs intended only for array synthesis with these microelectrodes under optimal gap distances could employ 40 μm stoppers attached to the microelectrode array and on which the substrate would rest during deprotection. Precise and accurate substrate alignment in this arrangement would be dependent only on the rigidity of the microscopic stoppers, rather than on the macroscopic fit of the piston and cylinder. In addition, these stoppers could also be configured to limit the horizontal deviation, if this indeed proved significant with smaller microelectrodes.

Simplicity of Fabrication. Though this work focuses on the scientific and technical aspects of electrochemical microarray fabrication, discussion is only complete with a brief prediction of difficulties in practice. Although technical advances are rapidly decreasing costs per microarray analysis, very significant research and development expenses and high equipment costs are typical of most means of making arrays at present. Costs are falling, but the high investment required in clean-room construction, mask development, and equipment upkeep and maintenance add significant expense to photolithographic arrays. The difficulty of maintaining the ink-jet head alignment and fluidics over many thousands of runs could likewise prove troublesome. The cDNA libraries or large sets of synthetic oligonucleotides used for spotting arrays are expensive to buy and maintain.

Costs for the electrochemical method could be less than others, given the mechanical simplicity of the fabrication apparatus. Reagents for *in situ* synthesis are relatively inexpensive and the main cost is in the solvents. The most expensive part of the apparatus is the microelectrode set, which can be reused to make many arrays (one electrode set was used in this work over 1,000 times with no sign of deterioration). The microelectrode sets, although customised designs in this work, could be produced more cheaply in higher volumes at conventional microfabrication facilities. The actual oligonucleotide synthesis lends itself to automation; equipment should be easy to maintain, given its simple design and few moving parts.

Customisation of Probe Design. Some array fabrication techniques are best suited to large-volume production of identical devices. For example, designing and making the many chrome masks used for photolithographic arrays represents a significant expense, so that the method is only reasonable when a known optimum, fixed set of probes is relevant to a wide range of different investigations. Such a prerequisite is not feasible in many applications. It may be necessary to change the probe set from one run to another, for example, in experiments to optimise probe design. Likewise, adding a probe for a newly discovered gene or a mutation in an existing set could require an entirely new set of expensive masks.

The ink-jet and micromirror methods, in contrast, allow complete flexibility; each array made may be different from any other. Changing the design is simply a matter of

changing the sequence data that drives the print head or micromirror driver. Conversely, these methods may also be used for large-volume production. However, as throughput is limited in the ink-jet method by the speed of the print head, and in the micromirror method by the total number of pixels on the micromirror device, these methods may not scale well to higher volumes of fabrication.

The electrochemical method is similar to the ink-jet and micromirror array methods in that the process is directed by a computer instruction set based on the sequence content of the array. A change in design is effected by simply changing the instruction set. This is not to say the method is unsuitable to large-volume processing with multiple electrode sets. In other words, it is as easy to make one thousand different, customised arrays as it is to make one thousand identical arrays.

Many types of surface patterning reactions (other than the conventional oligonucleotide synthesis presented here) could be controlled electrochemically. First, acid may be

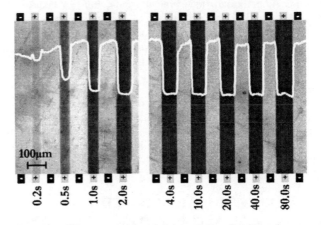

Figure 2 The extent of reaction at the surface may be controlled and investigated by delivering timed applications of current to the electrodes; the patterning reagent is confined over very long times. This image shows the patterns formed over a time course, using a fixed potential of 1.33 V applied to electrodes 20 μm from the surface. Diffusion does not blur the edges of the stripe formed after 80 s. Instead, the sharply defined stripes are limited to regions between the cathodes and show that patterning reagents may be directed to strictly confined areas (from *Anal Chem*).

involved in many other types of reaction on the substrate, for example eliminations, substitutions, rearrangements and chemical etching. Likewise, electrochemically-generated bases could also be used to eliminate base-labile moieties such as 9-fluorenylmethoxycarbonyl (Fmoc) or cyanoethyl groups, particularly relevant for peptide synthesis and peptide-oligonucleotide conjugation chemistry(28,29). Radicals, halogens, and other reactive intermediaries generated at microelectrodes could be used to elicit other exotic modifications. The large range of possible chemical modifications leaves much room for many types of probe synthesis and modification.

Range of Applications. As discussed in the context of synthesis fidelity and probe length, long probes are more applicable to expression profiling, while short probes are particularly suited to genotyping. Most purposes can be served through ink-jet *in situ* synthesis or spotting pre-synthesised probes at present. Photolithographic arrays are not particularly suited to expression profiling given short probe length, although they have successfully been used for this purpose(30), and higher coupling yields(31) could increase sensitivity and utility. One original disadvantage of the photolithographic method was that the photolabile protecting group could only be used to block 5′ hydroxyls, precluding reverse oligonucleotide synthesis useful for enzymatic extension-based detection methods, although new photocleavable groups(32) may permit such $5′ \rightarrow 3′$ syntheses.

The work described in this work does not push the electrochemical method to its physical limits, but we know probe length and fidelity depends on efficiency of chemical coupling at each base addition. The efficiency was demonstrated to be very high. As the feature sizes are small, the electrochemical method here could in theory combine the high density of photolithographic arrays with the high fidelity of ink-jet arrays. Furthermore, as conventional oligonucleotide synthesis reagents are used (with the exception of the electrochemical step which is only a substitution of one type of acid with another), non-standard or exotic syntheses (for example, reverse synthesis) will not require lengthy development of a new photocleavable linker for each application.

3 BIOLOGICAL AND MEDICAL APPLICATIONS

It is beyond the scope of this work to review in detail the immense range of biological and clinical microarray applications, as this manuscript and accompanying presentation focuses on device design.

Most arrays are used in research applications at present, but this will likely change as microfluidics capabilities improve(33-39). There are two types of clinical diagnostic applications used in research laboratories: detection of molecular lesions associated with disease states, such as cancer, and analysis of gene expression associated with pathological states. These studies can aid in disease classification, prognostic estimates, and help identify targets and biomarkers useful for guiding treatment decisions. Cancer provides a useful example: the tremendous breadth of somatic mutations possible in its initiation and progression can only be captured practically through highly parallel means; arrays are particularly suitable, and have been shown to greatly supplement differential diagnosis, tumour typing, and treatment planning in many human malignancies(40,41).

At present, however, the cost of each microarray analysis prohibits use in routine clinical diagnostic laboratories. With further development of fabrication methods, sample preparation techniques, analysis instrumentation, and data analysis algorithms, costs will fall to levels appropriate for widespread use of microarrays in an outpatient clinical setting. It is hoped the electrochemical method described in this work will become a useful tool in the future of genomic medicine.

REFERENCES

1. Duggan, D.J., Bittner, M., Chen, Y.D., Meltzer, P. and Trent, J.M. (1999) *Nature Genetics*, **21**, 10-14.
2. Schena, M., Shalon, D., Davis, R.W. and Brown, P.O. (1995) *Science*, **270**, 467-470.
3. Yang, Y.H., Dudoit, S., Luu, P., Lin, D.M., Peng, V., Ngai, J. and Speed, T.P. (2002) *Nucleic Acids Research*, **30**, art. no. e15.
4. Brown, C.S., Goodwin, P.C. and Sorger, P.K. (2001) *Proceedings of the National Academy of Sciences of the United States of America*, **98**, 8944-8949.
5. Freeman, T.C., Lee, K. and Richardson, P.J. (1999) *Current Opinion in Biotechnology*, **10**, 579-582.
6. Evertsz, E.M., Au-Young, J., Ruvolo, M.V., Lim, A.C. and Reynolds, M.A. (2001) *Biotechniques*, **31**, 1182,1184,1186.
7. Hoyt, P.R., Tack, L., Jones, B.H., Van Dinther, J., Staat, S. and Doktycz, M.J. (2003) *Biotechniques*, **34**, 402-407.
8. Clark, M.D., Panopoulou, G.D., Cahill, D.J., Bussow, K. and Lehrach, H. (1999), *Cdna Preparation and Characterization*. Academic Press Inc, San Diego, Vol. 303, pp. 205-233.
9. Drmanac, R. and Drmanac, S. (1999), *cDNA Preparation and Characterization*. Academic Press Inc, San Diego, Vol. 303, pp. 165-178.
10. Lockhart, D.J., Dong, H.L., Byrne, M.C., Follettie, M.T., Gallo, M.V., Chee, M.S., Mittmann, M., Wang, C.W., Kobayashi, M., Horton, H. *et al.* (1996) *Nature Biotechnology*, **14**, 1675-1680.
11. Jobs, M., Fredriksson, S., Brookes, A.J. and Landegren, U. (2002) *Analytical Chemistry*, **74**, 199-202.
12. McGall, G.H., Barone, A.D., Diggelmann, M., Fodor, S.P.A., Gentalen, E. and Ngo, N. (1997) *Journal of the American Chemical Society*, **119**, 5081-5090.
13. Giegrich, H., Eisele-Buhler, S., Hermann, C., Kvasyuk, E., Charubala, R. and Pfleiderer, W. (1998) *Nucleosides & Nucleotides*, **17**, 1987-1996.
14. Gao, X.L., Yu, P.L., LeProust, E., Sonigo, L., Pellois, J.P. and Zhang, H. (1998) *Journal of the American Chemical Society*, **120**, 12698-12699.
15. Beaucage, S.L. and Iyer, R.P. (1992) *Tetrahedron*, **48**, 2223-2311.

16. Hughes, T.R., Linsley, P., Marton, M., Roberts, C., Jones, A., Stoughton, R., Shoemaker, D., Blanchard, A., Phillips, J., Ziman, M. *et al.* (2000) *American Journal of Human Genetics*, **67**, 212.

17. Hughes, T.R., Mao, M., Jones, A.R., Burchard, J., Marton, M., Shannon, K.W., Lefkowitz, S.M., Ziman, M. and Schelter, J. (2001) *Nature Biotechnology*, **19**, 342-347.

18. Trabesinger, W., Schutz, G.J., Gruber, H.J., Schindler, H. and Schmidt, T. (1999) *Analytical Chemistry*, **71**, 279-283.

19. Vercoutere, W. and Akeson, M. (2002) *Current Opinion in Chemical Biology*, **6**, 816-822.

20. Selinger, D.W., Cheung, K.J., Mei, R., Johansson, E.M., Richmond, C.S., Blattner, F.R., Lockhart, D.J. and Church, G.M. (2000) *Nature Biotechnology*, **18**, 1262-1268.

21. Heller, M.J., Forster, A.H. and Tu, E. (2000) *Electrophoresis*, **21**, 157-164.

22. Swanson, P., Gelbart, R., Atlas, E., Yang, L., Grogan, T., Butler, W.F., Ackley, D.E. and Sheldon, E. (2000) *Sensors and Actuators B-Chemical*, **64**, 22-30.

23. Blanchard, A.P., Kaiser, R.J. and Hood, L.E. (1996) *Biosensors & Bioelectronics*, **11**, 687-690.

24. Pease, A.C., Solas, D., Sullivan, E.J., Cronin, M.T., Holmes, C.P. and Fodor, S.P.A. (1994) *Proceedings of the National Academy of Sciences of the United States of America*, **91**, 5022-5026.

25. Singh-Gasson, S., Green, R.D., Yue, Y.J., Nelson, C., Blattner, F., Sussman, M.R. and Cerrina, F. (1999) *Nature Biotechnology*, **17**, 974-978.

26. Cerrina, F., Blattner, F., Huang, W., Hue, Y., Green, R., Singh-Gasson, S. and Sussman, M. (2002) *Microelectronic Engineering*, **61-2**, 33-40.

27. Sampsell, J.B. (1994) *Journal of Vacuum Science & Technology B*, **12**, 3242-3246.

28. Stetsenko, D.A. and Gait, M.J. (2000) *Journal of Organic Chemistry*, **65**, 4900-4908.

29. Stetsenko, D.A., Malakhov, A.D. and Gait, M.J. (2002) *Organic Letters*, **4**, 3259-3262.

30. Nuwaysir, E.F., Huang, W., Albert, T.J., Singh, J., Nuwaysir, K., Pitas, A., Richmond, T., Gorski, T., Berg, J.P., Ballin, J. *et al.* (2002) *Genome Research*, **12**, 1749-1755.

31. Beier, M., Baum, M., Rebscher, H., Mauritz, R., Wixmerten, A., Stahler, C.F., Muller, M. and Stahler, P.F. (2002) *Biochemical Society Transactions*, **30**, 78-82.

32. Beier, M., Stephan, A. and Hoheisel, J.D. (2001) *Helvetica Chimica Acta*, **84**, 2089-2095.

33. Yang, J.N., Liu, Y.J., Rauch, C.B., Stevens, R.L., Liu, R.H., Lenigk, R. and Grodzinski, P. (2002) *Lab on a Chip*, **2**, 179-187.

34. Weston, D.F., Smekal, T., Rhine, D.B. and Blackwell, J. (2001) *Journal of Vacuum Science & Technology B*, **19**, 2846-2851.

35. Shi, W.M., Reed, M.R., Liu, Y.B., Cheung, L.N., Wu, R.J., Asatourians, A., Park, G. and Coty, W.A. (2002) *Clinical Chemistry*, **48**, 12.

36. Ramakrishnan, R., Dorris, D., Lublinsky, A., Nguyen, A., Domanus, M., Prokhorova, A., Gieser, L., Touma, E., Lockner, R., Tata, M. *et al.* (2002) *Nucleic Acids Research*, **30**, art. no.-e30.

37. Liu, Y.J., Rauch, C.B., Stevens, R.L., Lenigk, R., Yang, J.N., Rhine, D.B. and Grodzinski, P. (2002) *Analytical Chemistry*, **74**, 3063-3070.

38. Lenigk, R., Liu, R.H., Rhine, D., Singhal, P., Athavale, M., Chen, Z.J. and Grodzinski, P. (2002) *Abstracts of Papers of the American Chemical Society*, **224**, 027-BIOT.

39. Dorris, D.R., Ramakrishnan, R., Trakas, D., Dudzik, F., Belval, R., Zhao, C., Nguyen, A., Domanus, M. and Mazumder, A. (2002) *Genome Research*, **12**, 976-984.

40. Macoska, J.A. (2002) *Ca: a Cancer Journal for Clinicians.*, **52**, 50-59.

41. Van't Veer, L.J. and De Jong, D. (2002) *Nature Medicine*, **8**, 13-14.

Microfabricated Fluorescence-Activated Cell Sorter

with Hydrodynamic Flow Manipulation

Hyunwoo Bang*, Chanil Chung**, Jung Kyung Kim**, Seonghwan Kim**, Seok Chung**,
Dong-Chul Han*, and Jun Keun Chang**, ***

*School of Mechanical & Aerospace Eng., Seoul National University, Seoul, Korea, savoy@snu.ac.kr
**Digital Bio Technology Co., Seoul, Korea, jkchang@digital-bio.com
***School of Electrical Eng. & Computer Science, Seoul National University, Seoul, Korea
jkchang@digital-bio.com

ABSTRACT

Presented is a novel flow manipulation and particle detection method for the microfabricated fluorescence-activated cell sorter (µFACS). With hydrodynamic flow manipulation which includes passive flow channeling and hydrodynamic actuation with nozzle flow, we have developed a fast and robust method for utilizing the sorting function. These techniques are directly related to enabling the high throughput screening of the FACS machine. Moreover by detecting with synchronized imaging, we have also developed an on-line calibration technique for the accurate timing between detection and actuation. And this means that our system is very flexible that it can be tuned to any assay conditions with various flow speeds, particle densities and buffer or sample viscosities by real time calibration.

Keywords: micro-FACS, Lab-on-a-chip, hydrodynamic flow manipulation, plastic micromachining, PDMS

1 INTRODUCTION

Compared with conventional FACS machines, the µFACS provides higher sensitivity, lower cost, no cross-contamination and short flow path for the sorted cell viability. In this paper we present our latest results on the development of µFACS including the flow focusing and channeling concept, the design of the hydrodynamic flow manipulation, plastic microchip fabrication issues and the integration of the detection optics and signal processing system with real time system delay calibration.

For the microchip fabrication we applied plastic micromachining which is safer and easier to fabricate and suitable for mass production. And for the flow driving we chose simple negative pressure driving with vacuum suction. Actually vacuum driving setup requires only one peristaltic pump. Finally for the stream channeling and manipulation which is the main theme of this paper, we applied the passive channeling concept, hydrodynamic flow manipulation method and real time calibration with synchronized imaging technique.

The plastic microchip has three inlet ports for the sample and buffer, two outlet ports for sorted beads and waste. Between those inlet and outlet ports there is a single broad actuation channel that enables the 'timely hit' at the target cells. Simple principles of the hydrodynamic flow manipulation using off-chip valve switching technique was used to provide a much smaller size of the whole µFACS system, which resulted in a robust device design and highly reproducible focusing and sorting flow generations. Chips were evaluated by obtaining the substantial enrichments of the different fluorescent micro-beads in sorted micro chambers respect to their colors.

2 PRINCIPLE AND DESIGN

2.1 Hydrodynamic flow manipulation

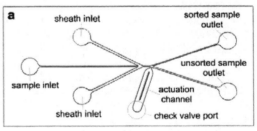

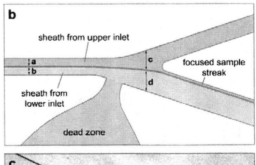

Figure 1: Channel configuration for the hydrodynamic flow switching. (a) Overall channel design with flow inlet and outlet ports (b) Passive flow channeling scheme (c) a CCD image of the focused Tryphan blue

Flow focusing is the core technology in flow cytometry. After flow focusing, the focused sample stream has to be channeled into desired outlet port for the sorting function. Then we should sort out specific particle by switching the focused stream into the sorting channel. Moreover for high throughput sorting flow switching should have high alternating frequency with good reproducibility. For this we designed a vacuum driving based fluidic network including passive flow channeling with hydrodynamic flow switching. And the device showed very fast switching ability with robust and reproducible operation.

The overall geometry and port configuration is shown in figure 1a. In figure 1a, there are two outlet channels which are connected to sort and unsorted sample outlet ports respectively. Those two channels have width of 100 μm and 150 μm respectively as shown in figure 1b. And all channels are 50 μm in depth. Because of the simple principle that streamline never crosses itself, the focused stream will roughly flow through the center line of the overall streamlines and as shown in figure 1b, the width c and d like a and b will roughly be the same. The difference in width between those two channels in the downstream makes the focused sample stream glance slightly downward the wedge structure at the diverging point.

At the diverging point it was a slight glance in flow path for the focused stream but in the end the focused stream follows through way down the lower channel and it results in drastic flow channeling. The fact that slight glance made the focused stream be channeled into sorting channel is a very important fact for flow switching because it means that a small disturbance can switch the stream up and down after the diverging point. In figure 1c you can see that focused stream of Tryphan blue is being channeled into lower channel. This concept is proved later in this paper by computer simulation. We didn't apply any active flow manipulation to channel the focused stream thus we call this concept passive flow channeling.

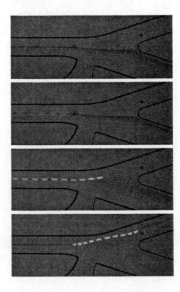

Figure 2: Focused streak of microbead being switched

As mentioned above if we just give slight disturbance toward upper channel at the diverging point, we can switch the focused stream upward. And after the disturbance has been disappeared the focused stream will rapidly switch back to its original state automatically. For this disturbance we designed an actuation channel (figure 1a) which is connected to piping from the buffer chamber which is normally blocked by check valve thus choked normally. By promptly opening and closing the check valve, the actuation channel acts as an instance disturbance source toward upper channel providing very high speed flow switching scheme.

Empirically, micro scale channel structures [1] can't afford this kind of rapid disturbance because of its laminar flow characteristics, so we designed the pressure channel to be sub-millimeter scale structure which is about 700 μm in width. The effectiveness of this millimeter scale channel structure is proved below by computer simulation. Figure 2 shows focused streak being switched by the instance disturbance method mentioned above. The application of this switching scheme and detailed explanation of the sorting function will be described with calibration method in real time sorting inspection section.

With this configuration sample stream does not run through the check valve [2], so there's no chance for the contamination or clogging inside the valve module. Other research groups and conventional FACS machines also use check valves, but they use check valves in the downstream and focused stream normally flows through it. The focused sample stream flow and blocking of it results in contamination and sample clogging inside the check valve which becomes one of main causes for system malfunction.

And our hydrodynamic flow manipulation scheme does not induce any surface modification [3] or on chip electrodes for flow manipulation [4]. And our sorting scheme occurs in a simple and continuous manner that any complicated flow driving schemes [5] are not required. This simplicity ensures the highly robust operation with easy and reproducible fabrication processes.

2.2 Simulation

Simulations were performed on a multiple-physics software package based on the computational fluid dynamics (CFD-ACE 2003, CFD Research Corporation, Huntsville Alabama, USA).

In figure 3 there are three pairs of triangular markers by the inlet and outlet of the channel. Three markers in the left inlet are used to trace streamlines in the upper, center and lower point of inlet channel respectively. On the right side of figure 3a you can see the green marker (trace 2) which traces focused sample stream of the inlet is finally in the upper side of lower channel. This proves that the channel network make the focused stream slightly glance at the diverging wedge.

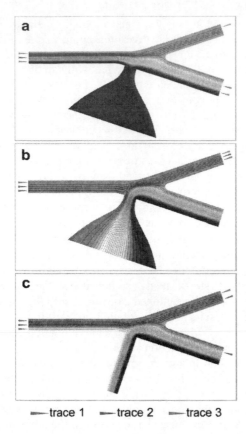

—— trace 1 —— trace 2 —— trace 3

Figure 3: Computer simulation of the flow switching scheme (a) Initial state : the focused stream is passively channeled into lower channel (b) Flow actuation by the actuation channel with nozzle structure (c) Flow actuation by the actuation channel without nozzle structure

In figure 3b and 3c you can find out the effectiveness of the millimeter scale actuation channel structure. In figure 3b all markers are finally in the upper channel. But in figure 3c only green marker (trace 2) is in the upper channel which means the actuation barely switched the focused stream to the upper channel. Considering the time until the flow switching is fairly accomplished, the channel configuration in figure 3b requires much shorter time to accomplish flow switching operation than the channel configuration in figure 3c. You can see that with the same width of the opening gate, channel configuration with millimeter scale actuation channel followed by nozzle structure exerts much faster switching ability.

2.3 Real-time Sorting Inspection

Figure 4 shows overall system setup with data signal and trigger signal wiring and flow driving configuration. The cell-sorting device was mounted on an inverted microscope (IX71, Olympus, USA). A green laser (GM32-10H, Intellite, USA) was used for epifluorescence excitation. And a (USH102D, Olympus, USA)W mercury lamp provided ambient fluorescence to aid capturing particle tracks after laser detection.

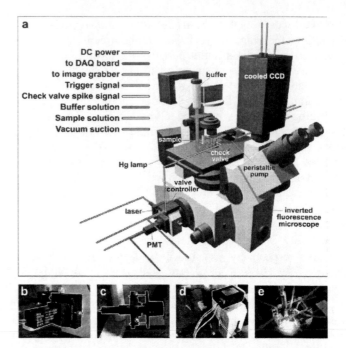

Figure 4: (a) System setup with signal and solution I/O of the μFACS system (b) XY stage and pin hole installed photo multiplier tube adaptor (c) XY stage and adjustable lens apparatus installed 532nm green laser adaptor (d) Spike signal generation kit for the check valve (e) Plastic microchip and its piping and a valve for the FACS operation

The fluorescence was collected through the objective and detected by a photo multiplier tube (PMT) (H5784-01, Hamamatus, Japan) with pinhole installed in its adaptor. The electrical signal from the PMT was sent via a preamplifier to a DAQ board (SD-104, COMIZOA, Korea) on PC. The signal was plotted on monitor and processed by appropriate threshold level to produce check valve trigger signal. Switching the flow switching valve (INKA2424212H, The Lee Company, USA) forced the beads of interest go to the collecting channel. And the same trigger signal activated cooled CCD (PCO, Kelheim, Germany) thus grabbing particle tracks after laser detection for on line sorting verification. A peristaltic pump (P625/10k.143, Instech, USA) were used for driving the sample and the sheathing buffer.

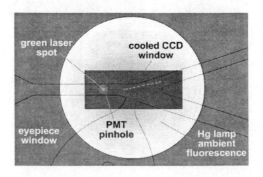

Figure 5: Light and window configuration

We focused on making a flexible detection and actuation system that can be tuned to any assay conditions with various flow speeds, particle densities and buffer or sample viscosities by real time calibration. There are lots of delay sources such as PC system delay by operating system, signal processing lag, valve actuation delay caused by mechanical characteristics of the inner valve structure and flow manipulation delay by tubing and chip (PDMS) elasticity. So if we can't perfectly predict and measure these delay sources and their degrees, we have to make flexible system that can be tune out any undesired delays.

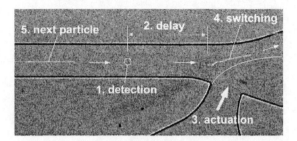

Figure 6: Sorting sequence with on line system delay calibration

Figure 6 shows how our system calibrates its delay sources. After fluorescence detection the valve trigger signal also triggers cooled CCD and the image is displayed on the monitor screen so we can inspect the verification image showing that the switching happened at appropriate moment or not. Adjusting the detection position by controlling XY stage of the microscope we can calibrate the distance between detection point and flow switching point. All the delay sources are implied in this distance, so by adjusting this distance real time sorting calibration can be accomplished. Actually it's like manipulating a jog dial watching monitor screen to calibrate the sorting operation.

3 FABRICATION

The plastic µFACS chip was fabricated in poly(dimethylsiloxane) (PDMS) using soft lithography [11] as is shown in Figure 7. The choice of the material PDMS depends primarily on its biocompatibility because this device is designed to dilute biological samples. Briefly, CAD file drawing has been patterned on a glass substrate with chrome masking. The chrome mask was used in 1:1 contact photolithography with SU-8 photoresist to generate a negative master mold, consisting of patterned photoresist on a Si wafer. Positive replicas with embossed channels were fabricated by molding PDMS against the master. Three inlet ports, two outlet ports and a actuation port (1-mm-diameter holes) for the fluids were punched out of the PDMS using a simple steel hand-puncher. The surface of the PDMS replica and a clean slide glass were activated in an oxygen plasma (2.666 x 10-2 Pa, 20 s, 25 W) and was brought together immediately after activation. An irreversible seal was formed between the PDMS and the slide glass. Finally, this assembly produced the required systems of microfluidic channels.

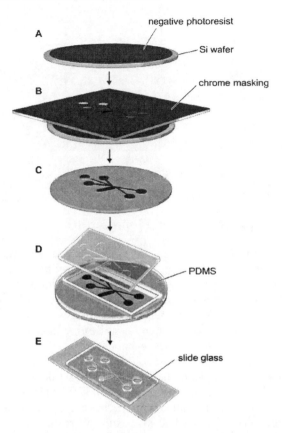

Figure 7: Scheme describing the steps involved in the fabrication process of PDMS-glass hybrid microchip

The fabricated microchip is smaller than a slide glass (14 mm x 29 mm x 4 mm, in width, length and thickness). The microchip has six holes five of which will latterly be connected to polyethylene tubing making inlet and outlet piping and one for the check valve connection. The check valve tip is very small thus directly inserted into the actuation hole. You can see fabricated plastic µFACS chip in figure 8.

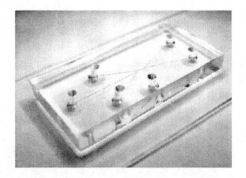

Figure 8: Fabricated plastic µFACS chip

4 RESULTS AND DISCUSSION

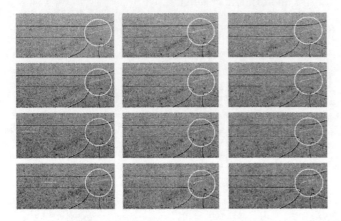

Figure 9: Micro beads being sorted out

Every time sorting happens the system leaves an image for the verification of its operation. Once the total system delay has been calibrated by the feedback inspection of the images, all detected particles are sorted out to the sorting channel. Figure 9 shows sequence of images after total system delay calibration, and you can see that all particles are sorted out to upper (sorting) channel.

Still for the on line calibration there's a limit in lateral flow speed. To take simultaneous images while sorting, there should be enough exposure time. So if the lateral flow speed is too fast particle streak images have to be taken in a very short time interval. And low exposure results in dark images which are so vague to be distinguished. But if we use conventional calibration method which is verifying the system delay by inspecting collected samples after each run or use fixed lateral flow speed, we can also calibrate the whole assay condition like other systems.

Our µFACS system is a very flexible one that it can be tuned to any assay conditions with various flow speeds, particle densities and buffer or sample viscosities by real time calibration. And its passive channeling and hydrodynamic flow manipulation concept ensures robust, fast and highly reproducible operation. We believe this kind of µFACS system will find a place next to conventional FACS machine as an essential facility of cell biology laboratory while cellomics is needed.

REFERENCES

[1] Jan Krüger, Kirat Singh, Alan O'Neill, Carl Jackson, Alan Morrison and Peter O'Brien, J. Micromech. Microeng., 12, 486–494, 2002.

[2] A. Wolff, I. R. Perch-Nielsen, U. D. Larsen, P. Friis, G. Goranovic, C. R. Poulsen, J. P. Kuttera and P. Tellemana, Lab Chip, 3, 22–27, 2003.

[3] Dongeun Huh, Yi-Chung Tung, Hsien-Hung Wei, James B. Grotberg, Steven J. Skerlos, Katsuo Kurabayashi and Shuichi Takayama, 4:2, 141-149, 2002.

[4] Petra S. Dittrich and Petra Schwille, Anal. Chem., 75, 5767-5774, 2003.

[5] Anne Y. Fu, Hou-Pu Chou, Charles Spence, Frances H. Arnold, and Stephen R. Quake, Anal. Chem., 74, 2451-2457, 2002.

Development of a Slug-Flow PCR Chip with Minimum Heating Cycle Times

S. Hardt, D. Dadic, F. Doffing, K. S. Drese, G. Münchow and O. Sörensen

Institut für Mikrotechnik Mainz (IMM)
Carl-Zeiss-Str. 18-20
D-55129 Mainz
Germany
hardt@imm-mainz.de

ABSTRACT

A microfluidic PCR device is presented capable of rapid temperature ramping and handling of sample volumes in the microliter and submicroliter range. The PCR chip comprises a straight micro channel in which a sample slug is periodically moved over three temperature zones. As part of the sample preparation a method for metering of the sample volume was developed. The PCR chip and the chips for sample preparation and fluidic actuation were fabricated by ultra-precision milling in polymer substrates. Computer simulations suggest that the sample slugs are heated or cooled on a time scale of some ten milliseconds when transported to a different temperature zone. The fluidic actuation based on a ferrofluid transducer is capable of positioning the sample volumes with a high accuracy after a large number of cycles. The design developed should be ideally suited for fast PCR of small sample amounts in a highly parallel manner.

Keywords: Microfluidics, PCR, heat transfer, slug flow.

1 INTRODUCTION

The polymerase chain reaction (PCR) is an enzyme catalyzed reaction for amplification of nucleic acids requiring three different temperature levels. At the highest temperature, a denaturation of double-stranded DNA takes place. The lowest temperature is required for annealing of primers being attached to the DNA strands, whereas at the intermediate temperature an extension step occurs using the existing strands as templates. In the past decades the PCR reaction has become one of the most important building blocks of nucleic acid analysis and has found a plethora of applications in such fields as medical diagnostics.

A number of systems allowing to perform the PCR on a microfluidic chip have been reported, among others the continuous flow chip developed by the group of Manz [1]. The solutions presented so far suffer from a number of disadvantages. As an example, a pressure-driven continuous flow device usually exhibits a considerable sample dispersion, thus preventing that all sample fractions can be processed at the same speed. Furthermore, many devices are not well suited for parallel, multiplexed PCR and require comparatively long heating cycle times, thus resulting in considerable time requirements for nucleic acid amplification. In addition, the processing of small samples in the micro- or submicroliter range is often not taken care of. The PCR system presented in this article was designed to eliminate some of these disadvantages.

2 DESIGN AND FABRICATION OF A TEST SYSTEM

In order to avoid cross-contamination of different samples, a PCR chip should ideally be designed as a disposable. Due to their low material and processing costs polymers are perfectly suited for corresponding Lab-ob-a-Chip applications. The feasibility of PCR amplification in polymer chips was proven earlier by Yang et al. [2]. The design concept which is studied in this article and was realized in polymers is sketched in Fig. 1.

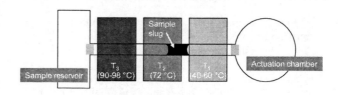

Fig. 1: Design principle of the slug-flow PCR chip.

A sample reservoir is connected to an actuation chamber via a micro channel in which a liquid slug is transported. The chip comprises three temperature zones corresponding to the temperatures required for the different steps of the PCR protocol. The sample slug is surrounded by air and stands in pneumatic contact with the actuation chamber. Through pneumatic transduction the slug is periodically transported between the temperature zones and is positioned in the center of each zone as appropriate. A similar principle based on a periodic motion of a sample slug between temperature zones has previously been reported [3,4]. However, in these studies no integrated microsystem was used and the slug was surrounded by an organic liquid phase.

For the experimental studies several test chips were realized in Poly (methyl methacrylate) (PMMA) and Cyclo olefin copolymer (COC) by ultra-precision milling

(Precitech Nanoform 350). In order to study different channel geometries, three straight channels with a depth of 200 μm and a width of 200, 600 and 1000 μm, respectively, were placed on each chip. At both ends of each channel ports for connection with a sample reservoir and an actuation chamber were provided. The polymer substrates were sealed with polymer sheets of 125 μm thickness by solvent bonding.

The PCR chip is coupled to a sample preparation chip and an actuation chip via fluidic interfaces. The complete setup of three chips in a row is placed on a metal rack which also provides the housing for three heating blocks. The heating blocks are connected with each other and comprise rectangular temperature zones on which the PCR chip is placed. In each of the blocks a heating cartridge and a thermocouple is inserted. A feedback control loop for the thermal management of the device was realized within the LabView (National Instruments) environment.

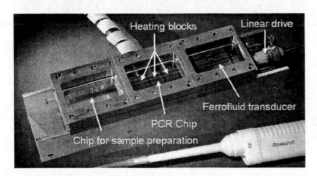

Fig. 2: Setup of a test system for on-chip PCR comprising a sample preparation and an actuation chip besides the actual PCR chip.

The complete test system comprising the three chips is shown in Fig. 2. One of the key components in this context is the actuation mechanism. The actuation chip contains a chamber partly filled with a ferrofluid (Ferrofluidics APG 047 n). Below this fluidic transducer a permanent magnet is moved by a linear drive. The ferrofluid follows the motion of the permanent magnet and displaces the sample slug in the PCR chip via pneumatic transduction. That way a periodic motion of the magnet is translated to a periodic motion of the slug which can be positioned within the temperature zones with a high accuracy.

3 SAMPLE PREPARATION

On of the basic steps performed in the sample preparation chip is the metering of a liquid slug from a larger volume. Typical sample volumes in which the PCR should be carried out are in the range of 1 μl. The series of steps allowing to create a sample slug of corresponding size is depicted in Fig. 3. The key element is a three way channel junction. In the channel extending to the right a ferrofluid volume is placed with the help of a permanent

magnet. The sample liquid is introduced from above, being transported through the channel extending to the left while the flow through the right channel is blocked by the ferrofluid. By subsequently moving the ferrofluid volume to the right a well defined amount of sample liquid is sucked into the channel provided for sample transfer. The actual metering occurs when the residual liquid volume is blown out with air. In this manner a well defined sample slug remains which can be transferred to the PCR chip.

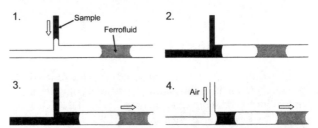

Fig. 3: Description of the steps employed for metering of the sample slug.

4 CFD SIMULATIONS

As noted by by Giordano et al. [5], among others, the time required for thermal ramping usually determines the time scale of a PCR cycle when the protocol is executed for DNA strands with a moderate number of base pairs. Under these conditions a reduction of the cycle times can only be achieved when the heating and cooling of the sample volume is speeded up. Typical heating and cooling rates of conventional thermocyclers are in the range of 2-10 K/s [6]. One of the drivers behind the development of the PCR system presented here is the reduction of the thermal ramping time scale.

As the temperature inside the sample slug is quite difficult to measure, simulations employing methods of computational fluid dynamics (CFD) were performed to study the flow and temperature distributions. In addition, such simulations allow for parameter studies which can not easily be conducted experimentally.

In order to study the heating of a liquid slug, the Navier-Stokes equation, the mass conservation equation and the enthalphy equation were solved on a structured hexahedral grid using the commercial finite-volume solver CFX5.6 (Ansys-CFX). For the free-surface flow problem the volume-of-fluid (VOF) method [7] was used assigning a "color" field to the different fluids, where a value of 1 indicates fluid 1 and a value of 0 fluid 2. Within the VOF method basically an advection equation for the color field is solved. A liquid slug with a length of 5 mm was initialized in a channel with a width of 1 mm and a depth of 200 μm (being the widest channel found on the test chip). The properties of water and air were chosen for the liquid and the gas phase, respectively. A value of 90° was assumed for the contact angle. In order to safe computational time, the problem was solved in the frame-of-reference co-moving with the slug. In that manner the number of computational cells required was reduced considerably. The channel walls were assumed as isothermal, having a temperature of 72 °C. The initial temperature of the fluids was set to 52 °C.

Simulation results for the temperature field as a function of time based on 46,000 grid cells are shown in Fig. 4. In these simulations a slug velocity of 2.5 cm/s was chosen, a realistic value easily achievable with the ferrofluid transducer. The figure shows the temperature field over a cut through the center of the channel for different times (in ms) indicated on the left. Each of the patches shown has a height of 200 μm corresponding to the channel height. The temperature range extends from 52 °C (white) to 72 °C (black).

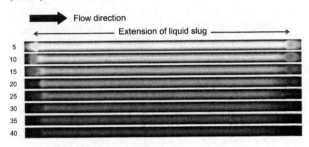

Fig. 4: Temperature field in the PCR channel as obtained from a CFD simulation.

As apparent from the figure, the temperature field is quite uniform over the slug, with gradients mainly occurring in the vertical direction. The figure also indicates heating times of some ten milliseconds. In order to quantify the heating time more precisely, the average temperature over a patch cutting through the center of the channel was determined by integration. This provides an upper bound to the heating time since liquid volumes located closer to the channel walls are disregarded in that way.

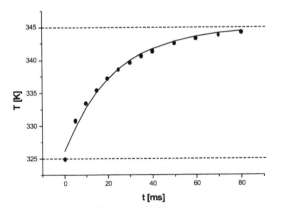

Fig. 5: Average temperature over a central plane cutting through the sample slug as a function of time.

In Fig. 5 the average temperature obtained in such a way is displayed as a function of time. The data points represent the CFD results and the solid line was obtained from a fit to the data using an exponential approach to the wall temperature. A temperature differing by less than 1 K from the wall temperature is reached after only 80 ms. The rapid heat transfer from the channel walls to the slug is due to two different effects. On the one hand, the small thermal diffusion path allows a rapid heating by heat conduction. On the other hand, the recirculation flow within the slug speeds up heat transfer additionally by convection.

5 EXPERIMENTAL TESTS

In a first experiment the temperature distribution on the PCR chip was measured using a infrared camera (FLIR ThermaCAM SC 500). The temperature along a PCR channel is displayed in Fig. 6. The three temperature zones are clearly visible as plateaus. Superposed to the temperature curve are three bands indicating the spatial extension of a typical sample slug. A worst-case estimate for the temperature variation over the slug can be obtained by evaluating the temperature span covered by the bands. Corresponding ΔT values are indicated in the figure. The real temperature variation over a sample slug is smaller than that, since owing to the higher thermal conductivity of water compared to the polymer material of the chip, a thermal equalization occurs.

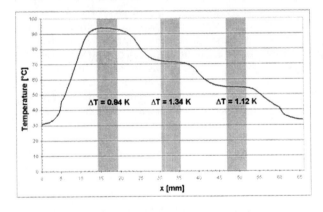

Fig. 6: Temperature distribution along a PCR channel.

The temperature distribution displayed in Fig. 6 suggests that for a rapid DNA amplification the sample slug needs to be positioned in the middle of the temperature plateaus with a sufficient accuracy. In order to study the sample transport, experiments with the ferrofluid drive were carried out for sample volumes ranging from 0.1-1.5 μl. A series of snapshots of such an experiment in the largest of the three micro channels is displayed in Fig. 7. A PCR cycle consists of successively positioning the slug in the middle of one of the temperature zones and moving it to the next temperature zone according to the protocol.

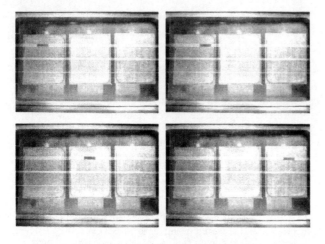

Fig. 7: Motion of a sample slug driven by pneumatic coupling to a ferrofluid transducer.

The experiments were carried out with the heating blocks brought to their appropriate temperature, so that possible problems with sample evaporation could be investigated. Besides evaporation, the focus of the experiments was on the positioning accuracy achievable with the ferrofluid transducer. It was found that problems due to evaporation mainly occurred in the smaller channels, whereas in the channel with a width of 1 mm evaporation is suppressed to a degree that enables a sufficiently large number of PCR cycles. More specifically, after 35 cycles the sample slug stayed intact, with no satellite droplets being observed, although some volume reduction due to evaporation was detected. In these experiments the position where a slug is deposited in the last cycle deviates from the corresponding position in the first cycle by less than 100 µm. Given a characteristic slug length of a few millimeters it can be concluded that the pumping mechanism by ferrofluidic transduction is suitable to carry out the PCR protocol by periodic slug motion. The simulation results presented in the previous section suggest that the sample virtually reaches its processing temperature already on the way to its "parking position".

In a second series of experiments the interaction of DNA strands contained in the sample slug with the walls of the micro channel was studied. Compared to conventional thermocyclers, the channels of microfluidic PCR chips are characterized by a significantly larger surface-to-volume ratio. For this reason, adsorption of DNA strands to the channel walls might drastically reduce the observed amplification efficiency. The corresponding samples were obtained from Evotec Technologies (Düsseldorf, Germany). Sample slugs with fluorescence-labelled human DNA strands of a molecular weight of 370 base pairs were moved through the PCR channel of 1 mm width over 30 cycles. The sample was then removed from the channel, analyzed by gel electrophoresis and compared with an unprocessed sample. The fluorescence intensity of the former was found to be only slightly reduced compared to the latter. Hence it

may be concluded that adsorption to the walls of the polymer channels should not pose a major problem for on-chip PCR.

6 CONCLUSIONS AND OUTLOOK

In conclusion it may be stated that a microfluidic PCR system has been established and its fluidic and thermal performance has been studied by experimental and simulation methods. The device allows for rapid temperature ramping on a time scale of some ten milliseconds and is able to process samples volumes in the micro- or submicroliter range. Important thermal and fluidic requirements such as comparatively small temperature variations over the sample volume and a reproducible and accurate positioning of the sample slug within the temperature zones are met. In addition, a sufficiently small wall adsorption of DNA strands was proven.

Experiments focusing at carrying out the PCR protocol inside the system are currently under way. The goal of these experiments is to maximize the observed amplification efficiency. First results indicate that nucleic acid amplification proceeds inside the micro channel. If necessary, the system bears a considerable potential for further optimization. As an example, the temperature profile shown in Fig. 6 could be further improved using a modified chip design. Owing to the principle of periodic slug motion and the very simple channel design derived from that, the system is ideally suited for parallelization: It is easily conceivable that the sample reservoir and the actuation chamber shown in Fig. 1 are not only connected by one channel, but rather many parallel reactions could be carried out in a multitude of channels.

ACKNOWLEDGEMENT

We are grateful to T. Hang for performing thermographic measurements. This work was supported by the German Ministry of Research and Education, grant number 16SV1355.

REFERENCES

[1] M. Kopp, A. de Mello and A. Manz, Science 280, 1046, 1998.
[2] J. Yang, Y. Liu, C.B. Rauch et al., Lab Chip 2, 179, 2002.
[3] M. Curcio and J. Roeraade, Anal. Chem., 75, 1, 2003.
[4] P.A. Auroux, P.J.R. Day, F. Niggli et al., Proceedings of Nanotech2003, 55, 2003.
[5] B.C. Giordano, J. Ferrance, S. Swedberg et al., Analytical Biochemistry 291, 124, 2001.
[6] I. Schneegaß and J.M. Köhler, Reviews in Molecular Biotechnology 82, 101, 2001.
[7] C.W. Hirt and B.D. Nichols, J. Comput. Phys. 39, 201, 1981.

COMPOSABLE SYSTEM SIMULATION OF DISPERSION IN COMPLEX ELECTROPHORETIC SEPARATION MICROCHIPS

Y. Wang, Q. Lin, T. Mukherjee

Carnegie Mellon University, Pittsburgh, PA, USA, yiw@andrew.cmu.edu

ABSTRACT

This paper presents a composable system simulation framework for electrophoretic separation microchips, using an analog hardware description language integrating analytical dispersion models that describe not only the behavior of individual components, but also the interactions between them. Both DC and transient analysis are performed in the framework. The accuracy (relative error less than 10%) and tremendous speedup (100~10,000×) of composable system simulations are verified by comparison to experiment and numerical studies.

Keywords: dispersion, system simulation, electrophoretic separation microchips

1 INTRODUCTION

While electrophoretic separation microchips have been widely studied in the past decade, their efficient simulation and design at the system-level continues to be a challenge. Experimental trial-and-error and numerical computation methods can lead to unacceptably long design cycles. Several analytical [1, 2] and semi-analytical [3] models and simulations of analyte band broadening (dispersion) have been proposed to speed up design at the component-level. A system simulation approach in which a design is decomposed into components has been presented [4], but involves over-simplified dispersion models that do not accurately account for component interactions and still require users with expert knowledge. To address these issues, this paper presents a composable system simulation framework using an analog hardware description language (Verilog-A) integrating analytical dispersion models [5] that are capable of capturing the interactions between components. The simulations illustrate the effect of chip topology, analyte properties and buffer properties on separation performance. Thus, it is generally applicable to the design of practical electrophoretic microchip devices.

2 COMPOSABLE SYSTEM SIMULATION

Our composable simulation framework consists of a model library and simulation engine. One major contribution over [5] is the development of a Verilog-A library consisting of parameterized behavioral models for commonly used components in separation microchips. Users can compose a complex design schematic by wiring these blocks for a fast and reliable top-down iterative approach to system-level design. Figure 1 illustrates two practical electrophoresis systems (serpentine and spiral) and their schematics. The systems are decomposed into a set of components including reservoirs, detector, injector, straight channels and turns, which are then linked via interface parameters according to the spatial layout and physics. Cadence is used to netlist the structure of the composable network and Spectre is employed as the simulator in this paper as in [6] (other schematic editors and simulators capable of simulating Verilog-A can also be used).

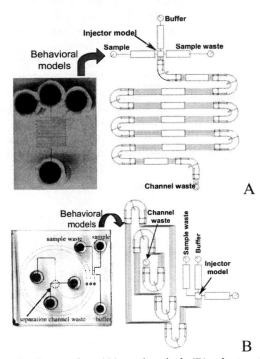

Figure 1: Serpentine (A) and spiral (B) electrophoretic separation microchips (Left) and their schematics (Right) with analog hardware description behavioral models

2.1 Interface parameters

The first step in creating a composable model of a system is to identify the parameters that will be communicated between neighboring components (interface parameters). There are two kinds of interface parameters involved in the network. One is the nodal voltage globally determined by Ohm's and Kirchhoff's laws. The other pertains to dispersion and includes variance (σ^2), the longitudinal standard deviation of the cross-sectional average concentration; skew coefficients (B_m), used to describe the skew caused by the turns; separation time (t), the moment the center of mass of the band reaches the component; and amplitude (A), the maximum concentration. Since the dispersion occurring in the downstream component doesn't affect the upstream, the parameters related to dispersion are calculated using a directional signal flow in which the output from one component is assigned as the input to the next, starting from an injector.

2.2 Behavioral Models

After selecting the interface parameters, the second step in developing a composable model library is selecting the list of composable elements, and deriving behavioral models for each element. Our composable library consists of seven basic models, which include turns (90° or 180°, clockwise or counter-clockwise), straight channel, injector and detector. The goal of each behavioral model is to capture the input-output signal flow relationship between the dispersion interface parameters and the equivalent Kirchhoffian electric network for the voltage interface parameter (see Figure 2). In a simulation scheme, the steady nodal voltage and the electric field through the component will be determined first. Users only need to input the voltages applied on all reservoirs as boundary conditions, and the simulator is able to calculate the nodal voltage distribution, which is another new feature beyond [4, 5]. Based on the computed electric field (E), the input-output functions of the dispersion interface parameters are then calculated. The inherent variable for the functions is the residence time Δt in a component.

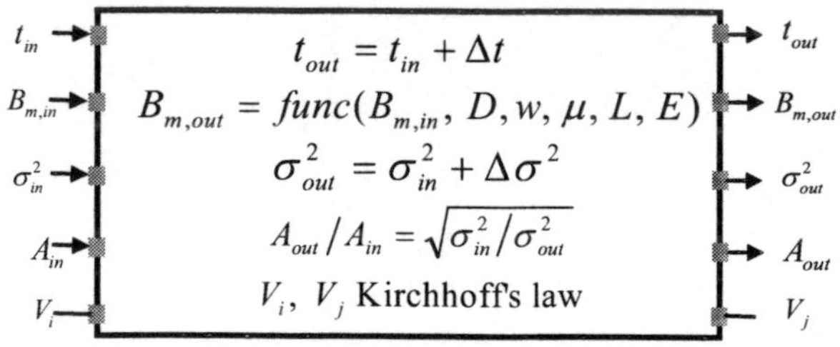

Figure 2: Structure of the behavioral model for composable system simulation. The arrow indicates the direction of signal flow for dispersion computation, and subscripts "in" and "out" represent the input and output to the model.

The residence time and amplitude ratio through a component is given as

$$\Delta t = L/E\mu \tag{1}$$

$$A_{out}/A_{in} = \sqrt{\sigma_{in}^2/\sigma_{out}^2} \tag{2}$$

where μ is the electrophoretic mobility of the species, L is the longitudinal length of the component; for a turn, $L=\theta R_c$ and θ is the angle of the turn. To obtain Eq (2), we always assume a Gaussian concentration distribution for the analyte band in the system.

The change of skew coefficients and variance depends on the specific component [5]. For a straight channel,

$$B_{m,out} = B_{m,in} \cdot e^{-(m\pi)^2 \Delta t D/w^2} \tag{3}$$

$$\Delta\sigma^2 = 2D \cdot \Delta t \tag{4}$$

For a turn,

$$B_{m,out} = \begin{cases} \pm\dfrac{4\theta\left(1-(-1)^m\right)\left(1-e^{-\lambda_m \Delta t D/w^2}\right)}{(m\pi)^4 \Delta t D/w^2} \\ \quad + B_{m,in}e^{-\lambda_m \Delta t D/w^2} \quad , \quad m \neq 0 \\ \\ \qquad\qquad B_{m,in} \qquad\qquad , \quad m = 0 \end{cases} \tag{5}$$

$$\Delta\sigma^2 = 2D\Delta t \pm \frac{8w^4\theta}{D\Delta t}\sum_{m=odd}^{\infty}\left(\frac{B_{m,in}\left(1-e^{-(m\pi)^2 D\Delta t/w^2}\right)}{(m\pi)^4}\right)$$

$$+ \frac{64w^6\theta^2}{(D\Delta t)^2}\sum_{m=odd}^{\infty}\frac{\left(-1+e^{-(m\pi)^2 D\Delta t/w^2}+(m\pi)^2 D\Delta t/w^2\right)}{(m\pi)^8} \tag{6}$$

where w is the channel width, D is the molecular diffusivity of the species, $m=0,1,2,3\ldots$ In Eq. (5) and (6), the plus sign is assigned to the first turn and any turn increasing the skew caused by the first; the minus sign is assigned to any undoing the skew.

To initiate and detect the dispersion, models of injector and detector are needed. Currently the injector model (see Figure 1) requires the user to specify its skew coefficients (B_m) and variance, which will be replaced with a detailed model from [7] when available.

In the detector model, the skew coefficient is transferred without change due to the small detection length L_{det}. The variance associated with the length is given as [8]

$$\Delta\sigma^2 = L_{det}^2/12 \tag{7}$$

The detector model generates an electropherogram (concentration vs. time) when the system is simulated using transient analysis. Spectre will first calculate DC operating points. Based on these points, the transient simulation is then performed by scanning the read-out time. Assuming the band doesn't substantially spread out during its passing through the detector, we approximate the cross sectional average concentration output (C_m) as

$$C_m = A_{out} \cdot e^{-(E\mu)^2(t-t_{out})^2/2\sigma_{out}^2} \tag{8}$$

where A_{out}, t_{out} and σ_{out}^2 are detector outputs from DC analysis; t is the actual read-out time.

3 RESULTS AND DISCUSSION

Our composable system simulation results are shown in Figures 3-6. In Figure 3A, an electrophoresis column of two complimentary turns used to separate TRITC-Arg is compared to our DC system simulations, showing excellent agreement. According to [5], the final variance normalized by w^2 depends on two dimensionless times τ_t and τ_s, the ratio of the time for an analyte molecule to advect through a channel to the time for it to diffuse across the channel width ($\tau = \Delta t D / w^2$, τ_t is the dimensionless time in the turn and τ_s in the inter-turn straight channel). In this simulation, $\tau_t = 0.068$ is relatively small, and transverse diffusion in the turn does not have enough time to remove all of the turn-induced skew, which accordingly incurs abrupt increase in variance (see the skewed band after the first turn in the numerical simulation plot in Figure 3A). During an analyte's migration in the long inter-turn straight channel, the transverse diffusion has adequate time (relatively high $\tau_s = 0.696$) to smear out most of the skew and presents a nearly uniform band before the second turn. The second turn then distorts the band again in the opposite direction, leading to another turn-induced variance equal to the one caused by the first turn. Figure 3B shows electropherograms from three detectors placed in the system. Respectively their positions are before the first turn, in the inter-turn straight channel, and after the second turn. Since both turns broaden the analyte band, the amplitude decreases consecutively and the band spreads out, where an initial band with variance $\sigma^2 = 100$ μm^2 and normalized amplitude $A=1$ was injected.

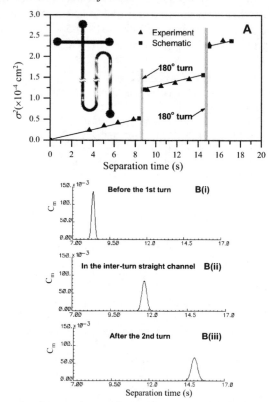

Figure 3: Comparison of experimental data [3] with DC system simulation (A) at $\tau_t = 0.068$ and $\tau_s = 0.696$. The mean width 50 mm [1] and diffusivity 3.12×10^{-10} m²/s were used in the simulation. Transient analysis (B) simulates three detectors' read-outs.

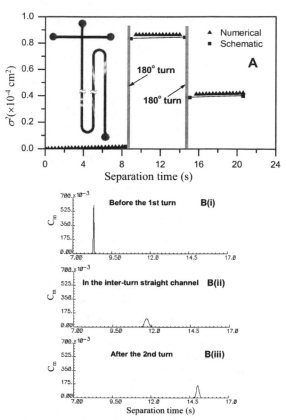

Figure 4: Comparison of DC system (A) simulation to numerical results at $\tau_t = 2.2 \times 10^{-3}$ and $\tau_s = 2.23 \times 10^{-2}$. Diffusivity of 1×10^{-11} m²/s was used in the simulation. Transient analysis (B) simulates three detectors' read-out.

In Figure 4, numerical and composable simulations are conducted, in which all parameters in Figure 3 are kept unchanged except for a very low diffusivity (common for DNA sequencing in gels or matrix materials), rendering extremely low $\tau_t = 2.2 \times 10^{-3}$ and $\tau_s = 2.23 \times 10^{-2}$. A maximum relative error of 5.6% is found. Netlisting and DC simulation take 50 seconds for the first time, and less than a second for subsequent iterations, leading to a 150~7,500× speedup. In this example, the longitudinal molecular diffusion is very small, inferred by the nearly zero change of variance in the straight channel; and the convection is significant in the total dispersion. Hence the analyte band keeps its sharp skew before arriving at the second turn that corrects most of the skew and leads to a reduced final variance. This indicates the importance of individual system level simulation and design for different species analysis. The transient simulation shows different concentration read-outs of the three detectors, compared to Figure 3B. Due to less dispersion of the analyte band, all detectors in Figure 4B show better performance compared to Figure 3B, as seen by the higher concentration output in Figure 4B. Because of the skew correction and dispersion reduction by the 2nd turn, the detector after the second turn shows a higher peak and narrower band than that in the inter-turn straight channel. This shows that for the microchip electrophoresis of low-diffusivity species, e.g.

the DNA in gels or other matrix materials that further reduce the diffusion coefficient, even numbers of turns should be applied to take advantage of the skew canceling interactions caused by complimentary turn pairs.

In Figure 5, a serpentine electrophoretic separation system (similar to Figure 1) involving six complimentary turns is simulated and compared to numerical results. The variance growth at the outlets of the second, the forth and the sixth turn are extracted. A worst-case error of 9.5% at the lowest τ_t and τ_s, and a 600~15,000× speedup are obtained. An observation can be made here that the variance increases with the even number of complimentary turns in a superlinear to sublinear manner depending on τ_t and τ_s [9].

In Figure 6, a complex spiral separation microchip of five turns to separate Dichlorofluoroscein is simulated and compared to experimental results. A worst-case error of 12% on plate number is found. The linear growth of plate number with electric field confirms that molecular diffusion is the major dispersion source in such a system [10].

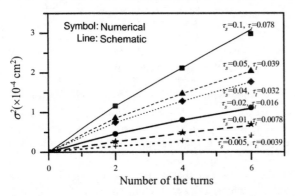

Figure 5: Comparison of composable serpentine system simulation (lines) to numerical results (symbols). τ_t and τ_s range from 0.078 to 0.0039 and from 0.1 to 0.005 respectively. The detections of the analyte band's variance are taken at the outlets of the second, the fourth and the sixth turns.

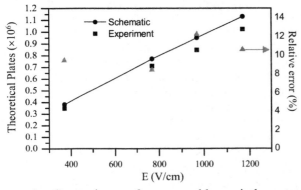

Figure 6: Comparison of composable spiral system simulation to experimental data on theoretical plate number vs. electric field. Right axis shows the relative error between simulation and experiments

4 CONCLUSION

A composable system simulation framework for complex electrophoretic separation microchips has been presented, in which a parameterized behavioral model library using an analog hardware description language (Verilog-A) has been developed. Kirchhoff's law and directional signal flow have been employed to solve the electric and dispersion network respectively. The system simulation results have been verified by numerical and experimental data. The proposed interface parameters and behavioral models are able to accurately capture the combined effects of system topology and parameters (such as analyte and buffer properties) on the separation performance. The transient analysis is also conducted to intuitively observe the actual detector read-out of the cross-sectional average concentration. Compared to numerical methods, a tremendous speedup (100~10,000×) can be achieved by the composable simulation, while still maintaining high accuracy (relative error less than 10%). This enables the sub-hour system-level synthesis and optimal design of electrophoretic separation microchips [11].

ACKNOWLEDGEMENT

This research is sponsored by the DARPA and the Air Force Research Laboratory, Air Force Material Command, USAF, under grant number F30602-01-2-0587, and the NSF ITR program under award number CCR-0325344.

REFERENCES

[1] S.K. Griffiths and R.H. Nilson, Anal. Chem., 72, 5473, 2000.
[2] J.I. Molho et al., Anal. Chem., 73, 1350, 2001.
[3] C.T. Culbertson, S.C. Jacobson and J.M. Ramsey, Anal. Chem., 70, 3781, 1998.
[4] Coventor Inc. "Behavioral Models for Microfluidics Simulations," CoventorWare, 2001.
[5] Y. Wang, Q. Lin, T. Mukherjee, MicroTAS'03, 135, 2003.
[6] Q. Jing, T. Mukherjee and G.K. Fedder. ICCAD'02, 10, 2002.
[7] R. Magargle, J.F. Hoburg, T. Mukherjee. NanoTech 2004, (accepted), 2004
[8] S.C. Jacobson, C.T. Culbertson, J.E. Daler and J.M. Ramsey. Anal. Chem. 70, 3476, 1998.
[9] R.M. Magargle, J.F. Hoburg and T. Mukherjee. MSM'03, 214, 2003.
[10] C.T. Culbertson, S.C. Jacobson and J.M. Ramsey. Anal. Chem., 72, 5814, 2000.
[11] A.J. Pfeiffer, T. Mukherjee, S. Hauan. IMECE'03, 2003.

Visualizing Individual Rhodopsin (a G Protein-Coupled Receptor) Molecules in Native Disk and Reconstituted Membranes via Atomic Force Microscopy

Eugene J. Choi*, Albert J. Jin*, Shui-Lin Niu**, Paul D. Smith*, and Burton J. Litman**

*IRDR, Division of Bioengineering and Physical Science, ORS, OD, National Institutes of Health, Bethesda, MD, 20892. choieug@ors.od.nih.gov
**Lab Membrane Biochemistry and Biophysics, National Institute of Alcohol Abuse and Alcoholism, National Institutes of Health, Rockville, MD, 20852. litman@helix.nih.gov

ABSTRACT

Individual rhodopsin molecules have been resolved with atomic force microscopy as both monomers and various oligomeric organizations that are sensitively dependent upon the physical state of membranes and environmental conditions. In intact native disk membranes, rhodopsin molecules are observed as randomly dispersed monomers and small oligomers. In reconstituted rhodopsin-DPPC (dipalmitoylphosphatidylcholine, di16:0-PC) membranes, phase separation of lipid and rhodopsin results in paracrystalline arrays of rhodopsin molecules. Since the coupling of rhodopsin and G proteins is an essential step in visual transduction, the results from this study could provide further insight into the structural relationship of rhodopsin to its function in vision and may have applications in nanotechnology.

Keywords: atomic force microscopy, rhodopsin, G-protein-coupled receptor, cell membrane.

1 INTRODUCTION

The rod outer segment (ROS) contains a stack of densely-packed, closed, flattened membrane disks [1]. These ROS disk membranes contain lipids and proteins, with docosahexaenoic acid (DHA) being the dominant phospholipid acyl chains and rhodopsin representing 95% of the protein content and acting as an important component for visual function [1]. Rhodopsin is a prototypical member of the heptahelical G-protein coupled receptor family, which is the largest family of cell surface receptors. The visual signal transduction pathway is initiated when a photon of light is absorbed by rhodopsin, resulting in the formation of metarhodopsin II, the G protein activating form of rhodopsin, from the inactive metarhodopsin I [1]. This transition promotes the binding and activation of the G protein transducin, which activates a phosphodiesterase, initiating the hydrolysis of cGMP and leading to Na^+/Ca^{+2} channel closures, resulting in a hyperpolarization of the plasma membrane transmitted to the synapse of the rod cell [1].

The structure and function of bovine rhodopsin are well characterized and studied [1-3]. However, the molecular organization of rhodopsin in disk membranes, which can have a critical impact on protein-protein interactions, was not directly observed until recently. Rhodopsin molecules from wild-type mouse ROS disk membranes were resolved as dimers organized in paracrystalline arrays on the cytoplasmic side of opened, single-layered disk membranes using atomic force microscopy [4].

Atomic force microscopy (AFM) is an innovative form of scanning probe microscopy with increasingly broad biological applications [e.g. 4, 5]. A sharp tip attached to the end of a cantilever is scanned across a surface permitting both atomic interactions between the probe and sample to be measured and sub-nanometer structures in biological samples *in situ* to be resolved. With specific advancements in AFM instrumentation, we have resolved bovine rhodopsin molecules on the top surface of intact ROS disks and have observed spontaneous phase separation of rhodopsin and lipid in reconstituted rhodopsin-DPPC membranes. In this manner, the organization of rhodopsin is captured in its most natural state, and its interplay with lipid environment is accessed.

2 EXPERIMENTAL

The ROS membranes were isolated from bovine retinas using the modified Shake-ate method [6]. Intact ROS disk membranes were isolated from ROS membranes using the Ficoll-floatation method [7]. Rhodopsin was purified on a conA affinity column [8]. Reconstituted rhodopsin-DPPC membranes were prepared using the dilution reconstitution method [9].

For AFM measurements, the ROS disk membranes and the reconstituted rhodopsin-DPPC membranes in large unilamellar vesicle (LUV) suspensions (c.a. 100 nm diameter via extrusion procedure) were adsorbed on mica in 5 mM Tris, 50 µM DTPA, 2mM DTT buffer and then washed with 20 mM Tris, 150 mM NaCl, 25 mM $MgCl_2$, 50 µM DTPA, 2 mM DTT. All imaging was performed under similar high ionic strength buffers and at room temperature, in both dark and light conditions, with the future capability of environmental control with argon gas. Both contact mode and tapping mode (TM) were utilized in this study with a PicoForce Multimode AFM under a Nanoscope IV controller and either a PicoForce or a type E scanner head (Veeco/Digital Instruments, CA) [5].

3 RESULTS & DISCUSSION

3.1 Rhodopsin in ROS Native Disk

This study shows, for the first time, individually resolved rhodopsin molecules on an intact, double-layered native membrane disk. AFM images (Figures 1A and 1B) reveal circular intact native disks with thicknesses between 13-18 nm and diameters of about 1 μm (Figure 1C). These thickness values are consistent with a height of two bilayer membranes stacked on top of each other, indicating that the disks are intact. Densely-packed rhodopsin molecules in close contact with each other can be seen on the surface of the disk membrane, and higher resolution low-pass filtered scans reveal that rhodopsin molecules are distributed in a random manner in the native disk as monomers and small oligomers (not shown). This random rhodopsin organization, as also evidenced by calculated power spectra (not shown), is contrary to the paracrystalline organization observed in single-layered disks from mice [4], but is consistent with historical biophysical and biochemical observations of random and diffusive rhodopsin dispersion in the fluid-like lipid environment of disk membranes provided by the high level of DHA phospholipids [10]. This discrepancy could be attributed to the effect of the mica substrate, which clearly caused lipid phase separation in the single-layered disks [4,10]. In our study, the images are obtained from the upper bilayer of the double-layered intact bovine disk, which is not in direct contact with the mica surface.

A typical AFM force spectroscopy curve (Figure 1D) shows that rhodopsin molecules in native disks can be stretched up to 120 nm, which is consistent with the full length of the extended rhodopsin peptide chain. The initial force needed to unfold and stretch the rhodopsin is about 500 pN. The cantilever then retracts and a smaller force of about 250 pN is sufficient to unfold the rest of the rhodopsin, comparable to other proteins [11].

The conditions necessary for optimal adsorption of the native disk to mica substrate required the use of high ionic strength buffer and the presence of monovalent and divalent ions. It is important to have native intact disk membranes adsorbed strongly enough to the mica support, but not so strong as to disrupt and burst the disks. This type of adsorption allows molecular rhodopsin to be visualized in the upper membrane by AFM. Our high ionic strength buffer conditions with divalent ions are similar to conditions used

by Fotiadis *et al.* [4] when they observed paracrystalline arrays of rhodopsin dimers in single-layered disks.

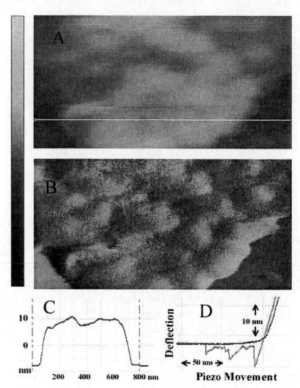

Figure 1: Examples of AFM measurements of native disk membranes with rhodopsin molecules in dark state at room temperature. **(A)** A contact-mode topographic image of a bovine ROS native disk membrane adsorbed on mica-supported surface (450 nm × 900 nm). **(B)** The deflection image corresponding to (A), showing both the densely-packed rhodopsin molecules and the smooth mica surface (on the left and right bottom corners of the figure). The brightness scales (top left) for height contrast in A and B is 60 nm and 1 nm, respectively. **(C)** A section profile corresponding to the white line in Figure 1A, confirming the presence of intact double bilayers in our disk preparations. **(D)** An example of an AFM force spectroscopy curve consistent with the unfolding of rhodopsin trans-membrane helices (blue curve for approaching; red curve for retraction and stretching). The nominal cantilever spring constant is 50 pN/nm.

3.2 Reconstituted Rhodopsin-DPPC

To address the issue of ROS disk adsorption and the interplay of rhodopsin organization with lipid phase separation on mica, this study now focuses on a reconstituted system with controlled rhodopsin to lipid ratios. Individual rhodopsin molecules were resolved by TM AFM in reconstituted rhodopsin-DPPC membranes with rhodopsin to DPPC ratios of 1:100 and 1:200. Under low ionic strength buffer conditions, the reconstituted membranes only weakly adsorb to the mica surface. However, as with the native disk membrane, the reconstituted rhodopsin-DPPC lipid membranes adsorbed more strongly to mica in the presence of high ionic strength buffer with both monovalent and divalent ions. At molar ratios of 1:100 and 1:200 [rhodopsin:DPPC] on mica (Figures 2 and 3), phase separation between rhodopsin-rich areas and rhodopsin-poor regions of the DPPC could be observed, with individual rhodopsin molecules being resolved. The rhodopsin-rich membrane is about 1-3 nm thicker than the surrounding membrane of thickness of 4-5 nm (Fig. 2A-C). Both rhodopsin monomers and oligomeric organizations occur in these rhodopsin-rich domains (Figure 3).

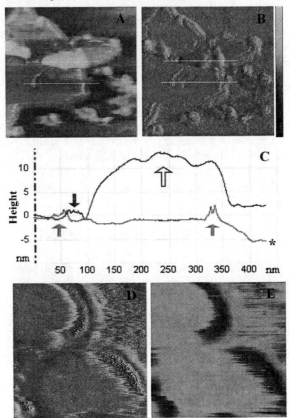

Figure 2: Phase separation in reconstituted rhodopsin-DPPC [1:100] membranes on mica substrates under light state at room temperature. **(A)** TM topographic image of a large rhodopsin-DPPC membrane region that contains elongated domains of rhodopsin monomers in DPPC bilayer as well as larger aggregates (0.75μm ×0.75μm scan). **(B)** TM amplitude map corresponding to (A). **(C)** Section profiles (lines in (A and B) with blue and red +) with filled arrows pointing to monomeric rhodopsin, and an open arrow pointing to larger multilayer membranes or rhodopsin aggregates. The mica substrate is indicated by *. **(D)** TM amplitude map of two un-fused liposome patches with phase separations (0.25μm ×0.25μm). **(E)** The corresponding phase image map showing huge AFM TM cantilever oscillation phase delay over rhodopsin-rich membrane, but not over rhodopsin-poor

membrane (upper left and center) or bare mica substrate (upper right). The brightness scale (top right) for (A), (B), (D), (E) is 20 nm, 300mV, 250mV, and 60 degrees.

The oscillation phase lag of the cantilever resonance relative to the signal sent to the cantilever's piezoelectric driver is an important measure of the viscoelastic and adhesion properties of the membrane. Figure 2E shows the first observations of membrane phase separation, into rhodopsin-rich areas and rhodopsin-poor areas, as measured with phase imaging using TM AFM. This technique provides clear resolution of membrane domains with substantial differences in material properties, and features that can be obscured by rough topography in height images. While structural details generally are revealed more clearly in error maps such as Figures 2B and 2D, TM AFM phase imaging is a powerful tool for displaying variations in sample properties both at very high resolution and in real time. At room temperature, the DPPC-rich phase is in the gel state, resulting in higher rhodopsin stability and lower rhodopsin mobility. This is in contrast to the native disk condition in which DHA phospholipids are highly unsaturated and more fluid, resulting in lower rhodopsin stability [3]. Tokumasu et al. have demonstrated that the phase behavior in supported reconstituted DPPC, dilauroylphosphatidylcholine, and cholesterol mixture membranes is modified compared with unsupported mixture membranes due to effects from the mica substrate. These mica effects depend on such factors as type and ratio of lipid mixtures, concentrations of monovalent and divalent ions, and buffer pH over the mica surface [5]. Our results show that rhodopsin-DPPC membrane phase separation occurs in single bilayers adsorbed to mica (Fig. 2).

In Figure 3A, regions of individual rhodopsin monomers can be clearly observed in the rhodopsin-rich phase at molar ratio [1:200]. Within this phase, individual rhodopsin molecules could be resolved with monomer peak spacing of about 3.5 nm (Figure 3B). Rhodopsin molecules packed together can be observed more clearly in Figures 3C and D, suggesting a structured organization. At a molar ratio of [1:200] of rhodopsin-DPPC membrane, two populations of rhodopsin molecules with different protrusion heights and shapes can be discerned in their respective arrays (Figures 3C). At the higher ratio of [1:100], two similar populations of individual rhodopsin molecules are evident although the structured arrays are less extensive (Figure 3D). This could reflect the non-vectorial incorporation of rhodopsin molecules in the LUV bilayers during reconstitution resulting in surface exposure of both the C- and N-terminal ends of the protein. The different extent of array formation may reflect a better membrane stress energy release in the lower rhodopsin composition. Indeed, topographic image maps corresponding to Figures 3D (not shown) reveal two protrusion classes of equal numbers of molecules with respective protrusion heights of about 0.9 nm and 1.9 nm over the surrounding membrane. The paracrystalline array structure observed here is similar to that observed in the single-layered disk studied by Fotiadis *et al.* [4], indicating that the crystalline structure may be a result of the above-mentioned phase separation [10].

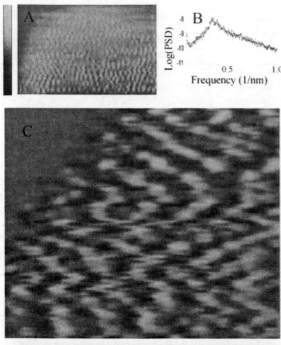

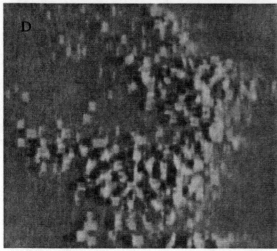

Figure 3: Molecular organization of rhodopsin molecules in reconstituted rhodopsin-DPPC membranes under light state at room temperature. **(A)** TM topographic image of reconstituted rhodopsin-DPPC [1:200] membranes (75 nm × 75 nm). **(B)** Horizontal 2D power spectrum of (A), showing rhodopsin spacing peaks at about 3.5 nm. **(C)** High-resolution low-pass filtered TM amplitude image of rhodopsin molecules in rhodopsin-DPPC [1:200] membrane (50 nm horizontally). Two different molecular shapes can be identified, and they each tend to form distinct arrays. **(D)** High resolution low-pass filtered TM amplitude images of rhodopsin molecules in rhodopsin-DPPC [1:100] membranes (50 nm horizontally). Molecular

arrays are less extensive, but two classes of the molecular protrusion height and shape can still be identified. The brightness scale (top left) for height contrast in A, C and D is 20 nm, 500 mV and 400 mV, respectively.

4 SUMMARY

For the first time, individual rhodopsin molecules were resolved in both the intact native disk membrane and the reconstituted rhodopsin-DPPC membranes using AFM. Unlike the random distribution in the intact native disk membrane with rhodopsin molecules in close yet transient associations, rhodopsin molecules in the reconstituted system are organized in a paracrystalline pattern and as monomers and oligomers that may depend on membrane phase separation, caused by direct interaction with the mica substrate. Our images obtained in native disk membranes arise from sampling the upper membrane surface of the disk, which is not in direct contact with the mica surface.

The results of this study should give further insight on the relationship between vision disk membrane structure and function. By manipulating the phase behavior between rhodopsin and membranes, one could control the lateral organization of rhodopsin for use in a smart device, molecular motor, or other bio- and nanotechnology applications.

5 ACKNOWLEDGEMENTS

We thank Sylvain Ho, Ed Wellner and Dr. Emilios Dimitriadis for contributions toward improving the AFM facility at DBEPS/ORS/OD/NIH.

6 REFERENCES

[1] Albert, A. D., Yeagle, P. L. (2002) *Biochim. Biophys. Acta* 1565, 183.
[2] Sakmar *et al.* (2002) *Annu. Rev. Biophys. Biomol. Struct.* 31, 443.
[3] Polozova, A., Litman, B. J. (2000) *Biophys. J.* 79, 2632.
[4] Fotiadis *et al.* (2003) *Nature* 421, 127.
[5] Tokumasu *et al.* (2003) *Biophys. J.* 84, 1.
[6] Miller *et al.* (1986) *Biochemistry* 25, 4983.
[7] Smith, H.G., Litman, B.J. (1982) *Methods Enzymol* 81, 57.
[8] Litman, B.J. (1982) *Methods Enzymol* 81, 150.
[9] Jackson, M.L., Litman, B.J. (1985) *Biochi.Biophys. Acta* 812, 369.
[10] Chabre *et al.* (2003) *Nature* 426, 30.
[11] Wang *et al.* (2001). *Prog. Biophys. Molec. Biol.* 77, 1.

Sample-Shunting Based PCR Microfluidic Device

P.-A. Auroux [*], P.J.R. Day [**], A. Manz [*, 1]

[*] Imperial College, London, UK, p.auroux@imperial.ac.uk
[**] CIGMR, University of Manchester, UK, philip.j.day@man.ac.uk

ABSTRACT

A new method to perform on-chip DNA amplification is presented in this abstract: we shunt the sample back and forth over temperature-controlled zones. Compared to existing systems, this approach offers, in addition to a greater flexibility in terms of cycling programme, an incomparable ease of parallelisation.

We report in this paper the first results obtained when using a sample-shunting approach to achieve on-chip PCR. An unoptimised chip displays an efficiency of 25% and is six times faster than a conventional thermo-cycler. To the extent of our knowledge no comparable results have been previously published.

Keywords: PCR, DNA, microfluidics, chip

1. INTRODUCTION

Since its discovery in 1983 by Kary Mullis et al., the polymerase chain reaction (PCR) has had a tremendous impact in the medical sciences and clinical applications. However, this technique suffers from critical drawbacks. It is for example prone to contamination and the cost associated with each reaction is not negligible. A lab-on-a-chip approach would address some of these issues. Indeed, as the sample is constantly enclosed in the micro-device, contamination risks are limited. In addition, the sample and reaction volumes are in the sub-microlitre range, thus reducing reactant consumption and costs.

To date, two approaches have been mainly studied to perform PCR on chip. The first one is based on a static-sample method and was first introduced in 1993 by Northrup et al. [1]. The device consisted of reaction chambers of 25 or 50 microliters, inside which the sample was subjected to temperature cycles by means of polysilicon heaters. An alternative approach based on a dynamic sample was proposed by Kopp and co-workers in 1998 [2]. In this continuous-flow system, the sample was pumped through a meandrous channel. Thermocycling was achieved by the PCR reactants passing several times over three fixed heating zones. Both methods have proved quite successful and several groups have since presented similar devices [3]. However, they also present some disadvantages. In the case of the static-sample approach for example, the devices present unnecessary inertia. In addition these designs take limited advantage of the flexibility of lithographic methods. Furthermore, with dynamic-sample based devices, it would be difficult to achieve a high throughput without a significant increase in the heater design complexity or in the microfluidic chip size.

A third approach was recently presented [4, 5]. It consists of shunting the sample back and forth over two/three heating zones in a single channel (see Figure 1). It elegantly associates the temperature cycling flexibility of the static-sample-chips to the quick temperature transition of the dynamic-sample microstructures. Indeed, the number of cycles can easily be modified by altering the pumping programme. Transition times from one temperature zone to another can also be optimised by adapting pump velocities. In addition, parallelisation is easily obtained due to the flexibility of lithographic methods. Finally, it is also possible to incorporate post-PCR applications (such as hybridization and/or optical based assays) at the end of each channel, providing unique capabilities to simultaneously apply different treatments to the same original sample. In this paper we present the first results obtained when using a sample-shunting approach to achieve on-chip PCR. In addition, the influence of the cycling conditions over the DNA amplification efficiency is studied.

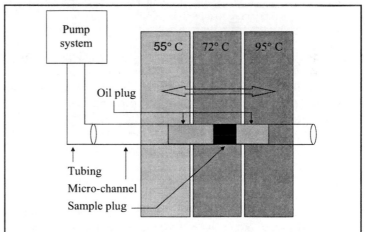

Figure 1. Schematic drawing of the shunting system. The thermocycling conditions are provided by shunting the sample back and forth over the temperature-controlled zones (the given temperatures are only examples).

2. APPARATUS

The device was designed using AutoCAD 2000. The length of the channels (4 cm) was determined by the heating system, whereas their width and depth (100 μm x 50 μm) were chosen to minimize risks of blockage from dust or PCR reagents (see Figure 2). The fabrication of the SU-8 chips was performed by Epigem Limited (Malmo Court, Kirkleatham Business Park, TS10 5SQ, Redcar, UK). It consisted in spin-coating and curing a thin layer of SU-8 (10 μm) on a PMMA substrate. An additional layer of SU-8 was subsequently spin-coated, the thickness of which was customized according to the application (50μm in this case). This layer was lithographically patterned prior to curing. A second PMMA substrate was also coated with a 10 μm SU-8 layer. The bonding of the substrates was achieved by bringing in contact the two SU-8 layers and applying high pressure and high temperature.

Figure 2. Top view of the PCR chip. The inlet is on the right. The three heating films are the dark-brown areas labeled A, B and C. The green wires link three thermo-couples recording the heater temperature to a monitor device.

The micro-fluidic inlet was linked to a Kloehn pump controlled by a home-written LabView programme. Three heating films were each connected to individual power supplies (RS, cat. #: 201-3446) and provided very stable temperatures (see Table 1). The test real-time PCR amplicon system was derived from Exon 2 of the protooncogene MYCN (forward primer: GCCGAGCTGC-TCCACGT; reverse primer: TCAAACTCGAGGTCTG-GGTTCT) and the PCR reagent kit was purchased from Roche Diagnostics (LightCycler DNA Master – SYBR Green I, cat. #: 2 015 099). The polymerase chain reaction was performed either with the micro-device previously described or with a Touchgene Gradient manufactured by Techne. The amplified samples were analysed using the Agilent 2100 Bioanalyzer with a 500 DNA kit (Agilent 500 DNA Array kit, cat. #: 5064-8284).

	Temp. A	Temp. B	Temp. C
average	95.91	72.18	60.33
stdev	0.11	0.09	0.12
% error	0.002	0.002	0.003

Table 1. Temperature stability study of the different heaters. Very stable temperatures were achieved giving percentage errors.

We performed two sets of experiments. The first one was conducted to assess the efficiency of a non-optimised system. The influence of the cycling programme over the PCR yield was estimated by the second set of experiments. To determine the effectiveness of the micro-device during PCR, a 1μL sample (template concentration: 179ng/μL) was first amplified using the microchip system. The amplified sample was then diluted 10 times and used as a 1μL-template for PCR using a conventional thermocycler. After increments of 5 cycles the amplified samples were analysed by gel electrophoresis using the Agilent 2100 Bioanalyzer. This set of samples was compared to a series directly amplified from the conventional thermocycler and using 17.9ng/μL concentrated DNA templates. The cycling programmes were: 40 cycles of 95°C for 15s and 60°C for 60s for the thermocycler amplifications; 40 cycles at 95°C for 8s and 60°C for 15s for the microchip amplifications.

Another set of experiments was performed in order to study the effect of the on-chip cycling programmes on the PCR efficiency. A template was first amplified with the micro-device following the procedure described above. A subsequent amplification was performed on a thermocycler in increments of 5 cycles. However, the on-chip thermocycling conditions differed. Three programmes, all performed for 40 cycles, were investigated: programme 1: 95°C for 8s, 60°C for 15s; programme 2: 95°C for 6s, 60°C for 6s; programme 3: 95°C for 2s, 60°C for 4s.

4. RESULTS

Table 2 presents the results of the efficiency assessment of the micro-fluidic PCR reactor. During this experiment a DNA sample was first amplified on-chip and was then used as a template for subsequent amplification on a conventional Techne thermocycler. The amplified product was then compared to a sample that had only undergone PCR on the Techne thermocycler (positive control). In this table, the correlation areas of the two series of amplified products are monitored after *n*-extra cycles on the thermocycler. As can be seen, the on-chip pre-amplified sample series displays a visible correlation area (amplified product) prior to the positive control series: amplified product from the micro-device series can be detected at 15 extra cycles, whereas the positive control series is still undetected. This result leads to a first conclusion: the micro-reactor produces some DNA amplification. The effect of the micro-device can be further assessed. Indeed, the correlation area of the on-chip pre-amplified series at 25-extra cycles corresponds to the correlation area of the positive control series at 35-extra cycles (correlation areas of 22.05 and 22.18 respectively). Consequently, the 40-cycle on-chip amplification is the equivalent of a 10-cycle thermocycler amplification. It can therefore be concluded that the micro-device presents a 25% overall efficiency.

In Figure 3, the effects of on-chip thermocycling programmes over the PCR efficiency are reported. Three programmes, all performed for 40 cycles, were investigated: programme 1: 95°C for 8s, 60°C for 15s; programme 2: 95°C for 6s, 60°C for 6s; programme 3: 95°C for 2s, 60°C for 4s. Although the three curves are comparable, the thermo-cycling programme 95°C for 6s, 60°C for 6s presents several advantages. First, it displays the highest amount of specifically amplified product. Furthermore, the efficiency/time-to-execute ratio for this programme is also excellent. The time to perform the polymerase chain reaction is then reduced by 84% when compared to a conventional thermocycler.

Extra cycles in the thermocycler	Corr. Area Microchip	Corr. Area Thermocycler
15	0.6	0
20	9.93	0.17
25	22.05	4.59
30	33.36	16.3
35	42.25	22.18
40	47.29	28.55

Table 2. Correlation areas of the specifically amplified product detected in the case of the microchip series and the thermocycler series after *n*-extra cycles in the thermocycler. The correlation areas underlined in the two series are comparable, which leads to an estimated 25% overall efficiency for the un-optimized microchip.

5. CONCLUSION

A sample-shunting based PCR microfluidic device has been successfully developed. An un-optimised chip presented a 25% overall efficiency and performed six times faster than a conventional thermocycler. Further optimisation is in progress.

Acknowledgements: P.-A. Auroux wishes to thank the Schweizer Forschungsstiftung Kind und Krebs for financial support.

6. REFERENCES

1. Northrup, M.A., et al., *DNA amplification with a microfabricated reaction chamber.* Transducers '93, 1993: p. 924-926.

2. Kopp, M.U., A.J. de Mello, and A. Manz, *Chemical amplification: Continuous-flow PCR on a chip.* Science, 1998. **280**(5366): p. 1046-1048.

3. Auroux, P.-A., et al., *Micro Total Analysis System 2. Analytical Standard Operations and Applications.* Analytical Chemistry, 2002. **74**(12): p. 2637-2652.

4. Auroux, P.-A., et al. *PCR Microfluidic Device for Detection of Low Copy Number Nucleic Acids.* in *Nanotech 2002.* 2002. Montreux.

5. Auroux, P.-A., et al. *PCR micro-volume device for detection of nucleic acids.* in *Nanotechnology conference and trade show.* 2003. San Francisco, USA: Computational Publications.

[1] Present address: ISAS, Dortmund, Germany, a.manz@isas-dortmund.de

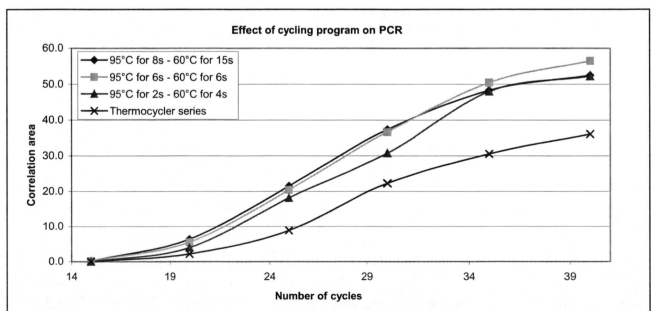

Figure 3. Curves representing the correlation area of the specifically amplified PCR products after *n*-extra cycles in a Techne thermocycler. The cycling programme providing the best efficiency/time-to-execute ratio comprised 95°C for 6s, 60°C for 6s. The time to perform PCR is then reduced by 84%.

Highly Efficient Electro-permeabilization of Mammalian Cells
Using Micro-electroporation Chip

Young Shik Shin[*], Jung Kyung Kim[**], Sun Hee Lim[**], Young Kyung Lee[**], Keunchang Cho[**], Yongku Lee[*], Seok Chung[**], Chanil Chung[**], Dong-Chul Han[*] and Jun Keun Chang[**, ***]

[*] School of Mechanical & Aerospace Engineering, Seoul National University, Seoul, 151-742, Korea
[**] Digital Bio Technology Co., Seoul, 151-742, Korea
[***] School of Electrical Engineering and Computer Science, Seoul National University, Seoul, 151-742, Korea

ABSTRACT

Electroporation, is widely used method that can be directly applied to the field of gene therapy. However, very little is known about the basic mechanisms of DNA transfer and cell response to the electric pulse. As a more convenient and effective tool for the investigation on the mechanisms of electroporation and the enhancement of the functional efficiency, we developed a micro-electroporation chip with polydimethylsiloxane (PDMS). Owing to the transparency of PDMS, we could observe the process of Propidium Iodide (PI) uptake in real-time, which shows promise in visualization of gene activity in living cells. Furthermore, the design of the electroporation chip as a microchannel offers advantages in terms of efficiency in permeabilization of PI into SK-OV-3 cells. We also noticed the geometric effect on the degree of electroporation in microchannels with diverse channel width, which shows that the geometry can be another parameter to be considered for the electroporation when it is performed in microchannels.

Keywords: electroporation, PDMS, microchannel, bio-MEMS, propidium iodide

1 INTRODUCTION

Electroporation is widely used method to introduce macromolecules into living cells by applying electric pulses. Under the high electric field, cell membranes are transiently rendered to porous and they become permeable to otherwise impermeable foreign materials. Electro-permeabilization depends on several factors: pulse amplitude, pulse duration, the number of pulses and other experimental conditions. With those parameters, many research groups have performed various theoretical as well as experimental studies about the electroporation to understand the mechanism and to enhance the efficiency of the transfection.

As the theory about the mechanisms of electroporation and gene transfer has not been developed well, the visualization is another important issue on the electroporation. Golzio *et al.* investigated the DNA transfer process at the single-cell level by using fluorescence microscopy digitized imaging [1]. Deng *et al.* assessed the effect of intense submicrosecond electric pulses on cells by means of temporally resolved fluorescence and light microscopy [2].

In common electroporation processes, a cuvette which is consisted of two parallel electrode plates is used to contain the mixture of cell suspension and genes. By applying a high electric field across those two electrodes, it is possible to deliver genes into cells. Aluminum is commonly-used material for disposable cuvettes. In this case, however, the release of Al^{3+} from the aluminum electrode, which has adverse effects on cells, has been observed by several research groups. In case of using aluminum electrodes, different field strength is expected because of the significant voltage drop across the oxide layer on the electrode surface. Therefore, although it is more expensive, platinum or gold electrodes are more advantageous [3].

In this work, we designed a micro-electroporation chip which has a microchannel and investigated the characteristics of electroporation induced by that specific design. The microchip was fabricated with polydimethylsiloxane (PDMS) by using MEMS techniques. PDMS has a great potential as a fabrication material. It is easy to fabricate, inexpensive, transparent, and biocompatible. Owing to the transparency of PDMS, it is possible to observe the process of transfection in real-time. Furthermore, PDMS chips can be easily integrated with other systems because of many preceded works using PDMS such as PCR, CE, mixer and filter. To show the advanced functionality of the micro-electroporation chip, we observed Propidium Iodide (PI) uptake into SK-OV-3 cells by means of fluorescence and light microscopy. To check additional effects induced by miniaturization, we performed experiments in different conditions by changing the width of microchannel.

2 MATERIALS AND METHODS

2.1 Fabrication of electroporation chips

The micro-electroporation chip has a simple design which contains a microchannel and two reservoirs for inlet and outlet. The channel height is 20 μm and the length is 2 cm. We varies the channel width as 100, 200, 300, 400, and 500 μm. Figure 1 shows the fabricated micro-electroporation chip.

Figure 1. Fabricated micro-electroporation chip.

The micro-electroporation chip was fabricated with the replica molding method. The microchannel patterns were fabricated by photolithography using chrome photo mask. Figure 2 shows the fabrication process of the micro-electroporation chip schematically. Negative photoresist (SU-8, MicroChem, MA, USA) was spin-coated onto a silicon wafer to create a mold master of 20 μm thick structure. After soft-bake, the pattern of the mask was transferred to the SU-8 coated silicon wafer by mask aligner (MA-6, Karl Suss GmbH, Germany). Post-exposure bake, developing and hard baking of the exposed SU-8 patterns are followed by pouring the mixture of PDMS and curing agent on the pattern (Sylgard® 184, Dow Corning Co., USA). The curing condition was 120 °C for 30 min. The PDMS replica was bonded with a glass substrate to form a microchannel by 25 W oxygen plasma treatments.

2.2 Cell preparation

SK-OV-3 cells were grown in Dulbecco's modified Eagle's medium (DMEM) supplemented with 10% (v/v) heat-inactivated fetal bovine serum (FBS, Sigma), penicillin (100 units/ml), streptomycin (100 μg/ml) and L-glutamine (4 mM) at 37 °C in a humidified 5% CO_2 incubator. Cells were dissociated from the 25 cm^2 tissue culture flask by using Trypsin-EDTA. The final cell suspension density was adjusted to 1×10^7 cells/ml.

Propidium Iodide (PI) was added to the cell media before applying an electric pulse. PI is a fluorescent marker that intercalates into nucleic acid molecules. When cell membrane becomes permeable, PI enters the cell and emits red fluorescence by binding to nucleic acids in the nucleus. Quantitative study is possible because the intensity of red fluorescence varies with the amount of PI bound to the nucleic acids. For the experiments, PI (1.0 mg/ml) was added to cell media as the ratio of 1:20 (v/v).

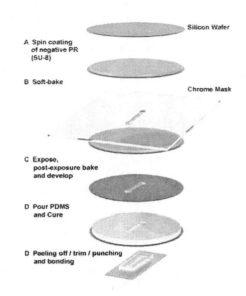

Figure 2. Schematic process for the fabrication of the micro-electroporation chip

2.3 Experimental setup

Experimental setup for the electroporation is consisted of a home made power supply, Pt electrodes and an electrode holder. As the electrode holder was mounted on a microscopy, we could observe the process of electro-permeabilization while applying the electric pulse. The power supply was connected to a computer via an analog output board (COMI-CP301, COMIZOA, Korea) and controlled with LabVIEW ver 6.1 (National Instruments, USA) program. To verify the performance of the micro-electroporation chip, a Square Wave Electroporation System (ECM® 830, BTX, USA) and a 2-mm gap cuvette with parallel-plate aluminum electrodes (BTX, USA) were used as a reference. To analyze the performance of two systems in the same electric field, we applied 200 V for cuvettes, and 2 kV for chips respectively. The electric field was 1 kV/cm and the pulse duration was 10 ms. We performed the experiments for five cases by changing the channel width: 100, 200, 300, 400, and 500 μm. Figure 3 shows the system setup for the experiment: (a) microchip and Pt electrode; (b) electrode holder on the microscopy.

Figure 3. System setup: (a) microchip and electrode; (b) electrode holder

To observe the PI uptake, we used an inverted fluorescence microscopy (IX70, Olympus, USA) equipped with a 100 W Mercury lamp and 20×/0.4 NA objective. The excitation light was filtered by a 510~550 nm bandpass filter and induced fluorescence from the electroporated cells was filtered with a 590 nm longpass filter. 640 × 480 pixel images were acquired by a 12 bit cooled CCD camera (PCO, Kelheim, Germany) at 10 Hz frequency.

3 RESULTS AND DISCUSSION

3.1 Electrolysis effect while applying electric pulse

During the electroporation within a cuvette, a two-phase layer of liquid and gas was created because bubbles were generated electrochemically at the surface of electrodes. Figure 4 shows the bubble formation on the surface of an electrode right after applying an electric pulse. The generated bubbles rose very fast and caused complex liquid motion. This bubble movement, combined with electrophoresis during pulse, constituted non-homogeneous condition of bulk media as well as cells. Our hypothesis for the mechanism of the bubble formation is that oxide layer on the aluminum electrodes acts as a high register layer. Aluminum is the material that forms oxide layer (Al_2O_3) very easily by intervening oxidizable electrolyte. Pliquett et al. also showed the same hypothesis with us and explained it very minutely [4].

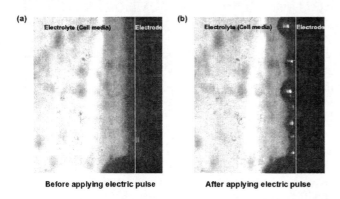

Figure 4. Bubble generation on the electrode of a cuvette: (a) before applying an electric pulse; (b) after applying an electric pulse. Applied voltage was 180 V and the pulse duration was 5 ms.

From the micro-electroporation chip, we couldn't observe bubble generation as well as complex movement. We may be able to explain this phenomenon in terms of the material. Pt electrodes that we use for the micro-electroporation experiments are very stable material. Therefore, it does not form oxide layer easily and even in case that oxide layer was formed, it can be removed with ease. As the motion of

the bulk media within a cuvette is very intense, the stable condition of the media within the micro-electroporation chip can be advantageous especially for the case of investigating the mechanisms of electroporation.

3.2 Transfection rate and detection of electro-permeabilization process

The transfection rates of SK-OV-3 cells in a cuvette and a microchip were compared under the same experimental conditions: 1 kV/cm electric field and 10 ms pulse duration. In a cuvette based system, the transfection rate was 34.5 %. The transfection rate was calculated by counting the number of cells with PI uptake using an Improved Neubauer Hemocytometer (Superior, Germany). In a microchip, the permeabilization of PI was detected from almost every cell in the microchannel even though the same experimental condition was applied.

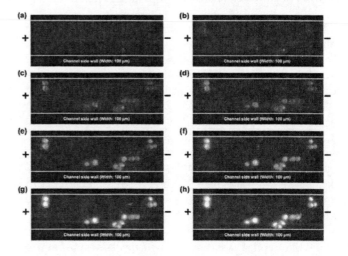

Figure 5. The PI inflow process within a 100 μm wide microchannel: before the pulsation (a) and after the pulsation (b) 1 ms; (c) 10 ms; (d) 5 s; (e) 10 s; (f) 15 s; (g) 25 s; (h) 35 s.

A localized inflow of PI in the mili-seconds was observed within microchannels after the pulse. Figure 5 shows the PI inflow process within a 100 μm wide microchannel. Right after the pulsation, the PI inflow was present only on the side facing the anode. This phenomenon shows that only the cell membrane on the left side was altered and the same observation was reported in the preceded work by Muriel Golzio et al. [5]. As time goes on, the fluorescence spreads into the whole cell interior (c-d) and after 10 s, the nucleus begins to emit fluorescence (e-h). This observation directly reflects the PI characteristics, binding to nucleic acids in the nucleus. The functionality of observing the real-time process in a single living cell is very advantageous because it can provide valuable information on the fundamental cellular processes.

3.3 The effect of channel width variation in the micro-electroporation chip

For microchips, the intensity of fluorescence by dye uptake was variable according to the channel width. As the channel width increases, the intensity of gray scale unit for cell area decreases even though the same electric pulses were applied. Figure 6 shows the representative microscopic images of cells for two different channels: 100 μm and 500 μm in width respectively. The electric field was 1 kV/cm and pulse duration was 10 ms. Pictures were taken at 30 s after the pulsation. It can be clearly noticed that PI uptake in a narrow microchannel (100 μm) is more than that in a wide microchannel (500 μm).

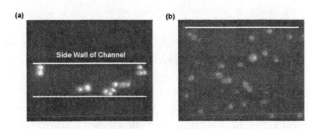

Figure 6. Fluorescence microscopic images of cells for two different channels: (a) 100 μm in width; (b) 500 μm in width. The images were taken at 30 s after the pulsation.

To compare the PI uptake for the five different cases of microchannels, images were acquired at 10 Hz and stored while performing the experiments. The image processing was performed for images at every 50 frames. The average intensity of gray scale unit for the background was subtracted from that for the cell area by using graphic software (Paint Shop Pro 7.0, Jasc Software, USA) and MATLAB program (MathWorks, Inc., USA). The comparative data for PI intensity are shown in figure 7. The effect of the channel width on PI uptake is obvious. As in a cuvette based system, the geometric parameters except for the electrode gap are not considered seriously, this unique phenomenon in a microchannel is noteworthy. Istávn P. Sugar et al. developed the low-voltage electroporator, which contains filter pores to increase local current density [6]. Owing to the large increase in current density in the filter pores, cells could be transfected by applying only 25 V. Their study was, however, based on local geometric changes for the reduction of current shunt pathways. So it can be differentiated from the phenomena that we observed in our microchips.

From this result, we could identify that we should consider the channel dimension when it comes to the electroporation in a microchannel. Further study is in progress to investigate the cause of the difference in a degree of electroporation within microchannels with various widths and the mechanism of it.

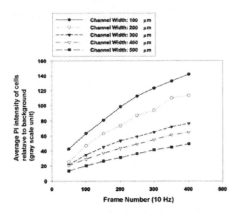

Figure 7. Average intensity of gray scale unit for cell area relative to the background for five different cases of microchannels. The relative intensity was calculated by image processing at every 50 frames. Images were acquired at 10 Hz.

4 CONCLUSIONS

We developed a micro device to perform electroporation in a microchannel. Owing to the advantageous material, PDMS, we could directly visualize the whole transfection process in real-time. In a microchannel, the bubble generation and complex motion of cell media as well as cells are not observed. This homogenious experimental condition led to the high transfection rate of PI into SK-OV-3 cells. The geometric effect on the degree of electroporation was identified through the experiments within microchannels with various channel width. This result reflects that the channel geometry can be another new parameter on the electropration when it is performed in a microchannel. Based on these experimental results, this new approach for the electroporation is expected to offer a lot of favorable possibilities for the investigation of the mechanism as well as the enhancement of efficiency.

REFERENCES

[1] M. Golzio, J. Teissié, and M.-P. Rols, *Proc. Natl. Acad. Sci.*, **2002**, 99 (3), 1292-1297.
[2] J. Deng, K. H. Schoenbach, E. S. Buescher, P. S. Hair, P. M. Fox, and S. J. Beebe, *Biophys. J.*, **2003**, 84, 2709-2714.
[3] U. Zimmernann and G. A. Davey, *Electroporation of Cells*, CRC Press, Boca Raton, FL, **1996.**
[4] U. Pliquett, E. A. Gift, and J. C. Weaver, *Bioelectrochem. Bioenerg.*, **1996**, 39, 39-53.
[5] M. Golzio, J. Teissié, and M.-P. Rols, *Proc. Natl. Acad. Sci.*, **2002**, 99 (3), 1292-1297.
[6] I. P. Sugar, J. Lindesay, and R. E. Schmukler, *J. Phys. Chem. B*, **2003**, 107, 2862-3870.

Effect of substratum morphology on cell behavior

Wen-Ta Su[*], I-Ming Chu[**], Jung-Yen Yang[***] and Chuan-Ding Lin[***]

[*]Dept. of Chemical Engineering, National Taipei University of Technology
[**]Dept. of Chemical Engineering, National Tsing-Hua University
[***]National Nano Device Laboratories, jyyang@ndl.gov.tw

ABSTRACT

Contact guidance induced by the surface topography of the substrata influences the direction of cells with artificial material. Mouse bone marrow stromal fibroblast was examined on grooved substrata of varying dimension (0.3 μm~1.2 μm groove width, 0.5 μm depth and 0.5 μm ridge). It was found that the repeat space had an effect in determining cell alignment. Measurement of cell alignment and examination by scanning electron microscopy showed the fibroblast interacted grooved substrata in monomodal responses and alignment to the direction of the grating.

Keywords: contact guidance, stromal, cytoskeletal, plasma etching, microfabrication

1 INTRODUCTION

Several in vitro studies have documented the ability of microgrooved substrata to produce cell orientation and directed migration. But the role of the texture of materials in inducing cell and tissues responses is still unclear. In spite of several publications and reviews have reported the effect of microtextured surface, little is known about the exact mechanism [1-5]. So controlling cell responses to a material is of interest with regard to tissue culture in the laboratory or implantable devices in body. Because the surface is in direct contact with cells, both chemical and topographical properties of the material surface can play a crucial role in determining cell responses. A number of physicochemical surface properties, including surface composition, surface charge, surface energy, surface oxidation, solidity, curvature, and surface morphology, have been shown to affect cell attachment and behavior [6].

Microfabrication technology employed to create substrates with topography is photolithography that produce features with controlled dimensions and specific shape [7]. Using this technique features in the range of micrometer could be etched precisely and uniformly on a variety of substrata such as silicon and glass, and this technique is flexible enough to allow a wide variety of textures and feature size. So this surface has facilitated systematic in vitro experiments to study influence of surface morphology on diverse cell physiological aspects such as adhesion, growth, and function. A common topographical feature investigated is a uniform multiple parallel grooves in the study of the effects of surface structure on cells. Typically, the grooves are often of the square ware, or V-shape. In general, grooved surface revealed that cells aligned to the long axis of the grooves, often with organization of actin, and other cytoskeletal elements in an orientation parallel to the grooves [8]. So we would examine how parallel grooved substrata with different space width affect the alignment to mouse bone marrow stromal fibroblast cell.

2 MATERIAL AND METHODS

2.1 Production of substrata

Oxide films were deposited at 700 ℃ to the thickness of 500 nm by LPCVD. Afterwards, trench-line patterns were defined on these oxide films by a g-line stepper (ASM PAS 2500). Subsequently, patterned films were etched in a CF_4/CHF_3 mixed plasma using a helicon wave plasma system (ANELVA ILD-4100). The line patterns are 40 μm in length, groove widths ranging from 0.3 μm to 1.2 μm and separated by 0.5 μm ridge.

2.2 Cell culture

Mouse bone marrow stromal fibroblast are performed in culture that is in a RPMI 1640 medium supplemented with 10 % fetal bovine serum (FBS), 1 % antibiotics (100 units/ml penicillin + 100 ug/ml streptomycin + 50 ug/ml nystatin). Then incubate at 37 ℃ 5 % CO_2/air for routine culture.

2.3 Cell seeding

The day before the test, all samples were immersed into 70 % ethanol and using an ultrasonic bath, subsequently, they were rinsed in double-distilled water and store overnight in cell culture medium at 4 ℃. Fibroblast cells, approximately a concentration 5×10^4 cell/ml were harvested from routine culture, were seed onto grooved substrata in 35 mm petri dish and incubated for 2 day at 37 ℃, 5 % CO_2/air.

2.4 Scanning electron microscopy

Cells on patterned substrata were fixed in 2.5%glutaraldehyde in phosphate-buffered saline (PBS) at 4 ℃ for 1 hr and two 15 min rinse in PBS at 4 ℃. Then samples were dehydrated in alcohol, dried by the critical point method using CO_2, sputtered with gold and viewed with JEOL JSM-5600 scanning electron microscopy.

3 RESULTS

Fig. 1 shows the microstructured substrata surface texture, SiO_2 on Si wafer. The depth of the grooved was 0.5 μ m with a serial of width (0.3 μ m,0.4 μ m,0.5 μ m,0.6 μ m,0.7 μ m,0.8 μ m,0.9 μ m,1.0 μ m,1.2 μ m), and 0.5 μ m ridge.

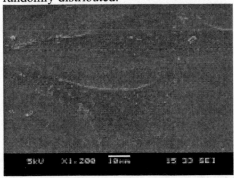

Figure 1: The scanning electron micrograph of microstructured surface, SiO_2 on silicon wafer.

Fig. 2 shows the cells on the unpatterned substrata (planar silicon dioxide surface) were extremely flattened and randomly distributed.

Figure 2: Mouse bone marrow fibroblast after 2 day incubated on the planar silicon oxide surface.

Fig. 3 and Fig. 4 were the cell growth on the grooved substrata, the cell was spindle shapes and both ends being drawing out into long, fine processes. The growth and alignment of fibroblast on grooved substrata was dependent on groove width, and being inversely proportional to groove width, the cells on the shorter grooved substrata were highly elongated and aligned but no cells on the wider parts.

Figure 3: The scanning electron micrograph of mouse bone marrow fibroblast after 2 day of incubated on a series of groove width, 0.5 μ m deep, 0.5 μ m ridge of SiO_2 on silicon wafer.

Figure 4: The scanning electron micrograph of mouse bone marrow fibroblast after 2 day of incubated on a series of groove width, 0.5 μ m deep, 0.5 μ m ridge of SiO_2 on silicon wafer.

The cells may have attempted to conform to the substrata, which their long axes cross and span the grooves and their lamellar regions cling intimately to the underlying planar substrata (Fig. 5).

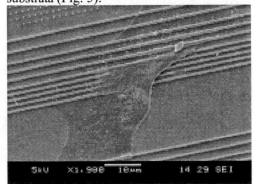

Figure 5: The scanning electron micrograph of mouse bone marrow fibroblast after 2 day of incubation on a series of groove width, 0.5 μ m deep, 0.5 μ m ridge and planar SiO_2 on silicon wafer.

must be straight, not bent. So that cells align to the substrata micromorphology to avoid a distortion of their cytoskeleton.

Further studies are required to clarify the ridge width and depth of grooves to cell behavior, and examined how various parallel grooved substrata affect the alignment of different cell type.

4 DISCUSSION

It has been known that photolithographic microfabrication techniques have been applied to address the problems to what extent and what mechanism topography influences animal cell behavior [5, 9, 10, 11, 12]. They examined some parameters of the grooved substrata to alter cell shape and guide cell movement. Most attention was the ridge width and depth of grooves. Dunn shown that the shape change and alignment of fibroblast on grooved substrata was most dependent on ridge width, alignment being inversely proportional to ridge width [11].

Our studied revealed that fibroblast migrated along the major axis of surfaces with parallel ridges/grooves of various line space, this phenomenon was called "contact guidance" by Weiss [13].

Contact guidance of fibroblasts on grooved substrata has reported by some laboratories. M. Abercrombie [14, 15] is the first to detect the fibroblast adhered to the substrata surface in specialized zone that was focal contact. They show a characteristic structure [16] with transmembrane integrin receptor, which are linked by several cytoskeletal proteins to intracellular fibers. Some experimental revealed that the stiffness of the cytoskeleton is probably a reason for cellular alignment [17] and their intracellular actin fibers must be straight, not bent. So that cells align to the substrata micromorphology to avoid a distortion of their cytoskeleton.

Further studies are required to clarify the ridge width and depth of grooves to cell behavior, and examined how various parallel grooved substrata affect the alignment of different cell type.

REFERENCES

[1] A. S. G. Curtis and P. Clark, *CRIT. Rev. Biocompat.*, 5, 344-362 (1990).

[2] R. Singhvi, G. Stephanopoulos, and D. I. C. Wang, *Biotechnol. Bioeng.*, 43, 764-771 (1994).

[3] E. T. den Braber, J. E. de Ruijter, H. T. J. Smits, L. A. Ginsel, A. F. von Recum, and J. A. Jansen, *J. Biomed. Mater. Res.* 29, 511-518 (1995).

[4] A. Curtis and C. Wilkinson, *Biomaterials*, 181573-1583 (1997).

[5] P. Clark, P. Connolly, A. S. G. Curtis, J. A. T. Dow, and C. D. W. Wilkinson, *Development*, 108, 635-644 (1990).

[6] R. G. Flemming, C. J. Murphy, G. A. Abrams, S. L. Goodman, P. F. Nealey, *Biomaterials*, 20, 573-588 (1999).

[7] J. A. Schmidt and A. F. von Recum, *Biomaterials*, 13, 1059-1069 (1992).

[8] C. D. W. Wilkinson, M. Riehle, M. Wood, J. Gallagher, A. S. G. Curtis, *Materials Science and Engineering*, 19, 263-269 (2002).

[9] D. M. Brunette, *Expl Cell Res.*, 164, 11-26 (1986a).

[10] D. M. Brunette, *Expl Cell Res.* 167, 203-217 (1986b).

[11] G. A. Duun, and A. F. Brown, *J. Cell Sci.* 83, 313-340 (1986).

[12] A. T. Wood, *J. Cell Sci*, 90, 667-681 (1988).

[13] P. Weiss, *Rev. Modern. Physics*, 31, 11-20 (1959).

[14] M. Abercrombie and E. J. Ambrose, *Exp. Cell Res.*, 15, 332-345 (1958).

[15] M. Abercrombie and G. A. Dunn, *Exp. Cell Res.*, 92, 57-62 (1975).

[16] K. Burridge, K. Fath, T. Kelly, G. Nuckolls, and C. Turner, *Abb. Rev. Cell Biol.* 4, 487-525 (1988).

[17] G. A. Dunn and J. P. Heath, *Exp. Cell Rev.*, 101, 1-14 (1976).

AN INJECTOR COMPONENT MODEL FOR COMPLETE MICROFLUIDIC ELECTROKINETIC SEPARATION SYSTEMS

R. Magargle[*], J.F. Hoburg[**] and T. Mukherjee[***]

Carnegie Mellon University, Pittsburgh, PA 15213, USA
*magargle@cmu.edu, **hoburg@ece.cmu.edu, ***tamal@ece.cmu.edu

ABSTRACT

This work presents a closed-form, numerically derived model of a key microfluidic electrokinetic separation system component, the cross injector. The model fits into a framework such that it can be combined with a network of other component models, such as a mixer, reactor, separation channel, and detector, to model a complete microfluidic separation system. The model is based on a reduction of the full 5-dimensional parameter space to a manageable 2-dimensional parameter space. The resultant model is valid for Peclet numbers ranging from 10 to 5000 and shows a worst case variance equal to the channel width squared.

Keywords: microfluidic, electrokinetic, cross, injector

1 INTRODUCTION

The injector is a very important component of the microfluidic separation system since it defines the shape and quantity of analyte that will be used for separation and analysis. Much work has been done previously to describe the various forms of microfluidic electrokinetic injectors, such as the tee, double-tee, cross, and double-cross [1-5]. This work presents a new form of the cross injector model that is compatible with a system network of component models, [6,7]. In 2000, Ermakov, et al., extensively modeled the effects of field ratios on the injected band for a cross injector [5]. The metrics used to measure the performance of the injector were based on the maximum concentration and the standard deviation of the transversely averaged concentration profile. The component injector model in this paper is based on similar metrics and uses similar notations to those in [5].

The methodology described here involves an exploration of a relevant portion of the injector parameter space using finite element, partial differential equation solutions of the convection diffusion equation. The local electrokinetic velocity is proportional to the local electric field. These solutions are then used to create fitted expressions for the desired input-output values, which are the input species concentration and the resulting output variance and maximum concentration. The output parameters are sufficient to construct a Gaussian distribution, approximating the width averaged plug concentrations as a function of longitudinal position, for the input to the separation channel.

2 INJECTOR DESIGN

A 2-stage cross injector with accessory fields is shown in Figure 1. The accessory fields in the loading stage confine the band within the injection chamber, reducing the variance of the band, and they provide temporal stability for analytes with components of varying mobilities. These loading stage accessory fields are also known as pinching fields and were recommended for use in cross topologies nearly a decade ago [1]. The accessory fields used in the dispensing stage pull the band out of the injection chamber to ensure a clean exit. The two-stage cross injector is a commonly implemented design due to its relative simplicity and good performance.

Some alternative injector topologies and control schemes are summarized in Table 1. This table provides a qualitative measure of the performance of each injector with regard to output band variance, maximum concentration, and uniformity of the band shape. It also gives an indication of the control complexity by listing the number of voltage control stages that are typically used for each injector type. The cross injector also sometimes employs 3-stage voltage control schemes. These include, the gated cross injection scheme [5] and a variation of the 2-stage injection that adds a pullback step before the dispensing stage [4]. Table 1 shows a qualitative assessment of the popularity of each topology in its final

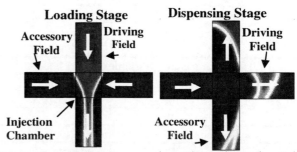

Figure 1 – The two stage cross injector. In the loading stage, the accessory electric fields horizontally pinch the analyte as it is driven vertically downward by the driving electric field. In the dispensing stage, the accessory electric fields vertically pull the analyte away from the intersection while the driving fields horizontally drive the plug into the separation channel.

Injector Type	Tee	Double Tee	Pi	Cross	Double Cross
	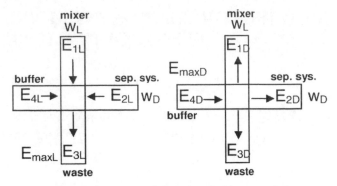 Tee	Double Tee	Pi	Cross	Double Cross
Common Injection Schemes	2 stage	2 stage	2 stage 3+ stage	2 stage 3+ stage	2 stage
σ^2	$> w^2$	$\sim \ell^2$	$\sim \ell^2$	$< w^2$	$< w^2$
C_{max}	H	H	H	M	M
Plug Uniformity	L	H	M	MH	H
Popularity	LM	MH	L	H	M

Table 1 – Injector topologies and control schemes. L, M, H refer to low, medium, high. w is the channel width, and ℓ is a channel separation length.

row, based upon instances of use in typical designs. Since the cross injector with 2-stage accessory fields combines a simple and popular topology with a reduced control scheme that is capable of good performance, it is pursued in detail here for incorporation into a component network model.

3 PROBLEM SPACE

The first step in the modeling procedure is identification of a complete set of non-dimensional variables that parameterize the problem space based on the known physical variables of the system. The physical variables governing the dynamics of the cross injector are summarized in Table 2(a,b), where μ is the electrokinetic mobility, κ is the species diffusivity, w_L is the width of the loading channels, w_D is the width of the dispensing channels, E_{iL} and E_{iD} represent the electric field at the injector inlet in the ith leg for the loading, L, and dispensing stages, D, respectively as shown in Figure 2.

The non-dimensional variables in Table 2(c) are

Loading Stage Variables – (a)		
Species	Geometry	Stimuli
μ	w_L	$E_{2,4L}$
κ	w_D	E_{1L}
Dispensing Stage Variables – (b)		
Species	Geometry	Stimuli
μ	w_L	E_{2D}
κ	w_D	$E_{1,3D}$
Non-Dimensionalized Variables – (c)		
$\omega = w_L/w_D$	$\varepsilon_L = E_{2,4L}/E_{1L}$	$\varepsilon_D = E_{1,3L}/E_{2L}$
$Pe_L = \mu E_{3L} w_L/\kappa$	$Pe_D = \mu E_{4D} w_D/\kappa$	

Table 2 – Cross injector physical variables for loading (a) and dispensing (b) stages. The resulting non-dimensional variables for both stages (c) reduce the number of degrees of freedom from 8 to 5.

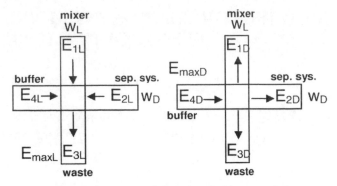

Figure 2 –The general layout of the cross injector with respect to the rest of the separation system for loading and dispensing stages.

obtained using the Buckingham-Pi Theorem [8]. Performance for each stage (loading and dispensing) is determined by 3 dimensionless parameters, the ratio of the channel widths, ω, the ratio of accessory to driving field, ε, and the Peclet number, Pe. The Peclet number is based upon the maximum electric field, E_{3L} in the loading stage and E_{4D} in the dispensing stage. The parameters ε_L and ε_D fix the ratio of accessory fields to driving fields in the respective stages.

In order to define the extent of the injector as a system component, the lengths of the legs of the cross are taken as four times the channel width, $\ell=4w$. This value, based upon numerical experiments, results in containment of the entire injection plug within the leg after injection, and isolates the injection chamber of the cross, within which the non-uniform electric fields are the dominant factor in the shaping of the injection plug.

3.1 Electric Fields

Four electric fields, measured at the injector inlets, are involved in each stage of injection, as seen in Figure 2. For the loading stage, E_{1L} represents the field that moves the fluid into the injection chamber and drives an inward current density. E_{2L} and E_{4L} represent the accessory fields that pinch the analyte fluid in the injection chamber, with inward current densities. For symmetric pinching: $E_{2L}=E_{4L}\equiv E_{2,4L}$. Finally, E_{3L} is the field with the largest magnitude, with an outward current density transporting the tail of the injection plug into the waste reservoir. Conservation of current for a surface that surrounds the injection chamber results in:

$$E_{3L} = E_{1L} + 2\frac{w_D}{w_L} E_{2,4L} \qquad (1)$$

For the dispensing stage, E_{2D} is the driving field that carries the plug into the separation channel and has an outward current density. E_{1D} and E_{3D} are the accessory fields with outward current densities that symmetrically pull the analyte out of the injection chamber, so that $E_{1D}=E_{3D}\equiv E_{1,3D}$. E_{4D} is the dispensing field with the largest magnitude; it flushes out the injection chamber pushing the plug into the separation channel. Again, conservation of current relates the dispensing stage inlet fields:

$$E_{4D} = E_{2D} + 2\frac{w_L}{w_D}E_{1,3D} \qquad (2)$$

Based upon eqs. 1 and 2, for any specific choice of the width ratio, ω, and field ratio, ε, all fields are determined in terms of the maximum field, E_{3L} or E_{4D}.

Finally, to determine the voltages that are necessary to achieve the desired fields for each stage, a linear system describes the relation between the applied voltage and electric field for any cross injector with legs of length $\ell=4w$. With electric fields normalized to the channel width, w, the linear relationships are:

$$\begin{bmatrix} \Phi_{1L} \\ \Phi_{2L} \\ \Phi_{4L} \end{bmatrix} = \begin{bmatrix} 0.1751 & -0.0604 & -0.0604 \\ -0.0604 & 0.1751 & -0.0543 \\ -0.0604 & -0.0543 & 0.1751 \end{bmatrix}^{-1} \begin{bmatrix} E_{1L} \\ E_{2L} \\ E_{4L} \end{bmatrix} \qquad (3)$$

$$\begin{bmatrix} \Phi_{1D} \\ \Phi_{3D} \\ \Phi_{4D} \end{bmatrix} = \begin{bmatrix} 0.1751 & -0.0543 & -0.0604 \\ -0.0543 & 0.1751 & -0.0604 \\ -0.0604 & -0.0604 & 0.1751 \end{bmatrix}^{-1} \begin{bmatrix} E_{1D} \\ E_{3D} \\ E_{4D} \end{bmatrix} ,(4)$$

where E_{mL} and E_{mD} are the desired field values with appropriate signs, and Φ_{mL} and Φ_{mD} are the resulting voltages to be applied to injector's inlets. The matrix coefficients are determined through numerical solutions of Laplace's equation for conduction problems within the injector. One of the node voltages is a reference ground node in each stage (Φ_{3L} and Φ_{2D}), so that a 3x3 matrix fully describes the relationship between applied voltages and fields. The voltages obtained in eqs. 3 and 4 are used solely for obtaining the fields in the injector. When connected to the rest of the network, these voltages are appropriately biased by the reference node voltage.

3.2 Reduced Problem Space

In order to create a simple analytic model for the output variance and peak concentration, the 5-dimensional parameter space must be considerably reduced. To do this the following assumptions are made:

- The channels widths are the same, $\omega=1$.
- The loading electric field ratio is fixed at $\varepsilon_L=1/2$.
- The dispensing electric field ratio is fixed at $\varepsilon_D=2$.

The majority of designs use channels with the same widths, justifying the first assumption. The second and third assumptions are based on a case study of optimal injector performance at $Pe_L=Pe_D=100$. For this study, numerical experiments were run on an injector with varying

values of ε_L and ε_D. As suggested by Ermakov, the performance was measured by the ratio of the maximum concentration to the standard deviation of the resulting plug after the dispensing stage [5]. This metric gives a measure of the quantity of the analyte injected and how dispersed it becomes. High performance results from large quantity and small dispersion, and reaches a maximum at $\varepsilon_L=0.5$ and $\varepsilon_D=2.0$. This agrees qualitatively with Ermakov's results for his case study at $Pe_L=Pe_D=250$, where he found $\varepsilon_L=0.33$, $\varepsilon_D=1.2$. Both cases suggest that the exit control field ratios should be higher than the pinching field ratios. After w, ε_L, and ε_D are fixed, the remaining non-dimensional parameters are Pe_L and Pe_D.

4 RESULTS

With the reduced parameter space, numerical simulations were run using Femlab for Pe_L and Pe_D ranging from 10 to 5000. Peclet numbers below this range for an injector lead to completely undesirable performance due to the large amount of diffusion that dominates each step. Peclet numbers above this range represent diffusion coefficients below 7×10^{12} m²/s for fields of 50kV/m and electrokinetic mobilities of 1.4×10^{-8} m²/Vs. Therefore, the range of Peclet number from 10 to 5000 encompasses the majority of practical system properties.

Simulations were run at 64 points in the parameter space. After the numerical results were obtained, they were converted to a closed-form analytical expression by fitting the numerical results with a two dimensional polynomial. The order of the polynomial in each dimension was chosen to minimize the error. The result was the product of a 7th order polynomial in $Log_{10}(Pe_L)$ and $Log_{10}(Pe_D)$ for describing the normalized variance (σ^2/w^2), and the product of a 6th order polynomial in $Log_{10}(Pe_L)$ and 7th order polynomial in $Log_{10}(Pe_D)$ for describing the normalized peak concentration (C_{max}/C_o, where C_o is the input concentration).

Figure 3 shows the plug variance and peak concentration as functions of Pe_L and Pe_d on logarithmic scales. These results indicate that the larger the Peclet numbers become, the better the injector performs. The worst expected performance for this injection scheme is approximately the width of the channel squared. The analytical model has a maximum error of less than 2% based upon the Femlab simulations. After the analytical model has been established through fitting to the numerical simulations, it takes less than ~0.5s to evaluate any point in the space, whereas the numerical simulations take ~4.5hrs per point, yielding a 32,000X increase in speed.

5 SYSTEM LEVEL INTEGRATION

Integration of the injector model with the other component models in a network requires a global determination of the node voltages. This is accomplished

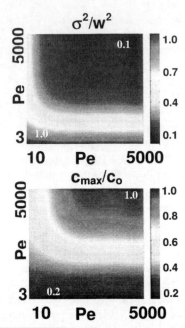

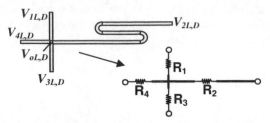

Figure 4 – Equivalent resistor network of injection and separation systems.

Figure 3 –Results of the injector model in the Pe_L and Pe_D problem space, showing the normalized variance and the normalized peak concentration.

through construction of an equivalent resistor network representing all legs of the system, as seen in Figure 4. The equivalent resistances are:

$$R_{1,3,4} = \frac{L_{1,3,4}}{\sigma wh} \qquad (5)$$

$$R_2 = \sum_{i=1}^{N} \frac{L_{2i}}{\sigma wh} + \sum_{j=1}^{M} \frac{\theta}{\sigma \ln(r_o / r_i)h}, \qquad (6)$$

where $L_{1,3,4}$ are the lengths of the channels connected to the injector, L_{2i} are the lengths of the separation channel segments, σ is the buffer conductance, w is the channel width, h is the channel depth, θ is the turn angle, r_o is the outer turn radius, r_i is the inner turn radius, N is the number of straight segments, and M is the number of turns. Eq. 6, accounts for the radially varying field structure in calculating the resistance of the curved separation channels.

The resulting network is then solved for the global node voltages (V_{1L}, V_{4L}, V_{oL}, V_{2L}, V_{1D}, V_{4D}, V_{oD}, V_{3D}), as seen in Figure 4. Then the internal separation channel node voltages are solved using a Kirchoffian linear system with the individual resistances. A complete description of the implementation in VerilogA is provided in [9].

6 CONCLUSION

In this paper a closed-form analytical model of a 2-stage cross injector is developed. The full 5-dimensional parameter space is reduced to only the Peclet numbers in the loading and dispensing stages for given field ratios and constant channel widths. The resulting polynomial fit model is 32,000 times faster than pure numerical simulation, and is integrated into a system network of

component models using the input concentrations and providing the plug variances and peak concentrations as outputs. The model predicts a worst case variance of $\sim w^2$ and has less than 2% maximum error relative to the original PDE solutions. The integration with the rest of the system is completed with a Kirchoffian network analysis to provide voltages for all ports and nodes in the network using equivalent resistor networks.

Future similar models for more complex injector types will permit system network representations of tradeoffs between alternative injector designs in interactions with other system components.

ACKNOWLEDGEMENTS

This research effort is sponsored by the Defense Advanced Research Projects Agency under the Air Force Research Laboratory, Air Force Material Command, USAF, under grant number F30602-01-2-0587 and the NSF ITR program under grant number CCR-0325344. Computing resources were made possible by NSF equipment grant CTS-0094407 and Intel Corporation.

REFERENCES

[1] Jacobson, S.C; Hergenröder, R.; Koutny, L.B.; et. al., Vol. 66, *Analytical Chemistry* 1994, pp. 1107-1113.

[2] Ermakov, S.V.; Jacobson, S.C.; Ramsey, J.M., Vol. 70, *Analytical Chemistry* 1998, pp. 4494-4504.

[3] Shultz-Lockyear, L.L.; Colyer, C.L.; et. al., Vol. 20, *Electrophoresis* 1999, pp. 529-538.

[4] Deshpande, M.; Greiner, K.B.; West, J.; Gilbert, J.R., *Micro Total Analysis Systems* 2000, pp. 339-342.

[5] Ermakov, S.V.; Jacobson, S.C.; Ramsey, J.M., Vol. 72, *Analytical Chemistry* 2000, pp. 3512-3517.

[6] Pfeiffer, A.J.; Mukherjee, T.; Hauan, S., *Proceedings of NanoTech* 2003, pp. 250-253.

[7] Wang, Y.; Lin, Q.; Mukherjee, T., *MicroTAS* 2003, Oct. 5-9, Squaw Valley, CA, USA.

[8] Hanche-Olsen, H., "Buckingham's π-Theorem", http://www.math.ntnu.no/~hanche/notes/ buckingham/buckingham-a4.pdf. 1998

[9] Wang, Y.; Lin, Q.; Mukherjee, T., *NanoTech* 2004, Mar. 7-11, Boston, MA, USA.

Dielectrophoresis (DEP) of Cells and Microparticle in PDMS Microfluidic Channels

Thirukumaran T. Kanagasabapathi [*], Christopher J. Backhouse [**] and Karan V.I.S. Kaler [*1]

[*] Department of Electrical and Computer Engineering, University of Calgary,
Alberta, Canada (T2N 1N4). kanagasa@enel.ucalgary.ca, kaler@enel.ucalgary.ca
[**] Department of Electrical and Computer Engineering, University of Alberta,
Alberta, Canada (T6G 2V4). chrisb@ece.ualberta.ca

ABSTRACT

Electromanipulation of microparticles utilizing microelectrodes has demonstrated considerable promise for the characterization, separation and handling of biological cells. Presently considerable research and development effort is directed towards the development of miniaturized fluidic systems with integrated dielectrophoresis (DEP) electrodes. The design, development and fabrication of a DEP microfluidic assembly with an in-built interdigitated microelectrode array is presented. Continuous fractionation of microparticles in a PDMS microfluidic channel is described. Experimental verification of positive and negative DEP of yeast cells and polystyrene latex beads is demonstrated.

Keywords: dielectrophoresis, pdms, microfluidics, continuous fractionation, microfabrication

1 INTRODUCTION

New and improved methods to characterize and sort cells and micron sized particles are in high demand for a wide range of applications in the areas of biomedical research, clinical diagnosis and environment analysis. With the use of technologies from the microelectronics process industry, fabrication of low cost microfluidic devices [1] has been made possible and conventional methods of fabricating microfluidic devices by etching glass or silicon are fast been replaced by soft lithography techniques [2,3]. Fabrication of microfluidic devices in Poly(dimethylsiloxane) [PDMS] is rapid and cost effective compared to conventional methods. The utilization of this soft lithography technique [3,4,9] for the implementation of DEP in a microfluidic system is demonstrated.

2 THEORY

Dielectric particles such as cells are electrically polarized when subjected to an alternating electric (A.C.) field. If the this field is furthermore inhomogeneous, then the cells will experience a dielectrophoretic (DEP) force [5] that can act to convey them toward strong or weak field regions, depending on the dielectric polarization of the cell and that of the suspending medium [6,7,13].

The time averaged DEP force $\left\langle \vec{F}_{DEP} \right\rangle$ exerted by a non-uniform field of peak strength E acting on a homogenous spherical particle of radius a, immersed in a medium is given by [7]:

$$\left\langle \vec{F}_{DEP} \right\rangle = 2\pi\varepsilon_m a^3 \, \mathrm{Re}[K_e] \nabla E_{rms}^2 \vec{r} \qquad (1)$$

Where ∇E_{rms}^2 is the square of the electric field gradient.

The DEP force may attract (positive DEP) or repel (negative DEP) particles from the regions of higher field. The DEP force determined by the sign of K_e, the real part of complex Claussius-Mosotti factor, which is dependent on the complex permittivity of the particle and medium respectively [5,6].

3 FABRICATION OF MICROFLUIDIC DEVICE

Poly(dimethylsiloxane) has been one of the most actively developed polymers for microfluidics, as it reduces the time, complexity and cost of prototyping and manufacturing [8].

Patterning a gold film on an insulating glass substrate forms the microelectrode array responsible for the synthesis of a periodic nonuniform electric field. Borofloat glass wafer [Micralyne Inc., Edmonton] with a chrome-gold layer of thickness 150nm was coated with HPR 504 [Microchem Co.,] and soft baked at 110° C for 30 min. After exposure to a 365 nm UV light source through a chrome mask [ABM Mask Aligner], the exposed wafer was developed [Dev-354 Microchem Co.,] The gold-chrome metallization layer in the exposed regions were etched by a wet chemical process. The resulting electrode structure is shown in figure 1, where the width and the spacing between the successive electrode fingers is 20 μm.

---

[1] Corresponding author.

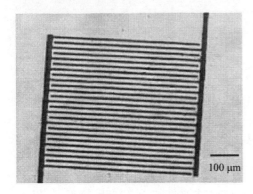

Figure 1. Interdigitated electrode structure

3.1 Molding PDMS microchannel

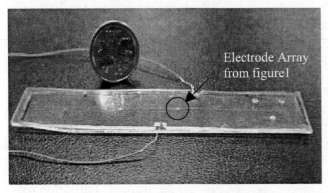

Electrode Array
from figure1

Figure 2. PDMS microfluidic assembly

Figure 2 shows the complete microfluidic device embedded with an electrode array as indicated by the arrow. The three walls of the channels are formed in PDMS and the cover plate, necessary for an enclosed channel, is formed by the glass substrate housing the interdigitated electrode array. The Si wafer with a sacrificial oxide layer of thickness 0.6 μm was coated with hexamethyldisiloxane (HMDS) at 150°C and a layer of photoresist, HPR504 [Microchem Co.] was spin coated on top and soft baked for 110°C. The wafer was then exposed to UV light through a chrome contact mask. The exposed areas were dissolved in Dev-354 [Microchem Co.,] and the sacrificial oxide layer removed using Buffer oxide Etch. Deep Reactive Ion Etching (DRIE) [Oxford Plasmalab-System100] of the exposed Si wafer resulted in a negative replica of the desired channel with a vertical sidewall as shown in Figure 3B.

PDMS prepolymer was prepared by mixing two commercially available components [Sylgard 184 Elastomer & Sylgard 184 Curing Agent, Dow Corning]. The prepolymer was mixed at 10:1 ratio by weight and subsequently poured onto the Si wafer and cured at 60° C for 1 hr [3].

The PDMS replica was then peeled from the master as shown in figure 3C. Access holes for reservoirs can be made by placing posts on the masks or punched out of the cured layer.

3.2 Surface Treatment

PDMS is hydrophobic due to the presence of negatively charges silanol groups on the surface which results in the absorption of hydrophobic species and can easily nucleate air bubbles. Exposing the cured PDMS layer to oxygen plasma at a pressure of 0.15 torr renders the surface hydrophilic [3]. This process creates ozonation on the surface and enables an irreversible bonding of the PDMS to the glass substrate. The glass substrate and the cured PDMS layer were exposed face-up to 80% oxygen plasma at a power of 45 watts for 90 secs. They are then sealed and placed on a hot plate at 60° C for 45 secs. This forms a permanent seal, attempting to break the seal can result in the failure of the bulk PDMS. The seal can withstand pressures ranging from 30-50 psi [8,9].

4 EXPERIMENTAL
4.1 Polystyrene Latex Beads

Polystyrene latex beads 6 μm in diameter with carboxyl (COOH⁻) or plain surface (NH⁺) were purchased from Interfacial Dynamics (Interfacial Dynamics Corp., USA). The beads were washed and suspended in deionised water. Dilute samples were injected into the fluidic device using a microfluidic syringe pump [Cole-Parmer Co., Cambridge]. A 5ml plastic syringe provides a continuous flow in the range of 0.001 μl/h – 14.33 ml/min [10]. Teflon tubing, tube end-fitting [Fisher Scientific Co.,] facilitates connection to the channel reservoir.

4.2 Yeast Cells

Yeast cells *(Saccharomyces Cerevisiae)* were cultured for 2 days at 30° C in a growth medium [1% yeast extract, 2% Glucose]. The samples are washed repeatedly with deionised water by centrifugation, the supernatant liquid decanted and the residual cells resuspended in fresh liquid. The final cells collected at the bottom after three successive centrifugations were diluted 1000-fold with deionised water prior to experimentation [11,12].

The dilute samples were injected into the microfluidic assembly as described and a function generator [Hewlett Packard, Model - 33120A] was used to supply a sinusoidal voltage required for the electrode array. The DEP induced cell motion was observed utilizing a optical microscope (Olympus, BH2)and the images captured by a video camera [Hitachi, VK-C350] coupled to the microscope station.

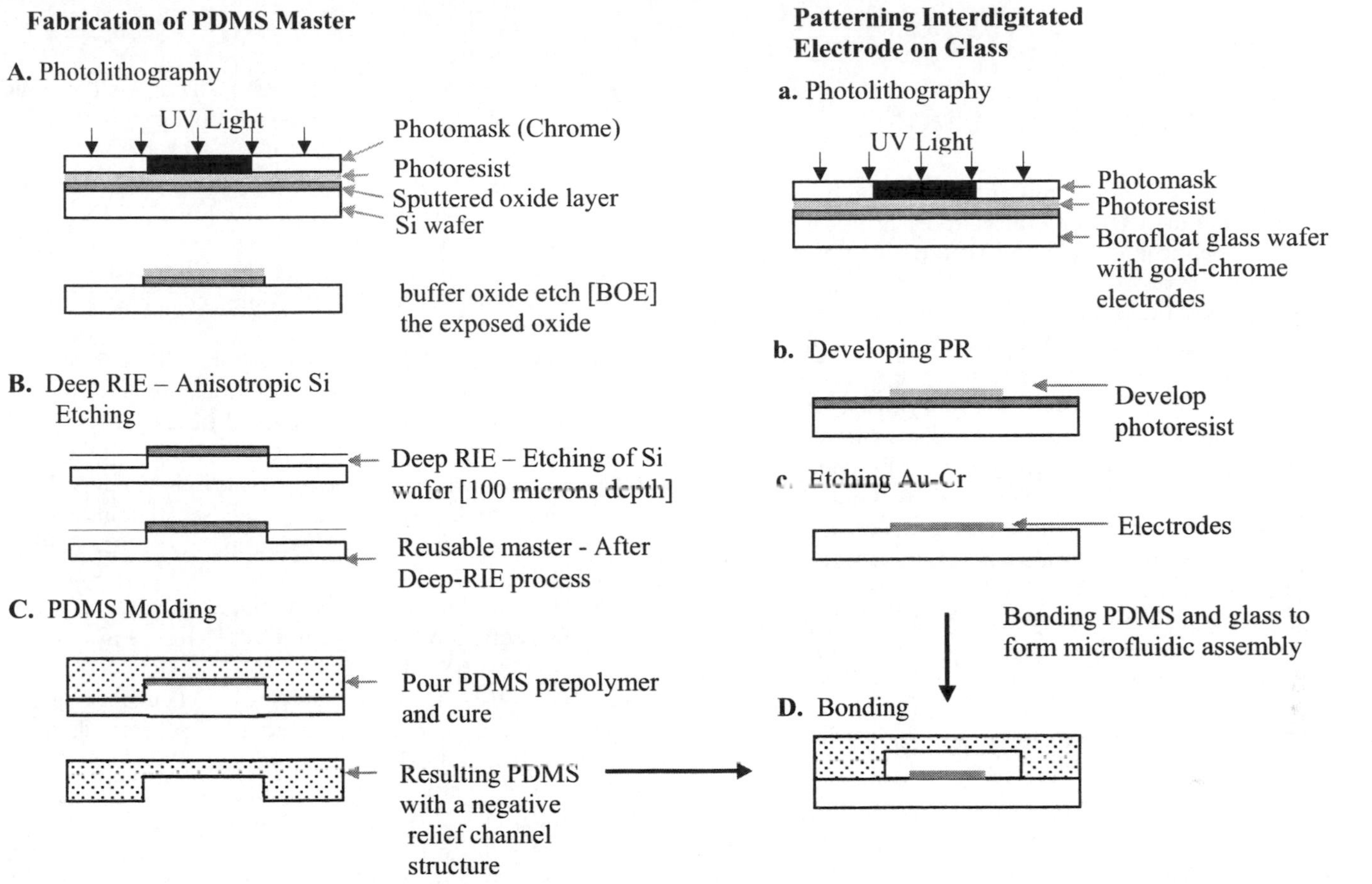

Fabrication of PDMS Master

A. Photolithography

UV Light

Photomask (Chrome)
Photoresist
Sputtered oxide layer
Si wafer

buffer oxide etch [BOE]
the exposed oxide

B. Deep RIE – Anisotropic Si Etching

Deep RIE – Etching of Si wafer [100 microns depth]

Reusable master - After Deep-RIE process

C. PDMS Molding

Pour PDMS prepolymer and cure

Resulting PDMS with a negative relief channel structure

Patterning Interdigitated Electrode on Glass

a. Photolithography

UV Light

Photomask
Photoresist
Borofloat glass wafer with gold-chrome electrodes

b. Developing PR

Develop photoresist

c. Etching Au-Cr

Electrodes

Bonding PDMS and glass to form microfluidic assembly

D. Bonding

Figure 3: Fabrication Process Sequence

5   RESULTS AND DISCUSSION

Figures 4A and 4B shows negative DEP of the polystyrene latex beads, observed when a voltage of 3.8 V_{p-p} of field frequency 480 kHz was applied to the chamber electrodes. The microbeads were observed to be levitated and formed 'pearl-chains' above the electrode. The difference in focus of the electrode array and the beads in figures 4A and 4B confirm levitation and 'pearl-chain' phenomena.

In contrast, yeast cells when subjected to a similar A.C voltage (3.7 V_{p-p}) at a field frequency of 580 kHz exhibited positive DEP and hence were attracted towards the region of maximum field intensity and collect at the electrode surface as shown in figures 4C and 4D. Figure 4D shows a higher concentration of yeast cells collected near the electrode edges at regions of field maxima and formed 'pearl-chains'.

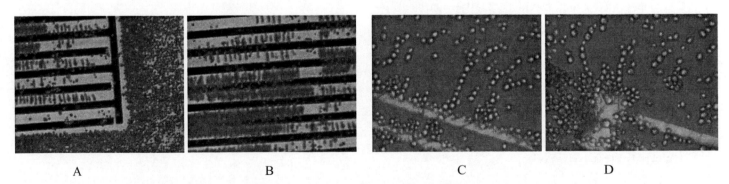

A   B   C   D

Figure 4: A - Negative DEP of polystyrene latex beads, cells were repelled from the electrodes and were levitated above them.
B - levitated beads forming 'pearl-chains' above the electrode surface.
C - Positive DEP of yeast cells; cells attached to the electrodes and formed 'pearl-chains'.
D – Higher population of yeast cells collected near the electrode edges at regions of field maxima.

6 CONCLUSION

A novel polymeric-glass microfluidic system with an integrated interdigitated microelectrode array has been described. Fabrication of such low cost, reusable microsystems capable of electro-manipulation of cells and experimental verification of positive and negative DEP has been demonstrated. The polystyrene beads levitated and confined above the electrode array were continuously removed by fluid flow. Thus, this non-invasive, easy to fabricate technique could be employed for the continuous fractionation of heterogenous mixture of cells. Since PDMS can be molded at low temperatures without elaborate fabrication requirements, the microfluidic device can be readily fabricated in a normal laboratory setting. Further, integration of this technology with on-chip imaging and control will provide a microsystem capable of quantitative and sensitive analysis of DEP signatures of various cancerous cells.

7 ACKNOWLEDGEMENTS

The work presented was supported by National Science and Engineering Research Council of Canada (NSERC) and in part by Microsystems Technology Research Institute (MSTRI) – University of Alberta. The authors also thank Nanofab staff for their guidance in fabrication and Colin Dalton for helpful discussions.

REFERENCES

[1] T.B. Jones et al, "Dielectrophoretic microfuidic devices," Proc. of IEJ/ESA Joint Symposium on Electrostatics, 78 – 87,2000.

[2] David C. Duffy, J. Cooper McDonald, Olivier J.A. Schueller and George M. Whitesides, "Rapid prototyping of Microfluidic Systems in Poy(dimethylsiloxane)," Anal. Chem., 70, 4974-4984, 1998.

[3] J. Cooper McDonald, David C. Duffy, Janella R. Anderson, Daniel T. Chiu, Hongkai Wu, Olivier J.A. Schuller and George M. Whitesides, "Fabrication of microfluidic systems in poly(dimethylsiloxane)," Electrophoresis, 27, 27 - 40, 2000.

[4] George M. Whitesides, Emanuele Ostuni, Shuichi takayama, Xingyu Jiang, and Donald E. Ingber, "Soft Lithography in Biology and Biochemistry," Annu. Rev. Biomed Eng., 3, 335-373, 2001.

[5] H.A Pohl, "Dielectrophoresis," Cambridge University Press, New York, 1978.

[6] K.V.I.S Kaler and H.A. Pohl, "Dynamic Dielectrophoretic levitation of living cells," IEEE Trans. Ind Apps., 1A-19, 1089 - 1093, 1983.

[7] Youlan Li and K.V.I.S. Kaler, "Dielectrophoretic fluidic cell fractionation system," Analytica Chimica Acta, 2003 [In press].

[8] Younan Xia and George M. Whitesides, "Soft Lithography," Annu. Rev. Mater. Sci., 28, 153 –184, 1998.

[9] J. Cooper McDonald and George M. Whitesides, "PDMS as a material for fabricating microfluidic devices," Accounts of Chemical Research, 35-7, 491 - 499, 2002.

[10] Li Cui, David Holmes and Hywel Morgan, "The dielectrophoretic levitation and separation of latex beads in microchips," Electrophoresis, 22, 3893 – 3901, 2001.

[11] W.M. Arnold, "Positioning, Levitation and Separation of Biological Cells," Inst. Phys. Conf. Ser. No 163, 28-31, 1999.

[12] Peter R.C. Gascoyne and Jody Vykoukal, "Particle separation by Dielectrophoresis", Electrophoresis, 23, 1973 – 1983, 2002.

[13] Ronald Pethig, Ying Haung, Xia-Bo Wang and Julian P H Burt, "Positive and negative dielectrophoretic collection of colloidal particles using interdigitated castellated micro-electrodes,"J. Phys. D: Appl. Phys.,25, 905-912, 1992.

Dry microarrays: a generalized data source for genome and proteome analysis

G. Sampath

Department of Computer Science, The College of New Jersey, Ewing, NJ 08628
sampath@tcnj.edu

ABSTRACT

Microarray analysis methods can be applied to characteristic data obtained from biological sequences in matrix form (which could include data from experiments or bioassays as well). The rows correspond to a genome region (intron or exon) or a protein, the columns to data obtained from the genome or proteins under selected conditions. In such a *dry microarray*, tasks like protein classification and construction of gene networks are aided by the use of various clustering methods applied to the heterogeneous data matrix. These methods, suitably modified, can also be used for simultaneous multiple alignment of all the sequences instead of pairwise or over small numbers of sequences at a time (as is the current practice). The feasibility of the concept is shown with results obtained from the construction and analysis of a dry microarray for protein classification.

Keywords: bioarrays, proteomics, computational methods

1 GENOMICS AND PROTEOMICS: MICROARRAY AND SEQUENCE DATA

Effective and efficient biological data analysis and visualization have become necessary with increasing amounts of data becoming available through the use of rapid sequencing and bioassay techniques. Thus a variety of sequence analysis methods have been developed to study higher-level properties of biological sequences, such as secondary and tertiary structure of proteins, homology, and phylogeny [11]. Similarly microarray technology [1] has made it possible to obtain expression data for large numbers of genes under a range of conditions (such as time evolution, samples from subjects both target and control, and differing environments), with the technology now being extended to proteins as well [6]. A large number of algorithms are now available to extract meaningful information from the data such as secondary and tertiary structure from sequences [13], relationships among the genes and conditions [2], classification of proteins [7, 9, 10], and construction of gene and protein networks [14]. Some of the methods [3, 6, 11] developed include multiple alignment, dendrogram analysis, principal component analysis, eigengene analysis, biclustering (patterned submatrices), support vector machines, singular value decomposition, hidden Markov models, and self-organizing maps, with statistics playing an important role in many of them [15].

By and large the analysis and modeling of biological data is currently centered on specific kinds of data and has specific objectives. Thus computational modeling in biology is usually driven by the kind of data it is based on and has concentrated on an individual goal such as sequence analysis at different levels, phylogeny or classification, or construction of networks. Recently this has given way to computational models that attempt to extract patterns from heterogeneous data using various probabilistic/statistical [4] and data fusion models [5, 8, 12] that may include phylogenetic profiles in the form of bitmaps. In this report, a general form of microarray is proposed in which heterogeneous data from diverse sources can be brought together in a single matrix to which a wide range of microarray-based computational techniques can be applied simultaneously at all levels: sequence, genome, proteome, and cell.

2 DRY MICROARRAYS AND THEIR POTENTIAL APPLICATIONS

If sequence level data from genomes and proteins are arranged in the form of a microarray-like matrix, many of the methods used in gene expression analysis can be applied to biological sequences as well. Such an array could lead to efficient ways of classification as well as identification of relationships that may be useful in constructing networks at the gene and protein levels. This suggests further that by combining heterogeneous data from a number of sources in the microarray, a microarray can serve as a framework for data fusion. In such a *dry microarray*, the columns can be any of a number of defining properties of the genome or protein, including, for example, 1) subsequences of given lengths, 2) subsequences that are similar in some sense, or 3) secondary structure motifs. These are in addition to the more conventional types of microarray data such as experimental, tissue, and cell cycle data. The wide range of statistical and computational methods used in microarray analysis can now be applied to the dry microarray. The columns can also be weighted to bias the analysis towards properties of interest. Potential applications include:

1) *Cluster analysis* of coding and non-coding regions of a genome, which can be extended to the study of relationships between specific sites and upstream regions,
2) *Protein classification*, which could lead to identification of protein families based on different criteria on the columns, and
3) *Secondary structure* (RNA and protein) and *tertiary structure identification* (with column properties based on motifs in the primary sequences).

3   FEASIBILITY: EXAMPLE AND COMPUTATIONAL RESULT

The feasibility of the approach is illustrated by constructing a dry microarray for a set of 200 proteins randomly selected from the Swiss-PROT database (which contains 6115 unique sequences after duplicates are removed). *Hierarchical clustering* [15] was used to group proteins in an unsupervised classification. Use of the CLUSTER software of Eisen (*http://rana.lbl.gov*, see also [15]) led to Figure 1, which shows a clustered display of the 200 proteins. The y axis corresponds to protein, the x axis to 7872 subsequences of 1 to 3 residues in the sample protein set. The proteins were randomly selected from out of 6115 unique proteins (obtained after removing duplicates from over 13000 listed). The selected ones were scanned to obtain all 1-, 2-, and 3-residue subsequences and their frequencies. Each frequency was divided by the length of the sequence and entered into the dry microarray. Thus an array entry is the ratio of subsequence frequency to length of primary sequence. Light bands (visible as vertical stripes) occur for certain 2- and 3-residue subsequences as well as for individual proteins over (ordered) subranges of the subsequences (visible as short dull horizontal stripes). The vertical bands are indicative of the presence of multiple alignment among the proteins mapped by the stripes, while the horizontal stripes are suggestive of groups of subsequences (typically of the same length) occurring in a single protein.

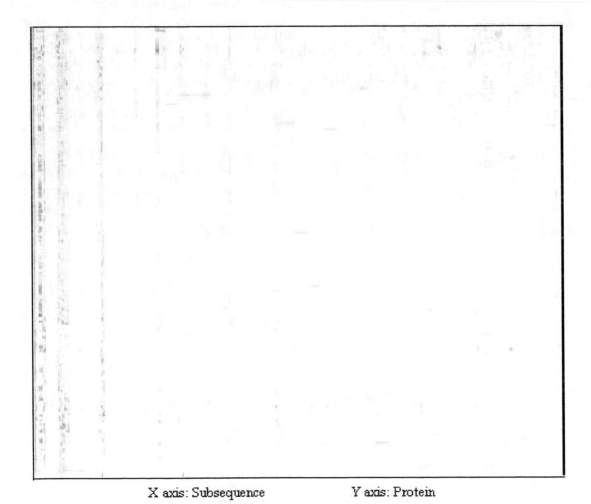

X axis: Subsequence   Y axis: Protein

Figure 1. Hierarchical clustered display of dry microarray of 200 proteins from Swiss-PROT

29	17	19	40	29	1	17	9	6	33

Table 1. Sizes of clusters obtained by K-means clustering of 200 proteins from Swiss-PROT

Clustering was also done using K-means clustering (using the same software). Table 1 shows the cluster sizes that were obtained with 10 clusters.

Work is ongoing on data analysis using a wide range of column definitions based on one or more of the following: sequence data (based partly on scoring matrices like PAM and BLOSUM), data based on DNA/protein motifs, expression data, and secondary structure data (proteins/RNA).

REFERENCES

[1] P. Baldi and G. W. Hatfield. *DNA Microarrays and Gene Expression.* Cambridge University Press, Cambridge (U.K.), 2002.

[2] A. Brazma and J. Vilo. "Gene expression and data analysis." FEBS Letters **480**, 17-24 (2000).

[3] Y. Cheng and G. M. Church. "Biclustering of expression data." Proceedings Intelligent Systems in Molecular Biology, 2000.

[4] M. Deng, S. Mehta, F. Sun, and T. Chen. "Inferring domain-domain interactions from protein-protein interactions." http://www.genome.org/cgi/doi/10-1101/gr.153002.

[5] R. Jansen, N. Lan, J. Qian, and M. Gerstein. "Integration of genomic datasets to predict protein complexes in yeast." J. Structural and Functional Genomics **90**, 1-11 (2002).

[6] I. S. Kohane, A. T. Kho, and A. J. Butte. *Microarrays for an Integrative Genomics.* MIT Press, Cambridge (Mass.), 2003.

[7] A. Kumar, B. Smith. "A framework for protein classification." Proceedings of the German Conference on Bioinformatics, volume 2, 55-57, 2003.

[8] G. R. G. Lanckriet, M. Deng, N. Cristianini, M. I. Jordan and W. S. Noble. "Kernel-based data fusion and its application to protein-protein function prediction in yeast." Proceedings of the Pacific Symposium on Biocomputing, January 3-8, 2004.

[9] A. H. Liu and A. Califano. "Functional classification of proteins by pattern discovery and top-down clustering of primary sequences. IBM Systems J. **40** (2), 379, 2001.

[10] L. Lo Conte, S. E. Brenner, T. J. P. Hubbard, C. Chothia, and A. G. Murzin. "SCOP database in 2002: refinements accommodate structural genomics." Nucleic Acids Research **30** (1), 264-267 (2002).

[11] R. Mount. *Bioinformatics*, Cold Spring Harbor Press, Cold Spring Harbor (NY), 2001.

[12] P. Pavlidis, J. Weston, J. Cai, and W. N. Grundy. "Gene functional classification from heterogeneous data." Procs. 5th International Conference on Computational Molecular Biology, April 21-24, 2001, pages 242-248.

[13] B. Rost. "Protein secondary structure prediction continues to rise." J. Structural Biology **134**, 204-218 (2001).

[14] B. Schwikowski, P. Uetz, and S. Fields. "A network of protein-protein interactions in yeast." Nature Biotech. **18** (12), 1257-1261 (2000).

[15] T. Speed (ed.). *Statistical Analysis of Gene Expression Microarray Data.* CRC Press, Boca Raton, 2003.

Molecular Simulation of DNA Microarrays
– an Application of 2D particle mesh Ewald algorithm –

T. Watanabe[*], M. Kawata[**], and U. Nagashima[*,**]

[*]Japan Science and Technology Corporation (JST), 4-1-8 Honcho,
Kawaguchi 332-0012, Japan, donjo@ni.aist.go.jp
[**]Grid Technology Research Center, National Institute of Advanced Industrial
Science and Technology, 1-1-1 Umezono, Tsukuba, 305-8568 Japan

ABSTRACT

We have developed the two-dimensional particle mesh Ewald (2D-PME) algorithm for molecular dynamics (MD) simulation, and applied it to have knowledge of the behavior of DNA and its interaction on microarrays, which crucial in the design of DNA microarrays. The 2D-PME algorithm can rapidly calculate the Coulomb interaction for a 3D system with 2D periodicity, which is the most time-consuming process in the MD simulation.

The MD simulation of DNA microarrays that contain a single DNA molecule in a unit-cell was carried out under the two types of the boundary conditions, 3D systems with the 2D and 3D periodicities, in which the Coulomb interactions were calculated by using 2D-PME and 3D-PME methods, respectively. Effect of the infinite replicas in the vertical direction to the microarray surface in 3D periodic system was examined based on the results of the 3D system with 2D periodicity. Comparison of the computational efficiency and accuracy between the 2D-PME and 3D-PME methods revealed the availability of the 2D-PME method.

Keywords: molecular dynamics, DNA microarrays, two-dimensional particle mesh ewald, 2D-PME

1 INTRODUCTION

DNA microarrays have attracted widespread attention in various areas such as drug design and genetic screening/sequencing. Although knowledge of the behavior of DNA and its interaction on microarrays is crucial in the design of DNA microarrays, such knowledge is still lacking.

Molecular dynamics (MD) simulation has been applied to have such knowledge at the molecular level. Wong and Pettitt, for example, executed MD simulation of DNA microarrays by using an all-atom model.[1] They applied novel periodic boundary condition, the glide-plane boundary condition.[2] Their system has the 3D periodicity, thus also had the lamellar structure. The effect of the infinite replicas in the direction vertical to the microarray surface is not negligiable. One of the solutions for this lamellar structure is the use of the 3D system with 2D periodicity. On the other hand, there were sophisticated methods to treat the Coulomb interaction in the 3D periodic system,[3,4] in contrast to the 3D system with the 2D periodicity.

Recently, Kawata et al[5,6] were developed computationally efficient method, two-dimensional particle mesh Ewald method (2D-PME), to calculate the Coulomb interaction in the 3D system with 2D periodicity. The 2D-PME method was not only computationally efficient but also gave the accrate energy and force[5,6]. There was only preliminary result but no application to the practical MD simulation.

The 2D-PME algorithm was applied to the MD simulation of DNA microarray systems to verify that the method can simulate efficiently the 3D systems with 2D periodicity. Effect of the infinite replicas in the vertical direction to the microarray surface in 3D periodic system was also examined based on the results of the 3D system with 2D periodicity.

2 METHOD

2.1 Two Dimensional Particle Mesh Ewald Algorithm

In 2D-PME method, the Coulomb potential energy, V, for a 3D system of N charged particles, $\mathbf{r}_1$, $\mathbf{r}_2,...,$ $\mathbf{r}_N$, satisfying the neutrality condition, $q_1+q_2+\cdots+q_N = 0$, with 2D periodicity can be expressed as

$$V = \frac{1}{2}\sum_{\mathbf{h}_x,\mathbf{h}_y}'\sum_{i=1}^{N}\sum_{j=1}^{N}\frac{q_i q_j}{\left|\mathbf{r}_i - \mathbf{r}_j + \mathbf{h}_x + \mathbf{h}_y\right|}$$
$$\cong V^r + V^k_{\mathbf{k}\neq 0} + V^k_{\mathbf{k}=0} + V^s, \tag{1}$$

where $\mathbf{h}_x$ and $\mathbf{h}_y$ are the 2D lattice vectors, the prime indicates that terms with $i=j$ are omitted when $\mathbf{h}_x + \mathbf{h}_y = \mathbf{0}$, and $\mathbf{k}$ is the reciprocal vector. V^r is the contribution from the real space sum, $V^k_{\mathbf{k}\neq 0}$ and $V^k_{\mathbf{k}=0}$ are the contributions from the reciprocal space sums with and without the $\mathbf{k}=0$, respectively, and V^s is the self-correction term. Here, a Gaussian charge distribution, such as $\rho(\mathbf{r}_i) = q_i \alpha^3 \pi^{-3/2} \exp(-\alpha^2 |\mathbf{r}_i|^2)$, is used in the charge decomposition, α is a screening parameter. These terms can be written as

$$V^r = \frac{1}{2} \sum_{\mathbf{h}_x, \mathbf{h}_y}' \sum_{i=1}^{N} \sum_{j=1}^{N} \frac{q_i q_j}{|\mathbf{r}_{ij} + \mathbf{h}_x + \mathbf{h}_y|} \times erfc\left(\alpha |\mathbf{r}_{ij} + \mathbf{h}_x + \mathbf{h}_y|\right) \quad (2)$$

where *erfc* represents the complementary error function.

$$V_{\mathbf{k} \neq 0}^k = \frac{1}{|\mathbf{h}_x \times \mathbf{h}_y|} \sum_{m_x = -G_x/2}^{G_x/2} \sum_{m_y = -G_y/2}^{G_y/2} \int_{-\infty}^{\infty} dh \left| C\left(\frac{2\pi m_x}{G_x}, n_x\right) \right|^2$$

$$\times \left| C\left(\frac{2\pi m_y}{G_y}, n_y\right) \right|^2 |C(h, n_z)|^2 \frac{|\tilde{Q}(m_x, m_y, h)|^2}{4\pi^2(m_x^2 + m_y^2) + h^2}$$

$$\times \exp\left\{ -\frac{4\pi^2(m_x^2 + m_y^2) + h^2}{4\alpha^2} \right\} \quad (3)$$

G_x and G_y are the number of grid points for the FFTs in the x- and y-direction, respectively, and n_λ is the order of the B-splines in the λ-direction. $C(v, w)$ is defined as

$$C(v, w) \equiv \frac{\exp[i(w-1)v]}{\sum_{t=0}^{w-2} M_w(t+1) \exp(ivt)} \quad (4)$$

where M_w is cardinal B-splines. $\tilde{Q}(m_x, m_y, h)$ is the Fourier integral of $Q'(m_x, m_y, t_z)$ with respect to t_z and $Q'(m_x, m_y, t_z)$ is the 2D FFTs of $Q(t_x, t_y, t_z)$ with respect to t_x and t_y. $Q(t_x, t_y, t_z)$ is a charge distribution calculated by using the B-splines.

$$V_{\mathbf{k}=0}^k = \sum_{i=1}^{N} q_i B\left[V_{\mathbf{k}=0\,g}^k, z_i \right] \quad (5)$$

where B$[V_{\mathbf{k}=0\,g}^k, z_i]$ is the Coulomb potential energy at position z_i, interpolated with a B-spline interpolation by using the values of $V_{\mathbf{k}=0\,g}^k V_{\mathbf{k}=0\,g}^k, \ldots, V_{\mathbf{k}=0\,Ng}^k$, which the Coulomb potential energy on the gth grid point ($g = 1, \ldots, N_g$) in the z direction.

$$V^s = -\frac{\alpha}{\sqrt{\pi}} \sum_{i=1}^{N} q_i^2 \quad (6)$$

These equations are same with the two-dimensional Ewald method (2D-EW) [7,8] except Eq (3). The overall computational efficiency of the 2D-EW method strongly depends on the computational efficiency of the reciprocal-space sum. In the 2D-PME method, the calculation of the reciprocal-space sum without the **k=0** term can be accelerated by using the B-spline interpolations and the FFT.

2.2 Molecular Dynamics Simulation

The 2D-PME algorithm was implemented into the NAMD Version 2.5 program package, [9] which is widely used. The NAMD have implemented both the

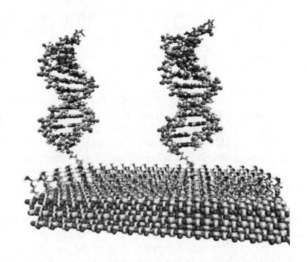

Figure 1. The system used in the MD simulation. This figure contains two unit-cells. Water molecules and the sodium and chloride ions were removed for visibility.

3D-PME method and the lattice for 3D system with 2D periodicity. In the implementation of the 2D-PME method, the classes and data structure for the 3D-PME in the NAMD program package were fully reused.

All calculation was carried out on the KOUME and UME PC-clusters at Grid Technology Research Center, National Institute of Advanced Industrial Science and Technology.

3 RESULTS AND DISCUSSION

The model system was essentially same with Ref.1 but used the pseudo-2D boundary condition instead of the glide-plane boundary condition. The system constructed on silica layer of β-cristobalite plate 1.2nm thick, of which a (1,1,1) surface was covered with sixty-four epoxides. A DNA strand d(CGTGTCCCTCTC) hybridized with its complement was tethered to the terminal of an epoxide by the amine linker. The empty space of the simulation box was filled with the water molecules and ions. This system is free from the infinite replicas in the vertical direction to the microarray surface, and has a gas-liquid interface in the non-periodic direction, z-direction. The water molecules which vapor into gas phase were inverted sign of z component of their velocity at the distance from a gas-liquid interface by 1.5 nm. The initial system without water molecules and ions was displayed in Figure 1. The potential parameters were used in the CHARMM22 proteins and CHARMM27 nucleic acids.

For purpose of comparison, the 3D periodic system of the DNA microarray was also constructed. The difference with the 3D system with 2D periodicity was presence of a silica-liquid interface instead of the gas-liquid interface and a lamellar structure. The total number of the molecules in the 3D periodic system was same with the 3D system with 2D periodicity. The calculation of the Coulomb interaction was carried out by the 3D-PME method implemented in the NAMD program package.

The main difference of the computational efficiency between the 3D system with 2D periodicity and the 3D periodic system was due to the structural difference described above, such as the lamellar structure and the silica-liquid interface, however, the total computational efficiencies of the 2D-PME and 3D-PME methods were about the same.

3 CONCLUSION

The two-dimensional particle mesh Ewald (2D-PME) algorithm was implemented into the NAMD program package for carrying out the practical MD simulation. The 2D-PME method implemented into the NAMD program package was applied to the MD simulation of the DNA microarrays. Two systems for the DNA microarrays were constructed for the 3D system with 2D periodicity and for the 3D periodic system. As these results, the 2D-PME method was free from the problem caused by the lamellar structure and the computational efficiency was comparable with the 3D-PME method. These results indicated the 2D-PME method had large possibility to be standard method to calculate the Coulomb interaction in the simulation for the interface, membrane, and surface.

Acknowledgements

This work was supported in part by the grant from ACT-JST ("Research and Development for Applying Advanced Computational Science and Technology" of Japan Science and Technology Corporation).

References

[1] K.-Y. Wong and B. M. Pettitt, *Theor. Chem. Acc.*, **106**, 233 (2001).

[2] K.-Y. Wong, B. M. Pettitt, *Chem. Phys. Lett.*, **326**, 193 (2000).

[3] T. A. Darden, D. M. York, L. G. Pedersen, *J. Chem. Phys.*, **98**, 10089 (1993).

[4] U. Essmann, L. Perera, M. L. Berkowitz, T. Darden, H. Lee, L. G. Pedersen, *J. Chem. Phys.*, **103**, 8577 (1995).

[5] M. Kawata and U. Nagashima, *Chem. Phys. Lett.*, **340**, 165 (2001).

[6] M. Kawata and U. Nagashima, *J. Chem. Phys.*, **116**, 3430 (2002).

[7] M. Kawata and M. Mikami, *Chem. Phys. Lett.*, **340**, 157 (2001).

[8] M. Kawata, M. Mikami, and U. Nagashima, *J. Chem. Phys.*, **115** 4457 (2001).

[9] L. Kale, R. Skeel, M. Bhandarkar, R. Brunner, A. Gursoy, N. Krawetz, J. Phillips, A. Shinozaki, K. Varadarajan, and K. Schulten, *J. Comp. Chem.*, **151**, 283 (1999). (NAMD)

[10] D. MacKerel Jr., D. Bashford, M. Bellott, R. L. Dumbrack Jr., J. D. Evanseck, M. J. Field, S. Fischer, J. Gao, H. Guo, S. Ha, D. Joseph-McCarthy, L. Kuchnir, K. Kuczera, F. T. K. Lau, C. Mattos, S. Michnick, T. Ngo, D. T. Nguyen, B. Prodhom, W. E. Reiher III, B. Roux, M. Schlenkrich, J. C. Smith, R. Stote, J. Straub, M. Watanabe, J. Wiorkiewicz-Kuczera, D. Yin, M. Karplus, *J. Phys. Chem. B*, **102**, 3586 (1998).

[11] N. Foloppe, A. D. MacKerell Jr., *J. Comput. Chem.*, **21**, 86 (2000).

Development and Characterization of a 2-D microchip-based Protein separation system.

J.J. Feng[2]*, V Olazábal[1]*, J.A. Olivares[1], S. Sundaram[2], S. Krishnamoorthy[2] & S. Iyer[1]

[1]B-4, Bioscience Division, Los Alamos National Laboratory, Los Alamos, NM 87545
[2]CFD Research Corporation, 215 Wynn Drive, Huntsville, Alabama 35805, USA
* these authors contributed equally
Correspondence to: siyer@lanl.gov and sk@cfdrc.com

ABSTRACT

We have developed a 2-D microchip based separation system that employs capillary isoelectric focusing (CIEF) coupled with capillary zone electrophoresis (CZE) and uses a real-time fluorescent imaging of the analytes as they are separated. Physics-based simulations have been developed to understand the fundamental physico-chemical processes occurring in this system. Here, we describe the design of the chip, instrument and detection system, characterize the operation and present results from preliminary simulations to understand and establish optimal separation conditions.

Keywords: capillary isoelectric focusing, capillary zone electrophoresis, chip, laser-induced fluorescence, and simulation.

1 INTRODUCTION

Traditional methods in proteomics have relied on gel-based separation of proteins followed by directed cleavage of these proteins and mass spectrometry to measure the products [1,2]. The emphasis on reliable, high-throughput analyses has led to the development of novel "pre-mass spec" separation approaches such as multidimensional liquid-chromatography (LC) of tryptic peptides [3]. While an effective technique, comparable expression levels cannot be determined using this approach unless some form of labeling is employed. Hence there is need for approaches that can suitably compare the effects of experimental treatment on intact proteins prior to peptide generation. The ideal separation system would allow real time imaging of the separation, use a labeling system that would enable a wide dynamic range of detection and facilitate interfacing with downstream procedures, such as mass spectrometry. To this end, we have developed a 2-D separation chip-based system, similar to that reported in [4], that employs capillary isoelectric focusing (CIEF) coupled with capillary zone electrophoresis (CZE) but uses real-time fluorescent imaging of the proteins over the whole separation area.

Design and development of such a system is a relatively complicated task. One needs to understand the fundamental physico-chemical processes, such as fluid flow, electric field and mass transport, occurring in this system before making any design changes. In this regard, physics-based simulations can play a vital role. Here we present the results of our preliminary simulations performed to characterize the separation conditions. Detailed models for the transport of protein mixtures in both porous and non-porous medium with associated electrokinetic phenomenon are under development. While, validation and design optimization of the 2D separation device using these models will be reported in a future communication, separations in individual dimensions were performed using a scanning LIF CE for cIEF [5] and a Beckman MDQ for CZE (data not shown) in gel filled media using neutral capillaries. Separation buffers and conditions were as per the manufacturer's instructions.

2 2-D SEPARATION SYSTEM

2-D Chip. A schematic of the 2-D chip is shown in Figure 1. The chip employs capillary isoelectric focusing (CIEF) in the first dimension coupled with capillary zone electrophoresis (CZE) in the second dimension. In this configuration, IEF separates proteins based on their isoelectric point. A fraction from the first capillary is transferred to a second capillary, where the contents are separated based on their size. In this fashion, the entire sample is subjected to two-dimensional separation.

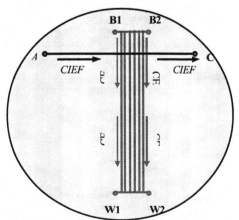

Figure 1. Schematic of 2D chip separation device. Reservoir A: Anolyte, Reservoir C: Catholyte, Reservoir B1 and B2: Buffer and Reservoir W1 and W2: Waste

This glass planar chip was designed in-house and fabricated by Motorola using standard photolithographic techniques and chemical wet etching [6]. It is placed in a holder with fluid reservoirs for loading sample, and where several platinum electrodes allow connection to the power supply.

Laser-induced fluorescence detection system (LIF). The laser-induced fluorescence detection system is shown in Figure 2. An epi-illumination configuration is used for the LIF detection. Light from a fiber-optic coupled argon ion laser, 488-nm emission line, is directed by a dichroic filter through lenses onto the 2-D chip surface. Emitted light passes through the dichroic, a bandpass, and a Holographic SuperNotch-Plus filter, and is focused onto a photomultiplier tube. The 2-D chip separation device is held in a stepper motor X-Y translation table. A Visual Basic program controls the X-Y table and acquires the photomultiplier output data from a 12-bit ADC board.

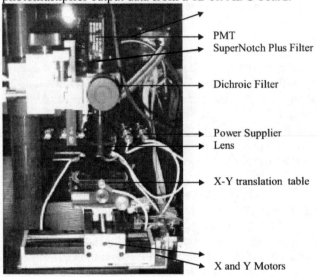

PMT
SuperNotch Plus Filter

Dichroic Filter

Power Supplier
Lens

X-Y translation table

X and Y Motors

Figure 2. *Laser-Induced fluorescence detection system for the 2-D.*

In order to design and develop an optimal and highly efficient separation system, the following design issues need to be addressed:

- What effect do gels have on separation?
- What are the optimal device dimensions to accomplish separation with minimal dispersion?
- How to generate required pH gradients within given space constraints of the CIEF channel and yet cover the range of pI points?
- What is the effect of geometry on leakage of buffer or current at CIEF-CZE channel intersection?

Multiphysics simulation of the combined CIEF-CZE phenomena can help answer the above questions. In this regard, an electrokinetic transport model in porous media has been implemented in CFD-ACE+[7], a commercial multiphysics simulation software developed and marketed by CFD Research Corporation. This, in conjunction with other electrokinetic transport models reported earlier [8], is used in the preliminary design analysis.

3 SIMULATION OF 2D SEPARATION

Simulation of 2D separation process is extremely complex due to nonlinear interactions among fluid flow, electric field, and sample transport. Full sets of governing nonlinear equations need to be solved. Besides, the presence of porous (gel) material in these systems will introduce additional complexities. In general the gel can be modeled as fibrous porous material that allows species smaller than the characteristic pore diameter to move through. The models developed are based on "effective medium approach" in which the effects of gel on flow, electric field and species transport are accounted for by introducing apparent parameters such as hydrodynamic permeability, effective electrical conductivity, hindered diffusion coefficient, and reduced electrophoretic mobility. Simulations that model hydrodynamic and transport processes at scales comparable to pore dimension, accounting for fiber orientation are infeasible. The governing equations are briefly summarized below.

Fluid mass Conservation:

$$\frac{\partial \varepsilon \rho}{\partial t} + \nabla \cdot \varepsilon \rho \mathbf{U} = 0 \qquad (1)$$

Momentum Conservation:

$$\frac{\partial \varepsilon \rho \mathbf{U}}{\partial t} + \nabla \cdot \varepsilon \rho \mathbf{U}\mathbf{U} = -\varepsilon \nabla p + \nabla \cdot \varepsilon \mathbf{t} + \varepsilon \mathbf{B} - \frac{\mu \varepsilon^2}{k_p} \mathbf{U} \qquad (2)$$

Analyte conservation:

$$\frac{\partial \varepsilon \rho c_i}{\partial t} + \nabla \cdot \mathbf{J} = R_i \qquad (3)$$

in which the flux is given by Nernst-Planck equation

$$\mathbf{J}_i = \varepsilon \rho \mathbf{U} c_i - \rho D_{i_{eff}} \nabla c_i + z_i c_i \omega_{i,eff} \nabla \phi \qquad (4)$$

Conservation of Electric Current:

$$\nabla \cdot \sigma_{eff} \nabla \phi = 0 \qquad (5)$$

Here ε is porosity of the fibrous medium and the other symbols are self-explanatory.

Of particular importance, the hindered diffusivity is calculated from [9]

$$\frac{D_{eff}}{D_0} = \left[1 + \frac{a}{\sqrt{k_p}} + \frac{a^2}{k_p^2} \right]^{-1} S(f), f = \left(1 + \frac{a}{R_s} \right)^2 (1 - \varepsilon) \qquad (6)$$

where a is species radius, R_s is the fiber radius and the steric exclusion is calculated as:

$$S = \frac{1}{1-f} \left[1 - 2f \left(1 + f - \frac{0.306 f^4}{1 - 1.403 f^8} - 0.0134 f^8 \right)^{-1} \right]$$

The electrophoretic mobility in gel is calculated based on Slater's model [10]

$$\frac{\omega_{eff}}{\omega_0} = \frac{1}{2 - \varepsilon}, 0.5 < \varepsilon < 1 \qquad (7)$$

for ordered fiber matrix and

$$\frac{\omega_{eff}}{\omega_0} = \frac{1}{3} + \frac{2}{3}\varepsilon, 0.5 < \varepsilon < 1 \qquad (8)$$

for a random fiber matrix. Table 1 summarizes models used in calculated apparent parameters:

Table 1: Summary of Effective Medium Models

Parameter	Model
Permeability k_p	Unit cell model [11,12]
Hindered Diffusivity D_{eff}	Brady's model [9]
Electrophoretic Mobility ω_{eff}	Slater's model [10]
Electric Conductivity σ_{eff}	Unit cell model [13]

The governing equations (1-5) are solved by an implicit finite-volume numerical scheme [7,8].

4 RESULTS AND DISCUSSIONS

The preliminary simulations are performed to analyze the impact of porosity of the gel matrix on sample separation and dispersion. Both separation (vertical) and IEF (horizontal) channels are assumed to be filled with gel matrix of porosity ε_1 and ε_2, respectively. In the first set of simulations, ε_1 was kept at a constant value of 0.9, while ε_2 was varied from 0.5 to 1.0.

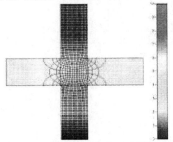

Figure 3. Potential and Field at Channel Junction

Electric field of 200 V/cm is applied in the separation channel, while IEF channel is floated. Governing equations (1-5) are solved satisfying appropriate boundary conditions. Figure 3 shows a contour plot of electric potential and iso-electric field lines. Significant distortion in the field is observed near channel intersection, which could be attributed to the fanning (leakage) effect.

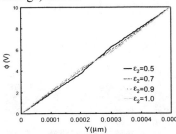

Figure 4. Potential Distribution Along the Separation Channel Various ε_2.

Figure 4 illustrates, variations in the electric potential along the axis of the separation channel as a function of porosity of gel in the IEF channel. The resulting electric potential is linear in the separation channel except near the junction of IEF and separation channels where distortions are observed due to the discontinuity in the gel porosity. The distortion in the contour lines is maximum for $\varepsilon_2 = 0.5$. The variations in the electric field strength in the separation channel, as well as in the center of the intersection are shown in Figure

5. As expected, only small variations are observed in the separation channel (which will ultimately vanish at locations far from intersection). However, the electric field strength increases with decrease in ε_2 since the fibers are assumed non-conductive and total current is conserved in the mathematical formulation. Changes in the field strength are attributed to both leakage current and changes in the area available for current conduction.

Next, the results from a sample separation study in gel and a non-gel media are presented. The purpose of this study is to understand the effect of gel matrix on separation and sample dispersion. The properties of species used in the simulation are listed in Table 2. The hydrodynamic permeability is assumed to be 1E11m². The fiber spacing is calculated from given porosity. Brady's model for diffusion coefficient and Slater's model for electrophoresis mobility, both valid for random oriented fibrous gel, are used. An electric field of 100 V/cm is applied to the separation channel. The detection point on the separation channel is located at 3cm from the intersection of CIEF and CZE channels.

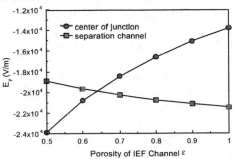

Figure 5. Electric Field at Separation Channel and At Channel Intersection for Various Porosities of IEF Gel.

Table 2: Species Properties

Species	Glutamic Acid	Histidine
Diffusivity (m²/s)	7.45E-10	7.316E-10
Mobility (m²/s/V)	2.9E-8	2.25E-8
Init. Conc. (μM)	1.0	1.5

Figure 6 shows an electropherogram predicted by the simulation at the detection point. The first plot shows concentration of glutamic acid and histidine when the separation channel is filled with the buffer solution only. In the second plot, concentration levels are shown when the separation channel is filled with a gel medium of porosity of 0.9. It is observed that both species are dispersed (largely due to molecular diffusion) as they move along the channel. The peak concentration is about 6% of their initial values. Also observed are time lags between the signals of the same species when it is run in the channel with and without a gel medium. This is attributed to hindered diffusion and reduced electrophoretic mobility due to the presence of gel medium. The signal peaks at the observation point are delayed by 7 and 8 seconds, respectively, for glutamic acid and histidine. These delays are approximately (inversely) proportional to reduced liquid volume fraction due to the presence of gel. On the other hand, it is also interesting to note that the dispersion of the sample profile is reduced (reduced band broadening) especially during the short separation time. This is illustrated in Figure 7 where concentration contour plots for glutamic acid and histidine are shown at T=80s after separation field is applied. This figure indicates that the species are focused into a tighter band due to the presence of porous medium, when compared with the situation when the channels are filled with only non-porous buffer solution.

5 CONCLUSIONS

Separation of species in a capillary electrophoretic system has been analyzed using computational models. Effective media approach based methodology has been implemented, where the transport coefficients and other parameters are computed using porosity-based "unit-cell" models. Computational tools have been applied to understand the influence of gel medium on electrophoretic separation of species. Preliminary results indicate that though the gel medium affected the transport (delay in peak signal), reduced band broadening was also observed due to reduced diffusion coefficient. Further model development is required that will account for transport of protein mixture coupled with ionization, biochemical reactions under the influence of electroosmotic flow in porous matrix. Development of these models, validation and design optimization of the 2D separation device are under progress and will be reported later.

ACKNOWLEDGEMENT

JJF, SK and SS acknowledge CFD Research Corporation for internal research support. VO, JO, SI acknowledge financial support from the LANL LDRD program. Los Alamos National Laboratory is operated by the University of California for the U.S. Department of Energy under contract no. W-7406-ENG-36.

REFERENCES

1. S-E, Ong, and A. Pandey, Biomol. Eng. 18, 195, 2001.
2. S. Beranova-Giorgianni. Trends Ana. Chem., 22, 273, 2003.
3. D.A. Wolters, M.P. Washburn. J.R. Yates, Anal. Chem. 73, 5683, 2001.
4. A.E. Herr, J.I. Molho, K.A. Drouvalaski, J.C. Mikkelsen, P.J. Utz, J.G. Santiago, and T.W. Kenny. Anal. Chem. 75, 1180, 2003.
5. JA Olivares, PC Stark, P Jackson. Anal Chem. 2002 May 1;74(9):2008-13.
6. C.S. Effenhauser, A. Manz, and H.M. Widmer, Anal. Chem. 65, 2637, 1993.
7. Anonymous, *CFD-ACE+ Users Manual*, CFD Research Corporation, Huntsville, AL, 2003.
8. Krishnamoorthy, J. J. Feng, and V. B. Makhijani, "Analysis of Sample Transport in Capillary Electrophoresis Microchip Using Full-Scale Numerical Analysis", *Proc. 4th Int. Conf. on Modeling and Simulation of Microsystems*, Hilton Head, SC, 2001.\
9. E. Johnson et. al., 1996 "Hindered Diffusion in Agarose Gels: Test of Effective Medium Model", *Biophy. J.* **70**, 1071.
10. J. Mercier & G. W. Slater 2001 "An Exactly Solvable Ogston Model of Gel Electrophoresis. 7. Diffusion and Mobility of Hard Spherical Particles in Three-Dimensional Gels", *Macromolecules*, **34**, 3437.
11. J. Happel 1959 "Viscous Flow Relative to Arrays of Cylinders", *AIChE J*, **5**, 174.
12. G. W. Jackson & D. F. James 1986 "The Permeability of Fibrous Porous Media", *Can. J. Chem. Eng.* **64**, 264.
13. J. Feng etc. "The Electric Conductivity of Fibrous Porous Media", To be published.

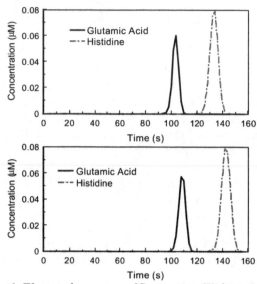

Figure 6. Electropherogram of Separation Without (top) and With Gel (bottom) in the Separation Channel

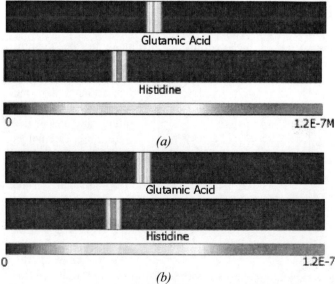

Figure 7. Comparison of Band Broadening in the (a) non-gel and (b) gel-based Medium at T=80s.

Amplification of protein adsorption on micro/nanostructures for microarray applications

E.P. Ivanova, Y.V. Alekseeva, D.K. Pham, L. Filipponi, D.V. Nicolau

Swinburne University of Technology, IRIS
PO Box 218, Hawthorn, VIC 3122, Australia, dnicolau@swin.edu.au

ABSTRACT

The fabrication and operation of biodevices require the accurate control of the concentration of the bioactive molecules on the surface, as well as the preservation of their bioactivity, in particular proteins that have a significant propensity for surface-induced denaturation. A method that allows the adsorption of proteins on 'combinatorial' micro/nano-surfaces fabricated via laser ablation of a thin metal layer deposited on a polymer that has been proposed allows for the up to 10-fold amplification of the protein adsorption on micro/nano-structures. This contribution studies the relationship between amplification of adsorption and protein molecular characteristics (total molecular surface; and charge- and hydrophobicity-specific surface).

Keywords: microarray, protein array, protein adsorption, microstructures, nanostructures

1 INTRODUCTION

Protein based microdevices have a long tradition, e.g. immunoassays, but the emergence of proteomics brought them in the center of attention of the scientific community. Compared with the classical protein-based devices, the new protein micro- and nano-arrays pose important technological challenges arising from (i) the large variety of proteins needed to be immobilized on the surface, in contrast with the rather similar antibodies; (ii) increased density of laterally defined areas with specific immobilization for specific proteins, required by high-throughput analysis; and (iii) the ever present complexities of protein-surface and protein-protein interactions, reflecting in complex fabrication and function, respectively. For all these reasons, the simple extrapolation of the technology responsible for the precursors of protein micro/nano-arrays, although attractive, is unwarranted or at least questionable.

Most technologies for the fabrication of protein chips have to ensure the confinement of molecularly different proteins in laterally-defined, either flat 2D; or profiled '2D+' micro-areas. The profiled features have the advantage of minimization of inter-spot contamination and the drawback of more difficult access of the recognition biomolecule (e.g. antigen for antibody microarrays) in a micro-confined area. Recently we have proposed a method for protein immobilization in micro/nano-channels fabricated via laser ablation of a thin metal layer deposited on a polymer immobilization substrate. The shallow profiles of the microfabricated features are expected to take advantage of the benefit of the flat areas and mitigate the drawback of the deep structures.

The present study investigates the adsorption of five, molecularly different proteins on micro/nano-structures fabricated via laser ablation, to probe the relationship between the amplification of the protein adsorption and their molecular characteristics (total molecular surface; and charge- and hydrophobicity-specific surface).

2 METHODS

The adsorption of five proteins with very different molecular characteristics, i.e. α-chymotrypsin, human serum albumin, human immunoglobulin, lysozyme, and myoglobin, has been characterized using quantitative fluorescence measurements and atomic force microscopy. The procedure for the fabrication of micro/nano-structures, which has been used for this study, has been described elsewhere [1]. Briefly, the process consists of (Figure 1) (i) the glass slides covered a thin polymer layer of Poly(methyl methacrylate) –PMMA- were covered with a thin (~50nm) layer of gold (sputtered); (ii) the gold-layered substrata were then incubated with bovine serum albumin (BSA); (iii) the gold and blocking protein layers were ablated (due to the opacity of the gold) with a laser (Cell Robotics workstation, 337nm, 20pulses/s, 10ns/pulse); and (iv) the proteins were deposited either by flooding the whole surface with a protein solution (protein attaches on PMMA-surfaces opened by ablation); or by spatially addressable deposition with a picoliter pipette mounted on a xy motorized table.

The molecular descriptors of the selected proteins have been calculated using an in-house developed program (freely downloadable [2]), that uses the Connolly algorithm beyond its original purpose (i.e. the calculation of molecular surface) for the calculation of the surface-related molecular properties (i.e. surface positive and negative charges; and surface hydrophobicity and hydrophilicity) as well as the molecular surfaces related to these properties.

Atomic Force Microscope (AFM) was used to characterize the topography and the relative hydrophobicity of the micro/nanostructures. The relative level of adsorption was quantified using a fluorescence reader.

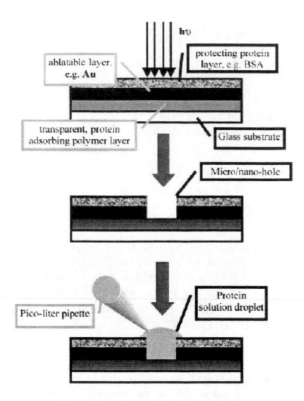

Fig 1: Fabrication of micro/nano-structures for protein arrays using microablation and directed deposition.

3 RESULTS AND DISCUSSION

The proposed technology produces surfaces that present to the proteins a large variation of the properties (in particular hydrophobicity and rugosity) of the surface concentrated in a small, micron-sized region. The ablation of the thin metallic layer induces the pyrolysis and partial 'sculpturing' of the polymer, with more hydrophilic surfaces towards the edges of the channel and a hydrophobic hump in the middle. The higher rugosity of the microstructures (Figure 2) translates in a 3 times more specific surface in than outside the channels.

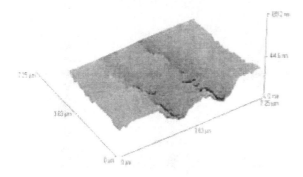

Fig 2: AFM mapping of the ablated micro-channels.

A more important feature than the increased rugosity in the microablated areas is the large variations in the relative hydrophobicity of different regions, as measured by AFM in lateral force mode. Fig 3 presents both the topography and the relative hydrophobicity of the surface of microablated lines.

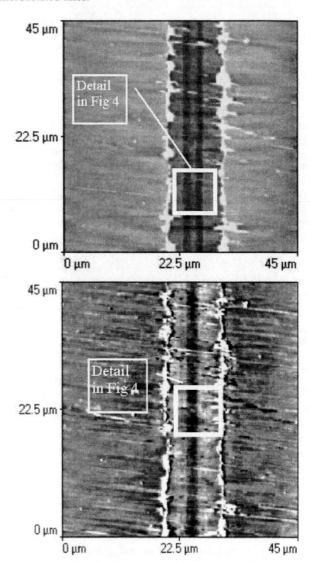

Fig 3: Topography (top) and relative hydrophobicity (bottom) of the surface of the microablated lines. Darker regions represent deeper levels and hydrophobic surfaces, on the topography and lateral force images, respectively.

These structures have micron-size dimensions laterally but tens of nanometers in depth. The latter dimensions make the structures comparable with medium to large proteins (Fig. 4).

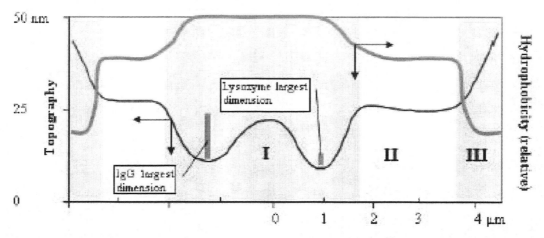

Fig. 4: Typical lateral distribution of topography and hydrophobicity in microablated channels. The largest dimensions of lysozyme and IgG are depicted for comparison.

Although there are variations of the dimensions and distribution of hydrophobicity vs. laser power, the general structure of the microablated channels remains the same. Although the process can use different polymers and metals, we found that PMMA (and gold) are so far optimum choices. In particular PMMA can offer large possibilities to 'combinatorialise' the chemistry of the surface upon thermolysis. Figure 5 depicts possible chemical pathways that would explain the AFM-measured variation in hydrophobicity.

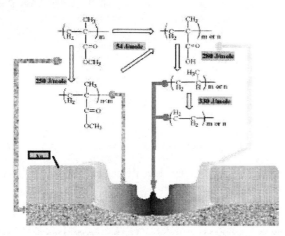

Figure 5: Possible pyrolysis pathways of PMMA localized in micro-regions leading to the observed lateral distribution of hydrophobicities.

The strategy behind this method was to allow different proteins, or different parts of the same large protein, to find the most appropriate –in terms of adsorption and preservation of bioactivity- surface. Because the different

micro/nano-surfaces are co-located in a small area (channel width around 10 µm or less) the florescence signal from the proteins would be perceived at the mm-range as being located in the same areas. Also it has been observed that the concentration of the proteins was apparently higher in the microstructures than on flat areas with similar material due to a higher specific surface and larger opportunities for attachment.

The combinatorial character of the surface would in principle also allow the probing of several patches on the molecular surface of the proteins, or modulate their bioactivity. Fig. 6 presents possible arrangements of IgG-like biomolecules on the combinatorial surfaces.

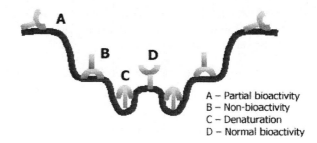

A – Partial bioactivity
B – Non-bioactivity
C – Denaturation
D – Normal bioactivity

Fig. 6: General concept of the probing of molecular surface of proteins.

The most important observation relates to the correlation between molecular properties of the adsorbed proteins and the amplification of their adsorption. For instance, it was also observed that the amplification of the fluorescence signal (due to increase of adsorption) is larger for smaller proteins (e.g. lysozyme) than for larger proteins (e.g. IgG). The proteins with a molecular surface in the order of tens on nm^2 adsorb up to 10-12 times more on microstructures than on flat surfaces, while large proteins with a molecular surface of few hundreds of nm^2 adsorb only 3 times more than on bare PMMA flat surface. Figure 7 presents synthetically this relationship.

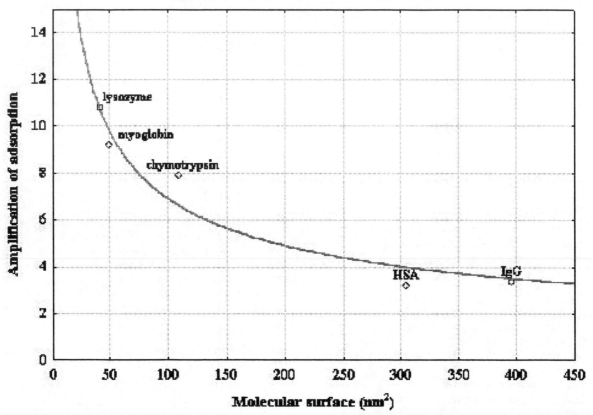

Fig. 7: Amplification of adsorption vs. molecular structure of the tested proteins, i.e. lysozyme, myoglobin, α-chymotrypsin, HAS and IgG.

The proposed technology for the fabrication of microarrays has the following potential advantages: (i) the combinatorial surfaces would improve the uniformity of biomolecule surface concentration, in particular for protein microarrays; (ii) the surface concentration of biomolecules (several very different proteins being tested) increases, and therefore the sensitivity increases accordingly, by 3-12 times, depending on the molecular properties of the biomolecules; (iii) it is possible -in principle- to probe different sides of the biomolecular surface and therefore the bioactivity;m and (iv) because the proposed method writes protein lines (that can encode information in a bar code manner) instead of dots, it can be used for the fabrication of informationally-addressable, as opposed to spatially-addressable microarrays.

4 CONCLUSION

We propose a method for the fabrication of random 'combinatorial' micro/nano-sized surfaces that allows a higher adsorption of proteins and possibly the immobilization of biomolecules on different sides of the molecular structure. The method, which is based on laser microablation of thin metal/blocking protein layers deposited on a polymer substrate, has proven to amplify the protein adsorption between 3 to 10 times.

ACKNOWLEDGEMENTS

The present study has been sponsored by Defense Advanced Research Projects Agency (DARPA) and Australian Research Council (ARC). LF is also the recipient of a Cooperative Research Centre (CRC) for Microtechnology fellowship.

REFERENCES

[1] E.P., Ivanova; J.P. Wright; D.K.Pham; L., Filipponi; A., Viezzoli; D.V. Nicolau. Polymer Micro-structures Fabricated via Laser Ablation for Multianalyte Microassays Langmuir, 18, 9539-9546, 2002.
[2] D. Nicolau, Jr.; D.V. Nicolau. Biomolecular Adsorption Database. URL: http://www.bionanoeng.com /BAD.

Evaluation of Electric Impedance Spectra for Single Bio-Cells with Micro-Fluidic Devices using Combined FEMLAB/HSPICE Simulated Models

V. Senez[*], T. Takatoki[**], B. Poussard[***], T. Fukuba[**], J.M. Capron[***] and T. Fujii[**]

[*]LIMMS/CNRS-IIS, [**]UTRC/IIS, Tokyo University,
4-6-1, Komaba, Meguro-ku, 153-8505, Tokyo, JAPAN, {vsenez,tfujii}@iis.u-tokyo.ac.jp
[***]IEMN/CNRS, Avenue Poincaré,
BP 69, 59652 Villeneuve d'Ascq, FRANCE, jean-marc.capron@isen.fr

ABSTRACT

This paper describes a simple method to predict the electrical impedance spectrum of single & cultured cells in micro-devices. It can be used for the rapid design of micro-sensors as well as for more fundamental studies about the interactions of electric fields with bio-cells. The finite element (FEMLAB) and the transport lattice (HSPICE) methods are coupled through the MATLAB environment for fast calculations of electrical impedance spectra (EIS). The impedances of simple test structures have been simulated and compared with their corresponding analytical solution. In the [10Hz-10GHz] frequency range, the maximum error was found to be lower than 6%. The geometry of our micro-device has been optimised with this method and the effects of various biological parameters have been analysed. This work has been performed in the framework of LIMMS, a joint laboratory between the CNRS (France) and the Institute of Industrial Science (University of Tokyo-Japan).

Keywords: lab-on-a-chip, impedance sensing, modeling, single cell analysis

1 INTRODUCTION

Recent advances in the understanding of cellular behavior in microenvironments have allowed the design and fabrication of micro-devices containing living organisms. Cell work has traditionally been done with equipment that is many orders of magnitude larger in size than a cell. As a consequence, it is difficult for a cell to maintain a microenvironment (i.e.: diffusion distance, stress, etc…) and it is assumed that its behavior in macro-scale devices is different to *in vivo* conditions.

On the contrary, micro-devices made from the proper materials (e.g.: gas-permeable polymers) allowing adequate oxygenation (without stirring or bubbling) and uniform nutrients diffusion are the perfect tools to conduct cell-based experiments at the micro-scale. The rich assortment of living micro-devices [1-5] that can be built using Bio-MEMS techniques have important applications in fundamental cell biological studies, tissue engineering, cell separation and culture devices, diagnostics, high-throughput drug screening tools, and cellular bio-detection.

Living cells have many interesting properties: they offer miniature size, biological specificity, signal amplification, surface binding capability, self-replication, multivariate detection, and other benefits. Their engineering requires rapid characterization methods. While current optical and chemical detection techniques [6-8] can effectively analyze biological systems, a number of disadvantages restrict their versatility. As examples: most samples must be chemically altered prior to analysis, and photobleaching can place a time limit on optically probing fluorophore-tagged samples. Purely electronic techniques provide solutions to many such problems, as they can probe a sample and its chemical environment directly over a range of time scales, without requiring chemical modifications [9-10].

One example of electronic detection is electrical impedance spectroscopy (EIS) [11-13]: examining permittivity as a function of frequency. EIS is a non-invasive method for characterizing cells suspensions or tissues. Because currents will pass either around or through cells depending on the frequency, EIS can be used to observe the structure and arrangement of cells.

In this paper, we described in the next section our micro-sensor, its fabrication process as well as our first prototypes. In section 3, our numerical model and its implementation are described. In section 4, we present several test cases to validate our modeling. Finally, we give our results concerning the optimization of the micro-sensor with this CAD tool.

2 MICRO-SENSOR

Micro-devices are typically fabricated in silicon or glass substrates. Techniques for processing these materials are well established and have been improved in the microelectronics industry. However, such manufacturing methods require clean room environments and can be very expensive. For this reason, there has been tremendous interest in the use of polymeric materials for device production. In addition to the lower costs associated with polymeric materials, micro-device processing involves reduced cycle times, thus further reducing the cost of a particular device. Polymers also offer a wide range of material properties, including hardness, hydrophobicity, and surface or bulk chemistries that can be adjusted to a specific applications. One important property is the biocompatibility of the material. Biocompatible materials are materials designed to exist and perform specific functions within living organisms.

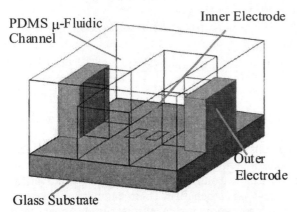

Figure 1: 3D schematic of the micro-EIS sensor showing the outer electrodes (3D) and inner electrodes (flat).

Our fabrication process includes the patterning of gold electrodes on a glass substrate and the molding of polydimethylsiloxane (PDMS) for the micro-fluidic parts. PDMS has been chosen since it is inexpensive, optically transparent, can be micro molded and has excellent O2 and CO2 permeability. It is also considered to be biocompatible and has been shown to be non-toxic to several cell types. The fabrication process has been detailed elsewhere [14] and allows the deposition of metal on PDMS [15]. The 3D schematic of the micro-sensor is given in Fig. 1.

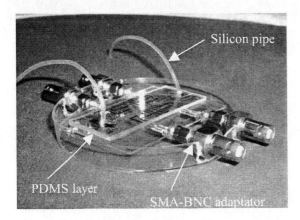

Figure 2: First prototype of the micro-fabricated EIS sensor.

One can observe that the outer electrodes are 3D electrodes in order to induce a uniform electric field in the sensing area. On the contrary, the inner electrodes are flat. Fig. 2 shows our first prototype. We can see the silicon pipes from where cells are introduced in the micro-device. The drive and receive cables of the impedance analyzer can

be directly connected to the glass substrate thanks to the micro-SMA/BNC connectors. The use of four-electrode set-up increases the linearity, accuracy and sensitivity of the measurement [16]. First, we want to know how to design the geometry of the sensor to obtain the highest sensitivity. More precisely, our purpose is to determine the size of the outer and inner electrodes and their location in the sensing area.

3 MODELING STRATEGY

Traditional approaches use Laplace equation in combination with the finite element (FE) [17]. Recently, this problem has been solved using Kirchhoff laws and transport lattice (TL) method [18].

In our approach, the advantages of these two methods are coupled. Indeed, the system being studied (i.e.: the micro-fluidic device) is divided into small finite elements (i.e.: triangles in 2D, tetrahedra in 3D) using the automatic mesh generator of FEMLAB (COMSOL). The initial mesh is refined around small geometrical features like the cell membrane (Fig. 3).

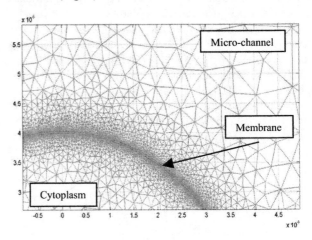

Figure 3: Triangular FEM mesh obtained with FEMLAB showing the automatic refinement of the cell membrane.

Then, the data structure is exported to the MATLAB (MATHWORKS) environment where the TL is defined. In our implementation, the nodes of the TL correspond to the nodes of the FE mesh (Fig. 4). Finally, the TL is solved by Kirchhoffs laws, using HSPICE (SYNOPSIS) giving electrical currents and voltages through the lattice. 2D results are presented in this paper and we are currently working at the implementation of the 3D models. The 2D TL is composed of basic electrical elements (Fig. 4), consisting of a resistance (R_i) in parallel with a capacitance (C_i) where the subscript "i" defined the region of the system (i.e.: PDMS, electrodes, micro-channel medium, membrane, cytoplasm). For a 3D implemetation, R_i and C_i are given by:

$$R_i = l^2/\sigma_i/V \text{ and } C_i = \varepsilon_i * V / l^2 \tag{1}$$

with l is thickness, V is the volume of the local ith element, σ_i and ε_i are the local electrical conductivity and permittivity. For a 2D space divided in triangles, their values are:

$$R_i = L/2/\Sigma_{k=1,2}[(A_k^j/L*1)*\sigma_k^j] \qquad (2)$$

$$C_i = 2*\Sigma_{k=1,2}[(A_k^j/L*1)*\varepsilon_k^j]/L \qquad (3)$$

with i the index on the edges of the FE mesh, A_k^j is the third of the surface of the jth FE triangle connected to the kth edges, $\Sigma_{k=1,2}[.]$ is the summation on the adjacent A_k^j defined for each FE edges (Fig. 3), L is the length of the impedance, the section is calculated with and equivalent width (A_k^j/L) and a thickness equal to one meter.

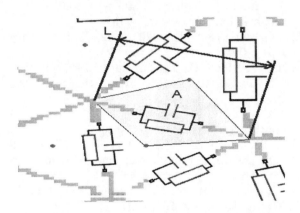

Figure 4: Details of the transport lattice. TL and FE nodes are located at the same position. Barycenters of triangles defined the surface represented by the local RC element.

4 VALIDATION AND OPTIMIZATION

First, we have compared the results (in terms of characteristic frequency or voltage) obtained by i) the FE method (FEMLAB), ii) the Weaver's approach [18] using HSPICE and iii) our own method with the corresponding analytical results. Plate capacitors with: i) one material, and ii) two different materials inserted as slab in series (Fig. 5a), or as a disk in a square (Fig. 5b) have been simulated.

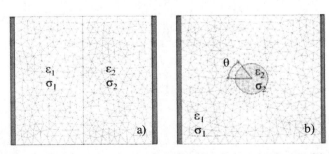

Figure 5: Test structures consisting in plate capacitors with different materials inserted as a) slab in series, or b) a disk in a square

We have studied the effect of the lattice density on the results. Table 1 gives the details of these validations. For the simulation of our sensor (130 000 RC elements), it takes about 12 minutes.

Structure	Error (%)
Plate Capacitor (1 material)	1.5
Plate Capacitor (2 slabs)	2.7
Plate Capacitor (1 disk in a square)	5.5
Simulation Type	Time (s.)
HSPICE	12
FEMLAB	1560

Table 1: Simulation errors compared to analytical models and typical simulation times (800 triangles or 2500 RC elements) using an Intel Celeron 1.8Ghz.

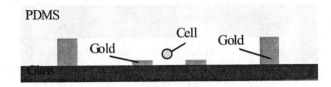

Figure 5: 2D cross-section of the simulated device. The outer electrodes are either planar or 3D.

Second, we have analysed the influence of several geometrical parameters on the variation of the impedance ($\Delta R/R = [Z_{without\ Cell} - Z_{with\ Cell}]/Z_{without\ Cell}$) due to one cell. Fig. 6 gives the 2D cross-section of the device used for the simulations. We have investigated: i) the height of the micro-channel (Fig. 7a), ii) the widths of the outer and inner electrodes and iii) the distance between the outer electrodes (Fig. 7b).

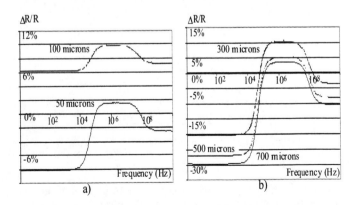

Figure 7: Calculated influence of: a) the height of the channel, b) the distance between the outer electrodes on the impedance spectrum from 1Hz to 10GHz.

In Fig. 7, one can see that the decrease of the micro-channel thickness and the increase of the distance between the outer flat electrodes improve the sensitivity at low frequency while they deteriorate this sensitivity at intermediate frequencies (Mhz). The simulation have also shown (not presented in this paper) that the width of the outer flat electrodes has no effect on the sensitivity at all frequencies; that the best sensitivity is obtained when the inner electrodes are close to each others; and that the widths of the inner electrodes has to be small at low frequencies and large at intermediate frequencies. As a result of this work, we have defined an optimised sensor.

For a geometrical configuration giving the highest $\Delta R/R$ at low and intermediate frequencies, we have studied the effects of: i) the radius of the cell, ii) the conductivity of the extra-cellular medium and cytoplasm (Fig. 7a), iii) the thickness of the membrane (Fig. 7b).

One can observe in Fig. 8a that the increase of the cell membrane thickness has no effect on the sensitivity at low and intermediate frequencies. It only shifts the sensitivity curve to higher frequency since the cell is less transparent to the electrical current. In Fig. 8b, one can observe that the increase on the conductivity of the cytoplasm has no effect at low frequencies and increase the sensitivity at intermediate frequencies.

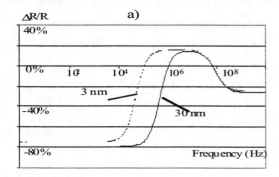

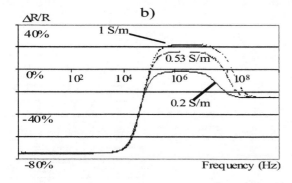

Figure 8: Calculated influence of: a) the thickness of the cell membrane, b) the conductivity of the cytoplasm on the impedance spectrum from 1Hz to 10GHz.

Our simulations show also (not presented in this paper) that the increase of the conductivity of the extra cellular medium increases the sensitivity at both low and intermediate frequencies. Finally, the increase of the radius of the cell increases the sensitivity more at low frequencies

than at intermediate frequencies. For these simulations, we have used the following values as default parameters: $\varepsilon_{PDMS}=2.65\varepsilon_0$; $\sigma_{PDMS}=0.5\times10^{-17}Sm^{-1}$; $\varepsilon_{Glass}=4.2\varepsilon_0$; $\sigma_{Glass}=10^{-14}Sm^{-1}$; $\varepsilon_{Electrode}=0Fm^{-1}$; $\sigma_{Electrode}=4.5\times10^{5}Sm^{-1}$; $\varepsilon_{Electrolyte}=80\varepsilon_0$; $\sigma_{Electrolyte}=0.12Sm^{-1}$; $\varepsilon_{Membrane}=9.04\varepsilon_0$; $\sigma_{Membrane}=10^{-6}Sm^{-1}$; $\varepsilon_{Cytoplasme}=50\varepsilon_0$; $\sigma_{Cytoplasme}=0.53Sm^{-1}$).

We are currently designing the electronic instrumentation to record the EIS (30kHz-30MHz) of these single or cultured cells in order to ensure a high accuracy of the experimental data and to provide the end-users with a compact, autonomous and inexpensive system.

REFERENCES

[1] J. Gimsa, T. Muller, T. Schnelle and G. Fuhr, Biophys. Journal, 71, 495, 1996.

[2] S. Gawad, L. Schild and Ph. Renaud, Lab on a Chip, 1, 76, 2001.

[3] H.E. Ayliffe, S.D. Brown, and R.D. Rabbitt, Proc. 2nd Joint EMBS/BMES Conf., Houston, 1692, 2002.

[4] H.E. Ayliffe, R.D. Rabbitt, P.A. Tresco, A. B. Frazier, Proc. Transducers, Chicago, 1307, 1997.

[5] R. Pethig, J.P.H. Burt, A. Parton, N. Rizvi, M.S. Talary, and J.A. Tame, J. Micromech. Microeng., 8, 57, 1998.

[6] S. Nie and R. N. Zare, Annu. Rev. Biophys. Biomol. Struct., 26, 567, 1997.

[7] S. Weiss, Science, 283, 1676, 1999.

[8] G. MacBeath and S. L. Schreiber, Science, 289, 1760, 2000.

[9] J. Viovy, Rev. Mod. Phys., 72, 813, 2000.

[10] L. L. Sohn, O. A. Saleh, G. R. Facer, A. J. Beavis, R. S. Allan, and D. A. Notterman, Proc. Natl. Acad. Sci. U.S.A., 97, 10687, 2000.

[11] K. Asami, E. Gheorghiu, and T. Yonezawa, Biophys. J., 76, 3345, 1999.

[12] C. Prodan and E. Prodan, J. Phys. D, 32, 335, 1999.

[13] G. Smith, A. P. Duffy, J. Shen, and C. J. Olliff, J. Pharm. Sci., 84, 1029, 1995.

[14] H. Becker and C. Gartner, Electrophoresis, 21, 12, 2000.

[15] M. Ishizuka, H. Hojo, M. Abe, K. Akahori, N. Honda, M. Mori, T. Sekiguti, and S. Shoji, Proc. μ-TAS, Nara, 413, 2002.

[16] F. Laugere et al., Sens. Actuators, A92, 109, 2001.

[17] S. Gawad et al., Lab on a Chip, 1, 76, 2001.

[18] T. Gowrishankar and C. Weaver, PNAS, 100, 3203, 2003.

Actin Adsorbed on HOPG and Mica Substrates: Characterization and Manipulation by Atomic Force Microscopy

D.V. Nicolau[*], G. S. Watson[**], C. Cahill[**], J. Blach[**] and S. Myhra[**]

[*]Swinburne University of Technology, PO Box 218, Hawthorn, VIC 3122, Australia
[**]School of Science, Griffith University, Nathan, QLD 4111, Australia

ABSTRACT

Actin deposited onto a hydrophobic graphite surface forms paracrystalline rafts consisting of F-actin filaments. Those filaments retain their internal coherence during manipulation by an atomic force microscope tip. However, the filaments can be 'herded' systematically in a direction perpendicular to their long axis, due to weak interaction with the substrate. Actin deposited onto a hydrophilic mica surface, on the other hand, tend to become immobilized due to strong out-of-plane interactions. Circular ordered actin features can be formed at preferred sites defined by the intersection of interfaces at micro-bubbles with the substrate. The observations may have applications for devices based on dynamics of molecular motors.

Keywords: actin, atomic force microscope

1 INTRODUTION

Actin is an essential constituent in systems and processes that are involved with muscle contraction, cell division and some pathological conditions (Somlyo et al. 2000) Accordingly the structure and function of actin has been receiving increasing attention due to its future nano-biotechnological roles in devices and processes, in particular those that aim to exploit the capabilities of molecular motors. The project has investigated 'paracrystalline rafts' of actin adsorbed onto graphite. As well, selective adsorption with nano-scale definition has been obtained on mica by taking advantage of the propensity of actin to seek sites of lowest energy configuration at interfaces.

2 METHODS

Frozen G-actin stock was thawed rapidly in DI H_2O at 37°C. Polymerizing was initiated in 5 mM potassium phosphate buffer at pH 7.5 with 100 mM KCl and 0.25 mM Mg ATP. The incubation took place at 37°C for 30 min. The incubated solution was spun at 45 Krpm for 60 min at RT, whereupon the supernatent was decanted. The actin pellet was then rinsed/washed in a buffer solution of 10 mM imidazole at pH 6.8, 40 mM KCl, 1 mM $MgCl_2$, and 0.5 mM DTT. Additional buffer was added for storage overnight at 4°C. The F-actin was then resuspended, resulting in a stock solution of 6 mg/mL.

Two dilutions were carried out, resulting in 0.06 mg/mL (S1) and $6x10^{-4}$ mg/mL (S2). Drops of 5 – 25 µL from S1 or S2 were placed on substrate surfaces and allowed to dry at RT (ambient humidity ca. 60%).

Two substrates were employed during the present study. Highly oriented pyrolytic graphite (HOPG) will present an atomically flat hydrophobic surface with sp^2 bonding. Clean surfaces were prepared by cleavage immediately before exposure to solution. Mica will likewise present an atomically flat surface composed of a silicate tetrahedral network. That surface is hydrophilic and will be covered with an adsorbed thin aqueous layer. Hydroxylation will then take place, thus generating a negative surface charge. However, ion exchange with the divalent cation ($K^+ \Leftrightarrow Mg^{2+}$) in the aqueous phase is likely to produce a positive surface charge density.

The AFM analysis and tip-induced manipulation were carried out with a JEOL JSPM-4200 multi-technique/multi-mode instrument under air ambient conditions. Imaging was performed in the constant force contact mode. Beam-shaped probes with nominal force constants of 0.06 and 0.58 N/m were used.

3 RESULTS AND DISCUSSION

Deposition of S1 actin on a hydrophobic HOPG substrate resulted in formation of paracrystalline rafts, Fig. 1, consisting of filamentary strands with longitudinal integrity being aligned at spacings of ca. 200 nm. The structures exhibited superficial resemblance to those reported in an earlier study by Shi et al. (2001). However, closer inspection in combination with the outcome of tip-induced manipulation has revealed distinct differences.

The spacing between the parallel rows was a factor of 4-5 greater than those of the earlier study. That study was based on transferring a self-assembled bi-layer consisting of a lipid monolayer and an F-actin layer onto a DPPC coated mica substrate (or an aged carbon-coated grid for EM analysis). The repeat distance perpendicular to the longitudinal structure was ascribed to the nodal distance, ca. 36 nm, along the double-stranded F-actin chain. The present ripple spacing must therefore have a different origin. The spacing between rows in Fig. 1 was found to be a function of interaction with the tip. The dependence is illustrated in Fig. 2 by the break in order at the edge of a second scan over a smaller field of view.

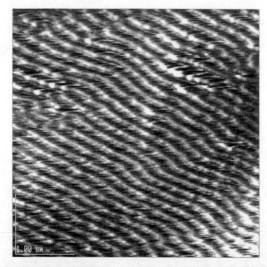

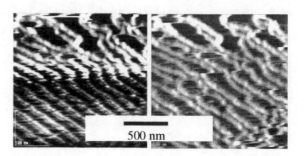

Fig. 2. The two contact mode images show: (left) the boundary between two successive scans over over-lapping fields of view, where the top half has been subjected to two raster scans, resulting in translation of every second row; (right) a third scan has been carried out over the full field of view, illustrating translation of every second row over the full field of view (note that the structure due to the original manipulation remained stable, top half of image).

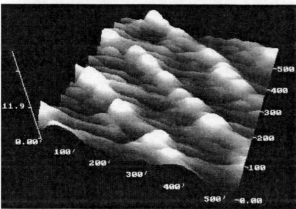

Fig. 1. Paracrystalline structure obtained from deposition of actin in S1 solution onto HOPG. The high-resolution image (right) reveals that the height (ca. 10 nm) and periodicity (ca. 40 nm) were consisted with SEM results for F-actin (Milligan et al. 1990).

Manipulation appeared to result in successive 'herding' of alternate strands, due to the delicate interplay of tip-induced shear stress, with in-plane and out-of-plane interactions within the raft and with the substrate.

Adsorption onto HOPG has the merit of effectively turning off double-layer interaction between the actin and the substrate. However, while HOPG is hydrophobic, the actin will be hydrophilic and will present a negative 'surface' charge. Thus ordering and spacings will be due to a subtle interplay between relatively weak and short-range out-of-plane van der Waals interaction with the substrate, longer-range in-plane repulsive electrostatic interaction, and attractive in-plane meniscus forces.

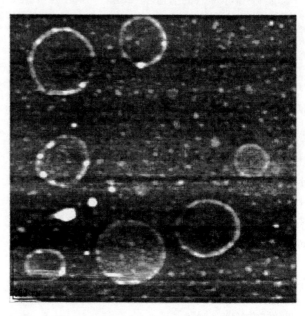

Fig. 3. Circular structures obtained from exposure of a hydrophilic mica surface to actin in S2 solution. The features are due to selective deposition of actin at preferred sites of low interface energy. Those sites are most likely located at the air/fluid/solid intersection where an adhering air bubble is attached to the mica surface. The existence of such bubbles is thought to account for anomalies in double-layer interactions (Considine et al. 1999). The features are stable against contact mode tip-manipulation, due to the double-layer force between a substrate with positive surface charge and a negative charge on the actin (St-Onge and Gicquaud, 1989).

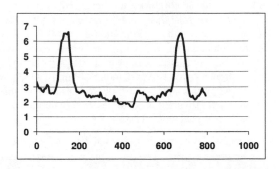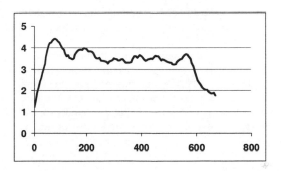

Fig. 4. Higher resolution images of circular features on mica shown in pseudo-3D representation. Most of the features exhibited adsorption along a 'triple-point' line (left). However, in some cases the circles were filled in (right), presumably due to availability of excess actin being forced into the circular confinement during evaporation. The contour lines are drawn through the center of the circle. The scales are in nm, and the heights of the features are consistent with that of a single layer of partially denatured actin.

4 CONCLUSIONS

Filamentary structures on a hydrophobic substrate form paracrystalline rafts that can readily be manipulated on the nano-scale, while adsorption on a hydrophilic surface suggests that tailored preferred adsorption sites at interfaces can be used to impose structural predictability. Both effects may have applications for nano-biotechnology.

REFERENCES

[1] Considine RF, Hayes RA, Horn RG, Langmuir, 15: 1657-1659, 1999

[2] Milligan RA, Whittaker M, Safer D, Nature, 348: 217-221, 1990

[3] Shi D, Somlyo A V, Somlyo A P, Shao Z, J. Microsc. 201: 377-382, 2001

[4] Somlyo AP, Somlyo AV, J. Physiol.-London 522: 177-185, 2000

[5] St-Onge D, Gicquaud C, Cell. Biol. 67: 297-300, 1989

A Molecular Motor that Links the Biological and Silicon Worlds

T. Robinson, Alex Blundell, C.F. Dutta, and K. Firman

IBBS Biophysics Laboratories, School of Biological Sciences, University of Portsmouth, King Henry
Building, King Henry I Street, Portsmouth PO1 2DY, keith.firman@port.ac.uk[1]

ABSTRACT

This project was funded by the European Community as part of the Future Emerging Technologies (FET-OPEN) initiative and represents an exciting and innovative approach to the use of biological systems in bionanotechnology.

We describe the initial stages in the development of a nanoactuator that is based on a molecular motor, which translocates DNA. The nanoactuator involves the movement of a magnetic bead, attached to the DNA, past a sensor/detector that is able to operate a silicon-based electronic device. Since the motor is biological in origin, using the biological 'fuel' ATP to drive the motor and, therefore, translocate the DNA, the nanoactuator links the biological and silicon worlds.

Keywords: Type I restriction-modification enzyme, endonuclease, DNA translocation, magnetic tweezers, atomic Force Microscopy (AFM)

1 INTRODUCTION

Type I Restriction-Modification (R-M) enzymes are the most complex of the many R-M systems known [for recent reviews see 1]. They are multisubunit, multifunctional enzymes composed of three separate subunits (HsdR – the restriction/motor subunit, HsdM – the methylation subunit, and HsdS – the DNA binding subunit). The active endonuclease (ENase) is composed of all three subunits in a ratio 2:2:1 ($HsdR_2$:$HsdM_2$:$HsdS_1$, or R_2-complex). The R_2-complex also functions as a DNA methyltransferase (MTase), an ATPase, and as a DNA pulling molecular motor [2, 3].

Unlike other restriction endonucleases, Type I R-M enzymes cut distal to the DNA recognition site to which they bind. DNA cleavage can occur many thousands of basepairs from the recognition site, using a process of DNA movement known as DNA translocation [2]. This ENase-based motor activity is driven by ATP hydrolysis [3], but unlike other DNA-based motors (e.g. DNA polymerase) it does not involve a linear tracking motion along the DNA; instead the motor remains bound at the recognition site and 'pulls' the adjacent DNA toward the bound enzyme (Figure 1). HsdM and HsdS alone are sufficient to assemble an independent MTase with a stoichiometry of M_2S_1 [4]. The R-M enzyme EcoR124I can be assembled *in vitro* from the core MTase by addition of the motor subunit HsdR [5]. However, the purified EcoR124I ENase exists as an equilibrium mixture of two species - $R_2M_2S_1$ and $R_1M_2S_1$ of which only the former is able to cleave DNA [5, 6]; although the R_1-complex is an ATPase and is able to translocate DNA [5, 7]. The R_2-complex is relatively weak, dissociating into free HsdR subunit and the restriction-deficient R_1-complex intermediate, under concentrations expected *in vivo* [5]. Therefore, to produce a more useful molecular motor, we engineered a special hybrid motor subunit (which we call HsdR(prrI)) that we were able to show was unique in forming exclusively a R_1-complex even when a vast excess of HsdR is mixed with MTase.

Figure 1   Motor activity of type I R M Enzyme

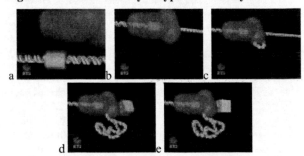

(a) The yellow block represents the DNA-binding (recognition) site of the enzyme, which is represented by the green object approaching from the top of the screen and about to dock onto the recognition sequence. (b) The motor is bound to the DNA at the recognition site and begins to attach to adjacent DNA sequences. (c) The motor begins to translocate the adjacent DNA sequences through the motor/DNA complex, which remains tightly bound to the recognition sequence. (d) Translocation produces an expanding loop of positively supercoiled DNA. The motor follows the helical thread of the DNA resulting in spinning of the DNA end (illustrated by the rotation of the yellow cube). (e) When translocation reaches the end of the linear DNA it stops, resets and then the process begins again.

The complex molecular motor function makes these enzymes particularly interesting. The restriction subunit (HsdR) contains a series of conserved amino acid motifs (DEAD box motifs), which are associated with helicase-like activities [8] including a Walker-type ATP binding site. They belong to a large superfamily (SF-II) of helicase-like enzymes [9] that also include type III R-M enzymes, chromatin remodeling factors and a few chimeric

[1] Corresponding author

enzymes. It has been suggested that chromatin-remodeling factors also make use of DNA translocation, in a similar mode as Type I restriction enzymes, which stresses the significance of a detailed analysis of the translocation process of the SF-II superfamily.

2 RESULTS

The use of the EcoR124I R-M enzyme as a molecular motor has been compounded by the existence of an equilibrium mixture between R_1- and R_2-complexes during purification of the restriction endonuclease [5, 6]. Therefore, we have produced a hybrid endonuclease that, as shown below, has a drastically increased dissociation constant for the R_2-complex, resulting in the production of R_1-complex only.

2.1 Production of a stable R_1-complex

The HsdR subunit of Ecoprrl differs from the HsdR subunit of EcoR124I only in the N-terminal region (Figure 2). The bulk of these differences are located within the first 150 residues. To engineer a system for overproduction of the HsdR(prrI) subunit we utilized the plasmid pACR124, which over-produces the HsdR(R124) subunit [10] and constructed a hybrid gene consisting of 600 bp from the 5'-end of the *hsdR* gene of Ecoprrl fused to the 3'-end of the *hsdR* gene of EcoR124I. To prepare this hybrid gene we produced a 600 bp PCR fragment of *hsdR*(prrI), produced by cleavage at the unique PstI site (which overlaps amino acid 200, and ligated this to the remaining *hsdR* gene of EcoR124I also cleaved at the unique PstI site.

Figure 2   Gel retardation using the hybrid ENase.

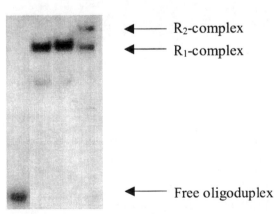

1 2 3 4

← — R_2-complex
← — R_1-complex

← — Free oligoduplex

10nM oligoduplex carrying a single s_{R124} recognition site was incubated with 200nM MTase. HsdR(prrI) was added at the following concentrations and the assembled products analyzed on a native 6% PAGE. **Lane 1**, *free oligoduplex;* **lanes 2-3**, *MTase, oligoduplex + 3200nM and 4000nM HsdR(prrI) respectively;* **lane 4**, *EcoR124I ENase produced from a 5:1 molar ratio of HsdR(R124):MTase, showing the location of R_1- and R_2-complexes.*

2.2   Structural analysis of motor subunit

The HsdR subunit of a type I R-M system is a large protein >100kDa, which means that to date there is no structural information for the protein (no crystals have been produced and it is too large for NMR). Therefore, an ideal approach to gain some information is to look for specific domains that might be isolated separately for NMR analysis. Several domains have been identified for the closely related EcoKI R-M system [11] using limited proteolysis to probe the structure. We have initiated a variety of structural studies of the HsdR(prrI) subunit including a similar analysis using limited proteolysis. We have shown that there is a similar number of proteolytic products from limited proteolysis of HsdR(R124I) as observed with EcoKI and these are very similar in HsdR(prrI). This suggests that the N-terminal region, used to make HsdR(prrI) may represent a domain in all HsdR subunits.

Figure 3   Motif X the site required for DNA cleavage

EcoKI (IA)	ADYVLFV--GLKPIAVVEAK	332
EcoAI (IB)	ADIVLYHKPGI-PLAVIEAK	78
EcoR124I (IC)	YDVTILVN-GL-PLVQIELK	167

This region includes the region responsible for DNA cleavage (motif X – Figure 3) that is another important structural feature of these proteins [11]. We have used protein-structure prediction algorithms to analyze the possible structure for this motif. Figure 4 shows the predicted structure.

Figure 4   Computer Generated Prediction (X-Motif)

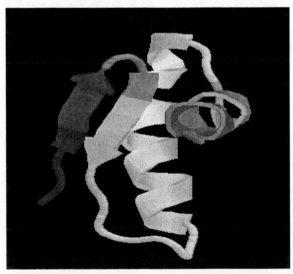

Predicted X-motif structure generated from superposed structural predictions of many aligned homologous sequences deposited on the protein sequence databank (http://protinfo.compbio.washington.edu/)

Figure 5   Limited Proteolysis of HsdRPrrI

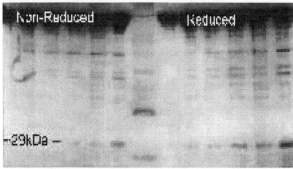

The above gel (13.5% SDS-PAGE) shows a limited proteolytic of the HsdR(prrI) subunit, using a modified trypsin protease. The proteolysis was performed on both the Non-Reduced and Reduced form of the subunit, respectively. The fragment of interest resides at ~29kDa and increases with intensity as proteolysis proceeds. This region has been mapped to the HsdR(prrI) sequence, suggesting a putative correlation with the N-terminal region of the subunit. This fragment will now N-terminal sequenced.

The hybrid HsdR(prrI) subunit was found to have an available cysteine at position three (as determined by chemical means and using cross-linking to a variety of ligands and polymers). This provides us with an ideal route for identification of the N-terminal proteolytic fragment (Figure 6), surface attachment of the motor, as well as a mechanism to study translocation through FRET. Since it has been previously shown that the R_1-complex is a unidirectional motor [7], it was important to show that this was also true for the hybrid R_1-complex.

Figure 6   HsdR N-Terminal Fluorescence Labeling

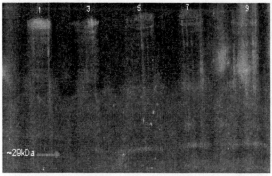

The above gel (13.5% SDS-PAGE) shows a limited proteolytic cleavage of the HsdR(prrI) subunit, carried as for Figure 5 and fluorescently labeled with Alexa-633 maleimide probe. Lanes 1,3,5,7,9 show a time-course trypsin digest (1, 5, 30, 60, & 90min, respectively). The fragment of interest resides at ~29kDa and increases with intensity the longer the proteolysis is allowed to continue.

We have initiated studies of the translocation process by measuring FRET between DNA-bound labeled motor protein and end-labeled DNA using Alexa 633 maleimide fluor (Molecular Probes) on the motor protein and Black-Hole Quencher-2 (BHQ-2) fluor on the DNA. A set of four individual 30-mer ssDNA oligonucleotide primers have been synthesized, of which two possess BHQ-2 (A_{max} 580nm) attached to the 5'-ends of two of the 30-mer's.

We obtained a successful FRET assay for the ENase complex when conjugated with Alexa-633 and BHQ2 (data not shown). However, further studies are underway to extend this work.

2.3   DNA translocation by hybrid R_1-complex

As a final check whether the hybrid HsdR(prrI)-MTase(R124I) R-M enzyme functions normally, its DNA translocation ability was tested. Figure 7 shows AFM images of typical hybrid R_1-complexes after incubation with 1 mM ATP and 724 bp DNA with a single recognition site. As anticipated, following the mechanism described in Figure 1, a DNA loop was formed by the hybrid R_1-complex. Only single loops were observed, consistent with the formation of R_1- rather than R_2-complexes. The size of the loops was comparable to those formed by wild-type enzymes at similar incubation times and under similar conditions (data not shown). With longer incubation times the average length of the loop increased. Both the presence of loops and the increase in size of such loops indicate regular translocation activity.

Figure 7   AFM images of DNA translocation

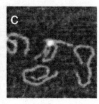

Typical images of translocation activity by HsdR(prrI)-MTase(R124I) after incubation for a) 10s, b) 30s, and c) 60s. in the presence of ATP and 724 bp DNA fragments, containing a single recognition site at 175 bp. At the position of the protein, which shows up as a white globular feature, one small DNA loop is observed that increases in size with longer incubation times. Scan range 250 nm, z range 3 nm.

2.4   Magnetic Tweezer studies

The study of these motor enzymes is encouraged by their unusual property of behaving as simple actuators – they move one part of the DNA relative to another section of the DNA and do not require immobilization on a surface to reproduce measurable motion (although the DNA needs to be surface attached for an observable displacement of a bead). This suggests a simple mechanism for detecting DNA movement through the use of a magnetic bead.

Figure 8   A simple Magnetic Tweezer setup

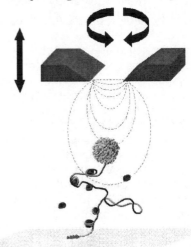

The simplest device for measuring such movement is a magnetic tweezer device (Figure 8). Magnetic tweezers allow real time monitoring of protein DNA interactions without surface interference and with femtonewton sensitivity. In addition, these systems can measure DNA displacements as low as 10nm as well as being able to produce negative, or positive, supercoils into the DNA, one turn at a time, through manipulation (spinning) of the magnetic bead [12, 13].

We have begun to measure the rate of translocation for the EcoR124I R-M enzyme in a single molecule situation and we expect this project to yield an abundance of interesting data over the next two years.

3 DISCUSSION

The EcoR124I R-M enzyme represents a very useful molecular motor that can be used as a nanoactuator. We have produced a derivative of the wild-type enzyme motor subunit, which only assembles as a R_1-complex that is also a unidirectional molecular motor, which can be used as a nanoactuator. We have initiated structural studies of this subunit and fluorescence analysis of DNA translocation. Single molecule studies have shown the motor is fully functional and we have initiated translocation studies using a magnetic tweezer setup.

Therefore, this motor has the potential to be a nanoactuator that will link the biological and silicon world through detection of the moving magnetic bead. Such a device could be silicon based with a variety of possible in built sensors for detecting the moving magnet from a simple magnetic force microscope cantilever etched into the device, to a Hall-type sensor built into the silicon.

4 ACKOWLEDGEMENTS

This project was funded by the European Community through a 5th Framework FET-OPEN project under the IST Thematic Programme (IST-2001-38036).

5   REFERENCES

1. Murray, N.E., *Type I restriction systems: sophisticated molecular machines (a legacy of Bertani and Weigle)*. Microbiology and Molecular Biology Reviews, 2000. **64**(2): p. 412-434.
2. Yuan, R., D.L. Hamilton, and J. Burckhardt, *DNA translocation by the restriction enzyme from E. coli K*. Cell, 1980. **20**: p. 237-244.
3. Endlich, B. and S. Linn, *The DNA restriction endonuclease of Escherichia coli B. I. Studies of the DNA translocation and the ATPase activities*. Journal of Biological Chemistry, 1985. **260**(9): p. 5720-5728.
4. Taylor, I., et al., *Purification and biochemical characterisation of the EcoR124 modification methylase*. Nucleic Acids Research, 1992. **20**(2): p. 179-186.
5. Janscák, P., D. Dryden, and K. Firman, *Analysis of the subunit assembly of the type IC restriction-modification enzyme EcoR124I*. Nucleic Acids Research, 1998. **26**(19): p. 4439-4445.
6. Janscák, P., A. Abadjieva, and K. Firman, *The type I restriction endonuclease R.EcoR124I: Over-production and biochemical properties*. Journal of Molecular Biology, 1996. **257**(5): p. 977-991.
7. Firman, K. and M. Szczelkun, *Measuring motion on DNA by the type I restriction endonuclease EcoR124I using triplex dissociation*. European Molecular Biology Organisation Journal, 2000. **19**(9): p. 2094-2102.
8. Gorbalenya, A.E. and E.V. Koonin, *Endonuclease (R) subunits of type I and type III restriction-modification enzymes contain a helicase-like domain*. FEBS Letts., 1991. **291**: p. 277-281.
9. Flaus, A. and T. Owen-Hughes, *Mechanisms for ATP-dependent chromatin remodelling*. Current Opinion in Genetics and Development, 2001. **11**(2): p. 148-154.
10. Zinkevich, V., et al., *The HsdR subunit of R EcoR124II: cloning and over-expression of the gene and unexpected properties of the subunit*. Nucleic Acids Research, 1997. **25**(3): p. 503-510.
11. Davies, G.P., et al., *On the structure and operation of type I DNA restriction enzymes*. Journal of Molecular Biology, 1999. **290**(2): p. 565-579.
12. Strick, T.R., V. Croquette, and D. Bensimon, *Single-molecule analysis of DNA uncoiling by a type II topoisomerase*. Nature, 2000. **404**(6780): p. 901-904.
13. Strick, T.R., D. Bensimon, and V. Croquette, *Micro-mechanical measurement of the torsional modulus of DNA*. Genetica, 1999. **106**(1-2): p. 57-62.

Viral Protein Linear (VPL) Nano-Actuators

A. Dubey[1], G. Sharma[1], C. Mavroidis[1, *], S. M. Tomassone[2], K. Nikitczuk[3], M.L. Yarmush[3]

[1]Department of Mechanical and Aerospace Engineering, Rutgers University, 98 Brett Road, NJ 08854
[2]Department Chemical and Biochemical Engineering, Rutgers University, 98 Brett Road, NJ 08854
[3]Department of Biomedical Engineering, Rutgers University, 98 Brett Road, NJ 08854
* Author for Correspondence, Tel 732-445-0732, Fax 732-445-3124, Email: mavro@jove.rutgers.edu

ABSTRACT

Computational studies are performed to predict and verify the performance of a new nanoscale biomolecular motor: The Viral Protein Linear (VPL) Motor. The motor is based on a conformational change observed in a family of viral envelope proteins when subjected to a changing pH enivronment. The conformational change produces a motion of about 10 nm, making the VPL a basic linear actuator which can be further interfaced with other organic/inorganic nanoscale components such as DNA actuators and carbon nanotubes.

1 INTRODUCTION

The recent explosion of research in nano-technology, combined with important discoveries in molecular biology have created a new interest in biomolecular machines and robots. The main goal in the field of biomolecular machines is to use various biological elements — whose function at the cellular level creates a motion, force or a signal — as machine components that perform the same function in response to the same biological stimuli but in an artificial setting. In this way proteins and DNA could act as motors, mechanical joints, transmission elements, or sensors. If all these different components were assembled together they could potentially form nanodevices with multiple degrees of freedom, able to apply forces and manipulate objects in the nanoscale world, transfer information from the nano- to the macroscale world and even travel in a nanoscale environment.

In this project, we are studying the development of Viral Protein Linear (VPL) nano-motors and their integration as actuators in bio-nano-robotic systems. The project consists of three research phases: 1) Development of concepts for novel bio-nano-motors and devices; 2) Performance of computational studies to develop models and design procedures that will predict and optimize the performance of the proposed bio-nano motors and systems; and 3) Execution of experimental studies to demonstrate the validity of the proposed concepts, models and design methodologies. In this paper we present the current activities and results for the first two phases. More specifically we will present the principle of operation of the VPL motor, the development of dynamic and kinematic models to study their performance and preliminary results obtained from the developed computational tools.

2 THE VPL MOTOR AND ITS TYPES

The basis of VPL motors is a pH dependant conformational change [1] observed in surface proteins (envelope glycoproteins) of certain retroviruses, such as the HIV1 virus. The change in the 3D structure and mechanical properties produces a linear-like motion. Envelope glycoproteins of various retroviruses play an important role in the process of membrane fusion, which is a process necessary for the virus to be able to infect a cell. During the process of membrane fusion, there is a distinct conformational change in the peptide on the viral surface as it 'readies' itself for infecting the cell. This change is due to the pH change associated with the vicinity of the cell. Given similar conditions, it is proposed to use this conformational change to produce VPL motors.

Peptides from different viruses can result in different VPL motors that can have different properties such as different weight, volume, range of motion, force and speed capabilities. However, the principle of actuation is the same. Studies have shown that the common characteristic in these viruses is the structure of a portion of the surface protein and the mode of infection. The envelope glycoproteins of these viruses can be divided into two subunits, the transmembrane (TM) and the surface subunit, which are a result of proteolytic cleavage of a common precursor protein. The surface subunit serves to recognize the cell to be infected when it comes in the vicinity of the virus with the help of receptors located on the cell surface. The TM subunits acquire an alpha-helical conformation when the virus is in its active or fusogenic state. The structure is like a hairpin composed of three coils, having one C terminal (carboxy- end) and the other N terminal (amino-end). This coiled coil structure undergoes a conformational change induced by mildly acidic conditions (i.e. pH around 5). This change is required for the process of membrane fusion, i.e. the fusion of viral and cellular membranes essential for infection of the cell. With the change in pH, the N-terminals pop out of the inner side and the peptide acquires a straightened position or the fusogenic state.

We have performed computational and experimental studies using the Influenza virus protein Hemagglutinin (HA) as the basis for forming a VPL motor. The reason for making this peptide selection is that based on current literature, this peptide seems to be able to perform repeatable motion controlled by variation of the pH.

The X-ray crystallographic structure of bromelain-released soluble ectodomain of Influenza envelope glycoprotein hemagglutinin (BHA) was solved in 1981 [2]. BHA and pure HA were shown to undergo similar pH-dependant conformational changes which lead to membrane fusion [3]. HA consists of two polypeptide chain subunits (HA1 and HA2) linked by a disulfide bond. HA1 contains sialic acid binding sites, which respond to the cell surface receptors of the target cells and hence help the virus to recognize a cell. There is a specific region (sequence) in HA2, which tends to form a coiled coil. In the original X-ray structure of native HA, this region is simply a random loop. A 36 amino-acid residue region, upon activation, makes a dramatic conformational change from a loop to a triple stranded extended coiled coil along with some residues of a short α-helix that precede it. This process relocates the hydrophobic fusion peptide (and the N-terminal of the peptide) by about 10 nm. In a sense of bio mimicking, we are engineering a peptide identical to the 36-residue long peptide mentioned above, which we call loop36. Cutting out the loop36 from the VPL motor, we obtain a peptide that has a closed length of about 4 nm and an extended length of about 6 nm, giving it an extension by two thirds of its length. Once characterized, the peptide will be subjected to conditions similar to what a virus experiences in the proximity of a cell, that is, a reduced pH. The resulting conformational change can be monitored by fluorescence tagging techniques and the forces can be measured using Atomic Force Microscope.

3 MOLECULAR DYNAMICS

To predict the dynamic performance of the proposed VPL motors (i.e. energy and force calculation) we are performing Molecular Dynamics (MD) Simulations that are based on the calculation of the energy released or absorbed during the transition from native to fusogenic state. We use the MD software called CHARMM (Chemistry at Harvard Molecular Mechanics) [4]. In MD, the feasibility of a particular conformation of the biomolecule in question is dictated by the energy constraints. Hence, a transition from one given state to another must be energetically favorable, unless there is an external impetus that helps the molecule overcome the energy barrier. When a macromolecule changes conformation, the interactions of its individual atoms with each other - as well as with the solvent - compose a very complex force system. With CHARMM, we can model a protein based on its amino acid sequence and allow a transition between two known states of the protein using Targeted Molecular Dynamics (TMD) [5]. TMD is used for approximate modeling of processes spanning long time-scales and relatively large displacements. Because the distance to be traveled by the N-terminal of the viral protein is relatively very large, we cannot let the protein unfold by itself. Instead of 'unfolding' we want it to undergo a large conformational change and 'open' up. To achieve this, the macromolecule will be 'forced' towards a final configuration from an initial configuration by applying constraints. The constraint is in the form of a bias in the force field.

In this project, the two known states are the native and the fusogenic states of the loop36. The structural data on these two states was obtained from Protein Data Bank (PDB) [6]. These PDB files contain the precise molecular make up of the proteins, including the size, shape, angle between bonds, and a variety of other aspects. We used the

PDB entries 1HGF and 1HTM respectively, as sources for initial and final states of the peptide.

By using TMD we have been able to prove that the final state of the VPL shown in Figure 2 is feasible, and it is the result of the action of the applied potentials. However, from the energy graph in Figure 3 we learn that even though the 36-residue long protein is forced to undergo a conformational change, it would require an energy jump to overcome the barrier that appears at approximately 4000 iterations. Unless an external force induces that jump, the protein would not go naturally to that state. The opening of the helical region followed by the adjustment of the remaining loop into an α-helical form is what requires more energy. This process occurs solely due to the pH drop in the natural setting of VPL since it is the rest of the large protein attached to the ends of this loop-36 that affects its behavior.

In a representative simulation, the "open" structure was generated arbitrarily by forcing the structure away from the native conformation with constrained high-temperature molecular dynamics. After a short equilibration, these two "closed" and "open" structures are then used as reference end-point states to study the transformation between the open and closed conformations. The transformation is enforced through a root mean square difference (RMSD) harmonic constraint in conjunction with molecular dynamics simulations. Both the forward (closed to open) and the reverse (open to closed) transformations are carried out. The RMSD between the two end-point structures is about 9Å, therefore the transformation is carried out in 91 intermediate steps or windows with a 0.1 Å RMSD spacing between each intermediate window. At each intermediate window, the structure is constrained to be at the required RMSD value away from the starting structure, it is minimized using 100 steps of Steepest Descent minimization, and then equilibrated with 0.5 picoseconds of Langevin dynamics with a friction coefficient of 25 ps on the non-hydrogen atoms. The harmonic RMSD constraint is mass-weighted and has a force constant of 500 kcal/mol/$Å^2$ applied only to the non-hydrogen atoms. The transformation is achieved by using the artificial RMSD constraint such that a conformation close to the final state is approached successfully. The energy plot using solvation function EEF1 [7] is shown in Figure 3 and the simulation snapshots for the intial and final states are shown in Figures 1 and 2.

FIGURE 1: Ribbon drawing of the closed conformation of 36-residue peptide as obtained from PDB entry 1HGF

FIGURE 2: Ribbon drawing of the open conformation as obtained by TMD simulations. There is a noticeable increase in alpha-helical content and the peptide opens.

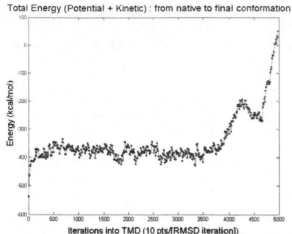

Total Energy (Potential + Kinetic) : from native to final conformation

Energy (kcal/mol)

Iterations into TMD (10 pts/[RMSD iteration])

FIGURE 3: Energy variation for LOOP-36 peptide with solvation model EEF1.

4 MOLECULAR KINEMATICS

Molecular kinematic simulations are being developed to study the geometric properties and conformational space of the VPL motors. The kinematic analysis is based on the development of direct and inverse kinematic models and their use towards the workpspace analysis of the VPL motors. In this section we present the derivation of the direct and inverse kinematic modules that have been incorporated into a MATLAB toolbox called BioKineLab that has been developed in our laboratory to study protein kinematics.

Proteins are macromolecules that are made up from 20 different types of amino acid residues. For kinematics purposes we consider these residues to be connected in a serial manner to create a serial manipulator. The "back bone" of the chain is a repeat of the Nitrogen - Alpha Carbon –Carbon (–N-Cα-C-) sequence. To the Cα atom is also attached a side-chain (R) which is different for each residue. These side-chains are passive 3-D structures with no revolute joints. Hydrogen atoms are neglected because of their small size and weight. The C-N bond joins two amino acid residues and has a partial double bond character and is thus non- rotatable. There are however two bonds which are free to rotate. These are the N-Cα and Cα-C bonds and the rotation angles around them are known as *phi* (φ) and *psi* (ψ) respectively. These angles determine the 3-D structure of the protein. Therefore a protein can be considered to be a serial linkage with K+1 solid links connected by K revolute joints. In case of loop36, K takes the value 72. In most kinematic studies, bond lengths and bond angles are considered constant, while the torsional angles (φ and ψ) are allowed to change [8].

4.1 Direct Kinematics

The direct kinematics problem calculates the VPL motor's final configuration when an initial configuration is given, all constant parameters of the chain are specified and a specific set of rotations for the torsional angles is defined. Frames are affixed at each backbone atom (Figure 4). Let b_i be a bond between atoms Q_i and Q_{i-1}. A local frame $F_{i-1} =$

$\{Q_{i-1}; \mathbf{x}_{i-1}, \mathbf{y}_{i-1}, \mathbf{z}_{i-1}\}$ is attached at bond b_{i-1} as follows: $\mathbf{z}_{i-1}$ has the direction of bond b_{i-1}; $\mathbf{x}_{i-1}$ is perpendicular to both b_{i-1} and b_i; and $\mathbf{y}_{i-1}$ is perpendicular to both $\mathbf{x}_{i-1}$ and $\mathbf{z}_{i-1}$. Similarly, a local frame $F_i = \{Q_i; \mathbf{x}_i, \mathbf{y}_i, \mathbf{z}_i\}$ is attached to bond b_i [9]. The Protein Denavit-Hartenberg (PDH) parameters are defined to facilitate the geometric representation of one frame to another [10] as follows: a_i is the distance from $\mathbf{z}_{i-1}$ to $\mathbf{z}_i$ measured along $\mathbf{x}_{i-1}$; α_i is the angle between $\mathbf{z}_{i-1}$ and $\mathbf{z}_i$ measured about $\mathbf{x}_{i-1}$; b_i is the distance from $\mathbf{x}_{i-1}$ to $\mathbf{x}_i$ measured along $\mathbf{z}_i$; and θ_i is the angle between $\mathbf{x}_{i-1}$ and $\mathbf{x}_i$ measured about $\mathbf{z}_i$. The coordinates of the origin and of the unit vectors of frame F_i with respect to frame F_{i-1} are represented using the following 4x4 homogeneous transformation matrix R_i [11] where $c\theta_i$ is $\cos(\theta_i)$, $s\theta_i$ is $\sin(\theta_i)$, and so on, l_i is the length of the bond b_i, θ_i is the torsional angle of b_i, and α_{i-1} is the bond angle between b_{i-1} and b_i:

$$R_i = \begin{bmatrix} c\theta_i & -s\theta_i & 0 & 0 \\ s\theta_i c\alpha_{i-1} & c\theta_i - c\alpha_{i-1} & -s\alpha_{i-1} & l_i s\alpha_{i-1} \\ s\theta_i s\alpha_{i-1} & c\theta_i s\alpha_{i-1} & -c\alpha_{i-1} & l_i c\alpha_{i-1} \\ 0 & 0 & 0 & 1 \end{bmatrix}$$

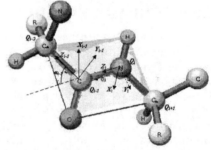

FIGURE 4: Frames F_{i-1} and F_i are attached to parent atom Q_{i-1} and Q_i and bond rotation angle is α_{i-1}.

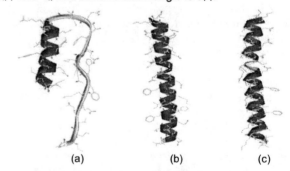

(a)　　　　(b)　　　　(c)

FIGURE 5: (a) loop36 protein in the native state, (b) open state generated by NMR experiments which is similar to that generated by MD, computation time for MD is about 2 hours, (c) open state generated by molecular kinematics, computation time is less than 40 seconds.

A representative result of the direct kinematics module of the BioKineLab Toolbox is shown in Figure 5. The results are obtained by running direct kinematics simulations on the native state of loop36 as shown in Fig. 5a. The final state of the same protein obtained from PDB is shown in Fig. 5b. Note that the random coil portion has turned into an α-helix after transformation giving us a linear motion of the end-effector. The goal was to achieve the final loop36 conformation using direct kinematics techniques. For this the torsional angles corresponding to the final state were determined using the Accerlys Viewer ActiveX software. These angles along with the initial state of loop36 were given as an input to the direct kinematics module of BioKineLab. Figure 5c shows the final structure generated by BioKineLab which gives a very good approximation of the actual output and clearly shows the relevance of using molecular kinematics for predicting and generating protein conformations.

4.2 Inverse Kinematics

The inverse kinematics problem calculates the VPL motor's torsional angles given an initial and final conformation and when all constant parameters of the chain are specified. A modified version of the Cyclic Coordinate Descent (CCD) method is used. The CCD algorithm was initially developed for the inverse kinematics applications in robotics [12]. For the inverse kinematics of protein chains, the torsional angles must be adjusted to move the C-terminal (end-effector) to a given desired position. The CCD method involves adjusting one torsional angle at a time to minimize the sum of the squared distances between the current and the desired end-effector positions. Hence, at each step in the CCD method the original n-dimensional minimization problem is reduced to a simple one-dimensional problem. The algorithm proceeds iteratively through all of the adjustable torsional angles from the C-terminal to the base N-terminal. At any given CCD step the bond around which the rotation is being performed is called the pivot bond and its preceding atom is called the pivot atom. The torsional angle corresponding to the pivot bond is to be determined. Figure 7 is the ball and stick model of a segment of the protein before and after the inverse kinematics simulation. Side chains are not shown for clarity.

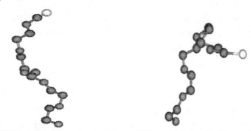

FIGURE 7: Initial conformation of the protein (left) and one of the solutions found by CCD simulations (right).

5 CONCLUSIONS

In this paper the concept of the Viral Protein Linear nanomotor was presented. Dynamic and kinematic analysis methods were described to calculate important properties of the motor. Preliminary results from the application of these computational methods in the VPL motor were shown. The dynamic analysis, though slower, attempts a more realistic representation of the system. Each intermediate conformation is energy minimized to make sure that it is stable and feasible. Targeted molecular dynamics studies show that a large impetus is needed to make the protein undergo the desired conformational change unless there are

other environmental factors present due to the presence of the remaining part of the protein not taken into account in this study. It however assures of the stability of the two end states of the system predicted by the kinematic analysis and experimental observations. Kinematics analysis can suggest the geometric paths that could be followed by the protein during the transition, while dynamics will narrow down the possibilities by pointing at the only energetically feasible paths. A combination of the two approaches – kinematics to give quick initial results and dynamics to corroborate and select the feasible solutions – can prove to be an indispensable tool in bio-nano-robotics.

6 ACKNOWLEDGEMENTS

This work was supported by the National Science Foundation (DMI-0228103 and DMI–0303950). Any opinions, findings, conclusions or recommendations expressed in this publication are those of the authors and do not necessarily reflect the views of the National Science Foundation.

7 REFERENCES

[1] Carr C. M. and Kim P., 1993, "A Spring-Loaded Mechanism for the Conformational Change of Influenza Hemagglutinin," *Cell*, **73**, 823-832.

[2] Wilson I.A., Skehel J.J., Wiley D.C., 1981, "Structure of the haemagglutinin membrane glycoprotein of influenza virus at 3 A resolution", *Nature*, **289**, 377-373.

[3] Skehel J.J., Bayley P.M., Brown E.B., Martin S.R., Waterfield M.D., White J.M., Wilson I.A., Wiley D.C., 1982, "Changes in the conformation of influenza virus hemagglutinin at the pH optimum of virus-mediated membrane fusion", *PNAS USA*, **79**, 968-972.

[4] Brooks R., Bruccoleri R.E., Olafson B.D., States D.J., Swaminathan S., and Karplus M., 1983, "CHARMM: A Program for Macromolecular Energy, Minimization, and Dynamics Calculations", *Journal of Computational Chemistry* **4**, 187-217.

[5] Schlitter J., 1994, "Targeted Molecular Dynamics: A New Approach for Searching Pathways of Conformational Transitions", *Journal of Molecular Graphics*, **12**,84-89.

[6] Berman H.M, Westbrook J., Feng Z., Gilliland G., Bhat T.N., Weissig H., Shindyalov I.N., and Bourne P.E., 2000, "The Protein Data Bank," *Nucleic Acid Research*, **28**, 235-242.

[7] Lazaridis T., and Karplus M., 1999, "Effective Energy Function for Proteins in Solution," *Proteins: Structure, Function and Genetics*, **35**,133-152.

[8] Gardiner E. J., Willett P. and Artymiuk P. J., "Graph-theoretic techniques for macromolecular docking", *Journal of Chemical Information and Computer Sciences*, 40, 273-279, 2000.

[9] Zhang M., Kavaraki L., 2002, "A new method for fast and accurate derivation of molecular conformations", *Journal of Chemical Information and Computer Science*, 42, 64-70.

[10] Denavit J., and Hartenberg S., 1955, "A Kinematic Notation for Lower Pair Mechanisms Based on Matrices", *ASME Journal of Applied Mechanics*, pp 215-221.

[11] McCarthy J., "Geometric Design of Linkages", *Springer Publications NY*, 2000.

[12] Wang L.C.T, Chen C.C, "A Combined Optimization Method for Solving the Inverse Kinematics Problem of Mechanical Manipulators", *IEEE Transactions on Robotics and Automation*, 7(4):489-499, 1991.

Biomimetic Molecules as Building Blocks for Synthetic Muscles – A Proposal

A.H. Flood, J. Badjic, J.-Y. Han, C. Pentecost, B.H. Northrop, S.J. Cantrill, K.N. Houk and J.F. Stoddart

Department of Chemistry and Biochemistry, and the California NanoSystems Institute,
University of California, Los Angeles, 405 Hilgard Avenue, Los Angeles, CA 90095-1569,
stoddart@chem.ucla.edu

ABSTRACT

The goal of this proposal is to cast biomolecular motors as the model for designing wholly artificial motor-molecules to serve as the building blocks for synthetic muscles. At the level of the single molecule, the artificial machines will be constructed from chemically powered motor-molecules with mechanically interlocked components. On the molecular ensemble level, the motor-molecules will be coherently self-organized on to surfaces in order to harness the cooperative operation of each motor-molecule in unison, consuming chemical energy to perform mechanical work up to the micro scale. These structures could be stacked in series to form artificial muscle tissue displaying large distance contraction and extension and to serve as a standard module for analogous applications on the meso scale. In the first phase of this work, the synthesis of the motor-molecules for attachment on surfaces is proposed, together with demonstrable nanoscale actuation as the target deliverable.

Keywords: interlocked molecules, motor-molecules, NEMS, self-organization, switches

1 INTRODUCTION

1.1 Biomimcry – Concept Transfer

Biomolecular motors (BMMs), such as the myosin-actin complex,[1] are nature's machines that convert chemical energy into mechanical work. The performance and scale of these systems are thought[2] to derive from their precise positioning and alignment into organized hierarchical levels beginning at the nanoscale. Moreover, this complex organization is one of the reasons why biomolecular motors are believed to outperform human-made machines.

Mechanistically, the BMMs couple the binding, hydrolysis and release of ATP with concomitant stepwise mechanical movements within an aqueous in vivo environment. Generally, the BMM's cycle of force production occurs through a sequence of alternating power and diffusive strokes.[3] We propose (Figure 1) artificial motor-molecules that operate in non-aqueous solutions, over a range of temperatures, by converting chemical fuel in the form of a proton gradient, into linear mechanical nanometer-scaled movements by a power-diffusive two-stroke cycle. Furthermore, we propose amplifying these

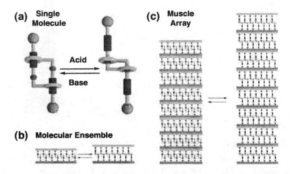

Figure 1. (a) The simple chemically powered contraction and extension of a linear molecular muscle could be (b) integrated into a monolayer assembly that will allow (c) higher order systems to amplify the nanoscale motion for macro scale applications.

motions by self-organizing the motor-molecules on solid surfaces so that they can coherently push and pull much larger objects, like the muscles in our bodies.

1.2 Linear Motor-Molecules

The motor-molecules of interest rely on chemical energy provided by base and acid to switch in a linear manner.[4] This type of switching derives from turning off and on the recognition between a dialkylammonium center (dark blue) and an aromatic crown ether (red). The function of a muscle is to contract and elongate; this relative movement in a molecular context can only be defined when the molecule itself is dimeric.[5] It is known, that when the molecule has both electron rich and poor moieties, the self-assembly of the dimeric [c2]daisy chain[6] is preferred over the entropically unfavorable infinite supramolecular arrays. The initial design is based on a hermaphroditically-shaped muscle motif (Figure 2) with either end differently modified for attachment to gold (disulfide) and silicon (hydroxyl) surfaces accordingly.

2 INNOVATIVE CLAIMS

Wholly artificial molecular motors are a unique class of compounds – only recently developed[2] – that perform relative mechanical movements on the nanometer scale. Their potential has not yet been fully realized, and so, acknowledging that these are "new systems for new applications" it is important to assess all classes of motor-molecules available for the most viable candidates.

Utilization of linear motor-molecules will provide a range of properties and enable a variety of auxiliary functions that will ultimately facilitate their inclusion into an integrated system.

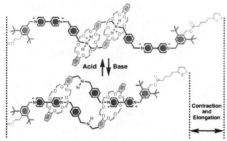

Figure 2. Chemical formulas of the hermaphroditic linear molecular muscle displaying the mechanical ring movement from one recognition station to another

(1) The integrated systems will be obtained by the organization at three hierarchical levels. *Single* linear motor-molecules are synthesized by employing self-organizing protocols. *Many* motor-molecules can be self-assembled into a molecular ensemble on a solid substrate for their coordinated movement and resulting cooperative force production. The force produced and strain observed can be *amplified* in the final level of organization by stacking the self-assembled monolayers on top of each other. The proposed hierarchical organization allows for optimization to be obtained at each level.

(2) The single motor-molecule – synthetically crafted – is tunable and customizable. Operationally, and as envisioned here, the molecules are predicted to produce (1) 10 pN of force, (2) 30% strain in the process of (3) repeatable (4) bistable mechanical switching, that is (5) chemically driven in a (6) "wet" non-aqueous environment, over (7) a range of temperatures (–38 to +62 C). Tunability is available for the strain, force, molecular rigidity and temperature ranges as directed from rational modifications to molecular structure. Customizability will allow (1) more than one length change, i.e., tristability, (2) for the system to be driven by light, such as a laser or sunlight, and (3) for attachment to a variety of substrates.

3 RELATED WORK

The proposed molecular design complements the work already undertaken[7] by Stoddart, Montemagno, Tai and Ho – however, the ammonium system presented here relies on a simple proton gradient or pH-switch, whereas the TTF systems are electrically actuated. Whereas the TTF system has a palindromic motif, the ammonium system is hermaphroditic, implying the optimal vectorial display of moving parts for linear movements. These molecular muscle designs are similar to those of Sauvage[8] but are more rigid, smaller, easier to make and rely on simpler chemical stimuli.

The nanoactuation of the proposed compounds relies on a uniquely different mechanism to other molecular systems. Liquid crystals operate[9] at ~micron scale by changes in the intermolecular alignment within molecular ensembles. Some electrically-activated polymers[10] and carbon nanotubes[11] rely on the diffusive migration of solvent (swelling) or ion migration (electrostatics) which are a phenomena associated with bulk materials rather than single molecular actuation. DNA hybridization[12] is neither reversible nor mechanistically well understood. An alternative contender to the molecular muscles is the photoactuation[13] of an azobenzene dye. The ammonium systems are ripe, however, for photoactivation utilizing excited-state acids and bases, such as $[Ru(bpy)_3]^{2+}$.[14]

4 TIMELINE AND BUDGET

4.1 Timeline

Phase I (18 Months)
(a) Synthesize prototype, characterize switching
(b) Synthesize tethered compounds and self-assemble
(c) Demonstrate surface switching
(d) Self-assemble 3D bulk network
(e) Demonstrate chemically-driven strain in 3D network
(f) Proof of principle light-driven motion
(g) Force spectroscopy and computational modeling
Phase II (18 Months)
(a) Interact with mechanical engineers
(b) Synthesize light-activated system and integrate like the chemically powered system (Phase I (a-d))
(c) Optimize strain
(d) Demonstrate force
(e) Demonstrate muscle module
(e) Demonstrate light-driven strain in 3D network
Phase III (12 Months)
Deliverable – Light-driven muscle module

4.2 Budget

Phase I   ($450K) – 2 Grad, 1 Postdoc, supplies
Phase II ($300K Chemistry) – 1 Grad student and 1 Postdoc plus supplies + $500K Engineering
Phase III ($300K Chemistry) – 1 Grad student and 1 Postdoc plus supplies + $500K Engineering
Total = $2.05M

5 TECHNICAL APPROACH

5.1 Synthesis of Linear Molecular Muscles

The molecular structure of the proposed molecular muscle consists of an electron rich ring in the shape of a dibenzo[24]crown-8 (DB24C8) macrocycle, and two electron poor sites – a secondary dialkylammonium (NH_2^+) center and a bipyridinium ($Bpym^{2+}$) unit – to which the electron rich ring can bind. A convergent synthetic route[4] (Scheme 1) was utilized to make the molecular muscle. Careful reaction conditions gave the DB24C8-based molecular muscle (DB-MM) in 14 % yield in five steps and

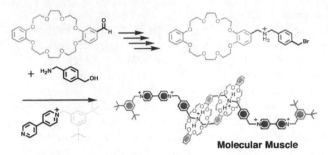

Scheme 1. Synthetic route used to obtain the molecular muscle in 14% yield.

following a chromatographic purification process. Compared to the molecular muscles made by Sauvage,[8] many fewer synthetic steps are required in order to assemble the final molecular muscle compound. Moreover, it still shows the same functions – contraction and elongation – upon the addition of stimulus, which, in this case, are acids and bases. Based on variable temperature ^{1}H-NMR spectroscopic studies, it has been concluded that the DB24C8 macroring is selectively bound to the NH_2^+ center from 235 to 335 K.

Fully reversible switching of the DB-MM dimer induces a length change, i.e., strain, of ~1 nm (29 % change).

The operation of this molecular muscle can be further extended into its self-assembly on solid surfaces by substituting the end groups, with disulfide or hydroxy substituted end groups.

5.2 Self-Assembly of Linear Molecular Muscles on Solid Surfaces

There is a growing range of molecular tethers for attachment[15] to solid substrates – disulfides on gold, polypodal carboxylates and phosphates on TiO_2 and SnO_2, and hydroxyl for silica (SiO_2) and chloro for silicon. Initially, disulfide-tethered molecular muscles will be utilized for self-assembly on gold using standard protocols.

5.3 Switching Analyses

Addition of a slight molar excess of a weak base, e.g., Et_3N or Bu_3N, which deprotonates the NH_2^+ center, causes the DB24C8 to move – i.e., the components slide with respect to each other – to the $Bpym^{2+}$ unit which can now provide stronger binding (Figure 2). This process can be observed either by UV-vis or ^{1}H-NMR spectroscopy. For example, the solution's appearance changed reversibly from clear to yellow when the base was added to the solution, indicating the creation of a charge-transfer interaction between the DB24C8 ring component and the $Bpym^{2+}$ unit.

Emission spectroscopy will used to probe the switching behavior of the system at the molecular level[16] on various surfaces. Fluorescein will be incorporated into the monomer unit to serve as both a stopper and fluorescent probe. A disulfide tether, allowing for the attachment to a gold surface, will be appended to this stopper at least six

carbons long in order to prevent quenching of the fluorescent probe by the gold.[17] The propensity of the aryl groups on the crown ether to π-π stack with any neighboring aryl groups, such as fluorescein, will facilitate the distinction between the extended and contracted states, as such π-π stacking interactions give rise to wavelength shifts in the emission spectra. The displacement incurred upon switching will be measured quantitatively using florescence resonance energy transfer (FRET) on the molecular muscle system, assymetrically stoppered with fluoroscein/rhodamine as the donor/acceptor pair. To monitor switching on a topographical level, characterization of the functionalized surfaces will be obtained using X-ray reflectivity, AFM, and SPR.[18]

5.4 Force Spectroscopy and Simulation

The force exerted by molecular muscles will be measured by AFM under contracted (base) and extended (acid) conditions. Dynamic force spectroscopy[19] will be utilized to probe host-guest interactions under thermodynamic equilibrium. Computations using the AMBER force-field and dynamics will assess free energy changes upon switching under acid/base conditions, as well as the details of molecular motions occurring in these processes.

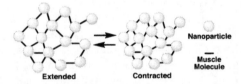

Figure 3. Extension and contraction of a bulk network of gold nanoparticles interconnected with muscle molecules.

5.5 Self-Assembly of Linear Molecular Muscles into a Random 3D Network

An interconnecting network (Figure 3) of muscle molecules and gold nanoparticles will be formed in order to provide a simple platform for the demonstration of strain in a bulk material. Two alternative protocols can be envisioned to self-assemble the network. The first relies on mixing bare gold nanoparticles with nanoparticles that are precoated with muscle molecules in which one of the two tethers is unbound. For example, a free disulfide could be displayed by utilizing either (1) two different tethers on the molecule, or (2) by optimizing the conditions (solvent, temperature, salt conc.) using a symmetrically tethered molecule so that one end is always free. Alternatively, stable solutions of nanoparticles coated with muscle molecules displaying hydroxyl groups may be esterified with disulfide tethers in situ and subsequently mixed with bare gold nanoparticles. Standard dynamic light scattering (DLS), TEM and UV-vis spectroscopic measurements will be employed to characterize the resulting assemblies.

5.6 Strain Demonstration in a Bulk Network

Changes in the size of the bulk network (Figure 3) will be determined by utilizing measurements during stimuli cycling that reflect (1) the overall size of the network, such as DLS or TEM samples, (2) the inter-nanoparticle distances, by UV-vis spectroscopy of the gold plasmon resonances, and (3) fluorescence to correlate the observed strain to molecular actuation.

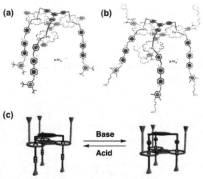

Figure 4. (a) Structural formulas of a molecular elevator,[20] (b) customized for (c) pulling two surfaces together.

5.7 Conceptual Extension

A second motif is conceptually formulated as a *molecular elevator* (**Figure 4a**).[20] The presence of three legs provides an internal amplification of the force produced from each extension and contraction cycle. The inclusion of the disulfide and hydroxyl units will again be utilized for surface attachment. In this way, not only is the force produced from a single motor enhanced.

Light-driven actuation is attractive for applications and so a series[14] of photogenerated acids and bases will be assessed for their facility to switch model systems, and then ultimately integrated into molecular muscles.

6 TECHNOLOGY TRANSFER

Success during Phase I anticipates the formation of a multidisciplinary team involving engineers and the emergence of commercial interests. These collaborations will rely upon constant feedback, allowing for an integrated systems-oriented approach to evolve an actuating muscle module for delivery into a niche product in the marketplace.

7 RELEVANCE TO DOD

In the first generation, pH-driven, single use modules with tunable force may deliver fast (ms) unparalleled, ultra-high density, light weight actuating material for deployment in the field. Recycling of the system will be possible upon addition of base. In the second generation, utilizing a photo-activated pH gradient, the contraction and extension could be cycled many times. These modules will serve niche or bulk deployment across a range of length scales.

REFERENCES

[1] D. S. Goodsell, Our Molecular Nature: The Body's Motors, Machines, and Messages; Copernicus: New York, 1996.

[2] V. Balzani, A. Credi, F. M. Raymo and J. F. Stoddart, Angew. Chem. Int. Ed. 39, 3349, 2000.

[3] C. Bustamante, D. Keller and G. Oster, Acc. Chem. Res. 34 412 2001.

[4] S. J. Cantrill, A. R. Pease and J. F. Stoddart, J. Chem. Soc., Dalton Trans. 3715 2000.

[5] M. C. Jiménez, C. Dietrich-Buchecker, J.-P. Sauvage and A. De Cian, Angew. Chem. Int. Ed. 39 1295 2000.

[6] S. J. Cantrill, G. J. Youn, J. F. Stoddart, D. J. Williams, J. Org. Chem. 66 6857 2001.

[7] Y.-C. Tai, Oral presentation, Biomolecular Motors Annual PI Conference, Aug. 2003, San Francisco.

[8] J. P. Collin, C. Dietrich-Buchecker, P. Gavina, M. C. Jimenez-Molero and J. P. Sauvage Acc. Chem. Res. 34 477 2001.

[9] J. Naciri, A. Srinivasan, H. Jeon, N. Nikolov, P. Keller and B. R. Ratna, Macromolecules 36 8499 2001.

[10] E. W. H. Jager, E. Smela and O. Inganas, Science 290 1540 2000.

[11] R. H. Baughman, C. X. Cui, A. A. Zakhidov, Z. Iqbal, J. N. Barisci, G. M. Spinks, G. G. Wallace, A. Mazzoldi, D. De Rossi, A. G. Rinzler, O. Jaschinski, S. Roth and M. Kertesz, Science 284 1340 1999.

[12] J. Fritz, M. K. Baller, H. P. Lang, H. Rothuizen, P. Vettiger, E. Meyer, H. J. Guntherodt, C. Gerber and J. K. Gimzewski, Science 288 316 2000.

[13] T. Hugel, N. B. Holland, A. Cattani, L. Moroder, M. Seitz and H. E. Gaub, Science 296 1103 2002.

[14] C. Hicks, G. Z. Ye, C. Levi, M. Gonzales, I. Rutenburg, J. W. Fan, R. Helmy, A. Kassis and H. D. Gafney, Coord. Chem. Rev. 211 207 2001.

[15] A. Ulman An Introduction to Ultrathin Organic Films from Langmuir-Blodgett to Self-Assembly; Academic Press, Inc: San Diego, 1991.

[16] P. Ashton, R. Ballardini, V. Balzani, M. Gomez-Lopez, S. Lawrence, M. Martinez-Diaz, M. Montalti, A. Piersanti, L. Prodi, J. F. Stoddart and D. Williams, J. Am. Chem. Soc. 119 10641 1997.

[17] K. Kittredge, M. Fox, and J. Whitesell, J. Phys. Chem. B. 105 10594 2001.

[18] K. Kim, W. Jeon, J. Kang, J. Lee, S. Jon, T. Kim and K. Kim, Angew. Chem. Int. Ed. 42 2293 2003.

[19] H. Schönherr, M. W. J. Beulen, J. Bügler, J. Huskens, F. C. J. M. van Veggel, D. Reinhoudt, and J. G. Vancso, J. Am. Chem. Soc. 122 4963 2000.

[20] J. Badjic, V. Balzani, A. Credi, and J. F. Stoddart, Science 2003 Submitted.

MEMS Biofluidic Device Concept Based on a Supramolecular Motor

M. J. Knieser[*], R. Pidaparti[*], F. Kadioglu[****], R. D. Harris[**], M. R. Knieser[***] and A. Hsu[*]

[*]Indiana University Purdue University Indianapolis,
Indianapolis, IN, USA, mknieser@iupui.edu, rpidapa2@iupui.edu, anihsu@iupui.edu
[**]Essential Research, Cleveland, OH, USA, harris@er.com
[***]Indiana Life Sciences, Indianapolis, IN, USA, drknieser1@knieser.com
[****]Istanbul Technical University, Istanbul, Turkey, fkadioglu@itu.edu.tr

ABSTRACT

The supramolecular machine, called the nuclear pore complex (NPC), controls the transport of all cellular material between the cytoplasm and the nucleus that occurs naturally in all biological cells. In the presence of appropriate chemical or geometrical stimuli, the NPC opens or closes, like a gate, and permits the flow of material into, and out of, the nucleus. Given the natural design of the nuclear pore complexes, their motor like function, and their direct engineering relevance to bio-molecular motors technology, our approach is to understand its design and mimic the supramolecular motor in an example of a biofluidic device through MEMS. A proof-of-concept based on a MEMS fluid pump will be designed and fabricated to demonstrate the applicability of the bio-inspired motor. The inspiration comes from the bio-inspired motor (Nuclear Pore Complex) which acts like a bi-directional pump for specific substances. While the NPC is about 200 nanometers in size, our proof-of-concept will be about 100,000 times larger in size. After fabrication of each MEMS component, process characterization, prototype evaluation and design evaluation will occur to iterate the modeling and simulation aspects of the project. The evaluations of the MEMS component will be done within a test fluid pumping environment.

Keywords: MEMS, motor, biofluidic, supramolecular

1 INTRODUCTION

The human body is a complex array of organs, each composed of a variety of tissues, composed of many cell types. It is the study of the individual cells which has led us and other researchers to be so marveled by the nature of cellular activity that we try to imitate what cells do. It is this impressive and efficient order of sub-cellular activities that has led to our study of the Nuclear Pore Complex (NPC) as a model of functionality that we wish to imitate.

This supra-molecular pump is located at the junction of the cytoplasm and the nucleoplasm [1]. It appears to represent a point of transition from the cytoplasmic cisterns, the smooth and rough endoplasmic reticulum, and the perinuclear cistern. It is the junction where nuclear material such as mRNA (messenger material produced by the nuclear code (DNA)) is sent as instructions to the rough endoplasmic reticulum to encode the production of specific molecules for use by the cell or distribution out of the cell for use elsewhere by the organ or organism. In the opposite direction material can be sent into the nucleus for messengering or wrapping and/or unwrapping DNA to allow for varied areas of DNA to be exposed for code production.

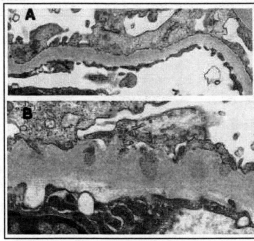

Figure 1: A: Normal peripheral capillary loop in kidney. B: Abnormal peripheral capillary loop with electron dense deposits in the basement membrane, representing chemical complexes trapped passing through the membrane.

Electron microscopic studies of the NPC has revealed an arrangement of filaments and ultrastructural forms which have been interpreted to be the backbone of the complex [2]. Our graphic representation of this suggests that the outer elements at both ends represent filaments and lipoprotein complexes used to anchor the core elements. It is these core elements, which appear to give specificity to the functionality of the NPC. Studies have shown that at different times the staining of the core at electron microscopy varies [1]. From studies of routine electron microscopy of normal and diseased kidney biopsies, we have demonstrated that at times of varied filtration by the lamina densa in electron micrographs it is possible to see materials in transition through the renal basement membrane [3][4]. (Figure 1) These materials in transition give darker and varied staining at electron microscopy.

Just as in the kidney biopsies shown, these pictures represent a dynamic liquid to semisolid material functioning as molecular elements and as supra-molecular structures. The kidney biopsies show immune complex disease frozen in a moment of time. The electron micrographs reveal the NPC when the complex is at a moment in time where molecular material may or may not be in the core area. How the material enters and transitions the core has been demonstrated using encapsulated viral material presented to the NPC [2]. Urs. F. Greber and Suomalainen demonstrated that the NPC has specific geometry, which must be met for the complex to allow transition through to the nucleus. It is unreasonable to assume that only geometry is in use to accomplish transition through the core area. It is reasonable from these studies to assume that when the geometry of the viral coat meets the requirements of the NPC, other chemistry takes place, allowing transition to the nuclear side. This process has been demonstrated to use APT as an on site energy source, consistent with most other supra-molecular reactions requiring energy input [2]. Reverse transitioning is most likely also based on some combination of geometry, chemical reactions and/or electro-mechanical activity, requiring energy.

Inspired by this complex and sophisticated supra-molecular activity, we have devised a MEMS pump, which will accomplish material transfer based on geometry (size) and electro-mechanical activity.

2 MEMS BIO-FLUIDIC DESIGN CONCEPT

The objective of developing a MEMS bio-fluidic proof-of-concept is to explore the design features of the NPC. For each of the components of the design its advantages, manufacturability and design features will be analyzed. Together, the proof-of-concept data and information obtained from the scaled prototype models will be used to refine the computational models for assessing the mechanical characteristics (assembly, weight, cost, stress concentrations, etc.) and scaling properties.

While the NPC is about 200 nanometers in size, our proof-of-concept will initially be about 10,000 times larger in size with the possibility 2 to 3 times larger in thickness and 1,000 times larger in width. The bio-inspired fluid pump initial concept is shown in Figure 2. The proof-of-concept NPC will be manufactured using MEMS related technologies. Specifically, the proof-of-concept will utilize electrostatic actuation, piezoelectric actuation and Silicon-On-Insulator (SOI) wafer processing.

Examining the NPC reveals the following characteristics. First, the NPC acts like a bi-directional pump. Second, the NPC has two types of sensors on each opening for the identification of specific substances: a geometric sensor and a chemical sensor. Third, the NPC has two controlling inputs from the nucleus: a controlling input that directs the NPC to pump a specific substance out of the cell and a controlling input that directs the NPC to pump a specific substance in to the cell. Fourth, the NPC contains all of the controlling logic to make the NPC pump function. These characteristics are summarized in a system schematic illustrated in Figure 3.

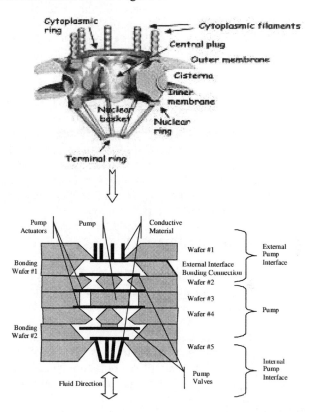

Figure 2: Bio-inspired MEMS fluid pump concept. [5]

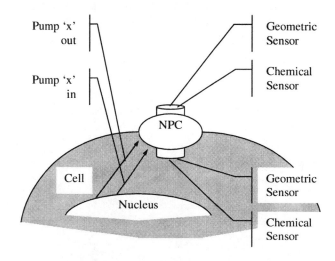

Figure 3: NPC system schematic.

Focusing on the proof-of-concept NPC pump, the bio-inspired design employs two actuation methods to achieve the pumping action. The bio-inspired pump consists of a moving central plug that is controlled by bimorph piezoelectric supports and two electrostatic actuated check values. The initial condition of the check values is that each are normally open. The basic pump sequence starts by the central plug actuated to the most outlet position and the outlet check value closed. With the outlet check valve closed, the central plug is actuated to the most inlet position pulling fluid into the void created between the outlet check valve and the central plug. Afterwards, the inlet check valve is closed and then the outlet check valve is opened. Then the central plug is actuated back to the most outlet position which would push the fluid through the outlet.

The MEMS manufacturing process for the proof-of-concept devices will encompass using SIO semiconductor wafers. These wafers provide many useful characteristics like etched cavities that follow crystal structures, doped silicon layers that can be used to create electrical conductors or possible active electronic components and proof-of-concept devices based on the deposited silicon layer thickness that can range from 50nm to 100um instead of wafer thickness of about 375um.

3 DESIGN ANALYSIS

The geometry considered for the bio-fluid pump proof-of-concept is shown Figure 4. Three computational models are considered using plane stress and beam type finite elements. Model-1 as shown Figure 4 is considered as a baseline with all the ports are closed so that no fluid flows through the pump. Model-2 and Model-3 consist of the same geometry as Model-1 but all the ports are completely open except the one at the bottom so that the fluid flows from top to bottom of the pump.

Two different material properties were used for the plane elements and beam elements. All beams have the following elastic properties in the finite element analysis: Young's modulus, E=70 GPa Poisson's ratio, ν=0.31 and density ρ=7600 kg/m^3. The materials properties for solid-planes are Silicon (E=180 GPa; ν=0.17, ρ=2330 kg/m^3).

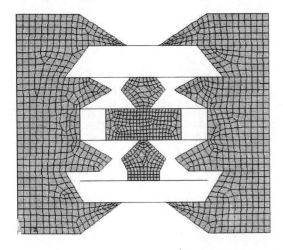

Figure 4: Bio-fluid pump, Model #1.

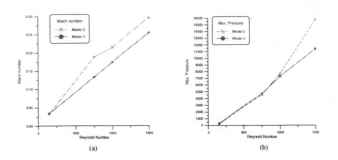

Figure 5: (a) Mach number and (b) pressure using different Reynolds numbers.

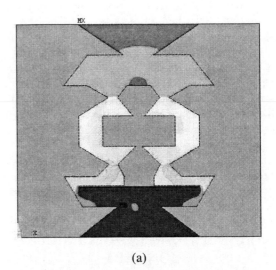

(a)

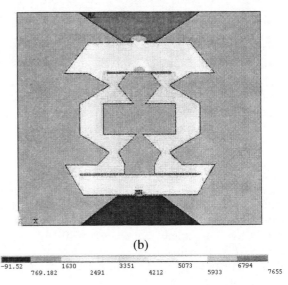

(b)

Figure 6: (a) Model-2 and (b) Model-3.

3.1 Fluid Flow Analysis

The material properties of water were used for the fluid. The material properties for the fluid used in the finite element analysis are viscosity, $\mu=1.14\times10^{-12}$ kg/nm-s and density $\rho=1000\times10^{-27}$ kg/m^3. The fluids are modeled using two-dimensional finite elements in ANSYS software. The PLANE141 element type is defined by four nodes and has two degrees of freedom in x and y directions at each node. Free mesh is used for creating the finite element model. The boundary conditions for the finite element models are assumed to be at inlet top line v_0 initial velocity, at the other zero and at the outlet bottom line zero pressure.

The results are obtained using the finite element meshes of 2045 for Model-2 and 1305 for Model-3 for the fluid flow with FLOTRAN/CFD, respectively. In addition, Model-2 and Model-3 were solved with using different Reynolds number during this analysis.

The maximum Mach number and maximum pressure for the Model-2 and Model-3 are shown in Figure 5 using FLOTRAN/CFD module in ANSYS software. For Model-2 and Model-3, the locations for maximum pressure are at the inlet, and locations for maximum velocity at the Model-2 are nearly at the bottom beam. The locations for maximum velocity at the Model-3 are nearly bottom beam for Re=150 and 750, and top beam for Re=1000 and 1500.

The pressure results using Reynolds number 1000 are presented in Figure 6 for Model-2 and Model-3.

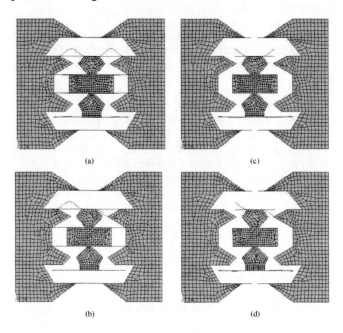

(a)   (c)

(b)   (d)

Figure 7: Model-1 and Model-3 Mode shapes.

3.2 Dynamic Analysis

Model-1 and Model-3 were solved for dynamic analysis to investigate the resonance characteristics of the electrostatic actuated check valve beams. The material properties of plane are Silicon. Polyimide and gold materials were used for all beams. The planes and beams are modeled using two-dimensional finite elements in ANSYS software. For planes, the PLANE42 element type

is defined by four nodes and has two degrees of freedom (translations in x and y directions) at each node. Free mesh is used for creating the finite element model. The boundary conditions for the finite element models are assumed to be at left and right ends fixed. The results are obtained using the total finite element meshes of 1519, 1406, for Model-2 and Model-3 for the dynamic analysis, respectively. Model-1 has 209 beam elements and Model-3 has 96 beam elements (BEAM3 in ANSYS Software).

Cases	First Mode	Second Mode
Model-1	42727 Hz	43149 Hz
Model-3	26882 Hz	27278 Hz

Table 1: Comparison of frequencies.

The fundamental mode shapes for the Model-1 are shown in Figure 7 using dynamic analysis. Table 1 shows the comparison of frequency values for the Model-1 and Model-3 model cases. It can be seen from Table 1, that first and second frequency values decrease about 37% for model-3 as compared to model-1. The third and fourth stiffnesses increase about 42% for Model-3.

4   SUMMARY

A process has been created and mechanical models explored that enable the creation of a proof-of-concept MEMS-based biofluidic device based on the supra molecular structure of the Nuclear Pore Complex. The device process has the capabilities of creating a 400 nano-meter thick device for testing in a fluid pumping environment.

REFERENCES

[1] D. Stooffler, B. Feja, B. Fahrenkrog, J. Walz, D. Typke, U. Aebi, "Cryo-electron Tomography Provides Novel Insight into Nuclear Pore Architecture: Implications for Nucleocytoplasmic Transport," Journal of Molecular Biology, 328, 119-130, 2003.
[2] F. Greber, M. Suomalainen, R. Stidwill, K. Boucke, M. Ebersold and AS. Helenius, "The role of the nuclear pore complex in adenovirus DNA entry." The EMBO Journal, 16:19, 5998-6007, 1997.
[3] MR. Knieser, EH. Jenis, DT. Lowenthal, "Pathogenesis of renal disease associated with viral hepatitis," Archives of Pathology, 97, 193-200, 1974.
[4] EH. Jenis, P. Sandler, GS. Hill, MR. Knieser, GE. Jensen, SD. Roskes, "Glomerulonephritis with basement membrane dense deposits," Archives of Pathology, 97(2), 84-91, 1974.
[5] http://physrev.physiology.org/content/vol81/ issue1/images/large/9j0110107002.jpeg

A Mechanistic Model of the Motility of Actin Filaments on Myosin

D.V. Nicolau, Jr. and D.V. Nicolau

Industrial Research Institute Swinburne, Swinburne University of Technology
P.O. Box 218, Hawthorn VICTORIA 3122 AUSTRALIA
sarmisegetusa@yahoo.com, dnicolau@swin.edu

ABSTRACT

The interaction of actin filaments with myosin is crucial to cell motility, muscular contraction, cell division and other processes. The *in vitro* motility assay involves the motion of actin filaments on a substrate coated with myosin, and is used extensively to investigate the dynamics of the actomyosin system. Following on from previous work, we propose a new mechanical model of actin motility on myosin, wherein a filament is modeled as a chain of beads connected by harmonic springs. This imposes a limitation on the 'stretching' of the filament. The rotation of one bead with respect to its neighbors is also constrained in similar way. We implemented this model and used Monte Carlo simulations to determine whether it can predict the directionality of filament motion. The principal advantages of this model are that we have removed the empirically correct but artificial assumption that the filament moves like a 'worm' i.e. the head determines the direction of movement and the rest of the filament 'follows' the head as well as the inclusion of dependencies on experimental rate constants (and so also on e.g. ATP concentration) via the cross-bridge cycle.

Keywords: Actin, myosin, motility, modeling

1 INTRODUCTION

Myosin is a type of molecular motor that brings about unidirectional movement of actin filaments using the energy produced by the hydrolysis of ATP. Muscular contraction occurs due to the sliding of actin filaments towards the center of bipolar myosin filaments. Additionally, in both muscular and non-muscular cells, the actin-myosin system is responsible for various cellular movement and shape-changing functions.

The *in vitro* motility assay is used to investigate the function of this and other molecular motor systems by arranging for actin filaments to slide along a surface coated with myosin, for myosin-coated microbeads to move along an actin-covered surface, or some variation on the above, in the presence of sufficient ATP. This assay is extensively used in assessing the impact of different chemical species (e.g. potential drugs, toxins, etc.) on the biomolecular motility which is involved in bioprocesses as varied as muscular contraction; neuroprocesses; and cell division.

In previous work [2], we proposed a model for actin motion on a myosin-coated surface that involved a myosin-actin filament power-stroke force that depended on the length of each filament as well as the density of myosin on the model surface. We made the artificial assumption that the "head" monomer of the actin controls the direction of the filament and all other monomers follow the head. Although this very simple model captures extremely well the observed sliding motion of filaments on myosin during motility assays, it is clear that it is inadequate for the purposes of predicting physical characteristics of the system *ab initio*.

In the present paper, we have made several important improvements to the model. Firstly, we have eliminated the assumption of "head controls direction" — instead, we model the filament as a chain of arbitrary-shaped beads connected by harmonic springs. Secondly, we have constrained the motion of the filament laterally, to account for its rigidity (also using a harmonic spring model). We have also introduced dependence upon ATP concentration, temperature and other environmental factors via the integration into the model of a cross-bridge cycle model. This is modeled as a Markov chain and evaluated stochastically over time in a Monte Carlo-like fashion, with arbitrary rate constants for each state transition. The present model begins with the meso-scale interaction of each monomer with a myosin head (rather than treating the filament as a whole) and thus is of a far greater predictive value as well as being physically meaningful. We envisage that it could be used not only for predicting the characteristics of actin-myosin interaction e.g. velocity versus force relationships for molecular motors but also for validation models of motility and thus muscular function as well as other critical biological processes.

2 MOTILITY MODEL

The model consists of three relatively independent components. The first of these is models the myosin molecules geometrically as planar structures equipped with two "arms" for binding to the actin filament. The second models these filaments as chains of beads connected by harmonic springs. Finally, the interaction of a model filament with a model myosin molecule is calculated under the framework of a stochastic Markov chain cross-bridge cycle model. In what follows we briefly describe each of these in turn.

2.1 Myosin Model

Myosin is a large protein molecule that uses chemical energy to perform directional biological motion. Myosin captures a molecule of ATP, the molecule used to transfer energy in cells and breaks it down, using the chemical energy produced to perform the so-called "power stroke" — pushing the actin filament. Myosin is composed of several protein chains: two large "heavy" chains and four small "light" chains and has a complex structure. For the purposes of motility, however, one can model the myosin as a "base" equipped with two "arms" that bind to the monomers in an actin filament and carry out a power-stroke.

Accordingly, the myosin head is modeled as a two-dimensional circle of radius r_{myo} oriented horizontally on an infinite (polymer) plane. A head has two "arms" and these act independently of each other. At any time an arm can be bound to an actin monomer or free. If bound, the actin monomer and myosin arm go through the cross-bridge cycle (see below). More precisely, within the "bound" state several other states exist — each corresponding to a stage of the cross-bridge cycle (the system traverses this cycle stochastically). When this is completed, the myosin arm becomes free to bind to another monomer. Arms can only bind to monomers that are "above" the circle representing the myosin head. The myosins are randomly distributed on the model surface with number density n_{myo}.

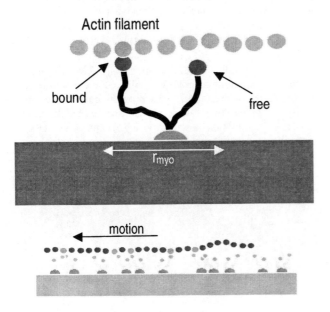

Figure 1: Interaction of an actin filament with a myosin-coated surface.

2.2 Actin Model

Actin filaments are modeled as chains of N arbitrary-shape monomers of mass m and nominal radius r, connected together by harmonic springs. Each monomer is connected to the next by a harmonic spring linking their centers in two dimensions (we assume the filament to remain horizontal and equidistant from the myosin-coated surface and thus ignore the z dimension) along the direction of an inbuilt "axis", $\mathbf{a}_i$ for the ith filament. All springs are identical with rest length l_{rest} and spring constant k. The force exerted on a monomer i from the two adjacent monomers is then given by

$$F_{spring}(i) = -k\left(\left|(\mathbf{p_{i-1}} - \mathbf{p_i})\right| + \left|(\mathbf{p_{i+1}} - \mathbf{p_i})\right| - 2l_{rest}\right) \quad (1)$$

where $\mathbf{p_i}$, $\mathbf{p_{i+1}}$ and $\mathbf{p_{i-1}}$ are the position vectors of monomers i, $i+1$ and $i-1$, respectively in 2-space.

Monomers are also constrained laterally. Let the angle between monomers $i-1$, i and $i+1$ be $\alpha_i(t)$ and let the straight line connecting monomers $i-1$ and $i+1$ be L. Then monomer i experiences a force proportional to $(\pi - \alpha_i(t))$ in the direction of a vector between the center of monomer i and the midpoint of L. The constant of proportionality is k_α. This then gives the angular response as

$$\frac{\partial^2 \alpha_i(t)}{\partial t^2} = -k_\alpha(\pi - \alpha_i(t)) \quad (2)$$

It is often the case that the behavior of the bonds connecting monomers in a polymer molecule is assumed to be harmonic. In this case we have also made this assumption for both types of response (positional and angular). It is probably safe to say that this is a good first approximation — however, this may be changed in a future model, depending on how well the present model captures the characteristics of a motility assay (and thus of the actin-myosin interaction in general).

Monomers experience three other forces. The first and most crucial of these is due to the power-stroke from the myosin arm. While the system is in the power-stroke state, the monomer that is bound to the myosin arm experiences a constant force, F_{myo}. This force is exerted in the direction of the monomer's axis — this assumption is physically meaningful since the binding of the myosin arm to the actin filament only happens in certain orientations of these and hence the power-stroke can only be applied in a given direction (possibly slightly perturbed due to thermal forces).

A frictional force proportional to the velocity of each monomer $|\mathbf{v}_i|$, with proportionality constant γ and opposed to the velocity vector is also experienced. This constant can either be estimated empirically or calculated under certain assumptions about the shape of each monomer (the assumption that monomers are spherical thus leads to the

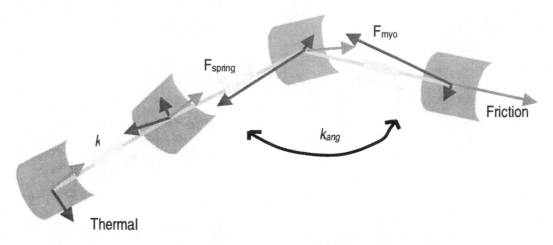

Figure 2: Mechanical model of actin filament.

Stokes' law constant). Finally, each monomer experiences a thermal force of normal strength (where the maximal strength depends on absolute temperature) distribution and random direction at each point in time. The sum of these forces on each monomer gives the total force experienced. At each time step, these are recomputed. The time step has length Δt.

2.3 Cross-Bridge Cycle Model

In order to model the interaction of a myosin head with an actin monomer in a physically meaningful way, we have introduced into our model a stochastic Markov chain cross-bridge cycle. The widely accepted scheme for this cycle was proposed by Lymn and Taylor [4] and is used here.

Initially, the myosin cross-bridge is unbound — in our model, the probability of binding at any time is $2p$ if a monomer is within the action radius of a myosin molecule and both arms are unbound, p if only one arm is bound and 0 if both arms are bound or no monomer is within the action radius. Upon binding, the myosin cross-bridge enters the weakly bound (non-force-producing) $(A) \cdot M \cdot ADP \cdot P_i$ state (A and M denote actin and myosin, respectively and the brackets indicate weak binding). Releasing the phosphate P_i leads to the next state — the strongly bound, force-producing $A \cdot M^* \cdot ADP$ state, leading to the execution of the power stroke. In our model, the length of this power stroke is a random variable which depends upon the rate constant for the transition to the next state. During the power stroke the monomer experiences a force F_{myo} along its axis direction. Thus the power stroke produces conformational changes in the myosin molecule and a translation of the monomer (and at a larger scale, the filament). The resultant $A \cdot M \cdot ADP$ state can no longer bind P_i and at the end of the stroke, ADP is released and the system enters the $A \cdot M$ state.

If ADP rebinds, then the system will continue to resist both motion and attachment. Eventually, ATP binds, leading to a detachment of the cross-bridge. Upon hydrolysis the cross-bridge enters the $(A) \cdot M \cdot ADP \cdot P_i$ state and can thus participate in a new working stroke [4]. The transition from each state to the next is controlled by the rate constants, which for different systems can be found in the literature and depend on ATP concentration, temperature and other factors. On the other hand, these can be estimated if other parameters (e.g. force) as well as motility characteristics are known in advance.

3 SIMULATION RESULTS

We have implemented our model into an in-house developed simulation program to facilitate the development and validation of the model. This program can be used to specify parameters (e.g. rate constants, filament length etc.) and extract simulation descriptors (e.g. filament velocity) under different conditions.

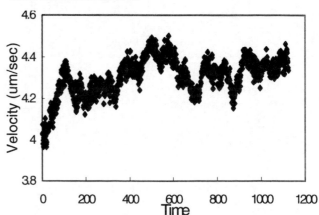

Figure 3: Velocity versus time in a typical simulation.

4 DISCUSSION

The motility of the protein molecular motors has been used in primitive dynamic nanodevices that have enormous potential in sensing, nano-level power generation and possibly computation [3]. It is difficult, however, to probe the feasibility of the usage of protein molecular motors in "real-device" conditions, that is, far from the conditions present in carefully optimized motility assays; thus, the simulation of motility assays with phenomenologically-relevant input parameters would constitute a very useful tool.

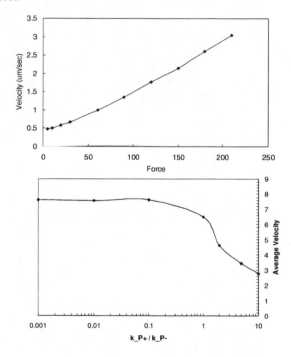

Figure 4. Velocity versus force (top) and forward/backwards rate constant ratio (bottom).

Biological molecular motors are generally thought to "rectify" Brownian motion — i.e. operate via a "Brownian ratchet" mechanism. The model proposed here also relies on a such an interaction between the actin filament and the myosin motor. The directionality of the motion (as shown by simulations) is guaranteed by the in-built orientation of actin monomers in a filament and the geometrical specificity of the myosin-actin binding. Finally, the quantitative nature of the interaction between the motor and the filament is controlled by the hydrolysis cycle rate constants. The qualitatively correct simulation results suggest that a Brownian ratchet mechanism combined with a stochastic hydrolysis cycle model can predict the nature (especially directionality) of the actin motility.

The combination of three models, while making for a fairly comprehensive overall description of actin-myosin based motility comes at the price of complexity — the model requires around 20 parameters (around 6 geometrical constants, 8 rate constants and 6 force/flexibility constants). Consequently, it cannot be considered analytically determinate in a practical sense — i.e. in order to predict one of the parameters (e.g. actin flexibility), it is necessary to estimate the others. For some parameters, e.g. filament-fluid drag coefficient, this can be done theoretically, while for others, e.g. rate constants, experimental data is available. The fully quantitative validation of the model using physical parameters is the subject of future work.

5 CONCLUSION

A comprehensive model for the motility of actin filaments on a myosin-coated substrate has been presented, comprising separate models of the hydrolysis cycle and mechanics of actin filaments and myosin motors. We have shown via simulation that the model is capable of predicting qualitatively the motility of actin filaments (especially directionality) and produces empirically credible relationships between parameters and observed motility descriptors (e.g. velocity). Further work will concentrate on the fully numerical evaluation of simulation results using empirically estimated biomolecular parameters.

6 ACKNOWLEDGMENTS

This study has been funded by a DARPA grant (no. 66001-00-1-8952). The authors would like to thank Kristi Hanson for useful discussions.

REFERENCES

[1]   Mahanivong C., Wright J.P., Kekic M., Pham D.K., Remedios C. and Nicolau D.V., Manipulation of the Motility of Protein Molecular Motors on Microfabricated Structures, Biomedical Microdevices, 4:2, 111-116 (2002)
[2]   Nicolau, D.V. Jr, Nicolau, D.V., Simulation of the motility of filaments on surfaces functionalised with molecular motors, Current Applied Physics (in print)
[3]   Nicolau, D.V. Jr, Nicolau, D.V., Computing with the actin-myosin molecular motor system, SPIE Proceedings, 4937, 219-226 (2002)
[4]   Lymn, R. W., Taylor, E. W.: Mechanism of adenosine triphosphate hydrolysis by actomyosin. Biochemistry 10, 4617-24 (1971).
[5]   Peskin, C., Odell, G., Oster, G.: Cellular motions and thermal fluctuations: the Brownian ratchet, Biophys. J., 65, 316-324 (1993).
[9]   Huxley, A.F.: Muscle structure and theories of contraction, Prog. Biophys. Biophys. Chem., 7, 255-318 (1957).
[10]   Julicher, F., Ajdari, A., Prost, J.: Modeling molecular motors, Rev. Mod. Phys., 69, 1269-1281 (1997).
[11]   Vale, R.D., Milligan, R.A.: The way things move: looking under the hood of molecular motor proteins. Science 288, 88-95 (2000).

A PROPOSAL CONCEPTUALIZATION OF A HYBRID BIO-MOTO-REDUCTOR ACTUATOR AND THE "VIRTUAL SHAFT" HYPOTHESIS

M.A.Venegas[*]

[*]Nano Science and Technology Institute affiliation.
[**]Texas Instruments de México, Sensors and Controls Division, Av. Aguascalientes Sur 401, Ex-Ejido Salto de Ojocaliente 20270, Aguascalientes, Ags. México, m-venegas1@ti.com
[***]Instituto Tecnológico de San Luis Potosí Campus Aguascalientes, Mechanical Engineering Department., Ags. México

ABSTRACT

Motors are one of the most used motion devices in so many places, with high impact in the industry; they provide rotary motions that generate useful work at almost any particular application. Most of the time, motors must need to be connected to the entire mechanical system through a transmission mechanism to give the optimal performance to the system, they tend do be generic Moto–Reductors or Speed-Increasers that can be used in certain applications in the macro world. This paper idealizes this concept at the nano world, with a different type of motor, a motor that has been created by Nature evolution and seems to be the tiniest motor known [16] with high efficiency performance [1][4], the F1-ATPase motor; and then, achieve some process or activity. We are conceptualizing a general idea for a prototype of a hybrid Bio-Moto-Reductor actuator using the F1-ATPase Motor Bio-Device as a main motion rotary producer. We also idealize the hypothesis of a new and novel concept for rotary motion transmissions, the "Virtual Shaft".

Keywords: Motor Reductor, Speed Increaser, F1-ATPase motor, Transmission.

1 INTRODUCTION

Mechanical engineering has created a solid theoretical and practical knowledge in transmissions and Power Adaptation for coupling motors to an entire mechanical system [6]. A very common device used in the industry is the Motor – Reductor, which sometimes is based on a Worm – Crown Gear type Transmission gearbox. These type of Motor – Reductors drastically increase the torque at the final shaft, and at lower speed, they also can be accurate for a certain application. However sometimes our needs will require speed instead of torque at the final shaft, from where a transmission of the type Spur Gears or Helical Gears can satisfy these needs. On the Other hand, the great advances achieved in Molecular Biology, have enabled us to explore more in detail the mechanical – chemical phenomena occurred in the biology systems down to the nano and meso scales world[1][2][8][14][17]. They seem to be Mechanical-Chemical models that convert Chemical energy into Mechanical useful work. Nature evolution has created molecular machines with high efficiency performance that has been the main focus of many studies.

The great advances in nanofabrication Technology and Bio- Chemical interfacing methods, [2][3][4][16], now enable us to "weld" or assemble organic subsystems with inorganic structures. All this knowledge converges in an interesting and new field that will allow for next generations of Machines and devices, "The Hybrid – Machines" that will bring an infinity of new fields of research and practical applications.

2 A THREE PHASE BIO-MOTOR MODEL

According to the Walker Structure and the research and experimentation done by Oster & Wang and Kinosita's Lab., the F1-ATPase motor sub-complex consists of three main subunits as minimal elements that work like a rotary and stator: α_3, β_3, and γ[1][4][7][10][11][12]. The rotation of the rotor, the γ subunit will be by done by the consecutives power strokes done at the βs subunits; when the ATP molecule binds to the Catalytic site near the βs subunits, producing a 30° bend on each β, and turning 120° the γ subunit rotor shaft at an acceptance concentration of ATP. There are several approaches that address the occupancy of the catalytic site in a uni-site or multi-site form. It is very important to conceptualize a three phase bio-motor device that can be controlled to turn in CW and CCW direction. Considering the switch-less F1 model proposed by Kinosita Jr.[4]. This paper proposes a three phase bio motor idea at first glance, figure 1.

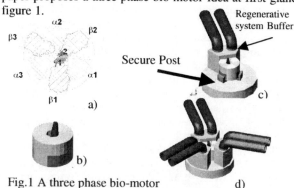

Fig.1 A three phase bio-motor conceptualization

In Figure 1, we conceptualize a three phase biomotor device, which has 3 Buffers filled with regenerative solution in order to avoid MgATP inhibition [14][18][19], and two input ports at every buffer from where the ATP nutrient is injected to each β subunit. Now we can think in a sequence of ATP injection as many authors have proposed [4], and depending on the sequence we may be able to turn either CW or CCW direction.

2 THE BIO-MOTO-REDUCTOR CONCEPTUALIZATION

The γ rotor shaft, has an asymmetric shape which turns in an asymmetric fashion, it is important to have a more concentric rotary motion before attaching it to the main shaft. An intermediate shaft, which consist of a conic Couple, a conic helical spring, the main base rotary assembly and a retainer, which helps to have either a concentric rotary motion but also to keep isolated the main cavity, filled with ultra clean water, and keeping the core structure protected from the environment is proposed. As shown in fig. 2

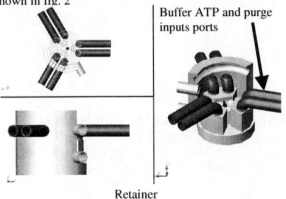

Buffer ATP and purge inputs ports

Retainer

Concentric rotary couple   Helical-Conical Spring

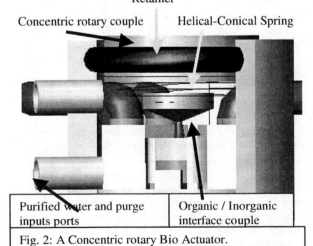

| Purified water and purge inputs ports | Organic / Inorganic interface couple |

Fig. 2: A Concentric rotary Bio Actuator.

Having a concentric rotary motion at the main structure assembled to the main gear transmission, a Worm gear type, with a mechanical base at the end and keeping the concentricity of the entire system. Fig. 3

Mechanical bushing

Worm gear   Mechanical Base

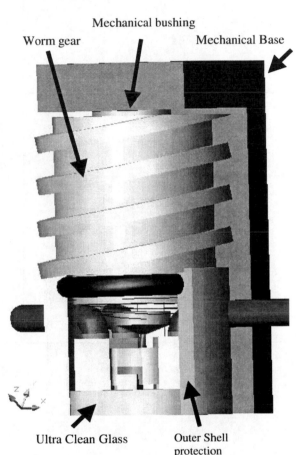

Ultra Clean Glass   Outer Shell protection

Fig. 3 The Worm gear assembly

These type of mechanical transmissions increase torque, but this conceptualization is restricted by so many factors, e.g. the total friction at the retainer and outer shell protection need to be lower than the permissible shear force of the γ rotor, but with a sufficient enclosure force to keep the core system hermetically assembled. Another important factor is the rigidity of the entire system, at this scale, matter seems to be soft, and a very mechanical rigid structure as well as a good assembly needs to be considered too. Other important parameters to be considered are: the Angular Speed, Angular displacement, torque and torsional moment at the γ rotor but also, this parameters differ at the main rotary concentric assembly, which will be the input of the Worm-Crown Shaft transmission and therefore, we must consider that this intermediate shaft, between the γ subunit and the Worm gear input will be affected by so many factors that will impact the rotary motion

The entire Bio-Moto-Reductor is shown in fig. 4.

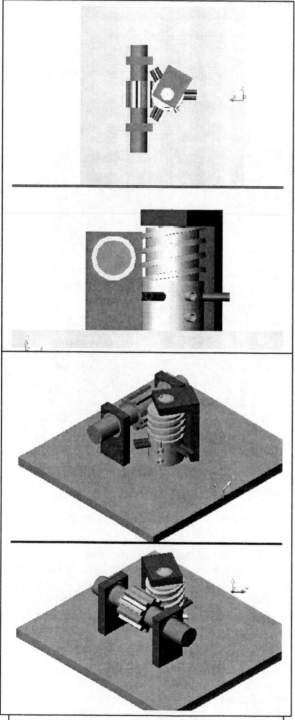

Fig. 4 The Hybrid Bio-Moto-Reductor Conceptualization.

The final shaft shown in figure 4 can also be connected to another type of reduction, we can also consider this shaft as input of a planetary gear box, and this will increase the torque.

Fig. 5 A planetary Gear Box at the final shaft.

3 THE HYPOTHESIS OF THE VIRTUAL SHAFT

The mechanical – bio-chemical model shown above is totally coupled in the entire transmission this means that there going to be so many restrictions that must be considered for a more formal design. A very important factor in Worm type transmissions is when changing the direction, this type of transmission tend to present a worm tied, and this is a very important feature that must be considered.

There are several types of transmission couples some of them are: Mechanical shaft transmission, Electrical Transmissions, hydrodynamic transmissions, etc. they are connected through mechanical systems, flow fluidics, or electrically, and so many other forms.

The virtual shaft consists in the use of the Magnetic force to transmit motion of a Magnet, increase speed and / or torque. This can give us two benefits: to amplify the γ subunit motion (either to increase speed or torque) but also to isolate the protein motor from the entire mechanical system. Fig 6 shows a general over view.

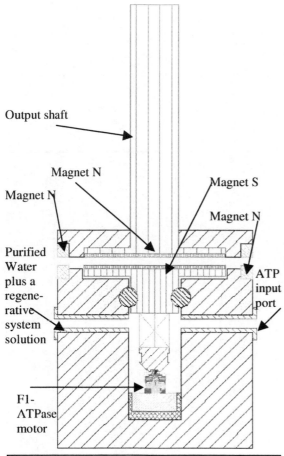

Output shaft

Magnet N

Magnet N

Magnet S

Magnet N

Purified
Water
plus a
regene-
rative
system
solution

ATP
input
port

F1-
ATPase
motor

Fig. 6 the Virtual Shaft powered with the F1-ATPase motor sketch.

3 CONCLUSIONS

Only a general overview of the hypothesis is presented. This research is only at the beginning steps and conceptualization ideas. The systems presented above require a lot of analysis and calculations for a more formal design, many restrictions and considerations must be taken into account. The research is however already started and the journey is underway.

4 REFERENCES

[1] George Oster and Hongyun Wang, "Reverse engineering a protein: the Mechanochemistry of ATP synthase", Biochimica et Biophysica Acta 1458, 29, 2000.

[2] Carlo Montemagno and George Bachand "Constructing nanomechanical devices powered by biomolecular motors", Nanotechnology, 10, 7, 1999.

[3] Ricky K. Soong,1,2 George D. Bachand,1,2 Hercules P. Neves,1,2 Anatoli G. Olkhovets,1,3 Harold G. Craighead,1,3 Carlo D. Montemagno1,2* "Powering an Inorganic Nanodevice with a Biomolecular Motor", Science, Vol 290, 4, 2000.

[4] Kazuhiko Kinosita Jr, Ryohei Yasuda, Hiroyuki Noji and Kengo Adachi, "A rotary molecular motor that can work at near 100% efficiency", The Royal Society, 355, 17, 2000.

[5] Raymundo Fernández Moreno, "Diseño de Maquinas I y II", Universidad Autónoma de San Luis Potosí, Facultad de Ingeniería, Área de Ingeniería Mecánica Eléctrica, Vol. I, II, 192, 2002, 2003.

[6] Raymundo Fernández Moreno, "Transmisión y adecuación de potencia", Universidad Autónoma de San Luis Potosí, Facultad de Ingeniería, Área de Ingeniería Mecánica Eléctrica, Vol. I, 170, 2000.

[7] George Oster and Hongyun Wang, "How Protein Motors Convert Chemical Energy into Mechanical Work", 22 pages.

[8] Carlos Bustamante, David Keller, And George Oster "The Physics of Molecular Motors" ACC Chem. Res., 34, 8, 2001.

[9] Kazuhiko Kinosita, Jr. "Single-Molecule Physiology ", Center for Integrative Bioscience, Okazaki National Research Institutes Higashiyama 5-1, Myodaiji, Okazaki 444-8585, Japan.

[10] Timothy Elston, Hongyun Wang & George Oster, "Energy transduction in ATP synthase", Nature, Vol. 391, 4, 1998.

[11] George Oster and Hongyun Wang, "Why is the mechanical efficiency of F1 ATPase so high? ", Departments of Molecular & Cellular Biology and ESPM, University of California, Berkeley, and School of Engineering, Dept. of Applied Mathematics & Statistics, University of California, Santa Cruz, California 95064, 26 pages, 2000.

[12] Kazuhiko Kinosita Jr., Ryohei Yasuda & Hiroyuki Noji "F1-ATPase: a highly efficient rotary ATP machine" Keio University, Essays in Biochemistry volume 35, 16, 2000.

[13] Carlo Montemagno, "Artificial Life and Nanobiotechnology; Foundations for Autonomous Robotic Systems" Presentation of the Workshop on Revolutionary Aerospace Systems Concepts for Human/Robotic Exploration of the Solar System Langley, VA November 6-7, 2001.

[14] Tomoko Masaike, Eiro Muneyuki, Hiroyuki Noji, Kazuhiko Kinosita Jr.,and Masasuke Yoshida, "F1-ATPase Changes Its Conformations upon Phosphate Release" THE JOURNAL OF BIOLOGICAL CHEMISTRY, Vol. 277, 7, 2002.

[15] Kengo Adachi, Ryohei Yasuda, Hiroyuki Noji, Hiroyasu Itoh, Yoshie Harada, Masasuke Yoshida, and Kazuhiko Kinosita Jr., "Stepping rotation of F1-ATPase visualized through angle-resolved single-fluorophore imaging" Department of Physics, Faculty of Science, Kanazawa University, Japan, vol. 97, 5, 2000

[16] Hiroyuki Noji, Ryohei Yasuda, Masasuke Yoshida, and Kazuhiko Kinosita Jr, "Ni-NTA immobilization of F1-ATPase — the World's mallest rotary motor, Qiagen News, 3, 3, 1998.

[17] George Oster and Hongyun Wang "ATP Synthase: two rotary molecular motors working together" University of California, Berkeley. Encyclopedia of Molecular Biology, Vol. 4, 19

[18] Yoko Hirono-Hara, Hiroyuki Noji, Masaya Nishiura, Eiro Muneyuki, Kiyotaka Y. Hara, Ryohei Yasuda, Kazuhiko Kinosita Jr., and Masasuke Yoshida, "Pause and rotation of F1-ATPase during catalysis", PNAS, Vol. 98, 6, 2001

[19] Ryohei Yasuda, Hiroyuki Noji, Masasuke Yoshida, Kazuhiko Kinosita Jr. & Hiroyasu Itohß, "Resolution of distinct rotational substeps by submillisecond kinetic", Nature, Vol. 410, 7, 2001.

[1] I would like to thank in a very special manner Dr. Christian Joachim, Head of the NanoScience Group CEMES/CNRS, Toulouse Cedex France; for his help, great contributions and orientation.

[2] Thanks to Dr. Masasuke Yoshida, Dr. Kazuhiko Kinosita Jr., Dr. Carlo Montemagno and Dr. George Oster for their permission to access the information used, product of their amazing and radically novel research in this extraordinary new field.

[3] **Alex Porter,** Texas Instruments SC Marketing, Thank you very much for spend some time in reading, reviewing and correcting this document.

[4] Thanks to Roberto Contreras Lizárraga, Head the Mechanization department at Texas Instruments de México, for his grate contribution and support.

[5] Thanks to John Birtwell, Head the Manufacturing Administration at TI S&C Division, for his help in the redaction of this document.

[6] Thanks to Dr. Sergio Rodríguez Quiñones and Dr. Raymundo Fernández Moreno, professors of the San Luis Potosí Technology Institute, Mechanical Department, for his valuable point of view and great orientation.

[7] Thanks to Engineer Carlos Javier Villagomez for his motivation and good advice.

[8] Thanks to Mr. Dennis Buss, SITD ADMIN of Texas Instruments, for sponsoring the Nanotech2004 congress and opening a new opportunity.

[9] Thanks to Texas Instruments for all the support.

[10] This work is dedicated to an incredible person that is not any more with us, **Laura Delgado Mata**, *"Rest peacefully"*.

The dielectric boundary force in ion channels: the case of gramicidin

B. Nadler*, U. Hollerbach** and B. Eisenberg**

* Department of Mathematics, Yale University, 10 Hillhouse Ave,
P.O. Box 208283, New Haven, CT 06520-8283, boaz.nadler@yale.edu
** Department of Biophysics and Physiology, Rush Medical Center, Chicago, IL, 60612

ABSTRACT

The dielectric boundary force (DBF) is the force that induced charges at dielectric boundaries exert on the charge that induced them. The DBF is present in any electrostatic problem with a non-uniform dielectric constant, and is important in the case of charged particles near dielectric boundaries. Yet, it is absent in common continuum theories of charge distribution and of charge transport, such as Poisson-Boltzmann and Poisson-Nernst-Planck. In this paper we define the DBF, formulate it as a solution of a partial differential equation, and present its crucial importance for ionic permeation through the gramicidin channel. The main permeation characteristics of gramicidin can be explained in terms of a delicate balance between the DBF and the force formed by the permanent charges of the gramicidin polypeptide. This balance, characteristic of a device, might suggest design principles for man-made nanopores.

Keywords: Dielectric forces, gramicidin, ion channels, continuum theories.

1 INTRODUCTION

Permeation through ion channels is one of the fundamental processes of life, and its understanding is one of the key problems in molecular biophysics and physiology of the cell [1]. At the crudest level of description, ionic permeation through a single protein channel embedded in a lipid membrane is an *electrostatic* problem, governed by the geometry of the pore, its permanent charge distribution and by the interactions of the mobile ions with the other mobile charges in the system.

One of the important features of this problem is the high spatial inhomogeneity of the dielectric coefficient. For an ion to pass through the channel, it has to move from an aqueous solution of high dielectric coefficient ($\varepsilon \approx 80$), through the narrow pore of the channel inside the membrane, which has a low dielectric coefficient ($\varepsilon \approx 2 - 5$).

In any electrostatic problem with a non-uniform dielectric constant, a charge q in one dielectric region induces surface charges at all boundaries separating regions of different dielectric values. These induced surface charges, in turn, exert an often neglected force on the charge q that induced them. Since most protein channels have a narrow pore or selectivity filter, this dielectric boundary force (DBF) is an important component in the overall forces acting on the mobile permeating ions, in addition to the familiar Coulombic interactions of the ions with all other fixed and mobile charges in the system.

In recent years, various continuum theories of charge distribution and of charge transport, such as Poisson-Boltzmann and Poisson-Nernst-Planck (PNP) have been applied to protein channels [2]–[9]. However, as shown both in simulations [8]–[13], and in theory [14], the DBF is not captured well in these theories. In [14], [15], we showed that a continuum description of charge transport in such nanoscale systems must include both the DBF and the excluded volume effects due to the finite size of ions. In [14], starting from a molecular model of interacting diffusing particles, an exact mathematical averaging procedure is presented that leads to a modified PNP type system of equations, explicitly containing the DBF and the finite size of ions.

In this paper we define the DBF and formulate it in a form useful for continuum theories, e.g., as a solution of a partial differential equation. We then present its crucial importance for ionic permeation through the gramicidin channel, widely studied in channology [16]. Analysis of the DBF provides a simple explanation of the main permeation characteristics of gramicidin in terms of a delicate balance between the DBF and the fixed charge force (FCF) formed by the permanent charges of the gramicidin polypeptide. For a monovalent positive ion, the DBF and the FCF almost cancel out, thus enabling permeation through the narrow pore of the gramicidin channel. For a negative ion or for a double charged positive ion, the two forces do not cancel out, yielding high insurmountable barriers that prevent these ions from crossing the channel. The remarkable fact that gramicidin, although overall neutral, produces an FCF that cancels out the DBF is characteristic of a device. This natural device might suggest design principles for man-made nanopores.

2 DIELECTRIC BOUNDARY FORCE

Consider a physical system consisting of charged particles in a non-homogeneous dielectric medium composed of an arbitrary number of regions of arbitrary shapes, denoted Ω_i. Further assume that in each region Ω_i the dielectric coefficient is constant with value ε_i. Now consider the electrostatic force on a single charge q_1 located, for example, at r_1 inside the region Ω_1. One component of this force includes all Coulombic interactions of q_1 with all other charges in the system. This force component, denoted F_c, is given by $-q_1 \nabla \phi_c(r_1)$, where $\phi_c(r)$ is the potential at r created by all *other* charges in the system, except q_1. It is the solution of Poisson's equation

$$\nabla \cdot [\varepsilon(r)\phi_c(r)] = -\frac{1}{\varepsilon_0} \sum_{j \neq 1} q_j \delta(r - r_j) \qquad (1)$$

with the standard jump conditions at dielectric boundaries,

$$[\varepsilon(r)\nabla\phi_c(r) \cdot n]\Big|_{\partial\Omega_i} = 0, \qquad (2)$$

where $\varepsilon(r)$ is the relative dielectric coefficient at r, n is a unit vector in the outer normal direction to a surface element on $\partial\Omega_i$, and square brackets denote the difference in the variable enclosed within them, between the value outside the region Ω_i and inside it.

Note that the potential ϕ_c does not include the charge q_1 explicitly, but rather treats it as an imaginary test charge. The physical presence of the charge q_1, however, induces surface charges at all boundaries between regions of different polarizability. Assuming a linear response of the dielectric mediums, this leads to additional induced surface charges, other than those induced surface charges produced by the other charges in the system, which are taken into account by the potential ϕ_c, through eqs. (1) and (2) .

While most standard textbooks on electrostatics explicitly write formulas for the induced surface charges due to a single charge, as gradients of the electrostatic potential, an important point usually not discussed explicitly, is that these induced surface charges, in turn, exert a force on the single charge itself. We denote this force by F_d, and refer to it as the *dielectric boundary force* (DBF). In an electrostatic problem with a non-homogeneous dielectric coefficient, the total electrostatic force on the charge q_1 is given by

$$F = F_c(r_1) + F_d(r_1)$$

while in a homogeneous system $F_d = 0$.

In [17] some properties concerning the DBF are proven.

Property 1: The DBF on a point charge q can be computed from the potential ϕ_0 created by the charge, by the following formula,

$$F_d(r_1) = -q \nabla_r \left(\phi_0(r) - \frac{q}{4\pi\varepsilon_0\varepsilon_1|r - r_1|}\right)\Big|_{r=r_1} \qquad (3)$$

In other words, the force acting on the point charge can be computed by subtracting from the electric potential produced by the point charge the singular homogeneous Coulombic term, and then computing the gradient of the resulting smooth potential at the charge location.

Property 2: The DBF is size-independent. The dielectric boundary force on a point charge of strength q is equal to the DBF on a uniform sphere of same strength and arbitrary radius a (as long as a is less than the distance from r_1 to the boundary $\partial\Omega_1$). Moreover, for a charged sphere, the DBF can simply be computed as

$$F_d(r_1) = -q \nabla \phi_a(r)\Big|_{r=r_1}. \qquad (4)$$

where ϕ_a is the electric potential created by a sphere of radius a. The potential of a uniform sphere is smooth everywhere, so there is no need to subtract a singular term as in the case of a point charge, eq. (3).

Property 3: The DBF is proportional to q^2. It has the same direction and magnitude for a positive or negative charge of same strength $\pm q$.

3 THE DBF IN GRAMICIDIN

We now consider the dielectric boundary force inside a gramicidin type channel geometry embedded in a membrane. Gramicidin is a small polypeptide (nearly a protein) widely used as a model of more complex natural channels [16]. Despite the absence of the DBF in continuum theories such as PNP, these theories have been used to study ionic permeation through gramicidin [4], [6]. Various authors noticed this error in the context of simulations [5], [11], [13]. The need for the inclusion of the DBF in a continuum description, derived from an underlying molecular model has been shown in a mathematical rigorous manner by Schuss & al. [14]. Recently various groups have studied the effects of the inclusion of the DBF in modified Poisson-Boltzmann and Poisson-Nernst-Planck equations applied to gramicidin and other channel-like geometries [7]–[9], [12], reporting that explicit inclusion of the dielectric boundary force in a continuum formulation yields better results than the standard theories that omit this force term.

In this paper we do not attempt to compute currents through gramicidin, but rather only present some interesting observations concenrning the role of the DBF in this channel. For the sake of our analysis, we assume that gramicidin is embedded in a lipid membrane with a uniform low dielectric constant of value $\varepsilon = 2$, while the pore of the channel and the surrounding aqueous baths have dielectric constant $\varepsilon = 80$. Since gramicidin is a long and narrow channel, we present the various forces only along the one dimensional channel axis, where the mobile ions are most likely to be. All numerical computations were performed with the "gramicidin" model described in [18].

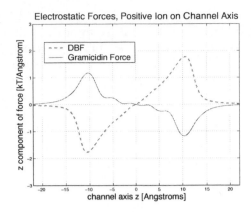

Figure 1: Gramicidin FCF (blue) and DBF (red) on a monovalent positive ion.

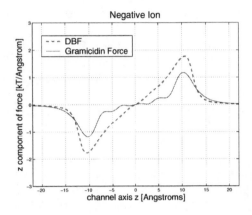

Figure 2: FCF and DBF on a monovalent negative ion.

In Fig. 1 the red dashed curve is the dielectric boundary force on a positive (or negative) ion as a function of position along the channel axis. Since the narrow pore is embedded in a region of low dielectric coefficient, surrounded by two aqueous solutions of high dielectric coefficient, the DBF is repulsive, and it creates a high potential barrier of more than $12kT$. Therefore, a simple hole with the same geometry as the gramicidin channel, i.e. length of about $25\mathring{A}$ and diameter of about $4\mathring{A}$, but with no protein with fixed charges in it, would be *impermeable* to passage of single positive ions. Since the dielectric boundary force is proportional to q^2 (see Section 2), such a non-charged pore opening in the membrane would be impermeable also to a negative ion or a double charged ion. Note that this analysis applies only to the movement of a single ion, and not to the coupled motion of a pair of say anion-cation, and also neglects the possible shielding of this force by mobile ions in the surrounding electrolytic solutions.

The gramicidin channel, however, differs from an idealized non-charged pore. Although gramicidin is overall neutral, there are non-vanishing partial charges along its atom groups, that create a non-vanishing electrostatic

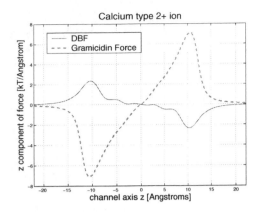

Figure 3: FCF and DBF on a Calcium type ion.

potential. The fixed charge force (FCF) on a positive ion, due to the gramicidin fixed charges is shown as the blue curve in Fig. 1. For a positive ion, the DBF and the FCF are nearly opposite, yielding a much smaller net force and a decreased potential barrier of about $5kT$.

For a negative ion, however, the situation is quite different. While the dielectric boundary force remains the same, the force due to the gramicidin channel is *inverted* with respect to the case of a positive ion because it is proportional to q. Now, the two forces do not cancel each other, but rather add up to produce a high insurmountable barrier of more than $20kT$ (see Fig. 2).

Finally, consider the case of a double charged ion, such as Ca^{2+}. Since the FCF is proportional to q while the DBF is proportional to q^2, the first is multiplied by two while the latter by four, in comparison to the case of a positive ion. The dielectric boundary force dominates, and it is not cancelled by the interaction with the gramicidin fixed charges (Fig. 3). This leads to a potential barrier more than $30kT$ high.

The cation selectivity of gramicidin can now be explained by a simple continuum type analysis of the balance between the different forces acting on an ion inside a rigid channel. Recently, similar results have been independently obtained by Edwards & al. [19], who also state that such a simplistic approach is not valid for quantitative results, such as computation of the net current through the channel. We stress that indeed, the development of a quantitative theory requires the computation of diffusion, friction and dielectric coefficients inside and near the channel, using more refined theories or molecular dynamics simulations that provide estimates of these parameters. In addition, the assumption that the channel is rigid needs to be reconsidered as well. For a study of the effective potential profile in a non-rigid gramicidin channel, see for example [7], [20].

It is instructive to have another look at the striking near-cancellation between the dielectric boundary force and the electrostatic interactions with the gramicidin fixed charges, for a positive ion (Fig. 1). It is our claim that the fact that these two forces nearly cancel each other cannot be purely coincidental. While the dielectric boundary force is a property of the geometry and dielectric coefficients of the problem, independent of the fixed charges of the protein, the electrostatic potential of the protein depends directly on its fixed charge distribution. The fact that the dielectric boundary force and the gramicidin force due to its fixed charges have extremal points (maxima and minima) at almost the same locations with almost the same heights (see Fig. 2), is characteristic of a *device*, designed or evolved to "have a purpose". It seems that the fixed charges of gramicidin have been *optimized*, by the course of evolution, to almost cancel out the dielectric boundary force, and thus allow the permeation of monovalent positive ions through the channel.

4 DISCUSSION

In an inhomogeneous system, charges *always* interact with (induced) dielectric boundary charges and with fixed charges if these are present near the boundary. In such cases, the induced surface charges and the corresponding DBF must be explicitly included in the computations.

In this paper we presented a simple example of the importance of the inclusion of the DBF in such computations, in the context of the gramicidin channel. We confined our analysis to the computation of the net force on a single mobile charge, due only to the dielectric boundary force and to the fixed charges of the protein, neglecting the effects of other mobile charges either inside the channel or in the surrounding electrolyte solutions.

Even though approximate and limited by our simplifying assumptions, our results show a striking cancellation between the DBF and the force due to the fixed charges of the gramicidin channel. This kind of cancellation is characteristic of a *device*, in which the free parameters, e.g., the fixed charges of the protein in our case, have been optimized to perform a certain function. This working hypothesis and the role of the dielectric boundary force should be investigated in other proteins as well.

While analysis of the DBF can explain why gramicidin is permeable mostly to monovalent positive ions, the DBF is not directly responsible for selectivity amongst different such ions, because the DBF on a spherical ion is independent of the radius of the ion [17].

In our analysis, we considered the force on a single ion, neglecting its interactions with the other mobile ions in the system. Obviously, in multi-particle systems, one has to consider the overall effect of all other mobile charges. A correct computation of these interactions is also required in continuum theories that attempt to compute currents, such as PNP. Standard PNP replaces discrete charges by continuum distributions composed of infinitely small charges. Thus, both the dielectric boundary force and the finite size of the ions are lost in this description. As shown in [14], a Brownian (Langevin) model for the motion of the mobile ions is equivalent to a hierarchy of Poisson-Nernst-Planck type equations containing conditional and unconditional densities, which explicitly contain the dielectric boundary force. Thus, to pursue further the analysis of shielding and the role of the dielectric boundary force from a theoretical approach, closure relations that are valid near dielectric interfaces need to be developed and checked against simulations.

REFERENCES

[1] Hille, B., *Ionic Channels of Excitable Membranes*, 2nd edition, Sinauer, 1992.

[2] Chen, D.P., Eisenberg, R.S., *Biophys. J.* 64:1405-1421, 1993.

[3] Chen, D.P., Lear, J., Eisenberg, R.S., *Biophys. J.* 72:97-116, 1997.

[4] A. Cardenas, R. Coalson and M. Kurnikova, *Biophys. J.* 79(1):80-93, 2000.

[5] M. Kurnikova, R. Coalson, P.Graf, and A.Nitzan, *Biophys. J.* 76:642-656, 1999.

[6] Hollerbach,U., Chen, D.P., Busath, D., Eisenberg, R.S., *Langmuir*, 16(13):5509-5514, 2000.

[7] A. Mamonov, R. Coalson, A. Nitzan and M. Kurnikova, *Biophys. J.*, 84:3646-3661, 2003.

[8] W. Im and B. Roux, *J. Mol. Biol.* 319(5):1177-97, 2002.

[9] W. Im, and B. Roux, *J. Mol. Biol.* 322(4):851-69, 2002.

[10] G. Moy, B. Corry, S. Kuyucak and S. Chung, *Biophys. J.* 78:2349-2363, 2000.

[11] B. Corry, S. Kuyucak and S. Chung, *Biophys. J.* 78:2364-2381, 2000.

[12] B. Corry, S. Kuyucak and S. Chung, *Biophys. J.*, 84:3594-3606, 2003.

[13] P. Graf P, A. Nitzan, M. Kurnikova and R. Coalson, *J. Phys. Chem. B*, 104:12324-12338 (2000).

[14] Z. Schuss, B. Nadler and R.S. Eisenberg, *Phys. Rev. E.* 64(3), 036116, 2001.

[15] B. Nadler, Z. Schuss, A. Singer and R.S. Eisenberg, *Nanotech 2003 technical conference proceedings,* Volume 3, pages 439 - 442.

[16] Wallace, B. A., (editor), *Gramicidin and related ion channel forming peptides*, Wiley, New York, 1999.

[17] B. Nadler, U. Hollerbach and R.S. Eisenberg, *Phys. Rev. E.* 68, 021905, 2003.

[18] Elber, R., D. Chen, D. Rojewska, and R. S. Eisenberg, *Biophys. J.*, 68:906-924, 1995.

[19] Edwards S., Corry B., Kuyucak S., Chung S., *Biophys. J.* 83:1348-1360, (2002).

[20] T. Allen, O. Andersen and B. Roux, *J. Am. Chem. Soc.* 125(32):9868-77, 2003.

Brownian Dynamics Simulation of Transport Properties in Potassium Ion Channels

S. Aboud*, D. Marreiro**, M. Saraniti** and R. Eisenberg*

*Molecular Biophysics Department,Rush University,Chicago, IL, USA
**Electrical and Computer Engineering Department,
Illinois institute of Technology, Chicago, IL, USA

ABSTRACT

In this work, a self-consistent Langevin dynamics-Poisson solver (LDPS) is used to model the ionic transport properties in ion channel systems. The charge transport behavior in the bulk electrolyte has been compared with the results of an analytic model. Within the LDPS framework, the molecular structure of both the protein and the lipid bilayer are explicitly accounted for, and the atomistic representation of the channel is embedded in the molecular structure of a patch of membrane, included in the computational domain.

Keywords: Brownian Dynamics, Langevin, Poisson equation, potassium channels, electrolyte solution

1 Introduction

Potassium (K) channels are a large class of transmembrane proteins that are probably present in all cells [1], and play a crucial role in stabilizing the membrane potential in excitable cells. They are characterized by an extreme selectivity (their permeability for K^+ is about 10^3 times larger than for the smaller Na^+ ions), and by a high diffusivity (comparable to bulk water).

The molecular structure of several K channels has been recently disclosed by means of X-ray spectroscopy. In particular, a 3.3 Å resolution mapping of the ligand-gated bacterial MthK channel structure, which opens in response to intracellular Ca^{2+}, has been successively disclosed by [2].

In this work, a Brownian dynamics simulator based on the self-consistent coupling of the Langevin equation with the P^3M force field scheme [3] [4], is used to model ionic transport. Within this approach, the molecular structure of both the protein and lipid bilayer are explicitly accounted for. Indeed, an atomistic representation of the channel is embedded in the molecular structure of a patch of membrane, and is included in the computational domain as described in [5] [6].

In the following section of this paper, the implementation of the LDPS simulation tool is discussed. In particular, the discretization scheme chosen for the numeric solution of the Langevin equation is presented along with the impact of the integration timestep on the simulated ionic dynamics. A calibration procedure for the bulk electrolyte solution is then presented and comparisons are made with the solution of an analytic integral expression. The implementation of the molecular structure of the lipid membrane and potassium channel protein is then described and the electrostatic behavior of the channel pore is shown. A discussion is carried out of the impact of the short-range field schemes on the simulated ion dynamics. Finally, the limitations of the Brownian dynamics approach will be discussed within the ion channel simulation framework.

2 Langevin Dynamics-Poisson Solver

The full Langevin equation is,

$$m_i \frac{d\mathbf{v}_i(t)}{dt} = -m_i\gamma\mathbf{v}_i(t) + \mathbf{F}_i(\mathbf{r}_i(t)) + \mathbf{R}_i(t) \qquad (1)$$

where m_i is the mass of the ith particle, and $\mathbf{v}_i$ is its velocity at time t, γ is the friction coefficient, $\mathbf{F}_i$ is the force on the particle i due to the presence of all other particles in the system and any external boundary conditions, including dielectric boundaries, and $\mathbf{R}_i$ is a fluctuating force due to the molecular bombardment of water on the ions, and is treated as a Markovian random variable with zero mean.

A third order algorithm is used for the calculation of the Brownian particle trajectories [7]. Within this approach the force is expanded in a power series,

$$F(t) \sim F(t_n) + \dot{F}(t_n)(t - t_n), \qquad (2)$$

where $\dot{F}$ denotes the time derivative, and is substituted back into Eq. 1, resulting in the following solution of the Langevin equation,

$$v(t) = v(t_n)e^{-\gamma\Delta t} + (m\gamma)^{-1}F(t_n)(1 - e^{-\gamma\Delta t}) \qquad (3)$$

$$+ (m\gamma^2)^{-1}\dot{F}(t_n)(\gamma\Delta t - (1 - e^{-\gamma\Delta t})) \qquad (4)$$

$$+ (m)^{-1}e^{-\gamma\Delta t}\int_{t_n}^{t} e^{-\gamma(t'-t_n)}R(t')dt', \qquad (5)$$

where $\Delta t = t_n - t$. The ionic positions are then calculated with the expression,

$$x(t_{n+1}) = 2x(t_n) - x(t_{n-1})e^{-\gamma(t_n - t_{n-1})} \qquad (6)$$

$$+ \int_{t_n}^{t_{n+1}} v(t')dt' + e^{-\gamma(t_n - t_{n-1})}\int_{t_{n-1}}^{t_n} v(t')dt'. \qquad (7)$$

The result is a set of equations for the particle trajectories that are dependent on a bivariant gaussian distribution, and which reduce to the Verlet [8] algorithm in the limit that the friction coefficient goes to zero. Because of this integration approach the timestep is not limited by the velocity relaxation time (i.e. the reciprical of the friction coefficient), and a long timestep (~20 fs) can be used.

A Poisson P³M algorithm [4] is implemented to resolve the electrostatic force fields. The long range interaction, which includes the external boundary conditions and the dielectric boundary interfaces is resolved with a multigrid [9] Poisson solver [10]. A nearest grid point scheme [3] is used for the charge assignment and force interpolation. The short range interaction for close particles is modeled with a Coulomb term and a van der Waals function which is either a Lennard-Jones [11] or and inverse power expression [12]. The short range Lennard-Jones potential between two particles is,

$$u(r_{ij}) = 4\epsilon \left[\left(\frac{\sigma}{r_{ij}} \right)^{12} - \left(\frac{\sigma}{r_{ij}} \right)^{6} \right] + \frac{q_i q_j}{4\pi \epsilon_r r_{ij}} \quad (8)$$

where σ and ϵ represent the zero of the the potential and the depth of the potential well, respectively, q_i is the charge of the ith particle and ϵ_r is the bulk dielectric constant. Another form of the short-range potential which can be used is,

$$u(r_{ij}) = \frac{\beta_{ij} |q_i q_j|}{4\pi \epsilon r_{ij}(p+1)} \left(\frac{s_i + s_j}{r_{ij}} \right)^{p} + \frac{q_i q_j}{4\pi \epsilon_r r_{ij}}, \quad (9)$$

where β_{ij} is an adjustable parameter, s_i is the size of the ith particle, and p is the hardness of the particle. A comparison of these two potentials is shown in Fig. 1, for β=1.

3 Bulk Electrolyte Solution

Several measurements are made to calibrate the proposed Brownian simulation approach for bulk electrolyte solutions. The simulation domain used is a homogeneous 20x20x20 tensor-product grid with mesh spacing 0.5 nm in all three directions. In this work, Dirichlet boundary conditions have been placed on opposite planes of the 3D volume, while Neumann conditions are imposed on the other 4 planes. Ions are specularly reflected from the Neumann boundaries, while the Dirichlet "electrodes" are treated as open boundaries and ions which traverse these regions exit from the simulation. In order to maintain the ion concentration ions must therefore be injected into the simulated volume. The computational domain is assumed to represent a small portion of a quasi-infinite electrolyte bath and the total molar concentration in the Dirichelt "electrodes" is conserved at each time step. The velocity of the injected ions is

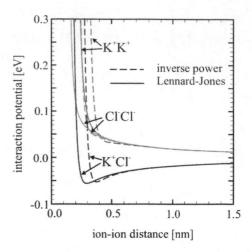

Figure 1: Comparison of short range Lennard-Jones and inverse power potential for K+ and Cl- in an aqueous solution.

calculated according to a Maxwell distribution in the directions parallel to the contacts and a half-Maxwellian in the normal direction.

As an initial test, the radial distribution function (RDF) of the electrolyte solution is calculated. The RDF is a measure of the liquid structure and is of central importance because it can be used to obtain a complete description of the thermodynamic state of the system [13]. To validate the approach, comparisons are made with the solution of the Ornstein-Zernike equation [13] which is combined with the closure relation given by the hypernetted chain approximation (HNC) [14]. The Ornstein-Zernike equation for a mixture is [13],

$$h_{ij}(r_{12}) = c_{ij}(r_{12}) + \sum_l \rho_l \int h_{il}(r_{13}) c_{lj}(r_{23}) dr_3, \quad (10)$$

where the RDF is $g_{ij}(r_{12}) = h_{ij}(r_{12}) + 1$, and h_{ij} is a measure of the total influence of particle 1 on particle 2 at a distance r_{12}. The direct correlation between particle 1 and particle 2 is expressed with $c_{ij}(r_{12})$ and ρ_l is the density of species l. To obtain the HNC closure relation, the direct correlation function is written as the difference between the total radial distribution and the radial distribution without any direct interactions included:

$$c(r) = e^{\beta w(r)} - e^{\beta[w(r)-u(r)]} \sim e^{\beta w(r)} - 1 + \beta[w(r)-u(r)], \quad (11)$$

where the total particle interaction is $w(r)$, $u(r)$ is the direct particle-particle interaction, and β is the thermal voltage $1/k_B T$. The radial distribution function is then calculated by iterating the following set of equations as suggested by [15],

$$\hat{y}(q) = \hat{c}(q)/(1 - \rho\hat{c}(q)) - \hat{c}(q) \qquad (12)$$

$$g(r) = exp[y(r) - u(r)] \qquad (13)$$

$$c(r) = g(r) - 1 - y(r) \qquad (14)$$

where $\hat{f}(q)$ denotes the three dimensional Fourier transform of $f(r)$ and

$$y(r) = e^{\beta u(r)} g(r). \qquad (15)$$

The RDF corresponding to aqueous KCl at a 150 mM concentration is shown in Fig. 2 for the analytic HNC approach and the LDPS. Both LDPS and HNC use the Lennard-Jones potential, with parameters taken from [16], and the integration timestep of the LDPS is 5 fs. Results show good agreement, although the peak is lower in the LDPS for the K^+Cl^- RDF. This may be due to spurious fluctuations of the ionic population present at low ion concentration. The impact of higher order charge assignment schemes in the Poisson solver is also being investigate.

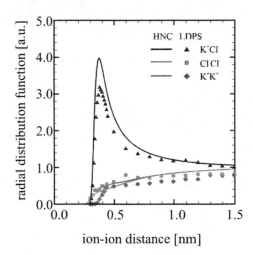

Figure 2: Radial distribution function in aqueous KCl at 298 K at a concentration of 150 mM calculated with HNC and LDPS

The equivalent conductivity for several aqueous KCl solutions under nonequilibrium conditions is also calculated with very good agreement with experimental values [6] [17], for several molar concentrations. The calculted conductivity tends to deviate from the experiemental values at molar concentrations higher than approximately 0.5 molar. This is due to an expected break-down in the validity of the primitive water model, occuring when the average distance between ions becomes comparable to the diameter of water.

4 Ion Channel Simulation

To include the channel protein and lipid membrane in the computational domain, the atomic coordinates and charge distributions are inserted explicitly using a combination of experimental data and simulation result, as explained in [6]. The atomic coordinates of the Mthk ion channel structure is obtained from X-ray diffraction experiments of crystallized proteins [2]. An energy minimization based approach [5] is applied to insert the channel into the interior of a phospholipid membrane, initially modelled with a molecular dynamics simulation [18]. A plot of the MthK K+ channel inserted into the lipid bilayer is shown in Fig. 3.

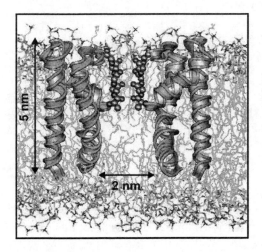

Figure 3: Two sides of the MthK K+ channel structure embedded in an explicit lipid bilayer.

Within the P^{3}M approach used for the force field calculation, the short-range component of the force, including both the Coulomb interaction and finite-size effects of the ions, is implemented by using a Lennard-Jones or an inverse power relation [3]. When applied to simple bulk solutions, this representation produces values for the equivalent conductivity that are in good agreement with experiment. However, the nature of the short-range interaction inside the ion channel is more complicated to model. As an initial probe into the electrostatic behavior of the channel pore, a test point charge is placed at various locations and the total force on the particle is calculated. This force is then integrated to obtain the potential profile along the channel. The potential profile is also calculted when a K+ ion is inserted in the center of the MthK channel, to investigate the effect of the short-range interaction. The short-range potential is given by the inverse power relation of Eq. 9. A comparison of the resulting potential profiles, both without the finite-size effects and with the short-range described by the inverse power relation are shown in Fig. 4. As can be seen, the inclusion of the finite size effect of the ion results in an increase of the electrostatic features due to the positions of the atoms in the selectivity filter. In addition the use of a non-homogeneous dielectric constant within the channel pore further alters the electrostatic profile.

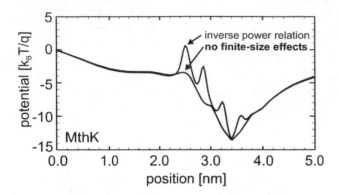

Figure 4: A comparison of the electrostatic potential acting on a K+ ion as a function of position along the center of the MthK channel for a short-range interaction that is based on an inverse power relation and when finite-size effects are neglected.

The atomic coordinates of both the protein and lipid bilayer have been treated as static quantities in this work. In reality, the motion can be significant and the result can drastically alter the shape of the potential, particularly in the selectivity filter. This motion must be accounted for in order to accurately describe the ionic transport. Further work needs to be done to investigate the influence of the motion of atoms on the electrostatic characteristics, as well as the correct form of the short-range electrostatic interaction within the cavity. Another important element of the simulation domain which must be included is a more accurate representation of the water inside the channel.

5 Conclusion and Discussion

The self-consistent LDPS has been used to model the transport behavior in ion channel systems. The first step is to calibrate the simulation approach for the bulk electrolyte solutions. An analytic integral equation describing the RDF was calclate with the HNC closure relation to validate the results of the LDPS in the equilibrium regime. The results show good agreement, although the LDPS peak is slightly lower. This may be due to fluctuations resulting from the sparsity of the ionic population.

To simulate the ion channel and lipid membrane system, their atomic coordinates and charge distribution have been explicately inserted in the compuational domain. Preliminary results of the distribution of the electrostatic potential along the channel show expected behavior, and demonstrate the sensitivity of the forces on both the short range interactions and the dielectric properties of the channel.

Acknowledgement

This work was supported in part by the grant T32 HL07692 of the National Institute of Health.

REFERENCES

[1] B. Hille, *Ionic Channels of Excitables Membranes*, Sinauer, Massachusetts, 1992.

[2] Y. Jiang, A. Lee, J. Chen, M. Cadene, B.T. Chait, and R. MacKinnon *Nature*, vol. 417, pp. 515–522, May 2002.

[3] R.W. Hockney and J.W. Eastwood, *Computer Simulation Using Particles*, Adam Hilger, Bristol, 1988.

[4] C.J. Wordelman and U. Ravaioli *IEEE Transaction on Electron Devices*, vol. 47, no. 2, pp. 410–416, Feb 2000.

[5] J.D.Faraldo-Gomez, G.R.Smith, and M. S. P. Sansom vol. 31, pp. 217–27, European Biophysics Journal.

[6] S. Aboud, M. Saraniti, and R. Eisenberg *Journal of Computational Electronics*, 2004, to be published.

[7] W.F. van Gunsteren and H.J.C. Berendsen *Molecular Physics*, vol. 45, no. 3, pp. 637–647, 1982.

[8] L. Verlet *Physical Review*, vol. 159, no. 1, pp. 159–103, July 1967.

[9] W. Hackbush, *Multi-Grid Methods and Applications*, Springer–Verlag, Berlin, 1985.

[10] S.J. Wigger, M. Saraniti, and S.M. Goodnick in *Proceedings of Second International Conference on Modeling and Simulation of Microsystems, MSM99*, Puerto Rico (PR), April 1999, pp. 415–418.

[11] R. Stephen Berry, Stuart Alan Rice, and John Ross, *Physical Chemistry*, Oxford University Press, II edition, May 2000.

[12] P. Turq, F. Lantelme, and H.L. Friedman *The Journal of Chemical Physics*, vol. 66, no. 7, pp. 3039–3044, April 1977.

[13] D.A. McQuarrie, *Statistical Mechanics*, University Science Books, Sausalito, CA, 2000.

[14] J. Barthel, H. Krienke, and W. Kunz, *Physical Chemistry of Electrolyte Solutions*, Springer, 1998.

[15] K.-C. Ng *Journal of Chemical Physics*, vol. 61, pp. 2680–2689, 1974.

[16] Wonpil Im, Stefan Seefeld, and Benoit Roux *Biophys. J.*, vol. 79, no. 2, pp. 788–801, 2000.

[17] V. M. M. Lobo, *Electrolyte solutions: literature data on thermodynamic and transport properties*, vol. 2, Coimbra, 1975.

[18] D. Peter Tieleman, Mark S. P. Sansom, and Herman J. C. Berendsen *Biophys. J.*, vol. 76, no. 1, pp. 40–49, 1999.

BioMOCA: a Transport Monte Carlo Approach to Ion Channel Simulation

T.A. van der Straaten, G. Kathawala, U. Ravaioli

Beckman Institute for Advanced Science and Technology,
University of Illinois, Urbana, IL 61801, USA
trudyv@uiuc.edu, gkathawa@uiuc.edu, ravaioli@uiuc.edu

ABSTRACT

Ion channels are highly charged proteins that reside in the membrane of all living cells to regulate charge transport to the cell. Most channels have device-like behaviors that are interesting to the electronics community for their possible application in bio-device design. Simulating ion channels over time-scales relevant to conduction present a difficult time and space multi-scale problem that is beyond the scope of the Molecular Dynamics approach. To address this problem we have developed BioMOCA, a 3-D reduced particle ion channel simulator based on the transport Monte Carlo methodology. BioMOCA has been described elsewhere; here we use it to study the link between the electrostatic properties of the protein and selectivity in the *gramicidin* channel. We also present representative simulations of Na^+ transport in *gramicidin*.

Keywords: ion channels, nanodevices, gramicidin, Monte Carlo simulations

1 INTRODUCTION

Ion channels are proteins that fold in such a way that they form natural conducting nanopores spanning the membranes of biological cells, thereby controlling the ion transport in and out of the cell. The side-chains of the amino acids are ionizable resulting in a strong permanent charge that varies sharply along the surface of the protein. Many channels can selectively transmit or block a particular ion species, and most have switching properties or perform specialized functions resembling those of complex electronic systems [1]. The device-like behaviors exhibited by ion channels intrigue the engineering community for many reasons. Ion channels have the distinct advantage of perfect structure duplication and self-assembly, while solid-state nanoscale device counterparts tend to be strongly affected by statistical fluctuations and defects. Channel mutation, which involves replacing or deleting one or more amino acids, allows the channel structure and charge distribution to be controlled in the laboratory with atomic resolution [2,3]. Engineering of natural channels, with specific conductance and selectivity properties, is thus conceivable.

Simulations can help provide a clearer understanding of channel operation. However, the simulation of ion channels embedded in a macroscopic membrane environment is a formidable multi-scale problem in both time and space.

Channel/membrane system dimensions range from ~10nm to ~1μm, while specific channel functions, such as switching or selectivity, are often controlled by sub-nm functional sub-units of the protein. Thus, at least part of the channel system needs to be represented with atomic resolution. In addition, ion traversal through the channel is a rare event, so ion motion must be resolved on a fs scale. On the other hand, reliable estimates of current require simulations lasting several ~μs. Realistic simulations are also difficult because at present the complete molecular structure and charge distribution is known from measurements for only a few channels. Even then, it is difficult to predict the structural and electrostatic changes that may occur when the channel is exposed to different physiological conditions.

With the advent of standardized software and massively parallel computing power, Molecular Dynamics (MD) [4] has become the most widely employed tool for studying ion dynamics in protein channels [5]. Although MD simulations can provide key information regarding the mechanisms of ion conduction in atomic detail, the resources required to compute steady-state channel currents directly using this approach well exceed the most powerful machines currently available. At the other end of the simulation hierarchy, continuum models based on drift-diffusion theory can be used to compute macroscopic currents quickly [6-8]. However, ion diffusivities and dielectric coefficients must be assigned with some care since the continuum paradigm deals with average densities rather than discrete particles of finite size, a premise that has been found to overestimate the ion concentration in very narrow channels.

In order to extend simulation times to ~μs, while still retaining some atomic level of detail, we have developed BioMOCA, a 3-D reduced particle ion channel simulator based on the Monte Carlo methodology developed for charge carrier transport in solid-state devices. Since BioMOCA has been described in detail elsewhere [9,10], we give only a brief overview. Water, protein and lipid membrane are treated as continuum dielectric media while ions are modeled as discrete finite-sized particles. Ion trajectories are traced in real space as sequences of free flights randomly interrupted by thermalizing scattering events that represent the interaction between ions and the surrounding channel/electrolyte environment. The duration of free flights is linked to the ion's diffusivity in the electrolyte. Ion size is modeled by dressing the ions with a pairwise Lennard-Jones potential and electrostatic forces are calculated self-consistently from the charged particles positions by solving Poisson's equation.

In the following section we discuss the electrostatic properties of ion channels, and the role this plays in the observed cation selectivity of the *gramicidin* (gA) channel. In section 3 we compare BioMOCA simulations of gramicidin with a 3-D continuum simulation. We also conducted BioMOCA simulations of the *porin* ion channel but because of space limitations these results are not discussed here.

2 GRAMICIDIN: PROPERTIES

Gramicidin A (gA) is a small channel-forming molecule consisting of 15 amino acids folded into a helical structure. When two gA molecules align end-to-end creating a dimer, a narrow channel (~25Å long, ~4Å in diameter) is formed spanning the membrane. Although it is charge-neutral overall, the charge distribution on the mouth of the gA channel and lining the pore imparts a signature double well electrostatic potential profile. gA only conducts small monovalent cations; anions are completely excluded from the channel

The issue of what permittivity values ε_p to assign to protein has been addressed in several recent articles (see [11] and references within). The protein environment can respond to an external field in several ways, each with its own relaxation time, e.g., field-induced dipoles, re-orientation of permanent dipoles and larger scale re-organization of ionized side-chains and water molecules, both within the interior and on the surface of the protein. In deciding what value to assign to ε_p one must consider how much physics is included explicitly. If all of the charges and dipoles are included explicitly, as in MD simulations, then clearly $\varepsilon = 1$ everywhere. If all but the field-induced atomic polarization is treated explicitly then $\varepsilon_p \sim 2$. However, when protein and water re-organization are included implicitly the dielectric coefficient is hard to define, particularly when ion motion takes place on the same time-scale as the protein's response to its presence. It has been suggested that for continuum models that seek to encapsulate the dielectric response of the water and protein side-chains with a dielectric coefficient, the value of ε_p should be significantly larger than 2 [11].

In gA the channel is also lined with dipoles formed by negatively charged carbonyls and positively charged amino groups, which hydrogen-bond to each other, helping to stabilize the helical structure. The dipoles are tilted with respected to the channel axis such that the carbonyls protrude slightly into the pore. MD simulations have shown that these carbonyls also form hydrogen bonds with water molecules that, due to the narrow width of the pore, must traverse the channel in single file [12]. As a result, water is highly ordered within the channel so the dielectric coefficient of the aqueous pore (ε_c) should really be much lower than the value of 80 used for water under bulk conditions. Nonetheless, in their attempts to explain gA's

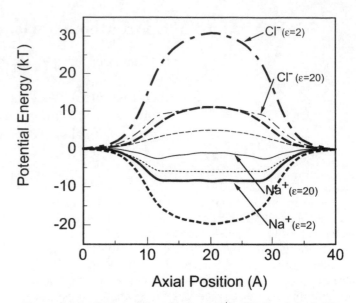

Figure 1 Net PE profiles seen by Na^+(solid lines) and Cl^- (dash-dot) ions traversing the center of gA. The individual contributions from the partial charges (short dash) and the dielectric barrier (long dash) are also shown. Thick lines indicate $\varepsilon_p = 2$, fine lines indicate $\varepsilon_p = 20$.

monovalent cation selectivity, several groups have assumed $\varepsilon_{ch} = 80$. For comparison, the results presented in this section were also obtained using $\varepsilon_{ch} = 80$.

Recently it has been proposed [13,14] that the mechanism allowing gA to conduct monovalent cations while blocking anions and divalent ions lies in the permanent charge being distributed over the protein in such a way that it creates a potential well for cations, that cancels the potential barrier created by the low dielectric ($\varepsilon_p = 2$) protein environment. Figure 1 shows the net potential energy (PE) profile seen by a single Na^+ ion traversing the center of a gA channel embedded in a low dielectric slab representative of the lipid membrane. The individual contributions from the partial charges and the dielectric barrier are also shown. These profiles were computed with BioMOCA, using the 1mag.pdb structure and the partial charges used by Edwards et al. [13] (PARAM22 version of CHARMM force field]). The dielectric coefficients of protein and lipid were set to 2, while the aqueous interior of the channel and electrolyte baths were given $\varepsilon_{water} = 80$. The large barrier $\sim 10 k_B T$ created by the protein dielectric is more than cancelled by the deep potential well $\sim 19 k_B T$ created by the permanent charge on the protein, allowing Na^+ conduction through the pore. In contrast, since the PE barrier caused by the dielectric is independent of the sign of the charge, and the permanent charges present an additional barrier to negative ions, the corresponding net PE profile for Cl^- shows a $\sim 30 k_B T$ barrier that would prevent Cl^- from entering the channel. Thus, if a low protein dielectric coefficient is used, the cancellation of permanent charge and dielectric barrier is a valid explanation for cation selectivity. However, as discussed above, 2 may not be a reasonable value for ε_p. Also shown in Figure 1 are the Na^+ and Cl^- profiles computed using $\varepsilon_p = 20$. In this case, the potential barrier due to the protein dielectric and the potential well due to the partial charges become much smaller. The cancellation of dielectric barrier and the potential well due to the partial charges still holds for Na^+ but the net barrier for Cl^- is now much lower (~$11k_BT$). Under an applied bias of 250 mV (~$10k_BT$) it may be possible for Cl^- to enter the channel. Thus it is not clear whether the explanation for gA cation selectivity in [13,14] holds unless a very low protein dielectric coefficient is assumed. Further, in the next section we show that alternative data for the permanent charge distribution can produce a sizeable potential barrier for the Na^+ that can still be overcome under an applied bias.

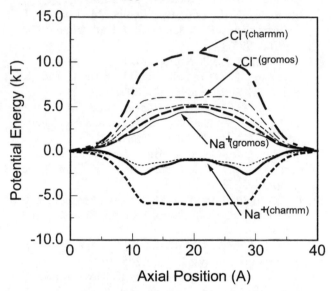

Figure 2 Net PE profiles seen by Na^+(solid lines) and Cl^- (dash-dot) ions traversing the center of gA. The individual contributions from the partial charges (short dash) and the dielectric barrier (long dash) are also shown. Thick lines indicate CHARMM partial charges, fine lines indicate GROMOS partial charges.

Although gA is charge neutral overall, the contribution of the permanent charge to the net potential energy profile is sensitive to the way in which this charged is distributed over the protein. There exist in the literature numerous different parameter sets for the distribution of charge on each amino acid. In Figure 2 we compare two different sets: the partial charges obtained with GROMOS [15] and with PARAM22/CHARMM. These two models present some difference in the charge distribution of the tryptophan side chains. We assume a dielectric coefficient of 20 for the protein, 2 for the lipid and 80 for the aqueous regions. The CHARMM partial charges result in a potential well almost 3 times deeper than that created by GROMOS. As a result, the net Na^+ PE profile for GROMOS is a barrier nearly $4k_BT$ high, while for CHARMM we have a double well of

about $3k_BT$. BioMOCA simulations show that under bias the channel described by GROMOS conducts Na^+ ions with a conductivity comparable to measurements. Indeed, the potential barrier for Cl^- is only ~$3k_BT$ higher and under an applied bias of ~250 mV one could expect Cl^- to also permeate the channel. In fact, our model of gA with $\varepsilon_p =20$ and GROMOS charges does indeed conduct Cl^- if the ionic radius is set equal to that of Na^+. Thus we purport that ion size also plays a crucial role in gA conductivity. Since the pore radius (~2Å) is comparable to the ionic radii ($r_+ = 0.95$ Å, $r_- = 1.8$ Å), the probability for Na^+ to enter the channel is very low, and effectively zero for Cl^- (in several longer simulations lasting ~μs no Cl^- ions were ever observed inside the channel).

3 GRAMICIDIN: SIMULATIONS

We now present two example simulations of the gA channel *in situ* in a neutral lipid membrane of thickness 22 Å, using the 1MAG.pdb structure and partial charges taken from the GROMOS force field. The lipid, protein and electrolyte regions were assigned values of $\varepsilon = 2$, 20 and 80, respectively. Planar contacts parallel to the X-Y plane are included at either end of the domain so that a fixed bias voltage can be applied across the system. Na^+ and Cl^- scattering rates were determined from published ion diffusivities in bulk electrolyte [16]. The scattering rates are high, $\lambda_+ = 8.1 \times 10^{13}$ s^{-1} and $\lambda_- = 5.3 \times 10^{13}$ s^{-1} necessitating small time steps of 10fs. However, since the ions diffuse very slowly, the charge distribution does not change appreciably during one time step. Benchmark simulations in bulk electrolyte have shown that for such high scattering rates the field can be updated as infrequently as every few ps without degrading the ion-ion pair correlation function [10]. Here we solve Poisson's equation every 100 time steps (1ps). Figure 3 shows the Na^+ and Cl^- concentrations and potential, averaged over a 200ns simulation of 1M NaCl, as a function of position Z between the contacts. Both contacts were held at ground potential (zero bias). The

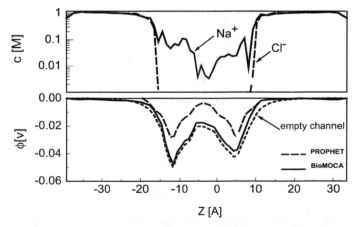

Figure 3. Time-averaged ion concentrations (top) and potential (bottom) as a function of z-position along the gA channel system, immersed in a 1M NaCl solution. No bias voltage was applied.

ion concentrations represent the average value in the X-Y plane, while the potential profile is taken along the center of the channel. The ion concentrations remain flat at ≈1M throughout the bath regions but inside the narrow pore region the Na^+ concentration drop by more than an order of magnitude while Cl^- is excluded altogether. The distribution of permanent fixed charge on the channel gives rise to an electrostatic potential distribution with two wells roughly $3k_BT/2$ deep, as shown by the solid line in the lower half of Figure 3. Also shown is the potential distribution in the absence of electrolyte (dotted line). The similarity between the two curves indicates that, at least over the period of observation, the fixed charge is not screened by the mobile ions. For comparison we have included the potential distribution calculated from steady-state continuum (drift-diffusion) theory on a finer (0.5 Å) mesh (dashed line). The assumption of average concentrations of point particles built into the continuum model allows a much higher concentration of ions of *both* species to enter the channel, resulting in partial screening of the fixed charge, as evidenced by the shallower potential wells.

The number of ions crossing the channel and the average Na^+ concentration inside the channel increases dramatically when a bias voltage is applied across the system. In Figure 4 we show the time-averaged Na^+ and Cl^- concentration and potential profiles with 250mV applied at the right contact. The simulation was run for 400 ns during which 10 Na^+ ions were detected crossing the channel from the right bath to the left. Though the applied field increases ion conduction, the average Na^+ concentration inside the channel is still an order of magnitude lower than in the baths − the channel presents a bottleneck to ion flow, causing almost all the applied potential to drop across the channel. In order to gauge the channel conductivity several channel simulations were performed with a 250mV bias and 1M NaCl in each bath. During the cumulative observation time of 26.3μs a total of 283 Na^+ ions crossed the channel, corresponding to a current of approximately

1.7 pA which compares well with single-channel current-voltage measurements [17].

ACKNOWLEDGMENTS

This work was partially supported by the National Science Foundation through the Network for Computational Nanotechnoogy Center, DARPA (contract SIMBIOSYS AF NA 0533) and the National Center for Supercomputing Applications. The authors are grateful to R. S. Eisenberg, B. Corry and E. Tajkhorshid for many useful discussions.

REFERENCES

[1] B. Hille, "Ionic Channels of Excitable Membranes" Sinauer Associates Inc., 1992.

[2] Saint, N., K.-L. Lou, C. Widmer, M. Luckey, T. Schirmer and J.P. Rosenbusch. J. Biol. Chem., **271**, 20676, 1996.

[3] Tajkhorshid, E., P. Nollert, M. Ø. Jensen, L. J. W. Miercke, J. O'Connell, R. M. Stroud, and K. Schulten. Science, **296**, 525, 2002.

[4] Ciccotti, G. and W.G. Hoover, eds. "Molecular-Dynamics Simulations of Statistical-Mechanical Systems", North Holland, 1986.

[5] B. Roux, Curr. Op. Struc. Biol.**12**, 182-189, 2002.

[6] Cardenas, A.E., R.D. Coalson and M.G. Kurnikova, Biophys. J., **79**, 80, 2000.

[7] T. A. van der Straaten, J. M. Tang, U. Ravaioli, R. S. Eisenberg and N. R. Aluru. J. Comp. Elec. **2**, 29-47, 2003.

[8] U. Hollerbach, D. Chen, W. Nonner and B. Eisenberg, Biophys. J., **76**, A205, 1999.

[9] T. A. van der Straaten, G. Kathawala and U. Ravaioli, J. Comp. Elec., in press, 2003.

[10] T.A. van der Straaten, G. Kathawala, Z. Kuang, D. Boda, D.P. Chen, U. Ravaioli, R.S. Eisenberg and D. Henderson, "Technical Proceedings of the 2003 Nanotechnology Conference and Trade Show", **3**, 447-451, 2003.

[11] C. N. Schutz and A. Warshel, Proteins: Structure, Function and Genetics, **44**,400, 2001.

[12] R. Pomes and B. Roux B, Biophys. J., **82**, 2304, 2002.

[13] S. Edwards, B. Corry, S. Kuyucak and S.-H. Chung, Biophys. J., **83**, 1348, 2002.

[14] B. Nadler, U. Hollerbach and R. S. Eisenberd, Phys. Rev. E., **68**, 21905, 2003.

[15] http://www.igc.ethz.ch/gromos/

[16] D.R. Lide, editor-in-chief, "CRC Handbook of Chemistry and Physics", CRC press, **5**-90, 1994.

[17] D. D. Busath, C. D. Thulin, R. W. Hendershot, L. R. Phillips, P.Maughan, C. D. Cole, N. C. Bingham, S. Morrison, L. C. Baird, R. J. Hendershot, M. Cotton, T. A. Cross, Biophys. J. , **75**, 2830, 1998.

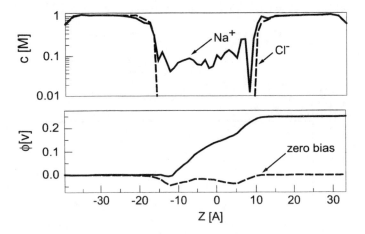

Figure 4. Time-averaged ion concentrations (top) and potential (bottom) as a function of z-position along the gA channel system, immersed in a 1M NaCl solution, under a bias voltage of 250 mV.

HYBRID MD-PNP SIMULATIONS OF THE α-HEMOLYSIN OPEN CHANNEL IONIC CURRENT

I. Cozmuta[*], J. T. O'Keeffe[**] and V. Stolc[**]

[*]Eloret[1] Corporation, NASA AMES Research Center,
Mail Stop 230-3, Moffett Field, CA 94035-1000, USA, ioana@nas.nasa.gov
[**]NASA AMES Research Center, Moffett Field, CA, USA,
jtok@nas.nasa.gov, vstolc@mail.arc.nasa.gov

ABSTRACT

Experiments have shown that single stranded nucleic acids polymers can be transported through the α-hemolysin ion channel under the action of an applied electric field. The translocation of the nucleic acid polymers causes transient blockade of the ionic current. The physical and chemical details of the interactions between polymer, channel and ionic solution that control the blockade events are not yet fully understood. To study such interactions, a hybrid MD-PNP model is proposed. The diffusion coefficient of the ions ($0.78 \cdot 10^{-9} m^2/s$) and pore wall charges are determined from MD simulations. These values are used in the PNP model to calculate open channel ionic currents. The calculated value for ionic current ($101\ pA$) is close to the experimental value ($120\ pA$) only when pore-wall charges are included in the model in addition to the pore geometry. This validates the present approach of bridging time scales by combining a microscopic and macroscopic model.

Keywords: ionic current, α-hemolysin, molecular dynamics, Poisson Nernst Planck theory, ion channel

1 INTRODUCTION

The study of biological systems at the molecular scale is useful for nanoscale engineering. One such example is the α-hemolysin protein (*33.2 kD*), which spontaneously self-assembles into a heptameric channel that inserts itself into cellular membranes causing leakage of ions and small organic molecules [1]. Recently, scientists used this protein as a model system for structural analysis of nucleic acids by deciphering ionic signature patterns [2-5].

Experimental studies show that single stranded nucleic acid polymers in extended linear configurations can be transported across the α-hemolysin pore under the action of an applied electric field. Translocation of such polymers causes partial blockades of ionic current through the channel. The level of fractional blockade differs for various nucleic acid homopolymers.

This result suggests the possibility for nanopore-based sequencing. However, the fast translocation rate (~*1-5 subunits per microsecond*) and small differences in levels of ionic current blockade presents a challenge in signal processing of the desired single nucleotide resolution. To design a nanopore that would achieve single nucleotide resolution, physical and chemical details of the interactions between polymer, channel and ionic solution need to be better understood.

Various computer simulations and theoretical models have been used to better understand structure-function relationships of ion channels. Continuum models such as the Poisson-Nernst-Planck and Eyring Rate Theory allow relatively quick computations of ionic fluxes and translocation times [6-9]. However, these continuum models do not account for fine structural details and depend on the choice of input values for the diffusion coefficients of the ions, system charges, energetic barriers and other parameters. On the other hand, it is presently not feasible with the present computation resources to reach time scales of physiological interest with full atomistic simulations[2].

For polymer translocation studies, the most important quantity to calculate and predict is the ionic current. Macroscopic scale events are influenced by the fine structural details and the time evolution of the model system (e.g. transport coefficients). The goal of the present work is to link atomistic (MD) and PNP simulations into a hybrid model to calculate the α-hemolysin open channel ionic current. An expression for the ionic flux at macro scale (electro-diffusion) is derived using the Poisson-Nernst-Planck theory that combines the solutions of both diffusion and Poisson-Boltzmann equations. Mezoscale models normally use a continuum representation for the system under study. However, in the present case information about the structural details and dynamics are implemented from MD simulations in which an atomistic representation of the system is employed. This hybrid model is thus more realistic than other mezoscale models in the ion channel literature and will be used in the future to calculate ionic currents for *organic and inorganic pores* in the presence of *pore-specific surface chemistries*.

---

[1] Eloret Corp, 690 W. Fremont Avenue, Suite 8, Sunnyvale, CA 94087-4202, USA

[2] The time scale of physiological ion permeation is in the order of ~100ns-1μs [15]. For a 100,000-atom system a reasonable upper limit with the present computer resources is about 10ns [8].

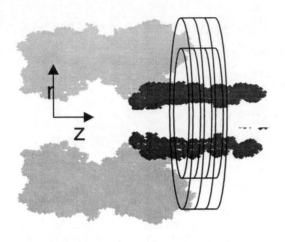

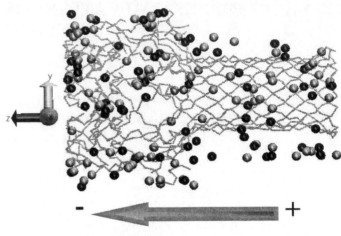

ELECTRIC FIELD

Figure 1: Schematic representation of α-hemolysin channel. In the PNP model a 2D grid (represented as concentric rings) corresponding to a cylindrical polar coordinate system is applied over the pore stem (dark area).

2 MODEL AND RESULTS

The X-ray diffraction structure of the α-hemolysin channel is used as model in the present calculations (a schematic representation is given in Fig. 1). The structure is available from the Protein Data Bank with a resolution of *1.9 Å*, determined at a temperature of *287 K* and *pH=6* [1]. The protein is a transmembrane homo-heptamer and contains 2051 residues (16,389 heavy atoms).

After replacing missing atoms in the PDB file, the structure was protonated in Amber7 [13] at neutral pH (total number of atoms 32,305) to reproduce the conditions in the polymer translocation experiments [2-4]. An NH_3^+ group at the amino terminus and a COO^- group at the carboxy terminus patched the terminal amino acids. The model was then aligned with the pore axis along the z direction of the coordinate system, with the origin at the center of the channel. Parameters for the potential functions and partial charges on the atoms were used from the Cornell 95 et al force field [12]. At neutral pH the total charge of α-hemolysin is $+7e^-$ and it was neutralized by adding Cl^- counter ions in Amber7. The energy of the structure was then minimized in Amber7 for 250 steps using the steepest descent method followed by 1250 steps of conjugated gradient method. First the hydrogen atoms were minimized, then the side chains, then all the atoms except the α-carbons and finally all the atoms. The RMSD value calculated for all the atoms was *0.8 Å*. Thus, no significant deviation during minimization was observed.

2.1 Molecular dynamics simulations

For MD simulations, the channel was inserted into a periodic unit cell (*95Åx95Åx125Å*) with electrolyte solution

Figure 2: Snapshot from an MD simulation of the α-hemolysin system in a 1M KCl solution and an external applied electric field. The K^+ (light colored spheres) and Cl^- ions (dark colored spheres) inside and within *5 Å* of the pore are shown using the van der Waals representation. The solvent (SPC/E water) is not displayed. This figure was generated using the VMD[3] software [14].

(*1M KCl*) and an electric field of *0.029 kcal/mol/Å/e⁻* (equivalent to the experimental value of *125 mV* over the *100 Å* long pore) is uniformly applied along the z axis (see Fig. 2). MD simulations were conducted using the NAMD[1] software [11] on NASA-AMES Research Center SGI supercomputers.

The coordinate and topology files are generated in Amber7. The model channel consists of a "reduced" atomistic representation of the α-hemolysin pore with polar walls, obtained by calculating the diameter of the pore corresponding to each position z along the channel axis and selecting the residues that lay within *25 Å* of the channel axis (see Fig. 3).

In all present MD simulations, a switching function was used to calculate the non-bonded interactions with a switch distance of *17.5 Å* and a cutoff of *18.5 Å*. To reduce the cost of calculating the non-bonded interactions, NAMD uses a non-bonded pair list that includes all pairs of atoms within a certain distance and may be updated periodically. The pair list distance was set in the present calculations to *20 Å* and updated every 20 MD integration steps. Also, to speed up the calculations, the RATTLE algorithm [16] was used to constrain all the bonds involving hydrogen atoms, thus allowing an increase in MD integration time to *2 fs*.

---

[1] NAMD and VMD are developed by the Theoretical Biophysics Group in the Beckman Institute for Advanced Science and Technology at the University of Illinois at Urbana-Champaign.

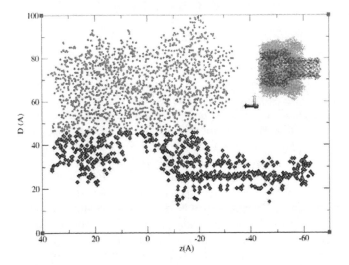

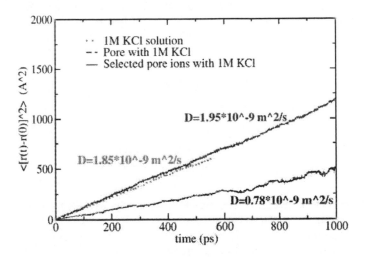

Figure 3: Projection in the axial plane of the α-hemolysin channel. The internal diameter of the pore corresponding to each position z along the channel axis is shown. The blue points represent the selection for the "reduced" representation of the pore used in the MD calculations of ion diffusion coefficients.

Figure 4: Mean square displacement versus time (MD calculations) for K$^+$ ions in: pure *1M KCl* solution (dotted line), *1M KCl* solution with α-hemolysin pore (dashed line), and *KCl* solution inside the α-hemolysin pore (solid line).

First, a periodic unit cell with ionic solution (*KCl*) was generated. The initial configuration of the K$^+$ and Cl$^-$ ions originated from the KCl crystal structure, and was solvated by adding water molecules (SPC/E model) in a proportion corresponding to a 1M concentration. The system was energy minimized over 1000 steps, heated from *50 K* up to *300 K* in steps of *50 K* every *10 ps*, and equilibrated for *200 ps* at *300 K* using the NPT ensemble with the Berendsen temperature and pressure coupling. This pre-equilibrated unit cell containing KCl solution was superposed over the "reduced" representation of the α-hemolysin pore. Overlapping ions and water molecules were removed and added back randomly to the unit cell to preserve the 1M concentration. The pore was fixed and the same procedure as above is repeated. To speed up the calculations[2] and obtain longer MD timescales, a multiple time step algorithm (MTS) was used where the non-bonded interactions are evaluated periodically. Within the 20 steps cycle, the van der Waals and electrostatic interactions were updated at intervals of 2 and 4 steps, respectively. After *700 ps* of MTS-NPT dynamics a production run of *1ns* of MTS-NVE dynamics was performed with the external electric field turned on. The diffusion coefficients, *D*, of the ions inside the protein channel are calculated for the 1ns production run using Einstein's equation of diffusion:

$$2Dt = \frac{1}{3} < \left| r_i(t) - r_i(0) \right|^2 >$$ (1)

with $r_i(t)$ the position vector of the particle at time t. If all the ions in the unit cell are selected, a value of *1.95·10^{-9}* *m^2/s* is determined (see Fig. 4, dashed line). This value is comparable to that determined from *550 ps* of MTS-NVE dynamics for pure *1M KCl* ionic solution (*1.85·10^{-9}* *m^2/s*, dotted line in Fig. 4). A smaller value (*0.78·10^{-9}* *m^2/s*, solid line in Fig. 4) is calculated if only the ions located inside the pore (and within *5 Å* of its boundaries) are selected and is used in the PNP calculations.

2.2 Poisson Nernst Planck calculation

In the PNP model a 2D grid (represented as concentric rings in Fig. 1) corresponding to a cylindrical polar coordinate system is applied over the pore stem (dark area). In contrast to the MD calculations where discrete ion positions are considered, the PNP approach assumes a continuum distribution of ions, where the ion concentration at a location is calculated as the time-averaged non-normalized probability of finding an ion at that location.

Following the formalism of Hollerbach et al. [10] a 2D-Poisson equation that relates the electrostatic potential ϕ to the fixed charge density ρ_F is used:

$$\nabla \varepsilon_r \nabla \phi = -\frac{q}{\varepsilon_0} (C_p - C_n + \rho_F)$$ (2)

with ε_r and ε_0 are the relative and vacuum permitivities. C_p and C_n are the concentrations of the positive (K$^+$) and negative (Cl$^-$) carriers, respectively, and are set initially at the bulk ion concentration (1M). In the present case, the only forces acting on the ions are the combined electrostatic forces $\nabla \phi$ resulting from the applied voltage V and the surrounding fixed (protein) and mobile (continuum KCl

---

[2] Using 200CPU's it takes 1.2 days to reach the time scale of 1ns. The modeled system contains ~120,000 atoms.

solution) charges. Under these conditions the ion flux, J, for each ion species (p for positive carriers and n for negative carriers) is approximated by the Nernst-Planck drift-diffusion equation:

$$J_i = e\mu_i C_i \left[-\nabla\phi \right] - eD_i \nabla C_i , \, i \in \{p, n\} \qquad (3)$$

$$\nabla \cdot J_i = 0 \qquad (4)$$

$$\mu = D \frac{e}{kT} \qquad (5)$$

with D (m^2/s) the ion species diffusion coefficient, μ (m^2/Vs) the corresponding mobility calculated from Eq. 5, k (J/K) the Boltzmann constant, T (K) the absolute temperature and e^- (C) the unit charge. Eq. 3 relates the flux for each ion species to the force on the ion (first term) and the concentration gradients of the ion species (second term). The conservation condition given by Eq. 4 simply states that the ionic current must be constant at steady state. Eq. 2 and 3 are solved iteratively for the potential profile ϕ and the equilibrium ion concentrations C_i. For each grid section of total area ∂A (Fig. 1), the ionic current I and the number of ions per unit time dN/dt are calculated by integrating the flux of positive and negative ions:

$$I = \int \left(J_n + J_p \right) \partial A = e \frac{dN}{dt} \qquad (6)$$

Values for the open channel ionic current and the ion flux are summarized in Table 1.

	I (pA)	dN/dt (ions/s)
Pore geometry	54	$3.37 \cdot 10^8$
Pore geometry and fixed charges	101	$6.28 \cdot 10^8$

Table 1 Calculated values for the open channel ionic current and the corresponding ion flux using the hybrid MD-PNP model. Diffusion coefficients and charges originate from MD simulations. The experimental determination for the open ion current is *120 pA*. [2].

In the 2D-PNP model, the ions have a diffusion coefficient $(0.78x10^{-9}$ $m^2/s)$ determined from MD simulations (section 2.1). The open channel ionic current is *54 pA* when only the channel geometry is considered. This means that the fine structural details of the stem (see Fig. 3) but not the partial charges on the atoms are included. By including the pore-atoms charges, this value becomes *101 pA*, comparable with the *120 pA* value measured in polymer translocation experiments [2].

3 DISCUSSION

The diffusion coefficient of potassium ions in the α-hemolysin channel is reduced relative to bulk solution by a factor of 2.5 as a result of the geometry and energetic interaction with the channel. The geometry of the channel seems to be responsible for about 50% of the total ionic current. However, it is yet unclear if the surface charges act as a mean field or if the local variations in the electrostatic potential play the leading role. Future work will attempt to refine this model to separate more clearly these influences. The MD-PNP model will also account for combined polymer-dynamics to predict blockade durations for various translocating polymers and will be used to calculate ionic currents for both *organic and inorganic pores* in the presence of *pore-specific surface chemistries*. Results could be used in the experimental design of a solid-state nanopore with single nucleotide resolution for sequencing purposes.

ACKNOWLEDGEMENTS

The present work was supported by DARPA and NASA Ames Research Center. I.C. was supported by NASA contract NAS2-99092 to Eloret and acknowledges the ASN administrators and the NAS support for technical assistance with NASA computers and the Amber, VMD and NAMD lists for software related support.

REFERENCES

[1] L. Song et al., Science, vol. 74, pp. 1859, 1996.
[2] J. J. Kasianowicz et al, Proc. Natl. Acad. Sci., vol. 93, pp. 13770, 1996.
[3] D. W. Deamer and M. Akeson, Trends in Biotechnology, vol. 18, pp. 147, 2000.
[4] M. Akeson et al, Biophys. J., vol. 77, pp. 3227, 1999.
[5] D. Branton and J. Golovchenko, Nature, vol. 398, pp. 660, 1999
[6] D. G. Levitt, J. Gen. Phys., vol. 113, pp. 789, 1999.
[7] D. K. Lubensky and D. R. Nelson, Biophy. J., vol. 77, pp. 1824, 1999.
[8] D. P. Tieleman et al., Qtly. Rev. Biophys., vol. 34, pp. 473, 2001.
[9] T. vd Straaten, et al., Proc. Second. Conf. Comp. Nanosc. and Nanotech., Puerto Rico, USA, April, 2002
[10] Hollerbach, U. et al., Biophysical Journal, vol. 76, A205, 1999
[11] L. Kale et al., J. Comp. Phys., vol. 151, pp. 283, 1999.
[12] W. D. Cornell et al., J. Am. Chem. Soc., vol. 117, pp. 5179, 1995.
[13] D. A. Pearlman et al., Comp. Phys. Commun., vol. 91, pp. 1, 1995.
[14] W. Humphrey et al, J. Molec. Graphics, vol. 14, pp. 33, 1996.
[15] G. R. Smith and M. S. P. Sansom, Biophys. Chem., vol. 79, pp. 129, 1999.
[16] H. C. Andersen, J. Comp. Phys., vol. 52, pp. 24, 1983.

Toward an Integrated Computational Environment for Multiscale Computational Design of Nanoscale Ion Channel Semiconductors

Shreedhar Natarajan[1,2], Sameer Varma[1,2], Yuzhou Tang[1,2], Scott Parker[1], Jay Mashl[1] and Eric Jakobsson[1,2*]

[1]NCSA and [2]Department of Biophysics and Computational Biology, University of Illinois.
*Contact: 4021 Beckman Institute, 405 North Mathews, Urbana, IL 61801, 217 244 2896 (voice) 217 244 2909 (fax), jake@ncsa.uiuc.edu

ABSTRACT

This paper describes the design and operation of an integrated multiscale computational environment for design of nanoscale ion channel semiconductors, the Ion Channel Workbench. The present work builds on an earlier multiscale calculation from our lab [1] in which we showed that this approach could provide a close correspondence to experimental electrophysiological data on potassium channels. The current paper advances the previous work by incorporation of multiscale into a single integrated computation, in which the results of calculation at one stage automatically feed as input to calculations at other stages. It also employs more advanced electrostatics and Brownian Dynamics techniques than the previous calculations. In addition, integration of molecular dynamics and transport Monte Carlo into the Ion Channel Workbench is being actively pursued.

Keywords: electrostatics, brownian dynamics, ion channels, nanotubes, CAD

1 OVERVIEW

Ion channels are natural nanotransistors; development of biomimetic devices using these nanotransistors would greatly benefit from a computer-aided design (CAD) environment. Compared to other gene products, data on ion channels is rich in precise quantitative functional data. This motivates the development of computational models and methods that can predict the electrophysiology of channels given their 3D structure. A challenge to computational models is the range of significant time scales from femtosecond for rapid atomic motions to microsecond for experimental observable like flux. Techniques that can capture all time scales in explicit detail within the scope of a single simulation would require a prohibitive amount of computer time. Multi-scale approaches follow a divide and conquer strategy. The simulation is split into several layers in a hierarchical fashion with the time scale of phenomena defining the layering and statistical mechanics providing the connectivity between the layers.

A multi-scale strategy is well suited for device simulation purposes. One of the requirements of a CAD system is to enable rapid prototyping. This drives the need to decompose the processes within a nanotransistor CAD system so that they can be computed efficiently. However, as we scale down in both size and time, the uncertainty inherent in the models increases and it is therefore necessary to have robust connections between these processes that capture the uncertainties effectively within a probabilistic framework. Our multi-scale approach to ion channel simulation uses well-studied statistical mechanical theory to connect across scales. This makes it a good candidate for a nanotransistor CAD system. Arguably, any biomimetic semiconductor CAD system could be modeled using this paradigm - process engineering with statistical mechanical theories providing the underlying conceptual connectivity.

2 SOFTWARE ENGINEERING

We designed our *in-silico* framework to address minimally the following issues:

- Predict accurately and efficiently electrophysiology of the channel being studied.
- Extract knowledge from these simulations and formulate strategies for designing novel channels with specific IV characteristics; as is often a requirement in device design (Reverse Engineering)
- Allow efficient pluggability of new theories and methodologies
- Allow for effective dissemination through a Portal environment

2.1 Requirements and Design

We created software modules to address vertical decomposition along guidelines specified by the hierarchical approach. Horizontal decomposition within layers is necessitated by variety of programs and theories that

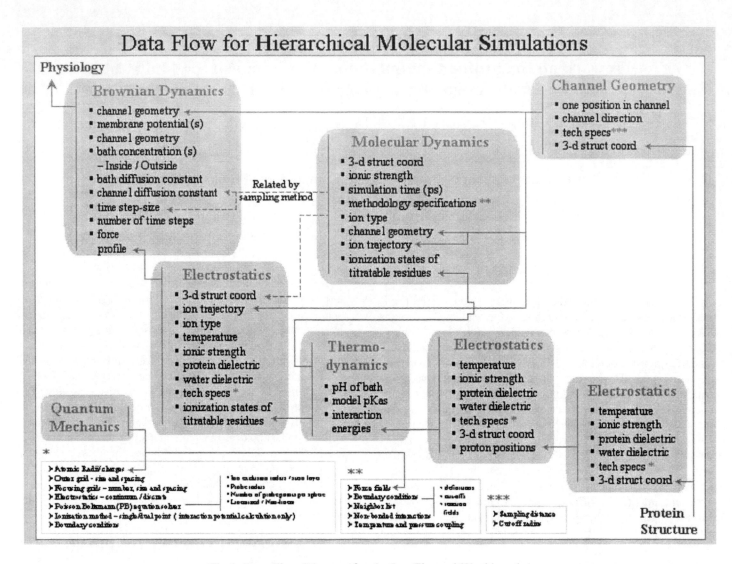

Fig 1: Data Flow Diagram for the Ion Channel Workbench.

contribute to it. E.g. Electrostatics. The workflow is shown in Fig. 1.

A module is characterized by its input and output. To ensure smooth connectivity across modules, it is necessary that the representation of information be uniform throughout the workflow. We have developed a simplistic model of module descriptors using XML [2] as the representation language. Using element overlap as the criterion enforces the connectivity between modules. These modules along with their XML metadata have been plugged and connected using a framework that we have developed locally. The framework includes a workflow interface, a workflow engine and a workflow metadata description also using XML.

Our approach to use a metadata based modular framework will have significant advantages in curating, archiving, searching, retrieving and analyzing simulation data in the future. A typical scenario in device design might be to "design a nanotube" with specific "ion flux" rates for a specific ion size and type. The first step would be to examine existing simulations and experiments for data that could

suggest a naturally occurring ion channel that had the required output characteristic. It is therefore necessary to have an automated analysis technique that can search through precompiled datasets. However, simulation data can run into gigabytes in size and are typically stored in Mass Storage systems. Hence it is necessary to create metadata archives that contain the essence of the simulation results. An advantage to using semi-structured metadata is that we can formulate precise queries using structured query languages to the existing datasets. Such query languages and optimizers for queries are being developed by a large and organized community of researchers and are readily accessible. This preempts the need to write specific programs tailored to address each and every analysis. Another potential advantage is that the CAD system can incorporate multiple theories and external programs into various "plug-points" within the framework in a seamless fashion. There is an increasing drive towards interoperability [3] in the digital biology community with increasing adoption of XML based representations of their outputs. The use of semi-structured data in the creation of open source NanoDevice simulation software can have a symbiotic and mutually beneficial relationship with the biological modeling community.

2.2 Implementation

The individual modules have been optimized to run in parallel if the problem allows for parallelization. The parallelized implementations of electrostatics and the Brownian dynamics components are described in Section 3 and Section 4. We discuss here the challenges involved in connecting the programs that reside on different computational clusters. Our implementation of the Ion Channel Portal addresses the issue of running a complete computational workflow on a computational grid seamlessly using a thin-web client. The web framework has been implemented mostly in Object-Oriented Perl. Each of the modules is wrapped with OO-Perl templates.

The modular metadata based framework discussed above lends itself to an implementation that utilizes the recently developed infrastructure of grid computing. Grid computing employs multiple, independent, distributed, heterogeneous, computational resources, coupled through a network and a common middleware standard, to perform complex computational tasks. In implementing the modular framework, the grid infrastructure allows a web server resource, which provides the web based workflow interface and the workflow engine, to be established and maintained separately from a set of remote high performance computational resources that are used to perform the computational tasks.

The workflow engine directs the execution of the ordered series of module calculations needed to characterize the electrophysiology of the channel being studied. These calculations may be performed on one or more remote computational resources, with the workflow manager assigning a module task to each resource and then receiving and managing the resulting output data. Tasks are assigned and data transferred using an asymmetric connection between resources. The connection from the web-server/workflow manager to the remote computational resource is made through the Globus grid middleware [4]. The connection from the remote compute resource back to the workflow manager is made through the web server and a custom remote http based client.

There are several advantages to an implementation based on the grid infrastructure. The coupling of distributed heterogeneous resources allows tasks to be performed on the resources best suited to their execution. Additionally, as more resources are made available on the grid, these resources can be utilized, extending the computation power of the portal. Finally, new modules can be integrated seamlessly into the framework, even those maintained on remote resources controlled by disparate members of the community.

3 ELECTROSTATICS

The long-range nature of Coulumbic forces makes electrostatics a vital player in both intra- and intermolecular interactions. Mathematically, the laws governing the distribution of potentials and electric fields are all encapsulated in the Poisson-Boltzmann equation (PBE). The energetics of conformational states, proton dissociation and diffusion; all have a strong electrostatic component. Within the context of our multiscale approach, we have decomposed electrostatic calculations into three modules.

Firstly, electrostatics is coupled with empirical equations (derived from quantum mechanics and probability density functions [5] to estimate and optimize coordinates of hydrogen atoms that are not resolved or ambiguous in protein structures [6].

Secondly, several functional groups in proteins (and some in synthetic nanotubes) have a tendency for proton dissociation. Proton dissociation alters the net charge of a functional group and thus plays a significant role in defining ion flux rates/ratios. Proton dissociation occurs with a probability (pKa) that is strongly influenced by environmental conditions. This probability is calculated from thermodynamic equations that require free energy functions calculable from PBE theory. We have recently formulated a systematic strategy that eliminates most parameter-dependent uncertainty in these calculations [7].

Finally, electrostatics has also been used to calculate the deterministic part of the force experienced by an ion in a Brownian dynamics simulation. We have also recently made several refinements to this methodology [8].

All the three modules in our hierarchy have been parallelized to efficiently run on multiple processors. Instead of parallelizing the PBE solver itself, each of these modules has been parallelized by decomposing the entire problem domain into a set of smaller domains. For instance, if one needs to determine electrostatic forces for a given set of trajectory points inside a nanotube (i.e. force profile), one can distribute these points across different processors. Each processor then calculates the forces for its own set of points. Such domain decomposition, as can be seen, is ingrained in the nature of the problem. We have used the same strategy to parallelize all the three modules [7, 8]. This context based parallelization strategy has two main advantages. Firstly, it does not depend upon the availability of an ingrained parallelization in the PBE solver. This allows the freedom to plug in any kind of a sequential PBE solver. And secondly, this strategy demands for a minimum possible inter-processor communication, which makes this parallelization ideal for grid-computing environments.

4　BROWNIAN DYNAMICS

Brownian motion is the random movement of solute molecules in dilute solutions that result from repeated collisions with solvent molecules. The basic principle behind Brownian dynamics (BD) is the same as molecular dynamics, but it introduces some new approximations that allow us to perform simulations at a microsecond time scale. We introduce implicit solvent description and ignore internal molecular motions that allow us to use a larger time step.

Our earlier implementation of BD simulation implemented this concept [9], along with the added assumption that diffusion of ion is along the channel axis; based on the observation that the pore region of ion channel is very narrow so that ions could only move in single-file style. These simplifications made it possible for the model system to be simulated for as long as 1ms within only 1 hr of CPU time; thus allowing us to retrieve statistically meaning results, such as current-voltage curves that are observable in experiments. We have made several improvements to the original algorithm [8]. It is now possible to do the simulation for multiple species simultaneously. Ion-ion pairwise interactions are pre-calculated by electrostatic calculations and converted into a look-up table so that this interaction could be more accurately represented during the simulation. The calculations have also been parallelized by temporal decomposition of the total simulation interval among multiple processors.

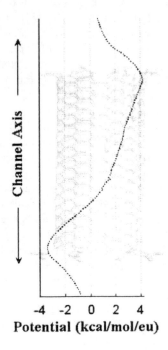

Fig 2: The Potential as seen by a K^+ ion as it moves along the permeation pathway of the Nanotube. The nanotube is shown as a watermark.

5　RESULTS AND DISCUSSION

Using the potential profile of nanotube showed in Fig 2 as input, the 1-D BD simulation generated the current-voltage (I-V) curve of K^+ that is shown in Fig 3. From this figure, it can be seen that there is a preference of influx than efflux of K^+ current because of the asymmetry of the potential profile. For this calculation, a total of 11 simulations were done with simulation time of 180 microseconds for each one. It cost 4 hour CPU time on NCSA IBM p690 machine.

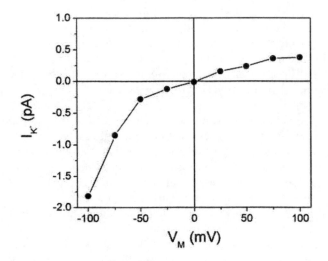

Fig 3: The I-V curve of a nanotube in a pseudo bilayer. [K^+] on both sides is 150mM

References

[1] Mashl, R.J., Y. Tang, J. Schnitzer, and E. Jakobsson. Biophysical Journal. **81**. 2473-2483. 2001.

[2] URL: http://www.w3.org/XML/.

[3] URL: http://www.sbml.org.

[4] URL: http://www.globus.org.

[5] Hooft, R.W., C. Sander, and G. Vriend. Proteins. **26**(4). 363-76. 1996.

[6] Malmberg, N.J., S. Varma, E. Jakobsson, and J.J. Falke. 2004.

[7] Varma, S. and E. Jakobsson. Biophysical Journal. 2004.

[8] Tang, Y., S. Varma, and E. Jakobsson. 2003.

[9] Bek, S. and E. Jakobsson. Biophysical Journal. **66**. 1028-1038. 1994.

Numerical modeling of diffusion in extracellular space of biological cell clusters and tumors.

J. Berthier, F. Rivera, P. Caillat

CEA-LETI, Département Systèmes pour l'Information et la Santé,
17, avenue des Martyrs, 38054 Grenoble, France ; jean.berthier@cea.fr

ABSTRACT

Biological cluster of cells may be seen as porous media, where the cells are the "solid grains" and the extra-cellular space (ECS) the "pores". It is often the case in dense clusters of cells that the flow rate of liquid inside the extracellular space is negligible in front of the molecular diffusion. In the particular case of tumoral cells, the extra-cellular path is the tumor interstitial matrix (IM) and the apparent (or effective) diffusion coefficient (ADC) determines the speed of delivery of drugs into the tumor. It is then of great importance to be able to estimate the value of the apparent diffusion coefficient. We present here a numerical approach, based on a Monte Carlo modeling to estimate the ADC in irregular, non repetitive morphologies of cell clusters.

Keywords: diffusion, cells, extracellular space, modeling, tumor.

INTRODUCTION

Porous media is composed of an arrangement of solid grains and interstitial fluid. Biological cluster of cells may be seen as porous media, where the cells are the "solid grains" and the extra-cellular space (ECS) the "pores" (fig 1 and 2). In fact, this is an approximation since some diffusing molecules can penetrate the cells (cellular uptake) and are thus removed from the extra-cellular space.

We place ourselves in the case where the fluid flow in the ECS is negligible in front of the molecular diffusion. One example is that of tumoral cells , the extra-cellular path is called the tumor interstitial matrix (IM) [2] and the apparent (or effective) diffusion coefficient (ADC) determines the speed of delivery of drugs into the tumor [3]. It is then of great importance in cancer treatment, and to any treatment targeted at dense cluster of cells, to be able to estimate the value of the apparent diffusion coefficient [4]. It can also be noted that any change of the ADC reflects a change in the cells shape and arrangement [5].

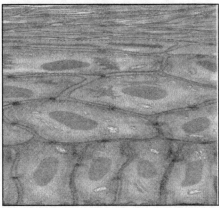

Fig. 2. Cell arrangement in the human skin from [6]. The ECS is small and the diffusion is close to the 1D case

Different types of numerical approaches have been already examined: homogenization theory [7] and Monte Carlo method [8] for regular repetitive patterns like squares or triangles. Regular patterns may seem sufficient to approximate an average ADC [9], however the recent use of microsystems to deliver drugs in-vivo requires the knowledge of diffusion in complex morphologies, specially if one wants to estimate the local uptake rates [10] or if any change in cell shape and arrangement takes place [7]. So far there have been very few investigations for irregular and disordered clusters, mostly because of the difficulty in describing the geometry. We propose here a new algorithm that includes three main features: (1) generation of a geometrical cell cluster by using the Evolver numerical program [11]; (2) a

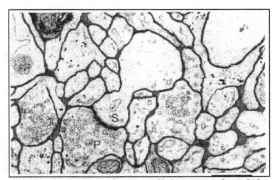

Fig. 1. Geometry of extra-cellular space from [1]. Electromicrograph of small region of rat cortex. The ECS is in dark on the picture. It can be seen at the bottom right "lakes" where the extracellular space widens.

Monte Carlo approach for calculating the random walk of the diffusing species; (3) a geometric tracking algorithm for constraining the diffusing molecules inside the ECS.

NUMERICAL MODEL

In order to investigate the diffusion process in irregular patterns of cells, we have set up a model based on a Monte Carlo approach. To the difference of the existing models, the geometry may be composed of different types of lattices of cells, not only regular but also irregular and anisotropic, mimicking real clusters of cells (fig. 3). These lattices of cells are obtained by making use of the Evolver code [11].

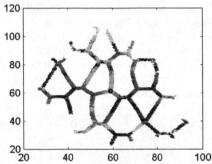

Fig. 3. A lattice of cell obtained with the Evolver code. The ECS is "painted" by the plot of the random walk of the particles.

A given set of points delimiting the cells is introduced in the Evolver numerical program in order to form an initial lattice of cells, each cell being bounded initially by linear segments. Depending on line tension and cell volume, the shape of the cells evolves until convergence to a minimum energy arrangement, mimicking real cell arrangement. It is assumed here that the cell membranes behave similarly to an interface having surface tension.

Particles are initially placed in a central micro-region, simulating the injection point at the tip of the micro-needle, the diffusion is then simulated by following the particles execute random walks inside the ECS or IM (fig.4 and 5).

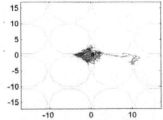

Fig. 4. Random walk of 2 particles inside the ECS of a regular lattice of rounded cells.

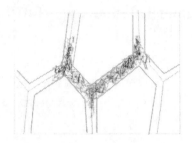

Fig.5. Random walk of 2 particles in the extra-cellular space of an irregular cluster.

The displacement (Δx, Δy) of a particle in the time step Δt is given by the relations:

$$\Delta x = \sqrt{4 D \Delta t} \cos(\alpha)$$
$$\Delta y = \sqrt{4 D \Delta t} \sin(\alpha) \qquad (1)$$
$$\alpha = random\ (0, 2\pi)$$

where D is the "free" diffusion coefficient , usually given by Einstein's law:

$$D = \frac{k_B T}{6 \pi \eta R_H} \qquad (2)$$

where k_B is the Boltzman constant ($1.38\ 10^{-23}$ J/K) , T the temperature (K), η the dynamic viscosity of the carrier fluid and R_H the hydraulic radius of the particle.

Particles location inside the cluster is permanently tracked and the particles are not allowed to cross solid boundary. To this point, the difficulty is to locate each particle and to detect if it numerically crosses a cell boundary; in such a case the particle is constrained inside the ECS.

RESULTS AND DISCUSSION

We distinguish two cases: that of a cluster with small to negligible extra-cellular space and that of a cluster with free extracellular spacing (like the lattice of fig 4 and the real cells of fig.1 bottom right).

In the case of a two-dimensional array of cells with negligible intercellular spacing, the results show that there is a direct relation between the ADC and the geometrical tortuosity τ of the extra-cellular space, whatever the arrangement of the cells (fig.6 and 7). In a porous media, the geometric tortuosity is the ratio between the shortest length joining one point to another one in the fluid and the straight line. In such a case, it has been theoretically shown for a regular cell arrangement [7] that the ratio of the apparent (effective) to the free diffusion coefficient is given by $D_{eff}/D = 1/\tau^2$ where $\tau = 2^{1/2}$.

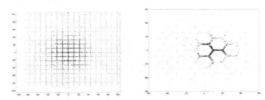

Fig. 6. (a) Diffusion in a cluster of square cells with small intercellular spacing, (b) in an hexagonal lattice.

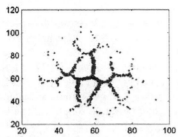

Fig. 7. Diffusion in an irregular cluster of cells with small intercellular spacing (scale in microns)

The numerical results of the present model are in agreement with this theoretical relation. In fig. 8, the normalized diffusion length

$$\beta = \frac{L}{\sqrt{4Dt}} \qquad (3)$$

(where L is obtained by averaging the distance of each particle between their location at time t and at time $t = 0$). is plotted versus time for the geometry of the cell cluster of fig. 3 and 7. At very small times, the diffusing particles execute random walk inside a small region of the ECS, so that the coefficient β is that of free diffusion: $\beta = 1$ at $t=0$. After a short time, the particles are constrained to 1D diffusion and β reaches a nearly constant value

$$\beta = \frac{L}{\sqrt{4Dt}} \approx 0.7 \approx \frac{1}{\sqrt{2}} \qquad (4)$$

By definition, the apparent diffusion coefficient satisfies

$$\frac{L}{\sqrt{4D_{eff}\,t}} \approx 1 \qquad (5)$$

From eq. (4) and (5), we find the relation

$$\frac{D_{eff}}{D} = \frac{1}{\tau^2} \approx \frac{1}{2} \qquad (6)$$

leading to the value $\tau=2^{1/2}$.

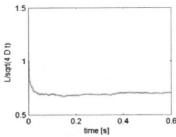

Fig. 8. Variation of the normalized diffusion distance β as a function of time, in the cluster geometry defined in figure 3 and 7.

In anisotropic clusters, tortuosity depends on the direction as shown in the figure 9 taken from [1].

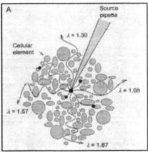

Fig. 9. Schematic view of diffusion paths in an anisotropic cell arrangement. Here λ is the tortuosity

It can be seen that even anisotropic clusters follow eq. (6) if averaging is done upon all the directions, but in such a case, it is shown that there exists a different ADC for each direction of the plane.
However, the real situation is often more complex (fig.1, bottom right) because the spacing of the cells lattice is not uniform and there are intercleft spaces. It is shown that diffusion speed may be reduced by entrapment when the dimensions of the residual spaces are large and the connecting exits sufficiently small. An idealized example is that of a cluster of round cells. If the dimension of the gaps between the cells is decreased, the apparent diffusion coefficient can be smaller than the value of eq. (6). In the case defined in fig 4., we obtain (fig. 10) $\dfrac{D_{eff}}{D} \approx \dfrac{1}{4}$.

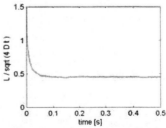

Fig. 10. Variation of the normalized diffusion distance β as a function of time, in a cluster geometry of rounded cells defined in figure 4.

A limiting case is that of a gap width smaller than the mean free path of the particle (percolation limit), in such a case, the particles are trapped inside the intercleft space.

Another example is given in fig. 11 where we have switched a cell to an intercleft space in order to compare the diffusion lengths.

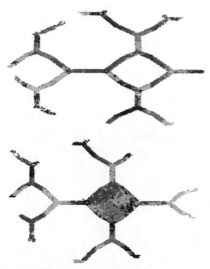

Fig 11. Random walk of particles in the ECS of a cluster of cells generated by Evolver. (a) no residual space between the cells, (b) after replacing a cell by a free space

CONCLUSION

Diffusion of biochemical species in a cluster of cells has be modelled by a three steps algorithm: (1) Evolver generation of cluster arrangement, (2) Monte Carlo random walk of the diffusing species, and (3) particle tracking to constrain the diffusing species inside the ECS.

The results of the model verify that the apparent diffusion coefficient in dense cell clusters with small extra-cellular spacing is that of the 1D case.

However the situation is much more complex in the extra-cellular space of irregular and anisotropic clusters of cells, specially if there exist intercleft spaces. Speed of diffusion can be considerably reduced by particle entrapment in the intercleft spaces.

In reality, two conditions are required for an efficient drug delivery: a diffusion speed sufficient to reach the ECS of all cells in the cluster in a reasonable time and a sufficient uptake rate to be sure that all cells are concerned by uptake.

Modeling the cellular uptake requires additional developments of the present algorithm. A very simple model for the uptake would be to introduce a probability of uptake as a function of the number of contact of the diffusing molecules on the cell membrane.

REFERENCES

[1] C. Nicholson and E. Sykova; "Extracellular space structure revealed by diffusion analysis," TINS; vol 21 n°5; 207-215; 1998.

[2] S. Ramanujan, A. Pluen, T. D. McKee, E. B. Brown, Y. Boucher, and R. K. Jain , "Diffusion and convection in collagen gels: implications for transport in the tumor interstitium," Biophys. J.; vol. 83; 1650-1660; 2002.

[3] J. Lankelma, R. F. Luque, H. Dekker, W. Schinkel, and H. M. Pinedo, "A mathematical model of drug transport in human breast cancer," Microvascular Research; vol. 59; 149-161; 2000.

[4] A. W. El-Kareh, S.L. Braunstein, T.W. Secomb. "Effect of cell arrangement and interstitial volume fraction on the diffusivity of monoclonal antibodies in tissue," Biophys. J.; vol. 64; 1638-1646; 1993.

[5] A. M. Herneth, S. Guccione, M. Bednarski, "Apparent diffusion coefficient: a quantitative parameter for in vivo tumor characterization, " European Journal of Radiology; vol 45, 208-213; 2003.

[6] M. Martin, "Conséquences d'une irradiation ionisante sur la peau humaine, " Clefs CEA, vol. 48, 53-55,2003.

[7] K. C. Chen, and C. Nicholson, "Changes in brain cell shape create residual extracellular space volume and explain tortuosity behavior during osmotic challenge, " Proc. Natl. Acad. Sci. USA; vol. 97, n° 15; 8306-8311; 1999.

[8] M. J. Saxton, "Lateral diffusion in an archipelago, the effect of mobile obstacles, " Biophys. J.; vol. 52; 989-997; 1987.

[9] J.J. Blum, G. Lawler, M. Reed, and I. Shin, "Effect of cytoskeletal geometry on intracellular diffusion, " Biophys. J.; vol. 56, 995-1005;1989.

[10] A. Szafer, J. Zhong, and J. C. Gore, "Theoretical model for water diffusion in tissues, " Magnetics Resonance in Medecine; vol. 33 n°5; 697-712; 1995.

[11] K. A. Brakke, "The surface evolver, " Experimental Mathematics, vol. 1, n°2; 141-165; 1992.

Mathematical Modeling for Immunocolloid Labeling

C. Raymond* and P. Milewski**

* New Jersey Institute of Technology, Newark, NJ, USA, raymond@njit.edu
** University of Wisconsin, Madison, WI, USA, milewski@math.wisc.edu

ABSTRACT

For many biological applications, it is becoming increasingly important to extend imaging capabilities to simultaneously identify multiple molecular species in order to gain insight into the structural and functional elements of cells. A key part of this drive is the ability to rationally optimize various labeling techniques, which in turn requires an understanding of the basic physical processes operating during the labeling process, and how these processes interact. The kinetics of the immunocolloid labeling process are typically modeled using a Langmuir kinetic equation [1]. However, very different behavior which can not be explained by a Langmuir-type model has been noted in certain experiments [2], [3]. We develop a mathematical model for the labeling process which exhibits these two behaviors in different parameter limits, gives some insight into how to optimize the labeling process, and can easily be extended to more complicated situations.

Keywords: immunocolloid labeling, immunogold labeling, mathematical model, reaction-diffusion model, surface-volume reaction

1 INTRODUCTION

Immunocolloid labeling is a technique for imaging molecular scale features in many biological applications. The method starts by conjugating colloidal heavy metal particles to ligand molecules which bind to the target molecules of interest. Gold is frequently used as the heavy metal, hence the process is often referred to as immunogold labeling. The specimen to be labeled is bathed in a suspension of the composite metal-ligand label particles, the label particles diffuse to and bind to the target molecules, and once the labeling process is complete, the labeling suspension is washed away and the specimen is prepared and viewed under an electron microscope. While the target molecules may be difficult to detect directly, the label particles are easily detected.

This process widely used, but in order to maximize its effectiveness and advance the development of multiple labeling techniques, in which several different target molecules are simultaneously marked with label particles which can be distinguished under the electron microscope by size, shape, or composition, a theoretical framework needs to be developed that will give some insight into how to choose the experimental parameters such as concentration of labeling particles, the time for which the specimen is exposed to the labeling suspension, and so on.

To this end, we have developed a mathematical model based on the two most basic physical mechanisms needed to describe the labeling process: diffusion of the label particles through the suspension to the immobile target sites, and the chemical reaction between the label particles and target molecules. In its simplest form, the model depends on only one nondimensional parameter, the Damköhler number Da, which can be thought of as the ratio of a time scale for the reaction to a time scale for diffusive transport of the label particles. In the limit of small Da, this model reproduces the results of the Langmuir-type model typically used to describe the labeling process [1], which assumes that the process is limited by the rate of the binding reaction.

On the other hand, it is well known that in certain situations the process is limited by diffusion of the label particles [1]. This can happen when the target molecules are densely packed on a surface, leading to depletion of the label particles close to the surface, as is the situation in certain experiments described in [2], [3]. This corresponds to the large Da limit of the model described herein.

The model has several nice features. In its simplest form, as presented here, it depends on only one nondimensional parameter, the Damköhler number Da. It is clear how all the experimentally accessible parameters enter into Da and hence easy to understand what happens when multiple experimental parameters are varied. Asymptotic solutions can be developed in the two limits $0 \ll \text{Da} \ll 1$ and $\text{Da} \gg 1$ which in fact turn out to be good approximations to the true solutions over almost the entire range of Da. The model is easily solved numerically, especially in the middle range when $\text{Da} \approx 1$ and neither of the asymptotic solutions does a good job. Finally, the model provides a good starting point. It can easily be extended to allow more complicated chemistry at the boundary, interactions between the label particles, complicated boundary geometry, multiple labeling, and so on.

2 MATHEMATICAL MODEL

As indicated above, our basic model consists of a diffusion equation for the concentration C of label particles in the suspension, coupled to an equation modeling the binding of the label particles to the target molecules on the boundary. The model assumes that there is a uniform surface concentration B at the boundary, which is appropriate for the experiments discussed in [2], [3], and hence ignores variations along the boundary. We have:

$$0 < x < \infty: \qquad C_t = C_{xx}, \tag{1}$$

where C is the nondimensionalized concentration, scaled by the dimensional initial concentration C_0, and x is the nondimensionalized distance from the boundary to which the target molecules are confined. We also need to keep track of the target molecules on the boundary, or equivalently, the surface concentration of bound label particles. This is done through the following two boundary conditions:

$$x = 0: \qquad B_t = C_x, \qquad \mathrm{Da}^{-1}B_t = C(1 - B). \tag{2}$$

The first of these states that the rate of change of the surface concentration B of bound label particles is equal to the net flux of label particles to the surface from the suspension. The second states that the labeling process follows mass action kinetics, with the reaction rate proportional to the product of the label particles in the suspension adjacent to the boundary and the surface concentration of free (unlabeled) target molecules. For simplicity, we ignore the dissociation reaction. The surface concentration is nondimensionalized with reference to B_0, the surface concentration of target molecules, or equivalently, the maximum surface concentration of bound label particles. We assume that the length scales associated with the problem are very small compared to the size of the experimental setup, and impose a no flux condition at infinity:

$$x = \infty: \qquad C_x = 0. \tag{3}$$

Finally, we have the initial conditions:

$$t = 0: \qquad C = 1, \qquad B = 0. \tag{4}$$

This is the form the equations take if we use diffusive length and time scales. The diffusive length scale is B_0/C_0, which can be thought of as the thickness of a layer of the initial suspension which contains exactly enough label particles to bind to all target molecules. The diffusive time scale is $B_0^2/(C_0^2 D)$, where D is the diffusion constant for the label particles in suspension. Since the particles are colloidal and subject to Brownian motion, D can be approximated by the Stokes-Einstein formula,

$$D = \frac{\kappa T}{6\pi\mu R}, \tag{5}$$

where κ is Boltzmann's constant, T is the temperature, μ is the dynamic viscosity of the fluid in which the particles are suspended, and R is the radius of the particle.

The only nondimensional parameter is the Damkhöler number,

$$\mathrm{Da} = \frac{k_A B_0^2}{DC_0} = \frac{B_0^2/(C_0^2 D)}{1/(k_A C_0)} = \frac{\text{diffusion time scale}}{\text{reaction time scale}} \tag{6}$$

The experimental parameter which may most easily be adjusted over a wide range is the initial concentration of labeling particles, C_0. Assuming that the labeling process is limited by the reaction, i.e. looking at the parameter regime $0 < \mathrm{Da} \ll 1$, really means that with everything else held fixed, we have chosen C_0 so that the process consumes only a small portion of the available supply of label particles. This may be difficult to arrange if the target molecules are densely packed, i.e. if B_0 is large. This complementary situation with $\mathrm{Da} \gg 1$ leads to a depletion of the supply of label particles adjacent to the boundary, so that the process is limited by the diffusive transport of labels to the boundary.

3 LARGE Da LIMIT

If we formally set $\mathrm{Da} = \infty$ in (2), we get a simplified solution with two possible solutions. The second condition in (2) reduces to two possibilities: either $B = 1$, or $C(x = 0, t) = 0$. The first of these leads to the very simple solution $B = 1$, $C = 1$, which corresponds to the long time limit in which the labeling process is complete. The second possibility also leads to a simplified problem, in which we can first solve for C, from (1), (3), (4) and $C(x = 0, t) = 0$. This problem has the solution

$$C = \mathrm{erf}\left(\frac{x}{2\sqrt{t}}\right), \tag{7}$$

where erf is the error function. We can then compute $C_x(x = 0, t) = 1/\sqrt{\pi t}$, and use the first equation in (2) to find that

$$B = \frac{2\sqrt{t}}{\sqrt{\pi}}. \tag{8}$$

This solution can only be valid for $0 \leq t \leq \pi/4$ at best, since we have scaled the problem so that $0 \leq B \leq 1$. In fact, an asymptotic analysis of the problem shows that in the limit $\mathrm{Da} \to \infty$, the solution to the problem converges to the limiting solution

$$B = \begin{cases} 2\sqrt{t/\pi} & 0 \leq t < \pi/4 \\ 1 & t \geq \pi/4 \end{cases}. \tag{9}$$

Corrections to this solution for large but finite Da can be obtained systematically, but are not reproduced here because they are somewhat complicated.

There are two points we want to make here about the limiting solution (9) and its refinements. The first is that that this simplified solution agrees with the experimental results and the theoretical analysis based on the Brownian transport of label particles reported in [2],[3]. The second is that including the first correction to (9) gives an approximate solution for B which agrees well with the solution obtained by numerically solving the problem for $2 < \mathrm{Da} < \infty$.

4 SMALL Da LIMIT

As noted above, in this limit we expect the solution to evolve on the reactive time scale. Since (1)–(4) were obtained using diffusive length and time scales, we expect that we will need to rescale the problem in this limit. In fact, naively setting $\mathrm{Da} = 0$ in (1)–(4) leads to the solution $C = 1$, $B = 0$, i.e. the solution is pinned at the initial conditions and does not evolve. Setting $\tau = \mathrm{Da}t$, $\xi = \sqrt{\mathrm{Da}}x$, we obtain

$$0 < \xi < \infty : \qquad C_\tau = C_{\xi\xi}, \tag{10}$$

$$\xi = 0 : \qquad C_\xi = \sqrt{\mathrm{Da}}B_\tau, \qquad B_\tau = C(1 - B), \tag{11}$$

$$\xi = \infty : \qquad C_\xi = 0, \tag{12}$$

$$\tau = 0 : \qquad C = 1, \qquad B = 0. \tag{13}$$

If we now set $\mathrm{Da} = 0$, the problems for C and B decouple. Solving for C using (10), $C_\xi(\xi = 0, t) = 0$ from (11), (12) and (13) gives $C = 1$. Therefore, we are left with a linear, constant coefficient, ordinary differential equation, in fact just the Langmuir kinetic equation, for B from (11), with solution $B = 1 - \exp(-\tau)$. If we go back to dimensional variables,

$$B = B_0 \left(1 - \exp(-k_A C_0 t)\right), \tag{14}$$

so that we can use a relatively simple curve fit to determine the association rate constant.

This can be corrected for small but nonzero Da by expanding both B and C as power series in $\sqrt{Da}$ and solving the resulting regular perturbation problem to the order desired. This leading order solution agrees well with the numerical solution of the full problem for $0 < \mathrm{Da} < 1/2$.

5 OUTLOOK AND CONCLUSIONS

We have presented a mathematical model for the immunocolloid labeling process. The model is consistent with existing theoretical results in the two limiting cases in which either the association reaction or the diffusive transport of label particles is the rate limiting step in the process. It allows the rational development of a sequence of approximate solutions valid in each of these limiting cases. In the intermediate range where the two subprocesses occur at commensurate rates, the model can easily be solved numerically.

The model is a first step toward a better understanding of the labeling process. More details of the surface chemistry can easily be included. For instance, dissociation of bound label particles from the targets can be included by a slight modification in the second equation of of (2). The Langmuir model can be modified to allow for crowding of the adsorbate when the target sites are densely packed [4], and the same can be done for this model in general (not just in the reaction limited case), again by a slight modification of the second equation in (2). The multiple labeling process can be approached by coupling together a number of these basic models, with separate variables for the concentrations of each of the label species. In actual applications of immunocolloid labeling, geometrical effects will potentially be an issue, for two reasons: the target molecules will generally not be homogeneously distributed on the boundary as assumed in the model as presented here, and the boundary on which the target molecules are located may itself be complicated. By making the obvious changes to pose the problem on a two or three dimensional spatial domain and allowing B_0 to vary with the location along the boundary, the model can be extended to cover these possibilities as well.

6 ACKNOWLEDGEMENTS

The authors thank Ralph Albrecht for bringing this problem to their attention, for providing experimental data, and for many stimulating discussions about immunocolloid labeling. The authors' work on this project is supported through NIH grant GM67244-01.

REFERENCES

[1] H. Nygren and M. Werthén, "Colloidal Gold Techniques and the Immunochemistry of Antigen-Antibody Reactions at Interfaces,", in Colloidal Gold: Principles, Methods, and Applications, volume 3, M.A. Hayat, editor, Academic Press, 307-319, 1991.

[2] K. Park, S. Simmons and R. Albrecht, "Surface characterization of biomaterials by immunogold staining – quantitative analysis," Scanning Microscopy 1, 339-350, 1987.

[3] K. Park, H. Park and R. Albrecht, "Factors Affecting the Staining with Colloidal Gold," in Colloidal Gold: Principles, Methods, and Applications, volume 1, M.A. Hayat, editor, Academic Press, 489-517, 1989.

[4] Adamson and Gast, "Physical Chemistry of Surfaces (6th ed.)," Wiley-Interscience, 603-615, 1997.

Crystallization of Membrane Protein Nanoparticles

Andrey Shiryayev*, Edward F. Holby** and J. D. Gunton*

* Department of Physics, Lehigh University
Bethlehem, PA 18015
** Carleton College, Northfield, MN 55057

ABSTRACT

The self-assembly of membrane proteins from solution into two-dimensional crystals is studied numerically using a continuum model proposed by Talanquer and Oxtoby (J. Chem. Phys. **109**, 223 (1998)). Although the mean field theory presented here is only qualitatively accurate in two dimensions, we obtain the main features of the crystal nucleation process, such as the nucleation free energy barrier and the number of particles and crystalline particles, respectively, in the critical nucleating droplet. Particular emphasis is given to the region near the metastable fluid-fluid coexistence curve, where the free energy barrier is small. The free energy barrier that separates a protein rich metastable fluid phase from its stable crystalline phase is studied for a variety of nucleation pathways in the metastable region. As in the three dimensional case of globular proteins, the nucleation barrier is smallest in the vicinity of the fluid-fluid critical point.

Keywords: nanoparticle, protein, crystallization, nucleation

1 INTRODUCTION

Membrane proteins are an important component of the cell. These proteins are associated with the cell membrane and serve a variety of important functions. They help in the transport of ions across the cell membrane, in cell-cell communication and as cell receptors. In order to determine the function of these membrane proteins, it is necessary to determine their structure, which traditionally is done via x-ray crystallography. This requires the growth of high quality membrane protein crystals, which is notoriously difficult and which is very sensitive to the initial conditions of the proteins in solution. While in the case of globular proteins three dimensional crystallization is a common step toward the determination of the atomic structure, success of growing three dimensional crystals for membrane proteins is still infrequent [1]. It is difficult to maintain a crystal lattice with the sole interactions between the hydrophilic domains of the proteins. A two dimensional crystallization in the presence of lipids, when the membrane proteins reconstruct into lipid membranes to form crystals, is an entire new way to determine the structure of the membrane proteins [2] [3].

Recently some attention has been given to understanding from a theoretical perspective the conditions for optimal crystal nucleation from solution, based on phenomenological microscopic models [8], [9]. In this work we present a brief summary of a recent, complementary approach, based on a continuum, density functional approach developed for the study of the crystallization of globular proteins from solution. As in the case of globular proteins, we focus on the region near the metastable two phase fluid-fluid coexistence, as this is presumably the region for optimal crystal nucleation. The two fluid phases correspond to protein-poor and protein-rich solutions. A prior simulation study of a two dimensional modified Lennard-Jones model of membrane proteins found that the lifetime of this two fluid metastable coexistence state was too short to be studied [9]. Hence it remains to be seen if this is also the case for the continuum model presented here.

2 MODEL

In this study we model membrane proteins as two dimensional nanoparticles, using a density functional model proposed by Talanquer and Oxtoby [4] for the study of globular proteins. Their model is in the class of the phase field models and is based on the well known van der Waals free energy density of the fluid phase and a corresponding phenomenological van der Waals like free energy density of the solid phase. The resulting free energy density is the minimum of these two free energy branches. The free energy functional is given by

$$\Omega[\rho, m] = \int d\vec{r} \Big[f(\rho, m) - \mu\rho + \frac{1}{2} K_\rho (\nabla \rho)^2 + \frac{1}{2} K_m \rho_s^2 (\nabla m)^2 \Big] \quad (1)$$

where f is the Helmholtz free energy density, μ is the chemical potential. The free energy depends on two order parameters: the (conserved) local density $\rho(\mathbf{r}, t)$ and a (non-conserved) local structural order parameter that shows whether the system is in a solid or fluid phase $m(\mathbf{r}, t)$. A critical cluster is a saddle-point of the functional (1); therefore in order to determine its size and profile we must solve the Euler-Lagrange equations with appropriate boundary conditions:

$$\frac{\delta\Omega}{\delta\rho} = 0 \quad \text{and} \quad \frac{\delta\Omega}{\delta m} = 0 \qquad (2)$$

Using the saddle-point solutions for $\overline{\rho}(\mathbf{r})$ and $\overline{m}(\mathbf{r})$ we can obtain such properties of the inhomogeneous system as the free energy barrier for nucleation, $\Delta\Omega = \Omega\{\overline{\rho}(\mathbf{r}), \overline{m}(\mathbf{r})\} - \Omega\{\rho_0, 0\}$, the surface tension of the nucleating droplet and the number of particles in the critical cluster.

3 RESULTS AND DISCUSSION

We solve the saddle-point equation (2) numerically using the shooting method. The boundary conditions correspond to the metastable disordered fluid state at infinity (an infinite distance from the center of the droplet) and an ordered solid state at the center of the droplet. We also assume that the system has a polar symmetry. The solutions to the equations (2) are very sensitive to the choice of shooting parameters and thus diverge easily. In order to avoid this divergence we use a shooting method with a fitting point which we choose to be at the intersection of the solid and the fluid branches of the free energy density. This allows us to decrease the numerical error. The starting point for shooting from infinity is chosen in agreement with the implementation of Sear's approximation [7] in two dimensions. Further details of this numerical approach are described in [5].

In figure 1 the density and order parameter profiles are shown for the case of the system close to the critical point, while in figure 2 the profiles are shown for the system further from the critical point. As we can see, the density profile near the critical point has a longer tail. This shows that the correlation length is larger as we approach the critical point and diverges at the critical point. This long tail is in agreement with Sear's results [7].

Next, we calculated the free energy barrier at constant supersaturation versus reduced temperature T/T_c. As in the three dimensional case [4], [6], [5], the free energy barrier has a minimum in the vicinity of the critical point. This can be explained in terms of the second derivative of the free energy density with respect to the number density of nanoparticles in the metastable disordered state $f_{\rho\rho}$. The free energy barrier decreases as one goes away from the liquidus line and increases as one approaches the liquidus line. The slope of the paths with constant supersaturation is proportional to $f_{\rho\rho}$, which vanishes at the critical point. Thus these paths become horizontal near the critical point [5] and the system changes its behavior from going away from the liquidus line to approaching it. Therefore the free energy barrier has a minimum near the critical point. Because the mean field theory breaks down near the

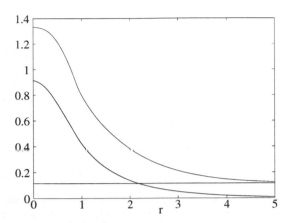

Figure 1: Density and order profiles of critical droplet close to the critical point. Higher profile is the dependence of the density on the distance from the center of the cluster. Lower profile is the dependence of the order parameter on the distance from the center of the cluster. The horizontal line shows the background fluid density.

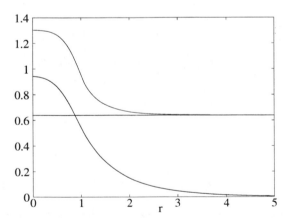

Figure 2: Density and order profiles of the critical droplet further from the critical point. Higher profile is the dependence of the density on the distance from the center of the cluster. Lower profile is the dependence of the order parameter on the distance from the center of the cluster. The horizontal line shows the background fluid density.

critical point, the details of the free energy barrier minimum can be explained only qualitatively by these calculations.

We also calculated the dependence of the number of particles in the critical cluster on the number of crystalline particles. One can see that as in the case of globular proteins, the number of particles in a two-dimensional critical cluster diverges as we approach the metastable critical point. This result is a mean field theory result. However ten Wolde and Frenkel [6] also obtained a similar liquid-like structure of the critical cluster in the vicinity of the critical point in their Monte-Carlo simulations. Further details of our results for the free energy barrier and the nature of the critical droplet will be published elsewhere.

4   CONCLUSION

Frenkel and Noro [9] were unable to determine the the fluid-fluid binodal for a modified Lennard-Jones model in two dimensions, as the barrier between the fluid-fluid metastable state and stable crystalline state was very small. However, they determined the metastable spinodal (defined as the locus of points where the barrier to self-assembly vanishes). There is thus a metastable critical point in the two-dimensional case. Although the decay time of the metastable binodal curve might be too short to be observed, the existence of the metastable critical point still affects the nucleation barrier and therefore the kinetics of self-assembly of the membrane proteins.

This work was supported by NSF grant DMR-0302598. One of us (EFH) was supported by a summer REU grant from Lehigh University.

REFERENCES

[1] C. Ostermeier and H. Michel, Curr. Opin. Struct. Biol. **7** 697 (1997).

[2] R.M. Garavito and S. Ferguson-Miller, J. Biol. Chem. **276**, 32403 (2001).

[3] D. Levy, M. Chami and J.-L. Rigaud, FEBS Letters. **504**, 187 (2001).

[4] V. Talanquer and D. Oxtoby, J. Chem. Phys. **109**, 223 (1998).

[5] A. Shiryayev and J.D. Gunton, J. Chem. Phys. to be published.

[6] P.R. ten Wolde and D. Frenkel, Science **277**, 1975 (1997).

[7] R. Sear, J. Chem. Phys. **114**, 3170 (2001).

[8] M. A. Bates, M. G. Noro and D. Frenkel, J. Chem. Phys. **116**, 7217 (2002)

[9] M.G. Noro and D. Frenkel J. Chem. Phys. **114**, 2477 (2001).

Monte Carlo Study of Models of Membrane Proteins

T. P. Connor, D. L. Pagan, A. Shiryayev, J. D. Gunton

Department of Physics
Lehigh University, Bethlehem, PA, USA

ABSTRACT

The crystallization of membrane proteins from so-lution is sensitive to the role of initial conditions. In order to understand this dependence, we report prelimi-nary results for two models for membrane proteins, each with short range attractive interactions. We determine the fluid-fluid coexistence curve for a square well attrac-tive model in two dimensions at short attractive interac-tion range $\lambda = 1.1\,\sigma$, where σ is the hard core diameter. We also extend an earlier study of a two dimensional modified Lennard-Jones model.

Keywords: membrane proteins, nano-particles, self-assembly

1 INTRODUCTION

Recent advances in decoding the human genome have identified a huge number of proteins to be investigated. While there are many classes of proteins, one important class is that of membrane proteins, which are confined in or near the cell membrane. High quality crystals, free from defects, are required to determine the structure of these proteins by crystallographic means. Obtaining these crystals depends on understanding the initial con-ditions that give rise to optimal crystallization. While the structure of many globular proteins has been ob-tained over the past decade, membrane proteins are no-toriously more difficult to crystallize, particularly so in three dimensions [1]. Difficulty arises in isolating inter-actions between the hydrophilic domains of the proteins in a lattice state. A two-dimensional crystallization in the presence of lipids, when the membrane proteins re-construct into lipid membranes to form crystals, is a whole new way to determine the structure of the mem-brane proteins [2], [3]. Many studies [4]–[6] have been aimed at determining the optimal initial conditions for growing high quality globular protein crystals from su-persaturated solutions. However, very few studies exist for membrane proteins.

Membrane proteins are nanoparticles that self assem-ble to form crystals. A typical membrane protein has three distinct domains: two hydrophilic extramembra-nous domains and the hydrophobic domain that spans the lipid bilayer. These nanoparticles regulate both the

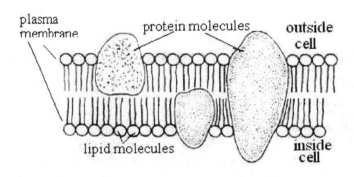

Figure 1: Illustration showing integral membrane pro-teins and the cell membrane. Not shown are peripheral membrane proteins which are associated with the sur-face of the plasma membrane.

flow of ions through the cell and the shape of the actual cell membrane itself. Figure 1 shows an illustration [7] of integral membrane proteins and the cell membrane. The former are characterized as having a small range over which they interact and exhibit a dependence on the strength of the interaction. Until recently no the-oretical studies of membrane proteins existed, with re-search focusing on three-dimensional globular proteins. For the latter, it is known that the details of the actual potential are unimportant, as long as the range of the at-tractive interaction is sufficiently short. As such, simula-tions of model globular proteins use different potentials such as the square-well [8], Yukawa [9], and modified Lennard-Jones (MLJ) [10] potentials and yield qualita-tively the same results; namely, a metastable fluid-fluid coexistence curve, and a fluid-solid coexistence curve. Recently, a two dimensional MLJ model [10] has been proposed as a generic model of membrane proteins. It was expected that for values of the attractive interaction small in comparison with the protein size, a metastable fluid-fluid region would exist, as has been found in the-oretical and simulation studies of globular proteins [4], [10]. Simulations of the two-dimensional model, how-ever, were unable to detect a metastable fluid-fluid co-existence, as the model yielded quite a rapid crystalliza-tion. This indicated that the free-energy barrier that has to be overcome to form a crystal nucleus from solution is quite small. The authors were, however, able to esti-

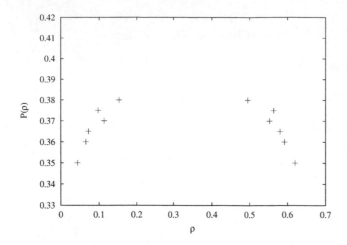

Figure 2: Fluid-fluid coexistence curve determined for the square well model.

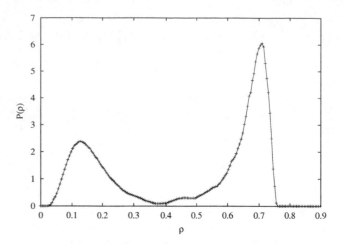

Figure 3: Density distribution obtained in a Monte Carlo simulation of membrane proteins along an isotherm.

mate the fluid-fluid spinodal curve – the locus of points in the phase diagram where the extrapolated value of the inverse isothermal compressibility vanishes.

In another study [11] the phase diagram of a model membrane protein, Annexin V, was determined, using a pairwise, directional interaction potential between proteins.

In this paper we extend these studies of models of membrane proteins by presenting preliminary results for two models with short-range attractions, in two space dimensions. The first is an attractive square well model; the second is the modified Lennard-Jones model, mentioned above.

2 ATTRACTIVE SQUARE WELL MODEL

Our first model is an attractive square well interaction model in two dimensions:

$$V_{SW}(r) = \begin{cases} \infty, & r \leq \sigma \\ -\epsilon, & \sigma < r \leq \lambda \sigma \\ 0, & r > \lambda \sigma \end{cases} \quad (1)$$

This model has been studied in three dimensions [8] as a model of globular proteins, but until now has not been studied in two dimensions. To look for coexistence between two phases, we use the Gibbs ensemble Monte Carlo method. Two simulation cells are thermodynamically connected and allowed to exchange volume and particles. To determine coexistence, the conditions $\mu_l = \mu_v$ and $P_v = P_l$ are imposed on both cells, where μ denotes the chemical potential and P denotes the pressure. We sampled the system at various cutoffs to see where coexisting phases could be found. We started with $\lambda = 1.25\,\sigma$, since this is known to be the value

in three-dimensions [8] above which no metastable coexistence curve is observed. We subsequently reduced the range of attractive interactions until coexistence was observed. The resulting coexistence curve is shown in Figure 2.

3 MODIFIED LENNARD-JONES MODEL

The second model [10] describes N particles in a two-dimensional box interacting via the modified Lennard-Jones (MLJ) potential

$$V(r) = \begin{cases} \infty, & r < \sigma \\ \frac{4\epsilon}{\alpha^2}\left(\frac{1}{[(r/\sigma)^2-1]^6} - \frac{\alpha}{[(r/\sigma)^2-1]^3}\right), & r \geq \sigma \end{cases} \quad (2)$$

where σ denotes the hard-core diameter of the particles, r is the interparticle distance, and ϵ is the well depth. The potential was unshifted and truncated at a cutoff distance $r_c = 3.5\,\sigma$, to compare with benchmarks in the literature [10]. The parameter α controls the range of attraction between particles. For $\alpha = 50$ the range of attraction is short compared to the diameter of the particles. In the grand canonical ensemble, the number of particles N is allowed to fluctuate while the chemical potential μ, temperature T, and volume V are fixed. After each Monte Carlo cycle, the joint probability distribution of energy and number density is recorded in the form of histograms. To test for coexistence, the density distribution was examined to determine if any bimodal peaks were present, indicating the presence of two phase at densities ρ_l and ρ_v.

Using these simulations, we attempted to locate a bimodal peak in the density distributions at values of temperature T where a metastable curve was expected

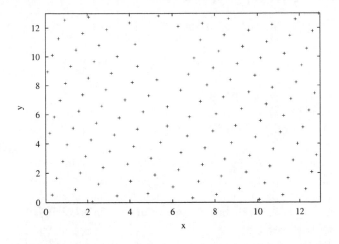

Figure 4: Configuration of the crystal state as obtained by starting the system in a vapor state at $T = .3124$ and $\mu = -2.43$. A regular lattice pattern evolves, along with some 'defects', during a single Monte Carlo simulation.

for the MLJ model. Figure 3 shows a density distribution at $T = .3125$ and $\mu = -2.43$. The bimodal peaks, however, do not correspond to fluid-fluid coexistence, but rather to fluid-solid coexistence. Further investigation reveals that this distribution does not correspond to an equilibrium state; allowing the simulation to continue reveals that the system starts in a fluid phase and subsequently 'crystallizes' into a single solid phase, as shown in Figure 4. A definite lattice structure is observed, confirming the previous findings that there is a low free-energy barrier in the two dimensional MLJ model, as opposed to the MLJ model of globular proteins in three dimensions [5], [12].

4 DISCUSSION

In this work, we report preliminary results for two models of membrane proteins. We were able to determine the fluid-fluid coexistence curve for an attractive square well potential, with $\lambda = 1.1\,\sigma$. Future work is needed to determine whether this is in a stable or metastable region of the phase diagram. We will also study the model for shorter range attractive interactions. Our results to date for the phase diagram of the MLJ model are consistent with the previous study [10].

ACKNOWLEDGEMENTS

This work was supported by a grant from the National Science Foundation, DMR-0302598. One of us (TPC) was a participant in a summer REU program at Lehigh University. He also wishes to acknowledge a travel grant from Lehigh.

REFERENCES

[1] C. Ostermeier and H. Michel, Curr. Opin. Struct. Biol. **7** 697 (1997).

[2] R.M. Garavito and S. Ferguson-Miller, J. Biol. Chem. **276**, 32403 (2001).

[3] D. Levy, M. Chami and J.-L. Rigaud, FEBS Letters. **504**, 187 (2001).

[4] D. Rosenbaum, P.C. Zamora, and C.F. Zukoski,Phys. Rev. Lett., **76**, 150 (1996)

[5] P. R. ten Wolde and D. Frenkel, Science, **277**, 1975 (1997)

[6] D. W. Oxtoby and V. Talanquer, J. Chem. Phys.,

[7] World Wibe Web, http://www.borg.com/ lubehawk/cell.htm

[8] Neer Asherie, et al., Phys. Rev. Lett., **77**, 4832 (1996)

[9] M. H. J. Hagen and D. Frenkel, J. Chem. Phys., **101**, 4093, (1994)

[10] Massimo G. Noro and Daan Frenkel, J. Chem. Phys., **114**, 2477(2001)

[11] M. A. Bates, M. G. Noro and D. Frenkel, J. Chem. Phys. **116**, 7217 (2002)

[12] D. L. Pagan, M. E. Gracheva and J. D. Gunton, submitted for publication, Los Alamos preprint 0311551

Biomolecule Electrostatic Optimization with an Implicit Hessian

J. P. Bardhan*, J. H. Lee*, M. D. Altman†, S. Leyffer⋆, S. Benson⋆, B. Tidor‡,*, J. K. White*

* Department of Electrical Engineering and Computer Science, † Department of Chemistry
‡ Biological Engineering Division
77 Massachusetts Avenue, Cambridge, MA 02139
⋆ Mathematics & Computer Science Division
Argonne National Laboratory, Argonne, IL 60439

ABSTRACT

Computational rational drug design is the application of computer simulation techniques to improve screening processes for new drugs or to design them *de novo*. The goal is to identify molecules that have high affinity and specificity for a target molecule. Optimizing the electrostatic binding free energy is tractable under certain assumptions as a quadratic optimization problem, but computationally expensive simulations have traditionally been required to determine the associated Hessian. Prior work showed that significant performance gains could be achieved by coupling physical simulation directly to the optimization process and avoiding the calculation of the Hessian. The present paper details recent improvements to this method and the implementation of a practical, full-scale code. Computational results demonstrate the code's efficiency on large problems and accuracy solving a realistic biomolecule optimization problem.

1 INTRODUCTION

Computational rational drug design is the application of computer simulation techniques to the problem of identifying new drug molecules that bind tightly and specifically to a given target molecule. The binding affinity is related to the change in free energy due to the binding of the drug molecule, or *ligand*, to the target molecule, or *receptor* [1]–[3]. Several types of atomic and molecular interactions contribute to this free energy change, including the hydrophobic effect, van der Waals interactions, and electrostatic interactions (including hydrogen bonding). Because electrostatic interactions are long range and contribute a significant component of the total binding free energy, it is important to understand how the partial atomic charges in the ligand and receptor interact with each other and with the surrounding solvent. To design optimal ligands, drug designers need to understand what set of ligand partial atomic charges optimize the electrostatic free energy of binding.

If continuum electrostatics are used to model the biomolecular interactions, then under certain assumptions [2] this set of optimal ligand partial charges can be determined by solving a convex quadratic program. Standard optimization techniques, however, typically require the expensive calculation of an explicit Hessian matrix to efficiently solve the program. We therefore introduced in [4] an optimization method that avoids the Hessian calculation; initial results showed promise

that the method might significantly accelerate solving these optimization problems. The present paper describes an improved formulation and an implementation that is capable of solving large, realistically sized optimization problems in biomolecule design.

Section 2 introduces the optimization problem, the electrostatic modeling techniques, and the Hessian-implicit primal-dual method of [4]. Section 3 describes recent improvements to the original method and important features of the full-scale implementation. Section 4 presents results that demonstrate the efficiency and accuracy of the new formulation. Section 5 summarizes the contributions of this work and suggests directions for further investigation.

2 BACKGROUND

2.1 Biomolecule Electrostatic Optimization

We present first the analysis of the electrostatic energy of the ligand alone in solution [1], [2]. A mixed discrete-continuum model is used: the ligand partial atomic charges are treated as point charges at discrete locations (the ligand atom centers) but with continuous value. The electrostatic potential in the interior satisfies the continuum Poisson equation with low dielectric constant; the potential in the solvent satisfies the linearized Poisson-Boltzmann equation and is characterized by a high dielectric constant and the inverse Debye screening length κ.

Each point charge polarizes the solvent; this polarization in turn creates a *reaction potential* field in the ligand. The reaction potential is a linear function of the vector x of point charges; since the sources are discrete points, the reaction potential must be found only at those points to calculate the electrostatic free energy. The mapping between x and φ_{rf}, the resulting vector of reaction potentials at the corresponding point locations, can therefore be represented by a matrix L_u, which is symmetric by reciprocity. The free energy due to x in the unbound ligand is then $\frac{1}{2}x^T\varphi_{rf} = \frac{1}{2}x^T L_u x$.

When analyzing the ligand–receptor complex in solution, the ligand charges again produce a field linear in x. The receptor charges, which are assumed to be fixed, produce an additional field c; the total free energy of the bound system is thus $\frac{1}{2}x^T L_b x + x^T c$. Improvements in affinity correspond to minimizing the change in free energy between the bound and unbound states, so the objective function for the optimization is the difference between these energies. Kangas and Tidor [2] showed that $L = L_b - L_u$ is symmetric positive semidefinite. Constraints are imposed on the charge vector x: first, sum of charge constraints on subsets of charges, and possibly on the entire set, are defined, which gives a set of equality constraints $Ax = b$. Physical and computational considerations lead to the imposition of box inequality constraints, $l_i \leq x_i \leq u_i \; \forall i$.

2.2 Biomolecule Electrostatic Modeling

We use the boundary element method for electrostatic simulation and address the free ligand; using the integral formulation from [5], the mapping L_u can be represented as a function of three integral operators $M_1, M_2,$ and M_3 [4]:

$$L_u = M_3 M_2^{-1} M_1. \qquad (1)$$

The fully dense matrices $M_1, M_2,$ and M_3 are too large for direct storage or inversion; instead, a fast method such as precorrected-FFT [6] is used to define functions that quickly perform matrix-vector multiplication by these matrices. The lack of explicit knowledge about M_2 necessitates the use of preconditioned Krylov iterative methods to apply the inverse M_2^{-1}. If there are n_c ligand charges, the matrix L_u can be calculated explicitly by solving, for $i = \{1 \ldots n_c\}$:

$$M_2 \varphi_i = M_1 e_i \qquad (2)$$

$$L_{u,i} = M_3 \varphi_i \qquad (3)$$

where e_i is the ith unit vector and $L_{u,i}$ is the ith column of L_u.

2.3 Hessian-Implicit Primal-Dual Method

The biomolecule electrostatic optimization problem can easily be transformed into standard form

$$\text{minimize} \quad \frac{1}{2} x^T L x + x^T c$$
$$\text{subj. to} \quad Ax = b \qquad (4)$$
$$\text{and} \quad x \geq 0$$

by introducing slack variables on both sides of the box constraints.

After introducing the Lagrange multiplier vectors λ and s, the Karush-Kuhn-Tucker optimality conditions for this problem can be written as a mildly nonlinear vector-valued function $F(x, \lambda, s)$ whose zeros are optimal solutions to (4) if $(x, s) \geq 0$: Primal-dual interior point methods [7] find an optimal solution to (4) by a modified Newton method: at iteration k, one linearizes F around the current iterate (x^k, λ^k, s^k) and biases the Newton step so that the pairwise products $x_i^{k+1} s_i^{k+1}$ are approximately equal. Define the average pairwise product $= (x^k)^T s^k / n$ where n is the size of x and s. The linearized system solved at each step is given by

$$\begin{bmatrix} L & -A^T & -I \\ A & 0 & 0 \\ S^k & 0 & X^k \end{bmatrix} \begin{bmatrix} \Delta x^k \\ \Delta \lambda^k \\ \Delta s^k \end{bmatrix} = -F(x^k, \lambda^k, s^k) + \begin{bmatrix} 0 \\ 0 \\ \sigma \, e \end{bmatrix}$$
$$= z(x^k, \lambda^k, s^k) = z^k, \qquad (5)$$

where X is the diagonal matrix with $X_{i,i} = x_i$, S is similarly defined, and e satisfies $e_i = 1 \; \forall i$. The second term on the right-hand side biases the Newton step as desired, where $0 \leq \sigma \leq 1$; the role of σ is discussed below. The calculated Newton update is then scaled by α^k, $0 < \alpha^k \leq 1$, to ensure that $(x^{k+1}, s^{k+1}) > 0$.

If L has the form $L = M_3 M_2^{-1} M_1$, one may view the Jacobian in (5) as the Schur complement of the system

$$\begin{bmatrix} 0 & -A^T & -I & M_3 \\ A & 0 & 0 & 0 \\ S^k & 0 & X^k & 0 \\ -M_1 & 0 & 0 & M_2 \end{bmatrix} \begin{bmatrix} \Delta x^k \\ \Delta \lambda^k \\ \Delta s^k \\ \Delta \varphi^k \end{bmatrix} = \begin{bmatrix} z^k \\ \\ 0 \end{bmatrix} \qquad (6)$$

where $\varphi^0 = M_2^{-1} M_1 x^0$. The Hessian-implicit method avoids direct calculation of L by solving (6) at each iteration using a preconditioned Krylov subspace method. To form the preconditioner we copy the expanded Jacobian from (6), set the M_1 and M_3 blocks to zero, and approximate M_2 by a matrix D_2 composed of the diagonals of the blocks of M_2.

3 IMPLEMENTATION

3.1 Designing an Aggressive Optimization Strategy

The parameter σ in (5) is called the centering parameter; it dictates how strongly the algorithm attempts to keep the pairwise products $x_i^k s_i^k$ equal. If σ is set close to unity, the algorithm makes slow progress towards an optimal solution but is robust and rarely stagnates. If instead σ is set very small, progress can be rapid but the optimization may stagnate; an iterate may approach the boundary of the feasible region $(x, s) > 0$, in which case the algorithm makes unacceptably slow progress. The original Hessian-implicit formulation set $\sigma = 0.4$ for all iterations, as suggested in [7], which balances robustness against the rate of convergence. The new formulation uses a simple rule to pick each σ^k independently, using the step multiplier α^{k-1} as the primary criterion:

Algorithm 1 *Choosing centering parameter* σ^k

$\sigma^k = 0.4$
$if \; \alpha^{k-1} > 0.7$
$\quad \sigma^k = 0.1$
$if \; \alpha^{k-1} > 0.95 \; and \; k > 8$
$\quad \sigma^k = 0.01$

This schedule was determined by practical experience with different model problems. The heuristic assumes that significant progress on the previous iteration has left the current iterate in a position to make good progress again. This assumption is generally good after a few iterations, and the two cases in which $\sigma^k < 0.4$ address its shortcomings.

3.2 Solving Equality Constrained Problems

The Hessian-implicit method was developed to solve programs with both equality and inequality constraints. The optimality conditions for an equality constrained program are linear, so only one linear system must be solved to find an optimal solution. The Hessian-implicit optimality conditions

$$
\begin{bmatrix} 0 & A^T & M_3 \\ A & 0 & 0 \\ -M_1 & 0 & M_2 \end{bmatrix} \begin{bmatrix} x^* \\ \lambda^* \\ \varphi^* \end{bmatrix} = \begin{bmatrix} -c \\ b \\ 0 \end{bmatrix} \qquad (7)
$$

can be solved quickly by using the preconditioner

$$
P_{eq} = \left[\begin{array}{cc|c} M_3 D_2^{-1} M_1 & A^T & 0 \\ A & 0 & 0 \\ \hline 0 & 0 & D_2 \end{array} \right]. \qquad (8)
$$

P_{eq} is effective because $M_3 D_2^{-1} M_1$ is an approximation to L; factoring P_{eq} is inexpensive because it is block diagonal as shown, and although the upper left block is dense, it is extremely small. In contrast to the preconditioner for (6), we include the $M_3 D_2^{-1} M_1$ block to prevent singularity of the preconditioner.

3.3 Full-scale Implementation

To implement the Hessian-implicit primal-dual method, we coupled the pFFT++ boundary element method code [6], [8] with the PETSc scientific library [9]. The pFFT++ code allows black-box multiplication by the integral operator matrices M_1, M_2, and M_3 to be done in $O(n \log n)$ time and space, where n is the number of panels used to discretize the surface. The PETSc library offers a variety of iterative linear solvers and preconditioners; each modified Newton update (6) is solved using GMRES and the LU factorized preconditioner. The preconditioner nonzero structure is fixed, so ordering need be performed only once.

4 RESULTS

4.1 Performance Comparisons

The calculations of M_2 matrix-vector (MV) products dominate the cost of the optimization process, so to assess the performance of different algorithms, the number of calculated M_2 MV products is used as the cost metric. Figure 1 illustrates the performance of the new and old formulations.

To compare the Hessian-implicit method's performance to standard algorithms, KNITRO, which implements a primal-dual interior-point nonlinear optimization algorithm [10], was used as a reference. KNITRO uses CG to calculate each Newton update; each CG iteration therefore performs one calculation of Lx if the system (5) is solved. Each Lx product requires one iterative solve to find $M_2^{-1} M_1 x$. We therefore estimate KNITRO's cost to solve (5) by multiplying the total number of CG iterations by the average number of M_2 MV products required to find a column of L. Figure 2 shows

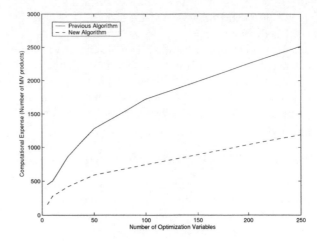

Figure 1: Performance of previous and new algorithms

that this alternative implicit scheme performs more poorly than the traditional explicit-Hessian method and much more poorly than the Hessian-implicit primal-dual method; we believe that the irregular behavior of KNITRO in Figure 2 results from particularly poor conditioning of one or more of the random test problems. When solving optimization prob-

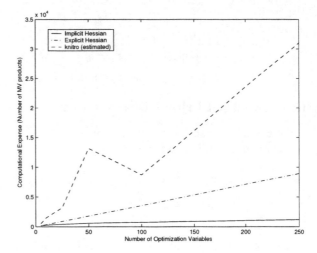

Figure 2: Performance of new and alternative methods

lems with only equality constraints, the computational advantage of using the Hessian-implicit method is even more pronounced, as illustrated in Figure 3.

4.2 Realistic Biomolecule Optimization

To assess the Hessian-implicit method's accuracy, we studied the ligand–receptor system of enzyme E. coli chorismate mutase (ECM) and an inhibitor transition-state analog (TSA) [3]. Plotted in Figure 4 are optimal charge distributions calculated by the Hessian-implicit method and an explicit Hessian method. The results agree very well. Inaccuracies are due largely to numerical Hessian asymmetry in the implicit method; when an explicit Hessian is formed, it can be symmetrized and nonphysical singular values can be removed [3],

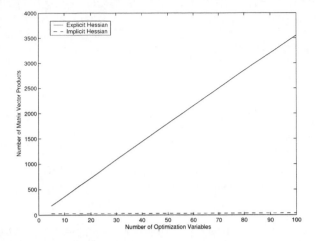

Figure 3: Performance of new algorithm on equality constrained problems

but these operations cannot be performed in the implicit optimization method. The optimization problem has 26 primary variables and the Hessian-implicit system solved at each iteration has approximately 130,000 variables.

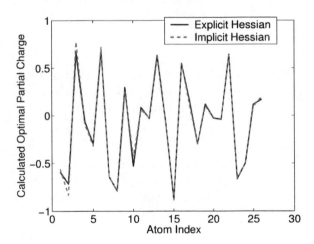

Figure 4: Accuracy of solution on realistic problem

5 DISCUSSION

This paper described improvements to the Hessian-implicit optimization method originally reported in [4] and a full-scale implementation of the new method. The implementation couples the pFFT++ fast boundary element method package and the PETSc scientific library. The recent improvements to the method improve performance by approximately a factor of two over the original formulation; in addition, the new full-scale code is capable of solving biologically relevant optimization problems. Simpler optimization problems with only equality constraints can be solved extremely rapidly using a newly designed preconditioner. Future work will investigate the sources of numerical Hessian asymmetry, and explore extending the formulation to allow convex constraints and the rapid update of a solution if the constraints are varied.

6 ACKNOWLEDGMENTS

This work was supported by the Singapore–MIT Alliance, the National Science Foundation, and the National Institutes of Health. J. Bardhan is supported by a Department of Energy Computational Science Graduate Fellowship. S. Leyffer and S. Benson are supported by the Mathematical, Information, and Computational Sciences Division subprogram of the Office of Advanced Scientific Computing Research under Contract W-31-109-ENG-38.

REFERENCES

[1] L.-P. Lee and B. Tidor. Optimization of electrostatic binding free energy. *Journal of Chemical Physics*, 106:8681–8690, 1997.

[2] E. Kangas and B. Tidor. Optimizing electrostatic affinity in ligand–receptor binding: Theory, computation, and ligand properties. *Journal of Chemical Physics*, 109:7522–7545, 1998.

[3] E. Kangas and B. Tidor. Electrostatic complementarity at ligand binding sites: Application to chorismate mutase. *Journal of Physical Chemistry*, 105:880–888, 2001.

[4] J. P. Bardhan, J. H. Lee, S. S. Kuo, M. D. Altman, B. Tidor, and J. K. White. Fast methods for biomolecule charge optimization. *Modeling and Simulation of Microsystems (MSM)*, 2003.

[5] S. S. Kuo, M. D. Altman, J. P. Bardhan, B. Tidor, and J. K. White. Fast methods for simulation of biomolecule electrostatics. *International Conference on Computer Aided Design (ICCAD)*, 2002.

[6] J. R. Phillips and J. K. White. A precorrected-FFT method for electrostatic analysis of complicated 3-D structures. *IEEE Transactions on Computer-Aided Design of Integrated Circuits and Systems*, 16:1059–1072, 1997.

[7] S. J. Wright. *Primal-Dual Interior Point Methods*. SIAM, 1997.

[8] Z. Zhu, B. Song, and J. White. Algorithms in FastImp: A fast and wideband impedance extraction program for complicated 3D geometries. *IEEE/ACM Design Automation Conference*, 2003.

[9] S. Balay, K. Buschelman, W. D. Gropp, D. Kaushik, M. Knepley, L. C. McInnes, B. F. Smith, and H. Zhang. PETSc home page. http://www.mcs.anl.gov/petsc, 2001.

[10] R. Byrd, M. E. Hribar, and J. Nocedal. An interior point method for large scale nonlinear programming. *SIAM J. Optimization*, 9:877–900, 1999.

Self Assembly of Globular Protein Nanoparticles

Andrey Shiryayev and James D. Gunton

Department of Physics, Lehigh University
Bethlehem, PA 18015

ABSTRACT

A continuum model of the self-assembly of globular protein nanoparticles proposed by Talanquer and Oxtoby (J. Chem. Phys. **109**, 223 (1998)) is investigated numerically, with particular emphasis on the region near the metastable fluid-fluid coexistence curve. The free energy barrier that separates a protein rich metastable fluid phase from its stable crystalline phase is studied for a variety of nucleation pathways in the metastable region. As shown earlier, the nucleation barrier is smallest in the vicinity of the fluid-fluid critical point. An approximate analytic solution is also presented for the shape and properties of the nucleating crystal droplet.

Keywords: nanoparticle, protein, crystallization, nucleation

1 INTRODUCTION

Globular protein molecules provide an excellent example of nanoparticles that can self-assemble to form crystal clusters. The formation of high quality protein crystals is a prerequisite in the determination of the protein structure, which in turn determines protein functions. As a consequence, there is great interest in determining the conditions necessary to form high quality protein crystals. Since nucleation of a crystalline droplet is the critical step toward the formation of the solid phase from the supersaturated solution, this is the focus of current studies. It has been shown that the crystallization of globular proteins can be explained as arising from attractive interactions whose range is small compared with the molecular diameter (corresponding to a narrow window of a small, negative value of the second virial coefficient B_2 [1],[2]). In this case the gas-fluid coexistence curve is in a **metastable** region below the liquidus-solidus coexistence lines, terminating in a metastable critical point (Figure 1). Self-assembling of the globular proteins into nanoscale clusters is a stochastic process which can be characterized by the nucleation rate; this rate is controlled by the free energy barrier between the supersaturated fluid solution and the crystalline state. Talanquer, Oxtoby [3] and ten Wolde and Frenkel [4] shows that the free energy barrier on the

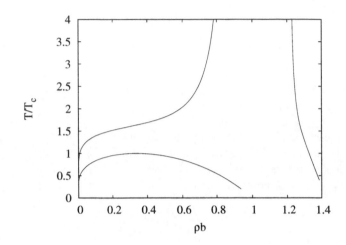

Figure 1: Phase diagram for the nanoparticles with the short-range attraction interactions. The fluid-fluid coexistence curve becomes metastable in this case.

path of constant supersaturation is the smallest in the vicinity of the critical point.

In this work we extend the numerical analysis of a continuum model of globular proteins studied by Talanquer and Oxtoby [3], with particular emphasis on the region near the metastable fluid-fluid coexistence curve.

2 MODEL

We use a model due to Talanquer and Oxtoby [3] to describe globular protein crystallization. Their phase field model is based on the following grand canonical free-energy functional:

$$\Omega\left[\rho, m\right] = \int d\vec{r}\Big[f(\rho, m) - \mu\rho + \\ + \frac{1}{2}K_\rho(\nabla\rho)^2 + \frac{1}{2}K_m\rho_s^2(\nabla m)^2\Big] \quad (1)$$

The free energy depends on two order parameters: the (conserved) local density $\rho(\mathbf{r}, t)$ and a (non-conserved) local structural order parameter that shows whether the system is in a solid or fluid phase $m(\mathbf{r}, t)$. Here $f(\rho, m)$ is the Helmholtz free-energy density and μ is the chemical potential. Talanquer and Oxtoby [3] use the van der Waals free energy density for the fluid branch:

$$f_f(\rho, m) = k_B T \rho \left[\ln \rho - 1 - \ln(1 - \rho b) \right]$$
$$- a\rho^2 + k_b T \alpha_1 m^2 \quad (2)$$

and a corresponding van der Waals free energy for the solid phase:

$$f_s(\rho, m) = k_B T \rho \left[\ln \rho - 1 - \ln(1 - \bar{\rho} b) \right]$$
$$- (a + a_m m_s^2)\rho^2 + k_b T (\alpha_1 m^2 + \alpha_2) \quad (3)$$

Here $\bar{\rho}$ plays a role of a weighted coarse grained density $\bar{\rho} = \rho(1 \quad \alpha_3 m(2 - m))$ where the term in square brackets implements a difference between the fluid and solid close-packing limits. For the fluid close-packing limit, $\rho b = 1$, while for the solid close-packing limit $\bar{\rho} b = 1$. The quantity $m_s(\rho)$ is the equilibrium value of the order parameter in the solid phase and hence is a solution of the equation $(\partial f_s / \partial m)_{m_s(\rho)} = 0$. Thus in the solid close-packing limit $(b\bar{\rho} = 1)$ $m_s = 1$. The parameter a_m in equation (3) has been introduced in order to change the range of the attractive interactions between molecules in the solid phase from that in the liquid phase (as described by the parameter a in equation (2)).

The chemical potential in the solid and fluid phases is the first derivative of the free energy with respect to density: $\mu = (\partial f / \partial \rho)_T$. In order to get coexisting densities ρ_α and ρ_β we have to solve the equations $\mu(\rho_\alpha) = \mu(\rho_\beta)$ and $\omega(\rho_\alpha) = \omega(\rho_\beta)$, where $\omega = f - \mu\rho$. Graphically this gives the well-known "common tangent" rule for coexistence. By repeating this for different temperatures we can obtain the entire phase diagram.

In order to obtain a critical droplet profile we must solve the Euler-Lagrange equations with appropriate boundary conditions:

$$\frac{\delta\Omega}{\delta\rho} = 0 \quad \text{and} \quad \frac{\delta\Omega}{\delta m} = 0$$

i.e.,

$$-K_\rho \nabla^2 \rho + \frac{\partial f}{\partial \rho} - \mu = 0 \quad (4)$$

$$-K_m \nabla^2 m + \frac{\partial f}{\partial m} = 0 \quad (5)$$

Using the solutions of (4) for $\bar{\rho}(\mathbf{r})$ and $\bar{m}(\mathbf{r})$ we can obtain such properties of the inhomogeneous system as the free energy barrier and the surface tension.

3 NUMERICAL RESULTS

Here we present some results, extending the work of [3], for the following choice of parameters: $a = 1$, $b = 1$,

$\alpha_1 = 0.25$, $\alpha_2 = 2$, $\alpha_3 = 0.3$, $a_m = 1$. The phase diagram for these values is shown in Figure 1, which shows in particular a metastable fluid-fluid coexistence curve. The existence of the metastable critical point affects the nucleation and growth processes in the vicinity of this point.

The main quantity characterizing the self-assembly process is a nucleation rate I, given by

$$I = I_0 e^{-\Delta\Omega/kT} \quad (6)$$

where I_0 is a prefactor given by the product of dynamical and statistical parts [6]. To be able to control the self-assembly process we need to understand how this metastable critical point affects nucleation. We calculate the free energy barrier for different thermodynamic paths. (Some results for this barrier were presented in [3] and in [5].) The barrier dependence on temperature and density in the metastable region between the liquidus (solubility curve) and solidus lines is quite straightforward. As we increase the temperature at the constant density of the disordered state the barrier increases and diverges at liquidus line. The same behavior is obtained [5] if we decrease the density of the disordered state an the constant temperature. This is because at any point on the solubility curve the free energy is infinite and decreases as one moves away from it and vanishes at the spinodal.

However, the existence of the metastable fluid-fluid coexistence curve changes the pathways of the constant free energy barrier and constant supersaturation lines as compared with the case in which the coexistence curve is not metastable. We can also see that along the lines with constant supersaturation the free energy barrier decreases as one approaches the critical point (Fig. 2), which is consistent with [3]. Figure 2 shows that the free energy barrier has a minimum near, but not at, the critical point. The location of this minimum changes from above the critical point for low supersaturation to below the critical point for high supersaturation. It also can be seen that the increase of the free energy barrier below the critical point becomes very sharp for the cases where the constant supersaturation lines intersect the fluid-fluid coexistence curve. Because there is a discontinuity in those lines of constant supersaturation which intersect the fluid-fluid coexistence curve, the free energy barrier has a corresponding discontinuity. this leads to the jumps in the dependence of the barrier on temperature (fig. 2 for $\Delta\mu = 0.8$ and $\Delta\mu = 0.9$.). As we can see from figure 3 the surface tension also has a discontinuity at $\Delta\mu = 0.8$ and $\Delta\mu = 0.9$. In order to understand the behavior of the free energy barrier along the constant supersaturation lines, we first consider their shape. At high temperatures these lines are more vertical (constant density lines) and the free energy barrier decreases with temperature. Near the critical

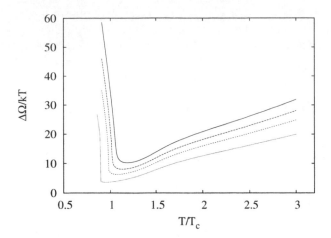

Figure 2: Dependence of the free energy barrier versus temperature at constant supersaturation. From top to bottom $\Delta\mu = 0.7, 0.75, 0.8, 0.9$.

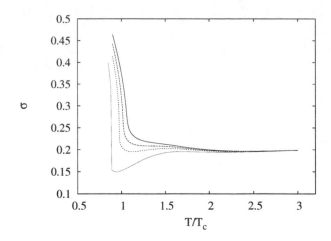

Figure 3: Dependence of the surface tension versus temperature at constant supersaturation. From top to bottom $\Delta\mu = 0.7, 0.75, 0.8, 0.9$.

point the curves become almost horizontal (isothermal lines) and the barrier starts to increase. Thus the shape of the constant supersaturation curve in some sense determines the behavior of the free energy barrier. To understand this in more detail, we note that the derivative of temperature with respect to density at constant supersaturation is

$$\left(\frac{\partial T}{\partial \rho_0}\right)_{\Delta\mu} = -\frac{\left(\frac{\partial \Delta\mu}{\partial \rho_0}\right)}{\left(\frac{\partial \Delta\mu}{\partial T}\right)} = -\left[\frac{f_{\rho\rho}^{(f)}}{\left(\frac{\partial \Delta\mu}{\partial T}\right)}\right]_{\rho=\rho_0} \quad (7)$$

where $f_{\rho\rho}$ is evaluated at the background fluid density. This derivative vanishes at the critical point, so that the constant supersaturation lines become horizontal in the vicinity of the critical point. On the other hand, at large temperatures the denominator vanishes, so that the constant supersaturation lines become vertical. Thus we see that the presence of the critical point changes the behavior of the free energy barrier from decreasing as we lower the temperature far from the critical point (the vertical part of the $\Delta\mu = const$ lines) to increasing as we lower the temperature to its critical value (the horizontal part of the $\Delta\mu = const$ lines). Therefore somewhere in between there is a minimum of the free energy barrier. Thus we can conclude that the free energy barrier is relatively unaffected by the existence of the critical point along paths of constant temperature or constant density, whereas it plays a crucial role along paths of constant supersaturation

One quantity of interest is the excess number of molecules, defined as the number of molecules in the presence of the droplet relative to the number of molecules in the spatially homogeneous metastable state:

$$N = \int_V (\rho(r) - \rho_0) d\vec{r} \quad (8)$$

We can calculate the number of molecules in the crystalline state using $m(r)$, since this phase field shows whether a point is in the solid or fluid state. Thus:

$$N_c = \int_V m(r)(\rho(r) - \rho_0) d\vec{r} \quad (9)$$

Mean field theory yields a divergence in the excess number of particles at the metastable critical point, whereas the number of particles in the crystalline state remains finite. We can check our numerical results with the nucleation theorem [7], [8]:

$$\frac{\partial \Delta\Omega}{\partial \Delta\mu} = -N \quad (10)$$

Thus we can take a derivative of the free energy barrier with respect to supersaturation and compare this with our results for the dependence of the excess number of particles on the supersaturation [5]. We see that the free energy barrier decreases rapidly near the critical point as a function of the supersaturation. This happens because the background fluid density as a function of the supersaturation becomes flat in the vicinity of the critical point.

4 DISCUSSION AND CONCLUSIONS

In this paper we have extended the numerical results [3] to obtain a better understanding of nucleation for their model. In particular, we have calculated the density and structure order parameter profiles of the critical nucleating droplet for different temperatures at constant supersaturation. This solution of the saddle

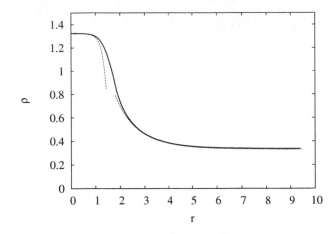

Figure 4: Comparison of the numerically obtained density profile (at $T/T_c = 1.2$ and $\rho_0 = 0.33$) with the tail and core approximations (dashed lines).

point equations then yields the free energy barrier to nucleation and the surface tension. As it can be seen the nucleation barrier has a minimum near the critical point. Also if the path with constant $\Delta\mu$ crosses fluid-fluid coexistence curve, then both free energy barrier and surface tension have discontinuity.

In our previous paper [5] we also presented analytical approximation for the core and the tail of the critical cluster. The tail solution is in agreement with Sear's approximation [9]. Figure 4 shows these approximation in comparison with numerical solution. We can see, that while the core approximation is very close to the numerical solution near the center of the droplet it diverges quickly as it approaches the interface region.

This work was supported by NSF grant DMR-0302598.

REFERENCES

[1] A. George and W. Wilson, Acta Cryst. D **50**, 361 (1994).

[2] D. Rosenbaum, P. C. Zamora, and C. F. Zukoski, Phys. Rev. Lett. **76**, 150 (1996).

[3] V. Talanquer and D. Oxtoby, J. Chem. Phys. **109**, 223 (1998).

[4] P.R. ten Wolde and D. Frenkel, Science **277**, 1975 (1997).

[5] Andrey Shiryayev and James D. Gunton J. Chem. Phys. to be published.

[6] J. Langer, Annals of Physics **54**, 258 (1969).

[7] D. Kashchiev, J. Chem. Phys. **76**, 5098 (1982).

[8] Y. Viisanen, R. Strey, and H. Reiss, J. Chem. Phys. **99**, 4680 (1993).

[9] R. Sear, J. Chem. Phys. **114**, 3170 (2001).

A Monte Carlo Study of a Model of Globular Protein Nanoparticles

D. L. Pagan*, M. E. Gracheva**, and J. D. Gunton*

* Department of Physics
Lehigh University, Bethlehem, PA, USA
** Department of Mathematics
University of Minnesota, Minneapolis, MN, USA

ABSTRACT

We explore the metastable fluid-fluid coexistence curve of the modified Lennard-Jones model of globular proteins of ten Wolde and Frenkel [4]. Using grand canonical Monte Carlo simulations along with mixed-field finite-size scaling and histogram rewighting, we estimate the critical temperature to be $T_c^*(\infty) = .4145(5)$ at infinite volume. The subcritical region of the metastable fluid-fluid coexistence curve is obtained using the hyper-parallel tempering method. In close proximity to the critical point, a region previously unattained by simulation, the phase diagram is calculated.

Keywords: nanoparticle, self-assembly, Monte Carlo, proteins

1 INTRODUCTION

Many diseases are known to be caused by the crystallization of certain globular proteins. For example, it has been shown that sickle cell anemia [1] begins with the onset of crystallization of the mutant hemoglobin-S molecule. Other diseases that exhibit this type of morphology include genetic cataracts [2] and Alzheimer's disease. Recent work [3] has focused on understanding the conditions which give rise to the crystallization of such proteins, in order to ultimately prevent the onset of such diseases.

Globular proteins are nanoparticles that self-assemble to form crystals. They are known to have a range of attractive interaction that is small in comparison with the size of the proteins. This is in sharp contrast to much smaller particles such as argon, whose properties have been more extensively studied. Both simulation [4] and theory [5] have predicted that the crystallization of these nanoparticles is optimized in the vicinity of a metastable, fluid-fluid critical point. An accurate determination of the critical point and the phase diagram in the nearby region, then, is needed to precisely identify the conditions of this optimization. We use a model of globular proteins [4] known to capture experimentally determined characteristics, the pairwise energy of which is shown in Eq. (1):

$$V(r) = \begin{cases} \infty, & r < \sigma \\ \frac{4\epsilon}{\alpha^2}\left(\frac{1}{[(r/\sigma)^2-1]^6} - \frac{\alpha}{[(r/\sigma)^2-1]^3}\right), & r \geq \sigma \end{cases} \quad (1)$$

where σ denotes the hard-core diameter of the particles, r is the interparticle distance, and ϵ is the well depth. The width of the attractive well can be adjusted by varying the parameter α; for $\alpha = 50$, the resulting phase diagram was found to map to that of globular proteins determined experimentally [3]. We determine the metastable fluid-fluid critical point of the above model using finite-size scaling techniques adapted for simple fluids by Bruce and Wilding [6] and determine the corresponding binodal curve near the critical point using the hyper-parallel tempering method [7].

2 MODEL

Grand canonical Monte Carlo simulates a system at constant volume $V = L^d$, chemical potential μ, and temperature T while the number of particles N is allowed to fluctuate. The joint probability distribution of the number-density $\rho = L^{-d}N\sigma^d$ and energy-density $u = UL^{-d}$ at a particular point in parameter space (β, μ) is analyzed, where U is the total pair-wise energy and $\beta = 1/k_B T$. We then obtain the joint probability distribution of mixed operators

$$P_L(\mathcal{M}, \mathcal{E}) = (1 - sr)\, P_L(\rho, u), \quad (2)$$

where

$$\mathcal{M} = \frac{1}{1 - sr}\,[\rho - su] \quad (3)$$

and

$$\mathcal{E} = \frac{1}{1 - sr}\,[u - r\rho] \quad (4)$$

are the ordering and energy-like operators, respectively, with the parameters s and r system-dependent field-mixing parameters. In simple fluids, or off-lattice models, there is an absence of particle-hole symmetry which gives rise to an asymmetrical density distribution. Analyzing the joint-operator distribution allows one to recover a symmetrical distribution, accounting for mixed-field scaling effects. At criticality, the ordering-operator distribution obeys a scaling relation [6] of the form

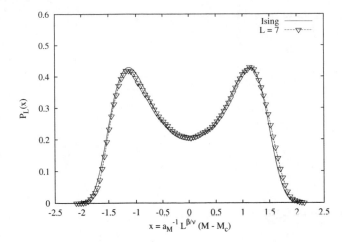

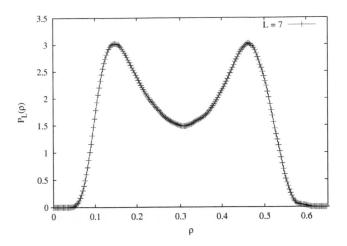

Figure 1: The form of the ordering operator distribution $P_L(\mathcal{M})$ for the $L = 7$ system size. Shown for comparison is the universal Ising fixed-point ordering operator distribution $\tilde{P}_{\mathcal{M}}^*(x)$ (solid). The data has been expressed in terms of the scaled variable $x = a_{\mathcal{M}}^{-1} L^{\beta/\nu}(\mathcal{M} - \mathcal{M}_c)$, where $a_{\mathcal{M}}^{-1}$ was chosen such that the distribution has unit variance.

$$P_L(\mathcal{M}) = \tilde{P}_{\mathcal{M}}^*(a_{\mathcal{M}}^{-1} L^{(d-1/\nu)}[\mathcal{M} - \langle\mathcal{M}\rangle_c]), \quad (5)$$

where $a_{\mathcal{M}}^{-1}$ is a non-universal scaling factor, d is the dimensionality of the system, ν is a critical exponent [8], and the symbol $\langle\rangle_c$ denotes a grand canonical average taken at criticality. It has been shown [9]–[11] that simple fluids belong to the Ising universality class. Matching the fluid fixed point distribution $P_L(\mathcal{M})$ to that of the Ising model at criticality, then, allows for an accurate determination of the critical point parameters.

In the subcritical region, sampling the joint distribution of coexisting phases becomes prohibitively difficult. In this region, a free-energy barrier exists that can no longer be overcome by the now low density fluctuations. We use hyper-parallel tempering Monte Carlo (HPTMC) [7] to tunnel through this barrier. Parallel-processing is used to set up a composite system of simulations (replicas) at different state points (β, μ), where the unnormalized probability density of the composite state $\mathbf{x}$ is given by

$$p(\mathbf{x}) = \prod_{i=1}^{M} exp[-\beta U(x_i) + \beta\mu N(x_i)], \quad (6)$$

with x_i denoting the state of the ith replica of the M replicas. Two types of trial moves are employed according to the above limiting function: 1) regular insertion/deletion trial moves are attempted within each replica and 2) configuration swaps are attempted between pairs of replicas. To enforce detailed-balance, the

Figure 2: The density distribution for the $L = 7$ system size shown at the critical point $T_c^*(L = 7) = .4130(2)$.

pair of replicas to be swapped are chosen at random and are accepted with the probability

$$p_{acc}(x_i \leftrightarrow x_{i+1}) = min[1, exp(\Delta\beta\Delta U - \Delta(\beta\mu)\Delta N)], \quad (7)$$

where $\Delta\beta = \beta_{i+1} - \beta_i$, $\Delta U = U(x_{i+1}) - U(x_i)$, $\Delta(\beta\mu) = \beta_{i+1}\mu_{i+1} - \beta_i\mu_i$, and $\Delta N = N(x_{i+1}) - N(x_i)$.

Both in the critical and subcritical regions, we employ histogram reweighting [12] to facilitate the analysis. Histogram reweighting allows one to extract accurate information about neighboring state points (β', μ') from one single simulation run at a specific set of state points (β, μ).

The potential in Eq. (1) was truncated at a cutoff distance $r_c = 2.0\sigma$ and left unshifted. Simulations were performed on system sizes of $L = 6\sigma, 7\sigma, 8\sigma$, and 10σ. The systems were equilibrated at two million steps and production runs ranged from 500 million steps to 1 billion steps for the smaller and higher system sizes, respectively. The chemical potential was tuned until a double-peaked structure was obtained in the density distribution in the vicinity of the critical point. We then matched $P_L(\mathcal{M})$ to the Ising distribution as described above. This procedure was repeated for all system sizes studied. In the subcritical region, six replicas were used to obtain coexistence for the $L = 7\sigma$ system. Lower temperatures were biased toward a high density distribution while those at higher temperatures were biased toward a low density distriubtion. This allowed for swapped configurations to 'melt' more easily. We chose our temperatuers and chemical potentials so that many swaps were realized; on average, swaps were accepted 25 percent of the time.

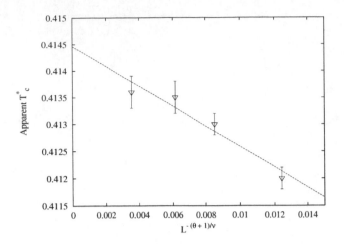

Figure 3: The apparent reduced critical temperature plotted as a function of $L^{-(\theta+1)/\nu}$, with $\theta = 0.54$ and $\nu = 0.629$. The extrapolation to infinite volume yields the estimate $T_c^*(\infty) = .4145(5)$.

3 RESULTS

In Figure 1, we match our order-operator distribution to that of the Ising model. For clarity, only the matching for the $L = 7$ system size is shown. As can be seen, the agreement between the two is good, giving us estimated values of the critical parameters for the $L = 7$ system size of $T_c^*(L) = .4130(2)$ and $\rho_c^*(L) = .312(2)$. In Figure 2, we show the corresponding density distribution for the $L = 7$ system at criticality. By carrying out our simulations at different system sizes, we are able to determine the infinite-volume critical temperature $T_c^*(\infty)$. Deviations from the true critical point are of the form

$$T_c^*(\infty) - T_c^*(L) \propto L^{-(\theta+1)/\nu}. \qquad (8)$$

In Figure 3, we plot the apparent critical temperature $T_c^*(L)$ versus $L^{-(\theta+1)/\nu}$. Extrapolation to infinite volume gives us the estimate for the true critical temperature $T_c^*(\infty) = .4145(5)$.

Using HPTMC, density distributions in the subcritical region in close proximity to the critical point were obtained for the $L = 7$ system size. In Figure 4, we plot the temperatures of our six replicas as a function of the average densities of the density distributions to obtain this portion of the phase diagram. Also shown is a power law fit of the form $\rho_\pm - \rho_c = A\,|T - T_c| \pm B\,|T - T_c|^\beta$, where T_c and ρ_c are are our extrapolated values as $L \to \infty$ and $\beta = .3258$ [8] is the Ising exponent, mapping out the rest of the metastable region. From our grand canonical Monte Carlo simulations, we were able to obtain the fluid coexistence curve; also shown is our extrapolated values for the solid coexistence curve. These results match well the previous results of ten Wolde and

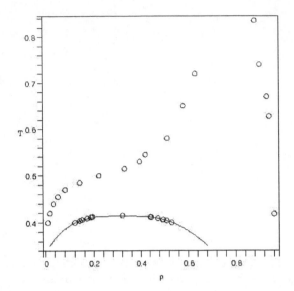

Figure 4: Phase diagram as obtained from both HPTMC and grand canonical Monte Carlo simulatons for the $L = 7$ system size shown with the critical point $T_c^*(\infty)$ and the corresponding critical density ρ_c. Also shown (solid line) is a fit to the data, with $\beta = .3258$, the fluid coexistence curve, and extrapolated values for the solid coexistence curve.

Frenekl [4]. The metastable data provide a reasonable match to the power-law fit, though we have not accounted for possible finite-size effects.

4 CONCLUSION

Mixed-field finite-size scaling techniques and histogram extrapolation methods were used to estimate the critical-point parameters of the modified Lennard-Jones model [4] of globular proteins. Our estimate for the infinite-volume temperature is $T_c^*(\infty) = .4145(5)$, slightly lower than a previous estimate. In the nearby subcritical region, we explored the phase diagram in close proximity to the critical point, a region previously unattained by simulation. Further details of this work are given in [13].

Many experimental studies have shown that nucleation of protein crystals is indeed enhanced near the critical point. They have also found, however, that globular proteins tend to form gels in part of this region. In a similar system of adhesive spheres, it was recently determined that the critical point lies within the percolation curve [14]. There is no pathway to form gels for the current model, though mode coupling theories describe the gelation region for other similar systems. Obtaining the 'gel-line' as described by mode coupling theory would resolve these differences.

ACKNOWLEDGEMENT

This work was supported by a grant from the National Science Foundation, DMR-0302598.

REFERENCES

[1] O. Galkin, et al., PNAS **99**, 8479 (2002)

[2] Ajaye Pande, et al., PNAS **98**, 6116 (2001)

[3] D. Rosenbaum, P.C. Zamora, and C.F. Zukoski,Phys. Rev. Lett., **76**, 150 (1996)

[4] P. R. ten Wolde and D. Frenkel, Science,**277**, 1975 (1997)

[5] D. W. Oxtoby and V. Talanquer, J. Chem. Phys., **101**, 223 (1998)

[6] N. B. Wilding, Phys. Rev. E, **52**,602 (1995)

[7] Qiliang Yan and Juan J. de Pablo, J. Chem. Phys.,**111**, 9509 (1999)

[8] A. M. Ferrenberg and D. P. Landau, Phys. Rev.B, **44**, 5081 (1991)

[9] M. W. Pestak and M. H. W. Chan, Phys. Rev. B,**30**, 274 (1984)

[10] U. Närger and D. A. Balzarani, Phys.Rev. B **42**, 6651 (1990); U. Närger, J. R. de Bruyn, M.Stein, and D. A. Balzarini, *ibid.* **39**, 11 914(1989)

[11] J. V. Sengers and J. M. H. Lefelt Sengers, Annu. Rev. Phys. Chem. **37**, 189 (1986)

[12] A. M. Ferrenberg and R. H. Swendsen, Phys. Rev. Lett., **61**, 2635 (1988)

[13] D. L. Pagan, M. E. Gracheva, and J. D. Gunton, submitted for publication, Los Alamos preprint 0311551

[14] Mark A. Miller and Daan Frenkel, Phys. Rev. Lett., **90**, 135702 (2003)

DNA and Estrogen Receptor Interaction Revealed by the Fragment Molecular Orbital Method
– Implementation of SCF Convergence –

T. Watanabe[*], Y. Inadomi[**], S. Tanaka[***], and U. Nagashima[*,**]

[*]Japan Science and Technology Corporation (JST), 4-1-8 Honcho,
Kawaguchi 332-0012, Japan, donjo@ni.aist.go.jp
[**]Grid Technology Research Center, National Institute of Advanced Industrial
Science and Technology, 1-1-1 Umezono, Tsukuba, 305-8568 Japan
[***]Toshiba Corporation, 1 Komukai Toshiba-cho, Saiwai-ku, Kawasaki, 212-8582, Japan

ABSTRACT

The fragment molecular orbital (FMO) method was applied to evaluate the interaction between estrogen response element (ERE) in DNA and DNA binding domain (DBD) of estrogen receptor (ER). In order to calculate the whole electronic state of DNA-ERE and ER-DBD, we developed the SCC-DIIS method and the field scaling method. The SCC-DIIS method is for acceleration of the self-consistent charge (SCC) convergence, and the field scaling method changes the field effect from other fragments for stable convergence. Hybridization of these two methods made it possible to calculate the whole electronic state of DNA-ERE and ER-DBD.

Keywords: estrogen receptor, DNA binding domain, zinc finger, fragment molecular orbital

1. INTRODUCTION

Estrogen receptor (ER) families belong to a superfamily of ligand-induced nuclear steroid hormone receptors. These families repress or enhance transcription when identifying and binding as a dimer to specific DNA target sequences, such as estrogen response element (ERE). In the DNA binding domain (DBD) of ER, Zinc finger (Zn-finger) motifs formed by eight cysteine residues and two zinc ions play a critical role in this specific binding between DNA-ERE and ER-DBD. There are many studies both experimental and theoretical studies about the specific binding.[1-4] The structural aspect of the binding between DNA-ERE and ER-DBD is revealed that one of the α-helix of Zn-finger interacts with primary groove of DNA-ERE and another α-helix positioned in the secondary groove specifies the sequences of DNA-ERE.[1] However the detailed structure and mechanism of the specific binding are not known. For example, the protonation state of cysteine residues in Zn-finger motifs are controversial.[2,3]

There were some molecular dynamics studies[4], importance of the water molecules was revealed. However the classical force field is insufficient for treatment of the highly charged motifs, such as DNA-ERE and Zn-finger. On the other hand, the studies using *ab initio* molecular orbital method could treat the highly charged motifs but only a small part of the whole system.[2]

The challenging studies by applying *ab initio* molecular orbital method to proteins and enzymes have been carried out,[5-7] but it's very time-consuming calculation even by using the most up to date computer. The fragment molecular orbital (FMO) method[8-12] is the approximation for the MO calculation of a huge molecule, for example protein and DNA, and surprisingly well reproduces the total energy of the MO calculation with a moderate computational effort. The FMO method divides a huge molecule into small fragments, and then calculates the MOs of each fragment and fragment pair. The computational efficiency of the FMO method is achieved by the escape from the hot spot of the conventional MO calculation, such as the two electron integrals (typical computational cost proportional to N^4, where N is the number of basis functions) and the large scale diagonalization (typically N^3). The FMO method gives not only total energy but also the inter fragment interaction energy and various physical variables, such as dipole moment and so on.[13] The whole molecular orbital is also able to obtain by using the FMO-MO scheme.[14]

Another advantage of the FMO method is the high parallel efficiency. Especially the MO calculation of each fragment pair can be carried out completely independent of other fragment pair. The MO calculation of the fragment, however, needs to repeat until the electron density of each fragment become self-consistent. For the efficient FMO calculation, the acceleration of this iteration is required. Sugiki et al[15] accelerated the convergence of this iterative procedure in the FMO-DFT method. Additionally the unstable MO state doesn't converge in this iterative procedure sometimes, the safety convergence is also greatly required. In this paper, the SCC-DIIS and field scaling methods were tested and implemented for the efficient FMO calculation of

2. METHOD

2.1 Fragment Molecular Orbital Method

In the FMO method, there are two procedures. At first, the MO calculations for each fragment are repeated until the electron density of each fragment become self-consistent. This converged state called as the self-consistent charge (SCC). After SCC achieved, the MO calculations for all possible combination of fragment pair are carried out and the total energy of calculation, E_N, is obtained using the equation as

$$E_N = \sum_I E_I + \sum_{I>J} (E_{IJ} - E_I - E_J)$$

where E_I and E_{IJ} denote the energies of the fragment I and the fragment pair IJ, respectively. The FMO method is free from the MO method used to calculate the E_I and E_{IJ}, however, we assume the HF based FMO calculation hereafter.

In the MO calculation of the fragment I, the effect of other fragments is introduced into Hamiltonian, H_I, as an electrostatic field from the electron density of other fragment J, ρ_J, as below,

$$H_I = \sum_{i \in I} \left\{ (-\frac{1}{2}\nabla^2) + \sum_A (-\frac{Z_A}{|r_i - A|}) + \sum_{J \neq I}^{N_J} \int \frac{\rho_J(r')}{|r_i - r'|} dr' \right\} + \sum_{i \in I} \sum_{i>j \in I} \frac{1}{|r_i - r_j|}$$

where A and r denote the coordinates of nuclear and electron, respectively, and Z is nuclear charge. The MO calculation of the fragment pair IJ is also represented by same Hamiltonian with the subscription IJ instead of subscription I.

There are several program packages of the FMO method, the ABINIT-MP program package[16] was used in the all calculation in this paper.

2.2 SCC-DIIS Method

The SCC-DIIS method is an application of the direct inversion of the iterative subspace (DIIS) method to the SCC problem. The DIIS methods[17,18] have been successfully applied for acceleration of convergence in various iterative problems such as the self-consistent field calculation of Hartree Fock. Sugiki et al[15] were implemented the DIIS method to the SCC problem and achieved the convergence acceleration of the FMO calculation of the glycine polymers. We implemented the modified DIIS method to the ABINIT-MP program package. In this method, the differences of the interfragment interaction energies were used as the error function of DIIS, and the SCC-DIIS method acts to minimize this error function. When SCC is achieved, this error function vanishes.

2.3 Field Scaling Method

If the target electronic state of the molecule is unstable, the FMO calculation sometimes diverges in the SCC procedure. For safety convergence, the field scaling method was developed. In this method, the field effect from other fragment was added stepwise to avoid the large noise which arise the divergence of the SCC procedure.

3. RESULTS AND DISCUSSION

3.1 Model Calculation of (gly)₁₅

The model calculation using the α-helix glycine 15-mer, (gly)$_{15}$, was carried out to test the convergence of SCC procedure. The FMO calculations were carried out with the STO-3G and 6-31G basis sets. Figure 1 was results of the SCC-DIIS method for the FMO/6-31G calculation. The thick and solid line was the calculation without the SCC-DIIS method, and other lines represented the SCC-DIIS methods with the various number of the histories of the electron density matrices in iterative procedure.

The SCC procedure without the SCC-DIIS method converged with 37 cycles. The SCC-DIIS calculations with more than 5 histories of the electron density matrices largely accelerated convergence, ca 23 cycles, although the SCC-DIIS calculations with 2-4 histories accelerated little. In the calculation of a huge molecule, the electron density matrix is very large, and a large number of the electron density matrices can't be saved on memory. This calculation indicated that 5 histories of the electron density matrix were sufficient for acceleration of the SCC convergence.

In the initial stage of the SCC procedure, all SCC-DIIS calculations deaccelerated comparison with the calculation without the SCC-DIIS method. In general, the DIIS method effectively accelerates the iterative procedure in the near region of the solution to be

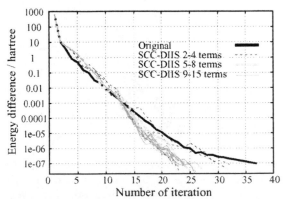

Figure 1. The convergence of the SCC procedure of the FMO/6-31G calculations of (gly)$_{15}$ with or without the SCC-DIIS method.

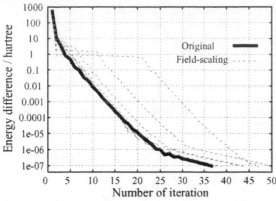

Figure 2. The convergence of the SCC procedure of the FMO/6-31G calculations of (gly)$_{15}$ with or without the field scaling method.

converged, although it doesn't work well and sometimes need other method in the far region from the solution. [19]

The field scaling method doesn't accelerate the SCC procedure of (gly)$_{15}$ as shown in Figure 2. The reason may be that (gly)$_{15}$ is homo-polymer and there is little noise for the SCC convergence.

3.2 Calculations of DNA-ERE and ER-DBD

The structure of DNA-ERE and ER-DBD was used in the reference 4, as shown in Figure 3. The calculation of DNA-ERE with no water molecule and no counter ion was shown in Figure 4. The thick and solid line which is the SCC procedure without the SCC-DIIS and field scaling methods didn't converge at all. The thick and dashed line which is the SCC-DIIS calculation also diverged. These behavior due to the unstable electronic state of DNA-ERE in the gas phase, because DNA has a large number of the anionic charged phosphoric acids.

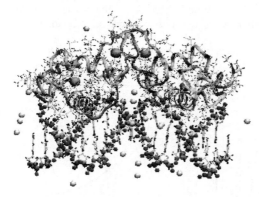

Figure 3. DNA estrogen response element and DNA binding domain of estrogen receptor. Zinc atoms are shown in orange and counter ions (sodium) in yellow. Water molecules surrounding these molecules are not shown. This structure used in this study was taken from reference 4.

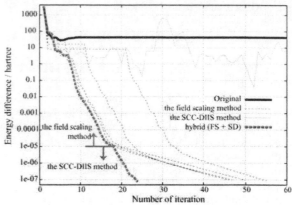

Figure 4. The convergence of the SCC procedure of the FMO/STO-3G calculations of DNA-ERE.

The calculations with the field scaling method shown as the thin and dashed lines in Figure 4 successfully converged, and indicated the effectiveness of the field scaling method for the instable electronic state. On the other hand, the final stage of the SCC procedure in the field scaling calculation was considerably deaccelerated. By the hybridization between the field scaling and SCC-DIIS methods shown as the thin and solid line, the SCC procedure was converged straightforwardly. In the near region of the solution to be converged, the SCC-DIIS method effectively accelerated. This hybridization was complementary and achieved the safety and rapid convergence.

4. Summary

For the interaction analysis between DNA-ERE and ER-DBD, the SCC procedure in the FMO method was developed by using the SCC-DIIS and field scaling methods. In the model calculation of (gly)$_{15}$, the SCC-DIIS method largely accelerated in the last stage of the SCC procedure but slightly deaccelerated in the initial stage. Five histories of the electron density matrices used in the SCC-DIIS method were sufficient.

In the FMO calculation of DNA-ERE, the SCC procedures with or without the SCC-DIIS method were diverged. The field scaling method was needed to converge but considerably deaccelerated in the last stage of the SCC procedure. Hybridization between the field scaling and SCC-DIIS methods was complimentary and achieved the safety and rapid convergence of the SCC procedure.

Acknowledgements

We are thankful to Dr T. Nakano and Dr K. Fukuzawa for helpful discussion. The DNA-ERE and ER-DBD structure[4] was obtained from Prof L. Nilsson. This work was supported in part by the grant from

ACT-JST ("Research and Development for Applying Advanced Computational Science and Technology" of Japan Science and Technology Corporation).

References

[1] J. W. R. Schwabe, L. Chapman J. T. Finch, D. Rhodes, *Cell,* **75**, 567 (1993).

[2] L. Y. Low, H. Hernandez, C. V. Robinson, R. O'Brien, J. G. Grossmann, J. E. Ladbury, B. Luisi, *J. Mol. Bio.,* **00**, 87 (2002).

[3] T. Dudev, C. Lim, *J. Am. Chem. Soc.,* **124**, 6759 (2002).

[4] M. A. L. Eriksson, L. Nilsson, *Eur. Biophys J.,* **28**, 102 (1999).

[5] M. Challacombe, E. Schwegler, *J. Chem. Phys.,* **106**, 5526 (1997).

[6] C. Van Alsenoy, C.-H. Yu, A. Peeters, J.M.L. Martin, L. Schafer, *J. Phys. Chem.,* **102**, 2246 (1998).

[7] F. Sato, T. Yoshihiro, M. Era, H. Kashiwagi, *Chem. Phys. Lett.* **341**, 645 (2001).

[8] K. Kitaura, T. Sawai, T. Asada, T. Nakano, M. Uebayashi, *Chem. Phys. Lett.,* **312**, 319 (1999).

[9] K. Kitaura, E. Ikeo, T. Asada, T. Nakano, M. Uebayashi, *Chem. Phys. Lett.,* **313**, 701 (1999).

[10] T. Nakano, T. Kaminuma, T. Sato, Y. Akiyama, M. Uebayashi, K. Kitaura, *Chem. Phys. Lett.,* **318**, 614 (2000).

[11] K. Kitaura, S.Sugiki, T. Nakano, Y. Komeji, M. Uebayashi, *Chem. Phys. Lett.,* **336**, 163 (2001).

[12] T. Nakano, T. Kaminuma, T. Sato, K. Fukuzawa, Y. Akiyama, M. Uebayashi, K. Kitaura, *Chem. Phys. Lett.,* **351**, 475 (2002).

[13] H. Sekino, Y. Sengoku, S. Sugiki, N. Kurita, *Chem. Phys. Lett.,* **378**, 589 (2003).

[14] Y. Inadomi, T. Nakano, K. Kitaura, U. Nagashima, *Chem. Phys. Lett.,* **364**, 139 (2002).

[15] S. Sugiki, N. Kurita, Y. Sengoku, H. Sekino, *Chem. Phys. Lett.,* **382**, 611 (2003).

[16] ABINIT-MP program package Ver.20021029.

[17] P. Pulay, *Chem. Phys. Lett.,* **73**, 393 (1980).

[18] P. Pulay, *J. Comput. Chem.,* **3**, 556 (1982).

[19] K. N. Kudin, G. E. Scuseria, E. Cances, *J. Chem. Phys.,* **116**, 8255 (2002).

Development of an Advanced Simulation System for the Analysis of Particle Dynamics in LASER based Protein Ion Sources

Andreas Hieke

Ciphergen Biosystems, Inc., 6611 Dumbarton Circle, Fremont, CA 94555 USA
ahi@ieee.org, ahieke@ciphergen.com

ABSTRACT

An advanced multi-physics simulation system henceforth referred to as GEMIOS (Gas and Electromagnetic Ion Optical Simulator) has been implemented which allows the computation of 3D trajectories and energy exchange of ion beams/clouds under the influence of 3D time-dependent external electromagnetic fields in the presence of 3D rarefied gas flow fields and has therewith capabilities beyond existing tools. GEMIOS achieves its unique capabilities by utilizing commercial (ANSYS) as well as special purpose codes (FORTRAN).

The fundamentally novel aspect of GEMIOS is that it combines electromagnetic field solutions and fluid dynamic field solutions obtained within a given domain to compute charged particle trajectories.

1. INTRODUCTION

The goal of clinical proteomics is to identify proteins which may be used as biomarkers and to correlate them with disease states. ProteinChip arrays serve to selectively capture proteins prior to further detailed analysis in specialized mass spectrometers (ProteinChip Readers). A principal and critical step in this process, performed in laser based systems called MALDI (matrix-assisted laser desorption/ionization) /SELDI (surface-enhanced laser desorption/ionization) sources or interfaces, is to generate, cool, and collimate ion beams of said captured proteins to allow injection into mass analyzing devices. Within these interfaces and devices protein ions are subjected to time dependent electromagnetic fields as well as collisions with intentionally introduced background gas molecules in order to reduce the ion temperature. The particle dynamics in such ion sources is extremely complex.

A simulation tool was needed which can provide insight into and a deeper understanding of ion extraction, collisional cooling and guidance processes in the presence of background gases. It also serves to optimize geometric shapes, electric potentials, frequencies, timing, gas flow rates and pressures.

2. METHODS

GEMIOS's capability to provide electromagnetic field solutions and fluid dynamic field solutions is essential to model ion optical systems operating at elevated pressures where electromagnetic forces and scattering effects are equally important. GEMIOS achieves this by utilizing a combination of codes (Fig.1) operating on the same input data set:

- a 3D Monte-Carlo Newtonian Motion and Collision module (MC-NMC), FORTRAN
- the FEM system ANSYS-Multiphysics,
- a semi-statistical 2D DSMC (Direct Simulation Monte Carlo) module for rarefied gas flow

ANSYS serves as fully parameterized 3D solid modeler and mesher and provides the 3D Laplace/Poisson solver for the computation of electromagnetic fields. According to the number of electrodes within a model and based on a set of canonical Dirichlet boundary conditions a data set of static, orthogonal base solution is computed from which any arbitrary static or dynamic field configuration is later obtained by superposition in the MC-NMC (Hieke [1]).

ANSYS also provides a 3D Navier-Stokes solver operating on the same mesh generated for the Lapace/Poisson solution and providing gas pressure, velocity distribution and temperature data of a background gas which is present in the considered ion-optical device. It is particularly advantageous for subsequent computations to obtain electromagnetic and gas flow solutions on the same mesh although optimal mesh distributions typically differ for both cases.

The DSMC module provides solutions for rarefied gas flow below the continuum range. Since DSMC is not based on macroscopic properties or derived from macroscopic behavior (as Navier-Stokes is) but rather simulates a gas by actually treating corpuscular collisions of a very large number of virtual super particles in a Monte Carlo algorithm it is capable of providing valid gas flow data at any pressure, Knudsen and Reynolds number (Bird[2]).

The 3D Monte-Carlo Newtonian Motion and Collision module (MC-NMC) is the core component of GEMIOS and computes trajectories and energy exchange of an ion beam/cloud under the influence of time dependent external electromagnetic fields in the presence of a background gas flow field. While particle tracing is possible in ANSYS using APDL (ANSYS Parametric Design Language) its execution speed is insufficient for Monte Carlo simulations (Hieke [3]). As a result, a compiled stand-alone code had to be used for the MC-NMC.

One of the challenges arising from the use of a FEM system is the computation of particle trajectories in arbitrarily meshes (Kenwright et al. [4]) which is far more complicated compared to equidistant orthogonal grids used by FDM, commonly found in particle tracing codes.

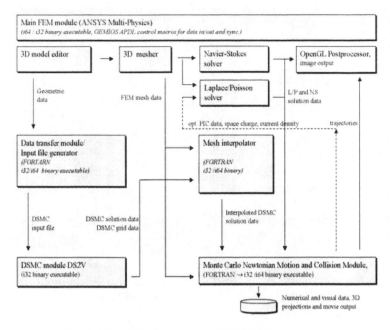

Fig.1: Simplified structure of GEMIOS

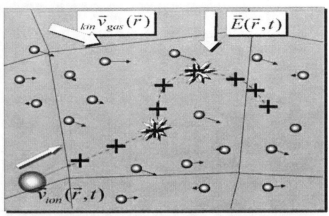

Fig.2: For ions (blue sphere) moving through the meshed domain under the influence of a time- dependent electric field a local collision probability is computed at each time step based on current ion velocity, ion-gas scattering cross section, local gas flow speed, pressure, and Monte Carlo sampled gas temperature. These quantities and the ion mass are also used to determine post-collisional velocity vectors if a collision event occurs (stars).

First, the MC-NMC reads in files containing information about the structure of the mesh such as node locations, connectivity, element to node correlation, element neighbors, and element center locations. It then builds various cross-reference and lookup tables and employs local search routines to avoid prohibitively expensive brute-force search routines during tracing. Thereby accurate first guess predictions for the element a particle is about to fly in with probabilities on the order of 80 to 90% are achieved.

Second, the MC-NMC reads the orthogonal Laplace base solutions for all nodes and performs a field superposition either globally, prior to the tracing for electrostatic cases, or locally in real time for time-dependent cases. Third, the MC-NMC module optionally reads in gas pressure, velocity and temperature data for element center locations if collisions are computed. In addition, information about particle masses and scattering cross sections must be provided.

The trajectory integration is based on a Runge-Kutta scheme with 2nd order accuracy in time and space with optionally automatic time stepping. Additionally, the MC-NNC employs local interpolation of electromagnetic fields within each element. Details are beyond the scope of this paper and will be reported elsewhere.

During tracing, and based on local gas density, (MC sampled) temperature, gas flow field velocity, scattering cross sections, actual ion velocity, and current time stepping a local, time-dependent probability for a collision event is computed. If in fact a collision event occurs, the post-collisional ion velocity is determined based on said local quantities and an additional MC sampling of angular velocity distributions as required for hard sphere models (Fig. 2). This procedure replicates aspects of the before mentioned DSMC approach. In future, additional collision models as well as rotational and internal degrees of freedom may be implemented. Finally, the MC-NMC module extracts time-dependent kinetic, translational, and thermal velocities. It also provides graphical output capabilities (2D projections) for very rapid animations and/or generation of movies of the computed events. Trajectory data can also be transferred to the ANSYS postprocessor for 3D visualization.

3. RESULTS

Trust in such a complex simulation system has to be established gradually by going from simpler to more complex models. The accuracy of the trajectory integration (without collisions) and field superposition has been verified for a number of cases with known analytical solution of trajectories and fields including long term dynamic stability in RF quadrupoles. The verification of the correct algorithmic approach of the collision treatment is more difficult. Appropriate test cases involve simple thermal equalization and/or constant gas flow speeds.

The application of electric RF fields in combination with background gas collisions introduces new phenomena such as collisional cooling and heating of ions for which analytical solutions are typically not available (Moore [5]).

Fig. 3 though Fig. 10 show some of such increasingly complex cases computed with GEMIOS ranging from simple ion plume expansion into resting background gas of constant pressure to RF quadrupole confinement and cooling as well as multi-electrode ion source systems.

The computational expense in terms of floating point performance and required memory is substantial. GEMIOS runs on a computer system comprising a dual Itanium2 64bit machine with 12Gbyte RAM, a dual Xeon 32bit machine with 4Gbyte RAM connected via Gigabit Ethernet and an external Arena 8-disk RAID-0 unit for temporary data storage with sustained I/O data rates on the order of 160Mbyte/s.

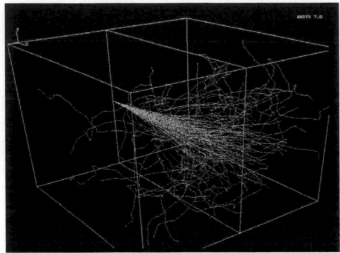

Fig.3: Simple test case: ion beam expanding into resting background gas, no external electromagnetic field, m_{ion}=1000u, d=1nm, v_{x0}=1000m/s, T_{gas}=300K, p_{gas}=0.1Pa, v_{gas}=0m/s

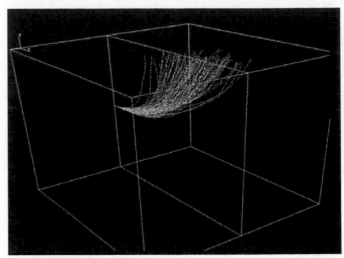

Fig.4: Test case: ion beam expanding into an upward streaming background gas, no external electromagnetic fields, m_{ion}=1000u, d=1nm, v_{x0}=1000m/s, T_{gas}=300K, p_{gas}=0.1Pa, $v_{y\ gas}$=500m/s

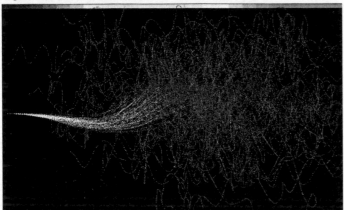

Fig.5: Test case: ion beam expanding into a resting background gas while subjected to a spatial uniform electric AC field; m_{ion}=1000u, d=1nm, v_{x0}=500m/s, T_{gas}=300K, p_{gas}=0.1Pa, v_{gas}=0m/s, Ey_{AC}=5V/m, f=500Hz, n_{coll}= $5 \cdot 10^4$, color indicates ion temperature

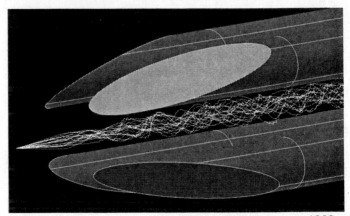

Fig.6: Ion injection into a RF quadrupole: m_{ion}=1000u, f=500kHz, U_{AC_eff}=500V; $\vec{v}_0 = \vec{e}_x \cdot 600\text{m/s} + \Re \cdot 100\text{m/s}$ some trajectories transferred back to the ANSYS post processor; notice the characteristic ion motion in the RF field

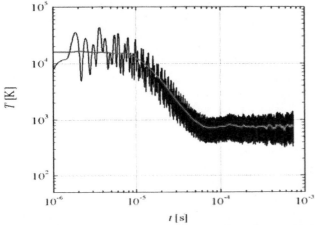

Fig.7: Thermal ion temperature as a function of time of a hot ion cloud injected into a gas filled RF quadrupole: m_{ion}=1000u, d=1nm, T_{gas}=300K, p_{gas}=10Pa, v_{gas}=0m/s, $|E_{max}| \approx 10^5$V/m, f=816kHz: cooling occurs within $t \leq 100\mu$s, final temperature higher than gas temperature due to collisional heating

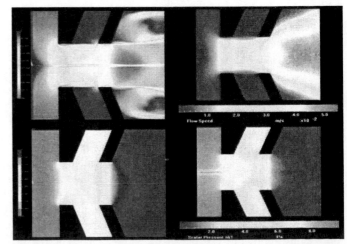

Fig.8: Comparison of Navier-Stokes (left) and DSMC solutions (right) for an axisymmetric gas expansion problem (system shown in Fig. 9): contour plots of flow velocity (top) and pressure (bottom)

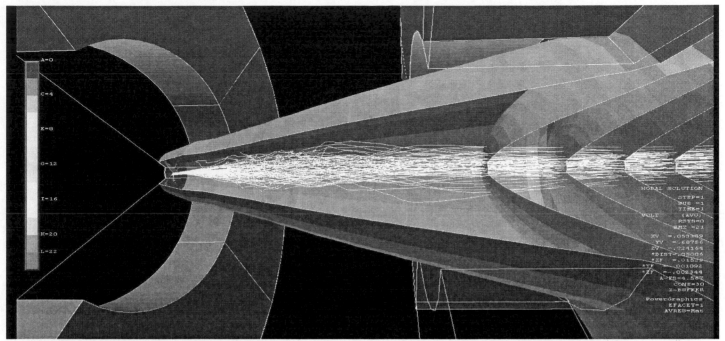

Fig.9: Early stage of an ion beam in a multi-electrode ion generation system, isosurface plot of electric potential, m_{ion}=1000u, $\vec{v}_0 = \vec{e}_x \cdot 1000$m/s $+ \vec{\Re} \cdot 200$m/s $\quad d$=1nm, p_{gas_max}=10Pa (computed gas flow field not displayed), n_{coll}= 5·10⁴

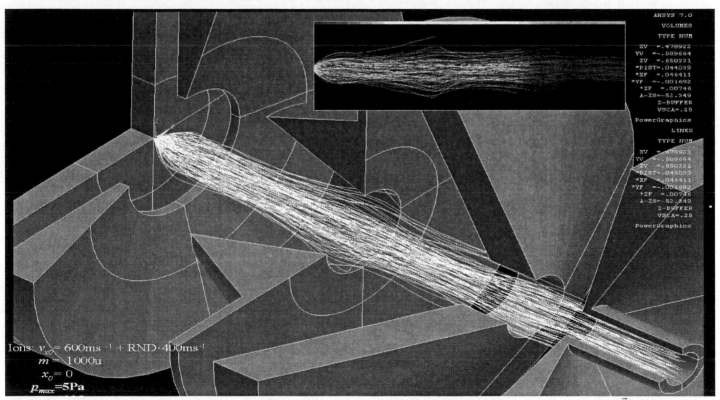

Fig.10: Ion beam in a multi-electrode ion generation system with background gas, m_{ion}=1000u, $\vec{v}_0 = \vec{e}_x \cdot 600$m/s $+ \vec{\Re} \cdot 400$m/s d=1nm, p_{gas_max}=5Pa, average number of collision per ion trajectory: 295, inserted picture shows trajectories colored according to local ion temperature; system exhibits ion losses on lenses under displayed conditions

References

[1] A. Hieke: "Simple ADPL Implementation of a 3D FEM Simulator for Mutual Capacitances of Arbitrarily Shaped Objects Like Interconnects", MSM '99, Puerto Rico, U.S.A. April 6-8, 1999

[2] Graham A. Bird: "Molecular Gas Dynamics and the Direct Simulation of Gas Flows", Clarendon Pr; ISBN: 0198561954; 2 edition (October 1994)

[3] A. Hieke: "Computation of a Magnetic Lens", 14th CADFEM Users' Meeting 1996, Bad Aibling, Germany

[4] David N. Kenwright and David A. Lane: "Interactive Time-Dependent Particle Tracing Using Tetrahedral Decomposition", IEEE Transactions on Visualization and Computer Graphics", 1996, vol. 2, no. 2, pp.120-129,

[5] R.B. Moore: "Ion Beam RFQ confinement and cooling", ISAC TRAP Workshop, Vancouver, April 2002

Resolved Particle Simulations of Microsphere/Cell-based Bioanalytic Systems

K. Pant, B.F. Romanowicz and S. Sundaram[*]

CFD Research Corporation
215 Wynn Drive, Huntsville, AL 35805
[*]Corresponding author, e-mail: sxs@cfdrc.com

ABSTRACT

Microsphere and cell-based assays are emerging as the platform of choice for rapid quantitation of biomolecular interactions in bioagent detection, clinical diagnosis, drug therapy, etc. [1,2] Traditional assay designs have been driven by expensive trial-and-error experimentation. The availability of high-fidelity simulation tools for these complex microsystems can help accelerate the design cycle significantly. However, current models rely on a point particle approximation and are incapable of describing large particle dynamics or biochemical coupling adequately. In a novel development, we have successfully developed computational models for finite-sized cells/bead assay design within a multiphysics biomicrofluidic simulation software suite, CFD-ACE+. The developed models fully integrate large bead dynamics, convective-diffusive analyte transport and biomolecular binding on bead surface. In this paper, we present the modeling framework, along with several validation and demonstration case studies.

Keywords: microsphere, cell, assay, design, resolved particle simulations

1 INTRODUCTION

The use of whole cells in microfluidic platforms to study biological interactions has been growing rapidly. Microspheres (or beads) are also used extensively in a variety of such applications as support structures for antibodies, proteins, DNA, etc. Microsphere- and cell-based assays provide several key benefits over conventional surface-derivatized assays including: (1) improved mixing and contacting, (2) ease of processing, e.g., cytometry, elution, etc., (3) ready multiplexing (4) continuous monitoring, and (5) enhanced functionality with magnetic/electric fields.

Traditionally, microsphere- and cell-based assay design development practices have been based on expensive, time-consuming, trial-and-error experimentation. The emergence of high fidelity, multiphysics simulations based analysis represents a shifting paradigm in bead/cell assay design [3]. A computational prototyping approach can be employed to screen design concepts, optimize performance and shorten the design cycle with tremendous savings in time and cost. However, widespread use of simulation-enabled design has been limited by two key technological deficiencies: (1)

inaccurate and inadequate representation of momentum and analyte exchange between beads and suspending fluid. This shortcoming is especially felt for large beads (relative to flow length scales) since conventional Lagrangian models rely upon a point particle approximation, where the volume of the buffer displaced by cell/bead is neglected. This clearly is not a valid assumption for microfluidic systems where bead/cell size is comparable to flow/system length scales, and (2) lack of bio-kinetic adsorption mechanisms and appropriate rate constants applicable to different classes of bio-molecular interactions.

In this paper, we present a recent development for high-fidelity design analysis of microsphere- and cell-based bioanalytic microsystems. The modeling framework couples an enhanced Lagrangian particle tracking methodology for finite-size particles with buffer/analyte transport and a suite of biochemistry models for surface binding. Flow features at bead/cell length scales are resolved using a novel surface marker point approach (termed *Resolved Particle Simulations* or *RPS*) to track the surface of the particle. Non-uniformities in surface coverage, shear stress, etc. on particle surface are computed implicitly in a finite volume formulation. The modeling approach is validated and demonstrated for different problems.

2 MODEL FORMULATION

Characterization of interactions between bulk (buffer/analyte) and bead/cell-surface immobilized biomolecules requires a fundamental understanding of (a) coupled fluid and bead motion (fluid-solid momentum transfer), and (b) analyte transport and binding (fluid-solid mass transfer). Mathematical models employed in RPS follow a generalized multiphase formalism [4] and are discussed next.

2.1 Buffer Transport

Buffer transport in the bulk is governed by the conservation of mass and momentum and is mathematically described by continuity and Naviér-Stokes equations as [5]:

$$\frac{\partial \alpha_c u_i}{\partial x_i} = 0 \tag{1}$$

$$\frac{\partial \alpha_c \rho_c u_i}{\partial t} + \frac{\partial \alpha_c \rho_c u_j u_i}{\partial x_j} = -\alpha_c \frac{\partial P}{\partial x_i} + \frac{\partial \alpha_c \tau_{ij}}{\partial x_j} - \quad (2)$$

$$\beta_V (u_i - v_i) + \alpha_c \rho_c g_i$$

where u, ρ_c, P and g are the buffer velocity, density, pressure and gravity respectively and v is bead/cell velocity. Note that the volume displaced by the particles is accounted for by including the buffer volume fraction or porosity, α_c. Fluid stresses (τ_{ij}) and interphase momentum transfer coefficient (β_V) are expressed in terms of basic variables using constitutive relations.

2.2 Analyte Transport

Analyte transport in the bulk (buffer) solution occurs due to convective (governed by convective flow rate) and diffusive mechanisms (determined by mass diffusivity of analyte). The mass conservation of analyte is written as [5]:

$$\frac{\partial \alpha_c C}{\partial t} + \frac{\partial \alpha_c u_j C}{\partial x_j} = D \frac{\partial^2 \alpha_c C}{\partial x_i \partial x_j} + S \quad (3)$$

where C, D and S are the analyte concentration, diffusivity and source/sink term (due to surface binding), respectively. The analyte transport equation is also modified to account for volume exclusion due to finite-sized particles.

2.3 Bead/Cell Motion

The equation of motion of a particle in a Lagrangian framework is represented as [6]:

$$\frac{\partial v_i}{\partial t} = A_D + g_i + \sum_{i \neq j} A_{c,ij} \quad (4)$$

where A_D is the bead acceleration due to fluidic drag, g_i is the body force and $A_{c,ij}$ is the collisional acceleration. In traditional trajectory methods, the drag force on a particle is obtained from a uniform flow over a sphere approach, which is invalid for large particles due to strong spatial variations in the flow field.

In RPS, the bead/cell is represented by a center of mass location and a pre-specified number of surface marker points as shown in figure 1. The region occupied by the bead is blocked off to the fluid using the porosity variable and tangential no-slip and zero normal flux boundary conditions are imposed on the surface of the particle. Drag force on the bead is computed by explicit numerical integration along the surface as:

$$F_D = \sum_{k=1}^{N} \left(-p_k + \tau_{ij,k} \cdot \mathbf{n}_k \right) \Delta S_k \quad (5)$$

where p_k is pressure at marker point cell k, $\tau_{ij,k}$ is stress tensor at marker point cell k, n_k is the outward normal at marker point k, ΔS_k is the surface area associated with marker point k, and N is the total number of marker points. The RPS approach is also different from the technique reported in [7], as it does not involve expensive remeshing.

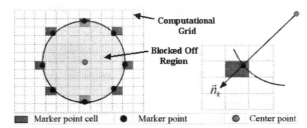

Figure 1. Drag calculation in Resolved Particle Simulations

2.4 Surface Binding

Analyte-receptor interaction takes place at the bead/cell surface and is modeled using a suite of biochemistry models including (a) enzyme kinetics, including Michaelis-Menten, substrate inhibition/activation, competitive/non-competitive inhibition and multiple substrate kinetics, (b) protein adsorption/DNA hybridization with arbitrary order kinetics, and (c) user-specified kinetics. Temporal evolution of bead/cell surface coverage is tracked. For pseudo first-order kinetics (Langmuir adsorption model), the evolution equation can be written as

$$\frac{d\theta}{dt} = K_{on} C_s (1 - \theta) - K_{off} \theta \quad (6)$$

where θ is the surface coverage fraction and K_{on} and K_{off} are the adsorption and desorption rate constants, respectively. A mass balance between the arrival flux at bead/cell surface (due to transport) and production/consumption rate (due to reaction) is imposed for each analyte in the system.

2.5 Particle Interactions

Particle-particle interactions are resolved using either (a) hard sphere collisions [5,8] or (b) inverse polynomial soft potential approach. The latter is based on Molecular Dynamics (MD) methodology and accounts for particle collisions using a short range repulsive force. The force on bead i due to bead j is given by

$$\vec{F}_{ij} = \begin{cases} \vec{0}, & if \ |\vec{r}_{ij}| > d + \rho \\ \varepsilon \left(\frac{d}{|\vec{r}_{ij}|} \right)^n \frac{\vec{r}_{ij}}{|\vec{r}_{ij}|}, & if \ |\vec{r}_{ij}| \leq d + \rho \end{cases} \quad (7)$$

where $\vec{r}_{ij}$ is the separation vector between i and j, ε is the force scaling (stiffness) parameter and ρ is the force range. The total net force acting on ith bead is obtained by summing up the contributions from all the other beads within the computational domain. The characteristics of the force are tailored such that it acts only over a short range to prevent bead overlap by acting continuously over time, which enhances model stability and robustness.

3 MODEL VALIDATION

The Resolved Particle Simulations modeling framework was validated using two different canonical flows: (1) Settling under gravity, and (2) Flow over a cylinder. The results from these studies are discussed next.

3.1 Settling Under Gravity

The first validation case was taken be that of a particle (diameter D) settling in a stagnant fluid under the influence of gravitational field as shown in figure 2. The two-dimensional planar computational domain was taken to be $15D$ in height and $10D$ in width, where D is the particle diameter. The sides of the computational domain were specified to be symmetry boundaries and the top and bottom were taken to be entrainment boundaries. The initial velocity of the particle was taken to be zero.

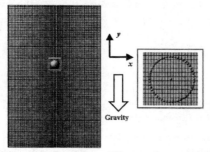

Figure 2. Particle settling under gravity

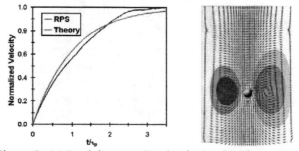

Figure 3. (a) Particle normalized velocity (b) Flow pattern

Particle velocity, normalized with respect to the predicted settling velocity, is presented in figure 3(a), along with an exponential rise curve, based on the aerodynamic response time (τ_p) of the particle. The computed velocity increases initially due to an imbalance in the forces acting on the particle. Particle location, streamfunction and velocity vectors at $t=1.91\,\tau_p$ is shown in figure 3(b). Note that as the particle pushes the fluid down, it sets up a recirculatory flow pattern within the computational domain.

3.2 Flow over a Cylinder

A second validation case for was taken to be flow over a cylinder. Both explicitly meshed (direct) and RPS calculations were carried out for a cylinder of diameter (D) exposed to a freestream velocity (U_0). The Reynolds number based on freestream velocity and cylinder diameter was taken to be 100. Flow velocities and pressure were monitored at three probes locations (shown in figure 4) downwind of the cylinder.

Figure 4. Flow over a cylinder (probe locations in red)

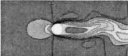

Figure 5. (a) RPS (b) Direct simulation

Planar visualizations of streamwise velocity contours are presented in figure 5. The results from RPS are in good qualitative agreement with direct calculations. To quantitatively compare the results, the shedding frequency for the two cases was computed and averaged over the three probe locations. The shedding frequency from RPS (0.182 s^{-1}) was found to be within 3% of the shedding frequency for direct calculations (0.177 s^{-1}).

4 SOFTWARE DEMONSTRATION

Resolved Particle Simulations provide for unprecedented fidelity for analysis of microsphere- and cell-based assays. As a demonstration case, we study here the evolution of surface coverage non-uniformities in a trapped bead binding platform. An axisymmetric planar representation (shown in figure 6) of a typical fluidization platform was selected for these demonstration calculations. The device consists of antibody-coated microspheres suspended in a chamber and trapped between two mechanical screens. Sample is injected through an inlet at the bottom of the device and is contacted with the beads in the chamber. The chamber was taken to be 16.6 mm in diameter and 80 mm long. Twenty (20) 3 mm diameter beads were seeded randomly throughout the domain between the two screens. Note that the beads are several times (6-8 times) bigger than the computational cell size. The assay of E. coli binding to its probe was studied for these calculations.

Time snapshots of bead locations and fluid streamlines are presented in figure 6. The prescribed fluid velocity is not large enough to adequately fluidize the bead bed and the particles exhibit hindered settling in the chamber. The primary advantage of RPS over traditional approaches lies in accurate and explicit resolution of flow around the particle. This is clearly demonstrated by the wake formation behind the beads and the deflection of fluid streamlines around the beads. This effect cannot be captured in such detail with traditional point particle methods.

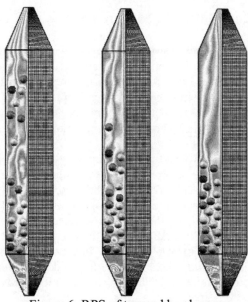

Figure 6. RPS of trapped bead assay
(a) t=0.5s, (b) t=1.0s, t=2.0s

Non-uniform binding on the bead surface is also captured effectively using RPS (see figure 7). The surface marker points are colored by surface coverage (θ) – for this scale, θ is defined as fractional coverage based on the surface area associated with the marker point. Analyte contours are plotted in the background with red indicating high analyte concentration and blue indicating low analyte concentration. The side of the microspheres exposed to higher sample concentrations displays higher surface coverage. Obviously, this has implications for signal to noise ratios during flow cytometry. Note that particle rotation, which would mitigate this phenomenon, is not included in these calculations. Such high-resolution analysis of microsphere/cell-based bioanalytic systems can be employed to evaluate design concepts and quantitatively explore design parameter space in a cost-effective manner.

5 CONCLUSIONS

Resolved Particle Simulations (RPS), a novel methodology for high-fidelity design of microsphere- and cell-based assays was developed and implemented within CFD-ACE+, a widely used general-purpose biomicrofluidic design software from CFDRC. The computational framework combines large particle dynamics with a suite of biochemistry models and buffer/analyte transport to provide for unprecedented details of flow features and surface coverage. The developed models were validated for different canonical flows and numerical predictions were found to be in good agreement with analytical results and direct computations. Demonstration calculations with a trapped bead binding platform were performed using RPS. The computational model was able to predict hindered settling of particles and non-uniform surface binding satisfactorily. In summary, CFDRC has successfully developed high-fidelity computational models for simulating biochemical reactions on flowing, large beads. Such an analysis tool can be used with great impact to develop a fundamental understanding of the biochemical and/or transport processes involved, optimize assay design, and screen new concepts for improvement.

ACKNOWLEDGEMENT

The authors gratefully acknowledge support for this work under NSF SBIR program (DMI-0216507, Program Monitor: Dr. Om Sahai).

REFERENCES

[1] Sundaram, S. & Makhijani, V.B., SIMBAD Phase I Final Report, Lockheed Martin/DARPA, 2000.
[2] Weimer, B. C., Walsh, M. K., Beer, C., Koka, R. & Wang, X., *Appl. Env. Microbiol.*, 67, 1300, 2001.
[3] Pandian, P.B., Shah, K.B., Sundaram, S. & Makhijani, V.B., *MSM International Conference*, 92, 2002.
[4] Duchanoy, C. & Jongen, T., *Comput. Fluids*, 32, 1453, 2003.
[5] Crowe, C.T., Sommerfeld, M., & Tsuji, Y., *Multiphase Flows with Droplets and Particles*, CRC Press, Boca Raton, FL, 1998.
[6] Riley, J. J. & Maxey, M. R., *Phys. Fluids*, 26, 883, 1983.
[7] Peskin, C.S. & McQueen, D.M., *J. Comput. Phys.*, 81, 372, 1989.
[8] Sundaram, S. & Collins, L. R., *J. Fluid Mech.*, 335, 75, 1997.

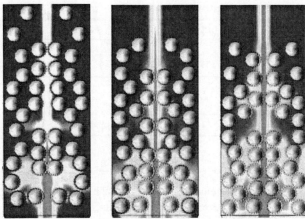

Figure 7. Non-uniform surface binding
(a) t=0.5s, (b) t=1.0s, t=2.0s

Modeling of the dielectrophoretic forces acting upon biological cells: A numerical comparison between Finite Element/Boundary Element Maxwell stress tensor methods and point–dipole approach

A.M. Benselama[*, **], P. Pham[**] and É. Canot[***]

[*]LEMD, CNRS of Grenoble, 25 Avenue des Martyrs, 38041, Grenoble, France, adel.benselama@cea.fr
[**]LETI, CEA–Technologies Avancées, 17 Avenue des Martyrs, 38054, Grenoble, France, pascale.pham@cea.fr
[***]IRISA, INRIA, Campus Universitaire de Beaulieu, 35042, Rennes, France, edouard.canot@irisa.fr

ABSTRACT

Maxwell Stress Tensor (MST) method is investigated in this study to quantify the degree of approximation made with the point–dipole method in respect to dielectrophoresis (DEP) in micro–devices. Latex particles and biological cells immersed in aqueous buffers of various conductivities are considered. The two methods (point–dipole and MST full approaches) are compared using analytical dipolar solutions and numerical ones obtained from the Finite Element (FEM) and the Boundary Element (BEM) methods. Particular emphasis is made on the insufficiency of the point–dipole method in micro–scale issues especially when particles are located near to planar electrode.

Keywords: Boundary Element Method, Finite Element Method, Maxwell stress tensor, point–dipole approach, Micro–devices

1 INTRODUCTION

When studying DEP, two fundamentally different methods are typically used to determine the resulting DEP forces applied upon particles. The point–dipole approach, which is by far the most used method in literature [1], assumes the particle as a point–dipole. Particle is assumed here not to disturb the electric field at its vicinity and its presence is neglected for computation. For biological cell DEP micro–systems, where particles size is of the same order than electrodes gap, those assumptions become to fail. The second method, the MST full method, makes no approximation at this level and the particle presence is considered for the electric field computation. The Maxwell Stress Tensor (MST) method is used to determine, in this case, the DEP force [2].

Point–dipole approximation has been used to model the electric polarization of materials for more than two centuries. Particles forming the matter can be identified as point–dipoles when macro–scales are involved. So neglecting the particle presence allows calculating the electric field with the Laplace equation. The DEP force is then obtained from the Clausius–Mossotti expression that assumes the particle shape to be spherical. As far as biological micro–devices are concerned, none of these

conditions is satisfied. However, a great deal of works done so far and connected to micro–DEP (where electrode–to–electrode distances are comparable to treated particles dimensions) do consider the point–dipole approximation as an established fact, perhaps for its easiness and straightforwardness, and hence make not enough critical of the obtained results [3].

This work is a tentative to quantifying the suitability of the point–dipole approximation to estimate the DEP forces for micro–scale devices. The dipolar, but also multipolar forces acting upon biological cells (or latex particles), deduced from the undisturbed electric field, are compared with a FEM/BEM MST force calculated by a former evaluation of the disturbed electric field. BEM technique is introduced here to make its validation in order to be able to simulate particle deformation afterwards not included in this paper.

The DEP micro–device considered in this work consists of a planar electrode near a hyperboloidal revolution electrode. The axisymmetry hence got allows studying only half of a meridian plan. The electrode gap distance equals 100μm. Latex particles experimented are of 20 μm diameter, and Chinese Hamster Ovocytes (CHO) are of 30 μm diameter. Both latex particles and CHO cells are immersed in Mannitol and in a very conductive physiological solution. The different electric properties are summarized in table 1.

Material	Relative electric permittivity, ε_r	Electric conductivity, σ S/m
Latex	2.55	2.379×10^{-3}
CHO, membrane	4	10^{-7}
CHO, cytoplasm	60	0.5
Mannitol	77.8	1.4×10^{-4}
Physiological solution	78.5	2.5

Table 1: Different electric parameters. The equivalent cellular complex permittivity formula used to compute CHO DEP is given by [4], Appendix C

2 THE POINT–DIPOLE APPROACH

This one is solely based on the electric behavior of a static configuration. Because of the skin thickness of the buffer (which is greater than the actual device size for the whole relevant frequency range), the electric field, $\mathbf{E}$, is supposed to be curl–free, *i.e.*, admits a scalar potential, ϕ, from which it derives. That quasi–static hypothesis writes down as

$$\mathbf{E} = -\nabla \phi \qquad (1)$$

When no free space charge density is present, the first Maxwell law gives

$$\nabla . \mathbf{E} = 0 \qquad (2)$$

Which leads then to

$$\nabla^2 \phi = 0 \qquad (3)$$

Taking into account the geometry of Figure 1, with no particle $\Omega^{(1)}$ for now, one can see that the problem is uniquely determined when considering the conditions, [5],

$$\begin{cases} \phi(\mathbf{x}) = V^1, & \text{for} \quad \mathbf{x} \in \Gamma_1 \\ \dfrac{\partial \phi}{\partial \mathbf{n}}(\mathbf{x}) = -E_n(\mathbf{x}) = 0 & \text{for} \quad \mathbf{x} \in \Gamma_2 \\ \phi(\mathbf{x}) = V^2, & \text{for} \quad \mathbf{x} \in \Gamma_3 \end{cases} \qquad (4)$$

For FEM simulation, the problem is solved using the FEMLAB package (version 2.3a). Once this done, the electric field and its derivatives are obtained in a straightforward manner. Then, as particles are supposed to be spherical, the DEP force is computed using the formula, [1][4],

$$F_q = 4\pi \varepsilon^2 R^3 \, \mathrm{Re}\left(\frac{\overline{\varepsilon}^1 - \overline{\varepsilon}^2}{\overline{\varepsilon}^1 + 2\overline{\varepsilon}^2} \right) E_k E_{k,q} \qquad (5)$$

Where R is the particle radius and $\overline{\varepsilon}^k$, $k = 1, 2$, the complex permittivity of the kth medium defined by

$$\overline{\varepsilon}^k = \varepsilon^k + \frac{\sigma^k}{i\omega} \qquad (6)$$

i is the square root of -1, ω the applied external field excitation pulsation. ε^k and σ^k are the permittivity and the conductivity of the kth medium, respectively. The expression between brackets in formula (5) is called the Clausius-Mossotti factor. This factor, which is based only upon different electric properties and frequency,

determines, separately from the electric field, the DEP regime as attracting to stronger electric field regions (positive DEP) or repelling from them (negative DEP).

For BEM simulation, equations are presented in section 3.2.

3 THE MST FULL APPROACH

An identical configuration as described in section 1 is involved by taking, however, the presence of the particle ($\Omega^{(1)}$) into account. That is by considering figure 1.

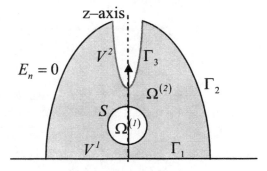

Figure 1: The configuration taken to evaluate the DEP forces by point–dipole approximation when particle is *ignored*, and by the full approach when the particle is considered.

Let us consider only a harmonic excitation, *i.e.*,

$$\begin{cases} V^1 = \overline{V}^1 e^{i\omega t} \\ V^2 = \overline{V}^2 e^{i(\omega t + \varphi)} \end{cases} \qquad (7)$$

Hence, all variables will be supposed to be a $e^{i\omega t}$ time dependent, which means that $X(\mathbf{x}, t) = \overline{X}(\mathbf{x})e^{i\omega t}$ for all variable X. Therefore, the local electric Gauss theorem and the conservativeness of the electric current density, give

$$\nabla . \left(\frac{1}{i\omega} \overline{\varepsilon}^{(k)} \nabla \overline{\phi}^{(k)} \right) = 0 \qquad (8)$$

3.1 FEM modeling

FEM results are obtained by solving equations (8) with FEMLAB package, which implicitly insures the continuity of the complex electric displacement vector, *i.e.*

$$\left(\overline{\varepsilon}^{(1)} \overline{\mathbf{E}}^{(1)} - \overline{\varepsilon}^{(2)} \overline{\mathbf{E}}^{(2)} \right) . \mathbf{n} \Big|_S = 0 \qquad (9)$$

through the interface S, in an intuitive direct manner. The other boundary conditions used are formally identical to equations (4)

3.2 BEM modeling

The following description concerns the BEM formulation. Into each medium k the complex electric potential $\overline{\phi}$ will satisfy, according to (8)

$$\nabla^2 \overline{\phi}^{(k)} = 0 \tag{10}$$

By applying the Green second identity to the later equation, the associated integral equation will be obtained, for $k = 1, 2$, as described by [6],

$$\alpha^{(k)}(\mathbf{x})\overline{\phi}^{(k)}(\mathbf{x}) = \int_{\partial\Omega^{(k)}} \left(G\frac{\partial\overline{\phi}^{(k)}}{\partial\mathbf{n}^{(k)}} - \overline{\phi}^{(k)}\frac{\partial G}{\partial\mathbf{n}^{(k)}} \right) dS'(\mathbf{x}') \tag{11}$$

$\mathbf{x}$ and $\mathbf{x}'$ are the position vectors of the observation point and the current integration one, respectively and G is the Laplace fundamental solution. $\alpha^{(k)}(\mathbf{x}) = 1$, if $\mathbf{x} \in \Omega^{(k)}$, $\alpha^{(k)}(\mathbf{x}) = \frac{1}{2}$, if $\mathbf{x} \in \partial\Omega^{(k)}$ and $\alpha^{(k)}(\mathbf{x}) = 0$, if $\mathbf{x} \notin \Omega^{(k)}$. Let us define

$$\overline{\varpi} = \mathbf{n}.\left(\overline{\mathbf{E}}^{(1)} - \overline{\mathbf{E}}^{(2)}\right)\Big|_S \tag{12}$$

$$\overline{\delta} = \left(\mathbf{n}.\overline{\mathbf{E}}^{(2)}\right)\Big|_\Gamma \tag{13}$$

If we take the derivative of equation (11) according to the observation point, $\mathbf{x}$, combine it to equation (9), we will obtain, after manipulation, for $\mathbf{x} \in \Gamma$ ($\Gamma = \Gamma_1 \cup \Gamma_2 \cup \Gamma_3$), the main integral equation to be solved

$$\frac{1}{2}\overline{\varpi} = \left(\frac{\overline{\varepsilon}^{(1)} - \overline{\varepsilon}^{(2)}}{\overline{\varepsilon}^{(1)} + \overline{\varepsilon}^{(2)}}\right) \times \left[\int_S (\mathbf{n}.\nabla'G)\overline{\varpi}dS' \right.$$
$$\left. - \int_\Gamma \left((\mathbf{n}.\nabla'G)\overline{\delta} + \overline{\phi}^{(2)}\left(\mathbf{n}.\nabla'\frac{\partial G}{\partial\mathbf{n}'}\right) \right) dS' \right] \tag{14}$$

This equation is solved using the CANARD BEM based programming kernel developed at IRISA, INRIA of Rennes, France, and written in FORTRAN90.

3.3 Evaluation of the DEP force by the MST

The computation of the applied DEP force upon the particle is led by using the MST method. The Maxwell tensor, for each medium k, writes as

$$T_{qj}^{(k)} = \varepsilon^{(k)}\left(E_q^{(k)}E_j^{(k)} - \frac{1}{2}\delta_{qj}E_l^{(k)}E_l^{(k)} \right) \tag{15}$$

(The sum for repeated indices is made).

The total force applied upon the $\Omega^{(1)}$ domain writes then down as [2]

$$F_q^{(1)} = \int_S T_{qj}^{(2)}n_j^{(1)}dS \tag{16}$$

4 RESULTS

For the point–dipole approach, BEM and FEM adequacy of the numerical obtained results is demonstrated through an analytical comparison not included in this paper. The electric potential map calculated by both methods in undisturbed configuration is shown in figure 2.

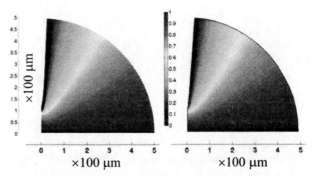

Figure 2: BEM, at left, and FEM, at right, electric potential computation mapping with no disturbing particle.

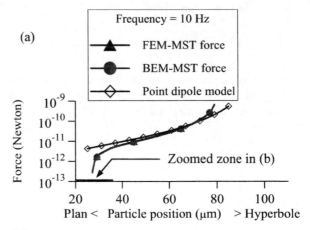

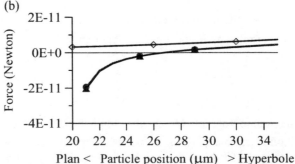

Figure 3: FEM/BEM MST force of a latex particle immersed in Mannitol at 10 Hz (radius equals 20 μm).

For the full approach, the computation is led, to evaluate the electric force applied upon latex spheres, by the MST with both BEM and FEM techniques, whilst the DEP force, deduced from the point–dipole approach, is evaluated only with FEM technique. Another computation is made on Chinese Hamster Ovary (CHO) cells with FEM technique by considering a double–layered particle model that matches up the shell like lipidic membrane surrounding the cytoplasm. Whereas BEM technique uses an equivalent homogeneous spheres model, as made by [4]. The main simulation results are shown here below.

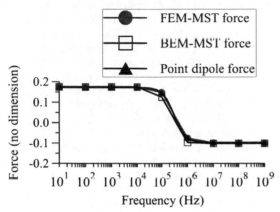

Figure 4: FEM/BEM MST and point–dipole forces at the gap center (z = 50 μm) for latex particle in Mannitol.

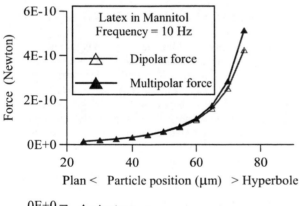

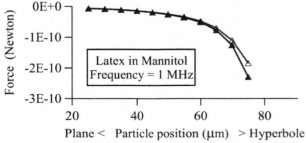

Figure 5: Multipolar forces sum (up to the fifth order), versus point–dipole force.

Latex particles and CHO cells do not perform significantly different patterns; hence for convenience only latex particle results are shown in this paper. As can be noticed by figure

3, MST FEM and BEM DEP forces exhibit satisfactory agreement. Point–dipole model shows, however, a quite different response for all frequencies and different combinatory particle–buffer experimented. The only exception (figures 3a and 4) to be mentioned is the innermost zone ($40 \leq z \leq 60 \, \mu m$) where point–dipole approach goes with MST one. Discrepancy of the right regime (positive or negative DEP) is also noticeable near the planar electrode (figure 3b), where positive DEP should occur according to point–dipole approach. This inconsistency continues to happen even for much smaller particles, provided that they are close enough, comparing to their radius, to the planar electrode. Point–dipole DEP model mismatch could partially be explained by the multipolar effect neglected near the sharp electrode, which has tendency to increase the DEP force as can be seen in figure 5. However, this is no more sufficient near the planar electrode.

5 CONCLUSION

A survey for micro–devices regarding DEP on biological cells and latex particles was made. The comparison between DEP forces estimated by both Boundary Element/Finite Element–Maxwell Stress Tensor method and by Finite Element–point–dipole approximation showed a fundamental statement. When particle–to–electrode distance is comparable to its radius, which is the most alike situation to happen as micro–scales are concerned, the point–dipole and even multipole approaches applied to micro–devices involving biological cells, as well as synthetic particles of similar size, are not always valid. They do not only fail to estimate the DEP force magnitude order but its regime as well (as positive or negative DEP). In such a configuration, the expensive, but unavoidable, method to properly describe the DEP is to consider the immersed cells (or particles) as part of the whole micro–device.

The authors would like to thank Pierre Atten, Alain Glière and Jean–Luc Achard for their helpful discussion about the Boundary Element technique and the physical consistency of the numerical simulation made in this project. This work was supported by the LEMD, CNRS of Grenoble, France.

REFERENCES

[1] Pohl, H.A., Cambridge University Press, 1978.
[2] Wang, X., Wang, X. -B. and Gascoyne, P.R.C., Journal of Electrostatics, **vol. 39**, 277-295, 1997.
[3] Markx, G.H. and Pethig, R., Journal of Biotechnology and Bioengineering, **vol. 45**, 337-343,1995.
[4] Jones, T.B., Cambridge University Press, 1995.
[5] Morse, P.M. and Feshback, H., **vol. 2**, McGraw-Hill, 1965.
[6] Wang, Q. and Onishi, K., VIIth Hydraulic and Engineering Software, Lisbon, 425–436, 2000.

An Approach for Fabrication of Polymer-based Bio-analytical Microfluidic Devices

Meiya Wang, Kuo-Yao Weng, Chien-Chih Huang, Wen-Pin Liu and Liang-Yu Yao

Industrial Technology Research Institute, ERSO
Hsinchu, Taiwan 310, R.O.C., meiya_wang@itri.org.tw

ABSTRACT

This research performed on the fabrication of thermal bonding of polymer microfluidic devices. Hot embossing process and thermal bonding were both achieved. Fine replication by micro hot embossing using polycarbonate substrate has been modified. Nanoscan machine using optical technique can be used to analyze the fabricated microstructures. Hot embossing results show that the near perfect replication of microstructure has been demonstrated, and the shrink degree of the geometry is less than 3%. The two stage process of thermal bonding was performed. The advantage of this method is to prevent the channel filled with the cover plate. The bond strength of the interface for the micro channel is greater than 17MPa using pull test method. The best results were able to seal the microstructure and drilled hole properly and allow for fluids to flow through channel from one reservoir to another one.

Keywords: polymer chip, micro mold, thermal bonding

1 INTRODUCTION

Microfabricated devices have been used in many areas of chemical and biological science. Various materials were and are used for fabrication of micro bio-analytical devices.[1-2] Common substrates are silicon, glass and quartz fabricated using conventional photolithography techniques. The most issues are the material and process costs and the long time to produce devices. Polymer is well suited as a substrate for bio-application devices due to its high dielectric constant, bio-compatible, low cost and ease of micro fabrication. Coupled with the hydrophobic nature and clear optical properties make it promising material for bio-analytical micro devices. The primary goal is the successful fabrication of micro channels for biochip commonly used in genomics and proteomics. A possible solution is polymer hot embossing and thermal bonding. [3]

Hot embossing is the technique used to fabricate high precision and quality plastics microstructures. Injection molding is normally achieved to make plastics components in industrial production. [4-6] A laboratory hot embossing process which used silicon mold made by MEMS technology has been developed when the polymer-based prototype chip of having bio-analytical function was researched.

The bonding technique for 3D microstructures has been described previously. General bonding technology can be divided into major branches, direct bonding and intermediate layer bonding. [7] For the bonding of polymer chip, there are some methods such as adhesive layer bonding and fusion bonding. Usually fusion bonding technology affects the bonded component yield. Even though adhesive layer bonding, it has the same problem to fill micro channels with intermediate layer. To prevent the problem above, the photo-definable material could be used. Because there are dissimilar materials in microchannels, the physical and chemical properties are a little different from similar materials in microchannels. In this work, fusion bonding for similar materials was performed.

The research is to develop a procedure for hot embossing of the microfluidic devices on PC substrate, and a method for thermal bonding to seal an embossed polymer substrate with a hole-drilled polymer plate.

2 MATERIALS AND METHODS

The masters were produced by three methods including wet anisotropic etching of Si, deep silicon RIE, and nickel electroplated through a photoresist mold. In this work, the micro channels for biochips were fabricated using a 60mm×50mm master, as shown in Fig. 1, which was fabricated from anisotropically KOH wet etched silicon. The master is bonded to a stainless steel support plate. In the embossing system, both master and polymer under vacuum are heated to a temperature above the glass transition point. Then the master is brought into contact with polymer substrate with the constant force applied and held for a certain time. Finally the master and the substrate are cooled down to the temperature less than the glass transition point, and then separated.[8]

Hot embossing condition, such as force, temperature, processing time and procedure, affects the polymer flow behavior and should be discussed. [9] A 500 μm thick polycarbonate (PC) is chosen as the raw substrate material in micro hot embossing. Differential scanning calorimeter is performed to investigate the glass transition temperature (Tg) of PC. The glass transition temperature of PC is around 140-160°C, above which PC exhibits higher fluidity. [10] The glass transition temperature of PC is higher than some polymers, such as PMMA or PS, it is suitable to apply for high temperature cycling requirement. When the temperature is above 160°C, loading force above 2000

Newton (N) and processing time including 120sec and 240sec can be selected as the embossing condition.

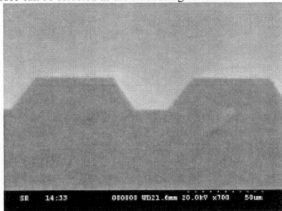

Figure 1: An example of silicon master with micro channels.

Different from other researchers, it's that thermal bonding was performed using two-stage process. The main reason to perform two-stage process is to prevent from micro channels filled due to higher temperature or to seal not properly due to lower temperature. The first stage is pre-bonding procedure using the hot-embosser as a heating and force source to perform bond the polymer wafers and to prevent from the micro channel filled by cover plate, so that the two wafers were combined together without the void between two wafers. The second stage is annealing process, which is to put the pre-bond wafer at the oven. The heating process mainly helps the interface between the channel plate and the cover plate seal stronger, not damages the structure of the micro device.

3 RESULTS AND DISCUSSION

The experimental setup of the hot embossing process is described at section two. The parameters of embossing process affect the processing condition of the hot embossing. The processing force ranging from 2000N to 20000N and the temperature set at the value, which is above 150°C and below 180°C, didn't get the better replication. When the temperature is maintained at 180°C, the micro channels structure transferred from the master looks no deformation as the processing force from 2000N to 20000N. The applied force helps the plastics material flowing into mold inserts. As the temperature higher, the viscosity and surface tension of polymer decrease. This may explain that the processing temperature is more dominant than the applied force. To get an accurate replication with micro features, the best process condition is that the temperature is raised to about 180°C, and a force of 20000 Newton is applied for 120sec. Figure 2 shows the embossed micro channels on PC substrate. The geometry of micro channel was measured with nanoscan, which doesn't

damage the microstructure. The shrinkage of microstructure is 2.4% at the width and 1.9% at the depth.

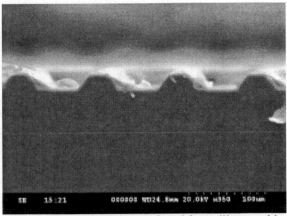

Figure 2: Micro channels transferred from silicon mold.

Microfluidic devices for PC substrates successfully pre-bond at a temperature of 150°C, force of 30000 Newton and time of 10 minutes. It is found that longer pre-bonding time (>10mins) and larger force (>30000N) didn't help the device seal properly. The second stage is annealing process, which is to put the pre-bond polymer device into the oven. This heating process mainly provides energy to bind the interface between the channel plate and the cover plate for sealing stronger, not damaging the micro device structure. In the annealing process, it doesn't need any force. The best condition is set up at an oven temperature 200°C and time larger than 4hrs. Figure 3 shows the SEM photograph of the micro channel after thermal bonding. The defects on the phptograph were resulted from the sample pre-treatment before using SEM. Bond strength could be measured by ROMULUS III. The bond strength is greater than 17MPa after thermal bonding time 4 hours. It was seen that the microstructure of micro channel didn't be filled with the cover plate. The cover plate didn't seem to be deformed.

Figure 3: The SEM photograph of the micro channel after thermal bonding.

4   CONCLUSIONS

It reported that the best process condition to get an accurate replication is that the temperature is raised to about 180°C, and a force of 20000 Newton is held for 120sec. The shrinkage of microstructure is 2.4% at the width and 1.9% at the depth. The pre-bond temperature isn't larger than glass transition temperature to prevent the channels from being filled by the drilled-hole plate, and the annealing temperature is larger than glass transition temperature for increasing the activation energy of the interface to let the fusion bond react rapidly. The pre-bonding process would be successful; so that the annealing process could help the polymer molecules in the interface have a stronger bond. The bond strength of the micro channel is greater than 17MPa. The results show that the fabrication method of polymer-based biochip could be performed successfully. The electroosmotic flow and flow field in PC microfluidic chip was determined using the current monitoring method and μPIV. It helps us to understand the surface characteristics in order to develop bio-analytical micro devices.

REFERENCES

[1] Y.C. Lin, H.C. Ho, et al., Transducers '01, 3D1.05P, 2001

[2] Y.C. Chan, R. Lenigk, et al., Transducers '01, 3D1.06P, 2001

[3] A. Vela and D.DeVoe, Undergraduate report-ISR

[4] L. Lin, Y.T. Cheng and C.J. Chiu, Microsystem Technologies, 4, 113-116, 1998.

[5] M. Heckele, W. Bacher and K.D. Muller, Microsystem Technologies, 4, 122-124, 1998.

[6] H. Becker, W. Dietz and P. Dannberg, Micro-TAS '98, 253-256.

[7] C.T. Pan, H. Yang, et al., J. Micromech. Microeng., 12, 611-615, 2002.

[8] Shan, X. C., R. Maeda, et al., Japanese Journal of Applied Physics, Part 1 42(6B): 3859-3862, 2003.

[9] X.J. Shen, L.W. Pan and L. lin, Sensors and Actuators A, 97-98, 428-433, 2002.

[10] Shan, X. C., R. Maeda, et al., Japanese Journal of Applied Physics, Part 1 42(6B), 3859-3862, 2003

A Generic Sensor for Ultra Low Analyte Concentration Detection

J. Hedley[*], C. J. McNeil[*], J. S. Burdess[*], A. J. Harris[*], M. G. Dobson[*], G. Suarez[*] and H. Berney[*]

[*]Institute for Nanoscale Science and Technology, Newcastle University,
Newcastle upon Tyne, NE1 7RU, England.
John.Hedley@ncl.ac.uk, Calum.McNeil@ncl.ac.uk, J.S.Burdess@ncl.ac.uk, Alun.J.Harris@ncl.ac.uk,
M.G.Dobson@newcastle.ac.uk, Guillaume.Suarez@newcastle.ac.uk, Helen.Berney@ncl.ac.uk

ABSTRACT

This paper reports on the development of a MEMS based generic mass sensor system (international patents pending) aimed at quantitative analyte measurement with a sensitivity of the order of 0.01 ng cm^{-2}. The sensor utilizes the property of cyclic symmetry to auto-compensate for any system drifts due to temperature fluctuations. Initial characterization of the sensor has demonstrated this auto-compensation mechanism. Immobilization protocols are currently being established using antibody immobilization onto a gold surface. The drive / sense electronics for the sensor system are under development.

Keywords: mass sensor, analyte detection, cyclic symmetry

1 INTRODUCTION

Measurement of mass using a resonant mechanical structure relies on added surface mass, produced by the immobilization of material onto the surface of the resonator, causing a shift in resonant frequency. Devices using this principle are referred to as micro-balances and are designed to use bulk, surface, or flexural acoustic waves to create resonant behaviour. Practical sensors have been manufactured using piezoelectric materials and well-known examples are the AT quartz shear wave resonator and the Rayleigh (SAW) resonator (see for example [1]-[4]). To achieve reliable measurements, particularly for low analyte concentrations, changes of the order of a few tens of hertz in a resonant frequency of 10~100 MHz must be detectable. Since these sensors rely on the unperturbed natural frequency of the resonator to provide an absolute reference it is important to have very stable operation or to account for parasitic frequency changes produced by temperature and the density and viscosity of any interfacial fluid. To overcome such problems adjacent matched resonator pairs can be used and the difference between their resonant frequencies taken as a measurement of added mass. This approach increases the cost and size of the basic system and there is the difficulty of producing two matched resonators.

In this project it was proposed to investigate the properties of cyclically symmetric structures and to explore how their special dynamic characteristics could be exploited to overcome the problems associated with the stability of the reference natural frequency. It is known that the natural frequencies of such structures can form sets of degenerate pairs — i.e. two independent modes of vibration share a common natural frequency. The intention of this research was to choose a structure and make selected regions of the structure chemically active and modified with suitable species to provide sites for the specific binding of the desired analyte. Mass added to the structure at these specific sites will disrupt the symmetry of the structure and thus break the degeneracy. The natural frequencies of the two modes will then have unique values and their separation will determine the amount of added surface mass. By using MEMS technology and with this inherent calibration, this sensor is theoretically 100 times more sensitive than conventional single frequency sensors.

2 THEORY

The sensor is essentially a circular plate which is clamped at its edges. A solution to the vibration problem of a clamped plate has been given by Wah [5] and this will be briefly described here. The equation of motion for a plate is given by:

$$\nabla^4(w) - \frac{T}{D}\nabla^2(w) + \frac{\rho_s}{D}\frac{\partial^2 w}{\partial t^2} = 0, \qquad (1)$$

where w is the deflection, ρ_s is the surface density and T is the tension (or compression) in the plate (force per unit length of edge). D is the flexural rigidity given by:

$$D = \frac{Eh^3}{12(1-\upsilon^2)} \qquad (2)$$

where E is modulus of elasticity and υ is Poisson's ratio.

The general solution of (1), using cylindrical coordinates (r, θ), is:

$$w = \left[A_n J_n\left(\frac{\alpha r}{a}\right) + B_n I_n\left(\frac{\beta r}{a}\right) \right] \\ \times (\cos n\theta + \lambda_n \sin n\theta)\sin \omega t \qquad (3)$$

where ω is the natural frequency of the plate, a is plate radius, n is the number of nodal diameters in the mode shape and:

$$\alpha^2 = \frac{Ta^2}{2D}\left[\left(1+\frac{4\omega^2\rho D}{T^2}\right)^{\frac{1}{2}}-1\right],$$

$$\beta^2 = \frac{Ta^2}{2D}\left[\left(1+\frac{4\omega^2\rho D}{T^2}\right)^{\frac{1}{2}}+1\right]. \qquad (4)$$

J_n, I_n are the Bessel and modified Bessel functions respectively. A_n, B_n and λ_n are arbitrary constants. For a clamped plate, the frequency equation becomes:

$$\alpha\frac{J_{n+1}(\alpha)}{J_n(\alpha)}+\beta\frac{I_{n+1}(\beta)}{I_n(\beta)} = 0 \qquad (5)$$

the solutions of which give the natural frequencies of the plate:

$$\omega_n = \frac{\alpha(\phi)\beta(\phi)}{a^2}\sqrt{\frac{D}{\rho h}}. \qquad (6)$$

Note that the solutions of equation (5), α and β, are functions of the tension, T, in the plate, defined by:

$$\phi = \frac{a^2 T}{14.68D}. \qquad (7)$$

Equation (7) has been defined such that $\phi = -1$ when T equals the buckling load of the plate.

Finite element analysis using the ANSYS package was performed to confirm the analytical model results. For this sensor, the two $n=2$ modes were chosen as the operational modes of the sensor, of which one is shown in figure 1.

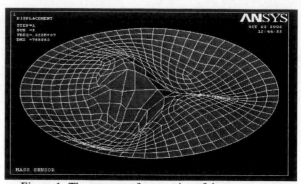

Figure 1: The resonant frequencies of the sensor were determined using an analytical model and finite element analysis.

Due to the cyclic symmetry of the structure, both the modes occur at the same frequency. However by selectively adding mass to specific areas (namely the anti-nodal positions corresponding to one mode and the nodal positions corresponding to the other mode) of the sensor, the frequency difference between the modes will become non-zero. The split Δω in the natural frequencies due to the addition of mass Δm is given by:

$$\frac{\Delta\omega}{\omega} = \frac{1}{2}\frac{\Delta m}{m}+\frac{1}{2}\frac{\Delta k}{k}. \qquad (8)$$

where m and k are the generalized mass and stiffness of the mode. Note that a change in stiffness Δk also contributes to this effect.

3   MANUFACTURE

An operational frequency of 5 MHz was chosen as this is well within the testing facilities available. To achieve this frequency with a 0.9 μm processing constraint on the sensor thickness a disc radius of around 50 μm was required. Stresses in the membrane layer were not expected however any present, because of the characteristics of the fabrication process, would increase the resonant frequency.

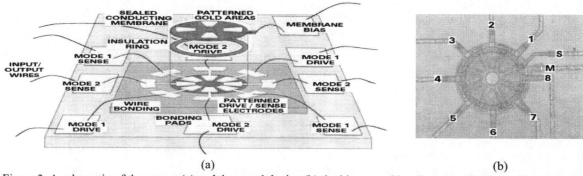

Figure 2: A schematic of the sensor (a) and the actual device (b), in this case with a diameter of 100 μm. Drive / sense electrodes are numbered '1' to '8', shielding ring connection 'S' and membrane connection 'M'.

A schematic diagram of the sensor is shown in figure 2a. A polysilicon membrane forms a sealed cavity over the drive / sense electrodes and gold is patterned onto the surface of the sensor at the specific binding sites.

The sensor, shown in figure 2b, was manufactured by a polysilicon deposition process by NMRC (National Microelectronics Research Centre, Ireland). The sensor was manufactured using 8 variations in diameter. Each of these had an optional shielding ring and shorting of opposite electrodes, giving a total of 32 design variations.

The process flow is as follows. A 450 nm polysilicon electrode layer was deposited onto a 1.25 μm field oxide and covered with a 300 nm protective nitride layer. Two sacrificial oxide layers were then deposited, of total thickness 600 nm, followed by the 0.9 μm thick polysilicon membrane layer. After a timed HF etch to remove the sacrificial oxide, an additional 150 nm polysilicon layer was deposited to seal the membrane cavity. A final 500 nm nitride passivation layer completed the process. Note that for mechanical characterization of the sensor, the final pattern gold areas for immobilization was not processed.

4 TESTING

The sensor was mechanically characterized using an optical workstation [6]. Both static (see figure 3) and dynamic (figure 4) tests were performed to ensure the cyclic symmetry of the device.

At this early stage of development, biological attachment protocols were not in place and to simulate the addition of mass onto the sensor, a d.c. voltage was applied across an electrode and the membrane. As shown by equation (8) this produced a frequency split between the two modes similar to the addition of mass. The measured frequency of one of the modes (taken over several days) for a range of applied voltages is given in figure 5(a). As with current sensor technology, the absolute value of the resonant frequency did drift over time. However the split between these modes (which is the operating principle of this sensor) remained constant for a given applied voltage, see figure 5(b). This initial testing clearly shows the virtue of using two degenerate vibrational modes to remove the effect of common mode noise sources, such as temperature.

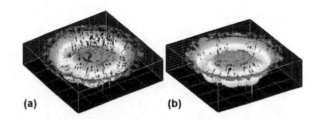

(a)   (b)

Figure (3): The deflection of the sensors under an applied voltage is used to confirm if cyclic symmetry has been achieved during manufacture:
sensor (a) no, sensor (b) yes.

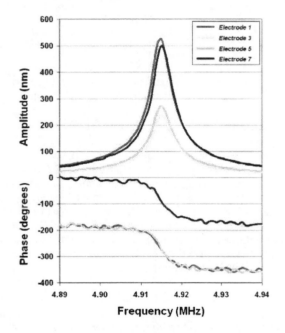

Figure (4): Vibrometer measurements are used to determine frequency and mode shape of the vibrating sensor. The phase information shows this to be the *n=2* mode. However the amplitude data shows a slightly asymmetric vibration which is responsible for the initial frequency split between the modes prior to the addition of mass.

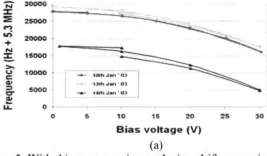

(a)

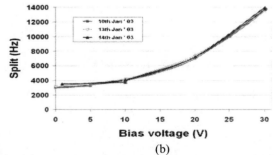

(b)

Figure 5: With this sensor, as in any device, drifts occur in the absolute value of the resonant frequency due to temperature variations (a) however the frequency split (which is this sensor's operational principle) remains consistent during testing (b).

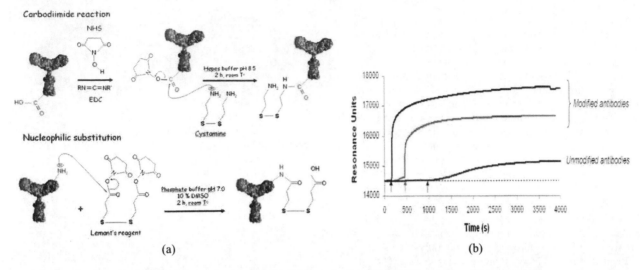

Figure 6: (a) Chemical modification of antibodies by the introduction of disulfide groups. (b) Surface plasmon resonance measurements indicate successful absorption of these modified antibodies onto a gold surface.

5 BIOLOGICAL PROTOCOLS

A study of the immobilization protocols has commenced. The introduction of disulfide groups onto the antibody structure, see figure 6(a), allows for a very strong adsorption process onto the gold surface. This feature has been characterized using surface plasmon resonance, the results of which are shown in figure 6(b). Note that the presence of these disulfide groups has increased the amount of antibody immobilized thereby increasing the sensitivity of the sensor to analyte detection.

6 CONCLUSIONS

The principle behind this mass sensor is to utilize the property of cyclic symmetry to give the device a surface mass sensitivity of the order of 0.01 ng cm^{-2}. Mechanical characterization of the sensor has been completed. Loading the sensor with an electrostatic force was used to show the measurement repeatability of the sensor over several days with these measurements being insensitive to environmental effects such as temperature variations. The biological attachment protocols are currently being investigated and the drive / sense circuitry for the sensor is under development.

7 ACKNOWLEDGEMENTS

The authors wish to acknowledge BBSRC for the research grant 13/ABY15453 and DTI for the research grant QCBC/C/002/00006.

REFERENCES

[1] J.W. Grate, S.J. Martin and R.M. White, "Acoustic wave microsensors: Part I & II", Analytical Chemistry 65, 940-948 & 987-996, 1993.

[2] S. Tombelli, M. Mascini, L. Braccini, M. Anichini and A.P.F. Turner, "Coupling of a DNA piezoelectric biosensor and polymerase chain reaction to detect apolipoprotein E polymorphisms", Biosensors and Bioelectronics 15, 363-370, 2000.

[3] E. Uttenthaler, M. Schraml, J. Mandel and S. Drost, "Ultrasensitive quartz crystal microbalance sensors for detection of M13-Phages in liquids", Biosensors and Bioelectronics 16, 735-743, 2001.

[4] N. Barie and M. Rapp, "Covalent bound sensing layers on surface acoustic wave (SAW) biosensors", Biosensors and Bioelectronics 16, 979-987, 2001.

[5] T. Wah, "Vibration of circular plates", Journal of the Acoustic Society of America 34, 275-281, 1962.

[6] J. Hedley, A. J. Harris and J. S. Burdess, "The development of a 'workstation' for optical testing and modifications of IMEMS on a wafer", Design, Test, Integration and Packaging of MEMS/MOEMS 2001 : Proceedings of SPIE 4408, 402-408, 2001.

Characterization of Micropumps for Biomedical Applications

Yousef Haik[1*], Jason Hendrix[1], Mohammed, Kilani[1], Paul Galambos[1], Ching-Jen Chen[1]
[1]Center for Nanomagnetics and Biotechnology, Tallahassee, FL 32310
*United Arab Emirates University, Department of Mechanical Engineering, Al Ain, UAE

ABSTRACT

A novel class of micro pump was developed at the Biomagnetic Engineering Lab at Florida State University. The pump operates by utilizing the viscous effects that are dominant the micro scale and by spinning a disk that contains a spiral and is in close proximity to a bottom plate. The pumps are highly compatible with biological applications as the system does not use thermal actuation which could denature proteins. Initial steps to characterize the pump consisted mainly of the 2D solution of the flow field. This paper outlines the secondary steps of characterization of this pump which includes internal resistance testing and CFD analysis.

Keywords: Micropumps, Surface micromachining

INTRODUCTION

MicroPumps were developed which utilize viscous drag. The pumps were fabricated using Sandia National Labs SUMMiT process which is a five layer polysilicon deposition process. One of the major benefits of this process is the all in one manufacturing. The entire pump and all actuation devices are fabricated at one time as a complete system. There is no assembly required. This technique allows for a high level of integratable systems to be used in a 'Lab-on-a-Chip' system. The major operating principle is the spinning of a top plate that contains a protruding spiral that is in close proximity to the bottom surface.

The internal resistance of the pump was an area of interest to determine and a developed priming procedure was utilized to determine this piece of data. The only

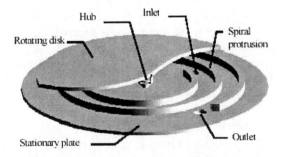

Figure 1. Three dimensional representation of spiral pump

modification to the procedure was introducing a hydrostatic head pressure. This was accomplished by raising the reservoir to a height higher than the outlet of the pump. The

pressure was determined by computing the pressure with the following equation:

$$P = \rho g h$$

In order to keep the pressure constant, a vessel of sufficient area was used as to not alter the height of the fluid inside the reservoir over time. The setup of the experiment appeared as in Figure 2. The height of the reservoir and therefore the pressure of each trial were altered and varying volumetric flow rates were measured. The flow rates vs. pressure data were then graphed and displayed in Figure 3. The graph was expected to be linear from the equation:

$$P = IR$$

where P is the driving gauge pressure, I is the mass flow rate, and R is the internal resistance.

The first step of characterization was a theoretical solution of the flow space performed by Mohammed Kilani et al. To further verify some of the assumptions made in that analysis and to further characterize the pump, a CFD model

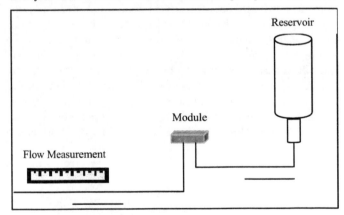

Figure 2: Internal Resistance Set-Up

of the device was developed. The geometry of the device was imported into the CFDRC program and meshed into three dimensions. Precise geometry measurements were determined by using a Fixed Ion Beam cut and scanning the device. To confirm and further validate the internal resistance tests, a CFD model of the pump was constructed[6]. After setting the boundary conditions to match the experimental trials, the mass flow rates were computed. The computations tend to support the experimental data. At low pressures, the resistance values are very similar, 9% different, and the graphs nearly overlap. However, as the pressure increases, the CFD model begins to give lower flow rates than were measured in the lab. This result tends to agree with a current hypothesis of an increased flow rate due to deflection in the channel geometry. One of the assumptions of the previous theoretical solution was to neglect centripetal effects. This assumption was verified as valid during the CFD analysis by plotting stream lines in the flow field a witnessing their position being fixed around the spiral.

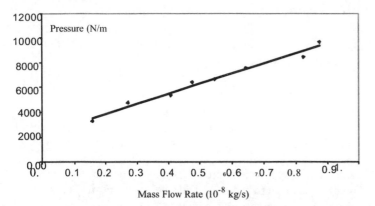

Figure 3: Head Pressure vs Flow Rate : No Pump Actuation

During the process of testing the devices some components of the pump tended to frequently fail. The first such failure was a combined effect of the torque of the TRA and the gear transmissions of the pump. Two style transmissions were used in the original design. One had a gear ratio of 0.36:1 and the other transmission had a ratio of 4.3:1. Both transmissions coupled with the capability of the TRA proved insufficient to turn the pump gear. To complicate matters further, the TRA very often ran sporadically and occasionally broke at the ratchet pawl head. Further frequent failures were found at the gears themselves. It was common to have a gear break away from the substrate at the hub. The most common gear to break free was the very small gear that is pointed out in Figure 4. This failure was due to the torque requirements of the pump.

The most common and by far the largest error with the

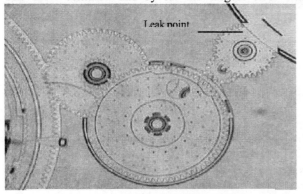

Figure 4: Gear Failure and Leak Point

pumps appears to have occurred during manufacturing. The poly-4 dimple used in the cover of the pump very often appears to have fused to the poly-3 top spinning plate of the pump. This idea was proven through the use of an FIB procedure which cut through the top surface and showed the dimple fused to the plate. This leaves the TRA and gear transmission trying to spin what is supposed to be a free-to-rotate plate in poly-3 that is fixed in place to poly-4 do to the

error. So instead of having five pumps per module, it was fortunate to find one pump per module that was free to rotate.

With torque requirements causing such large problems, other fluids than water were attempted as the pumping fluid. As the pumps operate on a viscous drag principle, using a fluid with lower viscosity should reduce the torque required to turn the pump at the cost of higher performance. Fluids that were investigated and used were silicon oil, methanol, 95:5 methanol: ethylene glycol, and 50:50 methanol: ethylene glycol. Silicon oil proved to have such low surface tension that it was able to leak out of the etch release hole. This rendered the fluid unusable. 100% methanol, which has a surface tension of about ¼ that of water, also proved to be unsuitable for pumping. It would leak out of the interface of the gear transmission and the top plate/housing of the pump which is pointed out in Figure 4. The final fluids investigated were mixtures of methanol and ethylene glycol. Both fluids proved sufficient for priming the device, but were unable to be pumped. An analysis on the viscosity of both of these mixtures showed that they were in fact at a higher viscosity than water. This could very well be the reason the pump was unable to move these liquids.

With so many errors in the design and fabrication, actual pump characterization curves could not be developed at this time.

2   DESIGN MODIFICATIONS

In order to overcome the before mentioned pump failing points, design modifications were submitted for production. The first failure point corrected was the problem of torque. Both the TRA and the gear transmissions were modified. The TRA used in the previous device was the first and oldest of the TRA's designed. A more recent TRA has been designed[4] which has higher torque outputs than the original and operates less sporadically. This higher output and more reliable TRA coupled with the new transmissions of 1000:1 and 2000:1 gear ratios should eliminate the error of low torque due to the driving mechanism. To further alleviate the problem of torque, each pump was reduced in size by 25% thus reducing the torque needed to turn the pump gear.

Modifications were also made with the TRA ratchet pawl. This very often broke in previous trials. The connection point and size of the head were increased in width to add more lateral support. Along the same idea, the small gear in the transmission would also commonly break and each hub in the new transmission was made larger to give a greater surface area contacting the substrate.

References:

1.   Mohamed Kilani, C. J. Chen, Yousef Haik; Florida State University
2.   *Priming procedure for spiral type MEMS pump*; Jason R Hendrix, procedure co-developed with Brian Sosnowchick
3.   Jeff Lantz, designer of the new TRA's
4.   Paul Galambos, Sandia National Labs
5.   Troy Lionberger, UC Berkley, CFD geometric

Modeling of Dielectrophoresis for Handling Bioparticles in Microdevices

G. F. Yao

Flow Science Inc.
683 Harkle Rd, Suite A
Santa Fe, NM 87505, USA
Email: gyao@flow3d.com

ABSTRACT

Dielectrophoresis can be used to trap and position bioparticles such as biological cells efficiently and successfully in BioMEMS devices. This paper presents a computational model to simulate this process. In particular, the model will solve for electric potential distribution, bioparticle movement in a Lagrangian frame, Joule heating, and medium flow due to electrothermal effects. As an illustration, the model was used to simulate a cell trapping process in a three-dimensional chamber.

Keywords: dielectrophoresis, Joule heating, cell traps, electrothermal effects

NOMENCLATURE

C_D	drag coefficient
$\mathbf{E}$	electric field intensity
$\mathbf{f}$	force due to electrothermal effect
$\mathbf{F}_{DEP}$	dielectrophoretic force
K	medium thermal conductivity
m_p	particle mass
P	pressure
r_p	particle radius
$\mathbf{U}$	velocity of suspending medium
$\mathbf{V}$	particle velocity
$\mathbf{X}$	particle location

Greek Symbols

ϵ	medium permittivity
ϵ^*	complex medium permittivity
ϵ_p^*	complex particle permittivity
μ	dynamic viscosity of medium
ω	angular frequency of electric field
ϕ	electric potential
ρ	medium density
σ	electric conductivity

1 INTRODUCTION

Dielectrophoresis (DEP) means the creation of forces on polarizable particles and the resulting movement of them in a nonuniform electric field (usually AC electric fields)[1-2]. Dielectrophoretic forces can be used as a mechanism in various BioMEMS devices for characterization, handling, and manipulation of microscale and nanoscale bioparticles including sorting, trapping, and separating of cells, viruses, bacteria, and DNA etc.[3-8]. For example, compared to traditional laboratory experiments, manipulating an individual cell or a group of cells in various BioMEMS devices based on dielectrophoretic force allows scientists to obtain new kinds of biological information, leading to new insights into how cells work. Therefore, dielectrophoresis, as well as MEMS devices based on it for operation, plays more and more important roles in the study of biology and medical sciences and it has a broad range of applications in drug discovery, diagnostics, and cell therapy. Developing a computer model to simulate a dielectrophoresis process is not only important and useful for design of new DEP devices for bioscience applications, but also for optimal operation of them. Motivated by these facts, a model for particle dielectrophoresis is developed and presented in this paper.

As mentioned above, the dielectrophoretic force is utilized in various BioMEMS devices for handling bioparticles. The magnitude and direction of DEP forces depend on the frequency of the AC electric field, conductivity and permittivity of both particles and the medium where particles are suspended, and the gradient of electric field. The gradient of electric field is dependent on the geometry and numbers of microelectrodes used. In addition, the strong electric fields adopted in dielectrophoresis generate a large power density (**Joule or Ohmic heating**) in the suspending medium. Due to high nonuniformity of electric field thereby power density, a temperature gradient yields, which results in gradients of conductivity and permittivity. The former produces a free volume charge and a Coulomb force while the latter creates a dielectric force. These two forces cause medium flow, called **electrothermal effect**, and give rise to an effect of dielectrophoresis of bioparticles[9]. In present work, a comprehensive model is developed. In addition to the calculation of electric field and dielectrophoretic force on bioparticles, Joule heating and electrothermal effects are also included in the model. In the following section, the related governing equations and numerical methods are described. To demonstrate the application of the developed model, a three-dimensional cell culture chamber is simulated and the results are presented in a separate section. Finally, the paper is ended by a brief conclusion.

2 GOVERNING EQUATIONS AND NUMERICAL METHODS

Flow of a suspending medium is described by the following incompressible mass and momentum conservation equations (The vector variables are represented in bold face in this paper.)

$$\nabla \cdot \mathbf{U} = 0 \tag{1}$$

$$\rho \left[\frac{\partial \mathbf{U}}{\partial t} + \mathbf{U}(\nabla \cdot \mathbf{U}) \right] = -\nabla P + \mu \nabla^2 \mathbf{U} + \mathbf{f} \tag{2}$$

where the force created due to aforementioned electrothermal effects can be calculated by

$$\mathbf{f} = -0.5 \left(\frac{\nabla \sigma}{\sigma} - \frac{\nabla \epsilon}{\epsilon} \right) \cdot \mathbf{E} \frac{\epsilon \mathbf{E}}{1 + (\omega \sigma / \epsilon)^2} - 0.25 E^2 \nabla \epsilon \tag{3}$$

with gradients of σ and ϵ determined in terms of temperature distribution[9]. Temperature distribution due to Joule heating is solved from

$$K \nabla^2 T + \sigma E^2 = 0 \tag{4}$$

while movement of bioparticles due to dielectrophoresis is given by

$$m_p \frac{d\mathbf{V}}{dt} = -\frac{1}{2} \pi r_p^2 \rho C_D |\mathbf{V} - \mathbf{U}|(\mathbf{V} - \mathbf{U}) + \mathbf{F}_{DEP} \tag{5}$$

$$\frac{d\mathbf{X}}{dt} = \mathbf{V} \tag{6}$$

with the dielectrophoretic force imposed on a bioparticle calculated by

$$\mathbf{F}_{DEP} = 2\pi \epsilon r_p^3 \Re \left[\frac{\epsilon_p^*(\omega) - \epsilon^*(\omega)}{\epsilon_p^*(\omega) + 2\epsilon^*(\omega)} \right] \nabla E^2 \tag{7}$$

where $\Re$ represents the real part of a complex number[2]. Electric potential is derived by solving

$$\nabla^2 \phi = 0 \tag{8}$$

followed by the calculation of the electric field distribution using

$$\mathbf{E} = -\nabla \phi \tag{9}$$

The model presented here is implemented in **FLOW-3D**®—a general purpose commercial CFD code. In particular, equations (1-4) and (8-9) are discretized in a set of fixed Eulerian grids or control volumes. Handling of a complicated computational domain is performed using a fraction- area- volume- obstacle- representation (FAVOR^{TM}) method[10]. The movement of bioparticles are tracked by solving Eqns.(5-6) in a Lagrangian frame. Detailed numerical methods adopted to solve for these governing equations can be found in [11].

Figure 1: Cell culture chamber geometry

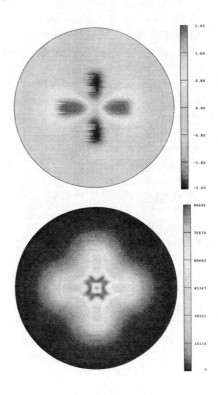

Figure 2: Top: electric potential; Bottom: electric field strength

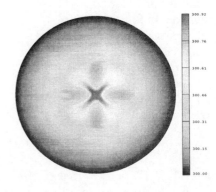

Figure 3: Temperature distribution

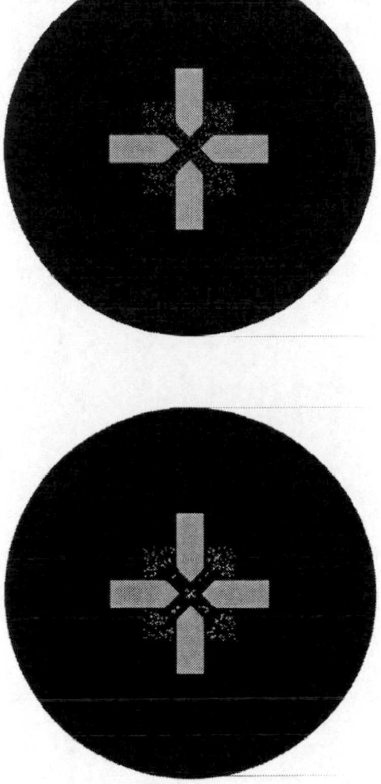

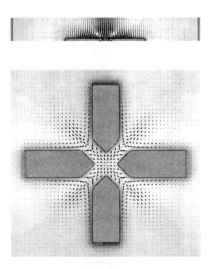

Figure 4: Particle distribution at two different times

Figure 5: Velocity distribution

3 NUMERICAL RESULTS

To demonstrate the application of the presented model, a cell culture process in a three-dimensional chamber is simulated. The chamber geometry is displayed in Fig.1 with a chamber diameter equal to 800 μm and a depth equal to 100 μm. Also shown in the same figure are four microscale electrodes deposited on the bottom of the chamber which provide the electric potential. Experimental studies of cell trapping in this kind of device was performed by Heida et al.[6-8]. The viscosity, density, permittivity, and electric conductivity of suspending medium used in the simulation are 8.55×10^{-4} $N \cdot s/m^2$, 1000 kg/m^3, 6.95×10^{-6} C^2/Jm, and 0.1 Sm^{-1}, respectively. The value of $\Re$ appearing in Eqn.(7) can be obtained from the related handbook based on the angular frequency ω for a specific medium-particle pair. In the simulation presented here, a value of -0.1 was used.

Cell culture operation involves particle movement (cell trapping) due to dielectrophoresis, Joule heating created by a strong electric field and electrothermal effects corresponding to Joule heating. The model presented here captures all these physical processes. The top of Fig.2 shows electric potential distribution created by four microscale electrodes on a cross section 5 μm above the chamber bottom. The magnitude of the rms (root mean square) value of electric potential on the electrode surface is 3 volts. The electric field strength distribution at the same cross section is plotted in the bottom of Fig.2. As expected, strong electric field strength was observed near the edges of electrodes. In particular, a "potential energy well" in the center of the chamber, with minimum electric field strength at its center, is formed where some cells are trapped. To see the trapping, particle distribution at beginning of the operation (top) and at a later time (bottom) are displayed in Fig.4. Even though it is not possible to make a qualitative comparison between numerical predictions and corresponding experimental measurements due to the lack of detailed information on experimental conditions, a quantitative comparison with the experiments performed by Heida et al.[6-8] indicates that particle distribution derived from the present model is consistent with the experimental observation. Due to the high electric field and sensitivity of bioparticles on temperature, Joule heating may be significant. The temperature distribution due to Joule heating is provided in Fig.3. About a 1K temperature increase was observed. Medium flow around the "potential energy well" due to electrothermal effects are illustrated in Fig.5 with the top figure representing flow in a vertical cross section passing through the middle of chamber and the bottom figure representing flow on a horizontal cross section immediately above the chamber bottom. It can be seen that medium flow could affect cell trapping depending on the operation conditions.

Here, we provide some results to illustrate the application of the presented model. For the design of a BioMEMS device, engineers need to know how to determine the number of microelectrodes and their geometry. For the use of these devices, it will be useful to know how to control different physical processes involved and get optimal operation. To this end, the present model will be useful for both design and operation of the related BioMEMS devices.

4   CONCLUSION

A computational model was developed and implemented in $FLOW$-$3D$®—a general purpose commercial CFD code—to solve for electric potential distribution, Joule heating, and medium flow created by electrothermal effects. Numerical results derived from the simulation of cell trapping in a three-dimensional chamber demonstrated that the present model could be useful for the design and operation of the related BioMEMS devices.

REFERENCES

[1] H. A. Pohe, Dielectrophoresis, London: Cambridge University Press, 1978.

[2] T. B. Jones, Electromechanics of particles, New York: Cambridge University Press, 1996.

[3] X. Wang, X. B. Wang, F. F. Becker, and P. R. C. Gascoyne, "A theoretical method of electric field analysis for dielectrophoretic electrode arrays using Green's theorem", J. Phys. D: Appl. Phys., 29, pp.1649-1660, 1996.

[4] G. H. Markx, P. A. Dyda, R. Pethig, "Dielectrophoretic separation of bacteria using a conductivity gradient", J. Biotechnology, 51, pp.175-180, 1996.

[5] N. G. Green and H. Morgan, "Dielectrophoretic separation of nano-particles", J. Phys. D: Appl. Phys., 30, pp.L41-L84, 1997.

[6] T. Heida, P. Vulto, W. L. C. Rutten, and E. Marani, "Viability of dielectrophoretically trapped neural cortical cells in culture", J. Neuroscience Methods, 110, pp37-44, 2001.

[7] T. Heida, W. L. C. Rutten, and E. Marani, "Dielectrophoretic trapping of dissociated fetal cortical rat neurons", IEEE Trans. on Biomedical Engng., 48, pp.921-930, 2001.

[8] T. Heida, W. L. C. Rutten, and E. Marani, "Understanding dielectrophoretic trapping of neuronal cells: modeling electric field, electrode-liquid interface and fluid flow", J. Phys. D: Appl. Phys., 35, pp.1592-1602, 2002.

[9] N. G. Green, A. Ramos, A. Gonzalez, A. Castellanos, H. Morgan, Electrothermally induced fluid flow on microelectrodes, J. of Electrostatics, 53, pp.71-87, 2001.

[10] C. W. Hirt and J. M. Sicilian, A porosity technique for the definition of obstacles in rectangular cell meshes, Proc. Fourth International Conference Ship Hydro., National Academy of Science, Washington, D. C., Sept., 1985.

[11] $FLOW$-$3D$® Theory Manual, Vol.1, Flow Science Inc., Santa Fe, NM, USA.

Quadruplets-Microneedle Array for Blood Extraction

S. Khumpuang, G. Kawaguchi and S. Sugiyama

Graduate School of Science and Engineering, Ritsumeikan University

1-1-1, Noji-higashi, Kusatsu, Shiga, Japan 525-8577

Tel.:+81(077)561-2775, Fax.:+81(077)561-3994, Email:gr017037@se.ritsumei.ac.jp

ABSTRACT

A novel fabrication of painless microneedle array using Synchrotron Radiation lithography with PMMA (Polymethylmethacrylate) is presented in this paper. The microneedles were shaped by Plain-pattern to Cross-section Technique (PCT). 4 patterns of X-ray mask were investigated and tested then the optimal one has been selected. The 1024(32x32) quadruplets-needles are consisted in 1x1 cm^2 chip. The crown-shaped structure with 4 tips microneedle can be usehd without the consideration of holes for blood extraction. The tip-size in nano-scale has also been approached. Fabrication of the PMMA microneedle is applicable for LIGA technique. The electroplating of metal and injection of a plastic material can be done for batch production.

Keywords: microneedle array, blood extraction, PMMA, X-ray lithography

1. INTRODUCTION

In biomedical application, the sharp tips, high strength and bio-compatible microneedles are required. PMMA is a material which can be safely used with human skin since it is not a brittle material, alike silicon. The High-Aspect-Ratio MicroStructure(HARMS) has been shaped by synchrotron radiation Lithography. Using the standard deep X-ray lithography exposing on an X-ray resist layer, the 2.5-dimension microstructure can be fabricated. However, the resulting side wall is usually vertical. In order to reach the 3-dimension microstructure, Plain-pattern to Cross-section Technique(PCT) was investigated. The structure has been resulted in the arbitrary shaped wall[1]. In addition, sharp micronneedle tip is the key of reducing the skin damage from piercing reaction[2,3]. Also the needle with sharp tip can smoothly penetrate to the skin at the correct position[4]. Nanoscale microneedle tip-size has also been achieved. Recently, a microneedle array has been fabricated using PCT method originated in Ritsumeikan University, Japan[5]. The previous work has presented single-tip microneedle arrays whilst the shape was simple and difficult to use for a blood extraction[6]. However, a method of approaching arbitrary structure was developed and now applied to the novel design in this work.

2. X-RAY MASK PATTERNS

Four patterns of X-ray mask have been investigated. The double right triangles, u-shape triangle, double trapezoids, and u-shape trapezoid as seen respectively in Fig. 1(a) to 1

(d), were used as X-ray absorber patterns for fabrication of PMMA microneedle array.

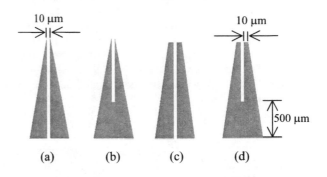

Fig.1 The four shapes of X-ray mask patterns used for X-ray lithography.

Figure 2(a) shows the dimension of a double right triangle pattern. In this figure, 32 patterns of 1500 μm-height with 300μm-pitch are consisted in a row. The base of each pattern is 100μm. Figure 2(b) shows the actual X-ray mask which is made by using UV lithography. X-ray absorber has been made of 3.5 μm-thick Au. The membrane is polyimide. Figure 2(c) is the closed up image of the tips of the mask pattern. For the other dimensions being unstated here are similar with the sample pattern.

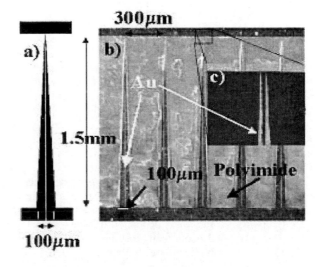

Fig. 2 The dimension of each right triangles pattern(a) the actual X-ray mask (b) and the closed up image of the tips.

3. FABRICATION TECHNIQUE

PCT Technique is a new method for fabricating 3-dimensional structures whose cross-sections are similar to the two-dimensional patterns on X-ray mask. A PMMA substrate is moving repeatedly with the scanning stage while X-ray is exposed through the X-ray mask. Figure 3 shows a simple schematic of PCT steps. The horizontal movement(x-direction) of the work stage forms the three-dimensional structures and gradually enlarges the exposed area of the PMMA. Since the X-ray absorption profiles in the PMMA depend on window spaces of absorber on the X-ray mask, the cross-section patterns in the PMMA forms similar structures to X-ray mask patterns whilst resulted in three-dimension. The exposed depth in the PMMA depends on X-ray dosage and the absorption coefficient of the PMMA mixture. Depth of the pattern after development can be controlled by exposure time and scanning speed. During the fabrication of microneedle, PMMA must be exposed twice by SR light source. The second X-ray exposure is done after PMMA substrate has been rotated 90° from the first exposure position.

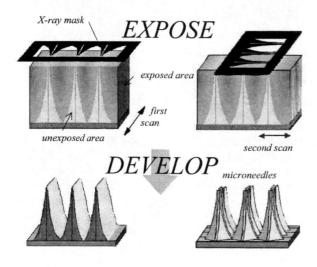

Fig.3 Twice scanning X-ray exposures by moving stage, so called PCT

Using four types of the masks suggested above, the PMMA was exposed through the masks with BL-13 of synchrotron radiation source AURORA at SR center, Ritsumeikan University, Japan. The wavelength of the SR light was between 0.15 to 0.73 nm. The applied electron energy and the maximum storage current of AURORA in the experiment were 575MeV and 300mA respectively. The exposure environment was fixed to Helium gas at 1 atm. The data of deposited doses against the processed depths on the resist collected 6 times in each experiment is shown in Fig. 4. The experiment started from a short exposure time at 60 min. At the lower deposited dosage (below 0.02 A.h.), the processed depth tends to increased uniformly to the dosage. However, at the higher doposited dosage (above 0.02 A.h.), the processed depth starts to be exponentially respecting to the dosage.The selected exposure dosage for one directional scan during the fabricate of micro needle array was about 0.06 A.h. The

exposed PMMA structures appear after developing by GG developer at 37°C for 3.5 hours.

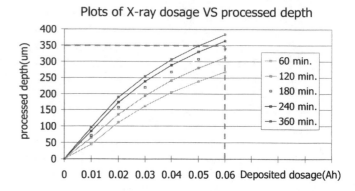

Fig. 4 The deposited dosage data against the processed depths of PMMA resist at the exposure times varied from 60 min. to 360 min.

4. PRINCIPLE OF THE MICRONEEDLE

The skin was flaring while piercing the microneedle. By the capillary force, blood can be extracted and kept after flowing to the grooves. Figure 5 shows the schematic diagram of blood extraction process. The quadruped grooves enhance the area of storing blood after penetrating the microneedle to skin.

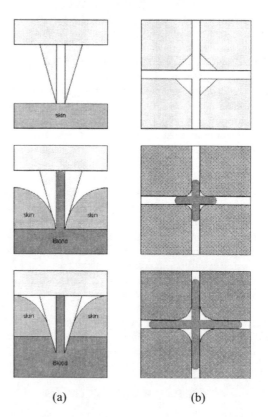

(a)   (b)

Fig.5. Schematic of blood extraction process. The cross section(a) and top view (b) of microneedle during the penetration

In order to consider the length dependent on the groove size the capillary equation is shown in equation (1).

Capillary effect $\qquad r = 2\gamma\cos\theta / \rho g h$ $\qquad$ (1)

r is the gap between parallel groves, h is the capillary height, θ is the angle of edge of blood surface against a groove wall and γ is the surface tension. Value of r must be as small as possible yet the volume of blood contained in the needle grooves is enough for a medical test.

5. EXPERIMENTAL RESULTS

The experiment has been done firstly with both trapezoid patterns. The results are shown in Fig. 6. The microneedles exposed from double trapezoids pattern(Fig. 6(a)) are more useful than one exposed from u-shaped trapezoid pattern (Fig. 6(b)) since flow channel in the grooves is connected to the middle of the quadruplets. The blood can be extracted and kept inside the groove. However, the structures were found that each spike was slim and weak.

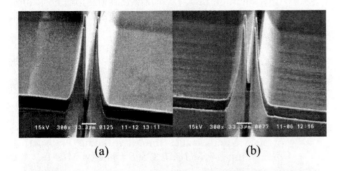

(a) $\qquad\qquad$ (b)

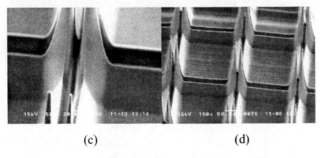

(c) $\qquad\qquad$ (d)

Fig. 6 SEM photographs of fabricated microneedles. The pattern used were, double right triangles(a) and u-shape triangle(b). The close up image of double right triangles(c) and array of u-shape triangle(d).

In order to be applicable for a safe use, the mask patterns, therefore, have been changed to be the triangle pattern. Using both double right triangles and u-shape triangle gave very sharp tips in nanoscale. Figure 7(a) is SEM photographs of microneedle array fabricated by triangle patterns and Fig. 7(c) shows the tip-size is about 300 nm. Thus, the best X-ray mask pattern for fabricating quadruplets microneedle array for safe use has been selected to be the double right triangle.

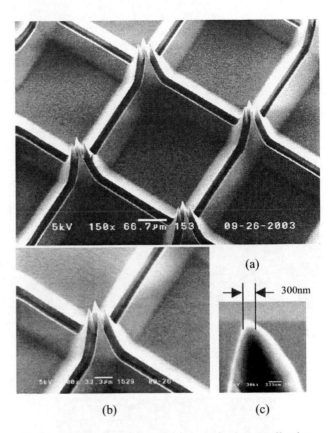

(a)

(b) $\qquad\qquad$ (c)

Fig. 7. SEM photographs of fabricated microneedles by . The triangle patterns. The array of microneedle(a), close-up image of a microneedle(b) and close-up image of the needle tip

6. DISCUSSION

A test of the microneedle array has been done with a thin piece of chicken meat. Aniline blue solution(Wako chemicals, Japan) was used as the test-liquid since it has the similar viscosity as blood and the blue color can be simply observed. The test was done by dripping several drops of Aniline to a container and put the piece of thin chicken meat(slices less than 0.3mm.) on the liquid surface. After that, press the microneedle array from the back so that the needles could penetrate the meat. The result of testing is shown in Fig. 8 and Fig. 9. In each microneedle, Aniline was kept in the middle, between each spike. However, liquid still did not flow along the grooves. The cause might be from the pressure of liquid that was free in lateral directions but the interface between the chicken meat and the container. After pressing the array, liquid was spread out of the testing area to the rest of the container. This problem may be solved by using a thin tube with applied pressure to the liquid flowing inside the tube or test with an alive animal by an assist of medical doctor or licensed person. Figure 9 shows the chicken meat after flared by microneedles. It could be seen that all the needles have been pierced through the meat without any part broken. The holes left on the meat were found to be as big as the microneedle base(100μm).

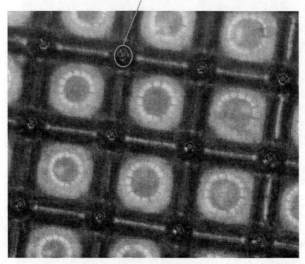

Liquid is kept between each spike

Fig 8. The array of microneedle after insertion through chicken meat. The liquid was kept between each spike by capillary force.

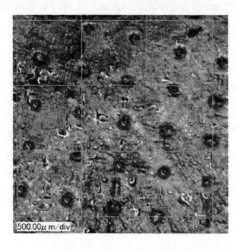

Fig.9. A photograph of chicken meat taken by an optical microscope after flaring by the microneedles.

7. CONCLUSION

The alternative way of using microneedles without holes to get the blood sample has been introduced. The optimum mask pattern was investigated. In this work, the double right triangle is suggested. The tip-size of less than 300nm of the microneedle was achieved. The PMMA microneedle structures can be used as molds for further LIGA process (electroplating and plastic injection) in order to reduce the cost of batch fabrication. Although the process is uncomplicated, the experience for mask design and dosage control for each experiment is necessary. The dosage chart has to be changed from time to time due to the life cycle of X-ray. Since the fabrication of each array is expensive, the number of sample arrays is limited. Only few tests have been done. Also the Aniline is a poison if the intake quantity is high. However, the test of microneedle on a tube of liquid with applied pressure will be tried in order to prove that quadruplets microneedle can be used in substitution of the microneedle with hole.

8. ACKNOWLEDGEMENT

This work was supported by 21st century COE(Center of Excellent) research program of Regional Science and Technology Promotion from the Ministry of Education, Culture, Sports, Science and Technology.

REFERENCES

[1] G. Kawaguchi, N. Samoto, H. Tosa and S. Sugiyama, "Investigation of X-ray Mask Patterns for Fabrication of Micro Needle Array Using SR Lithography", Proc. of Sensors Symposium, Tokyo, July, 2003

[2] S. Henry, D.V. Mcallister, M.G.Allen and M.R.Prausnitz, "Microfabricated Microneedle : A Novel Approach to Transdermal Drug Delivery", J. of Pharmaceutical Sciences, Vol.87, No.8, Aug. 1998

[3] P. Griss and G. Stemme, "Novel, Side Opened Out-of-plain Microneedles for Microfluidic Transdermal Interfacing", Proc.IEEE Conference of MEMS, Las Vegas, Jan., 2002

[4] B. Stoeber, and D. Liepmann, "Two-Dimensional Arrays of Out-of-Plane Needles", Proc. ASME 2000 International Mechanical Engineering Congress & Exposition, Orlando, Nov., 2000.

[5] S. Sugiyama and H. Ueno, "Novel Shaped Microstructures Processed by Deep X-ray Lithography, Digest of Technical Papers", The 11th International Conference on Solid-State Sensors and Actuators, Munich, Germany, June, 2001

A Disposable Glucose Biosensor Based on Diffusional Mediator Dispersed in Nanoparticulate Membrane on Screen-Printed Carbon Electrode

Z. Gao, F.Xie, M. S. M. Arshad and J. Y.Ying

Institute of Bioengineering and Nanotechnology, 51 Science Park Road,
Singapore 117386, Republic of Singapore, zqgao@ibn.a-star.edu.sg

ABSTRACT

A disposable glucose biosensor based on co-dispersion of a diffusional polymeric mediator and glucose oxidase (GOX) in nanoparticulate membrane on screen-printed carbon electrode was described. The biosensor exhibited classical features of a kinetically fast redox couple. At scan rates up to 200 mV/s, the peak-to-peak potential separation of the mediator was 59 mV at 25°C. In the presence of glucose, a typical catalytic oxidation current was observed, which reached a plateau at 0.30 V (*vs.* Ag/AgCl). Amperometric peak current at 5 s and the subsequent currents at longer test times were linear to the glucose concentration in the range of 5 to 600 mg/dl. By constructing a miniature channel on the screen-printed electrode and partially filling the channel with the nanoparticulate sensing membrane, glucose can be accurately determined in as little as 0.20 µl of sample. Successful attempts were made in blood sugar analysis.

Keywords: glucose, biosensors, nanoparticles, ferrocene, glucose oxidase

INTRODUCTION

Since Clark and Lyons developed the first oxygen electrode-based enzyme electrode, enzyme-based biosensors have been used in an increasing number of clinical, environmental, agricultural, and biotechnological applications.[1,2,3] To address the needs for frequent or continuous monitoring of glucose in diabetics, particularly in brittle diabetics, electrochemical glucose biosensors are, by far, the most widely employed. Since the two redox centers (FAD/FADH$_2$) of glucose oxidase (GOX) are prevented from transferring electrons to an electrode by an insulating glycoprotein shell,[4] the presence of a mediator is necessary to achieve direct electron exchange between the electrode and the redox centers of GOX. Electrochemical glucose biosensors normally employ two major groups of mediators, hydrogen peroxide oxidation mediating reagents and glucose oxidation mediating reagents. The former usually require high operating potentials where unwanted reactions of coexisting electroactive constituents in blood complicate the interpretation of the signal. Moreover, being limited by the least concentrated substrate, oxygen, in the enzymatic reaction, these biosensors suffer from low sensitivities. On the other hand, glucose oxidation mediating reagents offer a preferential catalytic oxidation of glucose at much lower potentials, and hence greatly enhance the selectivity of the corresponding blood sugar measurements. The sensitivity of the measurements is also greatly improved due to the replacement of oxygen by synthetic mediators that are usually in large excess in the systems. In these cases, the natural route of glucose oxidation catalyzed by GOX becomes a side-reaction, competing with the mediated one. Special attention must be paid to the effect of this reaction, and extensive effort must be made to minimize the degree of glucose consumption by dissolved oxygen in blood.

Unlike enzyme electrodes based on immobilized mediators, biosensors employing diffusional mediators are particularly attractive for developing disposable biosensors for diabetes home-care. Different forms of diffusional mediator-based systems have been employed by commercially available glucose biosensors. In most of the disposable glucose biosensors, the signal is a balance between the mediated reaction and the natural glucose oxidase catalyzed oxidation of glucose. Since each of the reactions has its own characteristics, the response of the biosensors is profoundly influenced by temperature, partial pressure of oxygen, pH, etc. A more challenging task in glucose biosensors development is to identify appropriate means of eliminating the interferences from blood cells and coexisting electroactive constituents. For example, high levels of red blood cells result in readings that are lower than the true values because of fouling of the sensing membrane and the increased magnitude of the side-reaction between oxygen and glucose. On the other hand, coexisting electroactive constituents such as ascorbic acid, uric acid, dopamine and acetaminophen all contribute to readings higher than the true values. Because of the importance of accurate blood sugar monitoring to the well-being of diabetic patients, it is highly desirable to have a sensing device that does not suffer from these drawbacks.

EXPERIMENTAL SECTION

Fig. 1 illustrates the fabrication of the disposable glucose biosensor. The nanocomposite membrane was screen-printed onto carbon strip using an aqueous slurry "ink" of poly(vinylferrocene-co-acrylamide) (PVFcAA), GOX, a poly(vinylpyridine-co-acrylic acid) (PVPAC) binder and alumina nanoparticles. A uniformed nanoparticulate membrane was printed on top of the working area and dried at 37°C in a controlled environment. The thickness of the nanoparticulate membrane was controlled by adjusting the total content in the ink while keeping a constant volume applied on the working area. After removing the protecting film of the double-sided adhesive tape, a Ag/AgCl coated polyester film was applied, forming a microchannel between the two films. Finally, individual sensors were cut off from the substrate.

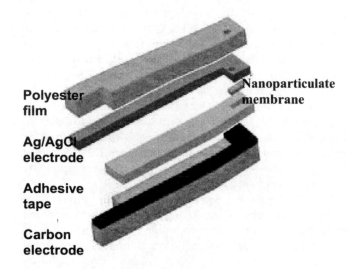

Fig. 1. Exploded view of the disposable glucose biosensor used in this work.

All electrochemical measurements were carried out with a model CHI 660A electrochemical workstation (CH Instruments, Austin, USA) at room temperature. Cyclic voltammetric measurements were performed using a conventional three-electrode system, consisting of a screen-printed carbon working electrode, a Ag/AgCl reference electrode (Cypress Systems, Lawrence, KS, USA) and a platinum wire counter electrode. To avoid the spreading of the printing ink beyond the 2-mm diameter working area, a patterned hydrophobic film was applied to the carbon electrode. To avoid electrode fouling and possible concentration changes in the ink, fresh electrode and ink were used for each voltammetric test. All glucose measurements were performed in a phosphate buffered saline (PBS) solution. In experiments where the pH was varied, 1.0 M HCl and 1.0 M NaOH solutions were used to adjust the pH of the PBS. In amperometric experiments, the working electrode was poised at 0.30 V (vs. Ag/AgCl).

RESULTS AND DISCUSSION

A typical cyclic voltammogram of the PVFcAA mediator in a plain nanocomposite ink is shown in Fig. 2. The electrode exhibited classical features of a diffusion-controlled kinetically fast redox couple. The peak current increased linearly with the square root of potential scan

rate, and the difference between the reduction and oxidation peak potential remained unchanged at 59 mV for scan rates up to 200 mV/s, showing that charge transfer from the mediator to the electrode is rapid.[5] Spiking this ink with glucose did not change the voltammogram at all, which suggests that there is no catalytic oxidation of glucose by the mediator alone. Practically identical voltammograms, as that shown in Fig. 2 trace a, were obtained in the presence of different amounts of GOX, ranging from 0.10 to 20 mg/ml, indicating that the enzyme does not appreciably affect the electrochemistry of the Fc^+/Fc redox couple in the ink. However, an addition of a very small amount of glucose to this ink resulted in an enhanced anodic current and a diminished cathodic current (Fig. 2 trace b). In addition, as can be seen in Fig. 2, the voltammogram was lifted up towards the anodic side around the redox potential of the mediator. Such changes are indicative of a typical chemically coupled electrode process (electrocatalysis).[5]

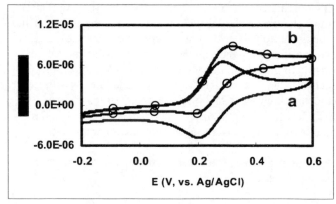

Fig. 2. Voltammograms of the nanoparticulate printing ink (a) and (b) with the addition of 100 mg/dl glucose. Potential scan rate = 100 mV/s.

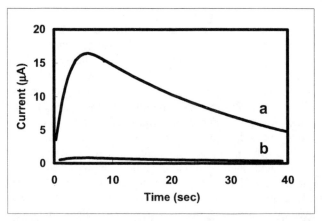

Fig. 3. Amperometric responses of the biosensor in PBS containing (a) 200 and (b) 0.0 mg/dl glucose. Poised potential = 0.30 V.

A typical amperometric response of glucose in an air-saturated PBS at the biosensor is shown in Fig. 3 trace a. Amperometric tests demonstrated that the biosensor has a rapid response time and high sensitivity to glucose. At 0.30 V, after spiking the glucose concentration, the oxidation current increased and reached the maximum very rapidly, within 5 s, followed by a gradual transient which maintains more than 60% of the peak current for a period of 20 s. No catalytic oxidation current was observed in a blank PBS buffer under identical experimental conditions (Fig. 3 trace b), but the presence of the nanoparticulate membrane did increase the background current and it took a considerably long time to drop to a minute level.

In order to obtain a satisfactory performance of the biosensor, the formulation of the nanocomposite ink was optimized. Since the catalytic reaction occurs between the mediator and GOX, the concentration of the mediator must be high enough to have a high sensitivity and linear relationship between the catalytic oxidation current and glucose concentration. Otherwise, the fraction of mediated glucose oxidation will be small and dependent on the amount of mediator in the membrane, instead of the glucose concentration. It was found that a mediator concentration of of 15 mg/ml is best for our purpose and the optimal concentration of GOX was found to be 0.20 mg/ml, taking into consideration of both sensitivity and biosensor economy. The poise potential is expected to affect the amperometric response of the biosensor; it was therefore examined in the range of 0.0 to 0.70 V. The current sensitivity increased with increasing poise potential and reached a plateau at 0.30 V. A slight decrease in sensitivity was observed when the poise potential became more positive than 0.50 V, presumably due to an increased background current. Moreover, too high a poise potential compromises the accuracy of glucose measurements owing to complications from both the much increased background current and possible direct oxidations of a number of electroactive species at the underlying electrode. For amperometric measurements of glucose, the potential of the biosensor was therefore poised at 0.30 V. The dependence of the catalytic oxidation current of glucose on the thickness of the nanoparticulate membrane was also investigated. The catalytic oxidation current reached maximum for nanoparticulate membranes with thickness of 250-500 μm. Insufficient materials in thinner membranes resulted in lower sensitivity and the disappearance of the current peak, On the other hand, further increase in membrane thickness beyond 500 μm could inversely affect the membrane permeability for glucose and the oxidation products of the GOX-catalyzed reaction. In addition, longer response times were noted for thicker membranes.

As mentioned earlier, oxygen affects the sensitivity of the glucose biosensor because glucose oxidation by dissolved oxygen occurs simultaneously as a side-reaction. Initial amperometric tests on thin nanoparticulate membranes employing a hydrophobic polyvinylpyridine (PVP) binder showed that the response was higher in the absence of oxygen than that with dissolved oxygen. At low glucose concentrations, e.g. 50 mg/dl, the competition with oxygen caused a significant decrease (~ 20%) in peak current. Hence, there was a need to suppress the oxygen interference in the system to achieve a highly selective and accurate biosensor. Introduction of acrylic acid units into the hydrophobic PVP resulted in a marked improvement of the biosensor performance. The resulting nanoparticulate membrane was highly hydrophilic, which improved the glucose/oxygen permeability ratio and optimized the accuracy and linearity of the biosensor response. The two amperometric graphs for 200 mg/dl glucose solutions bubbled with nitrogen and oxygen overlaid nicely with a difference of less than 5% in peak current. This illustrated that the biosensor was rather insensitive to the oxygen content in the samples

Unlike those utilizing surface-immobilized sensing membrane, the utilization of the non-conductive nanoparticulate sensing membrane offers great advantages over known disposable glucose biosensors in terms of selectivity. In the former systems, the sensing membrane is part of the electrode and is in direct contact with blood samples. Some constituents in blood, such blood cells, both red and white, proteins and ascorbic acid may interact with the sensing membrane and compromise the accuracy of blood sugar measurements. In this work, the nanoparticulate membrane is non-conductive, and therefore structurally and functionally is not part of the electrode. Catalytic oxidation of glucose only takes place at the electrode/nanoparticulate membrane interface. In other words, no electroactive species exchanges electrons with the electrode unless it passes through the nanoparticulate membrane to reach that interface. Thus, the nanoparticulate membrane provides a barrier to the passage of possible interferences of bulky species in blood such as cells and proteins. When this formulation was used to print the nanoparticulate membrane, the PVPAC binder serves a dual function in the sensing membrane: binding and analyte regulating. On rehydration, the membrane does not break up, but swells to form a gelled layer on the screen-printed carbon surface. Reactants, such as glucose and mediators move freely within this layer, whereas interfering species, such as red blood cells containing oxygenated hemoglobin are excluded. Anionic ascorbic acid and uric acid are expelled by the anionic PVAC polymer, and the partition of dissolved oxygen into the nanoparticulate membrane is largely minimized owing to the highly hydrophilic nature of this layer. This resulted in a sensing membrane whereby the amount of current generated in response to a given glucose concentration varied by less than 5.0% over a hematocrit range of 40-60% and in the presence of 0.20 mM ascorbic and 0.10 mM uric acid. Such desirable insensitivity towards the interfering constituents in blood was also observed in whole blood samples. Furthermore, the nanoparticulate membrane presented an analyte regulating layer for glucose too, significantly slowing down the transport of glucose so that the system was not kinetically controlled, thereby extending the linear domain through the entire physiologically relevant glucose concentration range of 40 to 540 mg/dl.

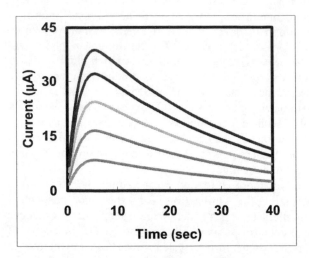

Fig. 4. Current responses to sequential increases of 100 mg/dl glucose from 100 to 500 mg/dl.

The sensitivity was ~76 nA/mg/dl for the peak current for glucose concentrations of ≤ 600 mg/dl. As seen in Fig. 4, the catalytic oxidation current was directly proportional to the glucose concentration up to 600 mg/dl, covering the entire physiologically relevant blood sugar levels. Interestingly, currents obtained at any point of time after the current peak was also linearly dependent on the glucose concentration, providing alternative sampling possibilities within the first 20 seconds of the amperometric tests. The precision was estimated from two series of 20 repetitive measurements of 40 and 300 mg/dl glucose solutions. The relative standard deviations were 4.0% and 8.6%, respectively. The detection limit, estimated from 3 times the standard deviation of repetitive measurements of 5.0 mg/dl glucose under optimal conditions, was found to be 1.8 mg/dl, which is limited by the charging current of the biosensor. More importantly, the blood sample volume needed for a single test was about 0.20 to 0.30 μl, the smallest sample volume amongst all the disposable glucose biosensors available on market. The stability tests were carried out at different temperatures. It was shown that the biosensor maintained 100% of its initial sensitivity for the first 180 days of storage at room temperature, lost 10% of its initial sensitivity after 60 min exposure at 50°C and about 50% of its initial sensitivity after 60 min at 60°C. This may be due to the loss of enzyme activity in the biosensor. The proposed method was successfully applied to the determination of glucose in whole blood (Table 1). The results were in good agreement with the reference values obtained with a yellow springs blood sugar analyzer (YSI Model 2300). The recoveries obtained were also good enough for practical use.

Table 1. Results of blood sugar analysis (average of 10 tests)

Sample	Glucose (mg/dl)	Reference Value (mg/dl)*	Recovery (%) (+ 50 mg/dl)
Blood 1	80	84	96.3
Blood 2	110	115	99.2
Blood 3	105	105	104
Blood 4	185	179	98.5
Blood 5	155	158	97.1

* Obtained with the YSI blood sugar analyzer.

In conclusion, we show here that glucose oxidase and PVFcAA can be readily and homogeneously dispersed into the nanoparticulate alumina together with the hydrophilic PVPAA binder, and the resulting membrane produced a typical catalytic oxidation current for glucose. The experimental results showed that the mediator retained its fast electron transfer properties and the GOX retained its catalytic activity after they were screen-printed onto the carbon electrode. They also demonstrated that this biosensor has good sensitivity and stability for blood sugar monitoring with a blood sample volume of as little as 0.20 μl. The use of screen-printing technique in the fabrication of the biosensor enables easy and low-cost mass production. These biosensor characteristics are promising for development of miniature glucose biosensors of high commercial values.

REFERENCES

1. L.C. Clark and C Lyons, Ann. N.Y. Acad. Sci. 102, 29, 1962.
2. A. Heller, Curr. Opin. Biotech. 7, 50, 1996.
3. H.A.O. Hill and G.S. Sanghera, in A.E.G. Cass (ed.)
4. Biosensors, a Practical Approach, IRL Press, 1990.
5. H.J Hecht, H.M Kalisz, J Hendle, R.D. Schmid, D.and Schomburg, J. Mol. Biol. 229, 3793, 1993.
6. A.J. Bard and L. R. Faulkner, Electrochemical Methods, John Wiley & Sons, New York, 2001.

A Microfluidic System with Nonenzymatic Glucose Sensor

Segyeong Joo[*], Sejin Park[**], Hyojin Lim[**], Teak Dong Chung[**] and Hee-Chan Kim[***]

[*]Interdisciplinary Program-Biomedical Engineering Major,
Seoul National University, Korea, sgjoo@melab.snu.ac.kr
[**]Department of Chemistry, Sungshin Women's University, Korea
[***]Department of Biomedical Engineering, College of Medicine and
Institute of Biomedical Engineering, Medical Research Center, Seoul National University, Korea

ABSTRACT

In this paper, we report an amperometric-type enzymeless glucose sensor system based on nanoporous platinum (Pt) and platinum/platinum oxide (Pt/PtO) reference electrode, which are integrated with microfluidic chip. This microchip system is comprised of microfluidic transport channels and a nonenzymatic glucose sensor system with nanoporous Pt electrode. From this microfluidic chip system, rinsing and reconditioning of sensor can be possible and multiple-use sensor system can be achieved. The sensitivity of developed system was 0.28 µA/mM when tested with glucose solution diluted in PBS buffer.

Keywords: microfluidic system, nonenzymatic glucose sensor, nanoporous electrode

1 INTRODUCTION

More reliable and cost-effective measurement of glucose is obviously one of the most important challenges in the industry of biomedical analysis system [1-2]. However, biological components, mostly the enzymes immobilized, in conventional glucose sensor cause intrinsic problems such as poor preservation and short lifetime, which ultimately lead to cost increase. In this respect, glucose sensor employing no enzyme implies a breakthrough in developing a new one with better reliability and longer shelf time.

Recently our research group reported that nanoporous Pt electrodes work as an excellent glucose detector, which substantially overcomes the critical drawbacks of the previously developed enzymeless sensors [3]. This novel invention is expected through further straightforward modification to be applied to the disposable strip-type glucose sensor which is now all enzyme-based sensor. For a multiple-use sensor system, however, it requires appropriate peripheral devices that rinses and reconditions the sensor in a programmed sequence. To address this matter, we integrated porous Pt films on patterned Pt substrate and Pt/PtO as a solid state reference electrode on a microfluidic chip including such peripheral devices.

2 MICROFLUIDIC CHIP STRUCTURE AND FABRICATION

Fig. 1 shows the schematic diagram of proposed microfluidic system. Pt electrodes for electroosmotic flow (EOF) and fore electrochemical detection were deposited on the bottom glass substrate. The upper layer of poly(dimethyl siloxane) (PDMS) was cast from a silicon wafer mold that has the pattern for microchannels and reservoirs. The mold is an etched silicon wafer with reactive ion etching (RIE) system. Both of the channel width and height are 30 µm.

Fig. 2 shows the overall fabrication process. First, a 2" × 3" slide glass (Model S, Matsunami, Japan) was cleaned with piranha solution and photoresist (AZ5214E, Clariant, Switzerland) was patterned on the top of the glass. Sputtering Ti/Pt (200/2000 Å) was followed by lift-off process to yield the metal electrode pattern, on which the nanoporous Pt working electrode and a Pt/PtO reference electrode were laid in the same way as previously reported [1]. On the other hand, PDMS on the prefabricated silicon mold was underwent additional treatment with O_2 plasma to promote adhesion with the bottom slide glass. Fig. 3 shows the complete system as made.

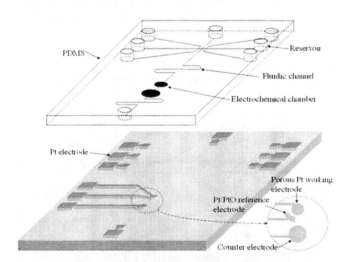

Figure 1: Schematic diagram of proposed microfluidic sysyem.

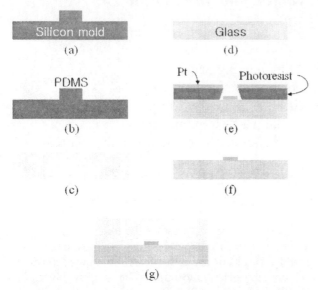

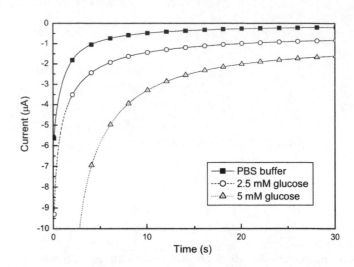

Figure 2: Fabrication process flow (a) Silicon mod (b) Curing PDMS on silicon mold (c) Peer off cured PDMS cover (d) Glass substrate (e) Pt deposited on patterned photoresist (f) Bottom glass after lift-off (g) PDMS bonded to bottom substrate.

Figure 4: Real-time measurement data of oxidation current of glucose with PBS buffer, 2.5 mM glucose and 5 mM glucose solution.

3 RESULTS

Fig. 4 shows the result of preliminary tests of the amperometric responses from the working electrode in real-time. Each glucose solution was diluted in PBS buffer (pH 7.2) and injected to each reservoir. Then, they were electroosmotically delivered to the electrochemical reaction (sensor) chamber by applying high voltage (~2 kV) to the electrode under the corresponding reservoir. Once the electrochemical reaction chamber filled with the glucose solution, EOF was stopped and the oxidation current of glucose from the working electrode was monitored upon applying 0 V to the Pt/PtO reference electrode for 30s. Repeating this process to other glucose solutions with different concentrations provided the data of oxidation current corresponding to the glucose concentration. The sensitivity of the glucose sensor system was 0.28 μA/mM.

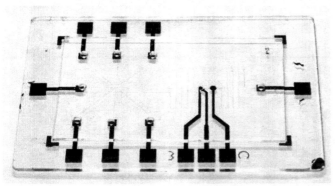

Figure 3: Fabricated microfluidic system with nanoporous Pt nonenzymatic glucose sensor.

4 CONCLUSION

We reported a multiple-use glucose detection system as a microfluidic system with an enzymeless glucose sensor. The developed system consists of a PDMS microchannel network and a glass substrate with Pt electrodes, nanoporous Pt working electrode and Pt/PtO reference electrode. We successfully observed the EOF transport of glucose solutions and verified the performance of the automatic glucose sensing system. We envision that this system will be a promising prototype of a portable micro glucose analysis system.

REFERENCES

[1] Frieder Scheller and Florian Schubert, "Biosensors," Elsevier, 85-125, 1992.
[2] George S. Wilson, "Bioelectrochemisty," Wiley-VCH, 438-458, 2002.
[3] S. Park, et al., "Nonenzymatic glucose detection using mesoporous platinum," Anal. Chem., 75, 3046, 2003.

Glucometers to Detect for Heart Attack

Yousef Haik[*1], Ahmad Qablan[1], Andrew Zowlinski[1], Ching-Jen Chen[1]
*United Arab Emirates University, Department of Mechanical Engineering, Al Ain, UAE
[1]Center of Nanomagnetics and Biotechnology, Tallahassee, FL 32310

ABSTRACT:

Acute Myocardial Infarction (AMI), commonly referred to as a heart attack, is a disorder in which damage to an area of heart muscle occurs because of an inadequate supply of oxygen to that area leading to damage or death of cells. The damaged tissue results in a permanent loss of contraction of this portion of the heart muscle. AMI occurs in approximately 2 out of 1,000 people per year. It is a major cause of sudden death in adults. Currently there is an interest for the earliest possible determination of AMI for patients admitted to the emergency room with chest pains, both to determine which patients can be safely sent home and which need further care, and because for those how suffered an AMI, treatment with thrombolytic therapy is most beneficial the earlier after an AMI the treatment is instituted.

INTRODUCTION

Development of a cost effective, simple and efficient assay determining weather or not an AMI has taken place will greatly assist health care providers in diagnosis. To further assist the medical community it would be beneficial to package the assay into a point of care (POC) device that may be utilized in route to an emergency department or at a patient's bedside. Of equal importance is for the device to be easy to use by nurses in the emergency room. Several proteins or markers such as Myoglobin (MB), Creatine Kinase (CK) and Fatty Acid Binding Protein (FABP), can result from the cardiac tissue injury and are released into the blood stream immediately after the cardiac injury. The increase of concentration of these markers is used to detect for AMI.

A device that measures the concentration of proteins (cardiac markers) that result from cardiac injury is developed. The device utilizes magnetic immunoassay to isolate the cardiac markers from the blood serum. Then with the aid of a glucometer, conventionally used to measure glucose concentration in the blood, the concentration of cardiac markers is measured.

MAGNETIC IMMUNOASSAY

The magnetic immunoassay utilizes a standard solid-phase enzyme linked immunoassay (ELISA). The ELISA takes advantage of the multiple epitopes that are found on the target protein, in our case the cardiac markers. A sandwich is formed by attaching two different antibodies to different epitopes on the same target antigen, which it is in our case, myoglobin (see figure 1). One antibody is attached to a solid surface of the magnetic particle, and the other is attached to a glucose molecule. The first antibodies is used for the separation of the Myoglobin from the blood sample, while the second antibody, attached a glucose molecule, is used to measure the relative concentration of myoglobin in

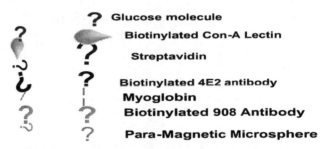

Figure 1. Schematic representation of a two-site immunoassay

blood stream. Attaching glucose molecules at the end of the anti-myoglobin antibody will facilitate the detection of AMI by regular glucometer.

Blood glucose concentration is measured by An amperometric system for point of care determination. In this system, the reagent strips contain a test area, which is impregnated with the enzyme glucose dehydrogenase, and two metal strips that serve as electrodes. Enzymatic action on glucose liberates an electron, which reacts with a mediator also present in the strip. The monitor in which the strip is placed applies a voltage across the electrodes, causing the mediator to be reconverted to its oxidized form. This reaction generates a current proportional to the amount of glucose present and is translated by the meter to a numeric glucose concentration.

Figure 2 shows a measurement of the electric resistance correlated with the glucose concentration. The glucose is attached to an antibody that is linked to the myoglobin. The resulted resistance from each concentration was different and inversely correlated with the concentration; the highest concentration had the lowest resistance.

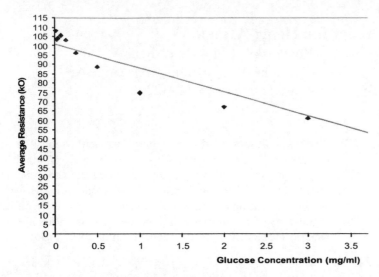

Figure 2: The relationship between each glucose concentrations and its average resistance

Optical Detection of Integrin $\alpha_v\beta_3$ Using a Nanostructured Lipid Bilayer Membrane Biosensor

O.Worsfold* and T. Nishiya

Frontier Research Division, Fujirebio Inc, Tokyo, Japan. ov-worsfold@fujirebio.co.jp

ABSTRACT

An ultra-sensitive optical bionanosensor platform for integrin $\alpha_v\beta_3$ loaded human umbilical vein endothelial cells (HUVECs) was fabricated. The bionanosensor design involved the incorporation of an RGD (arginine-glycine-aspartate) containing peptide conjugated BODIPY lipid soluble dye (donor) into a substrate supported bilayer lipid membrane (SBLM). The RGD peptide binds with the integrin $\alpha_v\beta_3$ via a multimerisation mechanism thus reducing the energy transfer from the donor dye moiety to a second higher excitation wavelength BODIPY lipid soluble dye (acceptor) moiety. The reduction in energy transfer is visualised by an increase in the fluorescence intensity of the donor and a decrease in the fluorescence intensity of the acceptor. The surface chemistry of the substrate has a significant effect on the fluidity of the lipid bilayer (measured by Fluorescence Recovery after Photobleaching (FRAP)) and on the subsequent biosensing response. Using this novel bionanosensor architecture a sensitivity of 25 HUVEC cells/ml was recorded.

Keywords: Integrin, optical, bionanosensor, lipid membrane

1. INTRODUCTION

A novel approach to next generation biosensor applications is to mimic the natural lipid bilayer membrane. Due to the high hydrophilicity of these membranes the degree of protein non-specific adsorption is significantly reduced. The Swanson group (Song, 1999) have proposed the use of lipid bilayers incorporating fluorescent dyes as the signal transducer and Forster Resonance Energy Transfer (FRET) as the interrogation technique. This type of optical system is only partially affected by pinholes/non-uniformity of the lipid bilayer and is therefore more robust than electrical interrogation systems. In the Swanson biosensor the lipid bilayer is either directly attached to microbeads (Song, 2000) or a glass substrate for optical waveguide applications. (Kelly, 1999) Therefore transmembrane proteins such as the biological important acetylcholine molecule cannot be used in these biosensors because they will have a reduced function in the lipid membrane due to the close proximity of the substrate surface disrupting the 3D structure of the transmembrane protein. A number of methods have been investigated to enable the incorporation of transmembrane proteins into substrate supported bilayer lipid membranes, including the use of aqueous polymers (Wong, 1999), unsymmetrical SAMs (Jenkins, 1998) and/or tethered bilayers (Naumann, 2002) to act as bridging links between the substrate and the bilayer lipid membrane. Our research proposes the use of a biotin/streptavidin "cushion" linker between the substrate and the bilayer. Other groups (Berquand, 2003; Delrouye, 2002) are also investigating similar systems using the streptavidin/biotin link, but these methods require the use of fusogens (i.e. PEG) to trigger the vesicle fusion of the streptavidin bilayer. Previous work has demonstrated that by manipulating the vesicle lipid composition and the small unilamellar vesicle preparation, it is possible to form bilayer lipid membranes on streptavidin without the use of external reagents. (Worsfold, 2003a)

Integrin $\alpha_v\beta_3$ is a cell surface bound adhesion receptor that is important in cell-cell and cell-extracellular matrix (ECM) interactions. (Hynes, 1992) Integrins are heterodimers composed of α and β subunits that have been shown to bind with RGD containing peptide ligands possibly via the β_3 chain (D'Souza, 1994). The primary mechanism of the integrin signalling pathway is their ability to recognise ligands from the ECM, anchor, then multimerise on the cell surface. (Stupack, 2002) The arraying (multimerisation) of the integrin $\alpha_v\beta_3$ will form the basis of the present biosensor signalling transducer. [The receptor RGD peptide was covalently attached to a BODIPY fluorescently labelled lipid molecule to act as a donor chromophore.] In the presence of integrin, the RGD attached donor molecule will array on the sensor surface thus affecting the ability of the donor to influence the spectral characteristics of an acceptor labelled lipid soluble material also present in the supported bilayer lipid membrane. (Figure 1)

Our previous work using substrate supported fluid bilayer lipid membranes supported on a streptavidin "cushion" layer demonstrated a sensitivity to integrin $\alpha_v\beta_3$ loaded HUVEC's of 1000cells ml^{-1}. (Worsfold, 2003a) The present work utilises the same biosensing mechanism but with different fluorescent dyes working as donor:acceptors pairs. We also achieved an improved methodology for the isolation and purification of the RGD peptide-dye conjugate (donor) thus enabling a greater sensitivity towards Integrin $\alpha_v\beta_3$ loaded HUVECs. The influence of the substrate surface chemistry is also investigated in terms of lipid bilayer membrane fluidity, biosensing response characteristics and long-term stability of the biosensor.

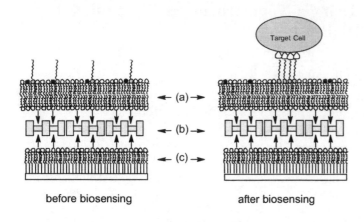

before biosensing   after biosensing

Figure 1: Schematic of the biosensor mechanism. (a)Lipid bilayer/RGDpeptide /BODIPY dye/Biotin lipid, (b) streptavidin, (c), EggPC/Biotin-PE LB monolayer. Substrate was HMDS treated glass or silicon.

2. EXPERIMENTAL

The materials used, synthetic methodology, substrate preparation and AFM characterisation, Langmuir monolayer studies and lipid bilayer membrane preparation will be described in full elsewhere (Worsfold, 2003b)

2.1 Lipid Bilayer Characterisation- FRAP

Fluorescence Recovery After Photobleaching (FRAP) measurements were performed using a Nikon ECLIPSE TS100-F microscope with an epi-fluorescent attachment. Suitable filters were used to allow broad-band excitation from 500-530nm and emission above 550nm. The sample was placed between two hematocytometer plate holders with a kalrez o-ring between the two plates. Photobleaching of the sample (r = 50μm at x20 magnification) with white light for 15-30secs was dependent on the sample. Images were recorded using a Hamamatsu C4742-95 digital camera with exposure times of 122ms – 1000ms depending on the sample. The images were analysed using SCION image beta 4.02 freeware.

2.2 Integrin $\alpha_v\beta_3$ Loaded HUVEC Biosensing

Fluorescence spectral analysis was performed using a Perkin-Elmer LS55 luminescence spectrometer. Scan speed = 25nm min^{-1}, Excitation slit = 2.5nm, Emission slit = 15nm. Excitation λ = 545nm (unless stated). The lipid bilayer samples were assembled/fabricated then inserted into a Perkin-Elmer cuvette slide holder. Biosensing experiments were performed by removing aliquots of the PBS solution and adding equal volumes of the HUVEC solution to the cuvette. The sensitivity of the biosensing response is determined from the following equation.

$$S = \frac{|F_0 - F_t|}{F_0} \times 100 \qquad (1)$$

Where S = Sensitivity of biosensing response, F_0 = Fluorescence (at a given wavelength) before and F_t = Fluorescence (at a given wavelength) after addition of cells respectively.

3. RESULTS AND DISCUSSION

The Langmuir monolayer and Langmuir-Blodgett film deposition characterisation are described elsewhere. (Worsfold, 2003b)

3.1 Fluorescence Recovery After Photobleaching (FRAP)

The fluidity of the lipid bilayer is described by the diffusion of fluorescently tagged lipids into an area, which was previously photobleached. The diffusion coefficient can be estimated from the following equation.

$$D = \left(\frac{r^2}{4t_{50}}\right) \qquad (2)$$

Where D is the diffusion coefficient, r = radius of the photobleached area and t_{50} = time taken to 50% fluorescent intensity recovery.

The lipid membrane fluidity of the ruptured vesicles was measured on three different surfaces (a) NaOH treated glass, (b) HMDS/streptavidin treated glass and (c) untreated glass. Figure 2 shows characteristic fluorescent micrographs produced in these FRAP studies for a lipid bilayer formed on a NaOH treated glass substrate. The t_{50} time is calculated from the fractional fluorescence recovery curve shown in figure 3 which gives a diffusion coefficient for this system D = 8.3x10^{-8}cm^{2}s^{-1}. Also shown in figure 3 are the fractional fluorescence recovery curves for the untreated glass substrates supported lipid bilayer and the HMDS/streptavidin supported lipid bilayer. The diffusion coefficients for the HMDS/streptavidin supported lipid bilayer (D = 2.6x10^{-8}cm^{2}s^{-1}) and the untreated glass substrate (D = 3.1x10^{-8}cm^{2}s^{-1}) indicate that the fluidity of the lipid bilayer on NaOH treated glass is greater than on either the untreated surface or the HMDS treated/streptavidin "cushion" substrates both of which have similar values. However, the long term stability of the lipid bilayer on NaOH treated glass is significantly less than on the other substrate surfaces.

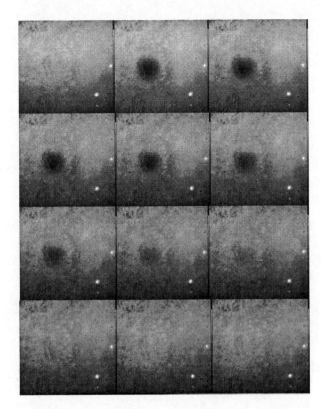

Figure 2: FRAP images of a lipid bilayer membrane on a NaOH treated glass substrate. From top left to right :- before photobleaching, after 1min photobleaching, 1min after photobleaching, 2mins after, 3mins after. 4mins after, 5mins after, 10mins after, 14mins after, 20mins after, 45mins after, 60mins after.

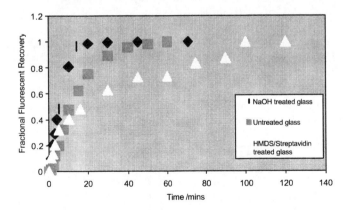

Figure 3. Fractional fluorescence recovery for lipid bilayer membranes on NaOH treated glass, untreated glass and HMDS/Streptavidin treated glass substrates.

3.2 Biosensing Experiments

All biosensing experiments were recorded on glass substrates etched for 24hrs, HMDS treated substrates and untreated substrates, at 298K. Figure 4 shows the effect of adding HUVEC cells, at a concentration of 25cells ml^{-1}, to a lipid bilayer formed on NaOH treated glass substrates. Here, there

is an increase in the intensity of the BODIPY 558/568-peptide donor moiety at 570nm and a decrease in the intensity of the BODIPY TR-X acceptor moiety at 625nm on exposure to HUVECs. These changes in the fluorescence are predicted from the detection mechanism, where the donor peptide moieties multimerise on the lipid membrane surface upon binding with the integrin $\alpha_v\beta_3$ proteins on the HUVEC surface. This causes a reduction of the energy transfer from the donor to the acceptor moiety. Figure 5 shows the effect of 5000 HUVECcells ml^{-1} on the streptavidin "cushion" lipid bilayer structure. A lower degree of response is observed for this system compared with the NaOH treated glass system at a concentration of 25 HUVECcells ml^{-1}.

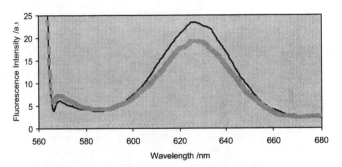

Figure 4. Effect of 25 HUVEC cells ml^{-1} (Thick line) to lipid bilayer membrane on NaOH treated glass substrate.

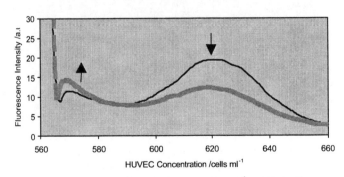

Figure 5. Effect of 5000 HUVECcells ml^{-1} (Thick line) to lipid bilayer membrane on HMDS/streptavidin treated glass substrate.

The effect of concentration on the sensor responses for both systems is shown in figure 6. In the lipid bilayer on NaOH treated glass substrate system (squares), there is a rapid increase in the response at low concentrations 0-500cells ml^{-1} followed by the onset of equilibrium saturation above this concentration. A lower degree of response is observed for the HMDS/streptavidin system compared with the NaOH treated glass system at a HUVEC concentration of 25cells ml^{-1}. However, the concentration plot of the HMDS/streptavidin supported system has a potentially greater dynamic range than the NaOH treated glass system. The long-term stability of the NaOH treated glass system is reduced (approx. 1 week) compared with the streptavidin "cushion" lipid structure which has demonstrated biosensing action after being stored for 1 month at 4°C. The untreated glass substrate bilayer system also exhibited response on exposure to HUVECs but at a significantly lower sensitivity.

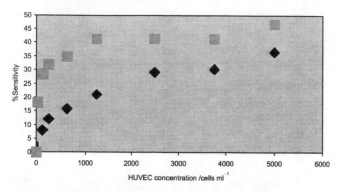

Figure 6. Effect of HUVEC concentration on response sensitivity for lipid bilayers on NaOH treated glass substrate (squares) and HMDS/streptavidin treated glass substra (triangles).

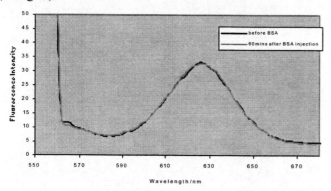

Figure 7. Effect of 63μm Bovine Serum Albumin on biosensor response.

Figure 7 shows the lack of any response of the HMDS/streptavidin system on exposure to 63μM Bovine Serum Albumin (BSA). This demonstrates the specificity of the proposed biosensor even on exposure to common interference proteins (BSA), which often cause false positive results in other optical biosensor systems.

4. CONCLUSIONS

In the present work we have changed the donor and acceptor dyes from the previous study to enable higher wavelength biosensor detection. Using this strategy, together with alterations in the donor:acceptor dye ratio, a higher sensitivity of 25 HUVEC cells ml[-1] was recorded. The importance of the substrate surface chemistry was highlighted by the increased bilayer fluidity on the NaOH treated substrates. One possible reason for this is the increased flatness of these substrates compared with the HMDS/streptavidin, or untreated glass substrate system. Interestingly however, the HMDS/streptavidin system provided the greatest long-term stability, as well as good biosensing sensitivity and dynamic sensing range.

5. REFERENCES

Berquand, A., Mazeran, P-E., Pantigny, J., Proux-Delrouyre, V., Laval. J-M., Bourdillon, C., (2003). *Langmuir*, **19**, 1700-1707.

Delrouye, V.P., Elie, C., Laval, J-M., Moiroux. J., Bourdillon, C., (2002), *Langmuir*, **18**, 3263-3272.

D'Souza, S.E., Haas, T.A., Piotrowicz, R.S., Byers-Ward, V., McGrath, D.E., Soule, H.R., Cierniewski,C., Plow, E.F., Smith, J.W., (1994), *Cell*, **79**, 659-667

Hynes, R.O, (1992), *Cell*, **69:1**, 11-25.

Jenkins, A.T.A., Evans, S.D., Knowles, P.F., Liu, Q, Miles, R.E.,Ogier, S.D, (1998), *Langmuir*, **14**, 4675-4678.

Kelly, D., Grace, K.M., Song, X., Swanson, B.I., Frayer, D., Mendes, S.B., Peyghambarian, N., (1999), *Optics Letters*, **24:23**, 1723-1725.

Naumann, C.A., Prucker. O., Lehmann, T., Ruhe, J., Knoll, W and Frank C.W., (2002) , *Biomacromolecules*, **3**, 27-35.

Song, X., Swanson, B.I., (1999), *Analytical Chemistry*, **71**, 2097-2107.

Song, X., Shi., J., Swanson, B., (2000), *Analytical Biochemistry*, **84**, 35-41.

Stupack, D.G., Cheresh, D.A., (2002), *Journal of Cell Science*, **115**, 3729-3738.

Wong, J.Y., Majewski, J., Seitz, M., Park, C.K., Israelachvili, J.N. and Smith G.S., (1999), *Biophysical Journal*, **77**, 1445-1457.

Worsfold, O., Toma, C., Nishiya, T., (2003a), *Biosensors and Bioelectronics*, accepted.

Worsfold, O and Nishiya, T (2003b), *Journal of Materials Chemistry*, submitted.

Contact Author
Oliver Worsfold,
Frontier Research Department,
Fujirebio Inc., 51Komiya-cho, Hachioji-shi,
Tokyo 192-0031, Japan
Email: ov-worsfold@fujirebio.co.jp

Micro-Patterning of Ionic Reservoirs within a Double Bilayer Lipid Membrane to Fabricate a 2D Array of Ion-Channel Switch Based Electrochemical Biosensors

José-María Sansiñena[*], Chanel K. Yee[**], Annapoorna Sapuri[**],
Basil Swanson[***], Antonio Redondo[*], Atul N. Parikh[**]

[*]T-12, Theoretical Division, Los Alamos National Laboratory, Los Alamos, NM 87545
[**]Department of Applied Science, University of California, Davis, CA 95616
[***]B-4, Bioscience Division, Los Alamos National Laboratory, Los Alamos, NM 87545

ABSTRACT

We present a new approach for the design of ionic reservoir arrays within a double phospholipid bilayer to ultimately develop a 2D array of ion-channel switch based electrochemical biosensors. As a first step, a primary bilayer lipid membrane (BLM) is deposited onto an array of electrodes patterned onto a substrate surface. Subsequently, an array of microvoids is created within the bilayer by a wet photolithographic patterning of phospholipid bilayers using a deep UV light source and a quartz/chrome photomask. The deposition of a secondary BLM onto the primary bilayer that spans across the patterned microvoids leads to the formation of an array of ionic reservoirs within the double phospholipid bilayer. This is accomplished using giant unilamellar vesicles and by exploiting membrane electrostatics. The eventual use of ligand-gated ion-channels incorporated into the secondary bilayer that covers the individual ionic reservoirs will lead to a 2D array of ion-channel switch based electrochemical biosensors.

Keywords: bilayer lipid membrane, ionic reservoir, ion channel, electrochemical biosensor

1 INTRODUCTION

Ion-channel based biological systems have not been extensively employed in current sensor technology because of the serious problems in reproducing a native-like microenvironment necessary for proper functioning. In fact, there are several problems in achieving this.

1. One needs to create a lipid bilayer, on a solid surface, that is sufficiently flexible and defect free such that ion-channel activity can actually be measured.
2. The lipid membrane must be in a physical state where biological molecules function as if in a natural system.
3. One needs to entrap an ionic medium between the membrane and the supporting electrode in order to facilitate the ion exchange through the ion-channels incorporated into the membrane.

A previous approach based on Gramicidin-A pairs, by Cornell *et al.* [1], exhibits limitations related to a choice of ion-channel switching mechanism that is different from any of those present in cell membranes. In addition, the preparation of the ionic reservoir structure between the membrane and the electrode evidences high experimental difficulties.

In contrast, surface micro-patterning methods in the solid state have been particularly useful for patterning supported phospholipid bilayers. These patterned phospholipid bilayers have played an important role to understand, emulate, pattern, and exploit many functions of cell membranes for fundamental biophysical research and many biomedical and sensing technologies [2-8].

The micro-patterning of supported membranes has been achieved by a variety of methods that fall into two broad categories:

1. Use of pre-patterned substrate surfaces that present chemical and/or electrostatic barriers to membrane formation [9-14].
2. Application of soft-lithographic methods of patterned deposition (stamping) or removal (blotting) using polymeric stamps [15, 16].

However, the methods requiring substrate pre-patterning depend on the prior deposition of exogenous materials on the substrate surface and form single patterns. While methods based on polymer stamps circumvent these issues, they require optimization of the physical contact, associated contact pressure, and deformability of the polymer stamps. Moreover, difficulties associated with achieving contact uniformity for large-area patterning remain.

In this approach we describe a simple light-directed method for patterning supported phospholipid membranes over large substrate areas in a non-contact manner and without requiring prior substrate patterning. The approach is generally analogous to solid-state lithographic methods developed for dry-state DNA and peptide arrays [17], but is applicable for the aqueous phase and fluid bilayer lipid membranes (BLM). It simply relies on the spatially-directed illumination of phospholipid bilayers submerged in an aqueous ambient to UV light. The pattern of exposure to light through a mask, or by other spatially addressable means, determines the relief pattern generated within the bilayer. Ultimately, an experimental procedure based on this approach has been developed to prepare ionic reservoirs within a double lipid bilayer with independent electrodes connected to individual ionic reservoirs.

2 APPROACH

The general procedure that we proposed to pattern supported membranes onto a solid substrate is schematically shown in Figure 1.

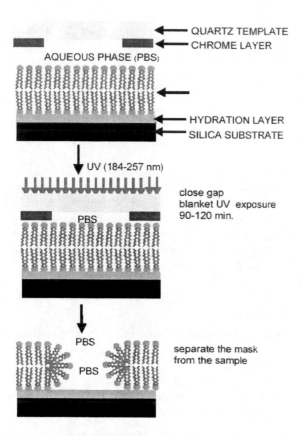

Figure 1. Direct patterning of voids within bilayer membranes using deep UV photolithography.

The process begins with the preparation of a continuous supported bilayer lipid membrane (sBLM) on a hydrophilic silica surface. To enable fluorescence measurements, vesicles are doped with appropriate concentrations of labeled lipids (e.g., 1 mol% 1,2-dihexadecanoyl-*sn*-glycero-3-phosphoethanolamine, triethylammonium salt abbreviated as Texas Red-DHPE). Next, a lithographically produced quartz mask (e.g., an array of square opaque (chrome) elements over a UV transparent quartz mask) is then brought in soft contact with the sBLM. Deep UV light in the 184–257 nm range, produced by low or medium pressure Hg lamp housed in a used quartz envelope (10-20 mW cm^{-2}), is then directed through the mask at the bilayer samples submerged in phosphate buffered saline (PBS) for ~120 min. Upon separation of the mask from the sample under the buffer, high-fidelity patterns of membrane bilayers comprising intact lipid patches in UV-protected areas and lipid-free regions in the UV-exposed areas are obtained. Using this approach, one can create both patterns of voids and isolated membrane islands.

The previous general procedure to pattern supported membranes onto a solid substrate has to be modified in order to prepare ionic reservoirs within a double lipid bilayer with independent electrodes connected to individual ionic reservoirs (Figure 2).

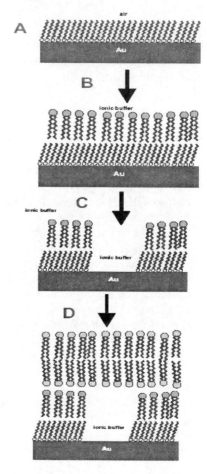

Figure 2. Schematic representation of an ionic reservoir formation within a double lipid bilayer constructed. **A)** Self assembly of alkanethiol (RSH) molecules on Au; **B)** Vesicle spreading on hydrophobized RSH/Au to produce a primary hybrid bilayer; **C)** Photochemical patterning of the hybrid bilayer; **D)** Deposition of the second bilayer that spans across the void in the primary bilayer.

First, a prior deposition of a metal onto the substrate with the same spatial distribution as the micro-patterned voids, using a customized mask with holes following the same distribution as in the previously mentioned quartz mask, is carried out. Next, the gold surface is hydrophobized by the self-assembly of alkanethiol molecules (Figure 2a) to allow the spreading of a vesicle to produced a primary hybrid bilayer (Figure 2b). The photochemical patterning of the hybrid bilayer leads to the formation of an array of voids within the bilayer (Figure 2c). Finally, the deposition of a second lipid bilayer that spans across the voids in the primary bilayer leads to the formation of a 2D array of ionic reservoirs (Figure 2d).

The eventual use of ligand-gated ion-channels incorporated into the secondary lipid bilayer that covers the individual ionic reservoirs will lead to a 2D array of ion-channel switch based electrochemical biosensors (Figure 3).

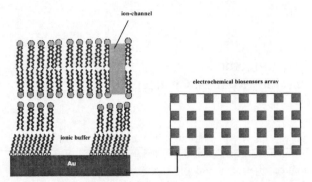

Figure 3. Electrochemical biosensors array based on ion-channels incorporated into the secondary lipid bilayer that covers the ionic reservoirs connected to individual electrodes.

In this kind of electrochemical biosensors the signaling occurs by electrochemical detection of changes in conductance when ion-channels open or close because of binding of the target-agent. Taking into account that each electrode is independently connected, one can get individual electric signals from each ionic reservoir.

3 CONCLUSIONS

In summary, the wet membrane photolithography approach presented here is simple, inexpensive, and an enabling procedure to create optically-defined patterns of fluid bilayer lipid membranes submerged in the aqueous phase. This approach allows the preparation of ionic reservoirs within a double bilayer with independent electrodes connected to individual ionic reservoirs for the development of a 2D array of ion-channel switch based electrochemical biosensors.

REFERENCES

[1] B. A. Cornell, V. L. B. Braach-Maksvytis, L. G. King, D. O. Osman, B. Raguse, L. Wieczorek, R. J. Pace, Nature, 387, 580-583, 1997.

[2] J. T. Groves and S. G. Boxer, Acc. Chem. Res. 35, 149-157, 2002.

[3] E. Sackmann, Science 271, 43-48, 1996.

[4] S. G. Boxer, Curr. Opin. Chem. Biol. 4, 704-709, 2000.

[5] Y. Fang, A. G. Frutos and J. Lahiri, J. Amer. Chem. Soc. 124, 2394-2395, 2002.

[6] C. M. Rosenberger, J. H. Brumell and B. B. Finlay, Curr. Biol. 10, R823-R825, 2000.

[7] A. Grakoui, S. K. Bromley, C. Sumen, M. M. Davis, A. S. Shaw, P. M. Allen, and M. L. Dustin, Science 285, 221-227, 1999.

[8] H. M. McConnell, T. H. Watts, R. M. Weis and A. A. Brian, Biochim. Biophys. Acta 864, 95-106, 1996.

[9] P. S. Cremer and T. L. Yang, J. Amer. Chem. Soc. 121, 8130-8131, 1999.

[10] J. T. Groves and S. G. Boxer, Biophys. J. 69, 1972-1975, 1995.

[11] J. T. Groves, N. Ulman and S. G. Boxer, Science 275, 651-653, 1997.

[12] J. Nissen, K. Jacobs and J. O. Radler, Phys. Rev. Lett. 86, 1904-1907, 2001.

[13] L. A. Kung, L. Kam, J. S. Hovis and S. G. Boxer, Langmuir 16, 6773-6776, 2000.

[14] K. Morigaki, T. Baumgart, A. Offenhausser and W. Knoll, Angew Chem. Intl. Ed. 40, 172-174, 2001.

[15] J. S. Hovis and S. G. Boxer, Langmuir 16, 894-897, 2000.

[16] J. S. Hovis and S. G. Boxer, Langmuir 17, 3400-3405, 2001.

[17] S. P. A. Fodor, J. L. Read, M. C. Pirrung, L. Stryer, A. T. Lu and D. Solas, Science 251, 767-773, 1991.

Potentiometric Biosensors Based on Silicon and Porous Silicon

Indranil Basu*, R.Venkata Subramanian**, Arun Mathew*,
Anju Chadha** and Enakshi Bhattacharya*

*Department of Electrical Engineering , Indian Institute of Technology Madras,
Chennai 600036, India, enakshi@ee.iitm.ernet.in
**Department of Chemistry, Indian Institute of Technology Madras, Chennai 600036,
anjuc@iitm.ac.in

ABSTRACT

We report fabrication of potentiometric biosensors with silicon for the estimation of triglycerides and urea based on enzymatic reactions. The sensor is an Electrolyte–Insulator-Semiconductor capacitor (EISCAP) that shows a shift in the measured CV with changes in the pH of the electrolyte. Enzyme mediated biological reactions involve changes in the pH of the electrolyte and an EISCAP can be effectively used for detection of biological compounds. Optimization of the conditions for the enzymatic reaction and calibration of the sensor are included. Effect of replacing silicon with porous silicon is discussed.

Keywords: biosensor, potentiometric, EISCAP, pH, enzymatic

1 INTRODUCTION

Detection and estimation of Triglycerides and Urea are clinically important, the first being correlated to heart disease while urea is an important analyte for the diagnosis of diseases such as renal malfunction. There is immense scope for improving sensitivity and ease of operation of commercial sensors presently used for estimating triglycerides and urea. We report fabrication of silicon and porous silicon based potentiometric biosensors for the estimation of tributyrin, a short chain triglyceride, and urea. The sensor is an Electrolyte–Insulator-Semiconductor capacitor (EISCAP) that shows a shift in the measured CV with changes in the pH of the electrolyte [1]. Many biological reactions, especially enzyme mediated ones, involve changes in the pH of the electrolyte. Hence an EISCAP can be effectively used for the detection of biological compounds through enzymatic reactions and can function as a biosensor.

2 PRINCIPLES

We can consider the basic EISCAP structure with silicon as the semiconductor, the oxide SiO_2 as the insulator and the electrolyte as the sample to be tested. When in contact with an aqueous solution, the SiO_2 surface is hydrolyzed to form silanol surface groups. These groups are positively charged, negatively charged, or neutral depending on the pH of the electrolyte[1]. The presence of this surface charge at the interface of SiO_2 and electrolyte affects the depletion layer within the silicon at the Silicon and SiO_2 interface, thereby producing a change in the flat band voltage of the electrolyte-insulator-semiconductor (EIS) system. Hence for solutions with different pH, C-V curves of the EIS system shift along the voltage axis with essentially no change in their shape.

The situation is similar if the insulator is silicon nitride[2,3]. Silicon nitride (Si_3N_4) has replaced SiO_2 as the sensing material in EISCAPs due to its superior sensing properties as compared to SiO_2. It not only offers a better pH sensitivity but is also stable in basic solutions in contrast to SiO_2, which dissolves in basic solutions [4]. Hence with Si_3N_4 based EISCAP, the detection can be extended to the basic range.

The pH sensitivity is determined by calculating the parallel shift in C-V curves in the depletion region along the voltage axis at a point of constant capacitance (usually at the midpoint of the C-V curve). The voltage value at that point is termed as U_{bias}. The EISCAPs, thus, can function as pH sensors and can be utilized for biosensing in case of enzyme mediated biological reactions resulting in the release of acid/base [5].

The amount of acid/base produced as a result of the enzymatic reaction is proportional to the amount of bioanalyte consumed. The pH of the electrolyte changes due to the production of acid/base that can be detected by the EISCAP. By correlating the change in pH of the electrolyte solution to the concentration of the bioanalyte, which produces this change, a pH sensing EISCAP can be used for detection of biological compounds.

When tributyrin, a short chain triglyceride, is hydrolysed by the enzyme lipase it results in the production of butyric acid as a product.

$$\text{Tributyrin} + \text{Water} \xrightarrow{\text{LIPASE}} \text{Glycerol} + \text{Butyric acid}$$

The pH of the solution changes with the butyric acid produced and the change is proportional to the concentration of tributyrin in the solution.

In the case of urea, the reaction is:

$$\text{Urea} + \text{Water} \xrightarrow{\text{UREASE}} \text{Ammonia} + \text{Carbon dioxide}$$

Ammonia and carbon dioxide liberated as a result of the enzymatic reaction dissolve in water thereby producing ammonium hydroxide and carbonic acid respectively. Ammonium hydroxide is a strong base while carbonic acid is a weak acid. Hence pH of the electrolyte solution after the enzymatic reaction shifts towards the basic range.

The change in pH is detected by CV measurements on the EISCAP and co-related to the concentration of tributyrin or urea[6].

3 EXPERIMENTAL DETAILS

Fig 1 shows the basic structure of the fabricated EISCAP. In order to obtain a good interface between the insulator layer and semiconductor in an EISCAP, Si_3N_4 is deposited on thermally oxidized silicon sample since the interface state density is negligible at the Si-SiO_2 interface. p-type(100) silicon wafers with a resistivity of 4 to 11 Ωcm were oxidized at 1000°C in dry oxygen for 2 hours. Si_3N_4 was then deposited by PECVD at 300°C for 7 minutes followed by a 10 min annealing at 800°C. The annealing step is essential because untreated Si_3N_4 when kept exposed to ambient conditions results in the oxidation of Si_3N_4 surface and degradation of pH sensitivity of Si_3N_4 [1,7].

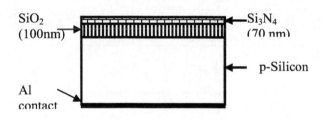

Fig 1: Cross-section of EISCAP

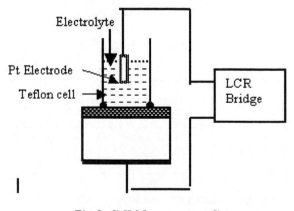

Fig.2: C-V Measurement Setup

The experimental setup used for characterizing the EISCAP is as shown in Fig. 2. The electrolyte is held with the help of a teflon cell and O-ring and the contact is taken from a platinum-wire dipped in the electrolyte. The semiconductor contact was taken from the back aluminium.

C-V measurements were done at room temperature with an HP 4275A LCR meter at 10kHz with 15mV signal amplitude by sweeping the dc bias from -10 to -3 V in steps of 100mV. Solutions of different pH were prepared by adding HCl and NaOH to 0.1M phosphate buffer with 0.5M KCl added as ionic strength adjuster for C-V measurements. After measurements, the device was thoroughly rinsed in deionised water. When the device is not in use it is stored in deionised water at 4^0C.

4 RESULTS AND DISCUSSIONS

The CV curves were normalized with respect to the accumulation capacitance value. With increasing pH of the electrolyte the C-V curves were found to shift towards right along the voltage axis as indicated in Fig. 3. U_{bias} is defined as the voltage where the capacitance falls to 70% of its maximum value. A plot of U_{bias} vs. pH shows a maximum slope of 55mV per unit change in pH, Fig. 4.

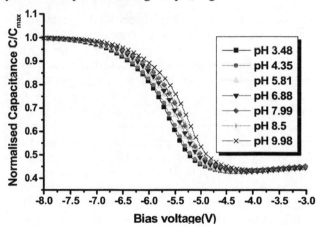

Fig.3: Normalized C-V curves for solutions of different pH

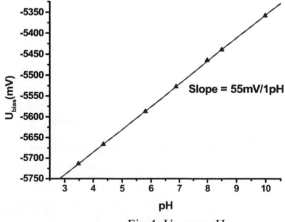

Fig.4: U_{bias} vs. pH

The rate of an enzymatic reaction is a function of the sample concentration and will be linear until the catalytic sites of the enzyme are saturated with the sample. At saturation, the rate of the reaction is independent of the sample concentration. It is important therefore to optimize various parameters of an enzyme catalyzed reaction in the linear region where the specificity of the enzyme to the sample is maximum. The parameters optimized for this study include enzyme concentration, sample concentration, time of the reaction, buffer concentration and its pH. For the detection of tributyrin, the amount of the enzyme, Lipase - with an activity of 40 units/mg - and the time of reaction were optimized to 12 mg for 30 ml of buffer solution and 30 min respectively, for the enzyme reaction to be in the linear region[8]. The buffer used was phosphate buffer, 0.01M, pH 7 and all reactions were done at room temperature.

The activity of the enzyme Urease (Jack beans, Fluka) was determined by the spectrophotometric method[3] before carrying out the optimisation studies and the activity was found to be 30 U mole/mg/min[9]. For the estimation of urea, the optimized values for a sample concentration of 150mM were 0.16mg of Urease in 0.5M phosphate buffer for 30 minutes.

The pH of the electrolyte changes after the enzymatic hydrolysis and is detected by CV measurements on the EISCAP. The final calibration plot of the sensor is a plot of U_{bias} vs. sample (tributyrin/urea) concentration. This is constructed by combining the two curves, namely: (1) sample concentration vs. pH which is obtained by carrying out the enzymatic reaction on known concentrations of the sample, monitoring the change in pH using a digital pH meter and noting down the pH value after the optimised interval of time (i.e. 30 min). (2) U_{bias} vs. pH which is obtained by carrying out the C-V measurements using known concentrations of the samples, after subjecting them to enzymatic reaction for the optimised interval of time. This calibration curve can be used to determine unknown concentrations of the sample. Figs 5 and 6 give the calibration plots for tributyrin and urea.

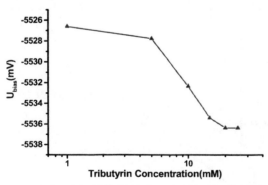

Fig.5: Calibration plot for sensing Tributyrin

Measurements were done with unknown concentrations of tributyrin and urea and the calibration plots, Figs 5 and 6, were used to determine the concentrations. For this, unknown concentrations of the sample to be estimated, tributyrin and urea, were prepared and the optimized value of the enzyme, lipase and urease , were added. After

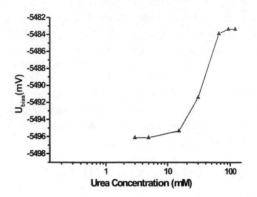

Fig.6: Calibration plot for sensing Urea

carrying out the enzymatic reaction for the optimized time, C-V measurements were taken and the corresponding values of U_{bias} were determined. From the calibration curve the tributyrin/urea concentration corresponding to the measured U_{bias} value was obtained and was compared with the actual value. The measured and actual values are given in Tables 1 and 2 and show good agreement.

Table 1: Unknown concentrations of Tributyrin as determined by the sensor

No.	U_{bias} (mV)	Concentration (mM)		% Error
		Measured	Actual	
1.	-5530.8	8.08	8.29	-2.53
2.	-5534.2	12.97	12.89	0.62

Table 2: Unknown concentrations of Urea as determined by the sensor

No.	U_{bias} (mV)	Concentration (mM)		% Error
		Measured	Actual	
1.	-5486.2	51.63	50	3.26
2.	-5488.3	41.72	40	4.3

4.1 Porous Silicon as the Substrate

Electrochemical dissolution of crystalline silicon in hydrofluoric acid (HF) based electrolyte solutions at constant current gives a sponge-like silicon matrix with pores, which is known as porous silicon [10]. Due to the large specific surface area ($> 200 \text{ m}^2 \text{ cm}^{-3}$) [11] and its

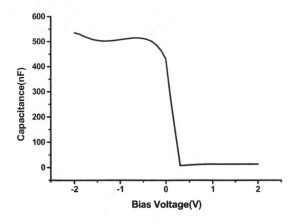

Fig.7: CV with porous silicon substrate

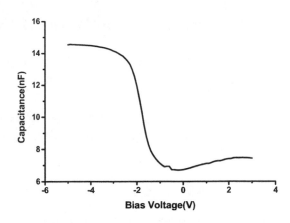

Fig.8 : CV with single crystalline Si substrate

porous structure, PS can adsorb molecules easily from the surroundings and can be used effectively as a sensor material. The pore size can be varied by varying the substrate doping, concentration of the etching solution, the anodisation current density and the time. Porous silicon was prepared under the following conditions:

Substrate – p-Si (ρ=4-11Ω-cm)
Etching solution – 1:2::HF:Iso-propanol
Anodisation current density – 10mA/cm^2.
Anodisation time- 10mins

The thickness of the porous silicon layer was found to be 5 μm and the average width of a silicon filament was about 10 μm while the average pore size was 2-3 μm. The porous silicon sample was subjected to dry oxidation at 400^{0}C followed by N_2 annealing after which 40nm of SiO_2 and 80nm of Si_3N_4 were deposited on it by PECVD technique. C-V measurements were then taken on the porous silicon sample using phosphate buffer solution (0.1M Na_2HPO_4/KH_2PO_4 in 0.5M KCl) at a signal frequency of 10kHz and is shown in Fig 7. C-V measurements were also performed using the phosphate buffer solution at a frequency of 10kHz on a similarly processed single crystalline silicon sample and is shown in Fig 8. As is observed, the porous silicon sample shows a 34.5 times increase in accumulation capacitance over the single crystalline silicon sample. This is due to the much larger surface area of the porous silicon sample and will be useful in miniaturization of the sensor.

5 CONCLUSIONS

A potentiometric sensor has been developed for detection and estimation of tributyrin and urea, which can be extended to detect other biological compounds as well. This sensor is an EISCAP with silicon as the substrate and Silicon Nitride as the sensing dielectric. Sensors using porous silicon as the substrate material have been shown to be more sensitive due to the increase in surface area as compared to the single crystal planar sensors. The sensors show good reproducibility.

ACKNOWLEDGEMENT

We thank the Department of Science and Technology, Government of India for funding this project.

REFERENCES

[1] William M. Siu and Richard S. C. Cobbold, IEEE Trans. on Electron Devices, ED-26, 1805, 1979
[2] D. L. Harame, L. J. Bousse, J. D. Shott, J. D. Meindl, IEEE Transactions on Electron Devices, ED-34, No.8, 1700,1987.
[3] C. Cańe, I. Grácia & A. Merlos, Microelectronics Journal 28, 389, 1997
[4] G. F. Eriksen & K. Dyrbye, Journal of Micromechanics and Microengineering, 6,55, 1996.
[5] Marion Thust, M. J. Schoning, S. Frohnhoff, R. Arens-Fischer, P. Kordos and H Luth, Meas. Sci. Technol., 7, 26, 1996
[6] R. R. K. Reddy, A. Chadha, E. Bhattacharya, Biosensors and Bioelectronics, 16, 313, 2001
[7] M. J. Schöning, A. Steffen, M. Sauke, P. Kordos, H. Lüth, A. Zundel, M. Müller-Veggian, Proceedings of the 3rd European Conference Sensors for the environment, Grenoble, 55, 30, March 1995.
[8] R. R. K. Reddy, I. Basu, E. Bhattacharya, A.Chadha, Current Applied Physics 3, 155, 2003
[9] R.V. Subramanian, I. Basu, A. Mathew, E. Bhattacharya, A.M. Kayastha and A. Chadha, Proc. 12th Intl. Workshop on Physics of Semiconductor Devices, IWPSD-2003, IIT Madras, Dec 2003, to be published.
[10] Joseph Fourier, and Jacques Derrien, "Porous Silicon Science and Technology", Winter School, Les Houches, 8 to 12 February, 1994.
[11] R. Herino, G. Bomchil, K. Barla and C. Bertarand, J. Electrochem. Soc., 134, 1994, 1987.

Bio-impedance Sensing Device (BISD) for Detection of Human CD4+ Cells

Nirankar N. Mishra[*], Scott Retterer[**], Thomas. J. Zieziulewicz[*], Mike Isaacson[**],
Donald Szarowski[*], Donald E. Mousseau[*], David A. Lawrence[*] and James N. Turner[*]

[*] Wadsworth Center, NYS Department of Health, Albany, NY, USA, nxm01@health.state.ny.us
[**] Cornell University, Ithaca, NY, USA, str8@cornell.edu

ABSTRACT

The aim of this work is to develop a small on-chip bio-impedance sensing device (BISD) to rapidly detect and quantify cells with a specific phenotype in a heterogeneous population of cells (e.g., human CD4+ cells in blood, for monitoring HIV-infected individuals) using a minimal sample volume and minimal preparation. The transducers, gold/titanium microelectrodes (100x100μm and 80x80μm), have been fabricated on glass substrates. The microelectrode surface is non-covalently modified sequentially with protein G', human albumin, monoclonal anti-human CD4 antibody, and mouse IgG. The anti-human CD4 antibody binds CD4+ cells present in human blood. The basic function of the microelectrodes was characterized using electrochemical cyclic voltammetry before and after protein G' deposition. The binding of biomolecules, protein G' and antibodies, as well as cells was detected by precisely measuring the electrical current, as a function of frequency (1-8 kHz), that flowed between the microelectrode and the much larger reference electrode. This current was measured using a specially designed very low-noise amplifier based on instrument-grade operational amplifiers. This measurement was plotted as impedance, using the constant-amplitude voltage applied between the two electrodes. We have conducted a series of AC impedance and output voltage measurements with various sizes of gold microelectrodes with adsorbed protein layers, as well as with the CD4+ cells. When a sample of human peripheral blood mononuclear cells (PBMCs) was incubated on the biosensor, an increase in impedance was observed; this increase was due to the presence of CD4+ cells, which cause a decrease in the current flow. Incubation of the captured cells with FITC-labeled anti-human CD4 antibody verified that all captured cells were CD4+, demonstrating the selectivity of the BISD system. The BISD system is a promising tool for the detection of CD4+ cells in HIV-infected individuals, and for the detection and quantification of antigen:antibody or receptor:ligand interactions.

Keywords: microelectrode, CD4 cell biosensor, impedance, nanofabrication, human blood

1 INTRODUCTION

Impedance measurement is not new to biological applications, as early as 1898. Stewart presented a paper at a meeting of the British Medical Association at Edinburgh, later published in the Journal of Experimental Medicine [1]. He observed that the electrical conductivity of culture media containing living bacteria changed during the growth of organisms. The electrical response curves that he presented were very similar to curves that are obtained from currently available impedance systems. The significant difference is that, today, impedance can constitute a rapid measure of microbial proliferation, whereas Stewart measured changes in conductance over periods of 30 days.

Following this pioneering work, some attempts to develop impedance systems [2-4] were made during the first part of 20th century. However, it was not until the mid 1970s that the technique began to receive the attention that it merits. This coincided with the introduction of dedicated impedance systems [5-8], notably by the Torry Research Station [9]. It was the work of the Torry group that paved the way for impedance microbiology, along with their use of the Malthus and Bactometer systems. Significant progress has been made since that time [10-14]. The effects of electric fields on cells have been extensively studied, since widespread applications exist in cell biology, for the monitoring of cell migration and morphological changes in real time on gold surfaces [15]. The main outcomes of the earlier studies were limited to the determination of the preliminary electrical response of a cell. It is now possible to make sensor devices that are targeted to particular cell types in blood and that can be diagnostic or prognostic for diseases such as cancer or AIDS. At least 40-100 million individuals are thought to have been infected with the human immunodeficiency virus (HIV) throughout the World, and more than one million of them are currently living in the United States. HIV was first identified in 1983, and in 1984 it was shown to be the cause of AIDS [16-18]. HIV infection is characterized by the depletion of the CD4+ helper/inducer subset of T-lymphocytes, leading to severe immunosuppression, neurologic disease, and opportunistic infections and neoplasms [19].

Dysfunction of CD4+ T cells may be the result of direct infection with HIV, but it also may be caused indirectly by exposure of infected cells to various viral proteins. Because CD4+ T cells play an important role in immune responses, it is not surprising that many of the immune defects that are observed during AIDS are secondary to the progressive decline in the number and function of CD4+ T cells [20]. Although flow cytometry is currently used to quantify CD4+ cells, it is expensive and time consuming, and it requires considerable expertise. To overcome this problem, we are developing a micro-biosensor to quantify CD4+ cells. It is a small, portable bio-impedance sensing device (BISD) based on gold surface coated with protein G' and mouse anti-human CD4 antibodies to capture CD4+ cells from heparinized whole blood on microelectrodes.

Our work suggests that the BISD has considerable potential as a biosensor that forms the basis of a rapid, sensitive system for monitoring and quantifying CD4+ cells. Initial results in our laboratory are promising. A further advantage of the BISD is the ability to collect data directly onto a PC, thereby allowing easy analysis. In summary, our BISD is a rapid, economical, and flexible tool, for use in detection of specific particles or cell types.

2 DEVICE DESIGN AND FABRICATION

A three-electrode device was designed and fabricated at the Cornell Nanoscale Facility. Four-inch glass wafers were prepared in a hexamethyldisilazane (HMDS) vapor prime oven for 34 min to improve the adhesion of the resist to the glass surface. Wafers were then spun with microposit S1818 photoresist (Shipley Company, Marlborough, MA) at 4000 rpm for 40 sec and soft-baked for 60 sec at 115°C.

The EV620 contact aligner was used to expose the resist for 7 sec. A dummy wafer was also used to check alignment prior to exposure of the glass wafers. The wafers were then put in a vapor priming oven and subjected to an ammonia bake. Following the image-reversal bake, the wafers were flood-exposed in the HTG System III contact aligner for 30 sec, and they were then developed in MF 321 developer (tetramethylammonium hydroxide) for 1-1.25 minutes, rinsed, and dried.

Titanium (150 nm) and gold (1200 nm) films were deposited on the wafers, using a CVC SC4500 combination thermal/E-gun evaporator via E-beam for the titanium and thermal for the gold. After deposition, the wafers were left in a microposit remover 1165 (N-methyl-2-pyrrolidine, Shipley company) bath. After 4 hr, the wafers were rinsed with a water spray and dipped in acetone, to remove any residual gold and/or resist.

Polyamide was spun on the patterned glass wafer and soft-baked for 1 min. The insulating layer was patterned to expose the electrodes and contact pads. Finally, wafers were baked for 8 hr in a 450PB high temperature vacuum oven using a programmable temperature controller and nitrogen flow cycles for curing of polyimide films. Details of the dimensions of the electrodes are shown (see Fig. 1).

3 SENSOR DEVELOPMENT AND MEASUREMENTS

3. 1 Chemicals and Instrumentation

Protein G' was obtained from Sigma (St Louis, MO). Mouse monoclonal anti-human CD4 antibody and fluorescein isothiocyanate (FITC)-conjugated anti-human CD4 antibody

was from Pharmingen USA. Human peripheral blood mononuclear cells (PBMCs) were obtained from volunteers in accordance with Institutional guidelines of the Wadsworth Center, Albany, NY. All other reagents and solvents were analytical grade.

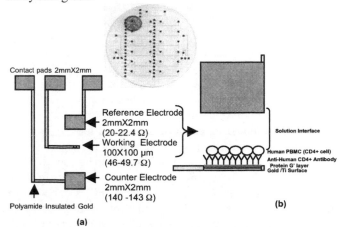

Fig. 1: BISD and its capture system(a) Electrode pattern and dimensions and (b) Modification of gold electrode surface by successive layer of protein, antibody and cells.

3.2 Procedure

The gold electrodes were first washed with nanopure water and dried under nitrogen. A PDMS well was applied on the 4 inch wafer, creating a well over the three electrode sets forming the biosensor. Prior to use, all electrodes were inspected under a microscope, and a blank impedance measurement was recorded with 1X PBS at pH 7.15. The protein was deposited on the working electrode (0.01mm^2) that was 400 times smaller than the gold reference electrode (4 mm^2). A small volume (10 µL) containing 1 µg of protein G' was incubated on the microelectrode surface overnight at 4°C in a water-saturated atmosphere. After incubation, the microelectrode was washed three times with 1 X PBS to remove any unbound protein. Protein G'-modified microelectrodes were then coated with human albumin as a blocking agent. The sensor's gold-protein G' electrode was further modified by incubating it in 1 µg/10µL of mouse monoclonal anti-human CD4 antibody in a wet chamber at 4°C for 2-4 hr, washed, and again blocked with 1µg/10µL mouse IgG. The microelectrode was washed after each coating, and impedance was recorded using gold spring contacts between the sensor's external pads and the external electronics. An AC voltage with an amplitude of 50.5mV was swept over a frequency range from 1 to 8 kHz. After the impedance measurements, the working electrode was washed. The microelectrode was then incubated in lysed whole blood containing CD4+ cells (10µL containing 3,000-100,000 total cells) for 1.5 hr in a wet chamber. After incubation, the working electrode was washed three times with 1X PBS to remove unbound cells. The impedance was measured again. To verify that the bound cells were CD4+ cells, the electrode was incubated for 1 hr with 10µL of anti-human FITC CD4 antibody. The electrode surface was washed with PBS, and images were recorded in a fluorescence microscope. To determine whether all the cells were CD4$^+$ cells, the electrode was also stained with 5µg/mL of 4', 6-diamidino-2-phenylindole (DAPI) in milliQ water.

4 RESULTS

4.1 Protein G' Adsorption on Microelectrode

The interaction between proteins and solid surfaces occurs widely in nature. There has been extensive research into the mechanisms intrinsic to protein adsorption, particularly in the fields of biomaterials and biocompatibility. It is well known that the adsorption of proteins onto electrode surfaces alters the electrochemical reaction at the electrode surface. Cyclic voltammetry is a very power tool for measuring electrode kinetics in the solid as well as the solution phase. In this work, the adsorption and subsequent effect of protein G' on the electrochemical properties of a gold electrode were observed, using cyclic voltammetry with a 0.05 mM potassium ferricyanide/0.01 mM sodium nitrate solution. The well-defined Fe^{2+}/Fe^{3+} redox couple peaks were observed in the potential range from -1.00 to 0.5V, at a scan rate of 100 mV/sec in the absence of protein G'. The same procedure had been applied to a gold electrode that was coated with 1 µg of protein G'. The Fe^{2+}/Fe^{3+} redox couple peaks were not observed in the presence of protein G' indicating that the protein was adsorbed to the gold electrode. The adsorbed protein G' layer acted as a blocking agent that hindered the electron transfer process at the electrode/solution interface (Fig. 2). Further, we verified that the optimal size ratio between the reference electrode and the working microelectrode is in the range of 300-600:1. In the present system, the reference electrode (4 mm^2) was 400 times larger than the working electrode (0.01 mm^2). This size ratio between two electrodes has little effect on the detection of the electrode's dynamics, and small changes in electrode behavior can be monitored.

Our BISD is capable of detecting the electrical response created by either single or multiple layers of protein deposited on the working microelectrode (see Fig 3.). Further, the signal resulting from two layers is significantly greater than that from a single layer. The protein layers include antibodies that serve as a molecular capture system for cells of the human immune system, in this case CD4$^+$ cells. The increase in impedance resulting from the capture of CD4$^+$ cells was significant with respect to the impedance produced by the capture system. Fig. 3 shows the increase in impedance produced by four cells in10 µL of PBMC's from previously frozen blood bound to the microelectrode. is The impedance change is due to presence of CD4$^+$ cells hindering the current flow from the small microelectrode. The number of cells on the electrode was confirmed by fluorescent light microscopy of FITC-labeled anti human CD4 antibody and DAPI (a stain for cell nuclei). Figure 4 shows that four CD4$^+$ cells were captured, which is in good agreement with the expected ~5-6 CD4$^+$ cells based on the microelectrode size and the number of cells in the preparation.

When the molecular capture system included blocking steps, (human albumin after protein G' and mouse IgG after the anti-human CD4 antibody), the change in impedance caused by a single cell captured from lysed blood (Figs. 5a & b) was less than that for the unblocked preparation (Fig. 5c). The smaller magnitude of the impedance change was probably due to differences in the electrical characteristics of the cells and to the presence of the additional layers of protein. Figure 5d also indicates that the output voltage may be a more sensitive measure of cell binding. The differences in output voltage before and after cell binding were due to a frequency-dependent change in capacitance.

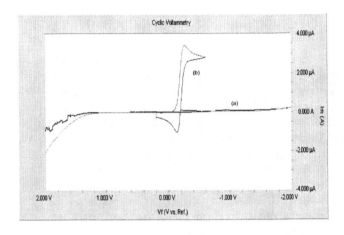

Fig. 2: Cyclic voltammogram of Fe $^{2+}$/Fe^{3+} redox couple in the presence (a) and absence (b) of 1µg protein G' on Microelectrode.

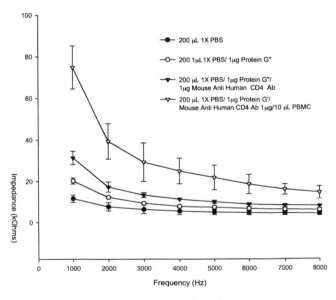

Fig. 3: Comparative impedance graph of successive layers of PBS, protein G', anti human CD4 Ab, and PBMCs on gold 100 X 100 µm electrode (no. of measurements/electrode = 3)

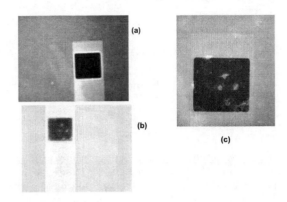

Fig. 4: Fluorescence image of blank electrode (a), FITC anti-human CD4 antibody-stained electrode showing 4 CD4$^+$ cells on the microelectrode (b), and image of DAPI stained biosensor, showing the same cells (c).

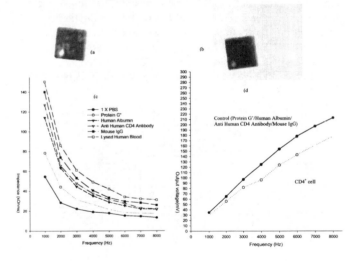

Fig. 5: Images and impedance/output voltage characteristics of a captured CD4$^+$ cell from 10µL of lysed whole blood. (a) UV image stained with DAPI, (b) fluorescence image of FITC anti-human CD4 antibody, (c) impedance of each successive layers, and (d) capacitance of modified electrode and the captured cell.

5 CONCLUSION

We are developing an on-chip BISD that makes use of a nanoscale cellular adhesion mechanism in an integrated electronic micro-system to capture CD4$^+$ cells from blood, and have demonstrated that all cells captured using CD4 antibodies were CD4$^+$. Future work will focus on quantitating the number of cells based on the electrical signal, as well as the possibility of quantitating specific protein levels in fluid samples, including serum proteins or toxic molecules such as botulinum toxin.

6 ACKNOWLEDGEMENT

This work was supported in part by the Nanobiotechnology Center (NBTC), an STC Program of the National Science Foundation (Agreement No. ECS-9876771), and the Cornell Nanoscale Facility (Agreement No. ECS-9731293).

REFERENCES

[1] G. N. Stewart, J. Exp. Med., 4, 235-243, 1899.

[2] M. Oker-Blom, Zbl. Bakteriol. Abs., 65, 382-389, 1912.

[3] L. B. Parsons and W. S. Sturges, J. Bacteriol. 11, 177-188, 1926.

[4] L. B. Parsons, E. T. Drake, and W. S. Sturges, J. Am. Chem. Soc., 51, 166-171, 1929.

[5] A. Ur, Nature, 226, 269-270, 1970.

[6] P. Cady and W. J. Welch, U. S. Patent No. 3, 743, 581, 1973.

[7] V. Pollak, Med. Biol. Eng., 12, 460-464, 1974.

[8] G. Eden and R. Eden, IEEE Trans. Biomed. Eng., 31, 193-198, 1984.

[9] J. C. S. Richards, A. C. Jason, G. Hobbs, D. M. Gibson, and R. H. Christie, J. Phys. E: Sci. Instrum., 11, 560-568, 1978.

[10] L. L. Hause, R. A. Komorowski, and F. Gayon, IEEE Trans. Biomed. Eng., 28, 403-410, 1981.

[11] M. Kowolenko, C. R. Kees, D. A. Lawrence, and I. Giaever, J. Immunol. Methods, 127, 71-77, 1990.

[12] Kevin S.-C. Ko, Chun-min Lo, J. Ferrier, P. Hannam, M. Tamura, B C. McBride, and R. P. Ellen, J. Microbiol. Meth., 34, 125-132, 1998.

[13] C. Xiao, B. Lachance, G. Sunahara, and H. T. Loung, Anal. Chem., 74, 1333-1339, 2002.

[14] Gi-Ho Kim, A. G. Rand, and S. V. Letcher, Biosensor & Bioelectronics, 18, 83-89, 2003.

[15] P. Mitra, C. R. Kees, and I. Giaever, Biotechniques, 11, 504-509, 1991.

[16] R. Gallo, S. Salahuddin, M. Popovic et al., Science, 224, 500-503, 1984.

[17] J. Levy, A. Hoffman, S. Kramer, J. Landis, J. Shimabukuro, and L. Oshiro, Science, 225, 840-842, 1984.

[18] F. Barre-Sinoussi, J. C. Chermann, F. Rey et al., Science, 220, 868-871, 1983.

[19] A. S. Fauci and H. C. Lane, HIV disease: AIDS and related disorder. In: A. S. Fauci, K. J. Isselbacher, J. D. Wilson et al., eds. Harrisons Principal of Internal Medicine, 14th ed New York: McGraw_hill, 1791-1856, 1998.

[20] Willium E. Paul, Fundamental of Immunology,. In: O. Cohen, D. Weissman, and A. S. Fauci, The Immunopathogenesis of HIV Infection, 4th ed Philadelphia: Lippincott-Raven, 1455-1481, 1999.

Behavior of Living Human Neural Networks on Microelectrode Array Support

R. Pizzi*, F. Gelain**, A. Fantasia*, D. Rossetti* and A. Vescovi**

*Department of Information Technologies, University of Milano
Via Bramante 65, 26013 Crema, Italy, pizzi@dti.unimi.it
** Stem Cells Research Institute, DIBIT S.Raffaele Milano, Italy

ABSTRACT

Researchers of the Department of Information Technologies of the University of Milano and of the Stem Cells Research Institute of the DIBIT-S. Raffaele Milano are experimenting the growth of human neural networks of stem cells on a MEA (Microelectrode Array) support.

The Microelectrode arrays (MEAs) are constituted by a glass support where a set of tungsten electrodes is inserted. We connected the microelectrodes following the architecture of classical artificial neural networks, in particular Kohonen and Hopfield networks.

The neurons are stimulated following digital patterns and the output signals are analysed to evaluate the possibility of organized reactions by the natural neurons.

The neurons reply selectively to different patterns and show similar reactions in front of the presentation of identical or similar patterns.

These results allow to design further experiments that improve the neural networks capabilities and to test the possibility of utilizing the organized answers of the neurons in several ways.

Keywords: neural networks, stem cells, microelectrode arrays, learning.

1 INTRODUCTION

During the past decade several experiments have been carried out on direct interfacing between electronics and biological neurons, in order to support neurophysiological research but also to pioneer future bioelectronic prostheses, bionic robotics and biological computation.

As microelectrodes implanted into brain give rise to rejection and infections, researches are under way, experimenting a direct adhesion between chip and neural tissue. Important results have been achieved at the Max Planck Institute [1], at the Georgia Tech [2], at the Northwestern and Genoa University [3], and at the Caltech[4].

Aim of our work is the development of architectures based on the Artificial Neural Network models on networks of human neural stem cells adhering to microelectrode arrays (MEAs). The MEAs are connected to a PC via an acquisition device that allows to stimulate the neurons with suitable inputs and to acquire the neuron signals in order to evaluate their reactions.

In this way we are investigating on the learning processes of biological neural networks and on the possible technological applications of hybrid biological/electronic networks.

2 MATERIALS AND METHODS

The problem of the junction between neuron and electrode is crucial: materials must be biocompatible with the culture environment, and neurons must firmly adhere to the electrodes in order to get maximum conductibility. Our support is constituted by a glass disk with tungsten $100 \times 100 \mu$ microelectrodes. The distance between the electrodes is around 20μ. (Fig. 1). These are connected to the PC via an USB acquisition card (IOTECH Personal Daq/56).

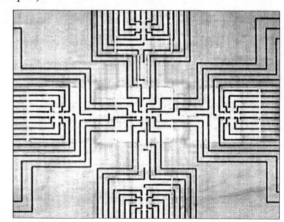

Fig. 1 Portion of the MEA support

Our neurons have been cultured starting from human neural stem cells, multi-potential undifferentiated cells whose main features are the ability of self-renewal and to differentiate into different adult cells[5].

On the MEAs we implemented two kinds of artificial architectures:

- A Kohonen Self-Organizing Map [6], composed by an input layer and an output layer connected in such a way as to classify the input by means of a competition between neurons, and
- A Hopfield network [7], fully interconnected network that stabilizes into equilibrium configurations that correspond to its memories.

We chose these models due to their straightforward architecture and their resemblance to some neurophysiological structures.

Then we realized two hardware/software models in order to discriminate simple patterns. The minimum software configurations able to recognize 8-bit patterns representing "zero" and "one" pure or affected by noise are 1) a Kohonen network with 8 input neurons and 3 output neurons and 2) a Hopfield network with 8 input/output neurons (Fig. 2) .

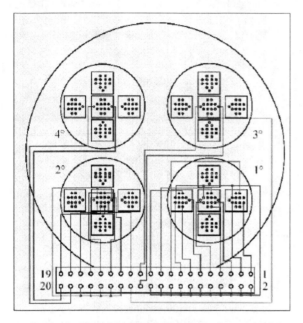

Fig. 2 Circuits of the hybrid neural networks
on the MEA

These networks have been implemented on the MEAs, culturing the stem cells on the connection sites. The software allows to design the input patterns, that are converted into suitable electrical stimuli (similar to the natural action potentials) at 40 Hz by a custom hardware device. The signals are also sampled at 40 Hz. This frequency has been chosen because of its probable involvement in the most advanced brain tasks [8][9].

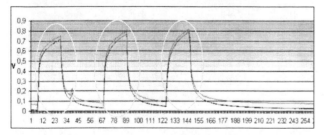

Fig. 3 Culture liquid output during stimulation
with "zero" patterns

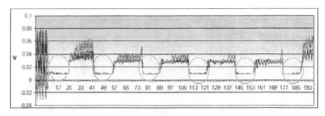

Fig.4 Network output during stimulation
with "zero" patterns

In order to check if the signals received by the acquisition device were actually coming from neurons, we measured the reactions of the only culture liquid, without cells (Fig. 3) , comparing them with those coming from the cells (Fig. 4).

You can see that the network reacts to the pattern "zero", constituted by the highest voltage ("11111111"), emitting the lowest voltages. The culture liquid, instead, answers to the "zero" pattern with a high voltage, as expected by a conductive medium .

In Fig. 5 you can see the reaction of the Kohonen network after stimulation with "zero" patterns, pure and affected by noise (green circles) , and with "one" patterns (red circles), pure and affected by noise.
Similar effects have been shown by the Hopfield network.

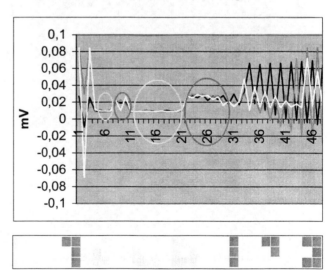

Fig. 5 Network output during training

At the end of the experiment we measured the network output in order to evaluate if the neurons had "stored" information in some way. Differently from the only culture liquid, that shows the same behaviour before and after experiment, the network output retains different voltages .

After the experiment, the output signals have been analysed using the Recurrent Quantification Analysis [10] .

This non-linear analysis tool processes the time series in a multi-dimensional space. In this phase space the time series represents the dynamical system to be analysed.

The Recurrent Plots show how the vectors in the reconstructed space are near or distant each other. The Euclidean distances between all the vector pairs are calculated and translated into colour bands. Hot colours (yellow, red, orange) are associated to short distances between vectors, cold colours (blue, black) show long distances. Signals repeating fixed distances between vectors are organized, signals without repeating distances are not. In this way we obtain uniform colour distribution for random signals, but the more deterministic and self-similar is the signal, the more structured is the plot.

The analysis of our data using the RQA method lead to interesting results. Signals coming from similar bitmaps gave rise to similar Recurrent Plots.
Moreover, in the following figures you can see the self-organization of a single output channel before stimulation, during the training, during the testing phase and after stimulation.
Fig. 6a shows one output channel of the Kohonen network before stimulation. Colours are cold and unstructured, showing lack of self-organization.

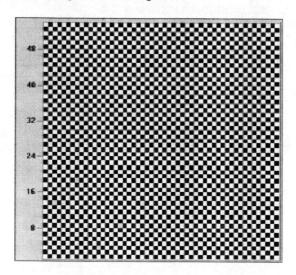

Fig. 6a   RQA plot of the Kohonen network
before training

The training phase shows a change in the structure. The Recurrent Plot of the channel during the testing phase shows wide uniform hot colour bands corresponding to a high organization. Fig. 6b is the plot of the output channel after the end of stimulations. In this case the uniform bands further widen, showing that the signal remains self-organized in time.

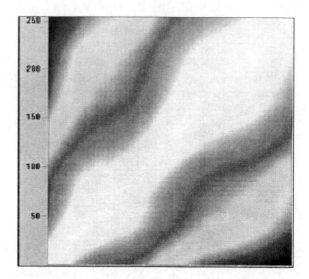

Fig. 6b   RQA plot of the Kohonen network
after stimulation

This analysis shows that introduction of organized stimuli modifies the network structure and increases the information content even after the end of stimulation, suggesting a form of learning and memorization..

We applied the same procedure to the output signals coming from the Hopfield network.
Fig. 7a shows one channel after stimulation with the pattern "zero": we see wide organized bands with peculiar features, different from the other channels. Fig. 7b shows the same channels after stimulation with pattern "one".

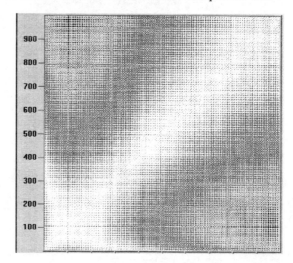

Fig. 7a   RQA plot of one channel of the Hopfield network
after stimulation with "zero" pattern

This analysis shows that the network behaves differently depending on the input signal and on the different channels.

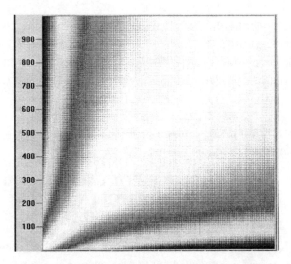

Fig. 7b   RQA plot of one channel of the Hopfield network after stimulation with "one" pattern

3   CONCLUSIONS

After a qualitative analysis of the output signals of the networks we can reasonably affirm that the networks show an organized behavior after the stimulation with patterns, and they are able to answer selectively to different patterns. The signal behavior changes depending on the network channels, and similar patterns give rise to similar answers.

Thus we can say the networks have shown a form of selective coding, highlighting a strong self-organization as a reply to stimulation.

Other experiments with more complex patterns and new output analyses are under way. In particular, we are using non-linear methods (including Artificial Neural Networks) to discriminate the output responses in form of dynamical attractors. Preliminary analyses give the hope that discriminant interpretation of the information content kept by the networks is actually possible.

In this case we would be able to use these outputs in several ways, ranging from robotics to biological computation to neuroelectronic prostheses.

In the future we will improve both the cell growing on the MEA supports and the measure methods and tools. We will also increase the number of connections in such a way as to build more complex networks and to test their learning abilities on more complex patterns.

ACKNOWLEDGEMENTS

We are indebted to Prof. G. Degli Antoni (University of Milano) for his valuable suggestions and encouragement and to ST Microelectronics for its important support.

REFERENCES

[1] Fromherz P., Schaden H., Vetter T., Neuroscience, 129,77-80 , 1991.

[2] Lindner J.F,. Ditto W., "Exploring the nonlinear dynamics of a physiologically viable model neuron", AIP Conf. Proc. 375(1), 709, 1996.

[3] Reger B., Fleming K.M., Sanguineti V., Simon Alford S., Mussa Ivaldi F.A., "Connecting Brains to Robots: The Development of a Hybrid System for the Study of Learning in Neural Tissues", Artificial Life VII, Portland, Oregon 2000.

[4] DeMarse T.B., Wagenaar D.A., Potter S.M., "The neurally-controlled artificial animal: a neural-computer interface between cultured neural networks and a robotic body", SFN 2002, Orlando, Florida 2002.

[5] Vescovi A.L., Parati E.A. Gritti A. Poulin P., Ferrario M., Wanke E., Frölichsthal-Schoeller P., Cova L., Arcellana-Panlilio M., Colombo A., and Galli R., "Isolation and cloning of multipotential stem cells from the embryonic human CNS and establishment of transplantable human neural stem cell lines by epigenetic stimulation. ", Exp. Neurol. 156; 71-83, 1999.

[6] Kohonen T., "Self-Organisation and Association Memory", Springer Verlag 1990.

[7] Hopfield J.J.," Neural Networks and Physical Systems with Emergent Collective Computational Abilities", Proc. Nat. Acad. Sci USA, 81,1984, 81..

[8] Menon V., Freeman W.J., "Spatio-temporal Correlations in Human Gamma Band Electrocortico-grams", Electroenc. and Clin.. Neurophys. 98, 89-102, 1996.

[9] Pizzi R., de Curtis M., Dickson C., "Evidence of Chaotic Attractors in Cortical Fast Oscillations Tested by an Artificial Neural Network", in: Advances in Soft Computing, J. Kacprzyk ed., Physica Verlag 2002.

[10] Zbilut J.P., Webber C.L., "Embeddings and delays as derived from quantification of recurrent plots", Phys. Lett. 171, 1992.

Glucose Sensors Based on Glucose Oxidase Immobilized in Polypyrrole-Polyacrylamide Microparticles

E. López-Cabarcos*, J. Rubio-Retama*, D. Mecerreyes**, B. López-Ruiz***
*Departamento de Química-Física Farmacéutica. Facultad de Farmacia. Universidad Complutense de Madrid, 28040 Madrid, Spain
** CIDETEC, Parque Tecnológico Miramon, Paseo Mikeletegui, 61-1º, 20009 San Sebastián, Spain
***Departamento de Química Analítica. Facultad de Farmacia. Universidad Complutense de Madrid, 28040 Madrid, Spain.

ABSTRACT

We have synthesized polyacrylamide-polypyrrole (PPy/PA) microparticles which present conductivity in the range of the semiconductor materials (around 10^{-5} S/cm). Glucose Oxidase was immobilized within microparticles with composition 50/50 (w/w) PPy/PA. These microparticles were used as biological component of an amperometric biosensor which shows similar glucose response in both aerobic and anaerobic conditions.

Keywords: polypyrrole-polyacrylamide microparticles, immobilization of glucose oxidase, glucose biosensor

1 INTRODUCTION

Intrinsically conducting polymers have become of great scientific and technological importance because of their electrical, electronic, magnetic and optical properties. Great effort has been dedicated to improve their processibility and nowadays the fabrication of films and fibbers of controlled thickness has been achieved [1-4]. However, to our knowledge, conducting polypyrrole microparticles have not yet been synthesized probably because colloidal polymer particles are produced by polymerization in the internal phase of emulsions and pyrrole is slightly soluble in water. Therefore, after polymerization the polymer would precipitate making difficult the formation of conducting microparticles. The approach that we present in this paper involves the immobilization of soluble polypyrrole into microscopic polyacrylamide particles using the concentrated emulsion polimerisation method, which has been recently employed in our laboratory to encapsulate enzymes [5-6]. The main novelty of this method in comparison with conventional emulsion polymerisation lies in the large volume fraction of the dispersed phase, which in the concentrated emulsion is larger than 74% . The aims of this contribution are twofold: First, to synthesise polyacrylamide-polypyrrole conducting microparticles following the W/O concentrated emulsion polymerisation method and to study their conductivity as a function of the polypyrrole content. The second objective is to encapsulate

GOx in these microparticles for its use as the biological component of a glucose sensor.

2 POLYPYRROLE-POLYACRYLAMIDE MICROPARTICLES WITH ENTRAPPED ENZYME

Soluble, doped polypyrrole has been synthesized following the procedure outlined in the literature [7]. The soluble polypyrrole was incorporated in the aqueous phase of a W/O concentrated emulsion. The content of polypyrrole in the concentrated emulsion was varied and micropaticles with 10/90, 15/85, 25/75, 30/70, 40/60 and 50/50 polypyrrole/acrylamide (w/w in the aqueous phase) were produced. Adding TEMED started the polymerization of the aqueous dispersed phase, during which temperature was controlled and kept below 35° C. After 1 h of reaction, the polymerization was complete and the microparticles were isolated by centrifugation (5000 rpm for 10 min at 10°C).

30 µm

Figure 1. Electron micrograph of polypyrrole-polyacrylamide (50/50 w/w) microparticles with entrapped Glucose Oxidase (polyacrylamide crosslinking is 3.2%)

Glucose Oxidase was immobilized in the PPy/PA microparticles by incorporating the enzyme in the aqueous phase (a solution of polypyrrole, monomer and cross-linker in phosphate buffer at pH 6.0) of the concentrated emulsion before beginning the polymerisation. As it is shown in Fig.1 redox polymerization of the gel-like emulsion produced polyacrylamide microparticles that contain polypyrrole and GOx entrapped inside them.

The electrical conductivity of the microparticles was measured by the four-point probe method at room temperature using disks (2mm thickness) prepared in a press under the application of $2 \cdot 10^7$ Pa during 1 min. The conductivity increases with the amount of PPy entrapped, from 10^{-12} S/cm for pure polyacrylamide, up to 10^{-5} S/cm (semiconductivity range) for the 50/50 PPy/PA microparticles. The conductivity was also investigated by IR spectrometry looking for polypyrrole conducting bands in the 10/90 and 50/50 PPy/PA samples. As is illustrated in Fig. 2, the sample with 50% PPy shows typical bands of polypyrrole such as: 1) polaron at 646 cm^{-1}, 2) bipolaron at 978 cm^{-1}, 3) and 4) vibration modes of the ring at 1236 and 1478 cm^{-1}, 5) and 6) stretching vibration of the C=C at 1565 and 1619 cm^{-1}, while the peaks associated to PPy segments are scarcely seen in samples with 10% PPy content.

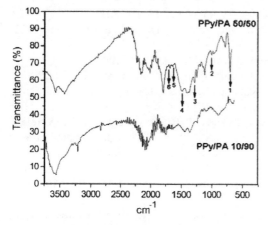

Figure 2. Infrared spectra of the 10/90 and 50/50 (% w/w) PPy/PA microparticles. The arrows indicate some IR typical bands of polypyrrole.

3 GLUCOSE BIOSENSOR

The microparticles with immobilized GOx were used as the biological component of an amperometric glucose sensor. To prepare the electrode, the 50/50 PPy/PA microparticles with GOx were selected because they showed the best conductivity properties. The enzyme electrode was prepared by depositing 3.0 mg of microparticles on the surface of a platinum electrode. The microparticles layer was covered and flattened around the platinum electrode surface using a dialysis membrane. The

resulting electrode was subsequently placed in a three-electrode cell as working electrode. The electrode was then washed with phosphate buffer and a potential of +400mV *vs* (SCE) was applied to the sensor until the background current had decreased to a constant level. More details about the electrode preparation are given elsewhere [5-6].

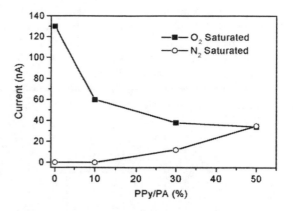

Figure 3. Biosensor current response to glucose as a function of the microparticle PPy content. The biosensor prepared with 50/50 PPy/PA microparticles gave the same response under both aerobic and anaerobic conditions.

The maximum current response to glucose of the 50/50 biosensor, measured in stirred solutions at 25ºC, under aerobic and anaerobic conditions is shown in Fig.3 as a function of the PPy content. To work in anaerobic conditions, oxygen was removed from the cell by purging with nitrogen during 20 min. The absence of oxygen was confirmed by using a biosensor prepared with PA microparticles with GOx but lacking pyrrole, which doesn't show response to glucose as is illustrated in Fig.3 (PPy/PA=0). In O_2 saturated solutions the biosensor current response decreased when the PPy content increased and the opposite bahaviour was observed in N_2 saturated solutions.

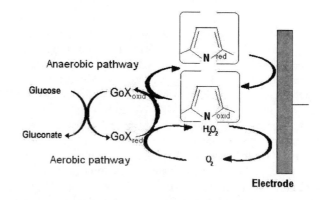

Scheme1. Proposed working mechanism of the 50/50 PPy/PA glucose biosensor when the electrode is polarized at 0.4V *vs* (SCE).

The 50/50 PPy/PA biosensor gave similar response in both cases (aerobic and anaerobic), which indicates that in anaerobic conditions electrons are transferred from the active site of the enzyme to the electrode *via* the polypyrrole, whereas in aerobic conditions the mediator is the O_2 molecule as in illustrated in the scheme 1.

The biosensor prepared with 50/50 PPy/PA microparticles shows a current intensity considerably smaller than that obtained when pure polyacrylamide with GOx microparticles were used in the biosensor. To improve the performance of the electrode, the cross-linking of polyacrylamide was reduced from 3.2% (selected as optimum in pure PA biosensors) to 2% with the result shown in Fig.4. As can be seen in Fig. 4 the reduction of cross-linking produced an increase of the current from 500nA up to 2000nA.

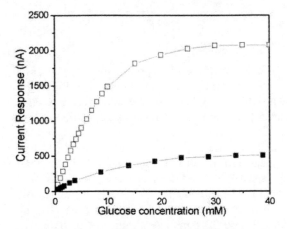

Figure 4 Response to glucose concentration of the 50/50 PPy/PA biosensor. Decreasing the polyacrylamide cross-linking from 3.2% to 2% increases the current intensity by a factor close to four.

The figures of merit of the 50/50 PPy/PA biosensor can be summarized as follows: I) The sensitivity to glucose is 189 nA/mM in aerobic and 92 nA/mM in anaerobic conditions. II) The limit of detection (signal/noise=3) is 0.03 mM in both cases (aerobic and anaerobic). III) The linear range of glucose concentration dependence (dynamic range) goes from 0.1mM to 10mM. IV) The average response time is 11s, which is smaller than the time of 32s obtained using glucose sensors based on GOx immobilized in PA microgels. Because the diffusion time of substrates is proportional to the square of the traveled length, the response time of this type of biosensors is similar to the response time obtained with biosensors prepared with enzymes entrapped in very thin membranes (10-15μm thick) which is around 30s [8] and is smaller than the response time commonly reported with polymeric membranes, which is around 2 min [9].

4 CONCLUSIONS

Using the concentrated emulsion polymerization method, we have synthesized PPy/PA microparticles which show conductivity in the range of the semiconductor materials. The conductivity increases seven orders of magnitude from 10^{-12} S/cm in pure PA microparticles up to 10^{-5} S/cm in 50/50 PPy/PA microparticles. Glucose Oxidase was immobilized in the 50/50 PPy/PA microparticles and an electrode was prepared having these microparticles as biological component. The 50/50 PPy/PA biosensor gave similar response in both, aerobic and anaerobic conditions which indicates that the electrons are transferred from the active site of the enzyme to the electrode *via* the polypyrrole.

5 ACKNOWLEDGEMENTS

The authors acknowledge financial support from the Spanish Science and Technology Ministry (BMF2000-0620, MAT 2002-10233-E and MAT-2003-03051-C03-03). We also thank A. Rodriguez (Electron Microscopy Center, UCM) for valuable technical assistance.

6 REFERENCES

1.- H.Shirakawa, Synth. Met. **125**, 3 (2001)

2.- D.Hohnholz and A.G.MacDiamid, Synth. Met., **121**, 1327 (2001)

3.- A. J. Heeger, Synth. Met. **125**, 23 (2001)

4.- A. G. MacDiamid, Current Appl. Phys., **1**, 269 (2001)

5.- J. Rubio Retama, E. Lopez Cabarcos, B. Lopez Ruiz, *Biomaterials,* **24**, 2965, 2003

6.- E. Lopez Cabarcos, J. Rubio Retama, D. Mecerreyes, B. Lopez Ruiz, *Nanotech2003,* Edt. Computational Publications, Cambridge, MA, USA, 59 (2003)

7.- D. Mecerreyes, R. Stevens, C. Nyuyen, J. A. Pomposo, M.Bengoetxea, H.Grande, *Synthetic Metals,* **126**, 173, 2002.

8.- C. Jiménez, J. Bartroli, N.F. Rooij, M Koudelka-Hep, Sensors & Actuators B, **26-27**, 421 (1995)

9.- A. Kross, S.W.F.M. van Howell, N.A.J.M. Sommerdijk, R.J.M. Nolte, Adv.Mater. **13**, 1555 (2001)

Nano-structure and phase transitions in the liquid phase of metallic and non-metallic elements

G.C. Vezzoli

Institute for the Basic Sciences
51 Park Drive Suite 22
Boston, MA 02215

ABSTRACT

The generally accepted model of the liquid phase is that of a single "homogeneous" phase, with a structure which is more or less homogeneous throughout the pressure-temperature field in which it is the stable phase. Herein we present data in the liquid state to argue that the liquid of any composition exists in several differently structured nano-phases. The local configurational variations that specify what can broadly be called a "phase" in the liquid include local coordination number, bond angle, coordination sphere and bond length. The elements that are studied herein include: Hg, Bi, Sn, S, Se, Te. The understanding of nano-phases in the liquid state can be a precursor to quenching persistently metastable solid phases which may have promise for applications in biotechnology.

Keywords: nanostructure, liquid phase, short range order.

1. Introduction

Herein is presented a coherent model and description of the behavior of the liquid state of elemental metals, and covalently-bonded non-metallic elements and binary molecules such as water. The data and interpretations specifically focus on metals, bismuth, tin, and mercury, and on group VI A elements, sulfur, selenium, and tellurium, with some reference to water. The interpretation given herein of over 30 years of data[1-7] proposes that the liquid state is composed of non-homogeneous local configurations with very short range nano-order (broadly referred to as 'phases') which undergo transitions from one to another, depending upon the thermodynamic environment. Thus the liquid state cannot be modeled as the conventionally /historically described disordered *random network* of atoms or molecules, but as a mixture of nano-heterogeneous regions in pressure-temperature-composition space (p-T-X) that are characterized by bond angle, bond strength, coordination number, coordination sphere, density, and chemical potential. Recent work on the polymorphic structures of ice and the local transitions in water [8] has reinforced the requirement and demand in modern phase equilibrium and the emergent field of nano-science that an overall model/theory of the liquid state must be forthcoming based on individual local structures that are favored at specific domains and ranges in a *p-T field*. The other thermodynamic variables in the equation of state that define the Gibbs function, electric and magnetic field, will also necessarily influence the nano configurations of liquid state species.

2. The Liquid State of Elemental Metals

We have conducted detailed studies of the elemental metals bismuth [1], tin [2], mercury [3], and lead [4], and interface studies with copper and platinum, using a wide variety of hydrostatic and quasi-hydrostatic equipment including argon pressure vessels, the Bridgman opposed anvil system, a diamond cell apparatus, and a hexahedral press. We measured *in-situ* optical, electrical conductance or resistance, specific volume, heat of formation, and magnetic properties over the course of 35 years of research. We also conducted detailed x-ray diffraction studies of solidified products quenched from p-T regions where the charge was held for various periods of time. All of these studies have consistently revealed second-order type transitions in the liquid state indicating that the liquid is *not* homogeneous and that liquids do *not* change structure monotonically as a function of pressure and temperature. *In situ* x-ray diffraction techniques, including angular dispersion analysis and energy dispersion analysis [5-7], also yielded a characteristic scattering pattern corresponding to a preferred probabilistic short-range liquid nano structure. The nano structure of the liquid state of metals is suggestive, of non homogeneity in the liquid state of semi-conductive materials and the liquid state of non metals because of the covalent bond vulnerability to the effects of changing pressure and temperature. These research findings give rise to a better understanding of nano-configurations in materials that have been either rapidly quenched, or slow-cooled from the liquid state, whether the resulting solid is non-crystalline such as glasses of whether polycrystalline or even single crystal materials (and defects therein). This implication is in keeping with the conclusions that the various "phases" and nano configurations of the liquid state are consistent with the character of the p-T melting curve (such as maxima and cusps) of the associated solid[1,2,9,10]. Implications from the research referenced and interpreted herin, are pertinent to precipitated phases and hence draw attention to biomedical applications related to blood flow through arteries, and to matter built up in the brain such as associated with brain diseases as Alzheimer's. Knowledge of the structure of precipitated and disproportionated phase will in due course lead to improved treatment of disorders that relate to material and chemical phenomena. For a comprehensive review of the specifics of the p-T-X characteristics of liquid-state nano-configurations, the reader is urged to study the original references [1-10] and the most recent review and interpretation given in Ref.11.

3. The Liquid State of Covalently-Bonded Non-Metals and Molecular Materials

Using the techniques described in the previous section detailed studies have been conducted on the GroupVI A elements, sulfur, selenium, and tellurium in the liquid state at atmospheric and the high pressure. Transitions between different liquid configurations, due to changes in pressure and temperature, were in general of a much larger magnitude than those associated with the low-melting metals, and were often more persistently metastable especially in molten sulfur This is because crystalline sulfur is based on an octameric puckered ring structure that upon melting transforms to an equilibrium mixture of rings, broken rings (ring scission), and polymeric chains, with the chain structure being favored above moderate pressures (in the tens of kbar range). The rather strong secondary bonding between neighboring chains (due in part to van der Waals interactions) favors and fosters persistently metastable phases that are insoluble in water and often insoluble in carbon disulfide as well[10]. The polymerization of liquid sulfur at 1atm at 159C, is a key liquid → liquid transition, and above about 0.8 kbar molten

sulfur melts *directly* to polymer [10]. Liquid sulfur undergoes chain scission and an accompanying decrease in viscosity at 1atm at about 188C, yet with increasing temperature the polymer concentration of the liquid continues to increase as the average chain length decreases from about ~10^6 atoms at maxium viscosity. The research has shown that sulfur in the crystalline and in the liquid state is an archetype material, and with increasing pressure and temperature shows a progression from an insulating ring structure, through a modestly semi-conductive chain structure, to the metallic state. In the liquid phases, the ring, the chain, and metallic species are associated with unique nanostructures rather than with a homogeneous liquid[10].

Selenium and tellurium, in their crystalline form at atmospheric pressure, are based on a helicoidal
A8 zigzag chain structure. At 1 atm selenium melts to a mixture of short chains and eight-membered rings. Liquid tellurium at atmospheric and at high pressure suffers a number of transitions based on changes in local coordination number and changes in bonding. Since group VI A elements are characterized by an s^2p^4 valence shell, the tendency toward forming closed rings (typically 8-memberend or 6-membered, but sometimes other variations) and long chains is favored with the typical primary coordination number (c.n) of 2, at least at 1 atm to the kbar range. The two available p-states to achieve a closed valence shell allow the liquid states to be very susceptible to structural transitions on a nano level caused by changes in pressure and temperature. It should be pointed out and emphasized that in the liquid state, the term "phase" is utilized in a loose or broad form, not in the sense of a true equilibrium phase in the crystalline form such as the diamond phase of carbon. Instead, the disordered character of the liquid state gives rise to a statistical property associated with the various local structural configurations, generally with a Maxwell Boltzmann distribution giving the probability of breaking chemical bonds. Therefore, only locally in short-range or in very short range (nano) scales, or neighborhoods, does a reconstructive-like or displacive-like "phase" transition in species configuration occur, depending upon p-T. Therefore in the liquid state a specific configuration or "phase" is favored thermodynamically within a domain of p-T-X,

rather than being a sole unique equilibrium form as in the crystalline state.

Water with its two covalent bonds and obtuse bond angle, and two lone pairs, and strong polar character has thus properties that render the molecule highly susceptible to structural changes caused by p-T. This delicate vulnerability to p-T can have profound biological implications in the immune system, and in genetics because of water molecules sheathing the DNA double helix.

4. Summary, Conclusion, and Inferences

In the liquid state of group VI elements, sulfur, selenium, and tellurium, and for the low-melting-point metals, bismuth, mercury, and tin, there is detailed reproducible evidence for non-homogeneity and unique nano-configurations at specific p-T(regions in p-T space or the projection of the Gibbs surface on the pressure-temperature plane). It is inferred that many other liquids behave in a similar manner, and possibly all liquids at appropriate values of p-T. It is concluded that the liquid state is well modeled by postulating a non-homogeneous mixture of nano-clusters as the general structural archetype for the liquid state of matter.

It is also proposed /inferred that unique liquid configurations can be useful in their quenched persistently metastable forms for specific applications potentially in a wide spectrum of areas such as electronics, optoelectronics and medical science/technology (to include diffusion barriers and thin film technology, particulate solvents in blood chemistry and in pancreatic and intestinal functions, and in catalysis such as in enzymatic action). Exploring for such possible applications is still in the area of very basic research, but the mechanistic phenomenology that is cornerstone to such feasible applications is the uniqueness of local short range order nanostructure regarding fundamental structure and chemical bonding properties. It should also be realized that useful properties of liquid nano-"phases" may be achieved by variations of applied electrical or magnetic fields, and even though modifying these variables (at relatively low to moderate magnitude) may not necessarily affect the strong covalent primary bonding or the electron delocalization associated with metallic bonding, the intervention of electromagnetic driving force may affect secondary bonding and non-satisfied valence states. The forced dielectric

breakdown of a nanophase, for example, can create a short or an electron (carrier) path that is needed for an alternate neuron communication pathway in the central or peripheral nervous system. Optical excitation of a liquid nanophase can also effectuate a property change that could be biochemically useful. Pressure-temperature induced influences on the water molecule can have very profound effects in living systems.

References

[1] H. Spetzler, M.D Neyers, T. Chan, High-Temp-High Pres 7, 481-496, 1975; GI Goryangcu Vestn.Mosk. Univ 1, 79-85, 1956; M. Gitis and I.Mikhailov, Sov. Phys. Acoust, 11, 372-380, 1996; Y. Dutchak, Phys. Met. Metallogr (USSR) 11, 133-143, 1961; G.C. Vezzoli, High-Temp-High Pres 10, 707-708, 1978; V. Ablordeppey, Phys. Rev. A3, 1680-1684,1971.

[2] G.C. Vezzoli, Mat. Res. Bull.23, 1-12 1988; J.P. Gabathuler and S.Z. Steeb, Naturforsch 34A(11), 1314-1318, 1979; G.C. Vezzoli, High-Temp- High Pres 20, 679-685,1988; E.S. Filippov, Izv. Vyssh. Uchebn. Zaved, chern Metall 11, 100-105, 1979.

[3] G.C. Vezzoli, phys. stat. sol (a) 46, k49, 1978.

[4] G.C. Vezzoli High-Temp-High Pres. 12, 195, 1980.

[5] K.R. Rao, Current Science 80(9), 1098-1100, 2001.

[6] P.H. Poole, et al. Science 275, 322-324, 1997.

[7] Y. Katayama et al. Nature 403, 170-172, 2000.

[8] R. Roy, Science 299, 5603, 2003.

[9] A. Epstein, H. Fritzcche, H. Lark-Horowitz, Phys. Rev. 107, 412-415, 1957; G.C.Vezzoli, J. Polym. Sci (Polymer Phys. Ed) 11, 1337, 1973; J.P. Gabathuler, Sci. (Polymer Ph Ed) 11, 1337-1345, 1973; S. Stishov, Zh Eksperim. Teore. Fiz 52, 1196-1198,1967; G. Tourand, B. Cabane, M. Breuil, J. Non Cryst Solids 8-10,676-680, 1972; N. Tikhomirova, S. Stishov, Zh. Ekisperim. Teor. Fiz 43, 2321-2324, 1962. G.C. Vezzoli, J. Amer. Ceram Soc 55(2), 65-67, 1973; K. Yaoita and K. Tsuji, J. Non. Cryst. Solids 156, 157-160, 1993.

[10] G. Gee, Trans. Faraday Soc. 48, 515-518, 1952; G.C. Vezzoli, J. Polym. Sci. A-1, 1587-1588, 1970; A. Tobolsky and A.

Eisenberg, J. Am. Chem. Soc. 81, 780-783, 1959; A.Eisenberg, Chem. Phys. 39, 1852-1855, 1963; G.C. Vezzoli, F. Dachille, and R. Roy, J. Polym. Sci. A-1, 7, 1557-1579, 1969; G.C. Vezzoli, P. Kisatsky, L.W. Doremus, P.J. Walsh, Applied Optics 15(2), 327-340, 1976; G.C. Vezzoli, F. Dachille, R. Roy, Inorg. Chem. 8, 2658-2661,1969.

[11] G.C. Vezzoli, Materials Research Innovations, April 2004 (in press).

The author wishes to acknowledge the collaboration and continuous encouragement provided by Professor Rustum Roy of the Pennsylvania State University, and the collaboration of the late Professor Frank Dachille (Also of The Pennsylvania State University), and the editorial assistance of Sandra Smalling.

Application of Local Radial Basis Function-based
Differential Quadrature Method in Micro Flows

D. Xu[*], H. Ding[**] and C. Shu[***]

[*]Institute of High Performance Computing, Singapore, xud@ihpc.a-star.edu.sg
[**]National University of Singapore, Singapore, mpedingh@nus.edu.sg
[***]National University of Singapore, Singapore, mpeshuc@nus.edu.sg

ABSTRACT

Numerical simulation of flows in micro geometry is one of the most appealing fields in computational fluid dynamics due to the advent of MEMS. In the present study, a local Radial Basis Function-based Differential Quadrature (RBF-DQ) method is applied to study the flow in micro cavity in the slip flow regime. RBF-DQ method is a natural mesh-free approach. The weighting coefficients of RBF-DQ method are determined by taking Radial Basis Functions (RBF) as test functions instead of using high order polynomials. RBF-DQ method has a unique feature of discretizing derivatives at a knot by a weighted linear sum of functional values at its neighboring knots, which may be distributed randomly. In the paper, two-dimensional incompressible flows in a micro lid-driven cavity are simulated with Knudsen number $Kn < 0.1$. Maxwell's first-order formula is adopted to treat slip boundary condition on solid walls.

Keywords: Micro flow, RBF-DQ method, meshless method, CFD

1. INTRODUCTION

In recent years, the theory of radial basis function has obtained intensive attention and enjoyed considerable success as a technique for interpolating multivariable data and functions. Its "true" mesh-free nature motivates researchers to use it to deal with partial differential equations. The first trial of exploration was made by Kansa [1]. Subsequently, Fornberg [2], Hon and Wu [3], Chen et al [4], Fasshauer [5], Chen and Tanaka [6] also made a great contribution in its development. It should be noted that most of above works related to the application of RBF for the numerical solution of PDE are actually based on the function approximation instead of derivative approximation. In other words, these works directly substitute the expression of function approximation by RBF into a PDE, and then change the dependent variables into the coefficients of function approximation. The process is very complicated, especially for non-linear problems. To overcome this difficulty, Wu and Shu [7], Shu et al. [8] recently developed a RBF Differential Quadrature (RBF-DQ) method. RBF-DQ method directly approximates derivatives of a PDE, which combines the mesh-free nature of RBF with the derivative approximation of Differential Quadrature (DQ) method. In this paper, we concentrate on the Multi Quadric (MQ) RBF to exploit its excellent performance in the function approximation. In a local MQ-DQ approach, a spatial derivative at a knot is approximated by a linear weighted sum of the functional values in the supporting region around the knot. The method shows its consistency to both linear and nonlinear problems. The weighting coefficients in derivative approximation are determined by MQ approximation of the function and linear vector space analysis. Some fundamental issues of this method are shown in [8]. This paper further explores the applicability of local MQ-DQ method for simulation of micro incompressible flows.

2. NUMERICAL METHODOLOGY

Following the work of Shu et al. [8], the nth order derivative of a smooth function $f(x, y)$ with respect to x, $f_x^{(n)}$, and its mth order derivative with respect to y, $f_y^{(m)}$, at (x_i, y_i) can be approximated by local MQ-DQ method as

$$f_x^{(n)}(x_i, y_i) = \sum_{k=1}^{N} w_{i,k}^{(n)} f(x_k, y_k) \qquad (1)$$

$$f_y^{(m)}(x_i, y_i) = \sum_{k=1}^{N} \overline{w}_{i,k}^{(m)} f(x_k, y_k) \qquad (2)$$

where N is the number of knots used in the supporting region, $w_{i,k}^{(n)}$, $\overline{w}_{i,k}^{(m)}$ are the DQ weighting coefficients in the x and y directions. The determination of weighting coefficients is based on the analysis of function approximation and the analysis of linear vector space.

In the local MQ-DQ method, MQ approximation is applied locally. At any knot, there is a circular supporting region, in which there are N knots randomly distributed. The function in this region can be locally approximated by MQ RBF as

$$f(x,y) = \sum_{j=1, j \neq i}^{N} \lambda_j g_j(x,y) + \lambda_i \qquad (3)$$

$$g_j(x,y) = \sqrt{(x-x_j)^2 + (y-y_j)^2 + c_j^2} \\ - \sqrt{(x-x_i)^2 + (y-y_i)^2 + c_i^2} \qquad (4)$$

where λ_j is a constant, c_j is the shape parameter. The weighting coefficient matrix of the x-derivative can be determined by

$$[G][W^n]^T = \{G_x\} \qquad (5)$$

where $[W^n]^T$ is the transpose of the weighting coefficient matrix $[W^n]$, and

$$[W^n] = \begin{bmatrix} w_{1,1}^{(n)} & w_{1,2}^{(n)} & \cdots & w_{1,N}^{(n)} \\ w_{2,1}^{(n)} & w_{2,2}^{(n)} & \cdots & w_{2,N}^{(n)} \\ \vdots & \vdots & \ddots & \vdots \\ w_{N,1}^{(n)} & w_{N,2}^{(n)} & \cdots & w_{N,N}^{(n)} \end{bmatrix},$$

$$[G] = \begin{bmatrix} 1 & 1 & \cdots & 1 \\ g_1(x_1,y_1) & g_1(x_2,y_2) & \cdots & g_1(x_N,y_N) \\ \vdots & \vdots & \ddots & \vdots \\ g_N(x_1,y_1) & g_N(x_2,y_2) & \cdots & g_N(x_N,y_N) \end{bmatrix}$$

$$[G_x] = \begin{bmatrix} 0 & 0 & \cdots & 0 \\ g_x^n(1,1) & g_x^n(1,2) & \cdots & g_x^n(1,N) \\ \vdots & \vdots & \ddots & \vdots \\ g_x^n(N,1) & g_x^n(N,2) & \cdots & g_x^n(N,N) \end{bmatrix}$$

The elements of matrix $[G]$ are given by equation (4). For matrix $[G_x]$, we can successively differentiate equation (4) to get its elements. For example, the first order derivative of $g_j(x,y)$ with respect to x can be written as

$$\frac{\partial g_j(x,y)}{\partial x} = \frac{x - x_j}{\sqrt{(x-x_j)^2 + (y-y_j)^2 + c_j^2}} \\ - \frac{x - x_i}{\sqrt{(x-x_i)^2 + (y-y_i)^2 + c_i^2}} \quad (6)$$

With the known matrices $[G]$ and $[G_x]$, the weighting coefficient matrix $[W^n]$ can be obtained by using a direct method of LU decomposition. The weighting coefficient matrix of the y-derivative can be obtained in a similar manner. Using these weighting coefficients, we can discretize the spatial derivatives, and transform the governing equations into a system of algebraic equations, which can be solved by iterative or direct method.

From the procedure of DQ approximation for derivatives, it can be observed that the weighting coefficients are only dependent on the selected RBF and the distribution of the supporting points in the support domain, Ω_{x_i}. In the process of numerical simulation, they are only computed once, and stored for future use. It should be noted that the computed coefficients can be applied to both linear and nonlinear problems.

The local MQ-DQ method provides an interesting and effective way to discretize the differential operators in the partial differential equations. However, to solve the Navier-Stokes equations in primitive-variable form, special splitting technique is required to deal with the difficulty arising from lack of an independent equation for the pressure, whose gradient contributed to the momentum equations. In this study, fractional step method is adopted to solve this difficulty.

3. FRACTIONAL STEP METHOD

Considering an unsteady, viscous, incompressible flow, the non-dimensional governing equations in terms of the primitive variables can be written as

Non-dimensional momentum equation:

High-speed flows:
$$\frac{\partial \mathbf{u}}{\partial t_h^*} + \mathbf{u} \cdot \nabla \mathbf{u} = -\nabla p + \frac{1}{Re}\Delta \mathbf{u} \quad (7)$$
where the convective time scale $t_h = L/U$ is used.

Low-speed flows:
$$\frac{\partial \mathbf{u}}{\partial t_l^*} + Re\,\mathbf{u} \cdot \nabla \mathbf{u} = -\nabla p + \Delta \mathbf{u} \quad (8)$$
where the convective time scale $t_l = L^2/\nu$ is used.

Continuity equation:

$$\nabla \cdot \mathbf{u} = 0 \quad (9)$$
where Re denotes the Reynolds number, defined as
$$\mathbf{Re} = \frac{UL}{\nu}$$
where L is the reference length, U the reference velocity and ν the kinematic viscosity. Solution of above Navier-Stokes equations confronts some difficulties like the lack of an independent equation for the pressure and nonexistence of a dominant variable in the continuity equation. One way to circumvent these difficulties is to uncouple the pressure computation from the momentum equations and then construct a pressure field so as to enforce the satisfaction of the continuity equation, which is named fractional step method or projection method.

In this paper, a two-step fractional formulation is applied with a collocated/non-staggered arrangement. For a multi-time-stepping scheme, the solution is advanced from time level "n" to "$n+1$" through a predicting advection-diffusion step where pressure term is dropped from the momentum equations. In the advection-diffusion equations, convective and diffusive terms are discretized by using Adam-Bashforth scheme and Crank-Nicolson scheme, respectively. It is known that the implicit treatment of diffusion term eliminates the viscous stability constraint which can be quite severe in numerical computations of viscous flow.

As an example, the following shows how the fractional step procedure is performed for high speed flows. For a time increment, $\Delta t = t^{n+1} - t^n$, the algorithm of fractional step method consists of two steps: Firstly, an intermediate velocity $\mathbf{u}^*$ is predicted by the advection-diffusion equation, which drops the pressure term. That is, for each interior node in the domain, the intermediate velocity $\mathbf{u}^*$ can be calculated by

$$\frac{\mathbf{u}^* - \mathbf{u}^n}{\Delta t} = -\left[\frac{3}{2}H(\mathbf{u}^n) - \frac{1}{2}H(\mathbf{u}^{n-1})\right]$$
$$+ \frac{1}{2\,\text{Re}}L(\mathbf{u}^* + \mathbf{u}^n) \tag{10}$$

where H denotes the discrete advection operator, L the discrete Laplace operator. Superscript $n-1$, n and $n+1$ denote the time levels. Secondly, the complete velocity $\mathbf{u}$ at t^{n+1} is corrected by including the pressure field,

$$\frac{\mathbf{u}^{n+1} - \mathbf{u}^*}{\Delta t} = -Gp^{n+1} \tag{11}$$

where G is the discrete gradient operator. The final velocity field should satisfy the continuity equation,

$$D\mathbf{u}^{n+1} = 0 \tag{12}$$

where D is the discrete divergence operator. Substituting Eq. (12) into Eq. (11) leads to the following Poisson equation for pressure

$$Lp^{n+1} = \frac{1}{\Delta t}(D\mathbf{u}^*) \tag{13}$$

Finally, the velocity $\mathbf{u}^{n+1}$ is updated by the solution of pressure equation (13). It should be emphasized that all the discrete operators mentioned above are discretized by the LMQDQ method.

Numerical experiments for micro flow in micro geometry

Flows in micro cavity in the slip regime are simulated in the study. The flows are considered to be incompressible, viscous, and driven by the moving upper wall. The flows therefore are described by those equations (7) - (9).

For the slip models on the solid walls, Maxwell's first-order formula is adopted, which is applicable for Knudsen number $Kn < 0.1$. Knudsen number is defined:

$$Kn = \frac{\lambda}{L} \tag{14}$$

where λ is mean free path of molecule, and L is characteristic geometry length. The physical boundary conditions of the problem can be specified as

$$u = 0, \ v = k_n\frac{\partial v}{\partial x}, \ \text{on left boundary } x = 0 \tag{15a}$$

$$u = 0, \ v = -k_n\frac{\partial v}{\partial x}, \ \text{on right boundary } x = 1 \tag{15b}$$

$$u = 1 - k_n\frac{\partial u}{\partial y}, \ v = 0, \ \text{on upper boundary } y = 1 \tag{15c}$$

$$u = k_n\frac{\partial u}{\partial y}, \ v = 0, \ \text{on lower boundary } y = 0 \tag{15d}$$

In the paper, numerical experiments are carried out at various Knudsen numbers. For convenience, uniform grid generation technique is adopted to generate the nodes in the domain. Based on time-dependent algorithm discussed in the previous section, numerical simulations for the micro lid-driven flows in a micro square cavity with Knudsen numbers of 0.005, 0.03 and 0.08 are carried out. The flow regime is visualized in the terms of pressure contours and streamlines which are shown in Figures 1 to 3. It can be observed that with the increasing of Kn number, the center of primary vortex still locates at the same position. The pressure contours are all more or less similar regardless of the changes of Knudsen numbers. However, due to lack of data from other researchers, the numerical results are not validated.

4. CONCLUTIONS

Flows in a micro square cavity with a moving lid in slip flow regime are numerically simulated using developed Local Radial Basis Function-based Differential Quadrature method with meshless approach. It demonstrates the broad applications of the method. Numerical results show that for slip flow in micro cavity, the Knudsen number does not have great effects to velocity and pressure contours.

REFERENCES

[1] E. J. Kansa. "Multiquadrics – a scattered data approximation scheme with applications to computational fluid-dynamics –I. Surface approximations and partial derivative estimates", Computers Mathematics application, 19, No (6-8) pp127-145 (1990).

[2] B. Forberg and T. A. Driscoll, "Interpolation in the limit of increasingly flat radial basis functions", Comput. Math. Appl., 43, 413-422 (2002).

[3] Y. C. Hon and Z. M. Wu, "A quasi-interpolation method for solving stiff ordinary differential equations", International Journal for Numerical Methods in Engineering, 48, 1187-1197 (2000).

[4] C.S. Chen, C. A. Brebbia and H. Power, "Dual reciprocity method using for Helmholtz-type operators", Boundary Elements, Vol. 20, pp. 495-504 (1998).

[5] G.E. Fasshauer, "Solving partial differential equations by collocation with radial basis functions", Surface fitting and multiresolution methods, eds. A. L. Mehaute, C. Rabut and L. L. Schumaker, pp.131-138 (1997).

[6] W. Chen and M. Tanaka, "A meshless, integration-free, and boundary-only RBF technique", Comput. Math. Appl., 43, 379-391 (2002).

[7] Y. L. Wu and C. Shu, "Development of RBF-DQ Method for Derivative Approximation and Its Application to Simulate Natural Convection in Concentric Annuli", Computational Mechanics, 29, 477-485 (2002).

[8] C. Shu, H. Ding and K.S. Yeo, "Local radial basis function-based differential quadrature method and its application to solve two-dimensional incompressible Navier–Stokes equations", Comput. Methods Appl. Mech. Engrg. 192, 941-954 (2003).

(a)

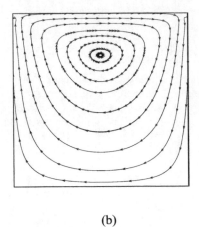

(b)

Fig.2 (a). Pressure contours; (b). Streamlines
Kn=0.03, Re=0.1, Vortex center at x=0.5, y=0.76

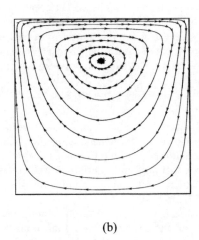

(b)

Fig.1 (a) Pressure contours; (b). Streamlines
Kn=0.005, Re=0.59, Vortex center at x=0.5, y=0.76

(a)

(a)

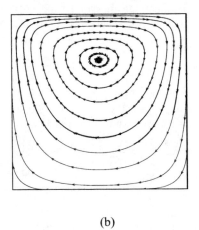

(b)

Fig.3 (a). Pressure contours; (b). Streamlines
Kn=0.08, Re=0.037, Vortex center at x=0.5, y=0.76

Numerical visualization and quantification of chaotic mixing for micromixers

Tae Gon Kang and Tai Hun Kwon

Department of Mechanical Engineering,
Pohang University of Science and Technology,
San 31 Hyoja-dong Nam-gu, Pohang, Kyungbuk, 790-784, Korea
E-mail: thkwon@postech.ac.kr

ABSTRACT

We present a systematic way of visualization and quantification of mixing in chaotic micromixers with a periodic mixing protocol. We name it colored particle tracking method, which consists of three steps. One is flow analysis step to obtain periodic velocity field of a periodic mixing protocol. The other is particle tracking step. At the inlet, depending on the species of the fluids, particles are labeled with a specific color. Then, using the initial distribution of colored particles, we repeat integration procedure up to the end period. Finally, from thus obtained particle distribution, we can evaluate mixing performance. In the present study, we propose a new measure of mixing by adopting the concept of the information entropy. We applied the developed method to three micromixers with patterned grooves and were able to evaluate mixing performance both qualitatively and quantitatively.

Keywords: microfluidics, micromixer, chaotic mixing, Galerkin/Least-squares (GLS) method, particle tracking

1 INTRODUCTION

Micromixer has a variety of applications in many biological processes [1]. It is well known that mixing in microchannel is difficult to achieve because most flows in micro scale belong to the regime of creeping flow, in which mixing is dominated by molecular diffusion.

Since early micromixer relies on only molecular diffusion, it takes very long mixing time to mix two different fluids [2]. But one can achieve enhanced mixing by incorporating chaotic advection [3,4]. In chaotic advection, fluid particles exhibit very complex trajectories even in creeping flow regime. Chaotic advection refers to kinematical phenomena in which motion of fluid particles is chaotic in Lagrangian sense even though the velocity field in Eulerian sense is simple. Since chaotic advection is possible to occur in micro scale, it must be a practically attractive mixing mechanism for micromixers. Many researchers have proposed various micromixers incorporating chaotic advection [5-8].

As far as mixing analysis of micromixers is concerned, most of numerical studies regarding micromixers are focused on solving mass transport equation in order to visualize and evaluate mixing performance of micromixers.

But there are several drawbacks in this approach. If the deformation of interface is too complicated or the length of channel is too long compared with the dimension of the height or width, the approach mentioned above may not catch the precise interface due to the numerical diffusion and insufficient resolution of discretization for computational domain. For three-dimensional numerical analysis of flow and mass transport, moreover, it needs enormous computational resources and CPU time to carry out mixing analysis for the entire micromixer. Therefore most of mixing analysis did not cover the entire micromixer but is limited to analyze the basic flow characteristics and mixing pattern in a single period or at most two periods.

Micromixers of interest in current work are patterned grooved micromixers. For this kind of micromixers, recently, there have been several theoretical and experimental works [7,8]. More recently, there has been a numerical study for the patterned groove micromixer by Wang *et al* [9]. To investigate the mixing performance of a micromixer clearly, it is required to carry out a numerical analysis for the whole domain. Therefore, it is essential to have an efficient numerical scheme for the entire domain and a quantitative measure of mixing for the evaluation of mixing performance.

The goal of the present work is to develop a systematic way of visualization and quantification of mixing in chaotic micromixers with a periodic mixing protocol. We name it "colored particle tracking method", which consists of three steps: the first is the periodic velocity field analysis; the second is the particle tracking using particles labeled with a specific color depending on the species; and the third is the quantitative measure of mixing from the particle distribution. Finally, we will present the results of the application of the developed numerical method to three patterned groove micromixers.

2 NUMERICAL METHODS

Colored particle tracking will be introduced in this section method as an efficient numerical method for analyzing micromixers with the help of a periodic mixing protocol. It consists of three steps. First, periodic velocity field of repeating mixing protocol is calculated via Galerkin/Least-squares (GLS) method [10]. Second step is the particle tracking. In this step, many particles are tracked using the velocity field obtained from the first step. The essence of this stage is the color labeled to particles according to their species. Since color information of a particle is an intrinsic property, it is possible to measure mixing from the distribution of colored particles at a specific cross section. The final step is the quantification of mixing with a proper measure. A measure of mixing is introduced by using the concept of information entropy. This overall approach is thought to be a general way of analyzing mixing in a micromixer composed of a periodic mixing protocol. The detailed method is described in the following sections.

2.1 Finite element formulation

A steady state, incompressible Stokes equation and boundary conditions are represented by

$$\nabla \cdot (2\mu \mathbf{D}) - \nabla p = 0 \qquad \text{in } \Omega \tag{1}$$

$$\nabla \cdot \mathbf{u} = 0 \qquad \text{in } \Omega \tag{2}$$

$$\mathbf{u} = 0 \qquad \text{on } \Gamma_\mathrm{w} \tag{3}$$

$$\mathbf{t} = -\Delta p \mathbf{n} + 2\mu \mathbf{D} \cdot \mathbf{n} \qquad \text{on } \Gamma_\mathrm{i} \tag{4}$$

$$\mathbf{t} = 2\mu \mathbf{D} \cdot \mathbf{n} \qquad \text{on } \Gamma_\mathrm{o} \tag{5}$$

where $\mathbf{u}$ is the velocity, p is the pressure, μ is the viscosity, $\mathbf{D}$ is the rate of deformation tensor, Δp is the pressure drop within one period, $\mathbf{n}$ is the outward normal vector to the boundary surface, and $\mathbf{t}$ is the traction force. In above equations, Ω, Γ_w, Γ_i, and Γ_o denote bounded domain, wall boundary, inlet boundary, and outlet boundary respectively. In addition to boundary conditions described above equations, periodic boundary condition is applied to the set of two corresponding nods, one for inlet and the other for outlet.

$$\mathbf{u}_\mathrm{i} = \mathbf{u}_\mathrm{o} \qquad \text{on } \Gamma_\mathrm{i} \text{ and } \Gamma_\mathrm{o} \tag{6}$$

Since periodic velocity condition is a kind of constraint to be satisfied by the velocity at the inlet and outlet surface, this condition is implemented via Lagrangian multiplier method as follows.

$$\sum \boldsymbol{\lambda} \cdot \left(\mathbf{u}_i - \mathbf{u}_o\right) = 0 \tag{7}$$

The final weak form incorporating periodic boundary condition is obtained by adding additional terms due to the Lagrangian multiplier to the original weak form of Galerkin/Least squares (GLS) method. In equation (8), φ is the weighting function for Lagrange multiplier, $\boldsymbol{\lambda}$. By employing GLS method, we can circumvent the compatibility condition of velocity and pressure interpolation function so that equal order interpolation function may be used [10]. In the present finite element formulation, we use tri-linear interpolation function for all variables, $\mathbf{u}$, p, and $\boldsymbol{\lambda}$.

$$\int_\Omega 2\mu \mathbf{D}(\mathbf{u}) : \mathbf{D}(\mathbf{v}) d\Omega$$

$$- \int_\Omega p(\nabla \cdot \mathbf{v}) d\Omega + \int_\Omega (\nabla \cdot \mathbf{u}) q d\Omega$$

$$+ \sum_k \int_{\Omega_k} (\nabla p - \nabla \cdot (2\mu \mathbf{D}(\mathbf{u}))) \cdot (\tau \nabla q) d\Omega_k \tag{8}$$

$$+ \sum \left[\varphi \cdot (\mathbf{u}_i - \mathbf{u}_o) + \boldsymbol{\lambda} \cdot (\mathbf{v}_i - \mathbf{v}_o)\right] = \int_{\Gamma_i \cup \Gamma_o} \mathbf{t} \cdot \mathbf{v} d\Gamma$$

2.2 Particle tracking

During the particle tracking procedure, one can identify the species of particles by their intrinsic color assigned initially. The problem of tracing the position of particles can be stated as integrating the following equation

$$\frac{d\mathbf{x}}{dt} = \mathbf{u} \tag{9}$$

where $\mathbf{x}$ is the particle position vector, $\mathbf{u}$ is the particle velocity, and t is the time. Instead of dealing with the original equation (9), it turns out to be helpful to modify it to a two-dimensional problem as

$$\frac{dx}{dz} = \frac{u}{w} = \tilde{u}$$

$$\frac{dy}{dz} = \frac{v}{w} = \tilde{v} \tag{10}$$

In equation (10), u, v, and w are the velocity component of x, y, and z coordinate, respectively. The 4th order Runge-Kutta method is used to integrate the set of ordinary differential equations.

What we need in integrating equation (10) is the information of velocity at a certain spatial location. In this particular case, velocity field is obtained from the solution of finite element formulation. Therefore velocity data is available at the nodal points. Thus, after searching the element in which a specific particle is located, spatial interpolation is required to find out the velocity at the desired location. Then a particle can be moved to a new position and at that location particle has the information of coordinates and species.

2.3 Measure of mixing

This section presents a way of characterizing mixing performance quantitatively by employing mixing entropy. First, a cross sectional area of micromixer is divided by finite number of cells. Then, for a certain particle configuration of multiple species, the mixing entropy is defined as a sum of the entropy of individual cells constituting the cross sectional area:

$$S = -\sum_{i=1}^{N_c} \left[w_i \sum_{k=1}^{N_s} \left(n_{i,k} \log n_{i,k} \right) \right] \qquad (11)$$

In the definition of mixing entropy, equation (11), i is the index for the cell, k is the index for the species, w_i is the weighting factor for the cell, N_c is the number of cells, Ns is the number of species to be mixed, and $n_{i,k}$ is the particle number fraction of the kth species in the ith cell. Since we are dealing with binary fluids mixing in the current study, N_s is 2. Weighting factor, w_i, is devised such that it is set to zero for the cell with no particle at all or having only single species.

Since the value of mixing entropy itself is not a practically meaningful quantity, degree of mixing (κ) was introduced as a measure of mixing. Degree of mixing (κ) is defined as a normalized entropy difference.

$$\kappa = \frac{S - S_0}{S_{max} - S_0} \qquad (12)$$

In equation (12), S is the mixing entropy of a cross section where κ is evaluated, S_0 is the mixing entropy of inlet particle distribution, and S_{max} is that of uniformly distributed particles. In the case of uniform particle distribution, which is the case of perfect mixing, κ is 1. At inlet, κ is fixed to be 0 by the definition of κ.

3   RESULTS

We applied the developed numerical method to three micromixers. Three micromixers of interest in this study, as depicted in figure 1, are slanted groove micromixer (SGM), staggered herringbone micromixer (SHM), and barrier embedded micromixer (BEM). These micromixers have patterned grooves on their bottom surface in order to induce rotational flow.

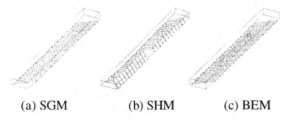

| (a) SGM | (b) SHM | (c) BEM |

Figure 1: Geometry of the periodic unit of three micromixers.

The geometrical parameter of SGM and SHM is from the paper by Stroock *et al* [7]. In the case of BEM [8], the height of barrier is 2/3 of the channel height, thickness of barrier is 20 μm, and the length of barrier is 720 μm, which corresponds to the length of six grooves.

The most important feature of a micromixer is its flow characteristics, which determines final mixing performance. Therefore, accurate velocity field is essential for the mixing analysis by particle tracking. In the particle tracking stage, particles are just the fluid particle; thus there is no interaction between particle and flow so that particles are just convected with the given velocity field of a micromixer.

Before mixing analysis via colored particle tracking, Poincaré sections are plotted to examine the dynamical system of three micromixers. As shown in figure 2, chaotic mixing takes place over the most cross section in SHM. In the case of BEM, mixing is also chaotic except two unmixed islands surrounded by KAM. However, in the case of SGM, one can expect that poor mixing will result in.

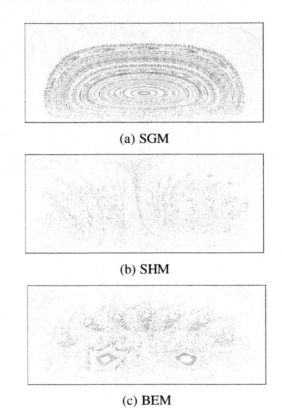

(a) SGM

(b) SHM

(c) BEM

Figure 2: Poincaré sections

We carried out colored particle tracking for three micromixers with the velocity solution obtained from finite element method mentioned in previous section. Particles, depending on their species, are labeled with a specific color and the color is inherent to each particle. Therefore, at a certain cross section of micromixer, the configuration of colored particles can reflect the status of mixing both qualitatively and quantitatively. From the distribution of colored particle in figure 3, one can easily figure out the progress of mixing through a micromixer.

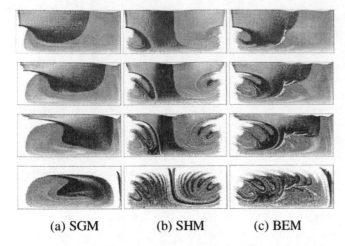

(a) SGM   (b) SHM   (c) BEM

Figure 3: Visualization of mixing; Distribution of colored particles at 1st, 2nd, 3rd, and 10th period. The number of initial particle is 80000 for three micromixers

Now we have the distribution of colored particles at all periods. Accordingly, we can calculate degree of mixing (κ) from equation (12). The change of κ along the down channel direction is plotted in figure 4. As expected from Poincaré sections and colored particle distribution, mixing performance of the SGM is the worst as expected and that of the SHM seems better than BEM of the current design. In the case of SHM and BEM, one can observe a rapid increase of κ and then convergence to an asymptotic value.

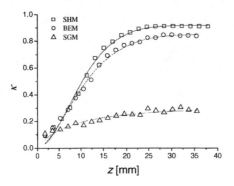

Figure 4: Degree of mixing (κ). κ was plotted to the 20th period for all micromixers.

4 CONCLUSIONS

A systematic way of mixing analysis was proposed to carry out direct numerical simulation of the chaotic mixing in micromixers with a periodic mixing protocol. Instead of solving mass transport equation, colored particle tracking was employed to visualize and quantify the mixing of two different fluids. This colored particle tracking method was applied to three micromixers. From the results of numerical simulation, we were able to obtain the flow characteristics of micromixers and the configuration of colored particles. From thus obtained particle distributions, we could evaluate the mixing performance quantitatively with the new measure of mixing. This method is not limited to a specific geometry but can be applied to general micromixers with a periodic mixing protocol.

Even though two fluids were involved in current study, colored particle tracking method could be extended to the mixing of three or more fluids without loss of generality. Furthermore, this numerical scheme is thought to be a useful tool in optimization of micromixers through design of experiment.

ACKNOWLEDGEMENTS

The authors are grateful to the Korean Ministry of Science and Technology for the financial support via the National Research Laboratory Program (2000-N-NL-01-C-148), and also to Mr. D.S. Kim for helpful discussions.

REFERENCES

[1] Tay F E H (ed) 2002 *Microfluidics and BioMEMS applications* (Boston: Kluwer Academic Publishers)

[2] Koch M, Chatelain D, Evans A G R and Brunnschweiller A 1998 Two simple micromixers based on silicon *J. Micromech. Microeng* **8** 123-6

[3] Aref H 1984 Stirring by chaotic advection *J. Fluid. Mech.* **143** 1-21

[4] Ottino J M 1989 *The kinematics of mixing: Stretching, Chaos, and Transport* (Cambridge: Cambridge University Press)

[5] Liu R H, Stremler M A, Sharp K V, Olsen M G, Santiago J G, Adrian R J, Aref H and Beebe D J 2000 Passive mixing in a three-dimensional serpentine microchannel *J. Microelectromech. Syst.* **9** 190-7

[6] Meneaud V, Josserand J and Girault H H 2002 Mixing processes in a zigzag microchannel: Finite element simulations and optical study *Anal. Chem.* **74** 4279-86

[7] Stroock A D, Dertinger S K, Ajdari A, Mezic I, Stone H A and Whitesides G M 2002 Chaotic mixer for microchannels *Science* **295** 647-51

[8] Kim D S, Lee S W, Kwon T H, and Lee S S 2002 Barrier Embedded Chaotic Micromixer *Proc. μTAS 2002 Symposium* Nara Japan 757-59

[9] Wang H, Ioventti P, Harbey E and Masood S 2003 Numerical investigation of mixing in microchannels with patterned grooves *J. Micromech. Microeng.* **13** 801-8

[10] Hughes T J R, Franca L P and Hubert G M 1989 A new finite element formulation for computational fluid mechanics: VII. The Galerkin/least-squares method for advective-diffusive equations *Comput. Methods Appl. Mech. Engrg.* **188** 235-55

A Low Cost Polymer Based
Piezo-actuated Micropump For Drug Delivery System

H.Liu*, Andrew A.O.Tay*, Stephen Y. M.Wan** and G.C. Lim**

*National University of Singapore, 10 Kent Ridge Crescent, Singapore 119260,
g0203811@nus.edu.sg
*National University of Singapore, mpetayao@nus.edu.sg
**Singapore Institute of Manufacturing Technology, Singapore.

ABSTRACT

This paper presents the design, realization and simulation of a novel polymer based check-valve micropump actuated by piezoelectric disc. Comparing with silicon substrate, polymer materials have such advantages as flexibility, chemical and biological compatibility, 3D fabrication possibility and low cost in materials and mass production. Laser micromachining technology and precision engineering techniques were used to fabricate the prototype with the dimension of $\Phi18mmx5.2mm$. Results of preliminary experiments on fusion bonding between polyimide and polycarbonate are also presented. Using DI water as the pumping medium, the presented micropump is expected to achieve self-priming, bubble tolerance and low power consumption and a flow rate of $30\mu l/min$ at the resonance frequency of 300Hz.

Keywords: polymer, laser microfabrication, fusion bonding, microvalve, flow rate.

1   INTORDUCTION

Interest in polymer materials has been motivated recently in the MEMS device for real world application purpose. Polymer materials, such as polycarbonate (PC), polyimide (PI), PMMA and polyester (PE) etc, have the potential abilities to be MEMS materials due to their properties such as flexibility, chemical and biological compatibility, 3D fabrication possibility and low cost in materials and mass production [1].

A micropump as one of the devices in the microfluidic system for bio-medical application is able to supply steady controllable fluid flow at the range from micro liter to milli liter per minute. There is a greatly growing market for the treatment of some diseases which need the micro dosing system [2].

This paper presents a low cost prototype of a polymer based self-priming micropump for drug delivery application. The micropump is actuated by the piezo-electrical disk while the pump chamber and the check-valves were machined in polycarbonate (PC) and polyimide respectively. The proposed micropump has the potentials to be fabricated by hot embossing technology for mass production purpose, which will reduce the fabrication cost. The actuation unit is driven by the controller unit which consists of an AC power supply and a function generator.

2   PUMP DESIGN

The micropump consists of a piezoelectric actuator, polycarbonate pump housing and two polyimide check valves. The working principle is shown in Figure 1[3].

When the actuator diaphragm is relaxed, both of the check valves are closed and there is no flow in the chamber. During the supply mode, the piezo disc is excited and the diaphragm is protruded caused by the deflection of the piezo disc. Thus, an under pressure is generated in the chamber which consequently generates a pressure difference across the inlet valve and the out valve. This pressure difference makes the inlet valve open and the outlet valve close

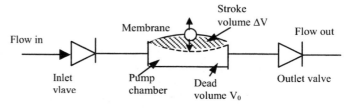

Figure1 The working principle of the check valve pump.

Hence, the flow (gas or liquid) comes into the pump chamber till the inlet valve closed by the inversed pressure difference generated by the actuator diaphragm. Then, the supply mode ends and the pumping mode starts. As the actuator diaphragm is depressed with the deflection of the piezo disc, an overpressure will be generated in the chamber. The flow goes out from the chamber through the opened outlet valve.

3 THE ACTUATOR UNIT

The actuator was made of a piezoelectric disc with the dimension of φ16mm × 0.2mm thick and a brass membrane with the dimension of φ16mm × 0.1mm thick. The piezo ceramic disc and the brass membrane were joined together by the Loctite epoxy adhesive (E-90FL).

When the piezo of the actuator is applied with an electric field of U/t, where U denotes the applied voltage and t the piezo disc thickness, the movement along the bonded surface of the piezo will be prohibited. Hence, it will generate forces and moments to lead to the curling of the brass membrane shown in Figure. 2(a). A longitudinal deflection larger than the horizontal change along the length of the piezo disc is obtained.

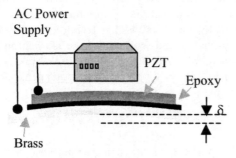

(a) Excitation schematics

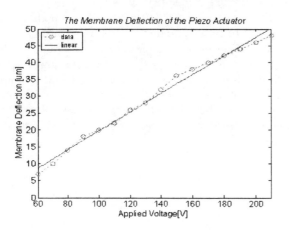

(b) The deflection variation with applied voltages

Figure.2. The piezo actuator working performance

From the plate and shell theory [4], the displacement $\delta(r)$ at any radial point r can be arrived at:

$$\delta_r = \lambda \frac{d_{31}(R^2 - r^2)}{2t^2} U \qquad (1)$$

where R denotes the radius of the actuator. The mechanical loading, the elastic properties of the brass membrane are assumed to be neglected for Equation (1). The constant λ is achieved empirically. From Equation (1), the maximum deflection is achievable at the centre of the disc and the deflection variation is linear with the applied voltage. The

testing result of the maximum deflection variations of the actuator diaphragm with applied was shown in Figure. 2(b). A linearity of the variation between the maximum deflection and the applied voltage (larger than 60V) was observed through a dial indicator with 2μm accuracy.

Integrating Equation (1), we can arrive the pump flow rate at:

$$Q = \lambda \cdot \frac{\pi}{4} \cdot \frac{d_{31}R^4}{t^2} U \cdot f \qquad (2)$$

where, f is the frequency of the excitation voltage. Therefore, the flow rate can be linearly controlled through changing the applied voltage and the excitation frequency. It offers the fundamental of the pumping working principle.

4 REALIZATION

The valve foil has a thickness of 50μm and it is fabricated using ultra-short pulse femtosecond laser micromachining technology. Femtosecond laser micromachining is a rapidly advancing area of ultra-short laser applications. It utilizes the ultra-short laser pulse properties to achieve an unprecedented degree of control in sculpting the desired microstructures internal to the materials without collateral damage to the surroundings. Because typical heat diffusion time is in the order of nanosecond to microsecond time scale whereas the electron-phonon coupling time of most materials are in the picosecond to nanosecond.

The reason for selecting polyimide as the material for the valve is because it has a relatively low Young's modulus (steel and silicon: 210GPa, polyimide: 30GPa). Hence, the critical pressure required to switch the valve flap is small and a larger gap (and flow rate) can be obtained. Therefore, it is not difficult to satisfy the basic operating conditions for the operation of the pump is easy to achieve. Also, polyimide has high resistance to corrosion. In addition, polyimide has good adherence with metals and has good sealing effect. All these mechanical and chemical properties make it as a suitable material.

For this micro-pump structure, the dead volume of the pump chamber comprises of the inlet and outlet and the small part encapsulating the inlet valve. A unique and distinctive feature of this micro-pump is the absence of a chamber immediately beneath the actuating diaphragm. This design yields a much smaller dead volume. Therefore, the compression ratio of the pump, which is the ratio of the stroke volume of the actuator unit and the dead volume will be improved. The higher compression ratio will help to achieve self-priming and bubble tolerance.

Polymer fusion bond technology was applied to assemble the pump structure. The bonding process was finished in a clean room oven with the applied pressure of 282.7 KPa, the temperature of 125°C for 80 minutes. Due to the small difference between their glass transition temperatures (T_g) [5], the polyimide and polycarbonate could be bond together relatively easily. A special fixture for the bonding process has been designed and made to control the lateral and vertical expansion less than 1.5%.

After that, the samples were tested with an Instron™ tensile testing equipment. The experiment setup is shown in Figure 3. Two roughened steel bars were fixed by the epoxy at the both sides of the sample as the holders for the equipment. It was observed that the fracture always occurred at the interface with epoxy. It means the bond strength is higher than the epoxy strength. The bonding interface was scanned by the Scanning Acoustic Microscopy (SAM). A bubble-free uniform bond interface is achieved. The bonding mechanism at the interface is still under study by the software of Cerius2®.

A piezoelectric actuator is laminated to the top of the pump housing to form a chamber. The check valves are fabricated in the polyimide membrane and bonded to the pump housing. The polyimide material has high glass transition temperature and flexibility, strong high anti corrosive coefficient, and high elasticity. Polycarbonate (thermo set) has been widely designed and used in the domestic and commercial market due to its good corrosion resistance and biocompatibility. And, using the polycarbonate as the pump housing material can improve the rigidity to increase the life time of the pump housing.

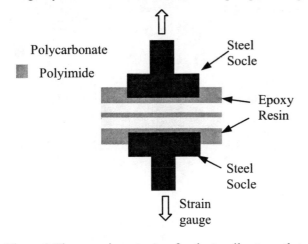

Figure 3 The experiment setup for the tensile strength test.

The pump structure, inlet and outlet were fabricated in polycarbonate by conventional precision engineering. The sample was successfully completed with the dimension of Φ18mmx5.2mm. A syringe was used to achieve a bubble-free chamber prior to the test. De-ionized water was used as the pump medium.

5 MODELING OF THE VALVE UNIT

A finite element analysis of the valve deflection was carried out using ANSYS™ to study the basic knowledge of the stress distribution and the static working condition.

Some assumptions were made to simplify the model. First, the fluid pressure was considered uniformly loading on the valve unit. And, it was modeled as a simply supported beam with the boundaries at the end of the beams. The following parameters were used for the analysis:

FEA Property:	3D
Analysis Method:	Static
Pressure Load Area:	Uniformity on Membrane
Material:	Polyimide
Young's Modulus :	2400MPa
Poisson Ratio:	0.35
Density:	1420 kg/m^3
Pressure:	10^4 Pa

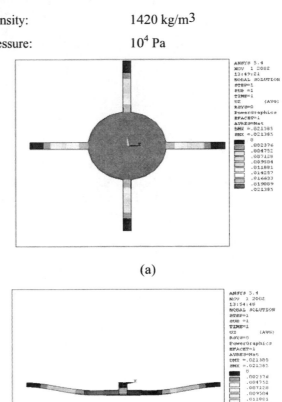

(a)

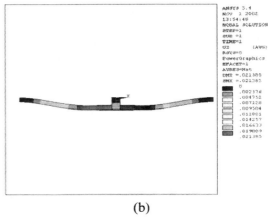

(b)

Figure 4. The modeling result of the valve deflection, (a) top view, (b) side view

The deflection result of the valve unit is shown in Figure 4. It was observed that the maximum deflection at the center point is about 21.4 μm under the pressure of 10^4 Pa. As the chamber height is about 500μm, for the safety consideration, the valve unit will not collapse with the actuators during the supply mode.

6 TESTING RESULTS

The sample was successfully completed with the dimension of Φ18mmx5.2mm. A syringe was used to achieve a bubble-free chamber prior to the test. De-ionized water was used as the pump medium.

When the pump was applied with a square pulse voltage (50V) at the pump head of (H=20mm shown in Figure 5). The flow rate rises almost linearly with the increase of the driving frequency till it reaches the maximum value of about 30.35µl/min. The resonance frequency was observed at 300Hz (Figure 5a). A wide working range from 200Hz to 450Hz was obtained to offer a flow rate of 25-30ul/min. The flow rate also increases very linearly with the variation of applied voltage in the range of less than 50v (yield voltage of piezo disc) (Figure 5b). A pumping head was achieved at 350mm with 150V and the power consumption of this pump was 15mW working at the 50V.

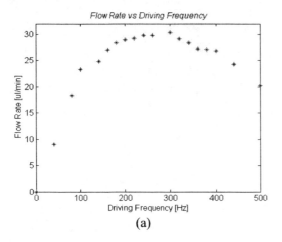

(a)

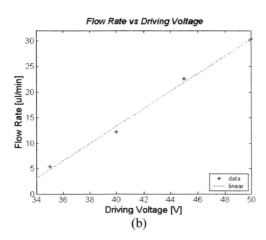

(b)

Figure 5. The testing result of the pump flow rate, (a) Flow rate variation with frequency, (b) Flow rate variation with driving voltage

7 CONCLUSION

The design, fabrication and testing of a polymer-based piezo-actuated micropump has been illustrated, which has the feasibility in the application of drug delivery system. Fusion bonding technology was tested successfully to bond the polymer materials at the appropriate temperature and pressure.

Using DI water as the pump medium, the maximum flow rate of 30.35µl/min has been achieved at the resonance frequency of 300Hz and the applied voltage only 50V. The pumping head can achieve 350mm and the power consumption is only about 15mW at 50V. The micropump's life time is more than 120hrs.

The micropump model would be developed further and the empirical constants determined at a later stage. The low production cost makes this micropump great potential to the disposable usage in the drug delivery systems.

ACKNOWLEDGEMENTS

The authors are grateful to thank Mr. Jeffrey Goh King Liang, Ms. Tan Joo Lett and Mr. Teh Kim Ming for their valuable efforts in the fabrication of this prototype. This work was sponsored by Agency for Science, Technology and Research, Singapore.

REFERENCES

[1] Gert Blankenstein, "Microfluidic Devices for Biomedical Applications", MSTNEWS, April, 2000.

[2] Angela Rasmussen, Mona E.Zaghloul, "In the Flow with MEMS", IEEE, Circuits and Devices, pp.12-25, July 1998.

[3] Nam-Trung Nguyen, Steven T. Wereley, "Fundamentals and Applications of Microfluidics", Artech House Inc., 2002.

[4] Stephen P. Timoshenko, S. Woinowsky-Krieger, "Theory of Plates And Shells", pp.5-pp400, McGraw-Hill, second edition, 1970.

[5] William D.Calister, Jr., "Materials Science and Engineering: An Introduction", New York, John Wiley & Sons Inc, 1997.

Galilean invariant viscosity term for an athermal integer lattice Boltzmann automaton in three dimensions

M. Geier, A. Greiner, and J. G. Korvink

Imtek, Albert-Ludwigs University of Freiburg
Georges-Köhler-Allee 103, 79110 Freiburg, Germany, geier@imtek.de

ABSTRACT

The athermal lattice Boltzmann automata (LBA) are promising replacements for Navier Stokes solvers in computational fluid dynamics (CFD). They are inherently parallel, scale exactly linear with the number of computational elements and can be applied to arbitrary geometries. The integer LBA presented here adds unconditional stability, roundoff-error freeness, and exact fulfillment of conservation laws to the list of benefits. The original LBA had artifacts not acceptable for industrial application. With ongoing research most of them could be removed. One of the artifacts was that the viscosity of the simulated fluid depended on the flow speed in an anisotropic manner. Previous work concentrated on removing this artifact by adding degrees of freedom to the unit cell. Here we present a method reducing the errors in the viscosity term without the need for new degrees of freedom. No computational overhead is introduced by this method.

Keywords: lattice Boltzmann, cellular automaton, central moments, Galilean invariance, viscosity

1 INTRODUCTION

The athermal LBA provide an alternative to solving the Navier Stokes equations for fluid dynamic problems [1]. The computational domain is divided into unit cells of equal shape and most often equal size. Each cell is either an ordinary fluid site or a boundary cell defining walls or inflows. Instead of obtaining a set of equations like in other numerical approaches, the LBA uses a cellular automaton as computational engine. Every cell is connected to a finite number of neighbors via links. The underling physics of a fluid is modeled as particles moving along these links in a synchronous streaming step and a subsequent mass and momentum conserving scattering step at the centers of the cells. The original LBA uses fuzzy particles to remove noise from the solution. With the applied floating point arithmetic it introduces numerical roundoff-errors that accumulate very fast due to the explicitness of the algorithm. They make ordinary LBA incapable of long time simulation. The cellular automaton description of the algorithm, however, makes it straightforward to get rid of the need for floating point arithmetic since no matrix has to be inverted and no equation needs to be solved. Instead, we allow a link to be occupied by a large number of particles modeled by an integer value. By further restricting the arithmetic to certain bounds we also exclude over- and underflow. The integer or digital LBA is hence unconditionally stable and fulfills conservation laws exactly (not only within machine precision). It comes with the additional benefit of linear complexity in space and time. Hence, it is obvious that the digital LBA will play an important role in the future when we encounter larger problems that have to be simulated for longer times. Only unconditionally stable and roundoff-error free methods, preventing us from the need of ever higher precision numbers for longer simulations, can survive in the long run. In order to prepare this method for its promising future ongoing research is yet to handle some teething problems.

2 ARTIFACTS

We simplify the collective behavior of all particles in a fluid to two basic operations: free streaming and rearrangement of particles according to entropy maximization. Particles cannot miss colliding with each other like in molecular dynamics since motion is restricted to links and scattering to sites. The computational efficiency with respect to memory consumption and speed is orders of magnitude better than for solving the Boltzmann Transport Equation directly. However, the restriction to a lattice introduces some unnatural artifacts, most notable the lack of Galilean invariance [2]. This means that some fluid properties depend on whether the frame of reverence is moving relative to the fixed lattice or not. In a preceding method, called the lattice gas automaton [3][4][5], this problem was so severe that Galilean invariance was associated with a Galilean breaking factor g in front of the convective term of the Navier Stokes equation derived from the lattice gas automaton. It took years of research to find a way to force $g = 1$. The result of this research was the LBA [1] that is hence said to be Galilean invariant. This is achieved by forcing all second order central moments (κ_{xx}, κ_{yy}, κ_{zz}, κ_{xy}, κ_{xz}, κ_{xy}) of the single site particle distribution function to isotropically approach a constant value. We define central moments as

$$\kappa_{x^m y^n z^l} = \sum_i (c_x^i - v_x)^m (c_y^i - v_y)^n (c_z^i - v_z)^l f_i / \rho \quad (1)$$

Here c_α^i is the velocity component of link i in direction α, f_i is the current occupation of link i, v_α is the macroscopic speed of the fluid in direction α, and ρ the number of all particles. This is unfortunately a very misleading definition of Galilean invariance, since it concerns only the convection term. Real Galilean invariance can only be obtained by forcing all central moments up to any order to constant values [9]. That is impossible, since the number of adjustable moments may never exceed the number of locally available degrees of freedom (number of links). We distinguish between athermal LBA, keeping all central moments up to second order constant, and thermal LBA, keeping all central moment to at least third order constant. Automata not able to adjust all third order moments cannot conserve energy, but they are significantly less computational intense (19 speeds compared to 49 in 3D) than thermal automata and are the method of choice for modeling incompressible flow with homogenous temperature distribution. Athermal LBA are only conform with the Navier Stokes equation to second order in velocity. That is typically taken to be sufficient, since the Navier Stokes equation itself is of second order only, but LBA introduce higher order coefficients not present in the Navier Stokes equation that alter the fluid viscosity at higher Mach numbers. This artifact can be reduced to insignificance by applying arbitrary fine grids. It is, however, our desire to use high Mach numbers and coarse grids for computational efficiency. Ironically, a lattice modifying only first and second order moments without introducing more degrees of freedom for higher moments has never been found. The original LBA sees them as unavoidable overhead. These degrees of freedom come at no additional cost. Our study reveals that a second and half order Galilean invariant method is better than the original automaton, while it introduces no overhead.

3 METHOD DESCRIPTION

The most popular lattice for the simulation of athermal flows in 3D is the so called D3Q19 lattice. Each site is a cube with 6 links pointing to the center of its faces and 12 to the middle of all edges [6]. One resting link is assumed to point onto the site itself. The site yields 19 degrees of freedom from which 4 are fixed in order to impose conservation of mass and momentum in each direction. By coordinate transformation we obtain a fifteen dimensional vector space perpendicular to the conserved quantities. Allowing only integer multiples of the transformed vectors to be added to the state vector of a single site during the scattering step, we avoid errors in the conserved quantities by design. Six degrees of

freedom are used to obtain second order Galilean invariance. From the nine remaining degrees of freedom only six can be used to achieve a further improvement by adjusting the third order central moments κ_{xxy}, κ_{xyy}, κ_{xzz}, κ_{yzz}, κ_{yyz}, κ_{xxz}. The other third order central moments (κ_{xyz}, κ_{xxx}, κ_{yyy}, κ_{zzz}) cannot be altered.

The links of the D3Q19 lattice are named intuitively after the directions they are pointing to: r, nw, w, sw, s, se, e, ne, n, nf, wf, sf, ef, f, nb, wb, sb, eb, and b. Here r refers to to the resting link f to front, b to back, w to west, s to south and so on. During the scattering step a new state vector $\vec{s'}$ is obtained from the incoming state $\vec{s}$. The process reads:

$$r' = r + 2(\iota_{nw} + \iota_{sw} + \iota_s + \iota_{ef} + \iota_{wf} + \iota_f + \iota_{nf} + \iota_{sf}) \quad (2)$$
$$nw' = nw + \iota_{yx} - \iota_{xy} - \iota_{nw} \quad (3)$$
$$w' = w + 2\iota_{xy} + 2\iota_{xz} - \iota_w \quad (4)$$
$$sw' = sw - \iota_{xy} - \iota_{yx} - \iota_{sw} \quad (5)$$
$$s' = s + 2\iota_{yx} + 2\iota_{yz} - \iota_s \quad (6)$$
$$se' = se - \iota_{yx} + \iota_{xy} - \iota_{nw} \quad (7)$$
$$e' = e - 2\iota_{xy} - 2\iota_{xz} - \iota_w \quad (8)$$
$$ne' = ne + \iota_{yx} + \iota_{xy} - \iota_{sw} \quad (9)$$
$$n' = n - 2\iota_{yx} - 2\iota_{yz} - \iota_s \quad (10)$$
$$nf' = nf + \iota_{yz} + \iota_{zy} - \iota_{nf} \quad (11)$$
$$wf' = wf + \iota_{zx} - \iota_{xz} - \iota_{wf} \quad (12)$$
$$sf' = sf - \iota_{yz} + \iota_{zy} - \iota_{sf} \quad (13)$$
$$ef' = ef + \iota_{xz} + \iota_{zx} - \iota_{ef} \quad (14)$$
$$f' = f - 2\iota_{zx} - 2\iota_{zy} - \iota_f \quad (15)$$
$$nb' = nb + \iota_{yz} - \iota_{zy} - \iota_{sf} \quad (16)$$
$$wb' = wb - \iota_{xz} - \iota_{zx} - \iota_{ef} \quad (17)$$
$$sb' = sb - \iota_{yz} - \iota_{zy} - \iota_{nf} \quad (18)$$
$$eb' = eb + \iota_{xz} - \iota_{zx} - \iota_{wf} \quad (19)$$
$$b' = b + 2\iota_{zx} + 2\iota_{zy} - \iota_f \quad (20)$$

The ι_β have to be integers. They are obtained from the incoming state of the cell. The algorithm performing this operations, while avoiding over- and underflow, is described in [6], [7]. A lengthy calculation yields:

$$\iota_{nw} = \lfloor \omega \left(\rho \left(\tfrac{-1}{12} \left(\tfrac{1}{3} + v_x^2 + v_y^2 \right) + \tfrac{v_x v_y}{4} \right) + \tfrac{nw+se}{2} \right) \rfloor \quad (21)$$
$$\iota_w = \lfloor \omega \left(\tfrac{-\rho}{6} \left(\tfrac{1}{3} + v_x^2 - v_y^2 - v_z^2 \right) + \tfrac{w+e}{2} \right) \rfloor \quad (22)$$
$$\iota_{sw} = \lfloor \omega \left(\rho \left(\tfrac{-1}{12} \left(\tfrac{1}{3} + v_x^2 + v_y^2 \right) - \tfrac{v_x v_y}{4} \right) + \tfrac{sw+ne}{2} \right) \rfloor \quad (23)$$
$$\iota_s = \lfloor \omega \left(\tfrac{-\rho}{6} \left(\tfrac{1}{3} + v_x^2 - v_y^2 - v_z^2 \right) + \tfrac{s+n}{2} \right) \rfloor \quad (24)$$
$$\iota_{ef} = \lfloor \omega \left(\rho \left(\tfrac{-1}{12} \left(\tfrac{1}{3} + v_x^2 + v_z^2 \right) - \tfrac{v_x v_z}{4} \right) + \tfrac{ef+wb}{2} \right) \rfloor \quad (25)$$
$$\iota_{wf} = \lfloor \omega \left(\rho \left(\tfrac{-1}{12} \left(\tfrac{1}{3} + v_x^2 + v_z^2 \right) + \tfrac{v_x v_z}{4} \right) + \tfrac{wf+eb}{2} \right) \rfloor \quad (26)$$
$$\iota_f = \lfloor \omega \left(\tfrac{-\rho}{6} \left(\tfrac{1}{3} - v_x^2 - v_y^2 + v_z^2 \right) + \tfrac{f+b}{2} \right) \rfloor \quad (27)$$
$$\iota_{nf} = \lfloor \omega \left(\rho \left(\tfrac{-1}{12} \left(\tfrac{1}{3} + v_y^2 + v_z^2 \right) - \tfrac{v_y v_z}{4} \right) + \tfrac{nf+sb}{2} \right) \rfloor \quad (28)$$

$$\iota_{sf} = \quad \left\lfloor \omega \left(\rho \left(\tfrac{-1}{12} \left(\tfrac{1}{3} + v_y^2 + v_z^2 \right) + \tfrac{v_y v_z}{4} \right) + \tfrac{sf+nb}{2} \right) \right\rfloor \quad (29)$$

Here $\lfloor \rfloor$ denotes the rounding down operation. The conserved quantities are given by

$$\rho = \qquad \sum_i f_i \qquad (30)$$

$$v_x = \frac{ne+se+eb+ef+e-nw-sw-wb-wf-w}{\rho} \qquad (31)$$

$$v_y = \frac{ne+nw+nf+nb+n-se-sw-sf-sb-s}{\rho} \qquad (32)$$

$$v_z = \frac{nf+sf+ef+wf+f-nb-sb-eb-wb-b}{\rho} \qquad (33)$$

where ω is the relaxation parameter which is restricted to $0 < \omega < 2$ by the second law of thermodynamics. Kinetic theory links the kinematic viscosity to ω through [1]

$$\nu = \frac{1}{3} \left(\frac{1}{\omega} - \frac{1}{2} \right) \qquad (34)$$

But this is only true if moments of higher than second order in Mach number are negligible. In order to achieve better results we use six remaining degrees of freedom according to the third order conditions [8].

$$\iota_{yx} = \lfloor \tfrac{1}{4} (se + sw - ne - nw - 2v_x^2 v_y \rho \quad (35)$$
$$+ v_y (\rho - r - n - sf - b - nf - sf - nb - sb))$$
$$- \frac{v_y}{2} \left(\iota_{nw} + \iota_w + \iota_{sw} + \iota_{ef} + \iota_{wf} \right)$$
$$+ v_x \left(\frac{ne - nw - se + sw}{2} + \iota_{nw} - \iota_{sw} \right) \rfloor$$

$$\iota_{xy} = \lfloor \tfrac{1}{4} (sw + nw - se - ne - 2v_y^2 v_x \rho \quad (36)$$
$$+ v_x (\rho - r - f - b - w - e - wf - ef - wb - eb))$$
$$- \frac{v_x}{2} \left(\iota_{nw} + \iota_{sw} + \iota_s + \iota_{nf} + \iota_{sf} \right)$$
$$+ v_y \left(\frac{ne + sw - se - nw}{2} + \iota_{nw} - \iota_{sw} \right) \rfloor$$

$$\iota_{yz} = \lfloor \tfrac{1}{4} (sf + sb - nf - nb - 2v_z^2 v_v y \rho \quad (37)$$
$$+ v_y (\rho - r - n - s - w - e - nw - ne - sw - nf))$$
$$- \frac{v_y}{2} \left(\iota_{ef} + \iota_{wf} + \iota_f + \iota_{nf} + \iota_{sf} \right)$$
$$+ v_z \left(\frac{nf + sb - nb - sf}{2} + \iota_{sf} - \iota_{nb} \right) \rfloor$$

$$\iota_{zy} = \lfloor \tfrac{1}{4} (sb + nb - sf - nf - 2v_y^2 v_z \rho \quad (38)$$
$$+ v_z (\rho - r - f - b - w - e - wf - wb - ef - eb))$$
$$- \frac{v_z}{2} \left(\iota_{nw} + \iota_{sw} + \iota_s + \iota_{nf} + \iota_{sf} \right)$$
$$+ v_y \left(\frac{nf + sb - nb - sf}{2} + \iota_{sf} - \iota_{nf} \right) \rfloor$$

$$\iota_{xz} = \lfloor \tfrac{1}{4} (wf + wb - ef - eb - 2v_z^2 v_x \rho \quad (39)$$
$$+ v_x (\rho - r - w - e - n - s - nw - ne - sw - se))$$
$$- \frac{v_x}{2} \left(\iota_{ef} + \iota_{wf} + \iota_f + \iota_{nf} + \iota_{sf} \right)$$

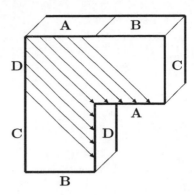

Figure 1: *A special geometry was used to measure viscosity in the shear wave decay experiment in the southeast direction. Letters indicate which faces are connected to each other. The domain is space filling and allows waves with constant wave number in southeast direction as indicated by the arrows. Previously, viscosity could only be measured in primary directions by simulating a periodic box.*

$$+ v_z \left(\frac{ef + wb - eb - wf}{2} + \iota_{wf} - \iota_{ef} \right) \rfloor$$

$$\iota_{zx} = \lfloor \tfrac{1}{4} (wb + eb - wf - ef - 2v_z^2 v_z \rho \quad (40)$$
$$+ v_z (\rho - r - f - b - n - s - nf - nb - sb - sf))$$
$$- \frac{v_z}{2} \left(\iota_{nw} + \iota_w + \iota_{sw} + \iota_{ef} + \iota_{wf} \right)$$
$$+ v_x \left(\frac{ef + wb - eb - wf}{2} + \iota_{wf} - \iota_{ef} \right) \rfloor$$

4 SIMULATION RESULTS

To measure viscosity values we simulate the decay of a sinusoidal shear wave in a box with periodic boundary conditions [5]. Here, we are particularly interested in the influence of a superimposed constant velocity perpendicular to the shear wave. Instead of computing the FFT [9] of the speed distribution, we correlate the speed only with the wave used for the initial condition. Mathematically identical, we call the new method lock-in shear wave decay (LSD) viscosimeter. It is computationally more efficient since it needs only $O(\#Sites)$ steps, while the FFT needs $O(\#Sites \log(\#Sites))$ and can only be applied if the length of the domain is a power of 2. A disadvantage of this method is that it can measure viscosity in the primary lattice directions (n, e, f) only and provides no information about the anisotropy of the viscosity term. Therefore, we developed an inclined shear wave decay viscosimeter using a special geometry with periodic boundary conditions to measure viscosity in the secondary lattice directions (see figure (1)).

Results for viscosity measured in the primary directions are shown in figure (2). The inclined viscosimeter was used to confirm that the new method introduces no new anisotropies.

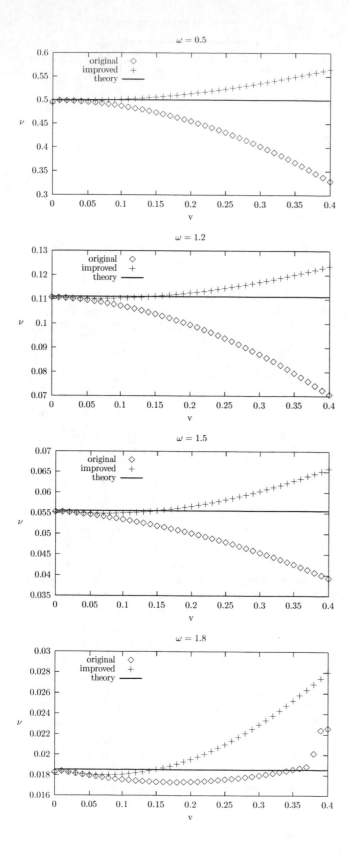

Figure 2: *Kinematic viscosity ν for different relaxation parameters ω versus flow speed v in lattice units for the original and the improved method compared with the theoretical value. The maximal possible speed is 1 (all particles moving in the same direction).*

5 DISCUSSION

We showed that the accuracy of the state of the art LBA can easily be improved by making use of spare degrees of freedom already present for isotropy reasons. The overhead introduced by a few additional local floating point operations is hardly noticeable. Still, athermal LBA are restricted to low Mach number computations by definition since all particles are assumed to move at the same speed. Neither the original nor the new method can be recommended for speeds $|v| > 0.2$ (in lattice units). Significant errors show up quickly in the original LBA, while the new method keeps viscosity nearly constant in the low Mach number regime. As accuracy increases, the method can be applied to higher Mach number problems without the need for grid refinement. Therefore, improvement of accuracy translates directly into higher computational efficiency. From the fifteen degrees of freedom only 12 could be fixed through second and third order moments. Trying to fit the remaining degrees of freedom to even higher moments results in contradictions. The benefit of using polynomials, as compared to transcendental functions, is their explicitness resulting in a much higher computational efficiency than entropically stabilized LBA. Stability concerns are absent due to the exact integer arithmetic.

6 ACKNOWLEDGEMENTS

We thank the Italian Research Association (CNR) and the Italian province Trento for financial support.

REFERENCES

[1] Succi, "The Lattice Boltzmann Equation for Fluid Dynamics and Beyond," New York, 2001.
[2] Wolfram, Journal of Statistical Physics, 45(3/4), 471-526, 1986.
[3] Frisch, Hasslacher, and Pomeau, Phys. Rev. Lett. 56(14), 1505-1509, 1986.
[4] Yepez, "Lattice-Gas Dynamics, Volume I Viscous Fluids", af.mil/Papers/pdf/LGD1.pdf, 1996.
[5] Rivet and Boon, "Lattice Gas Hydrodynamics", New York, 2001.
[6] Kauzlaric, Greiner, Geier, and Korvink, "Nanofluidic modeling and simulation", to be published in Advanced Micro and Nano Systems, Weinheim, 2004.
[7] Geier, Greiner, and Korvink, Proceedings of Polytronic 2003, 315-320, Montreux Switzerland, 2003.
[8] Qian and Zhou, Europhys. Lett., 42(4), 359-364, 1998.
[9] Teixeira, "Continuum Limit Of Lattice Gas Fluid Dynamics", Messachusetts Institute of Technology, 1992.

Liquid Disintegration: Investigation of Physics by Simulation

Jun Zeng and Tom Korsmeyer
Coventor, Inc.
625 Mount Auburn Street, Cambridge, MA 02138

ABSTRACT

Liquid disintegration includes the processes of formation of droplets and droplet fission. It is important in inkjet printing and in the miniaturization of biochemical analytical devices. This paper reports on an investigation of the mechanism of liquid disintegration via numerical simulation. The paper points out that the process is inherently multi-directional; droplet formation occurs only when the effect of the centripetal flow overcomes that of the axial flow.

A reduced-order model for droplet pinch-off, based on this insight, is derived and validated by simulations. Reduced-order models for droplet production have been derived previously from a phenomenological approach. This study shows the inherent limitation of this approach compared to that presented here, based on tracing the evolution of the multi-directional momentums.

Keywords: simulation, reduced-order modeling, droplet, liquid disintegration, inkjet

1 INTRODUCTION

Liquid disintegration includes the process of formation of droplets, or separating a liquid stream into discrete droplets; and the process of droplet fission, or dividing one droplet into smaller ones. Both processes share the same fundamental physics. From the energy perspective, both processes are nothing but energy transfer between surface potential and energy of other forms (thermal, electrical, kinetic, etc.). Disintegration is initiated by distortion of the fluid surface. When the surface area is increased, energy of other forms must be transduced into surface potential to make up the difference; when the surface area is reduced, the difference in surface potential is released.

The scientific study of liquid disintegration can be traced back to the early nineteenth century [1]. It has intensified since the 1970's, driven by the commercial success of the inkjet printing industry, whose core technology is a microelectromechanical (MEMS) device that produces discrete droplets or a stream of droplets [2].

Liquid disintegration is also important in the biotechnology sector. Droplets are the most natural vehicles to carry biochemical agents. Therefore, the control over their production and manipulation has attracted significant attention in the laboratory-automation community. Recently, a droplet-based programmable fluid processor, or PFP, has been developed [3]. The PFP is a centimeter-sized biochip with which general-purpose biochemical analyses may be conducted. In the operation of this chip, minute amounts (picoliter to nanoliter) of chemical sample are drawn from individual reservoirs in the form of droplets. These droplets are then delivered to a reaction chamber where multiple droplets may reside simultaneously. An individually addressable electrode array, embedded underneath the reaction chamber, generates dielectrophoretic forces that can translate droplets to pre-determined locations at pre-determined times. When droplets containing different chemical samples arrive at the same location, the droplets will merge into one droplet and a chemical reaction can occur. Chemical reactions can be detected, categorized, and reported. Hierarchical reactions can be achieved by merging droplets of intermediate reactions. A larger droplet may be split into smaller ones for more efficient manipulation or for detection (Figure 1). Other example applications include a droplet-based cytometer [4].

Understanding the mechanism of liquid disintegration is essential to the design of biochips. Methods that accomplish

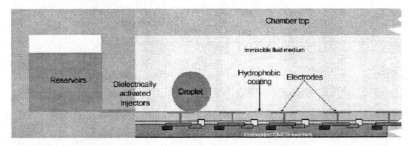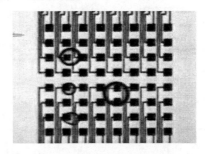

Figure 1. The Programmable Fluid Processor (PFP). Left shows an elevation view of the device. Right shows an experimental image of the reaction chamber with multiple droplets sitting on the top of electrodes in the array. Images courtesy of Professor P. R. C. Gascoyne of M. D. Anderson Cancer Center.

liquid disintegration may be classified into three categories: continuous jetting, drop-on-demand and those that utilize the Marangoni effect. In continuous jetting, a pressurized liquid flow from a nozzle breaks into a stream of droplets owing to the exponential growth of surface oscillations; a process that has been successfully explained by the linear stability analysis originated by Rayleigh [5]. This disintegration technique does not offer control over the event of droplet production. The Marangoni effect [6][7] creates an interfacial flow, which in turn induces liquid movement in the bulk due to viscosity. The movement in the bulk determines whether droplet pinch-off (fission) occurs, similar to the drop-on-demand method. The drop-on-demand method is the focus of this paper.

Numerical simulation is used throughout this paper and simulations of a drop-on-demand model problem are analyzed in detail. The model problem simulations show that, even though the axial flow along the injection direction is primarily responsible for the jet growth, the droplet formation process is largely determined by a secondary centripetal flow inside the jet. Therefore, the production of a droplet is a multi-directional rather than unidirectional process. This is the fundamental reason why the descriptive or phenomenological reduced-order modeling approaches [8] that treat droplet production as unidirectional can achieve only limited reusability. That is, a model designed for one device under one operating condition is usually not useful for other cases.

The model-problem simulations also reveal that the centripetal and axial flows work at cross-purposes. If the centripetal flow is sufficiently strong, droplet pinch-off will occur. Based on this understanding, a reduced-order model for droplet pinch-off is derived. Over a hundred simulations are carried out to adjust and validate the model. A discussion of these simulations provides a conclusion to the paper.

2 METHODS AND TOOLS

All of the simulation results presented in this paper were obtained by using the multi-physics simulation package CoventorWare[TM] (Coventor, Inc., Cambridge, Massachusetts).

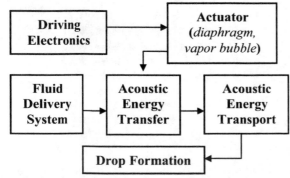

Figure 2. A schematic of the drop-on-demand process.

CoventorWare[TM] integrates FLOW-3D® (Flow Science Inc., Santa Fe, New Mexico) as its free-surface flow simulation engine. FLOW-3D® uses Volume-of-Fluid (VOF) methodology to handle problems with surfaces that undergo large-amplitude deformation and/or break-up. In contrast to other VOF software, FLOW-3D® applies true stress boundary conditions at free surface and eliminates the diffuse boundary layer between liquid and air that is commonly found in other VOF software. The dynamic contact angle is recovered from calculations accounting for effects including that of surface tension, wall adhesion, air pressure, liquid viscosity, and gravity.

3 MODEL PROBLEM

A typical drop-on-demand process may be described as in Figure 2. An electronically controlled actuator generates a large-amplitude fluid displacement. This disturbance travels in the fluid as acoustic energy and at the ejection nozzle generates one droplet. The model problem for this process is illustrated in Figure 3. A large-amplitude pressure pulse represents the acoustic signal delivered to the nozzle (distance l away in Figure 3). Without loss of generality, this model problem encapsulates all of the complex aspects of the actuation and acoustic propagation in the pressure pulse so that the numerical investigation can concentrate on the mechanism of droplet production.

Figures 4 and 5 show simulation results for nozzle radius $a=5\mu m$, and nozzle length $l=40\mu m$. The liquid has surface tension $\gamma=7.4e\text{-}2Kg/s^2$ and viscosity $\mu=1e\text{-}3Kg/m/s$. The actuation is represented by pressure pulse amplitude $P_0=0.7Mpa$ and pressure pulse period $T=4\mu sec$. For the first half of the period, the pressure is positive and this half-period is called the "acceleration stage". The second half-period, with negative pressure, is called the "deceleration stage".

Figure 4 shows the free surface profile. The necking, or reduction of the cross-section of the liquid jet at one point (the "pinch-off" point) occurs at the beginning of the deceleration stage. The necking continues throughout the deceleration stage until the cross-section converges to one point and the droplet breaks from the jet (t=18).

Simulation results show that the flow inside the nozzle

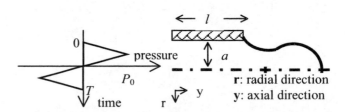

Figure 3. The model problem. The input pressure pulse is of triangular profile and is initiated in the liquid at $y=-l$. The nozzle is at $y=0$.

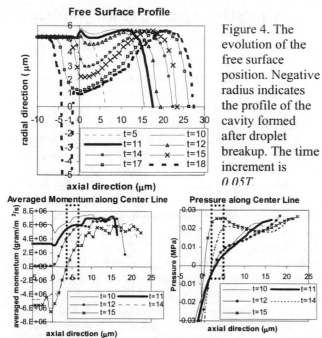

Free Surface Profile

radial direction (μm)

axial direction (μm)

t=5 --- t=10
t=11 — t=12 (△)
t=14 (•) t=15 (×)
t=17 (□) t=18 - - -

Figure 4. The evolution of the free surface position. Negative radius indicates the profile of the cavity formed after droplet breakup. The time increment is 0.05T.

Averaged Momentum along Center Line

averaged momentum (gram/m²/s)

axial direction (μm)

t=10 — t=11
t=12 (×) t=14 - - -
t=15 (×)

Pressure along Center Line

Pressure (MPa)

axial direction (μm)

t=10 t=11
t=12 (•) t=14 ·····
t=15 (×)

Figure 5. The evolution of the averaged momentum and pressure along the axial direction.

behaves as Poiseuille flow: the momentum has no spatial variation and the pressure varies linearly. In the acceleration stage, the liquid flow accelerates and the jet grows. In the deceleration stage, the flow inside the nozzle and that inside the jet are distinct from each other. The flow inside the nozzle slows down and eventually switches direction, following the negative gauge pressure at the actuation end. However, the flow inside the jet maintains more or less unchanged momentum (for instance, t>=11 in Figure 5). The disparity in momentum at both sides of the pinch-off point drains liquid out of its vicinity and eventually breaks the droplet from the jet body.

Such momentum disparity is a result of the dynamics of the pressure propagation and the presence of surface tension. During the acceleration stage, the positive portion of the pressure pulse propagates throughout the entire liquid body including the jet, where it linearly decays along the axial direction. However, during the deceleration stage, the negative portion of the pressure pulse appears to stop at the pinch-off section (dotted box in Figure 5) and does not propagate throughout the jet. Creation and maintenance of this signal blockage is the very reason that a successful droplet pinch-off is achieved.

Further understanding requires the Laplace-Young equation, which explains the role of surface tension. According to Laplace-Young, assuming the ambient air pressure is zero, the liquid pressure underneath the free surface may be expressed as $\gamma(1/R_1 + 1/R_2)$ where R_1 and R_2 are free-surface radii of curvature. At the tip of the jet, both R_1 and R_2 are of the order of a, thus the pressure P_t is approximately $2\gamma/a$. At the pinch-off section before necking occurs, R_1 is equal to a', the neck radius, and R_2 is infinite, thus the pressure P_p is equal to γ/a'.

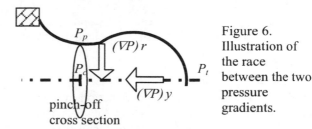

P_p $(\nabla P)r$ P_c P_t $(\nabla P)y$

pinch-off cross section

Figure 6. Illustration of the race between the two pressure gradients.

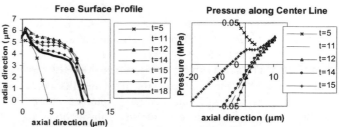

Free Surface Profile

radial direction (μm)

axial direction (μm)

t=5 (×)
t=11
t=12 (▲)
t=14 (•)
t=15 (•)
t=17 (•)
t=18 —

Pressure along Center Line

Pressure (MPa)

axial direction (μm)

t=5 (×)
t=11
t=12 (▲)
t=14 (•)
t=15 (+)

Figure 7. Evolution of the free surface profile and the pressure along the axial direction, for $P_0=0.4Mpa$ and time increment 0.05T.

During the deceleration stage, the reduction of P_c, the liquid pressure at the center of the pinch-off section, affects two pressure gradients: the axial pressure gradient $(\nabla P)y$ and the centripetal pressure gradient $(\nabla P)r$. $(\nabla P)y$ forces the liquid flow to respond to the pressure pulse and tends to slow down the flow inside the jet and eventually reverse the flow direction back towards the nozzle.

$(\nabla P)r$ induces a secondary centripetal flow. The centripetal flow shrinks the neck a' thus increasing γ/a' and consequently P_c. The rise of P_c can be clearly observed in the pressure plot in Figure 5. With more or less fixed P_t, the rise of P_c means $(\nabla P)y$ is reduced: from positive at the beginning of the deceleration stage, to approximately zero at t=14, and to negative at t=15. At this point, the flow inside the jet is accelerating rather than decelerating until the pinch-off.

It is the axial flow that contributes to the jet growth (and shrinkage). However, the secondary centripetal flow is the key to droplet detachment from the jet body. To prove this, Figure 7 shows another simulation result where P_0 is reduced to 0.4Mpa and the other parameters are kept the same. In this case, the secondary centripetal flow is too weak to raise P_c during the deceleration stage. Therefore, the pressure gradient is spatially uniform throughout the process. That is, the pressure pulse propagates throughout the entire liquid body without blockage. Hence, no momentum disparity is created and, of course, no droplet pinch-off occurs.

4 A REDUCED-ORDER MODEL FOR DROPLET PINCH-OFF

It is the effect of $(\nabla P)r$ overcoming that of $(\nabla P)y$ at the beginning of the deceleration stage that is the condition for successful droplet pinch-off. Recognizing that at the beginning of the deceleration stage $a'=a$, assuming a

Poiseuille flow profile, and accounting for the pressure jump at the jet tip due to the surface tension, such a condition can be expressed as $F>0$, with F defined as

$$F = \frac{aT}{64\pi\mu l}(p_0 - \frac{4\gamma}{a}) - 1 \qquad (1)$$

To validate this condition, a set of simulations has been carried out. The parameters are varied such that the surface tension coefficient $\gamma=0.02Kg/s^2$, $0.074Kg/s^2$ and $0.1Kg/s^2$; the viscosity $\mu=0.001Kg/m/s$, $0.002Kg/m/s$ and $0.005Kg/m/s$; and P_0 covers the range from $0.25MPa$ to $1MPa$ with a $0.05MPa$ increment. 144 data points have been collected and the results shown in Figure 8. Since Equation 1 predicts successful droplet pinch-off only for $F>0$, the differing simulation results indicate that the secondary flow effects not included in Equation 1 are indeed important. These effects are the transduction of acoustic energy to the kinetic energy and viscous dissipation of the secondary flow inside the jet that is induced by the surface tension. To compensate for this, the condition $F>0$ is modified to $G>0$, with G given by

$$G = \frac{aT}{64\pi\mu l}(p_0 - \frac{4A\gamma}{a}) + \frac{B}{\mu} + C \qquad (2)$$

where constants $A=5.2$ represents the additional pressure potential loss due to surface tension, $B=-2e-4Kg/m/s$

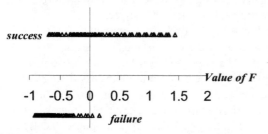

Figure 8. F values for 144 simulations. The two ordinate values are "success" indicating droplet formation is observed and "failure" indicating droplet formation is not observed.

represents additional viscous dissipation, and $C=-0.231$ is to reset the transition value of G to zero. These constants are determined

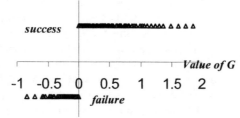

Figure 9. G values for 144 simulations.

by fitting the $G>0$ condition with the 144 data samples of Figure 8. Figure 9 shows the values of G for the 144 simulations. It is expected that since this model is derived from the underlying physical principals of droplet pinch-off, rather than a phenomenological approach, it is applicable to any drop-on-demand device driven by a large amplitude acoustic signal.

5 CONCLUSION AND FUTURE WORK

Numerical simulation is used to investigate the mechanism of drop-on-demand. The study shows that droplet pinch-off results from the centripetal flow effects overcoming those of the axial flow.

It is a rather broadly observed practice to adopt the so-called phenomenological approach to derive a reduced-order model for liquid disintegration (for a representative work see [8]). This method discards the secondary centripetal flow that is primarily responsible for the success or failure of the disintegration process. Hence this approach has achieved limited reusability. That is, a model derived for one device cannot be easily applied to describe the behavior of other devices that work under the same operating principle.

Recognizing the multi-directional nature of the liquid disintegration process, a novel reduced-order model for droplet formation is currently under development. Using the approach established here eliminates the reliance on phenomenological observation. Rather, this approach closely traces the evolution of the momentums of the centripetal flow and of the axial flow inside the jet; and so is expected to lead to a reduced-order model that will deliver accurate droplet pinch-off time and location, as well as the quality of the generated droplet (volume, terminal velocity, and sphericity). This model will be applicable to a broad range of liquid disintegration devices. The model and its performance will be presented in a future publication.

ACKNOWLEDGMENT

The authors acknowledge the support of The Defense Advanced Research Projects Agency (DARPA) under contract DAAD10-00-1-0515 from the ARO to the University of Texas M.D. Anderson Cancer Center.

REFERENCES

[1] Savart, F., Annal. Chim., Vol. 53, 337, 1833
[2] Le, H. P., J. Imaging Sci. & Tech., Vol. 42, No. 1, 49-62, 1998
[3] Vykoukal, J., Schwartz, J., Becker, F. and Gascoyne, P., Micro Total Analysis Systems 2001, 72, 2001
[4] Petersen, T. W. et al, Cytometry Part A, 56A:63-70, 2003
[5] Rayleigh, Lord, J. W. S., Proc. London Math. Soc., Vol. 10, 4, 1879
[6] Darhuber, A. A., Davis, J. M., Reisner, W. W. and Troian, S. M., Proc. Micro Total Analysis Systems 2001, 244, 2001
[7] Cho S. K., Moon H., and Kim C-J, Journal of Microelectromechanical Systems, Vol. 12, No. 1, 70, 2003
[8] Kyser E., Collins L. F. & Herbert N., Journal of Applied Photographic Engineering, Vol. 7, No. 3, 73, 1981

Capillary passive valve in microfluidic systems

Hansang Cho*, Ho-Young Kim[†], Ji Yoon Kang*, and Tae Song Kim*

*Microsystem Research Center, [†]Thermal/Flow Control Research Center
Korea Institute of Science and Technology, Seoul 136-791, Korea
Tel: 82-2-958-6708; Fax : 82-2-958-6910; E-mail: hscho@kist.re.kr

ABSTRACT

This work reports the theoretical and experimental investigations of capillary bust valves to regulate liquid flow in microchannels. The theoretical analysis uses the Young-Laplace equation and geometrical considerations to predict the pressure at the edge of the valve opening. Numerical simulations are employed to predict the meniscus shape evolution while the interface is pinned at the valve edge. Microchannel and valves are fabricated using the soft lithography. A wafer-rotating system, which can adjust a driving pressure by rotational speed, induces a liquid flow. Experimentally measured valve-bursting pressure agrees with theoretical predictions.

Keywords: capillary burst valve, microfluidcs, soft lithography, microchannel

1 INTRODUCTION

In microfluidic systems, valves are essential components to control the sequence of bio/chemical analyses. Among various valve schemes proposed to date [1, 2], a capillary passive valve is especially attractive because it is insensitive to physicochemical properties of the liquid and easy to fabricate by rapid prototyping. [2, 3] Although its basic concepts have been empirically verified before [2], its physical understanding is still far from complete. Recently delicate body forces generated in rotational motion are studied to drive microfluid. [1, 2, 4] This work rigorously analyzes the hydrodynamics of the capillary passive valve for the first time. Numerical simulations are conducted to predict meniscus shape evolution while the interface is pinned. Furthermore, the microfluidic valve and rotational wafer system is fabricated and driven by centrifugal force and its operation is visualized by a triggered video system.

2 THEORETICAL ANALYSIS

2.1 Pressure Difference across the Liquid Meniscus

The pressure difference across the liquid meniscus, ΔP_S is given by the Young-Laplace equation:

$$\Delta P_S = P_i - P_o = \sigma(1/R_1 + 1/R_2) \tag{1}$$

where P_i and P_o respectively denote the pressure inside and outside the liquid interface, and R_1 and R_2 are the principal radii of curvature. In a liquid channel, the radii of curvature are determined by the dimensions of the channel and the contact angle. Geometric considerations yield the following relation for the pressure difference in a straight channel:

$$\Delta P_S = -2\sigma \cos \theta_A (1/h + 1/w) \tag{2}$$

where σ is the surface tension, θ_A is the advancing contact angle, and w and h denote the width and the height of the channel, respectively. Here we assume that the advancing contact angle is determined by the liquid/solid combination and fairly insensitive to the speed of contact line. To drive liquid flow inside a hydrophobic microchannel, where ΔP_S positive, external pressure is should be applied to overcome such static pressure barrier and also viscous dissipation related to the liquid velocity.

A liquid flow through a microchannel can be regulated by devising an abruptly widening cross-section in the channel as shown in Fig. 1 (a). This area is called a capillary burst valve. The valve stops a liquid flow by pinning the contact line at the valve edge. The pinning takes place since the contact angle should increase until its value with respect to the new surface reaches the critical advancing angle. Then the pressure profile can be obtained by using a foregoing geometric considerations and the Young-Laplace relation. The analysis shows that the pressure difference occurring at the pinned state is

$$\Delta P_S = -2\sigma(\cos \theta_A / h + \cos \theta_I / w) \tag{3}$$

where θ_I is the intermediate contact angle which varies from θ_A to $\theta_A + \beta$ as the liquid meniscus bends toward the open area. Here β is the diverging angle of the valve with respect to the straight channel as indicated in Fig. 1 (a). When the driving pressure exceeds the value of Eq. (3), the valve bursts and the static analysis gives the following relation for ΔP_S :

$$\Delta P_S = -2\sigma\left[\cos\theta_A / h + \cos(\theta_A + \beta)/2b\right] \qquad (4)$$

Fig. 2 plots ΔP_S as the interface advances through the capillary burst valve, using Eqs. (2-4).

2.2 Pressure Difference in a Rotating System

In this work, we employ a rotating system to drive liquid flow. The driving pressure by a centrifugal force exerted on the liquid in a channel is written as

$$\Delta P_C = \frac{1}{2}\rho\omega^2(r_1^2 - r_2^2) \qquad (5)$$

where ρ is the liquid density, ω the constant angular velocity, and r_1 and r_2 are the radius as indicated in Fig. 2.

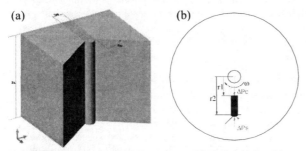

Fig. 1 (a) A schematic of a rectangular channel with a capillary burst valve. (b) Centrifugal force on a liquid in microchannel.

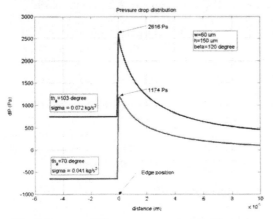

Fig. 2 Pressure difference across the liquid meniscus.

3 NUMERICAL SIMULATIONS

To investigate the temporal evolution of liquid interface at the valve area, a numerical analysis was performed for two-dimensional channels. The channels were composed of two straight lines, which met abruptly diverging area of the

valve. Numerical simulations were conducted using commercial software, CoventorWare™. The conditions of the simulations are summarized in Table 1. The simulation results, i.e. the temporal profiles of the advancing interface, are shown in Fig. 3, 4. The figure reveals that the meniscus shapes and wetting are different depending on the value of θ_A and β.

Table 1. Simulation conditions

θ_A		σ	ρ
120°	60°	7.2e-2 kg/s^2	998 kg/m^3

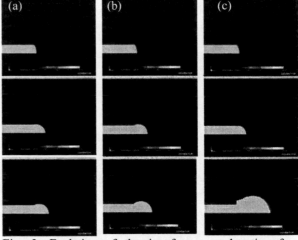

Fig. 3. Evolution of the interface near bursting for hydrophobic case. (a) $\beta = 30°$ (b) $\beta = 60°$ (c) $\beta = 90°$

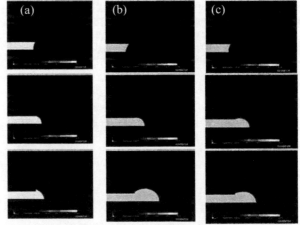

Fig. 4. Evolution of the interface near bursting for hydrophilic case. (a) $\beta = 90°$ (b) $\beta = 120°$ (c) $\beta = 150°$

4 EXPERIMENTAL STUDY

4.1 Design of Valves

The valve stops the liquid flow when the maximum value of ΔP_S given by Eq. (3) exceeds the driving pressure of ΔP_C given by Eq. (5). Therefore, the performance of the valve is determined by h, w, θ_A and β. The bursting pressure can be experimentally investigated by changing the rotational speed for a given liquid. The dimensions of the valves designed for this work are listed in Table 2.

Table 2. Dimensions of the valves used in this work

h	w	β		
150 μm	60 μm	60°	90°	120°

4.2 Fabrication Processes

The channels were fabricated by plastic molding process to decrease time and cost of fabrication. The mother mold was made of a photoresistor (SU-8). The photoresistor was spin-coated with the thickness of 150 μm and patterned using a conventional photolithography technique. A care was taken in baking processes to prevent the photoresitor from undergoing rapid temperature difference for minimizing thermal stress. The thermal stress tends to be concentrated at the opening edge of the valve, eventually deforming the microstructure. The opening edge of the valve was observed with a scanning electron microscope (S4200, Hitachi) to make sure no thermal stress deformation had occurred. The results were compared in fig. 5.

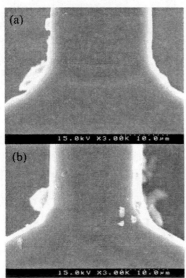

Fig. 5. Edge shape of SU-8 mold. (a) Normal heat treatment. (b) Careful heat treatment.

PDMS was then cast against the mold to yield an electrometric replica. The replication process is illustrated in Fig. 6 (a). The PDMS replica was bonded to a PC

(polycarbonate) substrate. For the bonding, the substrate was coated with 1 μm thick Primer 1200 (Dow Corning), which was then cured for 1 hour at room temperature. On the coating, a low-viscosity PDMS solution was spin-coated to a thickness of 10 μm. The replica and the coated substrate were bonded by curing the PDMS coating for 1 hour at 80°C. Such a process is shown in Fig. 6 (b). The image of the processed wafer is shown in Fig. 7.

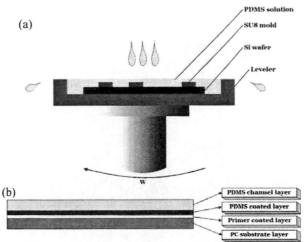

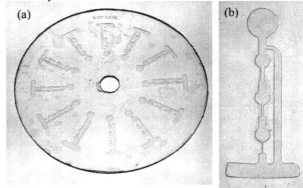

Fig. 6. Fabrication processes for (a) PDMS replica and (b) Wafer layers

Fig. 7. Images of the processed wafer. (a) The whole wafer. (b) Magnified section.

4.3 Experimental Procedure

The experimental apparatus is shown in Fig. 8. The system rotates a wafer with an AC servo motor controlled by an A/D converter, whose rotational speed ranges between 10 and 10,000 rpm. A CCD (charge coupled device) camera images the rotating wafer in situ using a trigger signal synchronized with a counter connected to a frame grabber. The frame rate ranges between 15 and 30 frames per second and the exposure time can be reduced to 1.25 μs.

A liquid used here is a water-dye (Phenol Red) mixture having physical properties as listed in Table 3. The microchannel wafer was loaded with the liquid of the volume of about 50 μl. The sample chamber was completely filled with 20 μl of the liquid, and the surplus was bypassed into a waste channel. By rotating the wafer, the liquid filling the sample chamber was driven to flow through a channel. At low rotating speeds, the flow was stopped at the valve. However, by gradually increasing the rotating speed, a critical pressure difference that caused the valve to burst was obtained.

Table 3. Physical properties of the liquid

θ_A	σ	ρ
95°	0.058 kg/s^2	1018.5 kg/m^3

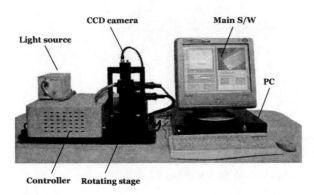

Fig. 8. A schematic of the centrifugal microfluidic system

4.4 Experimental Results

Experimental images of liquid flow around the valve are shown in Fig. 9. The images appear similar to the results obtained by the numerical simulations as discussed above. To compare the experimental measurements of the bursting pressure with our theoretical predictions, tests were performed using valves with different diverging angles, β. Fig. 10 reveals that the experimental results and the theoretical predictions are in a good agreement.

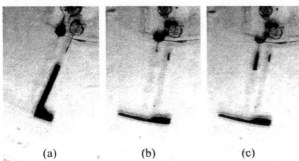

(a)   (b)   (c)

Fig. 9. Experimental images of the liquid around the valve. (a) Before pinning (b) Pinned state (c) After bursting

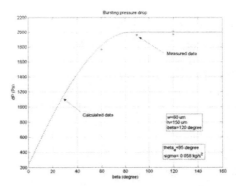

Fig. 10. Theoretical prediction of bursting pressure and experimental measurements.

5 CONCLUSIONS

Capillary burst valves provide attractive means to regulate microscale liquid channel flows since they involve no moving parts and are easy to fabricate. Especially, a wafer-rotation system can easily achieve flow regulation by adjusting rotation speed according to the critical bursting pressure of the valve. This work investigated the pressure difference that the capillary bust valve can withstand depending on the liquid properties and channel dimensions. We provided a rigorous theoretical formulation that predicts the pressure difference developing around the liquid meniscus. Numerical simulations were performed to show the meniscus shape evolution while the interface was pinned. Furthermore, the microfluidic valve and rotational wafer system were fabricated to test the performance of the capillary bust valves. The valve operation was visualized by a triggered video system. The experiments showed that the measurement results and the theoretical predictions were in a good agreement.

Acknowledgments

This research, under the contract project code MS-03-211-01, has been supported by the Intelligent Microsystem Center (IMC; http://www.microsystem.re.kr), which carries out one of the 21st century's Frontier R&D Projects sponsored by the Korea Ministry Of Science & Technology.

REFERENCES

[1] M. J. Madou, C.-H. Shih, *Biomed. Microdevices*, Vol.3, pp. 245~254, 2001

[2] D. C. Duffy, G. J. Kellogg, *Anal. Chem.*, Vol.71, pp. 4669 - 4678, 1999

[3] B.-H. Jo, D. J. Beebe, *J MEMS*, Vol.9, pp.76-81, 2000

[4] J. Ducree, R. Zengerle, *μTAS 2003*, pp. 606-606, 2003

Clogging pressure of bubbles in microchannel contractions: theory and experiments

Mads Jakob Jensen*, Xiaolin Wang**, Daniel Attinger**, and Henrik Bruus*

* MIC - Department of Micro and Nanotechnology, Technical University of Denmark
DK-2800 Kongens Lyngby, Denmark, mjj@mic.dtu.dk
** Department of Mechanical Engineering, SUNY at Stony Brook
NY 11794-2300, USA, daniel.attinger@sunysb.edu

Abstract

We present a theoretical, numerical, and experimental study of bubbles passing through a microchannel contraction. A certain pressure, the so-called clogging pressure, is needed to push a bubble through a microchannel contraction. We present the rst experimental results on such systems and compare them to results from a theoretical model rst presented at NanoTech 2003 [1]. The experiments and theory are seen to be in quite good agreement and we give possible explanations for some of the discrepancies.

Keywords: microchannel, contraction, bubble, clogging

1 Introduction

Many micro uidic systems on modern lab-on-a-chip systems contain channel contractions. These tend to become problematic if gas bubbles are present in the system. If the bubbles are large enough and span the entire microchannel they are prone to get stuck at the channel contractions. Here they can block and/or disrupt the ow, and disrupt measurements in an uncontrolled manner. To drive the bubbles out of the channels a large external pressure is needed to be applied, this is the so-called clogging pressure. These problems were already identi ed a decade ago [2], [3] but never really studied in depth. Our goal with this work is to contribute to the solution of this technological problem by means of basic research.

We start by presenting the experimental set-up and describe the measurement method. Secondly, we give a short review of the theoretical and numerical model used. At last we compare and discuss the results of the two methods, and we give some concluding remarks.

2 Experiments

For the experimental part of the work di erent capillary glass tubes of circular cross section were used. The tubes were manufactured in the glass shop of Stony Brook University. The tubes have typical internal diameters $D = 2$ mm and a contraction diameter d ranging between 40 m and 310 m. One outlet of a given glass

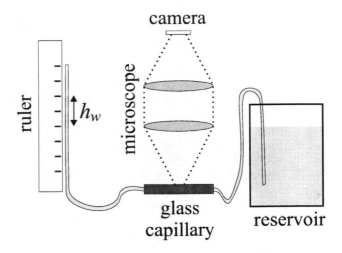

Figure 1: Experimental set-up used in the bubble clogging measurements. The height h_w of the water column is used to determine the pressure ΔP across the bubble. On Fig. 2 the detailed geometry of a bubble in the glass capillary is shown.

capillary is connected to a water reservoir and the other to a tube next to a scale, see Figs. 1 and 2. The connection tubes are made of Tygon. A hydrostatic pressure di erence can then easily be applied and measured. A bubble is introduced into the system by shortly disconnecting one Tygon tube from the glass capillary and letting in air. Through a microscope (Edmundoptics, ST-2-0099, USA) the position of the left and right meniscus, x_L and x_R, respectively, is measured by taking pictures with the camera (1.3 megapixel, C-920 Zoom, Olympus, Japan), see Fig. 1. The position is hence found as a function of the applied pressure di erence across the bubble $\Delta P = \ gh_w$, where is the density of water, g is the gravitational acceleration, and h_w is the water column height as sketched on Fig. 1.

While performing the experiments several precautions were taken as to ensure that the measurements were reproducible. Before any measurements: (1) distilled water is circulated in the tube in an ultrasonic bath for 5 minutes, (2) the water is then dried with clean compressed air (oil free). Neither solvents nor alcohol where put in the tubes as any residue might change

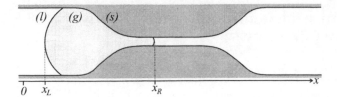

Figure 2: Schematic representation of the capillary where a gas bubble (g) in a liquid (l) is passing through the solid contraction (s). The left position x_L and right position x_R are marked as well as the zero point at the left of the bubble.

surface properties, e.g., the wetting angle.

As the measurements are to be compared with a theoretical model based on quasi-static assumptions, each measurement was performed when the system was in equilibrium. A 7 minutes delay between the setting of a new pressure point and the measurement allowed the bubble to reach its equilibrium position. Initial measurements with only tens of seconds of rest showed large discrepancies.

The experimental parameters that are used in the model are given in Table 1, together with estimates of the experimental error. Note that two different tubes have been used.

Table 1: Experimental values used in the theoretical model.

Variable	Value	Error		Tube Nr.
Contact angle	$\theta = 28°$	$\pm 4°$		All
Main	$D = 2.47$ mm	± 50	m	19
diameter	$D = 2.27$ mm	± 40	m	27
Restriction	$d = 110$ m	± 20	m	19
diameter	$d = 310$ m	± 20	m	27

3 Theory and Numerics

The theoretical and numerical part of the work is based upon an improved version of the model presented at NanoTech 2003 [1] (see also [4]). To apply and compare this model to the experiments one needs to: (a) to determine the shape $r(x)$ of the contraction (see Fig. 3) by fitting a curve to the contraction shape captured by the camera, (b) to scale the shape according to the measured values of D and d (see Table 1), and (c) to import this shape into the computer program. The computer program then determines the pressure ΔP, the bubble length $L_{\text{bubble}} = x_R - x_L$, and the corresponding positions x_L and x_R.

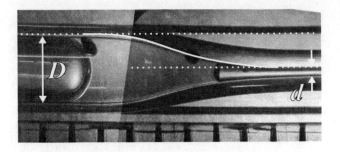

Figure 3: Images of the channel contraction taken by the camera depicted in Fig. 1. The images are used to make a curve fit and so determine the shape $r(x)$ of the contraction.

The model assumes quasi-static motion of the bubble. This implies that the contact angle θ between the liquid-gas interface and the solid-liquid interface is constant. An estimate of the maximal clogging pressure is given by the pressure required to move the bubble in a so-called sudden contraction

$$\Delta P_{\text{max}} = 4 \gamma \cos\theta \left(\frac{1}{d} - \frac{1}{D} \right), \qquad (1)$$

where γ is the liquid-gas surface tension. In the case of real physical systems as in our experiment the contact angle θ actually show some hysteresis. The contact angle has different values when receding θ_R and advancing θ_A. These different angles are due to different local wetting phenomena and contact line pinning [5].

4 Discussion and Conclusion

In the following we will only study two out the 27 glass tubes we have manufactured. They are referred to as tube 19 and tube 27.

On Figs. 4 and 5 the experimental data for tube 19 is compared to the results of the simulations. On these two first graphs the results are shown for a large bubble (full line) and a small bubble (dotted line), respectively. The experimental data is represented with error-bars.

Figure 4 depicts the pressure ΔP across the bubble as a function of the position of the left meniscus x_L. It should here be pointed out that the position is measured from different zero points, whence the different location of the two graphs. The pressure is seen to increase very rapidly. When the right side of the bubble x_R enters the contraction it advances rapidly compared to the left side x_L because of mass conservation. Moreover, most of the pressure drop is across the small meniscus having the largest curvature ($\propto 1/d$).

On Fig. 5 the length of the bubble L_{bubble} is shown as function of the pressure across the bubble. From this figure it is clear that the solid line represents the longer bubble.

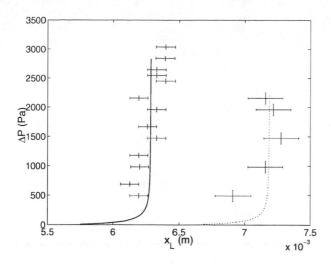

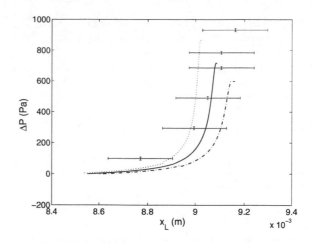

Figure 4: For tube 19: the pressure ΔP across the bubble as a function of the position of the left meniscus x_L. The gure includes: (a) the dots are experimental data, (b) the full line is the model calculation for a large bubble, and (c) the doted line similarly for a small bubble.

Figure 6: For tube 27: the pressure ΔP across the bubble as a function of the position of the left meniscus x_L. The gure includes: (a) experimental data (dots), and theoretical calculations using (b) the mean measurements of θ, D and d as input (full line) and (c) the largest deviation for $\theta \pm 4°$, $d \pm 20$ m, and $D \pm 40$ m (dotted lines).

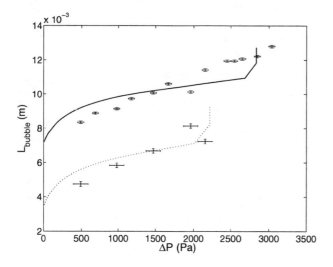

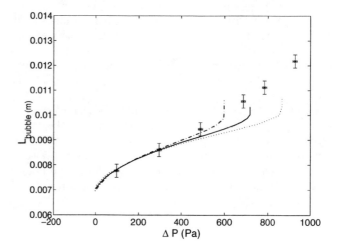

Figure 5: For tube 19: the length of the bubble $L_{\text{bubble}} = x_R$ x_L as a function of the pressure across the bubble ΔP. The gure includes: (a) dots representing experimental data, and model calculations for (b) a large bubble (full line) and (c) a small bubble (dotted line).

Figure 7: For tube 27: the length of the bubble $L_{\text{bubble}} = x_R$ x_L as a function of the pressure across the bubble ΔP. The gure includes: (a) dots (experimental data), and theoretical calculations using (b) the mean measurements of θ, D and d as input (full line) and (c) the largest deviation $\theta \pm 4°$, $d \pm 20$ m, and $D \pm 40$ m (dotted lines).

On Figs. 6 and 7 the experimental data for tube 27 is compared to the results of the simulations. For clarity we have chosen to present data only for a large bubble. In this case the dotted graphs represent numerical simulations where the parameters d, D, and θ have been chosen as the maximum and minimum of the deviations shown in Table 1.

As a general trend the data is seen to be in relatively good agreement with the numerical model. However, it is seen from Figs. 4 and 6 that the measured pressures generally are larger than the ones predicted by the model. This might be tentatively explained by the contact angle hysteresis mentioned earlier. On slightly contaminated surfaces the advancing angle θ_A is larger than the receding angle θ_R [5]. If we consider the measurements near the the maximum pressure, when the bubble is moving towards the right, the following might happen: the meniscus in the large part of the channel stops advancing with a contact angle slightly larger than θ, while the side in the restriction stops with a slightly smaller angle than θ. If we assume that $\theta_A = \theta + 10°$ and $\theta_R = \theta \quad 10°$ we may use the model equation Eq. (1) to give an estimate of the ratio between the experimental (with hysteresis) and the theoretical maximal pressures (without hysteresis):

$$\text{ratio} = \quad \frac{\cos\theta_R}{d} \quad \frac{\cos\theta_A}{D} \quad \frac{\cos\theta}{d} \quad \frac{\cos\theta}{D} \quad^{1} , \quad (2)$$

for typical values used in this project this ratio is about 1.1 meaning that the experimentally measured values are larger. This is exactly the trend seen.

Another source of possible error lies in the fact that the initial volume of the bubble is a parameter in the simulations. This initial value was not determined experimentally and a shooting method was used to t the correct initial volume. Finally, it should be mentioned that the glass capillaries were not perfectly symmetric which was assumed in the program. This last fact would however not have a vary large in uence on the results.

In conclusion it is obvious that although the model does not catch all the details of the experiments, like surface related e ects such as wetting and pinning of the contact line, it does in general yields good results which are within 10% of the measurements. As such the model is a good design tool that may be used in, e.g., the development of micro uidic systems. Most importantly this work is one of the rst combined experimental, theoretical, and numerical studies which is tackling the problem of bubble clogging in microchannel contraction.

REFERENCES

[1] M. J. Jensen, G. Goranović, and H. Bruus, NanoTech 2003, San Francisco, USA, February 2003, proc. vol. **1**, p. 258-261.

[2] P. Gravesen, J. Branebjerg, and O. Søndergård Jensen, J. Micromech. Microeng. **3**, 168 (1993).

[3] M. Elwenspoek, T. S. Kannerubj, R. Miyake, and J. H. J. Fluitman, J. Micromech. Microeng. **4**, 227 (1994).

[4] M. J. Jensen, G. Goranović, and H. Bruus, J. Micromech. Microeng., (submitted 2003), www.mic.dtu.dk/research/MIFTS/publications

[5] P. G. de Gennes, Rev. Mod. Phys. **57**, 827 (1985).

In-Plane Micropump: Design Optimization

J. Sin, W.H. Lee and H.E. Stephanou

Center for Automation Technologies
CII 8015, 110 8th Street
Troy, New York, USA, sinj@cat.rpi.edu

ABSTRACT

This paper presents the fabrication, analysis and optimization of a novel monolithically fabricated diaphragm micropump, referred to as an in-plane micropump. Compared to previous technology, the micropump in this paper is fabricated to have all necessary components in a single silicon layer. The actuation principle is electrothermal expansion of a silicon V-beam, and its expansion is amplified through a lever structure to create a greater diaphragm stroke. Either a check valve or a diffuser is incorporated with a diaphragm to direct fluid flow. The Deep RIE (Reactive Ion Etching) process is used to fabricate the pump structure on a SOI (Silicon On Insulator) wafer, and the electromechanical property of the pump structure is characterized to estimate the pump's performance. Considering that minimization of energy consumption is one of the critical design objectives in micropumps, the design parameters of the pump structure are optimized for minimal driving power.

Keywords: micropump, electrothermal, modeling, power, optimization

1 INTRODUCTION

Micropumps play a key role in microfluidic circuits where nano or pico-liter volumes of fluid are delivered and controlled. Low cost and low driving power are important characteristics in addition to precision and miniaturization in hand-held applications on biotechnology, drug discovery and drug delivery. In general, diaphragm driven pumps have self-priming advantages compared to electrically driven pumps, and a variety of actuation principles, valve mechanisms and fabrication processes have been used to build diaphragm pumps [1]-[4]. Most existing pump designs consist of multiple layers, where each layer has a unique function, such as a diaphragm, a valve, or an actuator. However, multiple layer structures require expensive wafer alignment and bonding as well as externally assembled actuators, for example, piezoelectric disks.

Monolithic design reduces fabrication complexity, and also allows a small diaphragm and a pump chamber therefore precision control of flow rate can be possible. In the silicon micropump presented here, electrothermal actuators are embedded. Electrothermal expansion provides higher force compared to electrostatic methods, but in practice the stroke is not sufficient to achieve the desired compression ratio. A high power source can generate larger stroke but the trade off is higher energy consumption. For this reason, we adopt a stroke amplification mechanism to design pump actuation structures. An important goal is to minimize the driving power, and design variables are geometrical parameters of the mechanism.

In this paper, we first describe a monolithic design of the micropump, then modeling and optimization processes are described.

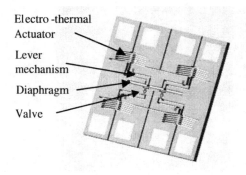

Figure 1: Illustration of an In-Plane Micropump.

2 In-Plane Micropump

2.1 Fabrication of Micropump

A monolithic silicon micropump is illustrated in Figure 1. The pump die is composed of electrothermal actuators, lever mechanisms, a pump diaphragm, and check valves. The actuator consists the ribs of thin silicon, so called V-beam, expands by means of Joule electrical resistance heating. When electrical voltage is applied to the actuator, the force caused by the rib's expansion pulls (or pushes) a rigid bar that is pivoted by a thin flexible bar. The bars act as a lever mechanism to amplify the motion of the actuator. Single or multiple actuators can be used to add force, and this figure shows the case of two thermal actuators for a diaphragm. The magnitude of the diaphragm deflection affects the intake volume and the compression ratio, therefore a flexible diaphragm is necessary in addition to a large stroke.

The pump die is covered with a glass plate to enclose fluidic flow. A few micron step structure is engraved on the cover plate to allow small gaps between the plates and moving parts of the pump layer. Through-holes in a cover plate are used for the electrical interconnections of the pump, and backside through-holes are used for fluidic interconnection between the pump chamber and the outside.

Either a check valve mechanism or a diffuser can be applied to this pump. Figure 2(a) shows a diffuser at an outlet side (outlet port is not shown). The flow path of an inlet port gradually increases, whereas, the one of outlet port decreases. The inlet flow path is designed to have a lower flow resistance when fluid flows into the pump chamber during the diaphragm expansion, but has higher flow resistance when fluid flows out of the chamber during the diaphragm compression. On the other hand, the output flow path has an opposite characteristic. Accordingly, fluid can be controlled to flow from an inlet port to an outlet port via the pump chamber with repeated diaphragm motions.

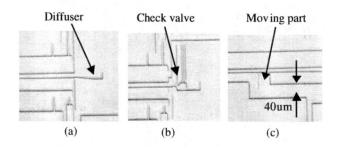

Figure 2: Fabircated diffuser and check valves. (a) Diffuser at an outlet port. (b) A cantilever check valve. (c) Active check valve (actuator is not shown).

Figure 2(b) shows a cantilever check valve mechanism that is passively operated by the pressure difference between the chamber and the ports. With a slender cantilever, each check valve opens and closes the inlet and outlet ports when the diaphragm expands and compresses. As an alternative, an active check valve was also considered as shown in Figure 2(c). It uses the same design of electrothermal actuation. The check valve of an inlet port is powered to open when the diaphragm expands to increase the volume of the pump chamber. The pressure drop due to the increased volume generates flow from the opened inlet port, which is located under the moving part of a valve. The check valve for the outlet port has the same configuration as the inlet check valve, but it opens the port when the pump chamber is compressed.

Figure 3 shows a fabricated micropump structure. In this fabrication, two diaphragms with four actuators

are used in symmetry to maximize the compression ratio (only one actuator and two diaphragms are shown).

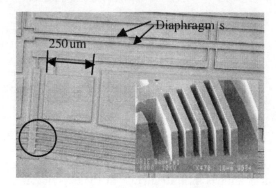

Figure 3: A fabricated micropump structure.

2.2 Structure Characterization

A preliminary experiment was performed to characterize the actuation mechanism of the pump as shown in Figure 4. Actuator dimensions are $12(W) \times 100(H) \times 1200(L)\mu m^3$, and the rib angle is 5.7. The power consumption is 2W for a serial connection of two actuators, and the deflection of the diaphragm is $30\mu m$ at $12V$. Figure 4(a) shows that the deflection of a diaphragm increases until 16 volts, and decreases afterward as silicon starts to melt at high temperature. In an effort to reduce the driving power, modeling and optimization are used to design the pump actuation structure in the following sections.

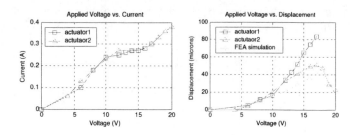

Figure 4: Characterization of pump actuation mechanism. (a) Energy consumption. (b) Displacement of diaphragm.

3 MODELING OF PUMP STRUCTURE

The electrothermal actuator of silicon U and V shape beams have been investigated in [5]–[7]. These models use a one-dimensional heat conduction equation (1),

$$k_s \frac{\partial^2 T(x,t)}{\partial x^2} + J^2\rho - Sk_a \frac{T(x,t) - T_\infty}{gh} = \frac{1}{\alpha}\frac{\partial T(x,t)}{\partial t} \quad (1)$$

where, J, ρ, S, k_s, k_a, g, and h are current density, resistivity, shape conduction factor, thermal conductivity of silicon, thermal conductivity of air, air gap, and structure thickness respectively. The radiation and convection effects are ignored. Maloney [7] applied the flexibility method [8] to solve the thermal expansion of the one-dimensional V-beam structure of Figure 5. Based upon his result, the thermal expansion δ_A of the V-beam actuator is the function of flexibility coefficient matrices f_{ij} and f_{0i}, where f_{ij} represents the displacement of a point i when a virtual force $X_j = 1$ is applied, and f_{0i} represents the displacement of a point i when all virtual forces are zero except the applied thermal load. Reproduction of this result [7] is summarized as

$$
\begin{cases}
\delta = \frac{L^2}{24EI}\left(\frac{X_1 L}{2} - 3X_3\right) \\
[X_i] = -[f_{ij}]^{-1}[f_{0i}] \\
f_{ij} = \sum_{k=1}^{n}\left[\int_{L_k}\frac{m_{ki}m_{kj}}{EI_k}dx + \frac{L_k n_{ki}n_{kj}}{EA_k}\right] \\
[f_{0i}] = \begin{bmatrix} L & -L & 0 \end{bmatrix}^T
\end{cases}
\quad (2)
$$

where, $[X_i]$ is a 3×1, and f_{ij} is a 3×3 matrix.

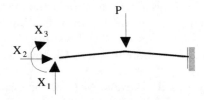

Figure 5: Silicon V-beam actuator with virtual force X_i.

The flexibility method [8] is applied to obtain the stiffness of a V-beam structure. When an external force P is applied to a V beam structure, the displacement due to this force is

$$
f_{0i}^P = \begin{bmatrix} \frac{5L^3}{48EI} + \frac{L\theta^2}{EA} & \frac{L^3\theta}{EI} - \frac{L\theta}{EA} & -\frac{L^2}{8EI} \end{bmatrix}^T
\quad (3)
$$

and the spring coefficient K_A of the V-beam is,

$$
K_A = \left[\frac{L^2}{24EI}\left(\frac{X_1^P L}{2} - 3X_3^P\right)\right]^{-1}
\quad (4)
$$

where, $X_i^P = [f_{ij}]^{-1}[f_{0i}^P]$.

Figure 6 illustrates the simplified beam model of a pump, and equations (5,6,7,8) describe corresponding spring constants of the mechanism. K_1 and K_3 relate the displacement of L_5 and L_1 respectively when force is applied at L_1. K_2 and K_4 relate the displacement of L_5 and L_1 respectively, when force is applied at L_5. When the beams L_2 and L_3 are designed to have a high moment of inertia, i.e. $I_2, I_3 \gg I_4$ deflection is mainly created by the bending of the beam L_4.

$$
K_1 = \left[\frac{L_2 L_3 L_4}{EI_4} - \frac{L_4}{EA_4}\right]^{-1}
\quad (5)
$$

$$
K_2 = \left[\frac{L_3^3}{3EI_3} + \frac{L_3^2 L_4}{EI_4} + \frac{L_4}{EA_4} + \frac{L_5}{EA_5}\right]^{-1}
\quad (6)
$$

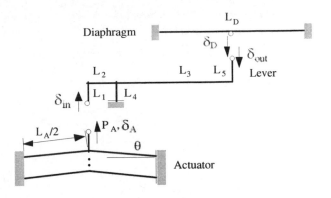

Figure 6: Illustration of pump mechanism as beam structures.

$$
K_3 = \left[\frac{L_2^2 L_4}{EI_4} + \frac{L_2^3}{3EI_2} + \frac{L_4}{EA_4} + \frac{L_1}{EA_1}\right]^{-1}
\quad (7)
$$

$$
K_4 = \left[\frac{L_2 L_3 L_4}{EI_4} + \frac{L_4}{EA_4}\right]^{-1}
\quad (8)
$$

When δ_{out} defines the displacement of output beam L_5 without the diaphragm having any effect, diaphragm deflection δ_D changes the output displacement $\delta_{out} = \delta_{out} - \delta_D = \frac{K_D}{K_2}\delta_D$, where $K_D = \left[\frac{L_D^3}{4 \cdot 48EI}\right]^{-1}$. By substitution of $\delta_{out} = \frac{P_A}{K_1}$, the diaphragm displacement due to the applied force P_A becomes

$$
\delta_D = \frac{K_2}{(K_2 + K_D)K_1}P_A.
\quad (9)
$$

Also, the force from the deflection of the diaphragm react against the actuation force P_A. When δ_{in} defines the displacement of the input beam L_1 without the diaphragm deflection, the actual displacement with the diaphragm deflection δ_{in}' is $\delta_{in} - \delta_{in} = \frac{P_A}{K_3} - \frac{K_D}{K_4}\delta_D$. Substitution of δ_D using equation (9) gives

$$
\delta_{in}' = \frac{P_A}{K_{eq}}
\quad (10)
$$

where, $K_{eq} = \left[\frac{1}{K_3} - \frac{K_D K_2}{K_4 K_1 (K_2 + K_D)}\right]^{-1}$.

From the force equilibrium condition at the input beam L_1, i.e. $K_{eq}\delta_{in}' = K_A \delta_{in}$, and the actuator displacement condition $\delta_A = \delta_{in}' + \delta_{in}$, the actuation force is determined as

$$
P_A = K_A \delta_{in} = K_A \frac{K_{eq}\delta_a}{K_{eq} + K_a}.
\quad (11)
$$

4 OPTIMIZATION

The optimization goal of this study is to minimize driving power, while achieving a desired pumping stroke and force. As shown in equation (13), the cost function

$f(\mathbf{x})$ is driving power, $V = voltage$, $I = current$ and the optimization variables are,

$$\mathbf{x} = [L_A, \ _A, n_A, V_A, L_2, L_3, L_D] \qquad (12)$$

where, n_A, and V_A are number of the V-beams and applied voltage respectively.

$$\underset{x}{\text{Minimize }} f(\mathbf{x})$$
$$f(\mathbf{x}) = \text{VI} \qquad (13)$$
$$\text{subject to}$$
$$\delta_{Diaphragm} \geq \delta_{desired}$$
$$P_{Diaphragm} \geq P_{desired}$$
$$T_{Actuator} \leq T_{limit}.$$

In this optimization, two parameters, $thickness = 100\mu m$ and $width = 10\mu m$ are fixed as fabrication constraints. Also, the L_4 is fixed to $100\mu m$, because ideally, a longer beam allows greater deflection in the boundary of yield strength. Desired displacement $\delta_{desired} = 30\mu m$, force $F_{desired} = 5mN$ and limit of temperature $T_{limit} = 800K$ are given as inequality conditions. The optimization was performed by Matlab's constrained optimization tool, and its result was shown in Figure 4.

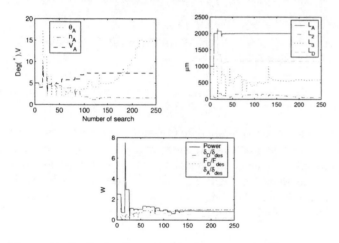

Figure 7: Optimization result of pump actuation structure. (a) Angle, number of ribs and applied voltage. (b) Length of beams. (c) Power, desired displacement and force.

Compared to typical V-beam actuators that use very small angle ($_A = 2 \sim 5$), the optimization result gives a larger angle $_A = 15$. The larger angle makes the actuator stiffer so that the actuator can withstand the reaction force of the lever mechanism. However, as the angle becomes larger, the actuator displacement δ_A becomes smaller. This problem can be solved by making L_2 smaller, while desired displacement is maintained. The driving power is optimized to $1W$ from an initial value $2W$, and it is regarded that multiple solutions exist since the power remained almost constant after 120 iterations.

5   CONCLUSION

A novel, monolithic micropump design was introduced. This approach offers the advantages of embedded actuation and reduced fabrication complexity. We formulated the design of pump actuation structure as an optimization problem by the use of the electrothermal V-beam actuator and lever amplifying mechanism. The result of this work shows that pump driving power can be reduced with optimized micropump structure design. Further study on optimization including fluid flow in the pump is on going, in addition to prototyping.

ACKNOWLEDGEMENT

This work was supported by the Infotonics Technology Center under grant No. DE-FG02-02ER63410.A000 from the US Department of Energy.

REFERENCES

[1] R. Linnemann, P. Woias, C.-D. Senfft, and J.A. Ditterich, "A self-priming and bubble-tolerant piezoelectric silicon micropump for liquids and gases," Proceedings on The Eleventh Annual International Workshop on Micro Electro Mechanical Systems, pp. 532-537, 1998.

[2] F.K. Forster, R.L. Bardell, M.A. Afromowitz, N.R. Sharma, and A. Blanchard, "Design, fabrication and testing of fixed-valve micro-pumps," Proceedings of the ASME Fluids Engineering Division, pp. 39-44, 1995.

[3] C. Grosjean and Y.-G. Tai, "A thermopneumatic peristaltic micropump," International Conference on Solid-State Sensors and Actuators, 1999.

[4] D. Maillefer, S. Gamper, B. Frehner, and P. Balmer, "A high-performance silicon micropump for disposable drug delivery systems," The 14th IEEE International Conference on MEMS, pp. 413-417, 2001

[5] L. Lin and M. Chiao, "Electrothermal responses of lineshape microstructures," Sensors and Actuators, 55, pp.35-41, 1996

[6] Q. Huang and N.K.S. Lee, "Analysis and design of polysilicon thermal flexure actuator," J. Micromech. Microeng., 9, pp.64-70, 1999.

[7] J.M. Maloney, D.L. DeVoe, and D.S. Schreiber, "Analysis and design of electrothermal actuators fabricated from single crystal silicon," in Proc. Micro-Electro-Mechanical-Systems (MEMS)-ASME 2000, pp.233-240, 2000.

[8] J. Oden, "Mechanics of Elastic Structures," McGraw-Hill, New York, 1967.

Analysis of Blood Extraction from a Lancet Puncture

Rex J. Kuriger[*] and Bart F. Romanowicz[**]

[*]Bayer HealthCare, Diagnostics Division
1884 Miles Avenue, P.O. Box 70, Elkhart, IN 46515, U.S.A., rex.kuriger.b@bayer.com
[**]CFD Research Corporation
215 Wynn Drive, Huntsville, AL 35805, U.S.A. bfr@cfdrc.com

ABSTRACT

A computational model was created utilizing the CFD-ACE+ software to model blood flow from a lancet puncture of the skin. A two-dimensional axisymmetric model was developed and the skin tissue was modeled as a porous media. The permeability of the porous media was adjusted to match the blood volume of an experimental end-cap for use with a lancet device. A parametric analysis of the end-cap radius shows that volume of blood collected increases as the opening increases, until a saturation point is reached for end-cap radii greater than 7 mm. A similar trend was found when compared with experimental data taken with the same constraints.

Keywords: lancet, blood extraction, CFD

1 INTRODUCTION

An important tool in maintaining tight control over blood glucose levels for people with diabetes is self-monitoring blood glucose (SMBG). The American Diabetes Association recommends that SMBG be performed several times daily for patients with diabetes [1]. Traditional SMBG requires lancing a fingertip with a device to acquire capillary blood. Most commercially available lancet devices utilize a stainless steel lancet that is spring loaded. When released, the lancet is launched forward and its tip momentarily protrudes through an aperture (end-cap) that is pressed flush against the user's skin surface. The lancet pierces the tissue, allowing a blood sample to form on the skin's surface. The patient typically squeezes or milks the finger near the puncture site until a sizable blood drop is obtained, which is then transferred to the testing device. This is generally considered a painful process, and repeated frequent sampling from the fingertip can generate tenderness. Pain is recognized as a major complaint of patients performing finger sticks [2] and is not conducive to testing in compliance with physician testing frequency recommendations.

The ability to perform SMBG on samples collected from sites other than the fingertip (forearms, thighs, palms, abdomen, etc.) may reduce pain and result in better testing compliance [3]. Such areas have a lower density of nerve endings [4]; therefore, blood collection is generally described as less painful. The ability to vary the site of testing can decrease the cumulative trauma to the fingertips. Some patients may also prefer to obtain samples in a less conspicuous area such as the forearm, thigh, or abdomen. One challenge to obtaining blood from these less painful sites is the lower capillary density of some alternative sites. The capillary density of the palm is 2.5 to 3.5 times higher than that of the forearm, abdomen, or thigh [5]. The fingertip appears to have an even higher capillary density than the palm [6].

2 PRIOR WORK

Several innovative techniques have been introduced to acquire blood samples from alternate sites for diagnostic purposes. The Microlet™ Vaculance™ Lancing Device developed by Bayer Corporation (Elkhart, IN) utilizes a plunger mechanism to actuate the lancet and create a vacuum around the puncture site. The device is pressed against the skin and the lancet is fired by completely depressing the plunger mechanism. An airtight seal is then formed by slowly releasing the plunger, which creates a vacuum. This causes the skin to bulge into a round transparent end-cap, dilating the puncture and increasing the flow of blood [7].

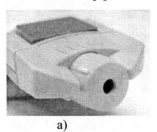 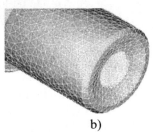

a) b)

Figure 1: a) Photograph of Microlet™ Lancing Device (Bayer Corporation, Elkhart, IN) end cap. b) Computational mesh of an end cap design candidate.

The AtLast™ Blood Glucose Monitoring System by Amira Medical (Scotts Valley, CA) was introduced to target alternate site testers. The device promotes blood flow with a stimulator ring that is pressed against the skin surrounding the puncture site. The initial depression of the device both lances the skin and kneads the surrounding skin tissue. It is sometimes necessary to pump the device multiple times in order to acquire an adequate sample size. The stimulator tip that interfaces with the skin's surface enhances blood flow by locating the skin for a consistent lancing depth. It also applies tension to the skin, which keeps the puncture site open for increased blood flow. And it kneads (or milks) the penetration site forcing blood to the skin's surface [8].

The MediSense™ SofTact™ Monitor by Abbott Laboratories (Bedford, MA) combined the lancing device and meter into one automated device intended for people who want alternative site testing with the touch of a button. To initiate a test, the user rests the device over the test site and the system automatically pricks the skin, draws the sample, and transfers it to the test strip. The monitor utilizes a vacuum-lancet nosepiece that interfaces with the skin's surface. A vacuum pump is used to draw the skin into the nosepiece and extract the blood from the puncture site after the lancing process [9].

In recent years, many technological advances have occurred that makes AST more feasible using sites with lower capillary density than the fingertip. In particular, the blood volume required for accurate testing with many meters has dropped to 1 μL or less over the past several years. For example, the Ascensia™ CONTOUR™ Blood Glucose Monitoring System (Bayer Corporation, Elkhart, IN) uses a test strip that requires a blood sample of only 0.6 μL [10]. This is a significant difference when compared to earlier AST devices such as the AtLast™ and the SofTact™ that required sample sizes of 2 μL and 2-3 μL, respectively [11]. In this realm, it may be possible to reliably acquire a sufficient blood sample from alternate sites without pumping or milking the site, or with a device that does not utilize a vacuum – maybe just some pressure for a couple seconds. This is extremely attractive, particularly when developing a cost-effective AST lancet device or an integrated system that automatically performs the SMBG process with a touch of the button.

Therefore it is of great interest to develop an end-cap or nosepiece design that interfaces with the skin tissue in a way to reliably extract a blood sample from a lancet puncture. Due to the complex nature of skin tissue and the numerous design possibilities, much experimental work must be performed to optimize a particular end-cap design, which is an expensive and time-consuming process. In order to minimize the development time and costs associated with such extensive studies, computer modeling of the system may prove beneficial. This is a first documented attempt at using computational fluid dynamics (CFD) to optimize the end-cap geometry of a lancing mechanism for maximizing blood flow from a lancet puncture. The CFD-ACE+ software package from CFD Research Corporation (Huntsville, AL) was employed for this purpose.

3 BACKGROUND

Given the geometric and material complexity of soft tissue structures and that they are subjected to complicated initial and boundary conditions, finite element models (FEMs) have been very useful for quantitative structural analyses. Recent applications of poroelastic and mixture-based theories and the associated FEMs for the study of the biomechanics of soft tissues, as well as future directions for research in this area have been surveyed [12].

Due to the great variability in the human physiology it is extremely difficult to actually quantify capillary blood flow from human tissue. Numerous factors can affect the capillary structure of the dermis layer such as aging, disease, temperature, etc. - all can contribute to person-to-person variability in blood flow from a puncture [13]. Therefore, the primary objective of this investigation is not to obtain quantitative results from the CFD analysis, but to acquire a qualitative sense of the system using a computational model for an engineering design analysis. Useful trends could be deduced from these results providing important design information that can be used for optimizing the device.

4 METHOD AND FLOW MODELING

All computational models of the skin simulated and presented here are time-efficient two-dimensional axisymmetric. The skin contains a slit (created by a lance) of known dimensions (0.38 mm diameter hole). The porous media feature of the CFD-ACE+ software was used to model the seeping of the blood from the puncture. This module allows the modeling of flow in materials consisting of a solid matrix with an interconnected void. In order to validate the feasibility of these simulations in obtaining useful design information, the modeling work will be divided into several parts. The first of which is presented in this paper where the skin is entirely modeled as a porous media and the flow rate of blood through the face of the slit is to be predicted and compared with experimental results.

4.1 Porous Media Model of Skin

Owing to the great difference in length scales between the micron-sized capillaries, and the end-cap diameter, it is computationally intractable to model the complete vasculature of the skin. Diffusion mechanisms of solutes within, across and between the intercellular lamellae of the mammalian stratum corneum may be modeled as a porous medium [14]. The skin is therefore modeled as a porous media, and taking advantage of the cylindrical symmetry involved, our problem is reduced to a two-dimensional axisymmetric model. A blood reservoir is introduced below and maintained at capillary pressure. The diameter of the end-cap defines the external model bounds.

Steady state simulations are performed using a laminar incompressible flow. The mass flow rate is multiplied by the blood collection time giving the blood volume obtained. The permeability of the porous media is adjusted to match experimental results for a nominal end-cap. Finally, parametric simulations are performed varying the end-cap radius to investigate how this parameter affects volume of blood collected.

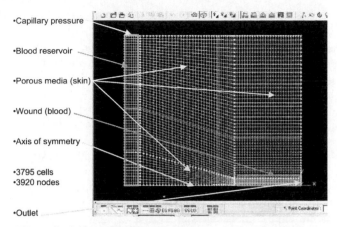

- Capillary pressure
- Blood reservoir
- Porous media (skin)
- Wound (blood)
- Axis of symmetry
- 3795 cells
- 3920 nodes
- Outlet

Figure 2. Axisymmetric grid of skin used for simulations.
The horizontal X-axis is axis of symmetry.

Fig. 2 shows the computational mesh used. The X-axis is the axis of symmetry. The blood reservoir (capillary pressure boundary condition at top) is at the left, and the skin surface is at the right. The lancet puncture is modeled as a blood-filled cavity, and is at the lower right. A natural pressure boundary condition is used to model the outlet of the puncture. Porous media regions model the epidermis, papillary dermis and reticular dermis of the skin.

4.2 Governing Equations

Transport through porous media is governed by the conservation equations of mass, momentum, energy and species. The pore dimensions may often be in the sub-micron range, and their overall effect is represented through volume-averaged quantities. Here, we present only the volume-averaged conservations for mass and momentum [15], as we are not solving for temperature, phase changes, species dilution, chemical reactions or electro-chemistry.

Mass conservation is given by the equation

$$\frac{\partial}{\partial t}(\varepsilon\rho) + \nabla \bullet (\varepsilon\rho\mathbf{U}) = 0 \qquad (1)$$

The conservation equation for momentum within the porous regions may be written as:

$$\frac{\partial}{\partial t}(\varepsilon\rho\mathbf{U}) + \nabla \bullet (\varepsilon\rho\mathbf{U}\mathbf{U}) =$$

$$-\varepsilon\nabla p + \nabla \bullet (\varepsilon\hat{\mathbf{o}}) + \varepsilon\mathbf{B} - \frac{\varepsilon^2\mu}{\kappa}\mathbf{U} - \qquad (2)$$

$$\frac{\varepsilon^3 C_F\rho}{\sqrt{\kappa}}|\mathbf{U}|\mathbf{U}$$

where ρ is the fluid density, p is the pressure, μ is the viscosity of the fluid, C_F is a quadratic drag factor, τ is the shear stress tensor, $\mathbf{B}$ is the body force vector, and $\mathbf{U}$ is the fluid velocity. ε is the porosity of the medium, and represents the volume occupied by the pores to the total volume of the porous solid, while the permeability κ is a

quantity representing the surface area to volume ratio of the porous matrix. The last two terms in Eq. (2) represent an additional drag force imposed by the pore walls on the fluid with the pores, and usually results in a significant pressure drop across the porous solid. In a purely fluid region, e _ 1 and k _ ∞, and the standard Navier-Stokes equation is recovered. In terms of boundary conditions, no special treatment to the momentum and pressure correction equations is required at fluid-porous solid interfaces.

In order to determine the validity of the simulation, results were compared to experimental values obtained from a test end-cap or nosepiece similar to the one used with the Microlet™ Lancing Device (Bayer Corporation, Elkhart, IN) [16]. Table 1 presents the dimensions, material properties and simulation parameters used in the computational model.

Dimension	Value
End-cap radius	4.72 mm
Depth of skin	5 mm
Depth of puncture	2.03 mm
Diameter of lance	0.38 mm
Depth of blood well	0.5 mm
Material property	
Blood viscosity	0.004 kg/m-s
Blood density	1060 kg/m^3
Skin porosity (porous medium)	0.3
Skin permeability (porous medium)	3.0e-14 m^2
Simulation parameter	
Blood capillary pressure	3684 Pa
Time of blood collection	10 s

Table 1: Dimensions, material properties and simulation parameters used for simulations.

First we adjust the permeability of the porous medium of the nominal end-cap design. We find that at a permeability of 3.0e-14 m^2 produces a volume of blood equivalent to that collected experimentally (Fig. 3.) [16].

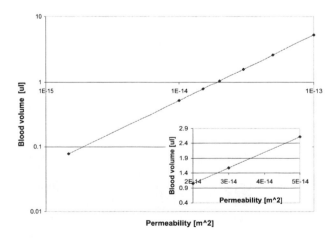

Figure 3. Volume of blood collected as a function of permeability for nominal end-cap radius of 4.72mm.

Simulation results for the nominal end-cap are shown in Fig. 4. Colors represent the pressure drop through the model. Arrows represent the direction and magnitude of the blood flow. Pressure drops off gradually through the porous medium, which offers greatest resistance to flow. Fluid velocities are greatest in the regions of the lancet puncture.

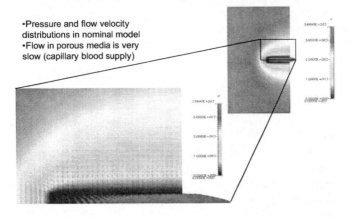

• Pressure and flow velocity distributions in nominal model
• Flow in porous media is very slow (capillary blood supply)

Figure 4. Simulated pressure distributions (colors) and flow velocity vectors (arrows). Inset shows model (top) and symmetry results rendered below.

5 RESULTS

Simulations were performed for end-caps with diameters ranging from 1.18 to 9.45 mm. Pressure distributions for these simulations are presented in Fig. 5. As the diameter of the porous region area is increased, a saturation point is reached beyond which no further benefit to blood collection volume is obtained. The incremental increase to blood flow from distant radial areas is insignificant compared to the flow from below.

For very small diameters of the end-cap, simulations indicate blood flows into the puncture from below. For larger end-cap diameters, simulations predict incrementally larger flow of blood will be supplied from the perimeter.

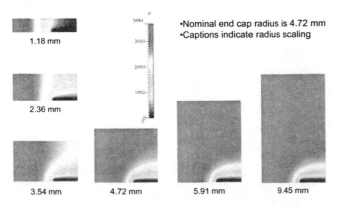

• Nominal end cap radius is 4.72 mm
• Captions indicate radius scaling

1.18 mm
2.36 mm
3.54 mm   4.72 mm   5.91 mm   9.45 mm

Figure 5. Simulated pressure distributions for various end-cap radii (vertical dimension of model bounds).

Fig. 6 illustrates a comparison of the blood volume collected from the simulation versus experimental values. The experimental results correlate with the computational model, and demonstrate an increase in blood volume with end-cap radius. However, it is difficult to quantify these values with a high degree of confidence due to the large standard deviations encountered during the experimental studies. For example, the average blood volumes obtained increased from 0.59 µL to 1.23 µL as the radius was enlarged from 1.52 mm to 2.8 mm, respectively, but with an average standard deviation 0.36. Other researchers have observed such variations when performing similar blood volume studies [13,17]. This is attributed to the vast variability in the human physiology.

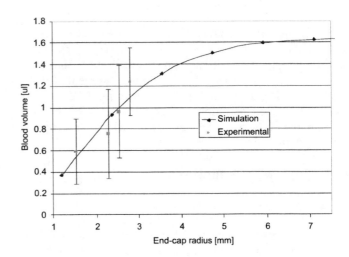

Figure 6. Comparison of blood sample volumes collected from experimental studies versus simulated results generated by the computational model.

5.1 Future work

Future work includes performing simulations where the lancing device is pressed down on the skin around a slit. Such simulations will consist of decoupled sequential elastic and flow simulations where the deformed computational grid from the elastic simulation will be used as a computational grid for the porous media flow simulations. The resulting model will provide a useful template for further investigations. And finally, an experimental coupled-elastic-flow model must be developed. This requires an experimental modification of the CFD-ACE+ solver. A user subroutine must be written to couple the results from elastic simulations with porous media material properties for the flow simulations in a similar sequential analysis procedure. An additional challenge for future work is finding a way of modeling the bulging/stretching of the skin/puncture and how it affects blood volume.

6 SUMMARY

A computational model was created utilizing the CFD-ACE+ software to model blood flow from a lancet puncture of the skin. A two-dimensional axisymmetric model was developed and the skin tissue was modeled as a porous media. The permeability of porous media was adjusted to match blood volume of experimental end-cap. A parametric analysis of the end-cap radius shows that volume of blood collected increases as the opening increases, until a saturation point is reached for end-cap radii greater than 7 mm. A similar trend was found when compared with experimental data taken with the same constraints.

REFERENCES

[1] American Diabetes Association, "Standards of medical care for patients with diabetes mellitus," Diabetes Care 25 (Suppl. 1): S33-S49, 2002.

[2] I. Berlin, J-C. Bisserbe, R. Eiber, N. Balssa, C. Sachon, F. Bosquet and A. Grimaldi, "Phobic Symptoms, Particularly the Fear of Blood and Injury, are Associated with Poor Glycemic Control in Type I Diabetic Adults," Diabetes Care, Vol. 20, 176-178, 1997.

[3] Y. Suzuki, "Painless Blood Sampling for Self Blood Glucose Measurement," Lancet, Vol. 339, 816-817, 1992.

[4] M. B. Carpenter and J. Sutin, "Receptors and Effectors," In: Human Neuroanatomy, Baltimore/London: William & Wilkins, 155-208, 1983.

[5] K. S. Pasyk, S. V. Thomas, C. A. Hassett, G. W. Cherry and R. Faller, "Regional Differences in Capillary Density of the Normal Human Dermis," Plast. Reconstr. Surg. Vol. 83, 939-45, 1989.

[6] S. I. Yum and J. Roe, "Capillary Blood Sampling for Self-Monitoring of Blood Glucose," Diabetes Technol. Ther., Vol. 1, No. 1, 29-37, 1999.

[7] Microlet™ Vaculance™ Lancing Device Product Literature, Bayer Corporation, Elkhart, IN.

[8] AtLast™ Blood Glucose Monitoring System Product Literature, Amira Medical, Scotts Valley, CA.

[9] The MediSense™ SofTact™ Monitor Product Literature, Abbott Laboratories, Bedford, MA.

[10] Ascensia™ CONTOUR™ Blood Glucose Monitoring System Product Literature, Bayer Corporation, Elkhart, IN.

[11] Children with Diabetes web-site, http://www.childrenwithdiabetes.com, 2003.

[12] Bruce R. Simon, Multiphase Poroelastic Finite Element Models for Soft Tissue Structures, pp. 191-218, Appl. Mech. Rev. Vol. 45, No. 6, June 1992.

[13] D. D. Cunningham, T. P. Henning, E. B. Shain, D. F. Young, T. A. Elstrom, E. J. Taylor, S. M. Schroder, P. M. Gatcomb and W. V. Tamborlane, "Vacuum-Assisted Lancing of the Forearm: an Effective and Less Painful Approach to Blood Glucose Monitoring," Diabetes Technol. Ther., Vol. 2, No. 4, 541-548, 2000.

[14] N. Kitson and J.L. Thewalt, "Hypothesis: The epidermal permeability barrier is a porous medium", Acta Derm Venereol 2000, Supp. 208: pp. 12-15.

[15] CFD-ACE+ User Manual, Chapter 8: Porous Media, CFD Research Corporation, Huntsville, AL, 2002.

[16] B. Flora and D. Hesser, "A Preliminary Report of Testing with a Lancing Fixture," Bayer Technical Report No. S-2002-6, Elkhart, IN, 2002.

[17] D. D. Cunningham, T. P. Henning, E. B. Shain, D. F. Young, J. Hannig, E. Barua, and R. C. Lee, "Blood Extraction from Lancet Wound using Vacuum Combined with Skin Stretching," J. Appl. Physiol., Vol. 92, No. 3, 1089-1096, 2002.

Polymeric micropump with SU-8 check valves assembled by lamination technique

T-Q. Truong and N-T. Nguyen

MPE School, Nanyang Technological University, Singapore, quangtt@pmail.ntu.edu.sg

ABSTRACT

This paper presents the first micropumps assembled using polymeric lamination technology. Each pump consists of two 100-μm-thick, 10-mm-diameter SU-8 discs; two 1.5-mm-thick, 15-mm-diameter polymethylmethacrylate (PMMA) discs; and one piezo disc. The SU-8 parts were fabricated by a two-mask polymeric surface micromachining process. The PMMA parts were machined from a PMMA sheet by CO_2-laser. The pump was assembled using adhesive bonding. The adhesive tapes were cut by the same laser system. With a drive voltage of $\pm$ 150 V the fabricated micropumps have been able to provide flow rates up to 2.9 ml/min and back pressures up to 1.6 m of water. The pump design and the polymeric technologies prove the feasibility of making more complex microfluidic systems based on the presented lamination approach.

Keywords: microfluidics, micropumps, SU-8, polymeric laminate technology

1 INTRODUCTION

In recent years, lab on a chips (LOC) has been attracting the interests of both industry and research community. Commercial LOCs are often made of polymers to achieve low cost, good biocompatibility, and good chemical resistance [1]. The transport of fluid in LOCs can be provided by micropumps. Therefore, there is a need for polymeric micropump designs which can easily be integrated in the polymeric fabrication process of LOCs. In [2], we reported such a micropump, which was fabricated with a layered concept. The pump consisted of SU-8 check valves fabricated by polymeric surface micromachining techniques, and polymethylmethacrylate (PMMA) plates machined by milling tools. Four bolts were used to fix the structure.

This paper presents an improved version of the above mentioned pump. First, three new valves with sealing ring have been designed in order to reduce leakage and to lower spring constants. Second, laser micromachining was used to fabricate PMMA plates and adhesive layers. Third, low-temperature adhesive bonding was used for laminating different polymer layers. The lamination technique is flexible and can be extended to more complex microfluidic systems.

2 MICROPUMP DESIGN

2.1 The pump structure

The pump consists of five structural layers: the piezo bimorph disc, the top PMMA plate, two SU-8 discs, and the bottom PMMA plate. These layers are bonded by four adhesive layers, Figure 1. The piezo bimorph disc is a brass plate 15 mm in diameter and 100 μm thick. A 90μm-thick layer of piezo ceramic is glued on top of the brass plate. The piezo bimorph disc works as both the actuator and the pump membrane. The top and bottom PMMA plates are identical. Each plate measures 15 mm in diameter and 1.5 mm thick. It has two through-holes for fluid access, and two other alignment holes. The fluid accesses are 0.6 mm in diameter. The alignment holes have a diameter of 1.6 mm. The space between the top PMMA plate and the piezo disc defines the pump chamber. Two fluid accesses of the bottom PMMA plate are fitted with two metal needles used as inlet and outlet. The two SU-8 discs are identical, 10 mm in diameter and 100 μm thick. The top SU-8 disc contains the outlet valve while the inlet valve is implemented in the bottom disc. Two alignment holes are incorporated in each of the PMMA plates and SU-8 discs.

2.2 Design of the SU-8 check valves

Each SU-8 disc consists of one check valve, one inlet hole and two alignment holes, Figure 2. The check valve consists of a 1-mm valve disc suspended on N folded beams, which have a cross section of 100 μm $\times$ 100 μm. The beams work as valve springs connected in parallel [1]. The valve disc has a sealing ring, which is 100 μm thick. Its outer diameter and inner diameter are 0.9 mm and 0.6 mm, respectively. The ring offsets the 50 μm gap, which was created by the adhesive layer, between the valve disc and the fluid access holes. The ring also creates a pre-loaded condition for the springs to reduce leakage. Figure 3 depicts the three valves implemented in the micropumps reported in this paper. The details on the designs and geometrical parameters are shown in Figure 3.

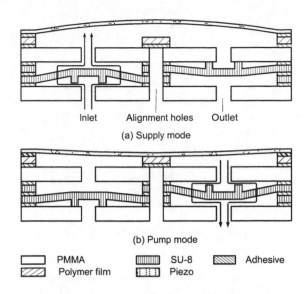

Inlet   Alignment holes   Outlet

(a) Supply mode

(b) Pump mode

☐ PMMA   ▥ SU-8   ▨ Adhesive
▨ Polymer film   ▦ Piezo

Figure 1: Schematic of the pump structure and is operation (not to scale). (a) Supply mode; (b) Pump mode.

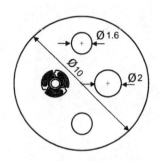

Ø1.6
Ø10
Ø2

Figure 2: The geometry of the SU-8 disc (dimensions are in mm).

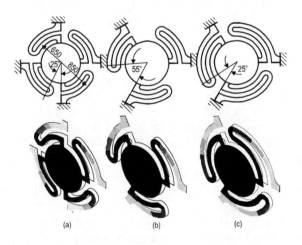

(a)   (b)   (c)

Figure 3: Three different valve designs result in three different spring constants (dimensions are in microns). Results obtained by FEM. (a) Valve 1, 4 springs, k = 927 N/m; (b) Valve 1, 3 springs, k = 691 N/m; (c) Valve 3, 3 extended springs, k = 286 N/m.

Numerical simulation with ANSYS was used to find the spring constants of the valves. A pressure of 3000 Pa was applied on the valve disc. The Young's modulus of SU-8 was assumed to be 4.02×10^9 Pa. The spring constant was determined by the force and the simulated displacement (1).

$$k = \frac{F}{\delta} = \frac{p\pi r^2}{\delta}, \qquad (1)$$

where k is the spring constant, F is the force applied on the valve disc, p is the pressure, r is the radius of the valve disc, and δ is the displacement. The results in Figure 3 show that Valve 3 (c) is three and two times softer than Valve 1 and Valve 2, respectively.

3 FABRICATION

3.1 SU-8 disc

The SU-8 discs were fabricated using a two-mask polymeric surface micromachining process, described in details in [1]. The process started with spin-coating SU-8 2100 photoresist (Microchem Corp., USA) on silicon, Figure 4a. This first SU-8 was then soft baked and exposed to UV light using the first mask containing all features of the SU-8 disc except the ring, Figure 4b. The intended thickness of this SU-8 film was 100 μm. After hard baking, a second 100-μm-thick SU-8 layer was spin-coated on top of the first layer. After a soft-baking step, the second mask containing the ring design was used for the second exposure, Figure 4c. The second mask was aligned to the first SU-8 layer using alignment marks. Even though the first SU-8 layer had not been developed, the alignment marks were still visible. The visible alignment marks were made possible by the different optical indices between cross-linked SU-8 and the unexposed area. With this two-mask process, we achieved an accuracy of 10μm, which satisfied our application. After the second UV exposure, the two SU-8 layers were hard baked and cooled down to room temperature. They were then developed using propylene glycol methyl ether (PGMEA), Figure 4d. In the final steps, the SU-8 discs were released in 30% KOH solution. Etch access created by many circular holes on the discs allows fast under etching. The circular shapes and the smooth designs circumvents the anisotropic characteristics of the KOH etch process. After releasing the SU-8 discs were cleaned with DI (deionized water) water and dried with nitrogen.

Figure 5 shows the SEM (scanning electron microscopy) images of Valve 2 and Valve 3.

3.2 PMMA and adhesive parts

The PMMA plates were designed in Autocad and sent to a laser cutter (Universal Laser Systems, M-300,

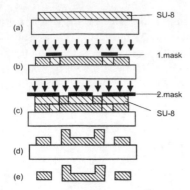

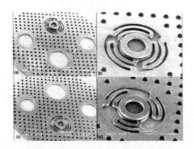

Figure 4: Process steps for the SU-8 valve disc

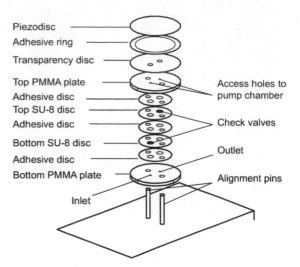

Figure 6: Lamination assembly concept of the micropump.

Figure 5: SEM picture of SU-8 check valves. (a) Top view, whole SU-8 disc, Valve 2; (b) Close-up of Valve 2; (c) Top view, whole SU-8 disc, Valve 3; (d) Close-up of Valve 3

25 W). The system can deliver a maximum laser power of 25 W and a maximum scanning velocity of 640 mm/s at a resolution of 1000 dpi. The power and the cutting speed for the PMMA parts were set at 9% and 1.1% of the maximum values, respectively. The material for the plates was a 1.5-mm-thick PMMA sheet (extrusion type).

The adhesive parts were cut out from a transfer adhesive tape (Adhesives Research, Inc, Arclad 8102 transfer adhesive). The transfer adhesive tape consists of a carrier and a 50-μm-thick adhesive layer. The laser power and the cutting speed for the adhesive tape were set at 1% and 1.1% of the maximum values respectively.

The polymer layer covering the alignment holes was cut out from a 100-μm-thick overhead projector transparency sheet (3M, PP2500) covered by the transfer adhesive tape. The laser power and cutting speed for these layers was set at 3% and 1.1% of the maximum values.

3.3 Assembly procedure

Figure 6 shows the assembly sequences of the micropump. First, the bottom PMMA plate was slotted into the alignment pins; followed by the first adhesive disc with the carrier facing up. The two layers were then pressed to bond. The carrier was then removed using two needles. Next, the bottom SU-8 disc with the sealing ring facing down was placed on the stack. Subsequently, the second adhesive disc was laminated on top

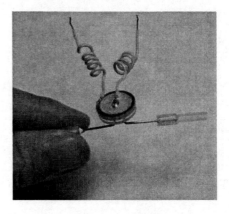

Figure 7: The assembled micropump.

the SU-8 disc; its carrier was then removed. In the next step, the top SU-8 disc with its ring facing up was assembled. The third adhesive later was laminated in the same manner as for the first two. The top PMMA plate was then placed on top of the stack. The whole stack was pressed to achieve good bonding. The bonded stack was then removed from the alignment pins. The carrier of the polymeric disc was removed. In the next step, the polymer layer covering the alignment holes was aligned to and pressed against the top PMMA plate. Next, the adhesive ring with the adhesive layer facing down was placed on top the polymer disc. The carrier of the ring was removed, ready to bond with the piezo disc. The piezo disc was aligned and pressed on the pump stack. Finally, two syringe needles, 0.65 mm in diameter, were attached to the inlet and outlet by epoxy glue (Araldite, Super rapid epoxy). The assembled pump, measuring 15 mm in diameter and 3.8 mm thick, is shown in Figure 7.

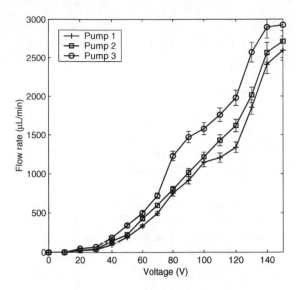

Figure 8: Flow rate versus voltage (frequency=100Hz).

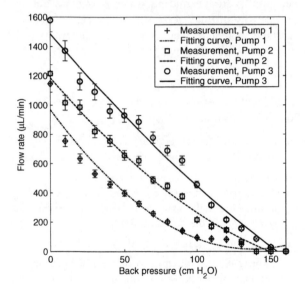

Figure 9: Flow rate versus back pressure (V=100± V, frequency=100Hz).

4   EXPERIMENTAL RESULTS

Three pumps were assembled by the lamination technique described in 3.3. Each micropump was equipped with a micro check valve discussed above. The pumps were characterized using a simple setup described in a previous paper [3]. The working fluid was DI water. The pumps needed to be primed using a syringe before testing. The measurement error is less than 5%. The measured flow rates versus actuating voltages characteristics at the frequency of 100 Hz are shown in Figure 8. Finally, Figure 9 shows the flow rates versus back pressures at ± 100 V and 100 Hz.

The three pumps show significant improvement over the pump reported in [2]. It is evident that the pump's performances are closely related to the spring constants

of the valves. Pump 3 has the best performance due to its lowest spring constant. Pump 3 achieved a flow rate up 2.9 ml/min and back pressures up to 160 cm of water (under a voltage of ± 150 V). That is nearly three times the flow rate (1 ml/min) and 8 times the back pressure (0.2 m) of the pump in [2]. Pump 3 delivers almost 24 times of the pumping power compared to the previous pump. The pumping power is estimated by (2).

$$P_{pump} = \frac{Q_{max} \times p_{max}}{2} \qquad (2)$$

where P_{pump} is the pumping power, p_{max} and Q_{max} are the maximum back pressure and the maximum flow rate, respectively.

The reasons for the improvements are: better alignment between the two valves and of the valves to the inlet and outlet; better seal between the valve disc and the inlet/outlet thanks to the pre-loaded spring; lower spring constant valve design.

5   CONCLUSION

We have designed, fabricated and characterized the first fully polymeric piezo-actuated micropumps assembled by lamination technique. The main advantages of the pumps are: modular design, low cost, simple fabrication and assembly processes. All fabrication and assembly processes, except for SU-8 parts, were carried out outside clean rooms. Moreover, the pump designs are easily integrated into complex microfluidic systems.

We are further improving the pump design to make it self-priming as well as to use thermal actuation force. Self-priming can be achieved by using thinner PMMA plates, which reducing the pump dead volume. Thermal actuator is being experimented; and shows promising results.

ACKNOWLEDGEMENT

This work was supported by the academic research fund of the Ministry of Education Singapore, contract number RG11/02.

REFERENCES

[1] NT. Nguyen, TQ. Truong, KK. Wong, SS. Ho and LN Low, "Micro check valves for integration into polymeric microfluidic devices," J. Micromech. Microeng., 14, 69–75, 2004.

[2] NT. Nguyen and TQ. Truong, "A fully polymeric micropump with piezoelectric actuator," Sensors and Actuators B, 97, 137–43, 2004.

[3] NT. Nguyen and XY. Huang, "Miniature valveless pumps based on printed circuit board technique," Sensors and Actuators A, 88, 104–11, 2001.

Electrokinetic instability mechanism of double-layered miscible fluids with different electrolyte concentrations in DC electric field

Jae W. Park[*], Kwan H. Kang[**] and Kang Y. Huh[***]

[*]Department of Mechanical Engineering,
Pohang University of Science and Technology, San 31, Hyoja-dong, Pohang 790-784, Republic of Korea,
[*]slam@postech.ac.kr, [**]khkang@postech.ac.kr, [***]huh@postech.ac.kr

ABSTRACT

Numerical investigation is performed on the electrokinetic instability mechanism of double-layered miscible fluids in DC electric field. Nonlinear interfacial waves as observed in flow visualization are successfully reproduced. The mechanism is explained on the basis of instantaneous velocity vectors, electric field, net charge distributions as well as concentration distributions. Counterclockwise vortical motion is produced according to the directions of concentration gradient and electric field. A strong vortex at the junction of a T-channel may allow fast mixing in high electric field. Simulation well reproduces the qualitative characteristics of instability, while the threshold instability criterion is underestimated.

Keywords: electrokinetic instability, electroosmotic flow

1 INTRODUCTION

Interfacial instability has recently been observed by Chen and Santiago [1] in electroosmotic flow of two miscible layers with different concentrations in a T-channel. Such instability needs to be suppressed in some micro-fluidic systems requiring stable transport with little species dispersion [2]. On the other hand, it can be utilized to enhance mixing of heterogeneous fluids against viscous dissipation at an extremely low Reynolds number. This instability was first found in a case with no mean flow motion by Hoberg and Melcher [3]. The difference is in the electroosmotic flow, which leads to convective instabilities. It has been suggested that the flow instability originates from the polarization due to concentration gradient in imposed electric field and the stability criterion is mapped for the first time through linear stability analysis [4]. The mechanism of the interfacial electrokinetic instability is elucidated numerically by the Nernst–Planck framework of transport equations in this study.

2 FORMULATION

We consider time-dependent mixing of the electrolyte layers flowing into a two-dimensional micro T-channel. The concentrations of the two inflowing streams are denoted as c_h and c_l, in which the subscripts h and l represent the high and low concentration. The electrolytes are free from chemical reaction. For simple analysis unit activity

coefficient and uniform dielectric constant of the solution are assumed throughout the domain. The unsteady Navier–Stokes equation with Coulombic body force for incompressible fluid is written as

$$\rho \frac{D\mathbf{u}}{dt} = -\nabla p + \mu \nabla^2 \mathbf{u} + \rho_e \mathbf{E} \tag{1}$$

where $D/Dt = \partial/\partial t + \mathbf{u} \cdot \nabla$, $\mathbf{u} = (u, v, w)$ is velocity vector, t is time, p is hydrodynamic pressure, ρ is mass density and μ is viscosity of fluid. $\mathbf{E} = -\nabla \phi$ is electric field and ϕ is electrostatic potential. The electrostatic potential ϕ is related with the charge density (ρ_e) as

$$\nabla^2 \phi = -\frac{\rho_e}{\varepsilon} \tag{2}$$

where ε is electric permittivity. The charge density is a sum of the concentrations of all species, i.e., $\rho_e = F \sum_i z_i c_i$, in which $i = 1$ and 2 respectively denote positive ions (cations) and negative ions (anions). c_i is molar concentration (mol/m^3).

The electroosmotic flow on the wall is characterized by the Helmholtz–Smoluchowsi slip velocity given as $u_w = (\varepsilon \zeta / \mu) E_t$. E_t is the tangential electric field on the wall and ζ is the zeta potential of the wall. Here, we exclude the flow in the electrical double layer while the osmotic velocity is imposed as a boundary condition on the wall ($u_t = u_w = -(\varepsilon \zeta / \mu)\nabla \phi$). Uniform average velocities are assumed at both inlets of the T-channel. Conservation of electrical charge can be written in the absence of chemical reaction as

$$\frac{\partial \rho_e}{\partial t} + \nabla \cdot \mathbf{J} = 0 \tag{3}$$

in which the current density is $\mathbf{J} = F \sum_i z_i \mathbf{N}_i$. Assuming Fickian diffusion, the flux density of each species $\mathbf{N}_i$ is expressed according to the following Nernst–Planck relation:

$$\mathbf{N}_i = D_i \nabla c_i + \frac{z_i F D_i}{RT} c_i \mathbf{E} + c_i \mathbf{u} \tag{4}$$

where D_i is diffusivity of ionic species (m^2/s), z_i is the valence of ionic species and F is the Faraday constant. Here we introduce the electroneutrality assumption as

$$\rho_e = F \sum_i z_i c_i \approx 0 \tag{5}$$

The first term in Eq. (3) is neglected on the basis of this assumption. Equation (5) has been validated for various electrochemical systems both numerically and experimentally. For instance, Riveros [5] showed its validity when $\kappa_l d > 20$, where κ_l^{-1} is the Debye length for the dilute solution. This condition is readily satisfied for the present system of interest, in which $\kappa_l d$ is normally greater than 10^2. Substituting Eq. (4) into Eq. (3), we obtain

$$\nabla \cdot (\sigma \nabla \phi) = -F \sum_i z_i \nabla \cdot (D_i \nabla c_i) \qquad (6)$$

where

$$\sigma = F^2 \sum_i z_i^2 \omega_i c_i \qquad (7)$$

Equation (6) can be rewritten as

$$\nabla^2 \phi = \nabla \sigma \cdot \mathbf{E} - F\left(\sum_i z_i D_i\right) \nabla^2 c \qquad (8)$$

Under the electroneutrality assumption, the transport equations for positive and negative species become

$$\frac{Dc}{Dt} = D_1 \nabla^2 c + \frac{zFD_1}{RT} \nabla \cdot (c \nabla \phi) \qquad (9a)$$

$$\frac{Dc}{Dt} = D_1 \nabla^2 c - \frac{zFD_2}{RT} \nabla \cdot (c \nabla \phi) \qquad (9b)$$

which yield the following convective-diffusion equation for c

$$\frac{Dc}{Dt} = D_{eq} \nabla^2 c \qquad (10)$$

D_{eq} is the equivalent diffusivity defined as

$$D_{eq} = \frac{\omega_1 D_2 + \omega_2 D_1}{\omega_1 + \omega_2} = \frac{2D_1 D_2}{D_1 + D_2}$$

We will solve Eq. (6) rather than the Poisson equation to obtain the charge density. In general it is not a simple problem to solve the transport equations for all species in combination with the Poisson equation. Here with 1:1 symmetric electrolytes, the conductivity in Eq. (7) becomes

$$\sigma = \frac{F^2}{RT}(D_1 + D_2)c \qquad (11)$$

If we replace the conductivity with the concentration by Eq. (11), all the equations may now be written as

$$\rho \frac{D\mathbf{u}}{Dt} = -\nabla p + \mu \nabla^2 \mathbf{u} + \varepsilon(\nabla^2 \phi)\nabla\phi \qquad (12)$$

$$\nabla \cdot (c\nabla\phi) = -\frac{RT}{F}\frac{(D_1 - D_2)}{(D_1 + D_2)}\nabla^2 c \qquad (13)$$

$$\frac{Dc}{Dt} = D_{eq}\nabla^2 c \qquad (14)$$

These governing equations are non-dimensionalized by introducing the following variables:

$$\bar{t} = \frac{t}{t_c}, \quad \bar{x} = \frac{x}{d}, \quad \bar{y} = \frac{y}{d}, \quad \bar{\mathbf{E}} = \frac{\mathbf{E}}{E_o}, \quad \bar{\phi} = \frac{\phi}{E_o d},$$

$$\bar{\mathbf{u}} = \frac{\mathbf{u}}{u_c}, \quad \bar{c} = \frac{c}{c_h - c_l}, \quad \bar{\rho}_e = \frac{\rho_e}{\varepsilon(E_o / d)} \qquad (15)$$

where $t_c = d / u_c$ and the characteristic velocity is taken as the electroosmotic Helmholtz–Smoluchowski slip velocity

($u_c = \varepsilon E_o \zeta / \mu$). Substituting the dimensionless variables in Eqs. (1), (6) and (10), we obtain the following non-dimensional equations. The overbar is dropped for brevity of the notations.

$$\frac{D\mathbf{u}}{Dt} = -\nabla p + \frac{1}{\text{Re}}\nabla^2 \mathbf{u} + \lambda(\nabla^2\phi)\nabla\phi$$

$$\nabla \cdot (c\nabla\phi) = G\nabla^2 c$$

$$\frac{Dc}{Dt} = \frac{1}{Pe}\nabla^2 c$$

The dimensionless parameters are defined as

$$\text{Re} = \frac{u_c d}{\nu} = Pe\frac{D_{eq}}{\nu} = \frac{Pe}{Sc}, \quad Sc = \frac{\nu}{D_{eq}}, \quad Pe = \frac{u_c d}{D_{eq}} = \frac{\varepsilon E_o \zeta d}{\mu D_{eq}}$$

$$G = \left(\frac{D_2 - D_1}{D_1 + D_2}\right)\frac{1}{\beta E_o d} = \frac{\alpha_2 - \alpha_1}{\beta E_o d}, \quad \lambda = \frac{\varepsilon E_o^2}{\rho u_c^2} = \frac{\mu^2}{\rho\varepsilon\zeta^2}$$

where $\beta = ze / kT$, which becomes $39V^{-1}$ for z = 1. κ is the inverse Debye length defined as

$$\kappa^2 = \frac{F^2(c_{h0} - c_{l0})}{\varepsilon RT} \qquad (16)$$

3 RESULT AND DISSCUSSION
3.1 Mechanism of instability

By substituting the Poisson equation into Eq. (6) the charge density is expressed as

$$\frac{\rho_e}{\varepsilon} = -\frac{\nabla\sigma \cdot \mathbf{E}}{\sigma} + \frac{F}{\sigma}\left(\sum_i z_i D_i\right)\nabla^2 c \qquad (17)$$

which describes how free charges are produced. The first term on the right-hand side is the contribution from concentration gradient. The strength of electric field decreases, if the current flows from high conductivity region to the region of low conductivity as in Fig. 1(a). Negative charges are generated to satisfy the Gauss' law in the domain, while the Coulombic force acts to the left in this case. The conductivity gradient is reversed in Fig. 1(b) with the electric field increasing in the current direction. Positive charges are produced to be forced to the right in Fig. 1(b). The second term on the right hand side of Eq. (17) represents the contribution from diffusivity difference of ionic species, which is related with the liquid junction potential. This term is usually much smaller than the first term.

Numerical simulation is performed for the case close to Chen and Santiago [1]. Let us consider the case of aqueous NaCl solution in a microchannel with the width of $d = 150$ μm. The electrolyte concentrations are chosen to be $c_l = 1$ mM and $c_h = 10$ mM, respectively. Then, the zeta potential becomes $\zeta_l = 55.2$ mV and $\zeta_h = 34.2$ mV. The diffusivities of Na$^+$ and Cl$^-$ ions in a dilute solution are 1.33×10^{-9} m^2/s and 2.03×10^{-9} m^2/s, respectively. Then it follows that $D_{eq} = 1.607 \times 10^{-9}$ and $\alpha_1 - \alpha_2 = -0.21$. In the experiment the applied potential difference is from 0.5 kV to 3.0 kV and the channel length is about 2.6 cm. The electric fields E_o in the unperturbed regions of the upper and lower branches are estimated to be from 16,025 V/m to 96,153 V/m according to the applied voltages. In the case of the threshold voltage 0.9 kV, the estimated Helmholtz–Smoluchowski slip velocity for each stream is u_h = 1.15 mm/s and u_l = 0.711 mm/s, thus u_c = 0.93 mm/s, Re = 0.14, Pe = 86.8, G = 1.1×10^{-3}, Sc = 598, and λ = 832.

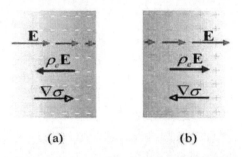

(a)   (b)

FIG. 1 Polarization due to conductivity gradient. Darkness level indicates the concentration.

Figure 2(a) shows the calculated concentration distribution for a typical case. The interface pattern is almost identical with that observed in experiment. Figure 2(b) shows the distribution of charge density together with the concentration distribution in solid lines. Darker regions have dominant positive charges, while brighter regions have dominant negative charges. The Coulombic force is proportional to $\rho_e \mathbf{E} \sim -(1/c)(\nabla c \cdot \mathbf{E}/)\mathbf{E}$. The growth of the interface pattern can be divided into two stages according to the charge production mechanism. In the first stage the species diffusion across the interface plays a crucial role. Molecular diffusion increases the thickness of the mixing layer in the flow direction. The gradient, $\partial c / \partial x$, is nonzero, so that ρ_e becomes nonzero in the thin mixing zone. $\partial c / \partial x > 0$ ($\partial c / \partial x < 0$) at the upper (lower) part of the mixing layer. Accordingly, ρ_e is negative (positive) above (below) the center region. Since the magnitude of ρ_e is proportional to $(1/c)$, the negative ρ_e in the upper part is more apparent in Fig. 2(b). Although not shown here, the vertical component of $\mathbf{E}$ is much smaller that the horizontal component, just after the junction region of the channel. Then, the Coulombic force generates counter-clockwise vortices at the mixing layer with subsequent development of the wavy interface pattern.

The later stage of growth originates from nonlinear deformation of the interface. Once the mixing layer forms a hump (which is convex upward), $\rho_e \sim -\nabla c \cdot \mathbf{E}/c < 0$ at the back of the hump while $\rho_e \sim \nabla c \cdot \mathbf{E}/c < 0$ at the fore of the hump. As a result negative and positive charges are induced periodically at the back and fore of each hump in Fig. 2(b). The flow is either decelerated or accelerated depending on the sign and magnitude of induced charges due to the Coulombic force. Downstream flow velocity is decreased at the back of each hump (P1 to P4), while increased at the

fore region (P5 to P8) in Figure 2(c). Decreasing flow velocity (due to the dragging effect of the Coulombic force) at P1 to P4 primarily contributes to formation of the unstable interface pattern. The spike pattern grows downstream, since the induced charge increases with deformation of the interface. The growth is limited by the stabilizing effect of molecular diffusion and the bounding effect of the channel wall.

Note that the electrokinetically generated vorticity is always in the counterclockwise direction. Consequently, the growing spikes form always in the backward direction, which is consistent for all experimental observations.

3.2 Mixing enhancement in high electric field

Diffusion is the only effective mixing mechanism for low Reynolds number laminar flow in a microchannel. There have been many trials to enhance mixing by complicated flow patterns of conventional passive mixers or external forces of active mixers. For instance, Hau et al. [6] showed that surface charges under steady electric field can drive unstable vortical motion enhancing mixing in electroosmotic flow. Here we consider similar, but much stronger and controllable vortical motion due to concentration gradients in applied electric field.

Figure 3(a) shows visualization in experiment at an applied electric field of 1.47×10^5 V/m. As the interface between two different concentration fluids is retreated to the direction of the lower entrance it seems that the stream of the higher concentration is choked at the lower corner of the junction. The concentration image reveals fast mixing of the two streams within a short distance. Figure 3(b) shows the calculated concentration profiles with instantaneous velocity vectors. It is clear that a strong counterclockwise vortex is located at the center of the T-junction, where the flow from the upper inlet meets that from the lower inlet. To satisfy the continuity relationship, a velocity vector around the vortex should have a considerable y-component yielding strong convective diffusion. The downstream concentration profile confirms fast and effective mixing, although weak molecular diffusion follows further downstream. To augment mixing, this instability mechanism may be combined with pressure-driven flow or the conductivity may be controlled by adding some salt to one of the fluids. This type of mixing mechanism can be beneficial in many practical application, since it does not involve any complex physical structures.

CONCLUDING REMARK

The detailed mechanism of electrokinetic instability for two miscible fluids of different concentrations is clarified from the calculation results of flow velocity, electric field and concentration distributions. It is suggested as an effective strategy for micro mixers to utilize the unstable vortex motion to enhance mixing of heterogeneous fluids with considerable concentration differences. Relevant experimental work is under progress.

REFERENCES

[1] C. H. Chen and J. G. Santiago, "Electrokinetic flow instability in high concentration gradient microflows," Proc. 2002 Int. Mech. Eng. Cong. and Exp., New Orleans, LA, CD vol.1, Paper No. 33563. 2002.

[2] S. C. Jacobson and J. M. Ramsey, "Electrokinetic focusing in microfabricated channel structures," Analytical Chemistry, vol. 69, No. 16, pp. 3212-3217, 1997.

[3] J. F. Hoberg and J. R. Melcher, "Internal electrohydrodynamic instability and mixing of fluids with orthogonal field and conductivity gradients," J. Fluid Mech. Vol. 73, pp. 333-351, 1976.

[4] C. H. Chen, H. Lin, S. K Lele and J. G. Santiago, "Electrokinetic microflow instability with conductivity gradients," 7th int. conf. on μTAS, Oct 5-9, Squaw Valley, California, USA, pp. 983-987, 2003.

[5] O. J. Riveros, "Numerical solutions for liquid-junction potentials," J. Phys. Chem., Vol. 96, pp. 6001-6004, 1992.

[6] W. L. W. Hau, L. M. Lee, Y.K. Lee and Y. Zohar, "Electrokinetically-driven vertical motion for mixing of liquids in a microchannel," 7th Int. Conf. on μTAS, Oct 5-9, Squaw Valley, California, USA, pp. 491-494, 2003.

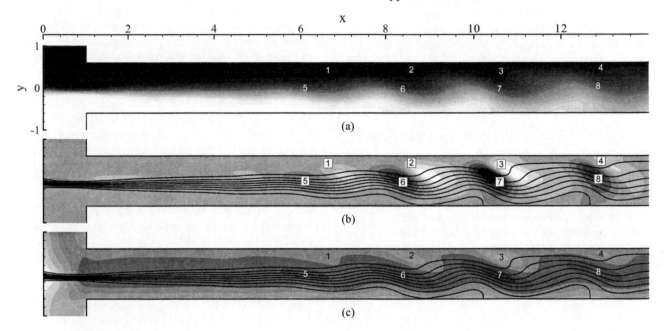

FIG. 2 Numerical results at an instant for $\lambda = 130$. (a) Concentration distribution. (b) Charge density with concentration distribution. (c) Total velocity (darkness level is proportional to velocity magnitude) and iso-concentration lines (line).

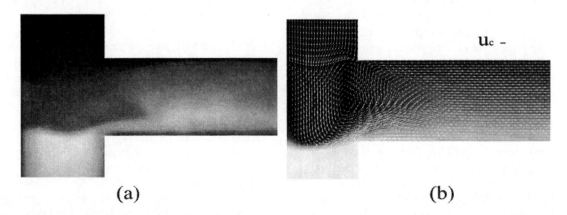

FIG. 3 Concentration profiles in case of high electro static field, (a) experimental result in case of 2.5 kV applied voltage; photo courtesy of S. M. Shin, (b) numerical result with instantaneous velocity vectors in case of λ=200.

Flow Simulation of a Microfluidic Analog of the Four-Roll Mill

F. R. Phelan Jr. and S. D. Hudson
Polymers Division
National Institute of Standards and Technology
Gaithersburg, MD 20899

ABSTRACT[1]

The fluid dynamics of channel geometries for liquid state materials characterization in microfluidic devices are investigated using finite element flow simulation and flow classification criteria. A pressure driven microchannel device is sought that has an adjustable flow type, approximating the function of the four-roll mill. Channel flow geometries are investigated in which the full range of linear flows (extension, shear and rotation) can be approximated in the neighborhood surrounding a stagnation point. A class of flow geometries is identified which makes use of opposing, laterally offset fluid streams that produce a stagnation point in the center of the geometry.

Keywords: microfluidics, four-roll mill, flow classification, finite element analysis, linear flow

1 INTRODUCTION

Fluid flow in microfluidic devices is becoming more and more common, driven by the miniaturization of many technologies [1-3]. Such devices afford a number of advantages over conventional methods, including greater speed, replacement of multiple manual steps, conservation of expensive and/or limited quantity reagents, and reduced human error.

The goal in this study is to develop a microfluidic device for liquid state materials characterization. This goal requires first and foremost, a flow geometry capable of generating flows suitable for materials characterization. The ideal types of flow for the characterization of fluids are linear flows which are described by the stream function and corresponding velocity vector

$$\psi = \frac{1}{2}\left(\xi\gamma x^2 - \gamma y^2\right) \qquad (1.1)$$

$$\underline{v} = \left(\gamma y, \xi\gamma x\right) \qquad (1.2)$$

The nature of these kinematics are characterized by the constants ξ and γ[4]. The quantity ξ is the flow type, and indicates the nature of the deformation experience by the fluid; the flow is extensional for $\xi = 1$, shear for $\xi = 0$, and pure rigid body rotation for $\xi = -1$. The quantity γ is the flow strength, and indicates the magnitude of the velocity gradient relative to the flow type.

In a linear flow, the velocity gradient tensor is constant and the flow type is uniform, which makes measurement of flow properties as a function of deformation rate unambiguous. In addition, linear flows contain a stagnation point at the origin point (0,0). The stagnation point serves a number of beneficial purposes in the present context. First, the material residence time in the neighborhood of the stagnation point is high, giving the material adequate time to respond to the applied deformation. Residence time is in effect, a third type of flow classification criteria [4]. In addition, the long residence time also provides a stable environment for material observation and measurement, and it allows one to ignore the inhomogeneities present in a complex flow, provided that the stagnation region is reasonably broad, and the flow gradients are suitably uniform.

A number of devices that approximate particular modes of linear flows at the macroscale are available [5-14]. The device that comes closest to the desired functionality is the four roll mill. While this is primarily thought of as a device for producing planar extension ($\xi = 1$), it can also be used to provide mixed flows with $0.2 < \xi < 1$ [7-8], and rotational flows where $\xi \sim -1$ [9]. However, the four roll geometry cannot produce simple shear ($\xi = 0$), and difficulties can be anticipated in trying to produce and control a miniature device with moving parts. This leads us to seek a solution based on channel flow.

In this study, a pressure driven microchannel device is sought that has an adjustable flow type, approximating the function of a four roll mill. In particular, we investigate classes of channel flows in which the full range of linear flows can be approximated in the neighborhood surrounding a stagnation point. To evaluate candidate geometries, finite element flow simulations in the zero Reynolds number limit were undertaken. Flow classification criteria [4] were then used to delineate combinations of geometry and boundary conditions for which the flow type can be adjusted between shear and extension in the neighborhood surrounding the stagnation point, while providing adequate flow strength. A class of flow geometries is identified which makes use of opposing, laterally offset fluid streams. Opposing fluid streams are

---

[1] Official contribution of the National Institute of Standards and Technology; not subject to copyright in the United States.

necessary to produce a stagnation point. Lateral offset is necessary in order to produce shear.

2 MODELING

2.1 Flow Simulation

To evaluate candidate geometries, flow was simulated using the finite element method. All flows were assumed to be steady, two-dimensional, and at the zero Reynolds number limit. Modeling was carried out using the commercial finite element packages[2] FIDAP [15] and FlexPDE [16]. The governing equations solved by FIDAP are the Navier-Stokes equations with the incompressible condition

$$-\nabla p + \eta \nabla^2 \underline{u} = 0 \tag{1.3}$$

$$\nabla \cdot \underline{u} = 0 \tag{1.4}$$

FlexPDE makes use of the alternative incompressible formulation

$$\nabla^2 p = -\nabla \cdot \nabla \cdot \left(\rho \underline{u}\underline{u} \right) + \beta(\nabla \cdot \underline{u}) \tag{1.5}$$

where β is a large value chosen such that $\nabla \cdot \underline{u}$ is sufficiently small.

2.2 Flow Classification

A suitable objective criterion for classifying the flow type in complex flows has been developed by Astarita [4]. For steady two-dimensional flows, where $\underline{u} = [u, v]$ and $\underline{\underline{D}} = \begin{bmatrix} \varepsilon & \gamma \\ \gamma & -\varepsilon \end{bmatrix}$, the flow type Ξ is given by the expression

$$\Xi = \frac{\varepsilon^2 + \gamma^2 - W^2}{\varepsilon^2 + \gamma^2 + W^2} \tag{1.6}$$

where $W = \omega - \Omega$, ω is the vorticity and

$$\Omega = \frac{\gamma\left(u\frac{\partial \varepsilon}{\partial x} + v\frac{\partial \varepsilon}{\partial y} \right) - \varepsilon\left(u\frac{\partial \gamma}{\partial x} + v\frac{\partial \gamma}{\partial y} \right)}{2\sqrt{\varepsilon^2 + \gamma^2}} \tag{1.7}$$

Like the constant ξ in linear flows, the variable Ξ has a value equal to 1 in purely extensional flows, 0 in shear flows, and -1 in a pure rigid body rotation.

For classifying flow strength, a relative flow strength can be defined as

$$S_r = \frac{\lambda}{\lambda_{ref}} \tag{1.8}$$

where $\lambda = \sqrt{\varepsilon^2 + \gamma^2}$ is an absolute measure of flow strength equal to the maximum eigenvalue of the stretching tensor,

---

[2] Identification of a commercial product is made only to facilitate reproducibility and to adequately describe procedure. In no case does it imply endorsement by NIST or imply that it is necessarily the best product for the procedure.

and λ_{ref} is the flow strength at a reference point in the flow. In what follows, we classify different flow fields using Eqs. (1.6) and (1.8), taking the reference flow strength to be an average flow strength at the inlet or outlet of a channel geometry. This enables comparison of the strength of deformation in the stagnation region with the deformation in the pressure driven channels that drive the flow.

3 SIMULATION RESULTS

In determining the parameters of a flow device that could be used to generate a range of approximately linear flows in the region surrounding a stagnation point, it was recognized that creating a flow with a stagnation point requires the use of oppositely directed fluid streams. While this is generally a means to produce extensional flow, by creating lateral offset between the opposing streams, shear and even rotation can be produced as well. We make use of this principle in our investigation.

The geometry together with boundary conditions studied here is depicted in Figure 1. Overall the geometry represents a modification of the basic cross flow device. Two important modifications are evident. First, the flow channels in the vertical direction are asymmetrically offset from the center of the channel on the top and bottom of the geometry by distances of $\pm H_2/2$, respectively. Second, the horizontal channels are split by a spacer plate of thickness h_{fin}. This provides two extra flow channels in the horizontal direction. The upper and lower spacer plates are offset horizontally by a distance d_{off}.

The flow geometry has three different boundary pressures, P_{in}, P_1, and P_2. The pressure P_{in} represents a pressure that remained fixed during a simulation, while the values of P_1 and P_2 represent pressures that were allowed to vary. Simulation results showed that the flow type at the stagnation point in this geometry is solely a function of the ratio $P_{rat} = \dfrac{\Delta P_1}{\Delta P_2}$, where $\Delta P_i = P_{in} - P_i$. By specifying values for P_{in}, P_{rat} and $P_{mag} = \Delta P_2$, values for P_1, and P_2 were set as $P_1 = P_{in} - P_{mag} \cdot P_{rat}$ and $P_2 = P_{in} - P_{mag}$. The relevant geometric and boundary condition parameters used in the simulations to be presented are listed in Table 1.

Plots of flow type and relative flow strength at the stagnation point as a function of P_{rat} are shown in Figures 2-3. Figure 2 shows that by cycling through values of P_{rat}, the entire range of flow types can be reached. For highly negative pressure ratios the flow type has a value that is negative, indicating a high degree of rotation. As P_{rat} decreases towards zero, the amount of rotation is continually reduced and the extensional flow type value of unity (extension) is reached at a pressure ratio of approximately -1. As the pressure ratio further increases and becomes positive, a shear flow type of zero at the stagnation point is eventually realized for a pressure ratio value of approximately 2. The relative flow strength depicted in Figure 3 is taken relative to the average flow strength at the inlet of the flow. The plot shows the flow strength continually rises (almost linearly) with increase in the pressure ratio. Thus, the flow strength value is relatively higher for shear (0.6), than for extension (0.23). Values are tabulated in Table 2.

4 DISCUSSION

In developing a microfluidic device for liquid state materials characterization, the most basic need is to have a suitable flow geometry for deforming the fluid in a controlled manner. We have sought to investigate classes of channel flows in which the full range of linear flows can be approximated in the neighborhood surrounding a stagnation point. Using a combination of finite element simulation and flow classification theory, a class of flow geometries that fit the above criteria has been determined. The full range of linear flow kinematics can be produced at the stagnation point, with varying, but in all cases acceptable degree of flow strength.

In the results presented, complete optimization of the flow geometry parameters has not been attempted as our primary goal in this initial study was simply to identify geometry and boundary condition combinations that allow us to achieve the basic goal of generating a full range of flow types. This, in coordination with experimental testing, is part of our ongoing work. There are a number of other items that will need to be examined in future work. First, our initial results presented here have considered the system to be 2-D in nature, as is done for example with the four roll mill. However, shear in the third direction is likely to have some effect and will be taken into account by means of 3-D flow simulation. In addition, non-Newtonian flow behavior will have to be addressed when working with polymer systems.

5 SUMMARY AND CONCLUSIONS

In this study, the fluid dynamics of pressure driven flow in microchannels was studied for the purpose of designing a liquid state, materials characterization device. A class of flow geometries was identified in which a full range of linear flows could be approximated in the neighborhood surrounding a stagnation point. The flow geometry that was identified makes use of opposing, laterally offset fluid streams. The opposing fluid streams are necessary to produce a stagnation point, and the lateral offset is necessary

in order to introduce rotation into the flow to produce shear. Geometry optimization in coordination with experimental testing is underway.

REFERENCES

1. Thorsen, T., S.J. Maerkl, and S.R. Quake, "Microfluidic Large-Scale Integration", Science, 298, p.580, (2002).
2. Ouellette, J., "A New Wave of Microfluidic Devices", The Industrial Physicist, p. 14, (August/September 2003).
3. Chow, A.W., "Lab-on-a-Chip: Opportunities for Chemical Engineering", AICHE Journal, 48(8), p. 1590, (2002).
4. Astarita, G., "Objective and generally applicable criteria for flow classification", Journal of Non-Newtonian Fluid Mechanics, 6, p. 69-76, (1979).
5. Taylor, G.I., "The formation of emulsions in definable fields of flow", Proc. Roy. Soc. London Ser. A, 146, p. 501, (1934).
6. Fuller, G.G. and L.G. Leal, "Flow birefringence of concentrated polymer solutions", J. Poly. Sci., Poly. Phys., 13, p. 43, (1981).
7. Bentley, B.J., and L.G. Leal, "A computer controlled four-roll mill for investigations of particle and rope dynamics in tow-dimensional linear shear flows", J. Fluid Mech., 167, p. 219, (1986).
8. Bentley, B.J., and L.G. Leal, "An experimental investigation of drop deformation and breakup in steady, two-dimensional flows", J. Fluid Mech., 167, p. 241, (1986).
9. Higdon, J.L.L., "The Kinematics of the Four Roll Mill," Phys. Fluids A, 5(1), p. 274, (1993).
10. Frank, F.C. and M.R. Mackley, "Localized flow birefringence of polyethylene oxide solutions in a two roll mill", Journal of Polymer Science A2, 14,1121-1131 (1976).
11. Hills, C.P., "Flow Patterns in a Two-Roll Mill", The Quarterly Journal of Mechanics and Applied Mathematics, 55(2), p. 273, (May 2002).
12. Berry, M.V. and M.R. Mackley, "The six roll mill: Unfolding an unstable persistently extensional flow," Phil. Trans. Royal Soc. (Lond.), 287, 1337, 1-16 (1977).
13. Keller, A. and J.A. Odell, Colloid Poly. Sci., 263, p. 181, (1985).
14. Bird, B., R.A. Armstrong, and O. Hassager, Dynamics of Polymeric Liquids, Volume 1, Wiley, (1982).
15. Fidap User's Manual, Fluent, Inc., Evanston, IL, (2003).
16. FlexPDE User's Manual, PDE Solutions, Inc. (2003).

Inlet Width (H$_{in}$)	Inlet/Outlet Width 1 (H$_1$)	Inlet/Outlet Width 2 (H$_2$)	Fin Width (H$_{fin}$)	Channel Length (L)	Offset Distance	Pressure Magnitude (P$_{mag}$)	Inlet Pressure (P$_{in}$)
2	1	2	0.05	5	0.5	1.0	0

Table 1. Geometry parameters and boundary conditions used for the parallel, laterally offset, channel geometry, with transverse flow channels. All units are relative.

Case	Pressure Ratio (P$_{rat}$)	Pressure 1 (P$_1$)	Pressure 2 (P$_2$)	Flow Type	Relative Flow Strength
Extension	-0.9	0.9	-1	0.9995	0.233
Shear	2.2	2.2	-1	-0.0044	0.596

Table 2. Values for flow type and flow strength at the stagnation point for the parallel, laterally offset, channel geometry, with transverse flow channels, at offset distances mostly closely corresponding to the cases of extension and shear.

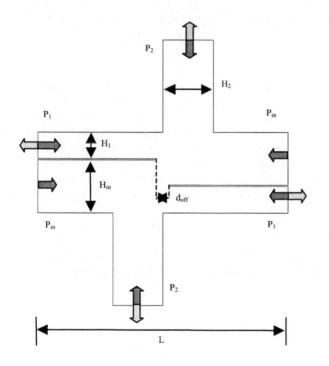

Figure 1. Flow geometry, parameters and boundary conditions. Geometry values used in the simulations are listed in Table 1.

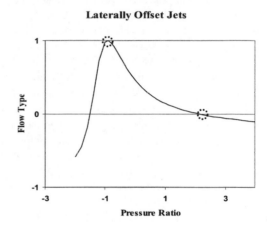

Figure 2. Flow type as a function of the pressure ratio.

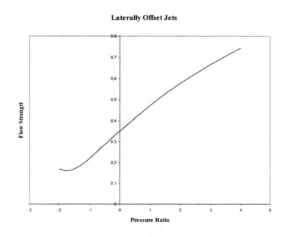

Figure 3. Relative flow strength as a function of pressure ratio.

SOI Processing of a Ring Electrokinetic Chaotic Micromixer

Y. T. Zhang[1], H. Chen[2], I. Mezic[1], C. D.Meinhart[1], L. Petzold[1, 2] and N.C. MacDonald[1, 3]

1. Department of Mechanical & Environmental Engineering, zhyt@engineering.ucsb.edu
2. Department of Computer Science
3. Materials Department
University of California, Santa Barbara, Eng.II, Room 2145, Santa Barbara, CA, 93106,USA

ABSTRACT

Micromixing has been an active research area in the past decade due to the rapid expanding application of the lab-on-a-chip system in life science. Using the silicon bulk micromachining technology, a ring electroosmotic micromixer, which uses a novel arrangement of electrodes and flow obstacles to induce chaotic mixing, has been designed and fabricated. The design idea uses heavily doped silicon as the fabrication material, and SOI wafer to ensure the electrical isolation of the device from the substrate. The isolation between the electrode and the non-conducting structures is achieved by thermally growing SiO_2 to fill the gaps between the two areas. The SOI processing and the later Focused Ion Beam (FIB) modification method produces integrated electrodes to the full depth of the microfluidic system. Other components of the lab-on-a-chip system can be fabricated following the same process flow. A 25-micron wide, 50-micron deep mixer has been fabricated and assembled for later testing.

Keywords: Silicon bulk micromachining, SOI, FIB, chaotic mixer, electroosmotic flow

1 INTRODUCTION

The technology of micro total analysis systems, or "lab-on-a-chip", has achieved a rapid development in the past ten years due to its wide application in life science and chemistry, such as DNA hybridization, PCR diagnostics, and drug discovery [1-4]. As an important component of the lab-on-a-chip system, micromixing has been given significant attention since most of the reagents need to be mixed with other reagents before the final analysis stage.

Most of the micromixers in the literature can be categorized as either passive mixers that use geometrical stirring [5-8], or active mixers that use movable parts or external forces such as pressure or electrical field [9-14] to achieve mixing. Current passive mixers still require mixing channels of considerable length, and the complicated asymmetric structure makes CFD simulation a challenging task. Active mixing is another approach for rapid mixing with either pressure disturbance [9], electrokinetic [11], magnetic [12] actuation or acoustic vibration [13].

The ring electroosmotic micromixer, which we present here, is a new type of electrokinetic micromixer. Four symmetrically located microelectrodes integrated at the wall of the central circular loop are used to apply a sinusoidal or other waveform electrical field to the fluid, thus inducing an oscillating electromosmotic flow. A two-dimensional FEMLAB simulation is performed to illustrate the exponential separation of the two neighboring particles , providing a proof of the chaotic nature of mixing.

2 PHYSICAL MODEL & DEVICE DESIGNING

The basic idea is to use electroosmotic force, which functions efficiently near a surface in a ring chamber, to induce chaotic mixing. Necessary conditions for chaotic motion require that (a) the system has three independent dynamic variables, and (b) the equations of motion contain a nonlinear term that couples several of the variables [15]. Although the Stokes equation for low Reynolds number flow is linear, the governing equation for the motion of the particle trajectories is nonlinear. Chaotic mixing can be achieved in laminar flows [16, 17]. Chaotic mixing in microchannels has been studied in a number of papers (Refer to the special issue of Proc. Roy. Soc. on microscale transport), such as using a 3-D serpentine pipe [5], side-channel pressure disturbance [9] [14], and ridged-floor with staggered herringbone stirring [7].

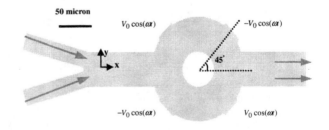

Fig. 1: Schematic of the new mixer with four symmetric electrodes equally distributed at the wall of the mixing chamber. Time-dependent electrical signals are given as shown in the picture.

A similar design for a pressure-driven micromixer appeared in [18]. However, the reported mixing rate was algebraic, compared to the exponential mixing rates achieved here. As shown in Fig. 1, two different fluid solutions from two individual reservoirs flow parallel to each other, then merge into one channel under a steady pressure gradient, and then enter the mixing chamber. Inside the mixing chamber, four electrodes are symmetrically positioned at the wall of the circular loop, so that a time-periodic electrical force can be applied transversally to the main stream. The middle circular island is designed to limit the transverse mixing length as the inlet mainstream transverse length.

Once the reagents enter the mixing chamber, the electrical double layer (EDL) or Debye layer is formed within a few nanometers from the SiO_2 surface, due to the spontaneous separation of charge generated at a liquid/solid interface. Net charge inside this EDL will migrate due to the Coulomb force. This leads to a flow profile that is a plug flow away from the Debye Layer. Voltage is applied as shown in Fig. 1, causing the complex instantaneous streamline pattern as shown in Fig. 2. The voltage is oscillated in time to provide breaks of separating surfaces in the streamline path. Continuing cycling of the voltage, flow at the rear of the mixer is separated into smaller fluid segments, thereby mixing the fluids.

3 NUMERICAL SIMULATION

The physical intuition, which led to the design of this device, has been verified by numerical simulation using the multiphysics modeling software FEMLAB (Comsol Inc.). The quasi-static electrical field is solved by the conductive media DC model with time-varying electrical potential at the 6-micron concave electrodes surface. The incompressible Navier-Stokes model is probed to determine the fluid velocity field. The two models are coupled together with slip boundary conditions [19] at the non-electrode area of the inner channel wall

$$ u_{slip} = -\frac{\varepsilon\zeta Ex}{\mu}, \qquad v_{slip} = -\frac{\varepsilon\zeta E_y}{\mu} $$

where ε is the permittivity of the liquid medium, μ is the dynamic viscosity, ζ is the local zeta potential with a typical value of $-0.1\,V$, E_x and E_y are the x, y components of the electrical field independently.

Snapshots of the electrical field and the fluid streamlines are given in Fig. 2 with maximum electrical potential on the electrodes surface of 1Volt. The medium fluid is water with conductivity 0.11845 S/m and relative permittivity 80.2. The inlet velocity is given as 0.1mm/s, and the mainstream channel width is 50 microns. Fig. 2 illustrates that the flow splits into domains with eddy rotation separated by separating streamlines. The time-dependent actuation causes the separatrix breakup and the associated chaotic advection. More detailed simulation analysis can be found in [20].

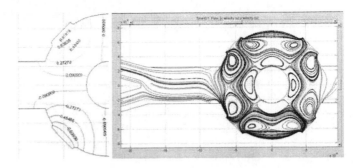

Fig. 2: Electrical potential contours (left) and fluid streamlines (right) distribution by 2-D FEMLAB simulation at time equals to 0.1s with v_0 1V and frequency 10Hz.

4 FABRICATION

RECM is a test device for micro/nano scale computational infrastructure. To efficiently compare 2-D simulation with experiments, it is required that the aspect ratio (channel depth over channel width) of the experimental device be as large as possible in order to minimize the floor effect on the flow profiles.

4.1 Principle

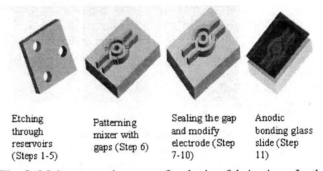

Etching through reservoirs (Steps 1-5)

Patterning mixer with gaps (Step 6)

Sealing the gap and modify electrode (Step 7-10)

Anodic bonding glass slide (Step 11)

Fig. 3: Main processing steps for device fabrication. In the processing, three reservoirs are etched-through after the first layer photolithography. Then the mixer and electrode networks are patterned and etched as the second layer.

Usually, metal sputtering or evaporating is used to deposit metal on the silicon chip as an electrode. But because of the difficulty in sidewall deposition (50 micron deep for our mixer) using current surface-based microfabrication technology, this method does not provide an electrode with a sufficiently high aspect ratio. To solve this problem, we use heavily boron-doped silicon with resistivity of 0.02ohm·cm (Ultrasil Corp.) as the fabrication material, and SOI (Silicon On Insulator) wafer to obtain isolation between the device and the substrate bulk material. The isolation between the electrode and the non-conducting structures is achieved by thermally growing SiO_2 to fill the gaps between the two areas. The 3 micron wide, 50 micron deep gaps are formed after the second photolithography.

4.2 SOI Processing

A 25-micron wide, 50-micron deep mixer (see Fig. 4) has been fabricated at our new MEMS/NEMS cleanroom and UCSB's nanotech research cleanroom. We also designed another two sizes (1 micron and 50 micron chamber size) to investigate length-scale dependence. Two photolithography steps are required to fabricate the mixer. One mask is used to create the fluid reservoirs and the second one is for patterning the mixer.

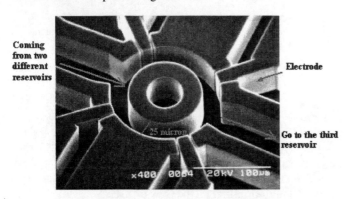

Fig. 4: SEM picture of the ring mixer (25micron wide, 50 micron deep) at the step 8 (before FIB characterization).

First, a 7.5-micron thick photoresist is spun (SPR220-7.0 spin at 3000 rpm for 30 seconds) onto the 4-inch SOI wafer (Step 1). This step is followed by photolithography patterning three reservoirs on the wafer by a GCA 6300 i-line wafer stepper (Step 2). Using a Bosch etcher (Plasma-Therm 770 SLR), the wafer is etched down 50 microns to expose the buried one-micron SiO2 layer (Step 3).

Then the buried SiO_2 layer in the reservoir area is removed. Li. et al [21] reported a 7 min dipping in saturated HF (49%) to remove the buried oxide. We developed a safer technology. After 15 seconds plasma O_2 cleaning, the wafer is immersed in improved buffer HF for 10 minutes, and then the procedure is repeated. RIE etching (O2 and CHF3) follows to totally strip the buried SiO2 layer inside the reservoir area (Step 4). Next, DRIE is performed to etch through the reservoirs (Step 5). By this step the first layer is done.

For the second layer, we use AZ-P4110 positive photoresist (spin at 4000 rpm for 30s, 1.1 micron in thickness). The pattern of the mixer (including mixer chamber, channels, integrated electrodes and contact pad) is defined (Step 6) and etched with DRIE down to 50 microns (Step 7). After cleaning (Nanostrip at 60 $^{\circ}C$ for 20 minutes), the wafer is loaded into the furnace (Tystar 8" oxidation furnace) to grow a 6-micron thick SiO2 layer (wet oxidation, 1050 $^{\circ}C$ for 96 hours) to enclose the gap between the electrode and non-electrodes area (Step 8). Although the gap is only 3 micron wide, since every 1micron growth of SiO2 consumes 0.44 micron silicon, we need to grow 6 microns of SiO2 to close the gap. RIE (Panasonic E640 etcher), at the etch rate of 2 microns per

10 minutes, is used to remove the surface oxides so that the silicon core of the contact pad is exposed for later probe testing (Step 9).

4.3 FIB Characterization

The FEI DB235 Dual-Beam Focus Ion Beam System is used for image and ion milling.

One measurement is to check the cross sectioning of the silicon-silicon dioxide isolation segment. The left image (Fig. 5) shows that the two silicon dioxide interfaces merge and close the gap after milling away half of the mixing chamber wall at the interface area.

Ion milling is used to modify the electrode surfaces. The design of the RECM requires that the electrodes must contact the fluid, and the remaining parts of the mixer are thermal oxide to achieve a higher slip velocity. We used a FIB with an ion source of $20\,nA$ for 3 minutes to selectively remove silicon dioxide from the electrode surfaces in contact with the fluid (Step 10), as shown in the right image of Fig. 5. Step 10 can be substituted by using a third mask to etch away SiO_2 on the electrodes sidewall.

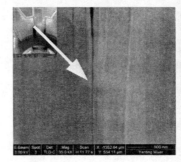 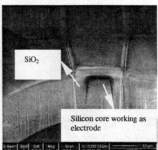

Fig. 5: Left: FIB characterization of the sidewall sealing. Right: FIB milled away 6-micron inner-sidewall silicon dioxide to expose the electrode to fluid.

4.4 Anodic Bonding

A 6 mm square, 170 micron thick Pyrex 7740 (Precision Glass & Optics) is anodically bonded to the chip by a self-made bonding apparatus in the air at 385$^{\circ}C$ (silicon and glass interface temperature) and –650V for 10 minutes (Step 11). Since our cover slip is quite thin, contrary to the widely believed 1000V voltage, our data shows that as low as 600V enables bonding. A high voltage can easily generate an electrical arc in air and thus damage the sample. Temperature is another key parameter. Bonding a SOI wafer requires a higher temperature than that used to bond a silicon wafer to the Pyrex glass slip when other parameters are kept the same.

5 MIXER TESTING SET-UP

A schematic diagram of the mixer testing is shown in Fig. 6. It is composed of three main parts: fluid supply system, electrical field supply system and video capture system. Three-micron diameter fluorescent polymer microspheres (Duke Scientific Corp.) are used for micro PIV analysis. Testing results will be presented in future publications.

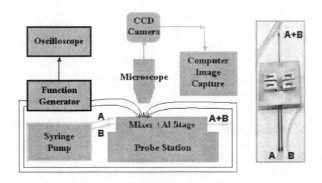

Fig. 6: Left: Schematic picture of the testing setup Right: Mixer on the Aluminum Stage with connecting tubings.

6 CONCLUSIONS

We designed and fabricated a new ring electrokinetic chaotic mixer (RECM) with 3-D integrated electrodes on a heavily doped single crystal silicon wafer. Simulation results indicate that chaotic mixing is possible with this design. The SOI processing and later FIB characterization provides high aspect ratio electrodes that can be fabricated to the full depth of the microfluidic system. Other components of the lab-on-a-chip system can be fabricated following the same process flow.

ACKNOWLEDGEMENTS

This work was supported under a grant from ITR/NSF (#ACI-0086061). The authors would like to thank Dr. Frederic Bottausci, Dr. Brian Thibeault and Dr. Jan P. Löfvander for technical support.

REFERENCES

[1] Manz, A., Graber, N., Widmer, H. M., "Miniatured total chemical analysis systems: a novel concept for chemical sensing", Sensors & Actuators B-Chemical, B1, 244–248,1990, Switzerland.
[2] Figeys, D. and Pinto, D., "Lab-on-a-chip: a revolution in biological and medical sciences", Analytical Chemistry-A page, 72(9), 330A-335A, 2000.
[3] Kopp, M. U., deMello, A. J., Manz, A., "Chemical amplification: continuous-flow PCR on a chip", Science, 280, 1046–1048,1998.
[4] Chow, A., Kopf-Sill A., Nikiforov T., Zhou, A., Coffin J., Wada, G., Alajoki, L., Spaid M., Yurkovetsky Y., Sunberg S. and Parce J.W., "High-throughput screening on microchips", Proceedings of the μ TAS 2000 Symposium, Enschede, the Netherlands, May, 489-492, 2000.

[5] Liu R.H., Stremler M.A., Sharp K.V., Olsen M.G., Santiago J.G., Adrian R.J., Aref H. and Beebe D.J., " Passive mixing in a three-dimensional serpentine microchannel", Journal of Microelectromechanical Systems, 9(2), 190-197,2000.
[6] Gobby, D., Angeli, P. and Gavriilidis, A., "Mixing characteristics of T-type microfluidic mixers", Journal of Micromechanics & Microengineering, 11(2), 126-132,2001.
[7] Stroock, A.D., Dertinger, S.K.W., Ajdari, A., Mezic, I., Stone, H.A. and Whitesides, G.M., "Chaotic mixer for microchannels", Science, 295, 647-651,2002.
[8] Jen, C.P., Wu, C.Y., Lin, Y.C. and Wu C.Y., "Design and simulation of the micromixer with chaotic advection in twisted microchannels", Lab on a Chip, 3, 77-81,2003
[9] Volpert, M., Meinhart, C.D., Mezic, I. and Dahelh, M., "An actively controlled micromixer", Proceedings of MEMS ASME IMECE, Nashville, Tennessee, November, 283-287,1999.
[10] Lee, Y.K., Deval, J., Tabeling P., Ho, C.M., "Chaotic mixing in electrokinetically and pressure driven micro flows", Proceedings of 14th IEEE Workshop on MEMS, Interlaken, Switzerland, January, 483-486, 2001.
[11] Oddy, M.H., Santiago J.G., and Mikkelsen J. C., "Electrokinetic instability micromixing", Analytical Chemistry, 73(24), 5822-5832,2001.
[12] Lu, L.H., Ryu, K.S. and Liu, C., "A magnetic microstirrer and array for microfluidic mixing", Journal of Microelectromechanical Systems, 11(5), 462-469,2002.
[13] Yang, Z., Goto, H., Matsumoto, M., and Maeda, R., "Active micromixer for microfluidic systems using lead-zirconate-titanate (PZT)-generated ultrasonic vibration", Electrophoresis, 21(1), 116-119, 2000.
[14] Niu, X. and Lee, Y. K. "Efficient spatial-temporal chaotic mixing in microchannels", Journal of Micromechanics and Microengineering, 13(3), 454–462,2003.
[15] Backer, G.L. and Gollub, J.P., "Chaotic Dynamics: an introduction", 2nd edition, Cambridge University Press, 1996.
[16] Aref, H., "Stirring by chaotic advection", Journal of fluid mechanics, 143, 1-21, 1984.
[17] Ottino, J.M., "The Kinematics of mixing: stretching, chaos and transport", Cambridge University Press, 1989.
[18] Chou, H.P., Unger, M.A. and Quake, S.R., " A microfabricated rotary pump", Biomedical Devices, 3(4), 323-330,2001.
[19] Probstein, R.F., "Physicochemical Hydrodynamics: an introduction", 2nd edition, John Wiley & Sons, Inc., 1994.
[20] Chen, H., Zhang, Y.T., Mezic, I., Meinhart, C.D. and Petzold, L., "Numerical simulation of an electroosmotic micromixer", Proceedings of Microfluidics 2003 ASME IMECE, Washington DC, November 2003.
[21] Li, Z.H., Hao, Y.L., Zhang, D.C., Li, T. and Wu, G.Y., "An SOI-MEMS technology using substrate layer and bonded glass as wafer-level package", Sensors and Actuators A-Physical, 96 (1), 34-42,2002.

SPECIFIC FLOW CONTROL SYSTEMS USING IR LASER INDUCED SOL-GEL TRANFER OF HYDROGEL

Jun-ichi Tanaka[1], Yoshitaka Shirasaki[2], Masayasu Tatsuoka[2], Takashi Funatsu[2], Shota Watabe[1], Shuichi Shoji[1] ,Tomohiko Edura[3] ,Jun Mizuno[3] Ken Tsutsui[3] ,Yasuo Wada[3]

[1]Department of Electric Engineering and Bioscience, [2]Department of Physics, [3]The Institute of Nanotechnology, Waseda University, 3-4-1 Ohkubo, Shinjuku, 169-8555 Tokyo Japan
E-mail: tunakaji@shoji.comm.waseda.ac.jp

ABSTRACT

IR laser induced sol-gel transfer of hydrogel is applied for the microfluidic controls. High performance biomolecules sorting systems using T-shaped microchannel including Ti absorbers to enhance the heat generation were realized. Mebiol Gel[TM] whose critical temperature 32 C was used as the carrier and high speed sorting of about 30 msec was achieved by optimizing the channel structure. We also apply sol-gel transfer to realize 2-D specific flow control in the microcavity by projecting near-IR light using a Digital Mirror Device (DMD).

Keywords: sol-gel transfer of hydrogel, biomolecules sorting system, 2-D specific flow control system

1 INTRODUCTION

Flow control in micro chemical/biochemical analysis systems is one of the key technologies. Various types of micropumps and microvalves have been studied and developed.[1] But most of the micro valves need multi step fabrication processes and have dead space due to their complicated structures. Even for the simple cell sorting systems including flow switches, the microchannel structures must be complicated when the microvalves are integrated. To avoid this problem, micro cell sorters using seath flow control were reported.[2,3] We proposed novel biomolecules sorting systems using sol-gel transfer of hydrogel.[4,5] Further studies on the switching characteristics, the sorting time is reduced by efficient IR absorbing procedure using metal thin film absorber as indicated below.

Flow control using sol-gel transfer of hydrogel can be applied to 2-D specific flow control in a free space glass microcavity, since the efficient heat generation is realized with metal thin film IR absorbers. First prototype of the 2-D specific flow control system by near-IR projection using DMD is also described.

2 BIOMOLECULES SORTING SYSTEM

2.1 Principle of Biomolecules Sorting System

Fig.1 shows the principle of the biomolecules sorting system using T-shaped microchannel, the sorting system has one inlet and two outlets of recover and waste ports. At initial state, IR focused laser (1480nm) is illuminated at the Ti absorber of the recover branch. Since the gelled layer is formed to block the channel, all molecules flow to the waste port. When the target molecule is detected, the illumination point is shifted to the Ti absorber of waste branch. Ti absorber and water are heated by the laser. Sol-Gel transfer of the Mebiol Gel[TM] occurs when the temperature exceeds the critical point (32 C). The formed gel blocks the flow. The target molecule is sorted to the recover port.

2.2 Design and Optical Setup

Fig.2 shows the structure of biomolecules sorting chip. The device consists of two quartz glass plates. A 5 μm wide and 5 μm deep channel is formed in the lower quartz plate. The channel is fabricated by femto-second laser ablation, and Ti thin films are formed close to the junction. The through holes of an inlet and two outlets are drilled with a diamond

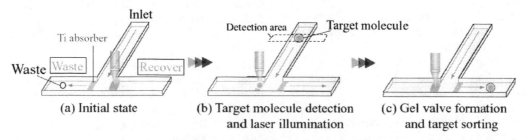

(a) Initial state   (b) Target molecule detection and laser illumination   (c) Gel valve formation and target sorting

Fig.1 Principle of the biomolecules sortingsystem

tool. Then the two plates are bonded by HF bonding[6].

Schematic of the optical setup is shown in Fig.3. The fluorescent signals are detected at 10um upstream from the junction. Wavelength and power of the used laser are 1480nm and 0.2W. PC controlled two scanner mirrors are used for switching of the focued IR beam (Mirror 1) and scanning (Mirror 2).

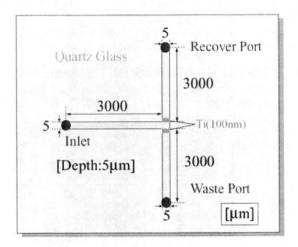

Fig.2 Structure of the T-shaped biomolecules sorting chip

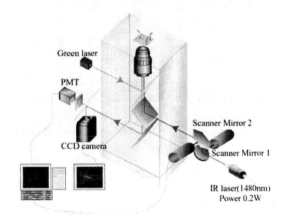

Fig.3 Optical setup of the T-shaped biomolecules sorting system

2.3 Experimental Results

We estimate the temperature distribution around the IR laser illumination point by monitoring the fluorescent intensity decrement of the Rhodamine due to the temperature increase. Fig.4 shows the temperature distribution around the near-IR laser illuminated, (a) water including Rhodamine on the glass and (b) that on the Ti (100nm) membrane formed on the glass. The IR energy necessary for elevating the certain temperature on the Ti

film is about 30 times smaller than that on the glass. The results show the Ti membrane is useful for the efficient IR absorber.

Fig.5 shows the sorting feature using fluorescent bead as the sample. Fig.5 (a) is streamline of the beam after detecting the target bead at the detection point and IR focused laser is illuminated at the left Ti absorber, while (b) is after switching the illumination point to the right absorber at 33 msec later. This result indicates that the sorting time is faster than 33 msec.

Fig.6 shows the relative velocity of the trapped beads and free beads around the sol-gel transfer measured by the high speed camera. By moving the illumination point, the trapped one is released while the free one is trapped. The transient time indicates the sol-gel transfer speed. The results indicate that the transfer is completed within 3 msec.

By employing the Ti IR absorbers, necessary IR power for sol-gel transfer is reduced about 1/30 and the sorting time was improved 4 times faster.

The proposed system is now going to apply actual sorting of biological cell, DNA and RNA. Recovery ratio of 95 % is already achieved in the case of λ–phage DNA sorting.

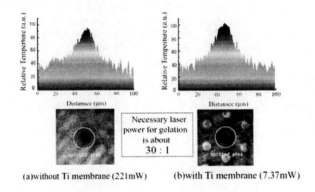

(a)without Ti membrane (221mW)   (b)with Ti membrane (7.37mW)

Fig.4 Temperature profile where the spotted IR laser illuminated

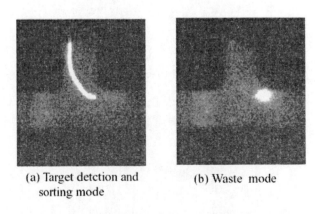

(a) Target detction and sorting mode   (b) Waste mode

Fig.5 Sorting feature using fluorescent beads

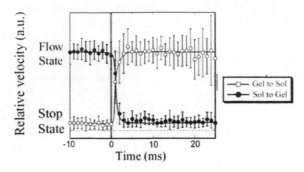

Fig.6 Relative velocity of the trapped beads and free beads around the sol-gel transfer area

3 2-D SPECIFIC FLOW CONTROL SYSTEM

3.1 Principle

The schematic principle of the 2-D specific flow control system is illustrated in Fig.7. A multi-inlets and outlets glass microcavity having the arrayed Pt IR absorbing spots on the bottom glass. At initial state, low viscosity Mebiol Gel™ is introduced into the inlets and flows out from all outlets. By projecting near-IR light patterns to the Pt absorbers using the PC controlled DMD. Mebiol Gel™ are heated and sol-gel transfer occurs when the temperature around absorbers exceeds the critical point (32 C). Then gelled patterns are formed and limited flow passes are formed in the microcavity (Fig.7(b)). The patterns of limited flow passes can be designed freely by PC programming.

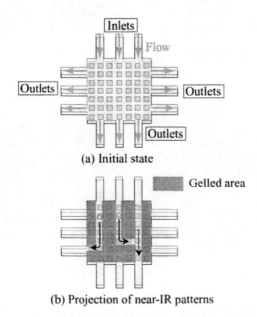

(a) Initial state

(b) Projection of near-IR patterns

Fig.7 Principle of the 2-D specific flow control system

3.2 Design and Optical Setup

Fig.8 shows the structure of the 2-D specific flow control chip. A microcavity and channels are fabricated on upper Pyrex glass substrate by the HF etching. Thickness of the upper substrate is 170μm. Cr/Pt near-IR absorbers are formed on the lower substrate. Size of the absorbers is 10 μm square and distance between the absorbers is 20 μm Thickness of the lower substrate is 500 μm. The through holes of inlets and outlets are drilled with a diamond tool. Both substrates are aligned and bonded by the thermal bonding.

Fig.9 shows the optical setup of the 2-D specific flow control system. Wavelength of the near-IR laser is 808nm. Maximum laser power irradiated to the DMD is 10W. Maximum incident power into the microcavity is 40mw. In the present stage, the IR projection area is limited within 100 μm x 80 μm.

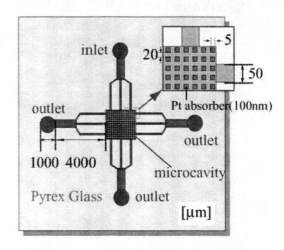

Fig.8 Structure of the 2-D specific flow control chip

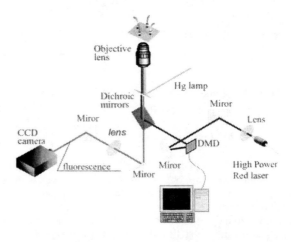

Fig.9 Optical setup of the 2-D specific flow control system

3.3 Experimental Results

Fig.10 shows the integration of the fluorescent beads streamlines (4.0sec) with and without IR projection to the microcavity. Fluorescent beads flow into the microcavity without control at initial state (a). By projecting near-IR light of two vertical patterns and one horizontal pattern with DMD, the beads flow passes are limited as shown in Fig.10(b). This result indicates that the 2-D specific flow control is available by this method. We are going to expand the projection area by improving the optical setup.

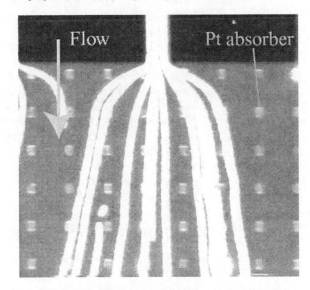

(a) Initial state

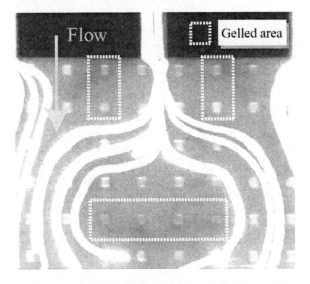

(b) Under gel walls formation by near-IR light projection

Fig.10 Fluorescent beads streamlines with and without IR projection

4 CONCLUSIONS

IR laser induced sol-gel transfer of hydrogel is applied for the microfluidic controls. High performance biomolecules sorting systems using T-shaped microchannel including Ti absorbers to enhance the heat generation were realized. Mebiol Gel^{TM} whose critical temperature 32 C was used as the carrier and high speed sorting of about 30 msec was achieved by optimizing the channel structure.

We also apply sol-gel transfer to realize 2-D specific flow control in the microcavity by projecting near-IR light using a Digital Mirror Device (DMD). Wide area 2-D specific control will be realized by improving the optical setup.

ACKNOWLEDGEMENTS

This research was partly supported by "Nanotechnology Support Project" of the Ministry of Education, Culture, Sports, Science and Technology (MEXT), Japan.

This research was partly supported by Grants-in-Aid for COE Research of Waseda University, Scientific Research Priority Area (B) No.13124209, Scientific research (B) No.12450167, and 13558088 from the Ministry of Education, Culture, Sports, Science and Technology of Japan

We would like to thank Prof. H. Yoshioka for providing Mebiol Gel^{TM}.

REFERENCES

[1] S. Shoji "Fluids for Sensor Systems", Topics in Current Chemistry, Vol. 194, pp164-188(1998)

[2] A. Wolff, U. Larsen, G. Blankstein, J. Philip, P. Telleman, "Rare event cell sorting in a Microfluidic system for application in prenatal", Proc. Micro Total Analysis Systems 1998, pp77-80

[3] T. Hara, T. Ichiki, Y. Horiike, K. Yasuda, "Fabrication of on-chip sorter devices with sub-micrometer scale channels and self-aligned microelectrodes", Proc. Micro Total Analysis Systems 2002, Vol. 1, pp124-126

[4] K Tashiro, S Ikeda, T Sekiguchi, H Sato, S Shoji, H Makazu, K Watanabe, T Funatsu, S Tsukita, "Micro flow switching using thermal gelation of methyl cellulose for biomolecules handling", Transducers'01,Vol. 2, pp932-935

[5] Y Shirasaki, H Makazu, K Tashiro, S Ikeda, T Sekiguchi, S Shoji, S Tsukita, T Funatsu,,"A novel biomolecules sorter using thermosensitive hydrogel in micro flow system", Proc. Micro Total Analysis Systems 2002,Vol. 2, pp925-927.

[6] H. Nakanishi, T. Nishimoto, R. Nakamura, A. Yoshida, S. Shoji, "Studies on SiO2-SiO2 bonding with hydrofluoric acid. Room temperature and low stress bonding technique for MEMS." Sensors & Actuators, Vol. 79 (2000), pp237-244

Optimization of Mixing in an Active Micromixing Device

George Mathew*, Igor Mezić*, Radu Serban** and Linda Petzold*

* Department of Mechanical and Environmental Engineering
University of California, Santa Barbara, CA 93106.
** Center for Applied Scientific Computing
Lawrence Livermore National Laboratory, Livermore, CA 94551.

ABSTRACT

Microscale mixers can be divided into two broad classifications - passive and active. Passive mixers rely on geometric properties of the channel shape whereas active mixers rely on time-dependent perturbation of the fluid flow to achieve mixing. In this work, we consider the problem of characterizing the mixing performance of an active micromixer which consists of a main mixing channel within which the flow is perturbed by pressure-driven flow from three pairs of orthogonal secondary channels. Using the newly developed measure of mixing, the so-called "Mix-Norm", we study optimal values for the frequency and amplitude of the oscillating flow in the secondary channels. It is shown that with one side channel operating regions of poor mixing parameter values interlay with good mixing parameter values and that for the second and third channels mixing is good for a large range of parameter values.

Keywords: micromixing, chaotic advection

1 Introduction

Mixing of fluids at the microscale has gained a lot of importance with the rapidly expanding use of micro fluidic systems in biology and biotechnology. Flow in microchannels typically have low Reynolds numbers and are therefore laminar. Molecular diffusion across the channels are too slow to cause mixing within reasonable time scales. These limitations make it necessary to design micromixers which efficiently stretch and fold fluid elements so that diffusion needs to act across only smaller length scales to achieve complete mixing. Microscale mixers can be divided into two broad classifications - passive and active. Passive mixers rely on geometric properties of the channel shape to induce complicated fluid particle trajectories and thus cause mixing ([1], [2]). Active mixers rely on time-dependent perturbation of the fluid flow to achieve mixing. In this work, we consider the problem of characterizing the mixing performance of an active micromixer which is based on the concept of chaotic advection. The concept of chaotic advection was introduced by Aref [3] - the basic idea being that even flow fields which have a simple structure from the Eulerian point of view could lead to complex Lagrangian fluid element trajectories, thereby causing efficient stretching and folding of fluid material resulting in the development of finer and finer structures in the advected passive scalar field. Books by Ottino [5] and Wiggins [4] address the problem of mixing using concepts and methods of dynamical systems theory.

To quantify the degree of mixing, we have developed a new measure of mixing called the Mix-Norm [9]. The Mix-Norm is able to capture the efficiency of the "stirring" stage of the mixing process accurately by probing the "mixedness" of the evolving density field at various scales. Previous approaches to this fundamental problem of measurement of mixing include using the entropy of the underlying dynamical system[7] as an objective for mixing and using the scalar variance of the density field which is being transported by a dynamical system[6].

2 Micro-Mixer Geometry and Models

We study an active micro-mixing device proposed in [8] and shown in Figure 1. It consists of a main mixing channel and three pairs of transverse secondary channels. The flow in the main channel is perturbed by a time-dependent pressure-driven flow from the secondary channels, thereby enhancing mixing. Two unmixed fluids, one at the upper half and the other at the lower half, enter the main channel. The two fluids referred to here are not two fluids with different properties, but can rather be thought of as the same fluid with different colored tracer particles in them.

For the purposes of our study, a simple analytical form for the flow field based upon superposition of elementary velocity profiles is assumed. The flow in the main channel follows a parabolic profile in the horizontal direction. The flow from the secondary channels consists of a vertical velocity with a parabolic profile that varies sinusoidally in time at different frequencies. The optimization problem here is to find the amplitude and frequency of oscillation in the secondary channels which give the best mixing. This micro-mixer has been built and studied experimentally [10]. In this analysis, diffusion is neglected and our objective is to optimize the "stirring" phase of the mixing process.

The micro-mixer is divided into three types of re-

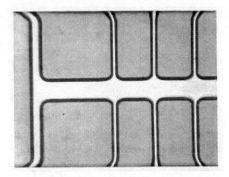

Figure 1: Micrograph of the micro-mixer device (device fabrication courtesy of K.S. Breuer, Brown University and R. Bayt, United Technologies)

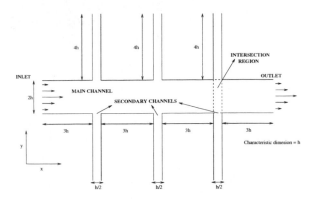

Figure 2: Dimensions of micro-mixer device

gions: the main channel (horizontal), the secondary channels (vertical), and the intersection regions. The dimensions of the mixer are as shown in Figure 2. A characteristic dimension of the mixer device is h, which is the half-width of the main channel, where $h = 100 \ \mu m$. In all our discussions, the origin of the x–y plane is assumed to be at the center of the inlet. The ordinary differential equations governing the motion of each particle are as follows:

$$\frac{dx}{dt} = \begin{cases} U_0 \left[1 - \left(\frac{y}{h}\right)^2\right], & |y| \le h \\ 0, & |y| > h \end{cases} \quad (1)$$

$$\frac{dy}{dt} = \begin{cases} 2\pi r_i f_i h \left[1 - \left(\frac{4\bar{x}}{h}\right)^2\right] \sin(2\pi f_i t), & |\bar{x}| \le 0.25h \\ 0, & |\bar{x}| > 0.25h \end{cases}$$

where $\bar{x} = x - (3.5i - 0.25)h$, for $i = 1, 2, 3$. The scaling constant $2\pi r_i f_i h$ is introduced so that in the absence of the main channel flow, a particle along the centerline of the secondary channel oscillates with amplitude $r_i h$. When $r_i = 1$, the amplitude of oscillation is exactly the half-width of the main channel. Therefore we need to optimize the amplitude ratios r_i and the pump frequencies f_i. Introducing a non-dimensional time $\tau = (U_0/h)t$ and using the characteristic dimension h as the unit for distance we get non-dimensional equations of the form:

$$\frac{dx}{d\tau} = \begin{cases} 1 - y^2, & |y| \le 1 \\ 0, & |y| > 1 \end{cases} \quad (2)$$

$$\frac{dy}{d\tau} = \begin{cases} 2\pi r_i F_i \left[1 - (4\bar{x})^2\right] \sin(2\pi F_i \tau), & |\bar{x}| \le 0.25 \\ 0, & |\bar{x}| > 0.25 \end{cases}$$

where $F_i = (h/U_0)f_i$ are the non-dimensional frequencies.

3 Simulation

To approximate the evolving density field at the outlet of the mixer a backward particle tracing method is employed here. An intial scalar density field is assumed all over the mixer as follows:

$$c_0(x, y) = \begin{cases} 1, & y > 0 \\ 0, & y < 0 \end{cases} \quad (3)$$

The incoming density field is also of the form (3). One can imagine a density of value 1 to represent a "blue" fluid and a density of value 0 to represent a "red" fluid. This density field is evolving under the dynamics of (2). We are interested only in the "mixedness" of the concentration field at the outlet of the mixer. To make the notation convenient, we define the space $C = \{X_o\} \times [-1, 1] \times [T, T + T_p]$ where X_o is the x-coordinate of the outlet, $[-1, 1]$ is the y-coordinate range at the outlet and $[T, T + T_p]$ is the time period within which we observe the outlet. T_p is chosen to be at least the period of the velocity field in (2) and $T > 0$ is some time beyond which we want to compute mixing. For each tracer particle at the outlet, we simulate the trajectories using (2), but backward in time. To be precise, for each initial condition $(X_o, y, \tau) \in C$, we solve the following ordinary differential equation

$$\frac{dx}{d\tau_b} = u(x, y, \tau) \quad (4)$$
$$\frac{dy}{d\tau_b} = v(x, y, \tau)$$
$$\frac{d\tau}{d\tau_b} = 1,$$

where τ_b is a dummy time variable and τ is time. Let $S : C \to \Re^2$ be the solution of (4) which gives the x, y-coordinates of the tracer particles when $\tau = 0$. Then the density field in the space C can be written as

$$c(X_o, y, \tau) = c_0(S(X_o, y, \tau)). \quad (5)$$

4 The Mix-Norm for Quantification of Mixing

In this section we introduce the Mix-Norm. In [9] we study the properties of the Mix-Norm from a theoretical perspective. Here, we define the Mix-Norm for a normalized 2-dimensional domain. Let $c : [0,1]^2 \to \Re$ be a scalar density field. The function c is assumed to be extended in each direction periodically, oddly or evenly based on the boundary conditions of the problem. All vectors are written in bold font and their respective elements are written in usual font with indices as subscripts and also for a given $\mathbf{s} \in (0,1)^2$ and $\mathbf{p} \in [0,1]^2$, $A_{[\mathbf{p},\mathbf{s}]} = [p_1 \quad s_1/2, p_1 + s_1/2] \quad [p_1 \quad s_2/2, p_2 + s_2/2]$. To define the Mix-Norm let

$$d(c,\mathbf{p},\mathbf{s}) = \frac{\int_{\mathbf{x} \in A_{[\mathbf{p},\mathbf{s}]}} c(\mathbf{x})d\mathbf{x}}{s_1 . s_2} \qquad (6)$$

for all $\mathbf{s} \in (0,1)^2$ and $\mathbf{p} \in [0,1]^2$. $d(c,\mathbf{p},\mathbf{s})$ is the mean value of the function c within the subset $A_{[\mathbf{p},\mathbf{s}]}$. Now define

$$(c,\mathbf{s}) = \left(\int_{[0,1]^2} [d(c,\mathbf{p},\mathbf{s})]^2 \, d\mathbf{p} \right)^{\frac{1}{2}}. \qquad (7)$$

$(c,\mathbf{s})$ is the L^2 norm of the averaged function $d(c,.,\mathbf{s})$ for a fixed scale $\mathbf{s} \in (0,1)^2$. Then the Mix-Norm of c is given by

$$\Phi(c) = \left(\int_{\mathbf{s} \in (0,1)^2} {}^2(c,\mathbf{s})d\mathbf{s} \right)^{\frac{1}{2}}. \qquad (8)$$

For our purposes, the density distribution $c : C \to \Re$ is periodically extended in the τ-direction and even-extended in the y-direction and we compute $\Phi(c \quad 0.5)$. For perfect mixing, any set with nonzero area within C should have an equal amout of "red" and "blue" fluid. The basic idea behind the Mix-Norm is to parametrize all rectangular sets within C and to measure the variance of the mean values of the function c within all these sets from the mean $c_m = 0.5$. For good mixing, $\Phi(c \quad 0.5)$ will be almost zero whereas for poor mixing, $\Phi(c \quad 0.5)$ will be close to 0.5. Also, if we know the function $c : [0,1]^2 \to \Re$, in terms of its Fourier expansion as follows

$$c(x_1,x_2) = \sum_{m=-\infty}^{m=\infty} \sum_{n=0}^{n=\infty} a_{m,n} e^{i2\pi m x_1} . (e^{i\pi n x_2} + e^{-i\pi n x_2})$$

$$= \sum_{m=-\infty}^{m=\infty} \sum_{n=0}^{n=\infty} a_{m,n} f_{m,n}(x_1,x_2), \qquad (9)$$

where $a_{m,n} = \langle c, f_{m,n} \rangle / \langle f_{m,n}, f_{m,n} \rangle$, then its Mix-Norm is given by

$$\Phi(c) = \left(\sum_{m=-\infty}^{m=\infty} \sum_{n=0}^{n=\infty} \lambda_m^p \lambda_n^e a_{m,n}^2 \langle f_{m,n}, f_{m,n} \rangle \right)^{1/2}, \qquad (10)$$

where

$$\lambda_m^p = \begin{cases} 1 \text{ if } m = 0 \\ \int_0^1 \frac{\sin^2(m\pi s)}{(m\pi s)^2} ds \text{ if } m \neq 0 \end{cases}$$

$$\lambda_n^e = \begin{cases} 1 \text{ if } n = 0 \\ \int_0^1 \frac{\sin^2(\frac{n}{2}\pi s)}{(\frac{n}{2}\pi s)^2} ds \text{ if } n \neq 0. \end{cases} \qquad (11)$$

Here, λ_m^p and λ_n^e are the eigenvalues of certain symmetric definite operators [9] and both of them are of $O(1/k)$ where k is the wavenumber.

5 Optimization of Mixing

By doing the simulations as in Section 3 and quantifying the mixing as in Section 4, we can study how the optimization parameters (amplitude ratios and frequencies of the pumps) influence the mixing. The space C defined in the above section is discretized uniformly in both directions and for each grid point, we solve (4) using a 4th order Runge-Kutta method. In all our computations, we set $T = 30$ and $T_p = 10$. Then the density field in the space C is approximated using (5).

First we discuss the selection of optimum parameters when only one pump is turned on, i.e, in (2) we set $F_2 = F_3 = 0$ and optimize for F_1 and r_1. Figure 3(a) shows the Mix-Norm as a function of the frequency, F_1 and amplitude ratio, r_1. There is an element of non-robustness as regions of poor mixing parameter values interlay with good mixing parameter values. The optimum parameters with lowest energy input to the mixer are $F_1 = 0.7$ and $r_1 = 1.0$. Note that these optimum parameters are also more robust when compared to other minimas. Even with high energy input to the micromixer, one can achieve very poor mixing as can be seen with parameter values $F_1 = 1.5$ and $r_1 = 1.0$. This shows how crucial the selection of the actuation frequencies are.

Next we discuss the variation of the Mix-Norm when the first two pumps are turned on. We set $F_3 = 0$, keep the first pump fixed at its optimum parameters ($F_1 = 0.7$, $r_1 = 1.0$) and vary F_2 and r_2. Figure 3(b) shows the Mix-Norm as a function of F_2 and r_2. It can be seen that reasonable mixing is achieved for a large range of frequency and amplitude values. Note that when $F_2 = 0.7$, for amplitude ratio values ranging from 0.5 to 1.0, there is a sharp decrease in the mixing performance and for amplitude ratio values ranging from 1.4 to 1.7, there is a sharp increase in the mixing performance. This clearly proves that keeping all the

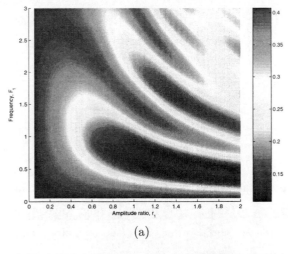

(a)

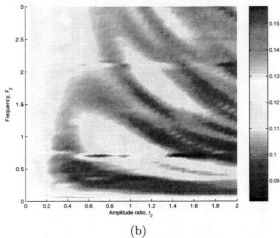

(b)

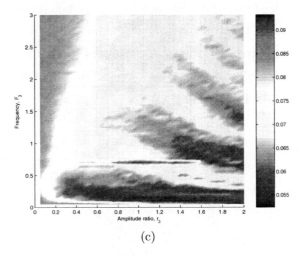

(c)

Figure 3: (a)$\Phi(c\ 0.5)$ as a function of F_1 and r_1 with 2nd and 3rd pumps o . (b))$\Phi(c\ 0.5)$ as a function of F_2 and r_2 with $F_1 = 0.7$, $r_1 = 1.0$ and $F_3 = 0$. (c))$\Phi(c\ 0.5)$ as a function of F_3 and r_3, with $F_1 = 0.7$, $r_1 = 1.0$, $F_2 = 0.7$ and $r_2 = 1.5$

pumps at the same parameters gives the worst mixing. Another interesting observation is that for the parameter space of the rst pump which gave reasonable mixing, the second pump gives poor mixing and vice versa. Also, for low amplitude values around $r_2 = 0.5$ we get good mixing, the reason being that the length scales which need to be mixed decreases after the rst pump. Finally, keeping the second pump parameters xed at $F_2 = 0.7$ and $r_2 = 1.5$, we optimize for the third pump. Figure 3(c) shows the Mix-Norm as a function of F_3 and r_3. The same observations as seen for the second pump optimization can be clearly seen for the third pump optimization.

REFERENCES

[1] R. H. Liu, K. V. Sharp, M. G. Olsen, M. A. Stremler, J. G. Santiago, R. J. Adrian, H. Aref, D. J. Beebe, *A passive micromixer: Three-dimensional serpentine microchannel.* Journal of MEMS, 9(2), 2000.

[2] A. D. Stroock, S. K. W. Dertinger. A. Ajdari, I. Mezić, H. A. Stone, G. M. Whitesides, *Chaotic mixer for microchannels.* Science, 295:647-651, 2002.

[3] H. Aref, *Stirring by chaotic advection.* Journal of Fluid Mechanics, 143, 1-21 (1984).

[4] S. Wiggins, *Chaotic Transport in Dynamical Systems.* Springer-Verlag Interdisciplinary Applied Mathematical Sciences Series, 1992.

[5] J.M.Ottino, *The Kinematics of Mixing: Stretching, Chaos and Transport.* Cambridge University Press, Cambridge, England 1989.

[6] D. Rothstein, E. Henry and and J. P. Gollub, *Persistent patterns in transient chaotic fluid mixing.* Nature, 401, 770-772 (21 Oct 1999) Letters to Nature.

[7] D. D'Alessandro, M. Dahleh and I. Mezić, *Control of Mixing: A Maximum Entropy Approach.* IEEE Transactions on Automatic Control 44(1999):1852-1864.

[8] M. Volpert, C. D. Meinhart, I. Mezić, M. Dahleh, *Modeling and numerical analysis of mixing in an actively controlled micromixer.* HEFAT 2002, 1st International Conference on Hear Transfer, Fluid Mechanics, and Thermodynamics, 8-10 April 2002, Kruger Park, South Africa.

[9] G. Mathew, I. Mezić, L. Petzold, *A Multiscale Measure for Mixing and its Applications.* Proceedings of the Conference on Decision and Control, Maui, Hawaii, USA. 2003.

[10] F. Bottausci, I, Mezić, C. D. Meinhart, C. Cardonne, *Mixing in the Shear Superposition Micromixer: 3-D Analysis.* Article submitted to Royal Society.

Droplet Movement Induced by Nano Assembled Molecules And Micro Textures

[1,2]Fan-Gang Tseng, and [2] Shu Chun Chien
[1]Engineering and System Science Dept., [2]MEMS Institute
National Tsing Hua University
101, Sec. 2, Kuang Fu Rd., Taiwan, R.O.C , [1]fangang@ess.nthu.edu.tw

ABSTRACT

This paper introduces a droplet movement method on top of a surface prepared by gradient nano-scale self-assembled molecules on micro sized textures. Experimental results demonstrate the contact angles on hydrophobic surfaces without/with micro posts increase from 110° to 143°, while on the hydrophilic surfaces decrease from 56.7° to 12.4°. As a result, the droplet moving speed is increased around twice on the gradient SAM surface with micro structures than that with only SAM gradient or only micro textures.

Keywords: hydrophobicity gradient, Droplet movement, surface tension, thiol SAM's, noano and micro Structures

1. INTRODUCTION

Droplet manipulation has been becoming an emerging alternative to continuous flow systems commonly employed in μTAS for its accurate dose control, less drag force from contact surface, and simpler flow system. Numerous approaches have been developed and tested to transport and manipulate μl-sized liquid droplets moving on the solid surface. The driving mechanisms for moving droplets on solid surfaces include electrostatic actuation [1], light-driven on photoisomerizable monolayer surface [2], the combination of Marogoni flow, capillary flow and phase change [3], and asymmetrically roughed surfaces [4], etc. All these driving forces are related to surface tension heterogeneity, and hysteresis of contact angles. This paper proposes another gentle approach, by employing surface hydrophobicity gradient with micro structured surfaces, to generate much larger surface tension gradient for droplet movement.

2. MATERIAL AND METHOD

2.1 Materials

To prepare gradient SAM's (Self Assembled Monolayers) for the generation of surface hydrophobicity gradient, agarose was botugh from Vegonia, USA, 98 % 1-octadecanethiol and 90% 16-mercaptohexadecanoic was bought from Aldrich, USA, ethanol was bought from Osaka, Japan, and De-ionized water was prepared in the lab.

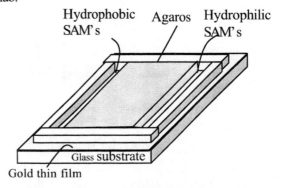

Fig. 1 Fabrication process of hyrophobicity gradient on Au surface by SAMs molecule diffusion

2.2 Gradient SAM's Preparation

The method for generating hydrophobicity

gradient on Au surface, as shown in Fig. 1, employs molecular diffusion inside Agarose gel on the top of Au surface with 1-octadecanethiol (hydrophobic) and 16-mercaptohexadecanoic (hydrophilic) molecules applied on the different side of the liquid reservoir. After 24 hours contact for molecule cross diffusion, the Agarose gel was peeled off from Au surface thus left a hydrophobicity gradient on the surface. [5]

2.3 Fabrication of Micro Structures

In the fabrication of micro structures, SU-8 thick negative tone resist was employed. SU-8 resist was first spun on silicon substrate for 20 μm thick, and then undergo lithography process to define the position and pitch for post arrays. Fabricated micro posts were then coated with 20nm Cr and 100nm Au. The fabricated posts were designed in different diameters (ranging from 30 μm to 100 μm) and pitches (ranging from 30 μm to 180 μm), as shown in Fig. 2. [6]

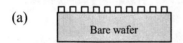

micro post structures were fabricated by thick film (20 μm thick SU-8 resist) lithography

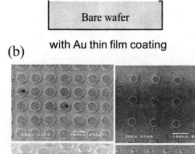

Fig. 2 Micro posts fabricated by thick film lithography(a) Fabrication process (b) SEM picture of Micro posts

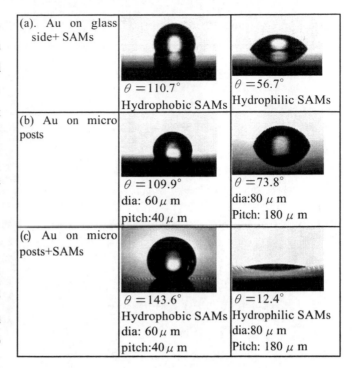

(a). Au on glass side+ SAMs	$\theta = 110.7°$ Hydrophobic SAMs	$\theta = 56.7°$ Hydrophilic SAMs
(b) Au on micro posts	$\theta = 109.9°$ dia: 60 μm pitch:40 μm	$\theta = 73.8°$ dia:80 μm Pitch: 180 μm
(c) Au on micro posts+SAMs	$\theta = 143.6°$ Hydrophobic SAMs dia: 60 μm pitch:40 μm	$\theta = 12.4°$ Hydrophilic SAMs dia:80 μm Pitch: 180 μm

3. RESULTS AND DISCUSSIONS

To understand contact angle variations for surfaces with different properties, six different surfaces have been prepared, including: (a) pure hydrophobicity gradient on Au surface, (b) micro post with different geometries, and (c) the combination of the above two. The contact angles were measured and tabulated in Fig. 3. When Au surface treated with only pure SAM's of 1-octadecanethiol (hydrophobic) molecules and 16-mercaptohexadecanoic (hydrophilic) molecules, the contact angle is about 110.7° and 56.7°, respectively, very typical for SAM's modified surface (Fig. 3a). In Au on post structures with different sizes and pitches, the maximum and minimum contact angles obtained in this experiment is 109.9° and 73.8°, respectively, not in a wide range as well (Fig. 3b). However in the combination of nano and micro structures, as shown in Fig. 3c, maximum and minimum contact angles can be changed to 143.6° and 12.4°, respectively. This demonstrates a contact angle gradient improvement from 56.7°-110.7° to 12.4°-143.6°, important for long range droplet transportation with high speed.

(a)

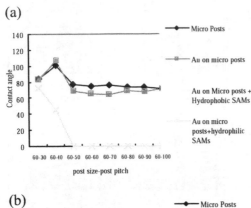

(b)

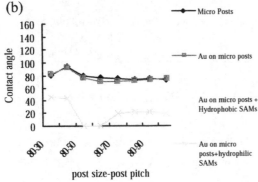

Fig.4 Contact angle variation versus pitch difference for (a) 60 µm posts, and (b) 80 µm posts in different SAM's coatings. Post pitches varied from 30 µm to 100 µm

To figure out the optimum post pitch for contact angle adjustment, 60 µm and 80 µm diameter posts with pitches varied from 30 µm to 100 µm were coated by different SAM's for contact angle measurement, and the results are shown in Fig. 4. In Fig. 4a, the maximum contact angle appears for the 60 µm posts at pitch of 50 µm, while the minimum contact angle happens interestingly at the similar angle. Similar results can be found in Fig. 4b for the 80 µm posts, the maximum and minimum contact angles appear at around 40-50 µm. This suggests that once the pitch is decided, the sizes of posts do not affect

contact angles very much. The pitch defines the roughness of the surface.

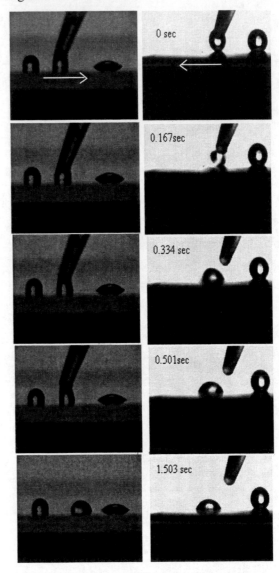

Fig.5 Droplet movement on SAMs/Au surface   Fig.6 Droplet movement on SAMs/Au/micro post surface

To compare the droplet movement speeds on the 3(a) and 3(c) surfaces, CCD image recording was applied for analyze droplet movement on different surfaces. The time sequences for these two cases are shown in Fig.5 and Fig.6, respectively, and the analyzed positions of droplet geometry center versus time are shown in Fig.7. It demonstrates the average droplet movement speed is around

1.88 mm/sec for the 3(a) surface while 3.26 mm/sec for the 3(c) surface under the same traveling time, almost as twice speed improvement by the combination of both surface hydrophobicity gradient and micro structures.

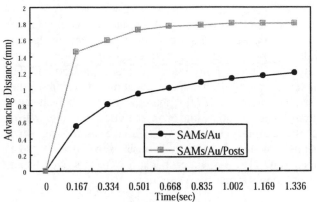

Fig. 7 Droplet movement distance versus time

4. DISCUSSIONS

In this experiment, the surface tension gradient built up by the gradient of SMA's coatings was employed to drive liquid droplet movement. However, most SAM's system can provide only about 60° variation in surface contact angle, not large enough for the droplet driving for long distance. With the assistance of micro structures, the apparent contact of nano SAM's molecules can be further expands to higher or lower level, thus greatly increases the contact angle gradients, 131.2° in this case, which help droplets more efficiently travel in long distance with higher speed.

5. CONCLSION

In this paper, we propose a novel method by employing both surface hydrophobicity gradient in nano scale, and post structures in micro scale, to enlarge the surface contact angle gradient from 60° to more than 130° for droplet actuation. Experiments have been conducted successfully in the fabrication of this micro/nano structured system, and the testing of the surface contact angles demonstrates a effective contact angle adjustment. Maximum and minimum contact angle conditions have also been found out and applicable to droplet movement. In the droplet actuation experiment, surfaces combined with SAM's coating and Micro posts carry out a droplet speed of 3.26 mm/sec, twice as large of that of the surfaces with only SAM's coatings. This gentle method can be applied to the transportation of bio-fluids, such as RNA or protein, without damage their bioactivities.

References

[1] Lee, J., Moon, H., Fowler, J., Scholllhammer, T. and Kim, CJ, "Electrowetting and Electrowetting -on-Dielectric for Microscale Liquid Handling", Sensors and Actuators A, Vol.95, pp.259-268, 2002

[2] K. Ichimura, S-K Oh, and M. Nakagawa, "Light-Driven Motion of Liquids on a Photorespon -sive Surface", vol.288, June 2, pp.1624-1626, 2000

[3] Daniel, D S., Chaudhury, M.K., and Chen, J.C., "Fast Drop Movements Resulting from the Phase Change on a Gradient Surface" Science, Vol. 291, pp 633-636, 2001.

[4] Bo He and Junghoon Lee, "Dynamic Wettability Switching By Surface Roughness Effect", MEMS'03, Kyoto, Japan(2003).

[5] Bo Liedberg and Pentti Tengvall, "Molecular Gradients of ω-Substituted Alkanethiols on Gold: Preparation and Characterization", Langmuir, 1995,11,3821-3827.

[6] Susan Daniel, Manoj K. Chaudhury, John C. Chen, "Fast Drop Movements Resulting from the Phase Change on a Gradient Surface", Science, 2001,291,633-636.

ANALYSIS OF AC-DRIVEN ELECTROOSMOTIC FLOW IN A MICROCHANNEL PACKED WITH MICROSPHERES

Y.J. Kang, C. Yang[*] and X.Y. Huang

School of Mechanical and Production Engineering
Nanyang Technological University, Singapore 639798, mcyang@ntu.edu. sg

ABSTRACT

This paper presents an analysis of the alternating current (AC) electroosmotic flow in open- and closed-end cylindrical microchanneles packed with microspheres. The time-periodic electroosmotic flow in an open capillary in response to the application of an AC electric field is obtained using the Green's function approach. The analysis is based on the capillary model. The backpressure associated with the counter-flow in a closed capillary is solved using the Brinkman's momentum equation. The results obtained from this study can be used for the development of novel electrokinetic microactuators.

Keywords: ac electroosmosis, porous media, closed channel, microchannel flow, green's function method

1 INTRODUCTION

Electroosmosis (EO) mostly is associated with a DC (direct current) electric field, which is used to generate a steady-state electroosmotic flow (EOF). A similar effect occurs with an AC (alternating current) electric field. In this situation, however, the electroosmotic flow becomes time and frequency dependent. Recently, the AC electroosmosis has drawn a wide attention by researchers [1-2].

With examination of the current literature known to the authors, no study on the AC electroosmotic flow in porous media has yet been reported. However, use of porous structures can enhance the pressure-building capacity of the AC electroosmotic flow. Paul *et al.* [3] proposed a method to generate high pressure using DC electroosmosis through a microchannel packed with microparticles. The pressure of 10 atm at 1.5 kV applied voltage have been achieved using fused-silica capillaries packed with charged 1.5 μm silica beads. In a recent work by Zeng *et al.* [4], an EO pump was fabricated to generate maximum pressures in excess of 20 atm or maximum flow rates of 3.6 μl/min for a 2 kV applied DC potential. Recently, a new separation mode for capillary electrochromatography using AC was proposed by Nakagawa *et al.* [5] as an application to control the chemical retention of a sample solute.

This paper presents an analysis of the AC electroosmotic flow in a cylindrical microcapillary packed with microspheres of uniform size. Two different conditions regarding the channel ends are considered.

When the channel is open, i.e., both ends are connected to the reservoirs, the oscillating EOF is analyzed using the capillary flow model. The corresponding Navier-Stokes equation is solved using Green's function method and the complete Poisson-Boltzmann equation governing the electric double layer (*EDL*) potential field is solved with arbitrary zeta potentials under an analytical scheme. When both the capillary ends are closed, the oscillating flow generated by the AC electroosmosis is balanced by the counter-flow. The associated backpressure is solved using the Brinkman's momentum equation.

2 PROBLEM FORMULATION

Consider a cylindrical microcapillary packed with charged spherical microparticles of diameter, d_p, as illustrated in Fig. 1. The zeta potential at the surface of the solid phase is ζ_p. The liquid through the porous structure is assumed to be an incompressible, Newtonian, symmetric monovalence electrolyte of density, ρ, and viscosity, μ. The porous medium is assumed to be homogeneous and the boundary effects are neglected. Two key parameters characterizing a porous medium are porosity and tortuosity. The porosity is defined as $\phi = V_f / V$, where V_f and V are the void and total volumes of the porous medium, respectively. The tortuosity is defined as $\tau = L_e / L$, where L_e is the effective length of travel for flow along the pore path and L is the physical length of the porous structure. The other two associated factors are pore size, $d_{pore} = 2R_{pore} = 2\phi d_p / 3(1-\phi)$,, and Darcy permeability, K, which are defined in following sections. These structure-related parameters can be determined by experiment. The AC electroosmotic flow in open and closed microcapillary will be discussed.

2.1 Flow in an open-end packed microchannel

In this situation, the flow is sorely electroosmosis-driven and no pressure gradient is present. We use so-called capillary model [6], in which the porous medium is assumed to be equivalent to a series of parallel tortuous tubules with effective pore size, d_{pore}. The liquid velocity field, $u(r, t)$ in the cylindrical tubules is governed by the Navier-Stokes equation [7]

$$\rho \frac{\partial u(r,t)}{\partial t} - \mu \frac{1}{r}\frac{\partial}{\partial r}\left[r\frac{\partial u(r,t)}{\partial r}\right] = E(t)\rho_e(r) \tag{1}$$

where $\rho_e(r)$ is the local volumetric net charge density of the electrolyte, and is governed by Boltzmann equation

$$\rho_e(r) = -2n_0 e_0 \sinh\left[\frac{e_0 \psi(r)}{k_b T}\right] \qquad (2)$$

Here $\psi(r)$ is the electrostatic potential of the *EDL*, and it is solved to follow an analytical scheme developed in ref. [2]. e_0 is the elementary charge, n_0 is the ionic concentration in the bulk liquid phase (i.e., far from the charged surfaces), k_b is the Boltzmann constant, T is the absolute temperature.

Introducing the following dimensionless parameters:

$$\bar{u} = \frac{u}{u_s} \quad \bar{t} = \frac{t}{t_{diff}} \quad \bar{r} = \frac{r}{R_{pore}} \quad \Psi = \frac{e_0 \psi}{k_b T} \quad u_s = -\frac{\varepsilon_r \varepsilon_0}{\mu} E_0 \zeta_p \qquad (3)$$

We can nondimensionlize Eq. (1) as

$$\frac{\partial \bar{u}}{\partial \bar{t}} - \frac{1}{\bar{r}}\frac{\partial}{\partial \bar{r}}\left(\bar{r}\frac{\partial \bar{u}}{\partial \bar{r}}\right) = \frac{(\kappa R_{pore})^2}{\Psi_s E_0}\sinh\left[\Psi(\bar{r})\right]E(\bar{t}) \qquad (4)$$

where the reference time $t_{diff} = \rho R^2_{pore}/\mu$ is the viscous diffusion time scale. $\psi_s = e_0 \zeta_p / k_b T$. κ is the Debye-Hückel parameter, defined as $\kappa = (2n_0 e_0^2 / \varepsilon_r \varepsilon_0 k_b T)^{1/2}$. ε_r is the dielectric constant of the electrolyte and ε_0 is the permittivity of vacuum. Eq. (4) is subject to the initial and boundary conditions

$$\bar{t} = 0 \qquad \bar{u} = 0 \qquad (5a)$$

$$\bar{r} = 0 \qquad \frac{\partial \bar{u}}{\partial \bar{r}} = 0 \qquad \bar{r} = 1 \qquad \bar{u} = 0 \qquad (5b)$$

Using the Green's function method, we can obtain an exact solution to Eq. (4) as

$$\bar{u}(\bar{r},\bar{t}) = \int_{\tau=0}^{\bar{t}}\int_{\xi=0}^{1} G(\bar{r},\bar{t}\,|\,\xi,\tau)\left\{\frac{(\kappa R_{pore})^2}{\Psi_s E_0}\sinh[\Psi(\xi)]E(\tau)\right\}d\xi d\tau \qquad (6)$$

where $G(\bar{x},\bar{t};\xi,\tau)$ is the Green's function expressed as [2]

$$G(\bar{r},\bar{t}\,|\,\xi,\tau)$$
$$= H(\bar{t}-\tau)\sum_{n=1}^{\infty}\frac{2 J_0(\lambda_n \bar{r})}{J_1^2(\lambda_n)}\xi J_0(\lambda_n \xi)\exp[-\lambda_n^2(\bar{t}-\tau)] \qquad (7)$$

Here λ_n are the positive roots of the zero-order Bessel function $J_0(\lambda_n)=0$, and $H(t-\tau)$ is the Heaviside step function.

Considering the application of a sinusoidally alternating electric field with an angle frequency ω, $E(t) = E_0 e^{i\omega t}$, we can show that Eq. (6) can be expressed as

$$\bar{u}(\bar{r},\bar{t}) = REAL\left[\frac{2(\kappa R_{pore})^2}{\Psi_s}\sum_{n=1}^{\infty}C_n\frac{J_0(\lambda_n \bar{r})}{J_1^2(\lambda_n)}\frac{\exp(i\beta^2 \bar{t})-\exp(-\lambda_n^2 \bar{t})}{\lambda_n^2 + i\beta^2}\right]$$
$$= \frac{2(\kappa R_{pore})^2}{\Psi_s}\sum_{n=1}^{\infty}C_n\frac{J_0(\lambda_n \bar{r})}{J_1^2(\lambda_n)}\frac{\lambda_n^2\cos(\beta^2 \bar{t})+\beta^2\sin(\beta^2 \bar{t})-\lambda_n^2\exp(-\lambda_n^2 \bar{t})}{\lambda_n^4 + \beta^4} \qquad (8)$$

where $C_n = \int_{\xi=0}^{1}\xi J_0(\lambda_n \xi)\sinh[\Psi(\xi)]\,d\xi$. "*REAL*" denotes the real part of the solution. i is the unit imaginary number. β is defined as $\beta = R_{pore}/\delta_s$, representing the aspect ratio of the tubule radius R_{pore} to the Stokes penetration depth δ_s, defined as $\delta_s = \sqrt{\mu/(2\pi\rho f)}$. (here $f=\omega/2\pi$ is the frequency of the applied electric field).

Integrating Eq. (8) along the radius of the cylindrical tubule gives the mean velocity,

$$\bar{u}_m(\bar{t}) = 2\int_{r=0}^{1}\bar{r}\bar{u}(\bar{r},\bar{t})d\bar{r}$$
$$= \frac{4(\kappa R_{pore})^2}{\Psi_s}\sum_{n=1}^{\infty}C_n\frac{1}{\lambda_n J_1(\lambda_n)}\frac{\lambda_n^2\cos(\beta^2 \bar{t})+\beta^2\sin(\beta^2 \bar{t})-\lambda_n^2\exp(-\lambda_n^2 \bar{t})}{\lambda_n^4 + \beta^4} \qquad (9)$$

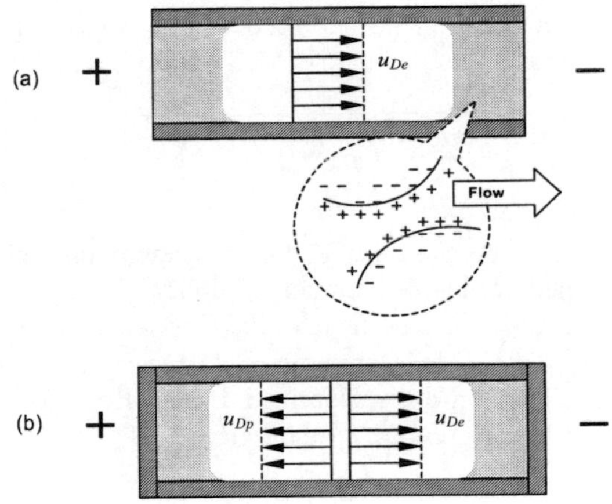

Fig. 1. Schematic illustrations of the AC electroosmosis in porous media. (a) flow in the packed microcapillary connected with two reservoirs. (b) flow in the packed microcapillary with two closed ends. Where u_{De} is the filter (Darcy) velocity due to electroosmosis, and u_{Dp} is the filter (Darcy) velocity due to backpressure.

Under the volume averaging method, the macroscopic filter velocity (Darcy velocity) of the fluid due to EO can be expressed as the mean velocity in each tubule taking into account of the tortuosity τ and the porosity ϕ, as

$$\bar{u}_{De}(\bar{t}) = \frac{\phi}{\tau}\bar{u}_m(\bar{t})$$
$$= \frac{\phi}{\tau}\frac{4(\kappa R_{pore})^2}{\Psi_s}\sum_{n=1}^{\infty}\frac{C_n}{\lambda_n J_1(\lambda_n)}\frac{\lambda_n^2\cos(\beta^2 \bar{t})+\beta^2\sin(\beta^2 \bar{t})-\lambda_n^2\exp(-\lambda_n^2 \bar{t})}{\lambda_n^4 + \beta^4} \qquad (10)$$

From Eq. (10) we can further show the expression of the frequency-dependent maximum Darcy velocity as

$$\left. \bar{u}_{De} \right|_{max} (f) = \frac{\phi}{\tau} \frac{4(\kappa R_{pore})^2}{\Psi_s} \sum_{n=1}^{\infty} \frac{C_n}{\lambda_n J_1(\lambda_n)} \frac{1}{\sqrt{\lambda_n^4 + \beta^4}} \quad (11)$$

2.2 Flow in a closed-end packed microchannel

In a microcapillary with closed ends, the electroosmotic flow induces a backpressure gradient. The backpressure gradient in turn generates a counter-flow to balance the electroosmotic volume flow, resulting in a zero flow rate. In porous media, no matter the flow is pressure-driven or electroosmosis-driven, the fluid velocity is represented by a uniform filter (Darcy) velocity u_D. Thus to balance the volumetric electroosmosis-driven flow, the back flow velocity, u_{Dp}, which is driven by the backpressure, must equal in magnitude to the flow velocity, u_{De} driven by electroosmosis (illustrated in Fig. 1(b)), i.e.,

$$\bar{u}_{Dp}(\bar{t}) = \bar{u}_{De}(\bar{t}) \quad (12)$$

Using the Brinkman's momentum equation for general pressure-driven flow in porous media, and neglecting the inertial forces and the boundary effect, we have

$$\frac{\rho}{\phi} \frac{du_{Dp}(t)}{dt} = -\frac{\partial p(t)}{\partial z} - \frac{\mu}{K} u_{Dp}(t) \quad (13)$$

Here $u_{Dp}(t)$ is the pressure-driven Darcy velocity through the porous medium inside the microcapillary. Where the Darcy permeability K is defined by the *Carman-Kozeny* equation [6], $K = \phi^3 d_p^2 / a(1-\phi)^2$, with $a = 150$ being the Ergun constant [8]. The non-dimensionlized form of Eq. (13), using the reference pressure p_0 and length L, is

$$\frac{\partial \bar{p}(\bar{t})}{\partial \bar{z}} = -\frac{L}{p_0} \frac{\mu}{K} u_s \left[\bar{u}_{Dp}(\bar{t}) + \frac{K}{\phi R_{pore}^2} \frac{d\bar{u}_{Dp}(\bar{t})}{d\bar{t}} \right] \quad (14)$$

Substitute Eq. (12) into Eq. (14), we obtain the expression for the periodically oscillating pressure gradient

$$\frac{\partial \bar{p}(\bar{t})}{\partial \bar{z}} = -\frac{L}{p_0} \frac{\phi}{\tau} \frac{\mu}{K} u_s \frac{4(\kappa R_{pore})^2}{\Psi_s}$$

$$\sum_{n=1}^{\infty} \frac{C_n}{\lambda_n J_1(\lambda_n)} \frac{\left(\lambda_n^2 + \frac{K\rho\omega}{\mu\phi} \beta^2 \right) \cos(\beta^2 \bar{t}) + \left(\beta^2 - \frac{K\rho\omega}{\mu\phi} \lambda_n^2 \right) \sin(\beta^2 \bar{t})}{\lambda_n^4 + \beta^4} \quad (15)$$

From Eq. (15), we obtain frequency-dependent maximum pressure gradient as

$$\left. \left| \frac{\partial \bar{p}}{\partial \bar{z}} \right| \right|_{max} (f) = -\frac{L}{p_0} \frac{\phi}{\tau} \frac{\mu}{K} u_s \frac{4(\kappa R_{pore})^2}{\Psi_s} \sum_{n=1}^{\infty} \frac{C_n}{\lambda_n J_1(\lambda_n)} \frac{\sqrt{1 + \left(\frac{K\rho\omega}{\mu\phi} \right)^2}}{\sqrt{\lambda_n^4 + \beta^4}} \quad (16)$$

3 RESULTS AND DISCUSSION

In the parametric study, the calculations are based on the values reported by Zeng *et al* [4]. The properties of the working solution were reported as: ionic strength 7.5 µM, zeta potential –95 mV, porosity 0.37, and tortuosity 1.5.

Several important parameters are also highlighted before discussion. The characteristic viscous diffusion time scale is denoted by the reference time $t_{diff} = \rho R^2_{pore}/\mu$. The time period for excitation electric field is denoted by $t_E = 1/f$. From the exponential term in Eq. (10), we can further estimate the characteristic time scale for the AC electroosmotic flow to reach its time-periodic steady state by choosing $\lambda_1^2 \mu t^*/\rho R^2_{pore} = 1$. Then $t^* = \rho R^2_{pore}/\mu \lambda_1^2$, where $\lambda_1 = 2.405$ determined form $J_0(\lambda_n) = 0$. The corresponding intrinsic frequency of the system is defined as $f^* = 1/t^* = \mu \lambda_1^2/\rho R^2_{pore}$. And we can infer that β represents the square root of the aspect ratio of the diffusion time scale to the period of electric field, i.e., $\beta = (2\pi t_{diff}/t_E)^{1/2}$.

3.1 Oscillating flow velocity in an open-end packed microchannel

The steadily time-periodic AC electroosmotic flow velocity for two-cycle periods in a packed microchannel with open ends are shown in Fig. 2. It is clearly shown that the flow presents a harmonic sinusoidal oscillation only with different magnitudes and phase lags. Fig. 2 gives a comparison of the effects of three different excitation frequencies for a fixed pore size. It is obvious that, for the same packing condition, using AC voltage of higher frequency decreases the EOF velocity. This scenario can be anticipated because under much high frequency, the electric field changes its direction so fast that the flow can never get developed across the bulk intraparticle domain. In case of very high frequency, corresponding to a large value of β, the diffusion time scale is much greater than the oscillation time period, i.e., $t_{diff} > t_E$. Therefore there is no sufficient time for the flow momentum to diffuse far into the intraparticle bulk phase. Thus the perturbed area is restricted only in a thin layer near the particle surface, while the fluid in the bulk phase remains stationary. This typical length scale of the oscillatory laminar viscous flow in response to a harmonic external excitation is represented by the frequency-dependent Stokes penetration depth, δ_s. Furthermore, the flow lags behind the excitation field with larger phase lag for higher frequency.

For a better understanding of the frequency-dependent velocity, we compare the magnitude of the oscillating velocity versus complete domain of frequency shown in Fig. 3, for three cases of different pore sizes. In accord with above analysis, for a fixed pore size, the magnitude of maximum velocity decreases with increasing excitation frequency. With increasing pore size, the magnitude of the maximum velocity gets higher in case of low frequency domain (lower than about 0.02f^*), and gets lower in case of high frequency domain (higher than about 0.2 f^*).

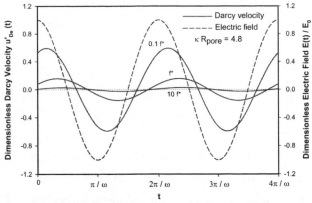

Fig. 2. Dimensionless time-periodic oscillating Darcy electroosmotic flow velocity versus time for three different excitation frequencies at a fixed pore size.

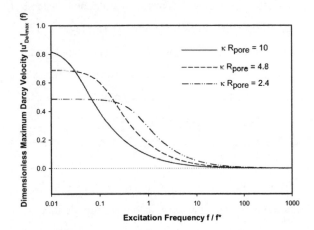

Fig. 3. Dimensionless maximum Darcy electroosmotic flow velocity versus excitation frequency, for three different pore sizes.

3.2 Oscillating backpressure in a closed-end packed microchannel

Fig. 4 presents the steadily time-periodic oscillation of the backpressure in a capillary of closed ends. It is shown that the backpressure associated with the counter-flow to balance the AC electroosmotic flow after application of the alternating voltage also presents a harmonic oscillation.

The magnitude of the maximum pressure drop versus the excitation frequency domain is shown in Fig. 5, for case study of three pore sizes. The maximum pressure drop generated by small pore size (corresponding to a smaller packing particle size) is always greater than that generated by large pore size (corresponding to a large packing particle size), for the entire domain of the excitation frequency. This result is in agreement with the experimental observations by Paul *et al* [3] who reported that the pressure generated per applied DC volt increases with smaller diameters of packing beads. It is also apparent in Fig. 5 that, for a fixed pore size, the backpressure decreases with increasing excitation frequency, until it reaches its system intrinsic

frequency f^*, and from which the backpressure increases with increasing excitation frequency.

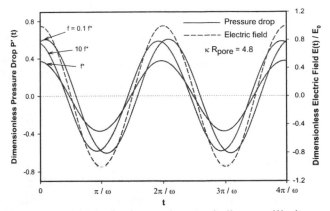

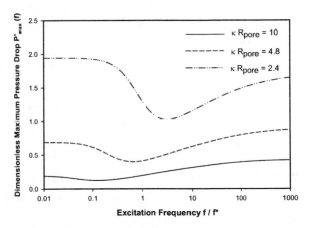

Fig. 4. Dimensionless time-periodic oscillating backpressure drop versus time for three different excitation frequencies at a fixed pore size.

Fig. 5. Dimensionless maximum backpressure drop versus excitation frequency for three different pore sizes.

REFERENCES

[1] P. Dutta and A. Beskok, Anal. Chem. 73, 5097, 2001.

[2] Y.J. Kang, C. Yang, and X.Y. Huang, Int. J. Eng. Sci. 40, 2203, 2002.

[3] P. H. Paul, D. W. Arnold, and D. J. Rakestraw, in: Proceedings of the Micro Total Analysis Systems '98 Workshop, Banff, Canada, 1998.

[4] S. Zeng, C. H. Chen, J. C. Jr Mikkelsen, and J. G. Santiago, Sensors and Actuators B, 79, 107, 2001.

[5] H. Nakagawa, M. Sato, S. Kitagawa, and T. Tsuda, Anal. Chem. 75, 3180, 2003.

[6] M. Kaviany, "Principles of Heat Transfer in Porous Media," Springer-Verlag, New York., 1995.

[7] R. F. Probstein, "Physicochemical Hydrodynamic," John Wiley & Sons, New York., 1994.

[8] C. T. Hsu, and P. Cheng, Int. J. Heat Mass Transfer 33, 1587, 1990.

An Effective Passive Micromixer Employing Herringbone Structure

Category: 4 Biomedical and Chemical Micro Sensors and Systems

[1]Chun-Fei Kung, [2]Chien-Fu Chen, [3]Chin-Chou Chu, and [4]Fan-Gang Tseng

[1,2,3]IAM, National Taiwan University, [4]ESS Dept., National Tsing Hua University, Taiwan

ABSTRACT

This paper proposes a high efficient micro mixer, passively employing surface tension force as driving power and Herringbone structures as vortex generator for fluid mixing. The fluidic channel was designed without sidewall and confined with only the bottom hydrophilic and top hydrophobic surface for later-on mixing process among channels. Besides, in order to increase the mixing efficiency, herringbone-like structures are arranged on the bottom of the channel to enforce liquids to produce three-dimensional flow automatically. The fabrication has been completed successfully, and the testing result demonstrated 5 times higher mixing rate in 15mm mixing range. This device is anticipated to be batch-fabricated and applied to power-free μ TAS or lab-on a-chip system in the future.

Keywords: hydrophilic, hydrophobic, herringbonelike

INTRODUCTION

Mixing in micro scale is a well known problem because Reynolds number is too low to generate turbulence. As a result, many mixing schemes, including zigzag channels [1], flow lamination [2], herringbone channel [3], or chaotic mixing [4], have been widely employed to solve this issue. Besides, there have been many actuation means to drive fluids in micro system, for example, by providing pressure gradient [5, 6], using thermal energies [7], or employing electrostatic force [8]. However, most of aforementioned mixers still required extra actuators like pumps [3] or external energies such as electrostatic or magnetic fields to drive the fluidic inside micro channel, and they usually need a large outside energy sources to support the desiring actuation, which greatly limit the capability of system integration. Therefore, for the sake of automatically producing mixing phenomenon without external driven means, capillary force is employed in this paper to drive fluid entering into micro channel with herringbonelike structures. These structures are the key factor to increase the mixing efficiency rapidly in the mixer because of the generation of three dimensional flow inside micro channel, greatly increasing fluid contact area for mixing.

As a result, a power-free device is proposed here employing only the surface tension energy defined by surface properties of the flow passages combined with herringbonelike concave structures to passively promote mixing effectively.

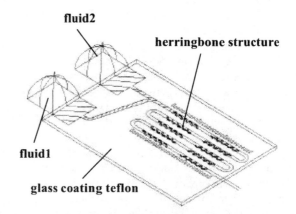

Fig 1. The surface-tension-driven micromixer with herringbone structures

DESIGN CONCEPTS

The design of the mixing device, as shown in Fig. 1, comprises a double fluidic entrance, and a winding channel confined by bottom hydrophilic patterns and a top hydrophobic surface employed to separate fluid in different channel temporarily, as the cross section shown in Fig. 2a.

When fluids are pipetted into the entrances, the pressure pushing-force established by the liquid puddle and the surface tension pulling-force generated by the hydrophilic side drag the fluids into the sidewall-free fluid passage, as shown in Fig. 2b. The advantage of that without sidewall is to reduce the frictional force on the interface between the fluid and the channel. Then, when two fluids are introduced into either liquid entrance, it will be driven automatically by surface tension into the winding channel and meet with each other for premixing.

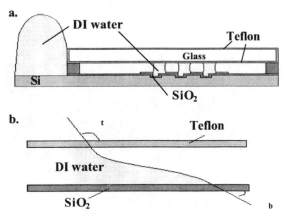

Fig 2. Cross section of the micro mixer and liquid driving method.

There are two different designs, one with smooth channel surface, and the other with herringbone structures, as shown in Fig. 3, are carried out for comparison. The purpose of the channel with herringbone structures is for rotating fluid in both spanwise and streamwise directions [3] to rapidly increase mixing surface by generating vortexes and decrease diffusion length by stretching fluid, thus

effectively improving mixing in micro scale. In the herringbone design, the depth of the concave structure is also important in promoting mixing process. When the ratio of the herringbone depth to the channel height is between 0.2 and 0.3, it will reach the most excellent mixing efficiency. Thus the depth of the herringbone structure is selected to be 3 μ m.

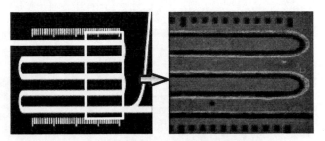

a. micro mixer with smooth channel designs.

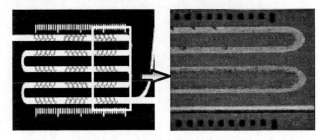

b. micro mixer channel with herringbone structures

Fig 3. Fluidic mixing effect in different designs

FABRICATION

Fabrication process, as shown in Fig. 4, is performed by patterning Teflon on silicon wafer with SiO_2 coating as the bottom hydrophilic channel guide, and the cover glass slide coated with Teflon as a hydrophobic liquid separation layer.

First, the fabrication starts from the silicon substrate etching by Deep Silicon Reactive Ion Etch (DSRIE) for 3 μ m, in order to define the position of the herringbonelike structures. Then Plasma Enhanced Chemical Vapor Deposition (PECVD) is used to deposit 1μm oxide (Fig. 4a) for hydrophilic surface property. And 1% liquid Teflon (Dupont, USA) is then spun on the oxide surface for 0.5μm

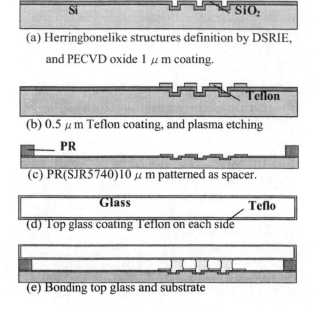

(a) Herringbonelike structures definition by DSRIE, and PECVD oxide 1 μ m coating.

(b) 0.5 μ m Teflon coating, and plasma etching

(c) PR(SJR5740)10 μ m patterned as spacer.

(d) Top glass coating Teflon on each side

(e) Bonding top glass and substrate

Fig 4. Fabrication process.

and patterned to form hydrophobic area (Fig. 4b). The process followed by Photoresist (SJR5740) patterning on silicon substrate as spacer for channel height definition (Fig. 4c), and Teflon coating on each side of the glass substrate (Fig. 4d) as the channel cover. Finally, The top and bottom substrates are bonded by photoresist with desired clearance, as shown in Fig. 4e.

EXPERIMENT RESULTS

To compare the ability of the mixing in the different designs, DI water and fluorescent particle mixture was pipetted into the micro channel, as shown in Fig. 3. First, the red fluorescent spheres (R900,Duke Scientific Corp.) were added into DI water to form working fluid. The concentration is 1% w/w, and diameter of the spheres is 0.93µm. Then the two work fluids were introduced into the entrance of the channel of 100 ?m wide and 2.5 cm long, and the liquid can be brought unto the end of the channel automatically without any other driving mean than surface tension. The velocity of the meniscus decreased from 1.7

cm/s to 0.5 cm/s due to the drag force increasing from the liquid-solid interface. The mixing process was visualized by a cooling CCD camera (DP50, OLYMPUS) under microscope (BX60M, OLYMPUS) , and the mixing process is shown in Fig. 3a.

The flow field in the smooth channel design, shown in Fig. 3a, formed only-laminar flow to the end of the channel. So the mixing effect is still not very conspicuous, it only has a little diffusion phenomena on the interface of the two liquids. However, another design with herringbone structures truly had very apparent mixing phenomena in the channels, as shown in Fig 3 b.

The detail mixing processes in different sections of the channels with herringbone structures are revealed in Fig. 5. The flow is laminar at the entrance region (Fig. 5a), but soon turn into vortex flow when passing through the herringbone region, and the particles start to scatter, as shown in Fig. 5b. The mixing rate reaches almost 45% when the flow distance approach 5 mm, as shown in Fig. 5c. Finally, it almost reaches fully mixing at the distance of 15 mm (Fig. 5d).

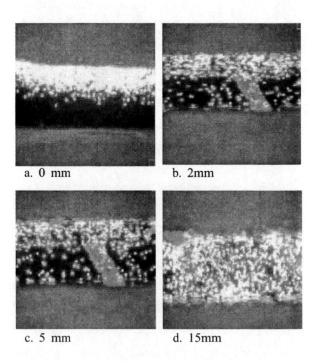

a. 0 mm   b. 2mm

c. 5 mm   d. 15mm

Fig 5. Photograph of the particle distribution on different positions of the herringbone channel.

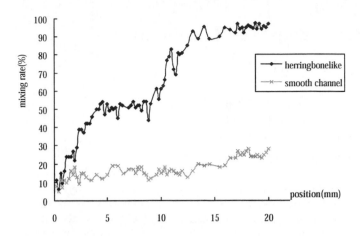

Fig 6. Mixing rate, on different positions for different channel designs.

The mixing rate for the mixers with or without herringbone structures was analyzed by image processing technique. The images were analyzed by ratio the area occupied by fluorescent particles to that of the whole channel to calculate the mixing effect, and the result is shown in Fig. 6. It's obvious that in the smooth channel, the mixing rate from the beginning to the end is always less than 20%, revealing a very poor mixing in this region. However, in the channels with herringbone structures, the mixing rate increases rapidly to about 100% at the end of channels because of the three-dimensional motion in particles. Thus it is almost 5 times better than that of smooth channel at the mixing distance of 15mm.

CONCLUSIONS

A power-free device is proposed in this paper employing only the surface tension energy defined by surface properties of the flow passages combined with herringbonelike concave structures to passively promote mixing effectively. Fabrication has been fully completely, and the designed hydrophilic channel area, hydrophobic spacer area, has also been demonstrated. Besides, the fluidic channel was designed without sidewall, it can increase the velocity of fluid flow in the channel. The testing results show that the mixing rate of the mixer with herringbonelike structures is 5 times better than that of smooth channel at the mixing distance of 15mm. This result is sufficient to prove that the herringbone structures in the micro channel can efficiently improve the mixing process efficiency. This device can be applied to power-free micro total analysis system or Lab-on-a-chip system in the future.

ACKNOWLEDGEMENTS

The work was supported by the National Science Council of Republic of China under the grant NSC-90-2212-E-002-204-.

REFERENCES

[1] Virginie Mengeaud, Jacques Josserand, and Hubert H. Girault, "Mixing Processes in a Zigzag Microchannel : Finite Element Simulations and Optical Study", *Analytical Chemistry*, Vol. 74, No. 16, August 15, 2002

[2] Jens Branebjerg, Peter Gravesen and Jens Peter Krog, "Fast Mixing by Lamination", *MEMS 1996*, pp.441-446

[3] Abraham D.Stroock, Stephan K. W. Dertinger, Armand Ajdari, Igor Mezic and Howard A.Stone, "Choatic Mixer for Microchannels", *SCIENCE, Vol. 295*, January 2002.

[4] Hiroaki.Suzuki, Chin-Ming Ho, "A Magnetic Force Driven Chaotic Micro-Mixer", *MEMS 2002*, pp. 40-43

[5] Branebjerg, J.; Gravesen, P.; Krog, J.P.; Nielsen, C.R.,"Fast mixing by lamination" 1996, Proceedings IEEE MEMS '96, pp. 441 -446. 1996.

[6] Deval, J. ;Tabeling, P.; C.M. Ho, "A dielectrophoretic chaotic mixer Deval" Proceedings IEEE MEMS' 02, pp. 36-39, 2002.

[7] Thomas K.Jun and C.J.Kim, " Valveless pumping using traversing vapor bubbles in microchannels", Journal of Applied Physics, Vol 83, No.11, 1 June 1998.

[8] Junghoon Lee and C.J. Kim," Liquid Micromotor Driven By Continuous Electrowetting", Proc. 1998 IEEE MEMS Conference, pp.538-543, Heidelberg, Germany, 1998.

Novel 2D Interfaces for Nanoelectrospray-Mass Spectrometry

S. Arscott,[1,2] S. Le Gac,[2] C. Druon,[1] P. Tabourier[1], C. Rolando.[2]

[1] IEMN, UMR CNRS 8520, 59652 Villeneuve d'Ascq, France
Steve.Arscott@iemn.univ-lille1.fr
[2] LCOM, UMR CNRS 8009, 59655 Villeneuve d'Ascq, France
Severine.Le-Gac@ed.univ-lille1.fr

ABSTRACT

We present the evolution of a novel two-dimensional 'nib-like' interface for nanoelectrospray-mass spectrometry (nanoESI-MS) applications. The design, fabrication and testing of three generations of such sources will be outlined here. The first generation of micro-nibs was fabricated using the negative photoresist (SU-8) by employing a simple 2.5D lithographic technique for rapid prototyping. The characteristic tip dimensions of such ESI-MS interfaces were 20×30 µm. The second generation of micro-nibs were also formed using SU-8 but using an improved 2D lithography and fabrication method. The smallest characteristic tip dimensions for these ESI-MS interfaces were 8×35 µm. A third generation of micro-nibs was fabricated from low-stress polycrystalline silicon (polysilicon) using silicon-based surface micromachining techniques. These micro-nibs were highly planar and had much smaller characteristic tip dimensions of 2×2 µm. MS testing on an ion trap mass spectrometer with standard peptides initially showed the feasibility of the 'nib-like' approach and the subsequent improvements obtained as the design and fabrication of the interfaces evolved.

Keywords: microfluidics, interface technology, micromachining, nanoelectrospray, mass spectrometry.

1 INTRODUCTION

Today, the exploding field of Proteomics provides an increasing amount of protein samples which need to be analyzed and identified. To these ends, Mass Spectrometry (MS) is a very powerful technique. This technique requires the test sample to be present in the machine as an ionic species. Electrospray ionization (ESI) [1-4] can achieve this and has the advantage of producing multi-charged ions which can bring high molecular weight molecules into the detection range [5]. However, even though MS equipment and techniques have highly evolved, ESI is itself still a relatively cumbersome technique. ESI is often performed using 'needle-like' capillary tubes manufactured from heating, pulling and breaking techniques. This fabrication method is inherently poorly controlled, resulting in the non-reproducibility of critical tip dimensions from tip to tip. This leads to poor reproducibility in nanoESI-MS analysis. The needle-like tip is manually loaded with the sample to be analyzed and then broken at the MS inlet resulting in a characteristic tip dimension (diameter) of 1-10 µm. It is clear that such a 3D needle-like topology prevents them from being compatible with the use of robotics for analysis

automation. In addition, a typical set-up to interface a microfluidic system to MS which employs such sources does not allow for optimizing the microsystem-MS coupling as the ionization tip is not part of the system. There is thus a current need for (i) stand-alone electrospray (ESI) sources with miniaturized, highly controlled tip dimensions which lead to nanoelectrospray (nanoESI) conditions, (ii) integrated nanoESI interfaces which provide world-to-chip interfaces to and from chip-based microfluidic systems and (iii) nanoESI interfaces which are compatible with robot automation for high-throughput and the inevitable reduced costs. Microtechnology can offer a solution to many of these requirements. The benefits of microtechnology for the productions of ESI sources are: (i) excellent control of critical dimensions, (ii) potential circuit-source integration, (iii) new device topologies, (iv) a greater choice of materials and (v) batch production.

The 'nib-like' approach which we present here borrows much from the topology of a simple fountain pen. Indeed, such a topology comprises the essential components required for an ESI tip: a reservoir, a capillary slot and a point-like feature. In addition to this, it is important to note that the fountain pen nib is in essence a quasi-2D object, this can be easily imagined if the nib is flattened. However, here the analogy will stop and we will proceed to present the design, fabrication and testing of three generations of novel nib-like sources intended for nanoESI-MS interfacing. The first two generations are fabricated using an epoxy-based negative photoresist (SU-8) and the third generation using Si-based microtechnology.

2 MICRO-NIB DESIGNS
2.1. First Generation Micro-nibs

Figure 1 shows a Scanning Electron Microscope (SEM) image of a first generation micro-nib nanoESI source. This first generation of sources was fabricated solely with the goal of testing the micro-nib concept. They were fabricated using the negative photoresist SU-8 2075 (Microchem, VA, USA) on a 3-inch silicon wafer. The essential features of the micro-nib are: (i) the main SU-8 support wafer, having a thickness of 400 µm, and (ii) a free-standing membrane-like point structure which contains the test liquid reservoir, a capillary slot and a nib tip. We estimated the characteristic tip dimensions to be of the order of 20 x 30 µm at the extremity of the tip. As the micro-nibs were fabricated using a simple 2.5D UV phtotolithographic method by varying the exposure doses for each feature, we have the advantage of the rapid production of prototypes.

Two simple photomasks were required for the fabrication. The process can also be defined a 'low temperature' as the maximum processing temperature was 95°C.

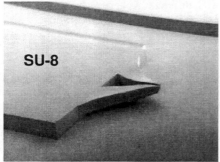

Figure 1 : SEM image of a first generation nanoESI-MS micro-nib. These sources were fabricated in SU-8 using a 2.5D photolithography The characteristic tip dimensions were 20 × 30 µm

In addition to this, device dicing was easy as the tip was not in proximity to the wafer surface following photoresist development. However, as the thickness of the membrane-like feature was difficult to control during UV dosage and photoresist development, the critical tip dimensions were thus difficult to control. Close inspection of tip dimensions proved that the effect of this was to reduce the device yield. In addition, the high tensile stress incurred whilst employing thick films of SU-8 tends to reduce the planarity of free-standing structures. The fabrication of these micro-nib prototypes has been described in detail elsewhere [6,7].

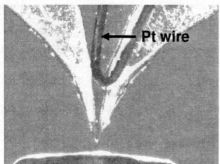

Figure 2 : Photograph taken during ESI-MS interfacing of a first generation micro-nib using a platinum wire to apply the high voltage (HV) to the test liquid.

The nib tips were tested in mass spectrometry on an ion trap mass spectrometer (LCQ Deca XP+, Thermo Finnigan). The nib tip was mounted on a tip holder which was then introduced in the mass spectrometer inlet. The high voltage (HV) was applied to the micro-nib via a platinum wire which was inserted in the reservoir feature of the nib tip, as shown on figure 2. MS tests were carried out using standard peptide samples which were prepared using a H2O/MeOH 50:50 solution acidified with 0.1 % HCOOH. Gramicidin S was mainly used as standard peptide for these experiments; it was tested at 50 down to 1µM. HV supply was in the 1.2-2.5 kV range.

Firstly, a camera allowed us to directly observe the tests; thus, the flowing of the liquid in the capillary slot could be observed as well as the formation of a Taylor cone (see figure 2) as soon as the voltage was applied to the liquid. Secondly, the signal (ionic current) in MS was observed to

be stable, reflecting thus the stability of the spray using these micro-nib tips (data not shown).

Figure 3 presents a typical mass spectrum obtained using one of first generation nib tips with two peaks at m/z 571.3 and 1141.7 corresponding to the Gramicidin S $(M+2H)^{2+}$ and $(M+H)^+$ species respectively. Nonetheless, it should be noted that the slightly more intense peak corresponds to the singly charged species.

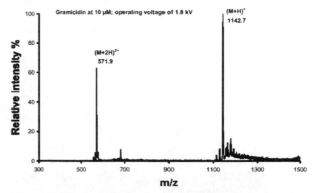

Figure 3: Mass spectrum obtained using a first generation nanoESI micro-nib having characteristic tip dimensions of 20 × 30 µm (Gramicidin S at 10 µM and an HV of 1.8 kV)

In conclusion, these MS tests using standard peptides allowed us to validate the idea of an ESI micro-nib tip, as these tips performed well also for MS/MS experiments. Nonetheless, the optimal functioning conditions for this first generation of micro-nibs did not satisfy us, as these were a 1.8 kV ionization voltage and a peptide sample concentration of around 10 µM. Thus, we proceeded to improve the design to produce a second generation of micro-nibs.

2.2 Second Generation Micro-nibs

Figure 4 shows an SEM image of a second generation nanoESI micro-nib.

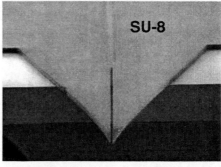

Figure 4 : SEM image of a second generation nanoESI-MS micro-nib. 2D photolithography was used to fabricate the micro-nibs in SU-8.

These micro-nibs were fabricated in a 2D topology using the negative photoresist SU-8 2035 on standard 3-inch silicon n-type substrates orientated (100). Firstly, a 200 nm nickel etch release layer was deposited onto the Si wafer surface. This nickel layer was then patterned using a photomask to form localized etch-release pads. Micro-nibs were then fabricated from SU-8 to have a thickness of 35 µm. Again, standard UV photolithographic techniques were employed in order to form the nibs in SU-8 which were composed a reservoir and a capillary slot, having an aspect ratio of >4, leading to the nib tip and microfluidic channels linking a main test liquid reservoir to the input of the capillary slot. The design contained a smallest characteristic tip dimensions of 8 µm × 35 µm. After development of the photoresist, the nickel etch-release layer was removed. Following this, the individual systems were diced using a technique which ensures that the nib tips remained unaffected by the wafer cleaving and thus gave a good fabrication yield. The fabrication details have been presented elsewhere [7,8]. It can be noted that the use of SU-8 gives rapid prototyping and low temperature processing. However, the main advantages of the second generation micro-nibs over the first generation are (i) improved control of critical tip dimensions due to the more 2D nature of the fabrication, (ii) smaller critical dimensions , i.e. 8 µm, (iii) improved cantilever planarity due to reduced stress of a thinner SU-8 film and (iv) improved yield due to improved fabrication process.

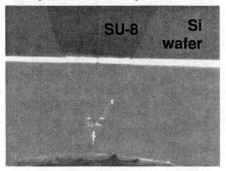

Figure 5 : Photograph of a second generation micro-nib during the nanoESI-MS tests. The high voltage (HV) was applied directly to the rear surface of the silicon.

The MS tests of these second generation nib tips were first carried out under the same conditions as described previously with the application of the ionization voltage on a Pt wire inserted in the nib reservoir, figure 5. The capillary slot width of the nib tip was either 16 µm or 8 µm. The test conditions consisted here of lower ionization voltage values, typically 0.8-1.5 kV and less concentrated peptide solutions, i.e. 1-5 µM.

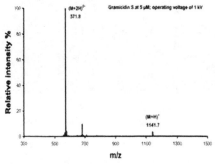

Figure 6 : . Mass spectrum obtained using a second generation nanoESI micro-nib having characteristic tip dimensions of 8 × 35 µm (HV supply of 1 kV, Gramicidin S at 5 µM).

Figure 6 presents the typical mass spectrum obtained with the 8 µm slot width nib tip with, as before, two peaks. However, it should be noted here that the mass spectrum pattern was different with a very intense peak for the doubly charged species and almost none for the singly charged one. Thus, the influence of the nib slot width was further investigated; 2 nib tips with a slot width of 8 or 16 µm were tested in the same HV and concentration sample conditions. This allowed us to observe how the slot width could influence the ionization phenomena and as a consequence, the intensity ratio for the singly and doubly charged species; the narrower the slot, the higher the intensity of the $(M+2H)^{2+}$. These MS results were compared to those obtained with a commercial tip (Proxeon). Using this latter tip, the mass spectrum pattern was seen to be the same as with a 8 µm slot width nib tip; this confirms that the ionization performances were enhanced using a nib tip having smaller dimensions.

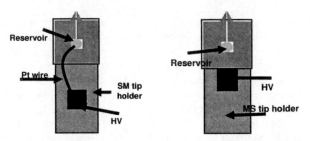

Figure 7: Set-ups used for testing the nib tips in mass spectrometry: (left) using a Pt wire that allows to apply the HV on the test liquid, (right) using the semi-conducting properties of the supporting Si wafer.

Lastly, we performed a last series of MS tests, whereby the ionization voltage was directly applied onto the supporting silicon wafer (see Figure 7). This greatly simplifies the test set-up as no manual step was required anymore to prepare the test. Once the tip mounted on the tip holder, the test liquid was simply dropped using a deposition robot into the reservoir feature and the tip introduced in the ion trap inlet. For this second series of tests, various samples were employed; standard peptides as before: Gramicidin S, Glu-Fibrinopeptide B and a Cytochrome C digest (figure 8).

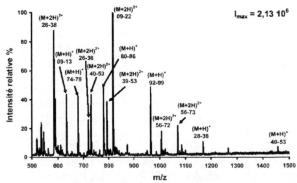

Figure 8: mass spectrum obtained using a 8 µm slot width nib tip for a biological sample, i.e. a Cytochome C digest (mixture of around 12 peptides); sample concentration of 1 µM and HV of 1.2 kV.

Irrespective of the nature of the sample, the nib tips were seen to perform even better than by applying the HV on a Pt wire. For example, in what we called 'limit conditions' using a Pt wire, the mass spectrum pattern did not look like the Proxeon one, i.e. there were two peaks with comparable intensities for both Gramicidin S ionic species. However, using this novel simplified set-up, better ionization performances could be yielded and the mass spectrum pattern was close to the Proxeon pattern.

As a conclusion, this second series of nib tip prototypes revealed that a decrease of the tip dimensions went along with better ionization performances. The nib tips with a 8 μm slot width performed as well as commercial nanotips. In addition, these nib tips were tested in a set-up fully compatible with analysis automation. Thus, we achieved our double goal with these nib tips; their geometry is fully appropriate for automation through the use of robotics and these microfabricated tips could be used in stand-alone conditions as well as being integrated as part of a microsystem. As the smallest critical dimensions for these micro-nibs were 8 × 35 μm, we have developed a process using silicon-based micromachining in order to fabricated micro-nib having critical dimension approaching 1μm.

2.3 Third Generation Micro-nibs

Figure 9 and 10 show SEM images of the polysilicon based micro-nibs produced for the third generation of nanoESI-MS interfaces. We fabricated this third generation of micro-nibs using Si-based microtechnology. The micro-nib cantilevers were made from polysilicon. A series of three photomasks were employed to produce: (i) pre-defined cleaving lines, (ii) localized sacrificial etch release pads (SiO$_2$) and (iii) micro-nib structures from polysilicon which contain: a reservoir, a capillary slot and a pointed cantilever. A very high planarity was achieved for the cantilevers as the deposition of the polysilicon was optimized in order to reduce the stresses to near-zero. In addition to this, a very high length/thickness ratio is obtainable due to the high Young's modulus of polysilicon compared to SU-8. The critical tip dimensions for this generation were 2 × 2 μm, much smaller than those obtainable with the second generation micro-nibs. Using Si-based microtechnology we achieved an excellent control of layer deposition. Figure 11 shows a global view of a stand-alone Si-based micro-nib (3rd generation). The details of the fabrication will be published elsewhere.

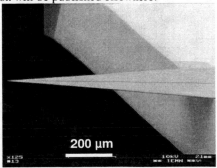

Figure 9 : Microscope photograph of a third generation of nanoESI-MS micro-nibs made from polysilicon. The characteristic tip dimensions were 2 × 2 μm

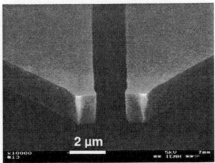

Figure 10: Close-up SEM image showing the critical dimensions (2 × 2 μm) of the capillary slot at the tip on the micro-nib made from polysilicon.

Figure 11: View of a third generation nanoESI-MS micro-nib fabricated from polysilicon using Si-based micromachining techniques.

3 CONCLUSIONS AND FUTURE WORK

We have presented a novel 2D electrospray source based on the idea of a nib. Such emitter tips have been successfully fabricated using the negative photoresist SU-8 and were seen to perform in nanoESI-MS mode. A third generation of micro-nibs has been successfully fabricated using Si-based microtechnology. The sources have several advantages over existing standard ESI sources: (i) excellent control of critical dimensions through microtechnology, (ii) simple easy-to-use stand-alone use (iii) integration with microfluidic circuitry and compatibility with the use of robotics for high throughput.

REFERENCES

[1] Chapman, S.; Physical Review; 10, **1937**, pp184-190

[2] Dole, M.; Mack, L.L.; Hines, R.L.; Journal of Chemical Physics, 49, **1968**, pp2240-2249

[3] Yamashita, M.; Fenn, J.B.; Journal of Physical Chemistry, 88, **1984**, pp4451-4459

[4] M.L. Alexandrov, L.N. Gall, N.V. Kravnov, V.A. Nicolaev, V.A. Pavlenko and V.A. Shkurov, Dokl. Akad. Nauk. SSSR 277, 379 (**1984**).

[5] Gieniec, J.; Mack, L.L.; Nakamae, K.; Gupta, C.; Kumar, V.; Dole, M.; Biomedical Mass Spectrometry, 11, **1984**, pp259-268

[6] Arscott, S.; Le Gac, S.; Druon C.; Tabourier P.; and Rolando C.;, J. Micromech. Microeng. 14, (**2004**) pp310-316

[7] Le Gac, S., Arscott S. and Rolando, C.; J. Am. Soc. Mass Spectrom. (in press)

[8] Arscott, S.; Le Gac, S.; Druon C.; Tabourier P.; and Rolando C.; Sensors and Actuators (in press)

[9] Le Gac, S., Arscott, S., and Rolando, C.; Electrophoresis 24, (**2003**) pp3640-3647

DEP Particle Dynamics and the Steady Drag Assumption

DEP Particle Dynamics and the Steady Drag Assumption
S.T. Wereley*, I. Whitacre*, R. Bashir** and H.B. Li**

*Purdue University, School of Mechanical Engineering,
West Lafayette, IN 47907-2088, USA, wereley@purdue.edu
**Purdue University, School of Electrical and Computer Engineering,
West Lafayette, IN 47907-2088, USA, bashir@purdue.edu

ABSTRACT

The interaction of fluid drag, dielectrophoretic forces, and Brownian motion on a nanoparticle's motion is studied using a microfluidic chip with interdigitated electrodes. The flow domain is a 11.6 μm deep by 350 μm wide channel with 23 μm wide electrodes located at the bottom surface of the channel and a glass top surface. The electrodes are covered by a thin layer of silicon dioxide to insulate them from the fluid medium, suppressing electrolysis and local Joule heating. Although these phenomena have been considered by other researchers, our experiments and modeling reveal it to be a considerably more complicated phenomenon than previously thought. Using an adapted micro Particle Image Velocimetry technique along with microscopic imaging, particle motion in three-dimensions is measured and compared to predicted results, showing not only the expected horizontal DEP retarding force but also a vertical force away from the electrodes. Further, because of the spatially varying nature of both the DEP force as well as the drag force, one of the main assumptions made in many previous DEP studies must be seriously questioned—whether steady low Reynolds number particle dynamics are insufficient to predict the particle behavior.

Keywords: dielectrophoresis, DEP, PIV, particle, velocity

1 INTRODUCTION AND SET UP

1.1 DEP Device

A dielectrophoretic device has been designed to trap, separate, and concentrate biological components carried in solution. This is done by creating a dielectrophoretic interaction between the spheres and the fluid. The device was designed and manufactured by Haibo Li at Purdue University, a student in Prof. Bashir's research group [1]. The device consists of a microchannel with a depth of 11.6 μm, width of 350 μm, and length of 3.3 mm. The channel was anisotropically etched in silicon to produce a trapezoidal cross-section. The channel was covered by a piece of anodically bonded glass. A schematic view and digital photo of the device are shown in Figure 1. Bright regions represent platinum electrodes and the dark regions represent the electrode gaps. The electrodes are covered by

a 0.3 μm thick layer of PECVD silicon dioxide, which insulates the electrodes from the liquid medium, suppressing electrolysis. The electrodes are arranged in interdigitated pairs so that the first and third electrodes from Figure 1 are always at the same potential. The second and fourth electrodes are also at the same potential, but can be at a different potential than the first and third electrodes. An alternating electric potential is applied to the interdigitated electrodes to create an electromagnetic field with steep spatial gradients. Particle motion through the resulting electric field gradients causes polarization of the suspended components, resulting in a body force that repels particle motion into increasing field gradients. In the experiments, sample solutions were injected into the chamber using a syringe pump (World Precision Instruments Inc., SP200i) and a 250μl gas-tight luer-lock syringe (ILS250TLL, World Precision Instruments Inc.). The flow rate could be adjusted and precautions were taken to avoid air bubbles. An HP 33120A arbitrary waveform generator was used as the AC signal source to produce sinusoidal signal with frequency specified at 1MHz.

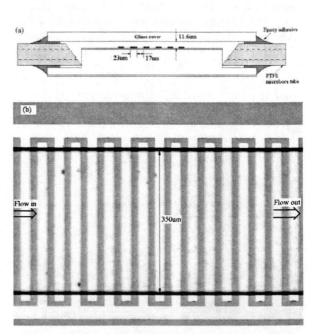

Figure 1. (a) Schematic view of experimental apparatus and (b) photo of apparatus.

1.2 Micro Particle Image Velocimetry

Particle image velocimetry (PIV) is a well accepted technique for acquiring full field velocity information from experimental fluid flows [2]. Micro particle image velocimetry (μPIV) has expanded conventional PIV to utilize microscopic imaging devices as developed by [3]. μPIV results have been used in a number of applications such as a detailed investigation of pressure driven flow through a microchannel [4], micron resolution temperature measurements [5], and investigations of dielectrophoretic particle trapping [6,7].

Figure 2 shows the typical layout of a μPIV system. The measurement volume can be illuminated by either a broad wavelength constant light source or a pulsed laser, such as a frequency doubled Nd:YAG. In these experiments a standard microscope mercury lamp is used. To ensure that all particles have the same properties, 0.7 μm polystyrene latex (PSL) microspheres (Duke Scientific) are suspended in de-ionized water in dilute concentrations (less than 0.1% by volume). Tracer particles in the measurement volume are coated with a red fluorescing dye (λ_{emit}=542 nm, λ_{emit}=612 nm). Images are captured with precise time delay separating exposures. The image sets are then divided into interrogation regions. The corresponding interrogation regions from consecutive images are cross-correlated to determine the most likely relative displacement of the particles in the interrogation regions in the form of a cross-correlation peak. Reducing the size of an interrogation region reduces the measurement volume and increases the spatial resolution of a measurement so long as displacements are less than half the size of an interrogation region, i.e. the Nyquist criterion.

The images were acquired using a Photometrics CoolSNAP HQ interline transfer monochrome camera (Roper Scientific). This camera is capable of 65% quantum efficiency around the 610 nm wavelength. The largest available image size that can be accommodated by the CCD array is 1392 by 1040 pixels, but the camera has the capability of pixel binning, which can drastically increase the acquisition frame rate by reducing the number of pixels that need to be digitized. A three-by-three pixel binning scheme was used in this experiment, producing images measuring 464 by 346 pixels, which were captured at a speed of 20 frames per second. The average focused particle diameter in the images was approximately 3 pixels.

Shallow Channel Considerations

When performing μPIV measurements on shallow microchannels, the depth of focus of the microscope can be comparable in size to the depth of the flow. A PIV cross-correlation peak, the location of which is the basis for PIV velocity measurements, is a combination of the velocity distribution in the interrogation region and some function of average particle shape. PIV velocity measurements containing velocity gradients can substantially deviate from the ideal case of depthwise uniform flow. Gradients in the in-plane direction have been addressed by image correction techniques [8], but gradients in the depthwise direction remain problematic. They can cause inaccurate velocity measurements due to the presence of multiple velocities within an interrogation region that are independent of mesh refinement. One problem with depthwise velocity gradients is cross-correlation peak deformation which reduces the signal to noise ratio of a PIV measurement [9]. Cross-correlation peak deformation can also reduce the effectiveness of subpixel peak fitting schemes which are based on a particular cross-correlation shape, such as a common five point Gaussian fit.

2 EXPERIMENTAL RESULTS

The experiments presented here designed to quantify the dielectrophoretic performance of the device. The experiments used six sets of 800 images each to analyze the effect of dielectrophoresis on particle motion in the test device. These images are high quality with low readout noise, as can be seen in the example fluorescent image of Figure 3. The top image demonstrates the many different particle intensity distributions which are typically present in a μPIV image in which the particles are distributed randomly within the focal plane. The bottom figure shows how, as the result of the DEP force, the particles migrate to the top of the channel and all have nearly identical images. The top image also shows that when a significant DEP force exists, the particles are trapped at the electrode locations by the increase in DEP force there. In general, the observed particle image shape is the convolution of the geometric particle image with the point response function of the imaging system. The point response function of a microscope is an Airy function when the point being imaged is located at the focal plane. When the point is displaced from the focal plane, the Airy function becomes a Lommel function [10]. Both Airy and Lommel functions feature concentric rings, although at different intensities

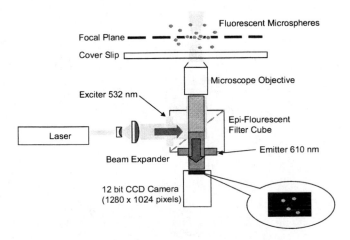

Figure 2. Diagram of typical μPIV system.

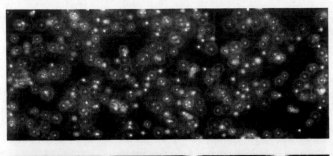

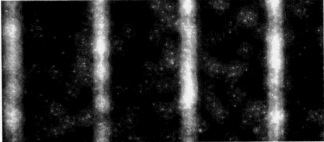

Figure 3. A photo of PSL particles with 0.0 volts (top) and 4.0 volts (bottom).

and diameters. For a standard microscope the diffraction limited spot size is given by

$$d_e = \frac{1.22 \cdot \lambda}{NA} \tag{1}$$

A numerical aperture (or NA) of 1.00 and an incident light wavelength λ of 540nm results in a diffraction limited spot size of 0.66μm, while the particles used are 0.69μm. Consequently the particle intensity distributions as recorded by the camera are partly due to the geometric image of the particle and partly due to diffraction effects. Hence, the distance of any particle from the focal plane can be determined by the size and shape of the diffraction rings.

A conventional μPIV analysis was performed to obtain an initial estimate of the average particle velocity field. Because the goal of the conventional μPIV analysis is not to extract velocity distribution, only the median velocity is reported. The spatial resolution can be higher than when using a velocity distribution analysis method such as the synthetic image method. In section 5.4, the velocity field will be extracted using the synthetic image approach. The PIV images were analyzed with *EDPIV*[1], an advanced μPIV interrogation package written in part at Purdue University by Lichuan Gui. An interrogation region of 32 by 32 pixels and a grid spacing of 8 by 8 pixels were used. Figure 4 shows vector plots for the experimental cases of 0.5 volts (top) and 4.0 volts (bottom). A uniform color scale was applied to these figures for representing velocity magnitude, such that velocity changes between cases can be

---

[1] Evaluation version available at
http://www.ecn.purdue.edu/microfluidics/Edpiv.zip

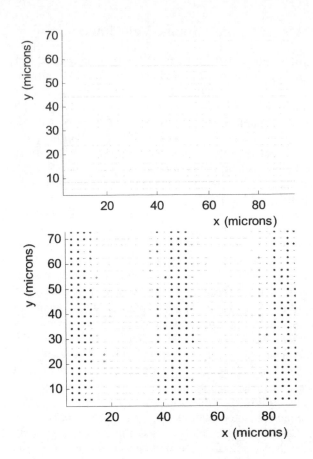

Figure 4. PIV vector plots for electrode voltages of 0.5 volts (top) and 4.0 volts (bottom).

more easily interpreted. In all cases, the electrode voltage signal frequency was set to 580 kHz. This was determined by sweeping between frequencies of 100 kHz to 10MHz and qualitatively determining the most effective trapping frequency.

The PIV results are summarized in Figure 5 which shows the average axial velocities for the six electrode

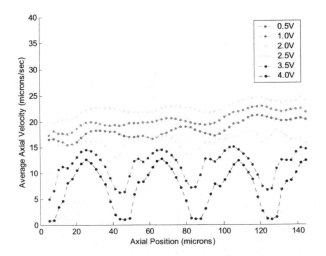

Figure 5. Average axial velocity from PIV results for all electrode voltage cases.

voltage cases. The three lowest voltages share a trend of decreasing particle velocity in the downstream direction. This phenomenon is evident in the average velocity curves in Figure 5. One explanation for this behavior may be that with each electrode a particle encounters, it lags the fluid velocity a little more. The cumulative effect results in a gradual slowing of the particle.

Another interesting result apparent from Figure 5 is that initially the average particle velocity increases as the voltage increases, 0.5 volts to 2.0 volts. This phenomenon is explained by particles being displaced from the channel bottom into faster areas of the fluid flow. This biases the velocity distribution toward higher velocities, altering the shape of the cross-correlation peak to favor higher velocities even though the fluid flow is constant. For higher voltages the effect of particles being hindered by axial field gradients is compounded by particles being forced beyond the high speed central portion of the flow profile by the DEP force. It can be qualitatively confirmed that particles migrate to the top of the channel by observing particle shapes in the images from the higher voltage cases, i.e. comparing the many particle shapes found in Figure 3 (top) which is acquired at 0.5 volts with the single particle shape found in Figure 3 (bottom) which is acquired at 4.0 volts. It is confirmed that the particles are indeed at the top of the channel by traversing the focal plane throughout the depth of the device.

The experimental results in Figure 5 were fit to the simple phenomenological model

$$V_{axial} = A \cdot x + B + C \cdot \sin\left(2\pi/D + E\right) \qquad (2)$$

by minimizing the root mean square difference between the fit function and the experimental data. The resulting constants capture important dynamics of the particle velocity. The A coefficient captures particles speeding up or slowing down as they pass through the device. The B coefficient captures the mean velocity of the particles within the device and the C coefficient captures the sinusoidal behavior of particles as they encounter each electrode. These results are shown graphically in Figure 6. The parameters for the cases of 0.5 to volts 2.5 volts are very similar, with the exception of a gradual increase in B. In the two highest voltage cases the linear slope A is reduced to nearly zero indicating that the particles do not speed up or slow down as they pass through the devices while the magnitude of the sinusoidal component C is greatly increased. From these results it is apparent that the particle velocities are reduced as particles travel downstream for low electrode voltages, and that the particles are dominated by a periodic trajectory for higher electrode voltages.

3 CONCLUSIONS

The dynamics of particles traveling through the device described in this paper are very complicated, exhibiting migration normal to the electrodes as well as trapping behavior in the plane of the electrode. Further work is needed to assess the accuracy of the steady drag assumption.

REFERENCES

[1] HB Li, Y Zheng, D Akin, R Bashir, "Characterization and Modeling of a Micro-Fluidic Dielectrophoresis Filter for Biological Species," *submitted* to J. Microelectromechanical Sys. (2004).

[2] R.J. Adrian, "Particle-imaging techniques for experimental fluid mechanics," Ann. Rev. Fluid Mech., Vol. 23, 261-304, 1991.

[3] J.G. Santiago, S.T. Wereley, C.D. Meinhart, D.J. Beebe, R.J. Adrian, "A particle image velocimetry system for microfluidics," Experiments in Fluids, Vol. 25, 316-319, 1998.

[4] C.D. Meinhart, S.T. Wereley, J.G. Santiago, "PIV measurements of a microchannel flow," Experiments in Fluids, Vol. 27, 414-419, 1999.

[5] V. Hohreiter, S.T. Wereley, M. Olsen, and J.Chung, "Cross-correlation analysis for temperature measurement," Meas. Sci. Tech., Vol. 13, 1072-1078, 2002.

[6] E.B. Cummings, "A comparison of theoretical and experimental electrokinetic and dielectrokinetic flow fields," AIAA 2002-3193, 2002.

[7] C.D. Meinhart, D. Wang, K. Turner, "Measurement of Ac Electrokinetic Flows," Biomedical Microdevices, Vol. 5, No. 2, 139-145, 2002.

[8] S.T. Wereley, L. Gui, C.D. Meinhart, "Advanced Algorithms for Microscale Velocimetry," AIAA J., Vol. 40, No. 6, 1047-1055, 2002.

[9] E.B. Cummings, "An image processing and optimal nonlinear filtering technique for PIV of microflows", Experiments in Fluids, Vol. 29, [Suppl.]:S42-50, 2001.

[10] M. Born, E. Wolf, Principles of optics, Oxford Press, Pergamon, 1997.

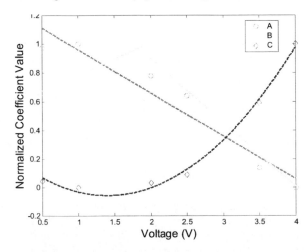

Figure 5. Curve fit parameters from Eq. 2 versus electrode voltage. The dashed lines represent linear, piecewise linear, and parabolic curve fits for parameters A, B, and C.

A New Method to Design Pressure Sensor Diaphragm

Xiaodong Wang*, Baoqing Li*, Sanghwi Lee*, Yan Sun*,
Harry T. Roman**, Ken Chin*, Kenneth R. Farmer*

*Microelectronics Center, Department of Physics, New Jersey Institute of Technology,
Newark, NJ USA, xw3@njit.edu
**PSE&G Company (T-10A) 80 Park Plaza, PO Box 570
Newark, NJ, USA, Harry.Roman@pseg.com

ABSTRACT

Over the last decade, silicon pressure sensors have undergone a significant growth. In most cases, these MEMS (Microelectromechanical System) devices are manufactured from rectangular or circular diaphragms whose thickness is constant and in the order of some microns. The development of high-performance diaphragm structure is of critical importance in the successful realization of the devices. In particular, diaphragms capable of linear deflection are needed in many pressure sensors. In order to increase the sensitivity, the diaphragm thickness should be thin to maximize the load-deflection responses. On the other hand, thin diaphragm under high pressure may result in large deflection and nonlinear effects that are not desirable. It is therefore important to characterize the relationship between diaphragm thickness, deflection, and sensitivity, both analytically and experimentally in order to establish the design guidelines for micro pressure sensors.

Keywords: resonant frequency, thickness, diaphragm, deflection

INTRODUCTION

The load-deflection method is a well-known method for the measurement of elastic properties of thin films. In this technique, the deflection of a suspended film is measured as a function of applied pressure. The load-deflection relation of a flat square diaphragm is given by [1]:

$$\frac{Pa^4}{Eh^4} = \frac{4.2}{(1-v^2)}\left[\frac{y}{h}\right] + \frac{1.58}{(1-v)}\left[\frac{y}{h}\right]^3 \tag{1}$$

Where
P applied pressure (Pascal),
y the center deflection of the diaphragm
a the half side length,
E young's modulus,
h the diaphragm thickness,
v Poisson's ratio of the diaphragm material.

The deflection range is divided into two regions: a small deflection region and (deflection less than 25% of the diaphragm thickness) described by the linear term in Equation 1 and a large deflection region described by the non-linear, cubic term in Equation 1.

As a general rule, the deflection of the diaphragm at the center must be no greater than the diaphragm thickness; and, for linearity in the order of 0.3%, should be limited to one quarter the diaphragm thickness.

For Linear Part

$$y = \frac{Pa^4(1-v^2)}{4.2Eh^3} \tag{2}$$

For a diaphragm clamped on its edges, from a mathematical point of view, the diaphragm can be viewed as a thin plate and the exact solution for the differential equation that describes its oscillation is given in [1]. Assuming that the deflection of the diaphragm is small compared to its thickness. The Reyleigh-Ritz method [2] was used to find the frequency of the lowest mode of vibration. The relationship for fundamental frequency for a square plate having density ρ_d and osculating in a medium (ρ_m) with was found to be:

$$f_1 = \frac{10.21}{a^2\sqrt{1+\beta}}\sqrt{\frac{gD}{\rho_d h}} \tag{3}$$

Where g is acceleration of gravity, and β and D is given by:

$$\beta = 0.6689\frac{\rho_m}{\rho_d}\frac{a}{h}$$

$$D = \frac{Eh^3}{12(1-\mu^2)}$$

SIMULATION

Figure 1 shows the relationship among side length, thickness, and natural frequency. With a given side length, the resonant frequency increases with thickness. With a given thickness of diaphragm, the frequency decreases with side length. Figure 1 also shows us that thickness has more effect on resonant frequency than side length.

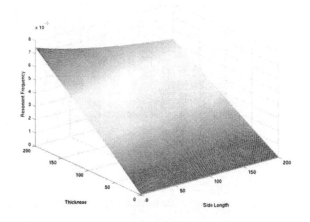

Figure 1 Relationships among side length, thickness and resonant frequency

According to Equation [2]

$$a = \left(\frac{4.2Eh^3 y}{(1-v^2)P} \right)^{1/4} = \left(\frac{50.4Dy}{P} \right)^{1/4}$$

Given E 150 GPa and v 0.25, Equation [3] becomes

$$f_1 = 4.5 \left(\frac{P}{y \left(\rho_d h + 0.6689 \rho_m \left(\frac{y}{P} \right)^{1/4} h^{3/4} \right)} \right)^{1/2}$$

Where ρ_d is 2330 kg/m^3 and ρ_m 1.25 kg/m^3.

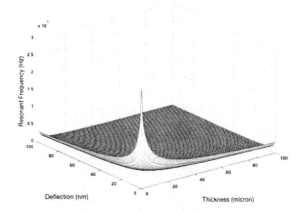

Figure 2 Relationships among deflection, thickness and resonant frequency

When the fixed pressure (1 Pa) is applied to the diaphragm, the relationships among natural frequency, deflection and thickness of diaphragm are shown in Figure 2. Figure 2 tell us that in order to get large natural frequency and large deflection at the given pressure, the thickness should be thin.

According to Equation [2],

$$h = \left(\frac{Pa^4(1-v^2)}{4.2Ey} \right)^{1/3}$$

Then Equation [3] becomes

$$f_1 = \frac{467.5P^{1/3}}{a^{2/3}y^{1/3} \left(\rho_d + 5863\rho_m \left(\frac{y}{aP} \right)^{1/3} \right)^{1/2}} \quad (4)$$

When the fixed pressure (1 Pa) is applied to the diaphragm, the relationship among natural frequency, deflection and diameter of diaphragm is shown in Figure 3. Figure 3 tells us that in order to get large natural frequency and large deflection of diaphragm at given load, the side length need to be small. Figure 2 and 3 tell us in order to get the large deflection and large resonant frequency of the diaphragm at the same times, both the side length and thickness of the diaphragm should be small.

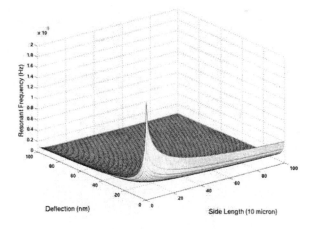

Figure 3 Relationships among deflection, side length and resonant frequency

Based on the understanding that partial discharge acoustic signal can occur at frequencies up to 100 kHz, and the fact that our proposed optical interrogation method should be capable of measuring displacements as low as 0.1 nm for 1 Pa sound pressure amplitudes, we found that many thickness/diameter combinations could be used to achieve a 200 kHz natural frequency, but the diameter must be less than ~1.4 mm in order to achieve the required displacement/pressure specifications according to equation 4. These results are illustrated in Figure 1, which shows that the curve denoted by triangles (1.4 mm diameter, 30 micron thickness) achieves 0.1 nm deflection for 1 Pa pressure. Smaller and thinner membranes can also be used, but these are less practical to fabricate.

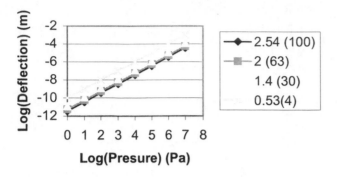

Figure 4 relationships among thickness, side length and deflection under same resonant frequency

EXPERIMENT RESULTS

A stainless steel housing for the sensor has been designed and fabricated. A finished system is shown in Figure 5. The sensor is placed at the end of a ~7" long tube that can be attached to a drain plug by a compression fitting connection. The end of the housing opposite the sensor has a stainless steel tee fitting for passing the optical fiber and allowing control of the sensor backside pressure and environment, as described above.

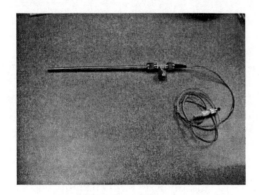

Figure 5 the optical fiber sensor

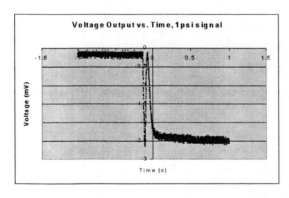

(I)

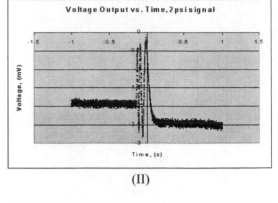

(II)

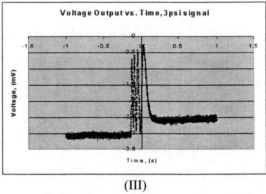

(III)

Figure 6 relationships between load and output

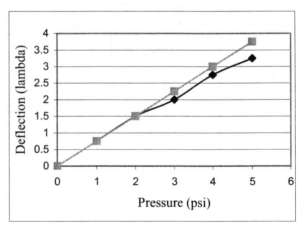

Figure 7 Comparison of the experimental and theoretical results

The frequency analysis of a sensor is important because we may get the resonant frequency of the membrane (actually also is sensor). Figure 8 shows sound response to different frequencies (below 100 kHz). It also shows that one of the resonant frequencies of the membrane is 88 kHz. It agrees with the simulation result.

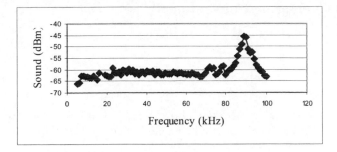

Figure 8 the frequency response of the sensor

CONCLUSIONS

The diaphragm is a very important part for MEMS sensor. The design of the diaphragm is critical to the output of the sensor. The relationships among side length, resonant frequency, deflection of the diaphragm are shown in the paper. According to the design rule, a optical fiber sensor was fabricated and tested. The calculated results agree with the experiments results.

REFERENCES

[1] Gregory T.A. Kovacs, "Micromachined Transducers Sourcebook", McGraw-Hill, 1998
[2] M. Di Giovanni, "Flat and corrugated diaphragm design Handbook", Mercel Dekker Inc., New York, 1982

Lattice Boltzmann Simulations Comparing Conventional
and a Heat Conduction Based Flow-Through PCR Micro-Devices

T. Mautner

SSC San Diego, San Diego, CA, USA, tom.mautner@navy.mil

ABSTRACT

This paper describes the lattice Boltzmann, numerical simulations used to evaluate a conduction-based, polymerase chain reaction (PCR) device for uniformity of the micro channel temperature profile and its velocity field distribution. For comparison purposes, numerical simulations were also performed using conventional PCR configurations. The simulation results demonstrate that in two-dimensions adequate performance of both the conventional and the heat conduction-based PCR systems. However, three-dimensional results indicate that nearly uniform wall temperatures (top, bottom and sides) are required to obtain uniform temperature fields in the PCR micro channels.

Keywords: lattice Boltzmann, PCR, microfluidics, simulations

1 INTRODUCTION

The current design considered for a bioagent detector includes a new heat conduction-based PCR [2] device requiring only two heaters. The two heaters will provide the temperatures required by denaturation and replication while heat conduction in the device substrate will provide the hybridization temperature. Prior to testing the new, heat conduction-based PCR device, numerical simulations were performed to evaluate the device and compare its performance to that of conventional, miniaturized flow-through PCR devices.

The fluid flow and temperature distributions in the PCR devices has been modeled using the lattice Boltzmann method. The lattice Boltzmann (LB) method [4] has been used as a computational fluid dynamics (CFD) tool for well over a decade. Unlike the lattice gas automata (LGA) and molecular dynamics approaches, the LB method simulates a flow system by tracking the evolution of particle distributions, not single particles. It differs from traditional CFD in that LB does not directly solve for the macroscopic variables found in the Navier-Stokes equations. The LB method, due to its kinetic nature, has a linear convection operator, and the local nature of the variables make it well suited for parallel computers. Additionally, the LB method is efficient, stable and allows "easy" implementation of boundary conditions and complicated geometries.

2 NUMERICAL METHOD

The technique used in computing incompressible flows with the LB method involves the time evolution of a density distribution function. The evolution of the Boltzmann equation has seen the development and adoption of the single relaxation Bhatnager, Gross and Krook (BGK) model for the collision operator which simplifies the operator and removes the Galilean invariance and dependence of pressure on velocity. It can be demonstrated that the Boltzmann equation recovers the Navier-Stokes equations via the Chapman-Enskog expansion. The LB method requires two steps, relaxation and streaming. During the relaxation step, distributions at the grid nodes relax to the equilibrium state according to the BGK rule. Then, in the streaming phase, the distribution advects freely at the characteristic velocity to the next node. It should be noted that the order of relaxation and streaming is not important. Once the distribution function is known, the macroscopic velocity and pressure can be calculated from its first two moments.

Building on previous work [3], the lattice Boltzmann BKG model in lattice units has the form

$$
\begin{aligned}
f_{\alpha,m}(\vec{x} + \vec{e}_\alpha \Delta t, t + \Delta t) - f_{\alpha,m}(\vec{x},t) = \\
(1/\tau_m)\left[(f_{\alpha,m}(\vec{x},t) - f_{\alpha,m}^{eq}(\vec{x},t)) \right] \\
+ F_T(\vec{x},t) \quad m = 1, 2
\end{aligned}
\tag{1}
$$

where f is the particle distribution and m=1,2 indicate equations for the density and temperature fields.

The relaxation coefficients are given by τ_m and the coupling between the velocity and temperature fields is achieved via the force coupling term F_T.

The numerical simulations used a 9 bit (D9Q2) for two-dimensions and a 19 bit (Q19D3) lattice BGK model in three-dimensions. In this formulation the density, ρ, momentum, ρu, temperature, T, pressure, p, and sound speed, c_s, are determined using the expressions

$$\rho(\vec{x},t) = \sum_{\alpha=0}^{18} f_\alpha \qquad \rho\vec{u}(\vec{x},t) = \sum_{\alpha=0}^{18} f_{\alpha,1}\vec{e}_\alpha \qquad (2)$$

$$T(\vec{x},t) = \sum_{\alpha=0}^{18} f_{\alpha,2} \quad p = c_s^2\rho \quad c_s = c/\sqrt{3} \quad c=1$$

The quantity c=1 indicates that in lattice Boltzmann units, the time and grid spacing are related by $\Delta x = c\Delta t$.

In the current formulation, the temperature, T, evolves according to

$$f_{\alpha_m,2}^{eq} = \frac{1}{6}\left[1 + 2\frac{(\vec{e}_\alpha \cdot \vec{u})}{c}\right]T(\vec{x},t) \quad \alpha_m = 1,....,6 \qquad (3)$$

Additionally, the viscosity and relaxation parameters for flow field, τ_v, and temperature, τ_T, have expressions which depend on the reference velocity and length, Reynolds number, viscosity and diffusion coefficient, D_T.

$$v = \frac{U_{in}L}{\text{Re}} \qquad \tau_v = \frac{6v+1}{2} \qquad \tau_T = \frac{6D_T+1}{2} \qquad (4)$$

3 BOUNDARY CONDITIONS

The boundary conditions used in the simulations specify the Reynolds number, fluid viscosity and diffusion coefficient. Equating the Reynolds number at full scale with the Reynolds number in lattice units provides the reference velocity, U_{in}, and the viscosity, v, which can then be computed using the equations given above.

A pressure driven flow can be achieved by specification of the inlet and outlet densities where $\Delta p = c_s^2\Delta\rho$. Alternatively, the flow may also be driven by application of a forcing term applied at each node whose magnitude is based on the pressure gradient in Poiseuille flow.

Additionally, the inlet and outlet velocities are specified. The inlet velocity components are (U_{in},0,0), and at the outlet the velocity is specified to have a zero gradient flow. The no-slip velocity boundary condition is applied to the channel walls using the bounce back technique in the LB formulation.

Finally, the wall temperature distribution is also specified on the boundary nodes, and the boundary temperature is obtained using the extrapolation procedure given by the equation

$$f_{\alpha,2}(x_b,t) = f_{\alpha,2}^{eq}(x_b,t) + f_{\alpha,2}(x_{b+1},t) - f_{\alpha,2}^{eq}(x_{b+1},t) \qquad (5)$$

4 TEMPERATURE-FLUID COUPLING

The forcing function, F_T, which provides the thermal-velocity coupling, can be computed using several parameters, the Rayleigh number, Ra, the viscosity, v, the Prandtl number, Pr and the reference length, L.

The magnitude of the forcing term will be determined using specific sets of variables depending upon if it is free or forced convection flow. Examination of the Nusselt number, Nu, indicates that for forced convection Nu=Nu(Re,Pr) and for free convection Nu=Nu(Re,Pr,Gr) where Re is the Reynolds number and Gr is the Grashof number. Typical values of constant portion of the forcing term range from $5.6 \times 10^{-6} \leq F_T \leq 5.6 \times 10^{-4}$ for z=25 grid cells and $5.6 \times 10^{-5} \leq F_T \leq 5.6 \times 10^{-3}$ for z=10 grid cells.

5 RESULTS

The lattice Boltzmann numerical scheme described above was used to compute the velocity and temperature fields in two- and three-dimensional geometries. The micro-channel geometry used is a representative section of a serpentine, continuous, flow-through PCR device having various aspect ratios and various corner configurations.

It was apparent early in the investigation that there were significant differences in the two- and three-dimensional simulation results. Straight and U-channels results demonstrate the effects of specified wall temperature distributions for both conventional and the conduction based flow-through PCR systems. The results, in the examples below, are given at time steps of 10K-20K.

5.1 Conventional PCR – Straight Channels

The first set of simulation results are presented in Figure 1 and illustrates the typical results obtained for three-dimensional simulations of conventional flow-through PCR. Conventional PCR in this context adheres to the notion of constant fluid heating in each temperature zone. The use of small thermal-fluid coupling, Figure 1a, results in nearly uniform temperature fields. However, specification of a larger thermal coupling force provides natural convection like effects resulting in significant non-uniform temperature fields as shown in Figure 1b. It should be noted that in these two cases, the bottom and side walls have the same temperature while the top wall has 0.95T of the bottom wall.

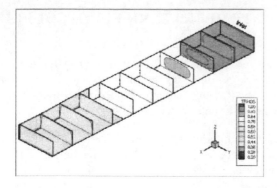

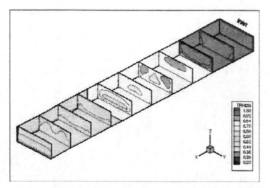

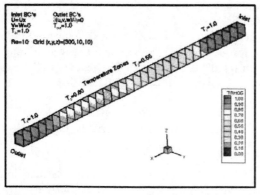

Figure 1. Straight channel section with aspect ratio of 3:1, Re=0.10 and grid (x,y,z)=200x32x10. From inlet to outlet, the specified wall temperatures are: bottom, T=1.0, 0.83, 0.63; top and sides, T=0.95T of bottom. (a) F_T=0.000056; (b) F_T=0.0056

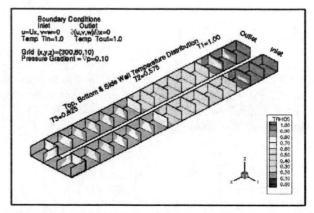

Figure 3. Temperature contours for a long U channel section with aspect ratio of 2:1, Re=0.1, grid (x,y,z)=300x60x10, and F_T=0.0056. From inlet to outlet, the specified wall temperatures are: bottom, top and sides, T=1.0, 0.575, 0.825, 0.575 and 1.0.

5.2 Heat Conduction Based PCR

The next two examples illustrate the effect of two- and three-dimensional geometries on the temperature field for the heat conduction-based PCR system considered for the biosensor design. First, the two-dimensional results are given in Figure 4, and indicate sharp transitions between the various temperature zones. Also, the computed velocity profiles have the expected shape characteristic of Poiseuille flow. These results are in agreement with other simulations found in the literature [1].

In contrast, the three-dimensional simulations using the same wall temperature distributions, shows the effects of volume (geometry) on the resulting flow characteristics, see Figure 5. The problems associated with non-uniform temperature distributions are more significant than those for the conventional PCR configurations. The concern here is that the temperature field will not be sufficient to maintain the sample temperature during each PCR stage during high volume processing. Also, the non-uniform temperature distributions may require longer channels (longer resident times) to achieve proper amplification of the sample.

Figure 2. Straight long channel section with aspect ratio of 1:1, Re=10, grid (x,y,z)=300x10x10, and F_T=0.000056. From inlet to outlet, the specified wall temperatures are: bottom and sides, T=1.0, 0.55, 0.80, T=1.0; top, T=0.95T of bottom.

The example, shown in Figure 2, illustrates the application of nearly uniform wall heating on a very long, square channel. The resulting fluid temperature is reasonably uniform except in the T=0.80 zone. This non-uniformity may be the result of having to abruptly transition from T=0.55 to T=0.80 in a short length of channel. Comparison of the results for Re=0.1-10 shows very little difference in the temperature contours over this range of Reynolds numbers.

As observed for straight channels, applying the same temperature on all walls of a U channel aids in providing a nearly uniform fluid temperature as demonstrated in Figure 3a. However, the problem of transitioning from the mid temperature T=0.575 to T=0.825 remains. The results also show the rotational flow due to the relatively small temperature gradient field in the T=0.825 region. The rotational flow does not help to completely mix the fluid temperature.

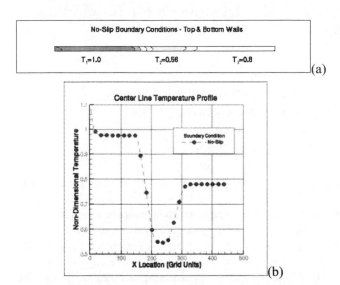

Figure 4. 2D simulation of the 2-heater PCR device, Re=0.10, time=20K, Grid 450x10, Forcing=0.00056: (a) temperature contours along the channel; (b) the computed centerline temperature distribution (similar to applied wall temperature boundary conditions)

6 SUMMARY

The two-dimensional flow channel results indicated that with nearly equal top and bottom wall temperatures, the PCR device would have essentially uniform temperature profiles across the channel. Thus, one would conclude that the PCR configuration would provide adequate sample amplification. However, the three-dimensional simulations present different findings. The three-dimensional simulations demonstrated that when the top and side wall temperatures were near ambient temperature there would be large temperature gradients and unsatisfactory temperature profiles across the channel. In an attempt to obtain uniform temperature profiles across the three-dimensional channel, various wall temperature distributions were tested. The most favorable temperature profiles consisted of specifying the top and side wall temperature to be on the order of 95-100% of the bottom temperature distribution.

ACKNOWLEDGMENTS

The author thanks the Joint Science & Technology Office for Chemical and Biological Detection for their support. Also, many thanks go to Dr. Peter Gascoyne, MD Anderson Cancer Center, U. Texas Houston and Dr. Shaochen Chen, U. Texas Austin for their continued discussions, encouragement and PCR device data.

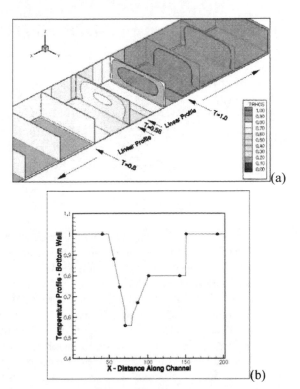

Figure 5. 3D simulations of the 2-heater PCR device, Re=0.10, time=10K, Grid 200x30x10, Forcing=0.000056: (a) temperature contours in a channel with 4 temperature zones, T=1.0, T=0.56, T=0.8 and T=1.0, close up view of the channel center region temperature distribution; and (b) wall temperature boundary condition.

REFERENCES

[1] Athavale, M., Chen, Z., Furmanczyk, M. and Przekwas, A., "Coupled multiphysics and chemistry simulations of PCR microreactors with active control," Modeling & Simulation Microsystems, 2001.

[2] Li, S. and Chen, S., "Design and analysis of a heat conduction-based continuous flow polymerase chain reaction system," ASME IMECE2002-33548, New Orleans, LA, 2002.

[3] Mautner, T. , "Investigation of enhanced mixing in microfluidic flows using lattice Boltzmann simulations of velocity and temperature fields," World Congress on Medical Physics and Biomedical Engineering, Sydney, AU, Aug., 2003.

[4] Succi, S., 2001, *The Lattice Boltzmann Equation for Fluid Dynamics and Beyond*, Clarendon Press, Oxford.

Geometrical Effects in Mechanical Characterizing of Microneedle for Biomedical Applications

Priyanka Aggarwal[*], C. R. Johnston[**]

[*]Department of Electrical and Computer Engineering, aggarwa@enel.ucalgary.ca
[**]Department of Mechanical Engineering, johnston@enme.ucalgary.ca
University of Calgary
2500, University Drive, Calgary, Alberta, Canada, T2N1N4

ABSTRACT

This paper discusses the design and implementation of MEMS microneedle with emphasis on integrated functionality for sensing the forces. The idea is based on the requirements of strength, robustness, and minimal insertion pain and tissue damage in patients The constraints of minimal microneedle dimension, force withstanding capabilities and their relations are the main basis for final design dimensions. Therefore, the design gives the efficient dimensions of the microneedle, while considering the influence of various mechanical forces and skin resistive forces. The design is extended by giving complete fabrication steps of the proposed microneedle design. Three piezoelectric sensors are fabricated on the microneedle, which reliable skin resistance measurement for different insertion depth. Performance analysis at the end proves the validity of the proposed design approach for different geometries and sizes.

Keywords: microneedles, mechanical forces, finite element modeling, skin resistance, piezoelectric sensing

1. INTRODUCTION

MicroElectroMechanical Systems (MEMS) is the integration of mechanical elements, sensors, actuators and electronics on a common silicon substrate through microfabrication technology [1]. The primary motivation for the development of microneedle is the need for needles that are economical to fabricate and which are small enough to reduce insertion pain and tissue trauma.

Today, the smallest conventional needle commercially available have a 305μm[2] outer diameter with a wall thickness of 76μm. Since traditional machining methods limit the diameter of the manufactured needles to 300μm, reductions in the size of the needles to ease the invasive effects of the needle are difficult to achieve with these methods. Further, it is extremely difficult to design and fabricate complex structures within a conventional needle.

On the other hand, microneedles are defined lithographically, so they can be designed for smaller size and geometry. Microneedles are extremely sharp, with sub-micron tip radii, as the fluid outlet is on the side of the needle. In addition, arrays of micro-needles with multiple outlet ports allow the injection of different fluids from the same device. Complex needle geometries such as 90-degree bends can be easily incorporated in microneedles [2].

In this work, performance analysis is made to validate the proposed design approach for different microneedle geometries and sizes. In addition, piezoelectric sensors are used to give reliable skin resistance measurement with the distance transversed into human skin.

This paper is organized into five Sections. Section 1 is the introduction. The various design issues related to selection of microneedle dimensions is covered in Section 2. The simulation results indicating the variation of forces are shown in Section 3. A brief overview of the fabrication steps are given in Section 4 and the paper is concluded in Section 5.

2. MICRONEEDLE DESIGN/FAILURE

Various factors are taken into account during the design of microneedles. Microneedles are far more reliable and efficient for transdermal drug delivery because of the small size. Also, microneedles may be used for obtaining precise blood samples, which are very helpful in the diagnosis of diseases. This paper presents the optimal design for the MEMS based microneedles. It takes into consideration the influence of mechanical forces (as compressive force, buckling force, free bending force, constrained bending and shear force) acting on the needle during insertion into human skin as well as skin resistive force. In addition, a comparison is drawn between several geometries and sizes of the microneedle.

The theoretical pressure required to pierce human skin has been given [3] as $3.183*10^6$ Pa ($P_{Piercing}$). This pressure is a major factor in determining the overall geometry of the MEMS microneedle. This pressure decreases drastically once the skin is punctured. The length of our MEMS based in-plane microneedle have been set at 600μm so as to withdraw blood samples from the capillaries in the dermis layer, which occurs at a distance of 500 to 2000 μm below the skin surface. These microneedles are long enough to withdraw sufficient bodily fluids from the dermis layer and yet small enough not to puncture pain causing nerves found in deeper tissues. Piezoelectric sensors are configured over the microneedle to detect the deforming forces acting on the microneedle as it penetrates through human body.

2.1 Modes of Loading

Analytically, microneedle dimensions can be calculated from the theoretical analysis for the compressive force, buckling, free bending, constrained bending and shear force, for a hollow beam with a fixed end. The tip force (exerted by the human skin) has to be smaller than any of these breaking forces. We have considered all of the above-mentioned forces and found that the forces with the most impact on the determination of the dimensions of the structure are the buckling and the bending forces as defined below.

The maximum Euler's buckling [2,4] force that a microneedle can withstand without breaking is given by the following equation:

$$F_{MaxBuck} = \frac{C\pi^2 EI}{L^2} \qquad (3.2)$$

where E =169 GPa represents the Young's Modulus of silicon, I (m^4) is the Moment of Inertia for different geometries as stated in Table 1, and L (m) is the length of the microneedle.

Type of Microneedle	Moment of Inertia
Circular Microneedle [1]	$I = \dfrac{\pi D^4 - \pi d^4}{64}$
Rectangular Microneedle [2]	$I = \dfrac{BH^3 - bh^3}{12}$
Square Microneedle [3]	$I = \dfrac{H^4 - h^4}{12}$

Table 1: Moment of Inertia for Type of Microneedle

In the computation, the base of the microneedle is modeled as a fixed joint giving end condition of C of 0.25.

The maximum free bending force that the microneedle can withstand is given by equation [2,4].

$$F_{MaxFreeBend} = \frac{\sigma_y I}{cL} \qquad (3.3)$$

where c (m) is the distance of the neutral axis to the outermost edge of the microneedle. For a circular microneedle, c is D/2 where D is the outer diameter while for rectangular microneedle c is H/2, H being the outer height. For square, c is H/2 where H is the outer dimension. Also, σ_y is the yield strength of the microneedle made of silicon i.e 7 GPa.

The resistance offered by skin before skin being punctured is given by the following equation:

---

[1] D is the outer diameter, d is inner diameter of microchannel running along length of microneedle

[2] B, H are the outer width and height while b, h are the inner width and height of the microchannel in the microneedle

[3] H is the outer dimension while h is the inner dimension of the microchannel

$$F_{Skin} = P_{Piercing} A \qquad (3.6)$$

where A is the cross sectional area of the microneedle. This force falls drastically once the skin is punctured [5, 6]. After the skin is pierced, the only force that acts on the microneedle is the frictional force due to tissue clamping the needle. This post-puncturing reduces the pressure $P_{Post-Punctured}$ to 1.6 MPa [5, 6] from the piercing pressure of 3.18 MPa.

2.2 Piezoelectric Sensing

High precision in the manual control of needles and biopsy probes in medical treatment requires high level of skill and dexterity. In anaesthesia, force sensation is an important feedback mechanism [5] and the practitioner needs to refresh or develop skills to improve the interpretation of the needles progress towards the target site. To have an idea of the force resisting progress of the microneedle into human skin, piezoelectric sensors have been included on the microneedle design. The progress of the microneedle in the body can be judged from the proportion of the microneedle remaining outside and force experienced during injection.

We know that, when a piezoelectric material is stressed mechanically by a force or moment, an electrical charge/voltage is generated. This is the basic sensing principle used in this design.

By expressing charges in terms of voltages, we get

$$V_{12} = \frac{d_{33} F_3 l_1}{\varepsilon \varepsilon_o A_{12}} \qquad (3.8)$$

where ε is the permittivity of air, ε_o is the relative permittivity of AlN and A is the area of the xy plane of the microneedle.

Here in this design, the force is being applied along z axis and the charge is measured along the z direction. Hence d_{33} coefficient is being used.

The sensor design consists of three layers. A single, unimorph 1 µm AlN layer is sandwiched between two electrodes layers each 0.8 µm thick. The top layer consists of three strips of chromium electrode and the bottom layer has a single common aluminum electrode covering the entire area occupied by the top three electrode strips, which has been distributed along the top surface area of the microneedle. The area of intersection between the top and bottom electrodes is where the piezosensitive material is located. The AlN layer consists of three strips of 20 µm width at interval of 300 µm. Three output voltages as given in Table 2 are obtained from this configuration of the sensing material. Here we are taking the length as 600 µm, cross sectional area of the microneedle to be 9575µm². Also to calculate voltage for sensor 1, $P_{Piercing}$ is used while to calculate voltage for sensor 2 and sensor 3, $P_{Post-Punctured}$ is being used. Fig 1 shows the variation of the three sensor voltages with the length of the microneedle.

Sensor	Sensing area (μm^2)	Position of sensor (μm)	Voltage (V)
Sensor 1	14500	1	0.1109
Sensor 2	8500	300	0.0952
Sensor 3	2900	600	0.2789

Table 2: Voltages Obtained at Various Depths

As the length of the microneedle increases, the voltages decrease. Voltage V1 remains same as it is independent of length of the microneedle.

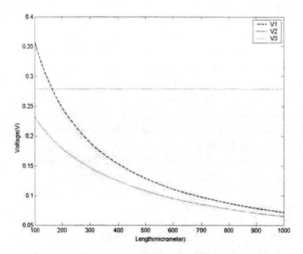

Figure 1: Variation of voltages

3. SIMULATION RESULTS

A cut-away view (showing the embedded channel) of a microneedle with a channel running along its length is shown in Fig 2. The dimensions of the microneedle are a width of 90 μm, height of 120μm, and length of 600μm, while the channel has a width and height of 35 μm. On substituting these values into the above mentioned formulas for rectangular microneedle, we get the maximum bending force 2.49N; buckling force of 14.851 N while the skin

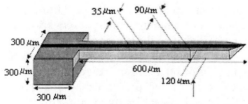

Figure 2: Microneedle dimensions

resistive force is 0.0305 N, making the design sufficiently strong.

The same results are confirmed using ANSYS. These stresses are obtained when a bending force of 2.1 N and buckling force of 11.135 N is being applied on rectangular, square and circular microneedle along the Y axis, taking the yield strength of silicon to be 7 GPa and young's modulus of 169 GPa. In Fig 3, the obtained yield stresses are plotted

along the Y axis, keeping X coordinate at 0 and Z coordinate at 52 μm (for Square microneedle), 45 μm (for Rectangular microneedle) and 65 μm (for Circular microneedle). These values are obtained on application of bending force of 2.1 N on the microneedles when the cross sectional area of the microneedles are kept same.

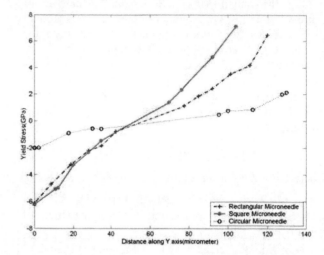

Figure 3: Yield Stresses for different shapes of microneedle

As can be seen in Fig 3, the values of stresses obtained for circular microneedle are less than those obtained for rectangular and square microneedles, indicating that circular microneedles can withstand more forces than rectangular or square microneedles.

In Fig 4, the Young's modulus of different shapes of the microneedle is plotted for varying distance along the Y axis, keeping X coordinate at 0 and Z coordinate at 52 μm (for Square microneedle), 45 μm (for Rectangular microneedle) and 65 μm (for Circular microneedle). From Fig 4, it is seen that the Young's modulus for the circular microneedle is the highest among the three. This shows that circular microneedle has more strength when cross sectional area of the microneedles is kept same.

Although it is established that the circular microneedles can withstand more forces than either rectangular or square microneedle, we are fabricating rectangular microneedles, the next preferred shape because of the fabrication limitations.

4. FABRICATION

This section covers the steps necessary to fabricate the microneedle in order to verify the theoretical performance shown above. In the first step of fabrication, a <100> silicon wafer is taken and cleaned with piranha. A layer of silicon dioxide of thickness 7500 ^{o}A [7] is thermally grown at 900 ^{o}C for 4 h over the wafer. Then, the wafer is cleaned in a solution of H_2SO_4/H_2O_2 and NH_4OH/H_2O_2 and dried in an oven at 150°C for 10 min. The oxide serves as an insulating layer. A layer of photosensitive material is spun onto the thermally deposited Si_2O layer.

A standard lithographic mask bearing the appropriate pattern to form the microchannel running along the length of the microneedle is positioned on top of the photoresist

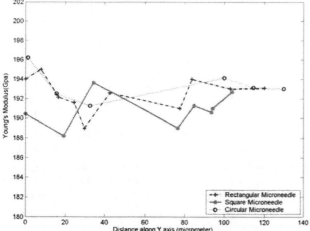

Figure 4: Young's Modulus for different shapes

layer. The wafer and photo-resist were then exposed to ultraviolet (UV) layer through the mask by means of optical mask aligner. The exposed photo-resist was removed by soaking the wafers in a liquid developer leaving the desired pattern of photo-resist. Now etching of the patterned oxide by DRIE [7] was done. This process forms the channel with the desired width 35μm and height 35μm in the bottom silicon wafer. The residual silicon dioxide is removed, followed by the necessary cleaning of the wafers before actual bonding takes place. The resulting wafer is fusion bonded with another silicon wafer [7]. The steps for the process include surface preparation, contact and finally annealing at 800 °C. The Si fusion bonding [7] is based on tendency of smooth surfaces to adhere. It utilizes thermal cycling after contacting to increase bond strength. Once the two wafers are bonded, the top wafer is subsequently thinned down, to achieve the desired top wafer thickness of 40 μm.

The sensor structure consists of three layers. The fabrication starts with sputtering of aluminum electrode over the entire bonded wafer surface to achieve Al layer thickness of 8000 °A. This forms the common bottom electrode layer for piezoelectric sensing of the forces as mentioned before. Magnetron sputtering of Al gives a deposition rate of 1μm/min using the following specifications: pressure of 3 mT, power 9.5KW, Al target and using argon at 30 sccms. Next, 1 μm thick layer of AlN is deposited by magnetron sputtering. This layer is then patterned and etched using 1(conc.) HF acid: 1 water because this mixture etches the underlying Al at negligible rate as compared to AlN layer. Approximately 8000 °A thick chromium layer is deposited onto the AlN layer using DC sputtering to form the top electrode for the sensor. The sputtering parameters are: base pressure 1.2x 10^{-6} Torr, Ar [7] pressure 5m Torr, substrate temperature 300 °C and sputtering power 400W. This layer is patterned and etched

to form the top three electrodes using (1:3 [50 g NaOH + 100 ml H_2 O]: [30 g K_3 Fe (CN) $_6$ + 100ml H_2 O]). This particular etchant is used in order to prevent etching of bottom AlN or Al layers. Now etching of unprotected bottom AlN and Al layers are carried out to form the desired shape. DRIE of the patterned silicon wafer is carried out using C_4F_8, SF_6, and Ar at etch rate of 2 μm/min, all the way down the wafer thickness to achieve vertical side walls. Finally, patterned backside etching. of the microneedle is carried out, to achieve the desired thickness of 120 um for the length of 600 um.

5. CONCLUSION

The design/dimension criteria of in-plane micro-needles have been discussed in this paper. It is established that the circular microneedle can withstand more amount of force than either rectangular or square microneedles. The importance of the skin resistance on the dimensions (length, cross sectional area) of the microneedle is a novel approach not generally considered in microneedles design. Integrated sensor design for sensing the skin resistance at different depths inside the tissue has been presented.

ACKNOWLEDGMENT

The authors would like to thank K.V.I.S Kaler, Department of Electrical Engineering at University of Calgary and Natural Sciences and Engineering Research Council of Canada (NSERC) for their valuable guidance and support.

REFERENCES

[1] William S Trimmer, "Micromechanics and MEMS: Classic and Seminal Papers to 1990", IEEE press, 1997
[2] J.D.Zahn, N.H.Talbot, D.Liepmann, A.P.Pisano, "Micro-fabricated Polysilicon Microneedles for Minimally Invasive Biomedical Devices Biomedical Microdevices" Vol 2, (2000), pp. 295-303.
[3] S. Henry, D.V. McAllister, M. G. Allen, M. R. Prausnitz, "Microfabricated microneedles: A novel approach to transdermal drug delivery", Journal of Pharmaceutical Sciences, Vol. 87, No. 8, August, (1998), pp. 922 – 925.
[4] J.Gere, S. Timoshenko, "Mechanics of Materials", fourth ed, May 1997
[5] Shankar Chandrasekhar, A Bruno Frazier, "Mechanical Characterization of Surface Micromachined Hollow Metallic Microneedles", Microtechnologies in Medicine & Biology 2nd Annual International IEEE-EMB Special Topic Conference, 2-4 May 2002, pp. 94 –98
[6] PN Brett, T J Parker, A J Harrison, T A Thomas and A Carr, "Simulation of Resistance Forces Acting on the Surgical Needles" 2nd International Workshop on Mechatronics in Medicine and Surgery, Bristol, 1997
[7] Marc Madou, "Fundamentals of Microfabrication", CRC Press, 1997

Design and Manufacturing of a Miniaturized 2.45GHz Bulk Acoustic Filter by High Aspect Ratio Etching Process

C.-H. Lin[*], H.-R. Chen[**], C.-H. Du[**] and W. Fang[*]

[*]National TsingHua University, Hsinchu, Taiwan, fang@pme.nthu.edu.tw
[**]Asia Pacific Microsystems, Inc., Hsinchu, Taiwan, chdu@apmsinc.com

ABSTRACT

A 2.45GHz filter is fabricated by thin film bulk acoustic wave resonator (FBAR) technology. Especially, it is designed to minimize die size and to simplify fabrication process simultaneously by using inductive coupling plasma etching. The quality factor of fabricated single resonator is 1567 and electromechanical coupling coefficient is 5.7%. The fabricated filter has minimum insertion loss around 1.5dB and maximum insertion loss 3.6dB in 83.5MHz bandwidth, which is suitable for WLAN and Bluetooth applications.

Keywords: bulk acoustic device, ICP, MEMS resonator

1 INTRODUCTION

The requirements of filters become more and more significant as wireless communication systems develop into higher frequency region. FBAR is a promising technology for these high frequency filters, because it has high power handling capability and it can precisely controls the resonance frequency by thin film thickness. Presently, the fabrication method for FBARs can be classified into three categories: back-side wet etching processes [1,2], front-side etching process [3,4], and multi-layer reflection method [5,6]. However, the multi-layer reflection method requires carefully controlled thickness for each layer. The front-side etching process face obstacles at the releasing step when chemicals have low selectivity between surface devices and sacrificial layer. And the back-side wet etching process has drawbacks of large device size and also has difficulties in front-side protection. In this study, Inductive Coupling Plasma (ICP) back-side etching process was used. Since ICP provided high aspect ratio etching, small device size can be fabricated. Moreover, the dry etching and high material etching selectivity characteristics of ICP process solved the front-side protection problem.

2 DESIGN AND SIMULATION

Thin film bulk acoustic resonators operate like quartz resonators, except that they are fabricated by thin film technology, therefore the operation frequency is much higher than quartz resonators. A basic FBAR structure includes a piezoelectric layer between two electrode layers.

And the stacked structure is suspended so that both the top and bottom surface are free to deform in out-of surface direction. Since longitudinal acoustic wave perfectly reflect at free surfaces, e.g. solid-air interface, standing waves can be created in the FBAR structure. Piezoelectric film then transfers energy from acoustic standing waves to electrical resonance.

An accurate model is required to simulate the complex behavior of FBAR. In this work, acoustic transmission line model was used. This model treats every acoustic layer segments as microwave transmission lines using force-voltage analogy. In addition, conventional Mason equivalent circuit model was used for piezoelectric material, which comprised two acoustic ports and one electrical port [7]. Connecting Mason's piezoelectric model and acoustic transmission line models in series, frequency response of the stacked structure was simulated. Figure 1 demonstrated the simulated frequency response of the proposed FBAR. The relationship between resonance frequency and structure thickness was analyzed after many times of simulation. And a favorable thickness set, included bottom electrode layer, piezoelectric layer and top electrode layer was selected for the following fabrication processes.

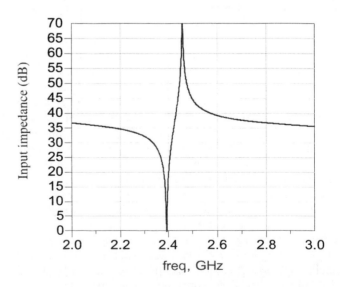

Figure 1: Simulation result of a bulk acoustic resonator using transmission line model.

3 PROCESS DESIGN AND RESULTS

Fabrication process is summarized in Figure 2. First, Silicon dioxide and nitride layers was grown and deposited as insulating layers and etching stop layers for ICP process (Fig. 2a). Next, bottom electrode was deposited and patterned (Fig. 2b). In this study, Ti-Pt was used, because Aluminum nitride (AlN) was selected to be the piezoelectric layer and AlN(002) structure can be better grown on properly oriented Pt. For the AlN deposition step, DC sputter process was used and the deposition temperature was below 400 °C (Fig. 2c). AlN was then etched by chemical solution to provide connection via between bottom electrode and top electrode. Then, the top electrode was deposited and the geometry was defined by lift-off process (Fig. 2d). The material of top electrode was Aluminum with low resistively and low density. Additional thin metal layer was deposited on selective area to make thicker resonator devices, which had lower resonance frequencies. Afterward, backside ICP etching was used to remove silicon substrate and formed thin membranes (Fig. 2e). These membranes were free at top and bottom surfaces, therefore resonators can achieve high quality factor. Finally, insulating layers under membrane was removed to increase quality factor and fine tune resonance frequency (Fig. 2f).

The fabricated devices was designed to have compact size smaller than 1mm×1mm, therefore the total number of devices on the six-inch wafer was about 10,000. Figure 3 shows the cross section view of one suspended FBAR filter, the cavity size was less than 600 um×600 um×600um, and it had vertical walls perpendicular to wafer surface. This characteristic of ICP etching process largely reduced the device size as expected. Figure 4 shows the membrane structure. The micrometer-thick AlN film had grown well on the Ti-Pt film, and collimate structure throughout the layer can be observed.

4 MEASUREMENT RESULTS

To demonstrate the performance of this design, two types of acoustic devices were measured. First, a resonator test key was employed to identify the quality factor. Next, filters were measured and compared to required specifications. The devices were characterized by network analyzer and on-wafer probe.

Resonator test keys were tested by one-port test method. Moreover, in order to demonstrate the overall performance of the resonator device, the parasitic resistance was not de-embedded in measurement. Test results of resonators are shown in Figure 5 and 6. Figure 5 shows the resonator S11 in the form of smith chart. Very small effective series resistance presented and shrunk the measured S11 circle. However, this small amount of series resistance was acceptable for further filter design.

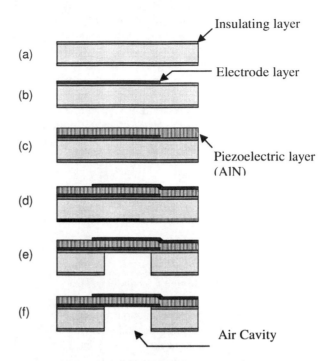

Figure 2: FBAR fabrication process using backside ICP.

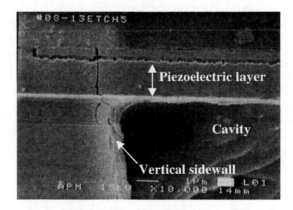

Figure 3: Cross section view of the fabricated device.

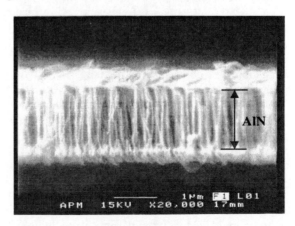

Figure 4: Collimated structure of AlN can be identified in the suspended membrane.

Figure 6(a) shows the magnitude of input impedance. One resonance point and one anti-resonance point could be determined by the minimum and maximum input impedance respectively. A good resonator for filter application should have wide separation between these two points, which represents the electromechanical coupling coefficient (K^2). Higher K^2 facilitates a wide band filter design and fabrication. From the resonance frequency (f_r) and anti-resonance frequency (f_a), the K^2 was calculated by:

$$K^2 = \frac{\dfrac{\pi}{2}\dfrac{f_r}{f_a}}{\tan\left(\dfrac{\pi}{2}\dfrac{f_r}{f_a}\right)} \tag{1}$$

The measured K^2 of this device was about 5.7%.

Figure 6(b) shows the phase of input impedance. For an ideal resonator, the phase should change from -90 to +90 degree in near vertical slope at resonance point. However, if some loss occurred in the resonance, the phase change became gentle. For this device, the phase change was quite steep, therefore the resonance had very high performance. The quality factor (Q) was calculated by the rate of phase change with frequency:

$$Q_r = \frac{f_r}{2}\left|\frac{d\angle Z_{in}}{df}\right| \tag{2}$$

The quality factor was calculated to be 1567 for this device. The electrical performances had shown that the ICP process was promising for the fabrication of bulk acoustic resonators.

Temperature characteristic was also tested with thermal chuck and wafer probing method. The temperature was controlled to rise from 25 °C to 80 °C and then cool down from 80 °C to 5 °C. Resonance frequencies were recorded and plot in Figure 7. The measured relationship between temperature and resonance frequency was linear and had temperature coefficient of resonance frequency about -27.4 ppm/°C, which was consistent with studies done by other research groups. After the device temperature cooled from high temperature to room temperature, the resonance frequency returned to its original value at room temperature. This results indicated that although bulk acoustic wave devices using ICP method was suspend as a membrane, it still hold good thermal conductivity, and no heat pile up in the device during temperature raise and fall period.

Next, the filter performance was measured. This filter was a traditional ladder type design, it included three serially connected resonators and four shunt resonators. Figure 8 shows the typical measurement result. The

minimum insertion loss was about 1.5 dB, maximum insertion loss in 83.5MHz bandwidth was 3.6 dB, and minimum rejection was 25dB. According to the test results, this high frequency filter had good electric performance and meets the required specification. Other frequency bands range from 1.5GHz to 2.2GHz also could be redesigned and fabricated by changing the electrode layout and thickness of suspended membrane.

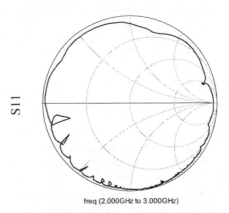

Figure 5: Measured resonator characteristic in smith chart form.

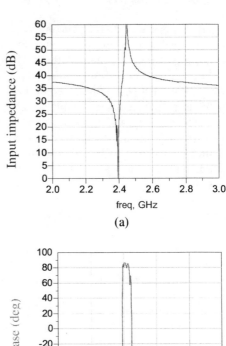

(a)

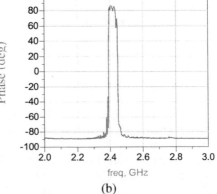

(b)

Figure 6: Measured input impedance, (a) magnitude, (b) phase.

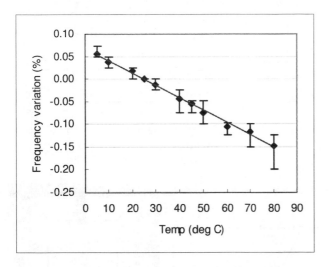

Figure 7: Relationship between temperature and variation of resonance frequency.

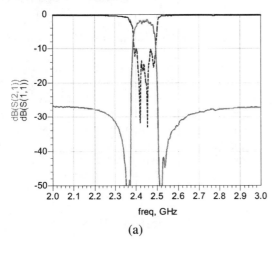

(a)

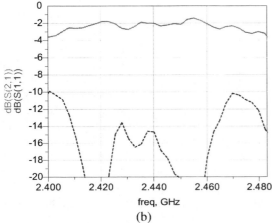

(b)

Figure 8: Filter performance (a) wide band, (b) narrow band.

5 CONCLUSION

Thin film bulk acoustic resonators and filters were fabricated using ICP process to reduce device size and facilitate fabrication process. Transmission line model was used to simulate the final resonator performance. And a preferable structure thickness was designed according to the simulation. Fabrication results demonstrated a small size device with near vertical cavity walls. The devices were tested by network analyzer and probe station. Quality factor and electromechanical coupling coefficient of a single resonator were 1567 and 5.7% respectively. Designed filter had maximum 3.6 dB insertion loss in 83.5 MHz band, rejection was more than 25dB. These fabrication results and test results confirmed that high quality thin film bulk acoustic resonators and filters can be designed and fabricated using ICP process.

REFERENCES

[1] Krishnaswamy, S.V., Rosenbaum, J., Horwitz, S., Vale, C. and Moore, R.A. "Film bulk acoustic wave resonator technology," Proceedings, IEEE Ultrasonics symposium, Honolulu, HI, December, 1990, pp 529-536.

[2] Park, J.Y., Lee, H.C., Lee, K.H., Lee, H.M. and Ko, Y.J. "Micromachined FBAR RF filters for advanced handset applications," IEEE International conference on solid state sensors, actuators and microsystems, Boston, MA, June, 2003, pp 911-914.

[3] Ruby, R., Bradley, P., Larson III, J.D. and Oshmyansky, Y., "PCS 1900MHz duplexer using thin film bulk acoustic resonator (FBARS) ," Electronics Letters, 35, pp 794-795, 1999.

[4] Larson III, J. D., Ruby, R. C., Bradley, P. D., Wen, J., Kok, S-L, and Chien, A., 2000, "Power handling and temperature coefficient studies in FBAR duplexers for the 1900 MHz PCS band," 2000 IEEE Ultrasonics Symposium, San Juan, Puerto Rico, Vol. 1, pp. 869-874.

[5] Lakin, K.M., McCarron, K.T. and Rose, R.E. "Solidly mounted resonators and filters," Proceedings, IEEE Ultrasonics Symposium, Seattle, WA, November, 1995, pp 905-908.

[6] Yoon, G. and Park, J.D. "Fabrication of ZnO-based film bulk acoustic resonator devices using W/SiO2 multilayer reflector," Electronics Letters, 36, pp 1435-1437, 2000.

[7] Auld, B. A., 1973, Acoustic fields and waves in solids, John Wiley and Sons, New York.

A MEMS Tunable Inductor

I. Zine-El-Abidine[*], M. Okoniewski[*] and J. G. McRory[*]

[*]University of Calgary and TRLabs
280, 3553 – 31 Street N.W. Calgary, AB, Canada
imed@cal.trlabs.ca

ABSTRACT

This paper reports a single-wafer micromachined radio-frequency (RF) inductor that can be integrated in a conventional RFIC device. The inductor achieved a quality factor greater than 9 at 5Ghz and a self-resonance frequency well above 15 GHz. The inductor is tunable and the inductance variation is greater than 8%.

Keywords: Radio frequency, MEMS, inductor, tunable.

1 INTRODUCTION

RF inductors are needed in any wireless front-end circuitry; the performance of both transceivers and receivers depends heavily on this component. High-Q inductors reduce the phase noise and the power consumption of VCO's and amplifiers and reduce the return loss of matching networks and filters. The quality factor of inductors can be increased by using a thick metal layer [1], [5] and by isolating the inductor from the substrate [2]-[6]. To isolate the inductor, bulk micromachining [2]-[5] or self-assembly [7], [8] can be used. Further, tunable inductors allow for performance optimization of RF front-end circuits. To date, there has not been a practical implementation of a high-Q tunable inductor. Most of the reported MEMS inductors are static fixed-value inductors. One variable inductor using MEMS switches has been reported [9]. However, it provides only discrete values of inductance depending on the ON/OFF configuration of the switches. A tunable inductor using self-assembly technique [8] has been reported. The drawback of that device is that such inductor, are lifted from the substrate, thus becoming prone to electromagnetic interference in the transceiver system.

2 DESIGN

The inductor reported here is formed of two layers, the bottom layer is silicon nitride and the top layer is gold. To increase the quality factor at the high frequencies, the silicon underneath the inductor is etched away. To achieve the tunability, two inductors are connected in parallel. The outer loop inductor is held to the substrate with nitride beams while the inner loop inductor is left free as shown in Figure1.

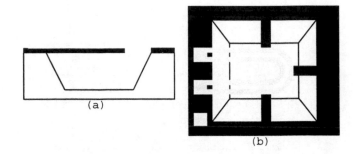

Figure 1: (a) side view, (b) top view of the inductor.

Since the inductors are made of two layers of different thermal expansion coefficients, they form a thermal bimorph [9], [10]. When a current is drawn through the inductors, it causes the inner inductor to move downwards into the pyramidal pit, while the nitride beams prevent the outer inductor from moving. The change of the angle between the two inductors results in a change in the mutual inductance and therefore the overall inductance.

3 FABRICATION

The fabrication process of the tunable inductor is shown in Figure 2. The process starts with depositing of a 600 nm nitride layer on the surface of the silicon wafer. The nitride layer is then etched using RIE (Reactive Ion Etching) to form the bottom layer of the inductor, the beams holding the outer loop and the cavity edges. Then, a 100 nm layer of chrome is deposited to serve as an adhesion layer for gold. The gold layer is 1μm thick. Gold and chrome are etched using iodine and nitric acid respectively. The final step is the bulk anisotropic etching using KOH to release the inductors and form the pyramidal pit. The silicon nitride layer played the role of the mask during the anisotropic etch step. The cavity depth is 300 μm. A scanning electron microscopy (SEM) photograph of the tunable inductor structure is shown in Figure 3.

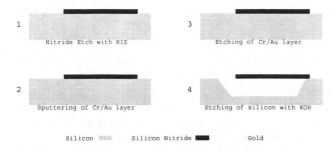

1 Nitride Etch with RIE

2 Sputtering of Cr/Au layer

3 Etching of Cr/Au layer

4 Etching of silicon with KOH

Silicon   Silicon Nitride   Gold

Figure 2: The fabrication process.

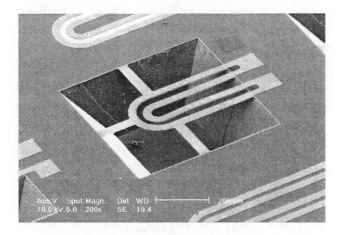

Figure 3: A SEM photograph of the inductor.

4 SIMULATION RESULTS

Figure 4 shows the simulated displacement versus the temperature difference using Coventorware [11]. The module used is MemETherm. Figure 5 shows the shape of the curved inner inductor.

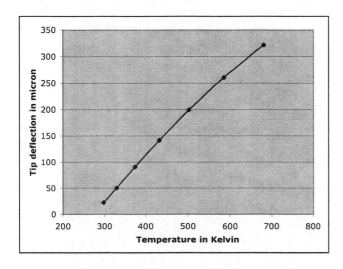

Figure 4: Simulated tip deflection versus temperature.

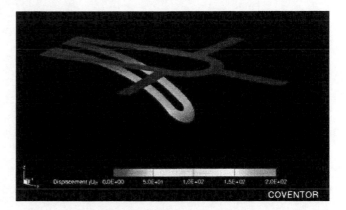

Figure 5: The simulated deflection of the inner inductor.

5 MEASURED RESULTS

The inductor was measured using 100-μm pitch ground-signal-ground (GSG) microwave probes connected to an HP8510 network analyzer with frequencies ranging from 1 GHz to 15 GHz. Probe calibration was done using Cascade CS-5 calibration substrate. Figure 6 illustrates the quality factor of the tunable inductor.

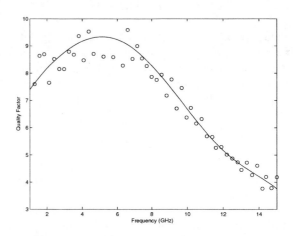

Figure 6: The quality factor of the inductor.

Figure 7 shows the tuning range of the inductor. The maximum value corresponds to the original state where the two inductors are at the same level. When the current flows through the two loops, the inner inductor deflects down as illustrated in Figure 8. By increasing the angle between the two loops, the mutual inductance decreases and therefore the overall inductance. The tuning range is 8%.

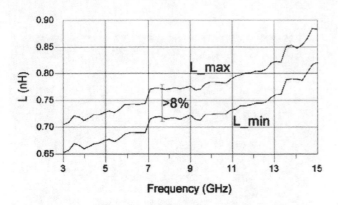

Figure 7: The inductance variation.

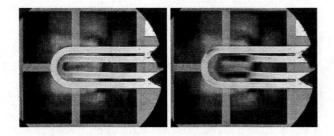

Figure 8: The deflection of the inner inductor under DC current.

6 CONCLUSION

A tunable MEMS inductor has been demonstrated with a tuning range of 8%. The inductor can operate beyond 15 GHz. The tunability is achieved using the principle of the bimorph actuation. Further improvements are underway to increase the quality factor and the tuning range.

7 ACKNOWLEDGMENT

The authors wish to thank Dr. Ken Westra, Stephanie Bozic, Dione Chambers and Hue Nguyen from the Nanofab of the University of Alberta for their assistance during the fabrication and Richard Humphrey from the Microscopy and Imaging Facility for his assistance with the SEM photographs caption. The authors wish to thank also NSERC for their support.

REFERENCES

[1] S. Pinel, F. Cros, S. Nuttinck, S-W. Yoon, M.G. Allen, and J. Laskar, "Very high-Q inductors using RF-MEMS technology for System-On-Package wireless communication integrated module," *IEEE MTT-S Digest*, 2003, pp. 1497-1500.

[2] J. Y.-C. Chang, A. A. Abidi, and M. Gaitan, "Large Suspended Inductors and Their Use in a 2-μm CMOS RF Amplifier," *IEEE Electron Device Lett.*, Vol. 14, May 1993, pp. 246-248.

[3] H. Lakdawala, X. Zhu, H. Luo, S. Santhanam, L. R. Carley, and G. K. Fedder, "Micromachined High-Q Inductors in a 0.18-μm Copper Interconnect Low-K Dielectric CMOS Process," *IEEE J. Solid-State Circuits*, Vol. 37, No. 3, March 2002, pp. 394-403.

[4] H. Jiang, Y. Wang, J-L. A. Yeh, and N. C. Tien, "On-Chip Spiral Inductors Suspended over Deep Copper-Lined Cavities," *IEEE Trans. Microwave Theory and Techniques*, Vol. 48, No. 12, Dec 2000, pp. 2415-2422.

[5] R. P. Ribas, J. Lescot, J-L. Leclercq. J. M. Karam, and F. Ndagijimana, "Micromachined Microwave planar Spiral Inductors and Transformers," *IEEE Trans. Microwave Theory and Techniques*, Vol. 48, No. 8, pp. 1326-1335, August 2000, pp. 1326-1335.

[6] J-B. Yoon, B-I. Kim, Y-S. Choi, and E. Yoon, "3-D Construction of Monolithic Passive Components for RF and Microwave ICs Using Thick-Metal Surface Micromachining Technology," *IEEE Trans. Microwave Theory and Techniques*, Vol. 51, No. 1, January 2003, pp. 279-288.

[7] G. W. Dahlmann, E. M. Yeatman, P. R. Young, I. D. Robertson, and S. Lucyszyn, "MEMS High Q Microwave Inductors Using Solder Surface Tension Self-Assembly," *IEEE MTT-S International Microwave Symposium Digest*, June 2001, pp. 394-403.

[8] V. M. Lubecke, B. Barber, E. Chan, D. Lopez, M. E. Gross, and P. Gammel, "Self-Assembling MEMS Variable and Fixed RF Inductors," *IEEE Trans. Microwave Theory and Techniques*, Vol. 49, No. 11, November 2001, pp. 2093-2098.

[9] W-H. Chu, M. Mehregany, and R. L. Mullen, "Analysis of Tip Deflection and Force of a Bimetallic Cantilever Microactuator," *J. Micromech. Microeng.*, Vol. 3, 1993, pp. 4-7.

[10] W. Peng, Z. Xiao, and K. R. Farmer, "Optimization of Thermally Actuated Bimorph Cantilevers for Maximum Deflection," Nanotech Proceedings, Vol. 1, pp. 376-379, 2003.

[11] MEMCAD User's Manual. Coventor, Inc., Cary, NC. [Online]. Available: http://www.coventor.com.

Design & Simulation of a Mechanical Amplifier for Inertial Sensing Applications

M. N. Kham, R. Houlihan and M. Kraft

University of Southampton, Microelectronics Centre, Highfield, Southampton SO17 1BJ

Email: mnk03r@ecs.soton.ac.uk

ABSTRACT

This paper describes the design and simulation of a mechanical amplifier used to improve the performance of a capacitive accelerometer. It comprises a long silicon beam attached to the proof mass and it amplifies the sensed motion based on the leverage system [1]. Finite element simulations have been used to validate the concept and facilitate the design. Although increasing the overall accelerometer dimensions, a mechanical amplifier provides amplification without adding significant noise to the system, unlike its electrical counterpart. Moreover, the increased amplification of the motion obtained with longer beam length results in higher sensitivity.

Keywords: capacitive accelerometer, mechanical leverage system, sensitivity.

1 INTRODUCTION

The magnitude of the signal in capacitive inertial sensing devices and particularly in differential sensors is often the limiting factor to sensor sensitivity and resolution. The signal is usually degraded by various noise sources. Due to the requirement of interface circuitry in capacitive sensing devices and therefore to the presence of inherent electronic noise sources in the circuit, noise can be minimised but never completely eliminated. In order to further improve the performance of a sensor, non-electrical circuit related alternatives need to be considered.

In this paper, a mechanical amplifier is presented which improves the sensitivity of a capacitive accelerometer by amplifying the sensed signal, as depicted in Fig.1. A schematic of the mechanical amplifier is shown in Fig.2. An advantage of using the mechanical amplifier is that it does not add significant noise to the system unlike its electrical counterpart. Modelling results for the amplifier are presented. Simulations were carried out using the MEMS dedicated finite element tool, Coventorware.

2 THEORY

The mechanical amplifier presented is based on the mechanical leverage system [1], which can be used as a way of increasing the scale factor of the sensor. An ideal pivot has the torsional stiffness, k_θ, which tends to zero and a vertical stiffness, k_v, which tends to infinity. Since true pivots cannot be realised using micromachining processes, they are approximated by flexural pivots. The model can be represented by Fig. 3 and described by Equation (1).

$$x1 = \frac{L2}{L1} x1 \qquad (1)$$

where L1, L2 are the beam lengths at either side of the pivot and x1, x2 are the deflections at the end of L1 and L2 respectively.

3 DESIGN

The design of the proof mass has been performed first and was followed by the addition of the mechanical amplifier. The two electrodes that are normally placed on either side of the proof mass in a conventional capacitive accelerometer, are placed at the end of the beam to detect the magnified deflection, x2, in this model.

Preliminary calculations have been performed to estimate the device characteristics and the values were verified using finite element analysis.

The proof mass measures 1×1mm^2 with a thickness of 50μm. Fabrication of such a high thickness is enabled by the use of SOI-MEMS technology and deep reactive ion etching (RIE) process. $10\times10\mu$m^2 etching holes are included in the proof mass to allow undercutting and release of the large structure during fabrication. In the simulation, the etch holes are lumped together and modelled as one large hole in the middle of the proof mass to increase simulation speed [2].

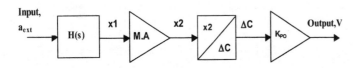

Fig. 1. An open loop, capacitive accelerometer including the mechanical amplifier (MA)

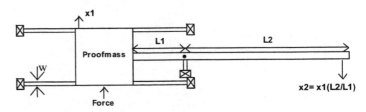

Fig. 2. Accelerometer with the mechanical amplifier

The volume of the proof mass is calculated to be 3.75×10^{-11} m^3 and the mass, about 100µg. The proof mass is supported by four suspension beams, each 1mm in length and 20µm in width, w. The depth of the beam is the same as the thickness of the proof mass.

To calculate the spring constant of the proof mass, the beam equation for a clamped-clamped beam with a point load at the centre (Fig. 4) is used [3]. For x less than or equal to L/2 away from a support, the deflection, y(x), corresponding to a point load, F, is given by:

$$y(x) = \frac{Fx}{48EI}(3Lx - 4x^2) \qquad (2)$$

where E is Young's modulus, L is the length of the beam, t is the thickness of the beam and I is the bending moment of inertia. For a beam of a rectangular cross section, the bending moment of inertia is given by,

$$I = \frac{1}{12}wt^3 \qquad (3)$$

where w is the width of the beam and t is the thickness.

Thus, in our model, the deflection of the beam, y(0.5L) is given by:

$$y(0.5L) = \frac{F(0.25L^3)}{48EI} \qquad (4)$$

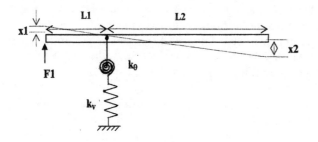

Fig. 3. Model used to predict leverage motion magnification

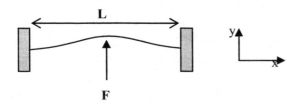

Fig. 4. Clamped-clamped beam [3]

The spring constant of a beam, k, can then be expressed as,

$$k = \frac{F}{y(0.5L)} = \frac{192EI}{L^3} \qquad (5)$$

The effective spring constant of the proof mass was obtained by multiplying k by 2 as there are two suspension beam pairs on the proof mass. The resonant frequency can be calculated using the equation below:

$$\omega_r = \sqrt{\frac{k}{m}} \qquad (6)$$

Several beam and pivot models have been evaluated in an effort to optimize the amplifier performance. To this end, finite element simulations were carried out using Coventorware. Optimized conditions for the accelerometer were determined based on the beam deflection, x2, which is sensed capacitively in the pick-off circuit.

During operation, the long beam, which acts as an amplifier, has to remain rigid while allowing maximum deflection. Parameters such as the beam width and the width of anchor (pivot) were evaluated. The beam thickness or depth is not a variable as it is decided by the fabrication process. In ref. [4], for the design of a silicon micromotion amplifier, silicon hinge-like systems were used to amplify motion. The same concept is applied here when various pivot structures, as shown in Fig. 5, were evaluated. The pivots are realised by etching the silicon material near the anchor in order to reduce torsional stiffness, k$_\theta$, and allow larger beam deflection. Fabrication of these structures is limited by the specification of the minimum etching width.

Having optimised the beam and pivot parameters, the length, L2 is varied and the resulting structure simulated to explore the behaviour of the mechanical amplifier. A static input acceleration of 1g is assumed in the simulations throughout these experiments.

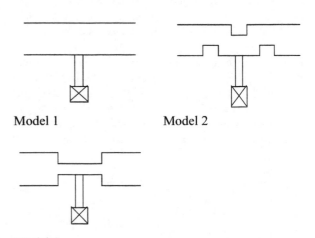

Fig. 5. Different pivot models

4 SIMULATION RESULTS AND ANALYSIS

Preliminary calculation shows that the proof mass has an effective spring constant of 34N/m and a fundamental resonant frequency of 2.93 kHz. These are comparable to the finite element simulation results, as shown in Fig. 6, which show the fundamental resonant mode occurring at 3.14 kHz (y-direction). The second and third resonant modes occur at 14.5 kHz (z-direction) and 22.6 kHz (torsional) respectively.

The simulation results showing optimization of the beam and pivot are presented in Figures 7 and 8. The figures illustrate the beam deflection, x2, for a range of beam/pivot parameters. Fig. 7(a) shows an optimum beam width of 30μm. For beams narrower than 30μm, the torsional vibration mode becomes the fundamental mode.

The length L1 was then varied for an assumed beam width of 30μm and shown in Fig. 7(b). Although increase in L1 results in greater beam deflection and therefore signal amplification, it increases the size of the device. Moreover, the dependence of x2 on L1 is less than on L2. For this reason, a value of 1mm was chosen for L1. Fig. 7(c) shows the variation in x2 with the pivot width. A pivot width of 5μm was chosen as it gives the highest beam deflection and is well above the specification for minimum etching width.

The beam length parameter, L2, was varied between 3 and 7mm and using the optimized conditions mentioned above and pivot model 1, the result is depicted 'model 1' in Fig. 8. It is found that the theoretical condition stated in Equation (1) is met when length of L2 is 6mm. Similarly, evaluations were carried out for pivot models 2 and 3. The figure shows that the improved pivot structures effectively reduce the length L2 to 5mm in model 2 and to 4mm in model 3.

In the simulation of these models, the 10x10μm² etching holes were not included in the long beam. As they are required for fabrication, they are incorporated both in the beam and in the proof mass. The modified structure yields a greater beam deflection denoted 'modified' in Fig. 8. The Coventorware model is shown in Fig. 9. The modal simulation for an input acceleration of 1g is shown in Fig. 10.

The simulation results can be summarized that L2 has to have a minimum length to act as a mechanical amplifier. Furthermore, the beam deflection is rapidly increased when length L2 is further increased. This property results in the rapid increase of the pick-off voltage signal and thus a higher resolution of the device is achieved.

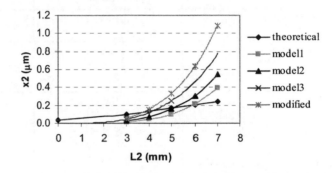

Fig. 8. Deflection of beam at various L2 lengths and different pivot models at input acceleration of 1g

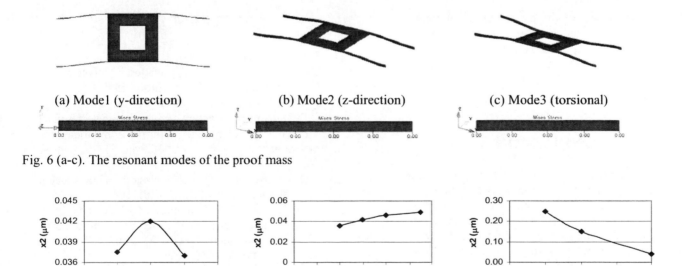

(a) Mode1 (y-direction)   (b) Mode2 (z-direction)   (c) Mode3 (torsional)

Fig. 6 (a-c). The resonant modes of the proof mass

Fig. 7(a-c). Simulation results of optimization of beam and the pivot parameters

 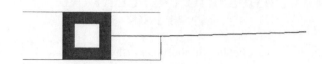

 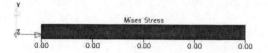

Fig. 9. Final simulation model Fig. 10. Model under static input acceleration 1g

Device Parameter	Without MA	With MA (L2=4mm, minimum length)	With MA (L2= 7mm)
Mass	~100 μg	~100 μg	~100 μg
Fundamental res. Freq.	3.14 kHz	1.2 kHz	0.7 kHz
Area (Dimensions)	$3\ \mu m^2$ (3mm x 1mm) (x 1)	$7\ \mu m^2$ (3mm x 1mm) (x 2.5)	$10\ \mu m^2$ (10mm x 1mm) (x 3.3)
Deflection	$x1 = 0.034\ \mu m$	$x2 = 0.15\ \mu m$	$x2 = 0.632\ \mu m$
Sensitivity = $\Delta C/G$ where $\Delta C = 2\varepsilon Ax/(d_o^2 - x^2)$	7.5 fF/G (x 1)	30 fF/G (x 4)	233 fF/G (x 30)

Table 1: Comparison of the performance of accelerometer with and without the mechanical amplifier

Finally, the parameters of the capacitive accelerometer with and without the mechanical amplifier are summarized in Table 1. The effect of exponential beam deflection when L2=7mm is evident. Although the total area occupied by the accelerometer is tripled, a 30-fold increase in sensitivity can be achieved.

5 CONCLUSION

Simulation results and calculation show that the concept of mechanical amplification based on a leverage system is feasible and it does not add significant noise to the system. Thus, incorporation of the amplifier into an inertial sensing system potentially can increase sensitivity. Currently we are extending the analysis to time varying input accelerations.

REFERENCES

[1] T. A. Roessig, "Integrated MEMS Tuning Fork Oscillators for Sensor application" Univeristy of California, Berkeley, Ph.D, Thesis, 1998.
[2] J. W. Weigold, K. Najafi, S. W. Pang, "Design and Fabrication of Submicrometer, Single Crystal Si Accelerometer", Journal of microelectro-mechanical systems, Vol. 10, pp 518-524, Dec 2001.
[3] H. Kovacs, "Micromachined Transducers Sourcebook," GTA McGraw-Hill, 1998.
[4] X. T. Huang, M. T. Saif and N. C. MacDonald, "A micromotion amplifier", Proc. 9th International Workshop on Micro Electro Mechanical Systems, San Diego, pp 424-428, 1996.

Micromachined Thermal Multimorph Actuators
Fabricated by Bulk-etched MUMPs Process

Adisorn Tuantranont[1,2] and V. M. Bright[3]

[1]National Electronics and Computer Technology Center (NECTEC)
[2]National Nanotechnology Center (NANOTEC)
112 Thailand Science Park, Paholyothin Rd., Klong Laung, Pathumthani 12120 Thailand
Tel: 662-564-6900, Fax: 662-564-6771, Email: adisorn.tuantranont@nectec.or.th
[3]CAMPmode, Dept. of Mechanical Engineering, University of Colorado at Boulder, USA
Tel: 303-735-1763, Fax: 303-492-3498, Email: brightv@colorado.edu

ABSTRACT

Micromachined thermal multimorph actuators for out-of-plane displacements have been designed and fabricated by Multi-Users MEMS Process (MUMPs) with a post-processing bulk etching step. Micromachined actuators have potential applications in micromanipulation and nanoposition such as manipulation of scanning tunneling microscope (STM) tips, nanopositioner for nano-assembly, micro/nanorobots and cantilever-based nanosensors. The actuator consists of a multilayer micromachined beam constructed of various combinations of polysilicon, oxide and metal layers. The multimorph is heat by applying current to an embedded polysilicon wire to achieve beam bending The maximum tip deflection of 2.5 µm with an input power of 40 mW is achieved. The maximum operating frequency of 2.7 kHz is measured by focus laser reflecting method. A design methodology, device fabrication, and device characterization are presented.

Keywords: Microactuator, Multimorph, MEMS, MUMPs, nanopositioner, Scanning Tunneling Microscope

1. INTRODUCTION

Microactuators found many applications in nanotechnology such as nanopositioner and nanoassembly. Presently, microactuators have been used in various applications such as micropump, microvalves, relays, switches, microrobots, and micro-optical elements etc [1]. Different principles such as electrostatic, piezoelectric, electromagnetic and thermal have been used in the microactuators. But thermal actuators are usually simpler, more reliable and easier to construct. Thermal actuation is generally based on either the linear/volume expansion or phase transformation of the materials [2]. The thermal actuators can exhibit large forces and good linearity relative to input power. Compared to electrostatic actuator, however, disadvantages of thermal actuators include high power requirements and lower bandwidths as a result of thermal time constant [1]. But the power consumption and thermal loss can be reduced through elaborated geometric and structural designs of the device and appropriate choice of materials, which can enhance the efficiency of thermal actuators.

The thermal actuators can be categorized based on their motion as in-plane and out-of-plane thermal actuators. In-plane thermal actuator consists of one layer of material with two arms. One arm that has a higher resistance and more power are dissipated as heat, is narrower than the other arm. This causes the bending motion in the direction parallel to the substrate [3,4]. Meanwhile, the out-of-plane thermal actuator mostly bases on bimorph effects and consists of two layer of material with different coefficient of thermal expansion (CTE) The bimorph beam will bend due to mismatch of CTE of various layers when there is temperature variation on the actuators. Out-of plane thermal actuators based on bimorph effects have been demonstrated by Baltes et al. [5] and Riethmuller and Benecke [6]. In previous work, thermal actuators were fabricated by a CMOS process [7]. In this paper, we present design, model, fabrication, and performance test of a micromachined thermal multimorph actuators fabricated by standard surface micromachining foundry called Multi-Users MUMPs Process (MUMPs) with a post-processing bulk-etching step [8].

2. DEVICE DESIGN

Figure 1 shows the tested thermal multimorph actuators with various material combination. All actuators have the identical geometry of 15 µm wide and 100 µm long with U-shape polysilicon heating wire of 3 µm wide and 200 µm long.

The actuators are designed with a combination of polysilicon and silicondioxide (SiO_2) layers with the top metal layer. The thermal actuator consists of a polysilicon resistor encapsulated in SiO_2 and a gold layer as shown in Fig. 2. Due to the different coefficients of thermal expansion of multi-layer sandwich of different materials, the actuator flexure curl when ohmic heating is generated by applying electrical power to the resisters, thus causing the high-resolution deflection of actuator tip. The achieved tip displacement is a function of the actuator design parameters, such as beam length, layer thickness and beam material composition.

An array of actuators is designed for performance test. The 12 actuators as shown in Figure 1 has been fabricated using different combinations of layers available in MUMPs as listed in Table.1.

Figure 1: Array of thermal multimorph actuators with various material combinations.

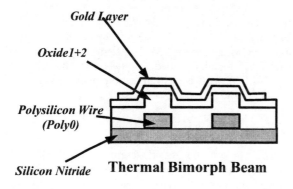

Figure 2: The cross section of the thermal bimorph actuator.

Beam	Layer Combination	Beam	Layer Combination
1	*P0*, M	7	P0, *P1*, M
2	*P0*, P1, M	8	P0, *P1*, P2, M
3	*P0*, P1, P2, M	9	*P2*, M
4	*P0*, P2, M	10	P1, *P2*, M
5	*P1*, M	11	P0, *P2*, M
6	*P1*, P2, M	12	P0, P1, *P2*, M

where *P0*, *P1*, *P2* is Poly0, Poly1, Poly2 respectively, and *M* is Metal. The italic layer is used as a heating wire.

Table 1: Combination layer of thermal multimorph actuators available through MUMPs.

3. DESIGN MODELING

Multimorph beams consist of layers of material sandwiched together to form a composite beam. In case where there are only two layers, the beam is called a bimorph. Since the thickness of the heating resistor is small compared to the thickness of the encapsulated oxide, the thermal actuation can be simply modeled as a bimorph [9]. The tip displacement for each type of actuators is a function of temperature and beam material composition. The material properties for each layer are assumed to be uniform throughout the layer and constant with temperature and no internal friction between the bimorph layers. The temperature distribution within the beam is assumed to be uniform since an embedded heating resistor that span the entire length of the beam is typically used for heating. The relationship between tip deflection d of the bimorph actuator and temperature variation is given by [10]:

$$d = \frac{kL^2}{2} = \frac{3}{4}L^2 \frac{\Delta T}{t_1 + t_2} \frac{\alpha_2 - \alpha_1}{1 + \frac{\left(E_1 t_1^2 - E_2 t_2^2\right)^2}{4 E_1 t_1 E_2 t_2 (t_1 + t_2)^2}} \quad (1)$$

where r is radius of curvature $=1/k$, k is the beam curvature, L is the beam length, α_1 and α_2 are the CTEs of two bimorph layers, E_1 and E_2 are their Young's modulus, t_1 and t_2 are their thickness, and ΔT is the temperature difference between the operating and initial temperatures. For a bimorph actuator, the force generated is distributed linearly along the beam length. From the beam theory, the force F_d generated at the end of the tip deflection d is given by [11]:

$$F_d = \frac{3EId}{L^3} \quad (2)$$

where EI is the flexural rigidity of the composite beam. Since the bimorph beam consists of two materials, the neutral axis of the beam may no longer in the middle of the beam cross-section. Either composite beam or transformed-section method must be use to calculate the flexural rigidity EI of the bimorph beams.

The commercial finite element package MARC with MENTAT was used to perform the analysis. The modeling consisted of an electro-thermal analysis to obtain the temperature distribution resulting from an input current as shown in Fig. 3. This was then coupled to a mechanical analysis in which the temperature distribution was used to determine the static deflections resulting from thermal expansion mismatch.

4. DEVICE FABRICATION

Micromachined thermal multimorph actuators have been fabricated through commercially available Multi-User MEMS Process (MUMPs) [12].

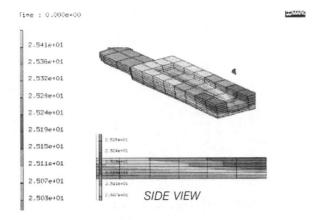

SIDE VIEW

Figure 3: Thermal distribution in a multimorph beam
when current is applied and its side view.

To suspend multimorph microactuators over cavities,
the bare single-crystal silicon substrate is exposed to
chemical etchant using the superimposition of
"Anchor1", "Anchor2", Poly1-poly2 via", "Dimple",
"Hole1" and "Hole2" layers on top of each other as
shown in Fig. 4.

By this method, the nitride and all the oxide layers
above nitride are opened through during regular MUMPs
fabrication, leaving exposed bare silicon while the other
area of the chip is covered with oxide and nitride layers
[8]. The silicon substrate is anisotropically etched by
EDP or KOH etchant or isotropically etching by xenon
difluoride (XeF₂). The polysilicon structures are not
effected by EDP etching because the etching rate of
polysilicon in these chemical etchant is much slower than
single-crystal silicon and can be negligible. By this
method, the surface and bulk micromachining can be
done and integrated on the same chip with only a
maskless post-processing step required on MUMPs chip.

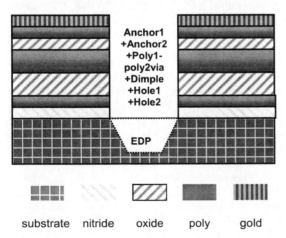

Figure 4: Schematic diagram of MUMPs and "cut
through substrate" mask for bulk etching post-processing
step with EDP.

5. DEVICE CHARACTERIZATION

Static deflections of the actuators are measured over
a range of input powers using a Zygo interferometric
microscope. When the beam is actuated, the tip of
actuator bends downward with different amount of
deflection depending on each beam structures. The
profile of actuated beam is shown in Fig. 5. Figure 6
shows the experimental plot of the actuator deflection as
a function of input power. Thermal actuator beam #1 has
a largest deflection compared to the other beams because
this beam is the thinnest beam with thickness of 3.87 μm
consisted of the layer of Oxide1 and Oxide2 between
gold and Poly0 wire. Thus the spring constant is the
lowest compared to the other beams. And with the
thermal expansion coefficient mismatch between silicon
dioxide and gold layer of 40:1 (CTE of gold is 14.2
ppm/K and CTE of SiO₂ is 0.35 ppm/K), the maximum
deflection of 2.5 μm is achieved using $P0$M actuator
(beam #1). Focused laser reflecting test setup is used to
determine the dynamic behavior of the actuators. Laser
beam is focused onto the tip of actuator and reflected to a
photodetector to detect the frequency of an operating
actuator. Rise time and fall time of reflected signal is
measured to determine the minimum cycle time of the
actuators. The rise time is measured as 150 μs and fall
time is 220 μs, thus minimum thermal time constant or
the time required to completely heat and cool is 0.37 ms.
Therefore, the maximum operating frequency of the
actuators is 2.7 kHz when the actuators are driven by a
square wave as shown in Fig. 7.

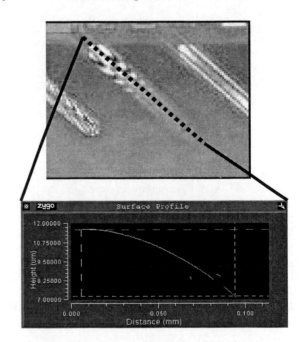

Figure 5: Actuator deflection measurement using
white light interferometric microscope.

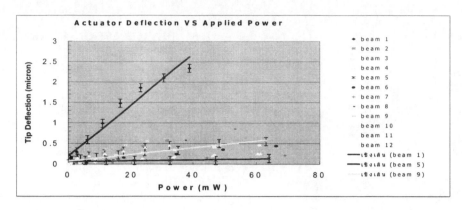

Figure 6: Measured actuator deflection versus applied power.

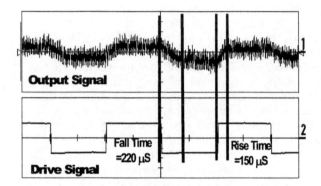

Figure 7: The operating frequency measurement of *P0*M actuator.

6. CONCLUSIONS

We present a thermal multimorph microactuator based on MEMS technology. The actuator is fabricated by standard MEMS foundry with additional post processing step of substrate etching. A design methodology used various combinations of available layers to determine the best actuator with a largest deflection is presented. Device characterization on both static and dynamic behaviors is experimentally determined. The maximum deflection of 2.5 μm with maximum operating frequency of 2.7 kHz is achieved. Micromachined thermal multimorph actuators found many useful applications such as microactuators in microrobotics and micromanipulators in a precise assembly. The use of these actuators in various applications is in study.

7. ACKNOWLEDGEMENT

This work was supported by the Air Force Office of Scientific Research (AFOSR), Grant No. F49620-98-1-0291. NECTEC Grant# 64041. Special thanks to Don Klaitubtim at AIT for system modeling discussion.

REFERENCES

[1] N. C. Tien and D. T. McCormick, "MEMS actuators for silicon micro-optical elements", *SPIE* vol. 4178, pp. 256-269, 2000.

[2] C. P. Hsu and W. Y. Hsu, "A two-way membrane-type micro-actuator with continuous deflections", *J. Micromech. Microeng.*, 10, pp. 387-394, 2000.

[3] H. Guckle, J. Klein, T. Christenson, K. Skrobis, M. Laudon and E. Lovell, "Thermo-magnetic metal flexure actuators", *Proc. IEEE Solid State Sensors and Actuators Workshops*, pp.73-75, 1992.

[4] J. Comtois and V. M. Bright, "Applications for surface micromachined polysilicon thermal actuators and arrays", *Sensors and Actuators*, A58, 99.19-25, 1997.

[5] H. Baltes, D. Moser, R. Lenggenhager, O. Brand, and G. Waschutka, "Thermomechanical microtransducers by CMOS technology combined with microchining", *Proc. Conf. MOEMS*, Berlin, pp. 98-103, 1991.

[6] W. Riethmuller and W. Benecke, "Thermally excited silicon microactuators", *IEEE Transactions on Electron Devices*, Vol. 35, No. 6, pp. 758-763, 1988.

[7] S. Eagle, H. Lakdawala, and G. Fedder, "Design and simulation of thermal actuators for STM applications in a standard CMOS process," *SPIE* vol. 3875, pp. 32-39, 1999.

[8] A. Tuantranont, L.A. Liew, V. M. Bright, J. L. Zhang, W. Zhang and Y. C. Lee, "Bulk-etched surface micromachined and flip-chip integrated micromirror array for infrared applications," *IEEE/LEOS International conference on Optical MEMS proceeding*, pp. 71-72, 2000.

[9] B. C. Read, V. M. Bright, and J. H. Comtois, "Mechanical and optical characterization of thermal microactuators fabricated in a CMOS process", *SPIE* vol. 2642, pp. 22-32, 1995.

[10] C. C. Tu, C. H. Liu, C. H.Du, J. J. Tsaur, and C. Lee, "A large-angle and large-mirror microscanner based on thermal actuators", *Proc. ASME Mech. Eng. Congr. Expo. (IMECE)*, New York, MEMS-23848, 2001.

[11] W. H Chu, M. Mehregany, and R. L. Mullen, "Analysis of tip deflection and force of a bimetallic cantilever microactuator", *J. Micromech. Microeng.* 3, pp. 4-7, 1993.

[12] D. A. Koester, R. Mahadevan, A. Shishkoff, and K. W. Markus, "Multi-User MEMS Processes (MUMPs): Design handbook, Rev. 4," *Cronos Integrated Microsystems*, 3021 Cornwallis Road, Research Triangle Park, NC 27709, 1999.

Micromachined Piezoresistive Tactile Sensor Array

Fabricated by Bulk-etched MUMPs Process

Adisorn Tuantranont[1,2], Tanom Lomas[1], and V. M. Bright[3]

[1]National Electronics and Computer Technology Center (NECTEC)

[2]National Nanotechnology Center (NANOTEC)
112 Thailand Science Park, Pahol Yothin Rd., Klong Laung, Pathumthani 12120 Thailand
Tel: +662-564-6900, Fax: +662-564-6771, Email: adisorn.tuantranont@nectec.or.th
[3]CAMPmode, Dept. of Mechanical Engineering, University of Colorado at Boulder, USA
Tel: 303-735-1763, Fax: 303-492-3498, Email: brightv@colorado.edu

ABSTRACT

The design, fabrication and testing of a 5×5 micromachined tactile sensor array for the detection of an extremely small force (micrometer-Newton range) has been discussed. An anisotropic etching of silicon substrate of a MUMPs process chip forms a central contacting pads that are trampoline-shape suspended structures and sensor beams. A piezoresistive layer of polysilicon embedded in sensor beams is used to detect the displacement of the suspended contacting pad. Each square tactile has dimension of 200 µm × 200 µm with 250 µm center-to-center spacing. The entire sensor area is 1.25 mm × 1.25 mm. The device was tested characterized under various normal force loads using weight microneedles. The individual sensor element shows the linear response to normal force with good repeatability.

Keywords: Tactile Sensors, Piezoresistive, MEMS, MUMPs, Force Sensor Array, bulk-etching

1. INTRODUCTION

Micro-Electro-Mechanical Systems (MEMS) technologies have found the broad field in intelligent solid-state microsensors, the sensing devices that are batch-fabricated by micromachining techniques and integrated with electrical circuits on the same chip. A micromachined tactile sensor is a promising area in the field of physical MEMS sensors. It has a function similar to the surface of a human fingertip. The measurement and processing of a contact stress found useful in robotic dexterous manipulation applications. When a robot grasps object, information on contact, shear force and torque determination are needed for feedback control of robot. Other potential applications of this sensor would be such as sensing of organic tissue on a small scale at the end of catheter or on the fingers of an endoscopic-surgery telemanipulator [1]. Recently, varieties of micromachined tactile sensors capable of measuring normal and shear stress are individually developed by many research groups [2,3] but none of them are foundry-fabricated. In this paper, we present a novel piezoresistive tactile sensor array designed for measurement in medium-density sub-millimeter tactile sensing and fabricated by a commercial available Multi-Users MEMS Process (MUMPs) with bulk etching in post processing step.

2. SENSOR DESIGN

2.1 Sensor Configuration

The individual sensing element in the array is a trampoline-shape structure composed of a central contacting plate suspended by four beams over an etched pit. Embedded in each of beams is a polysilicon piezoresistor (2.8 µm wide and 168 µm long). Each tactile sensor element has dimension of 200 µm × 200 µm with 250 µm center-to-center spacing. Each of four sensor beams has geometry of 90 µm long and 10 µm wide. The central contacting plate is a square plate of 40µm× 40 µm. The entire sensor area is 1.25 mm × 1.25 mm. Figure 1 shows the scanning electron micrograph (SEM) of the micromachined piezoresistive tactile sensor array and the configuration of an individual sensor element.

2.2 Piezoresistive Sensor

An applied normal force causes the deflection of the central plate in a direction normal to the plane of substrate and induces equal tension and axial elongation in each of the four beam elements. For a small deflection of the central plate, the beams can be modeled as solid bars with pin joints at both points where the beam joins the central plate and the substrate [3].

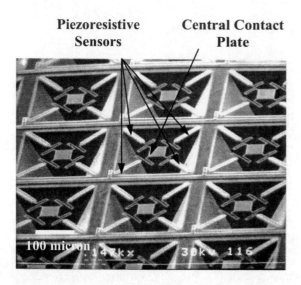

Piezoresistive Sensors　　Central Contact Plate

Figure 1: Scanning electron micrograph of the micromachined piezoresistive tactile sensor array and configuration of a sensor element.

We assume that the applied normal force is uniformly distributed on the central plate of each element.

With this simplified model, the strain induced in each beam due to applied normal force is characterized as:

$$\varepsilon_n = \frac{L}{2(EA)d} F \qquad (1)$$

where
ε_n　　strain induced in a beam due to an applied normal force

L　　the length of the sensor beam

F　　the applied normal force

d　　the normal deflection of the central plate or of the tip of the beam

E　　Young's modulus of beam composite

A　　cross-section area of the sensor beam

The working principle of the piezoresistive tactile sensor is piezoresistivity, which is the property of materials to change their resistivity under strain. The sensitivity to strain of a certain material is referred to as the Gauge factor (G). The gauge factor is defined as the ratio between the fractional change in resistance ($\Delta R/R$) and the strain (ε_n) induced in the resistor by an applied stress [4]. Longitudinal gauge factors (G_l) in which the direction of an electrical current flow is parallel to the applied strain is of interest here. The change of resistance due to strain in the beams is given by:

$$\frac{\Delta R}{R} = \varepsilon_n \times G_l \qquad (2)$$

From the mechanical beam theory, the deflection d at the tip of the cantilever beam is given by:

$$d = \frac{1}{2} kL^2 \qquad (3)$$

where k is the beam's spring constant and L is beam length. Thus, combining Eq.(1)-(3), the change of resistance as a function of the applied normal force is given by:

$$\frac{\Delta R}{R} = \left[\frac{G_l}{k(AE)L} \right] \cdot F = S \times F \qquad (4)$$

where S is sensitivity of the piezoresistive beam. The linear relationship of resistivity change and applied force is proven.

A commercially finite element analysis tool, MARC, was used to simulate the deflection of the structure under applied normal force [5]. When normal force is applied on the central plate, the sensor beams bend toward the pit as shown in Fig. 2 and the maximum stress is at the bases of the beams where piezoresistors are embedded. Figure 3 shows the stress distribution of the piezoresistive beam. The response of the sensor structure to pure normal stress loading was evaluated.

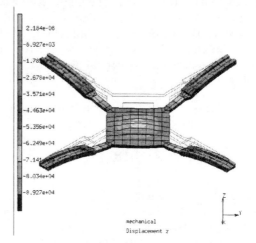

Figure 2: FEM result showing the deflection of the structure under normal force load.

3. DEVICE FABRICATION

3.1 Bulk-Etched MUMPs Process

The tactile sensors have been fabricated through the commercially available Multi-Users MEMS Process (MUMPs) [6]. This is a three-layer surface micromachining polycrystalline silicon (polysilicon) process.

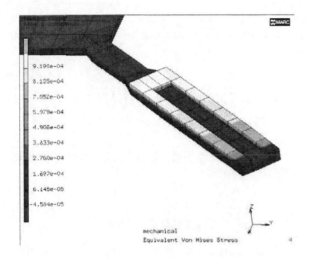

Figure 3: FEM result showing the stress distribution on the piezoresistive beam.

To suspend trampoline-shape structures over cavities, the bare single-crystal silicon substrate is exposed to chemical etchant using the superimposition of "Anchor1", "Anchor2", Poly1-poly2 via", "Dimple", "Hole1" and "Hole2" layers on top of each other to create "cut through substrate" mask.

By this method, the nitride and all the oxide layers above nitride are thoroughly opened during regular MUMPs process, leaving exposed bare silicon while the other area of the chip has covered with oxide and nitride layers [7]. The silicon substrate is selectively etched in-house using the method of anisotropic etching by EDP (ethylene diamine pyrochatechol and water). The polysilicon structures are not effected by EDP etching because the etching rate of polysilicon in these chemical etchants is much slower than single-crystal silicon. By this method, the bulk micromachining can be done on the surface micromachining chip (MUMPs) using only a maskless post-processing etching step.

3.2 Tactile Sensor Array Fabrication

The sensor array was fabricated using a bulk-etched MUMPs process as discussed in 3.1. Silicon nitride layer comprised the base layer of the free-standing sensor structures. The thinnest polysilicon, POLY0, is used to construct the piezoresistive structures and embedded in silicon dioxide layer of OXIDE1 and OXIDE2. The piezoresistor is located at the base of all four beams where the induced strain due to beam bending is at maximum. On the top layer, polysilicon, POLY2, was used to encapsulate silicon dioxide and increase the total thickness of the beams (total beam thickness = 5.35 μm).

Figure 4 shows the cross-section of the piezoresistive sensor beam. The central plate was constructed with the same structure as the beam except there was no piezoresistors embedded. The sensing structures were then released from the underlying silicon substrate using bulk silicon wet etching with an EDP solution at temperature of 90 °C for 30 minutes. The single crystal silicon is anisotropically etched in the area opened by the combination of "cut through substrate " layer (as discussed in section 3.1) until the central plate was undercut. The pits are formed underneath structures and their depths were controlled by etching time. The device was packaged in a ceramic DIP package and the read-out signal is electrically connected by gold wire bonding. A thin protective layer of polyvinyl chloride (PVC) or elastomer rubber can be adhered to the polysilicon surface of the sensor to provide interpolation of normal load between elements and protect the sensor array.

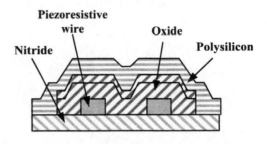

Figure 4: The cross section of the piezoresistive sensor beam.

4. TEST AND CHARACTERIZATION

First, High-resolution X-Y-Z micromanipulator is used to position the pre-known weight microneedle on the bare sensor (without protective layer) directly for force testing. The entire setup was placed under a microscope for visual inspection. The tactile sensor array was connected to the external circuit on a board to measure the resistance and voltage change when force is applied. Figure 5 shows the experimental setup of piezoresistive sensitivity measurement with the applied normal force.

A wheatstone bridge circuit is used to measure the extremely small change in the resistivity of piezoresistors. The resistivities of R_1, R_2, R_3 are adjusted evenly until equal to resistivity of the sensor, R_M, of 9.4kΩ. Therefore read-out voltage, V_0, is equal to zero before normal force is applied. Plot of the change in read-out voltage correspond to applied normal force is shown in Fig. 6. Sensor response shows linearity behavior with 0.02 mV/μN sensitivity gain.

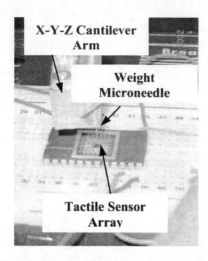

X-Y-Z Cantilever Arm

Weight Microneedle

Tactile Sensor Array

Figure 5: Experimental setup of piezoresistive sensitivity measurement with applied normal force by using pre-known weight microneedle.

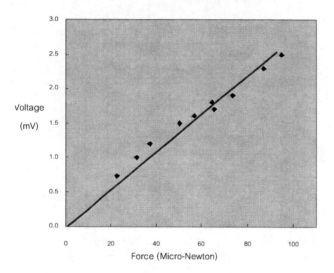

Figure 6: Plot of read-out voltage versus applied normal force.

5. CONCLUSIONS

The micromachined piezoresistive tactile sensor array has been successfully fabricated by the commercial available surface micromachining foundry (MUMPs) with bulk-etching post-processing technique. The bulk etching of silicon substrate through "cut through substrate" mask is successfully done with a complete undercut of a central contacting plate and create a suspended structures. The response of the sensor structure to pure normal stress loading was experimentally evaluated. Linear sensitivity of the sensors with normal force load was approximately 0.02 mV/μN as tested with pre-known weight

microneedle. More tests on operating temperature dependence and hysteresis behavior when the various protective layers are applied need to be studied further. Moreover, the packaging issues of the sensor must be addressed before the sensor can be considered useful for practical applications.

ACKNOWLEDGEMENT

This work was supported by NECTEC Grant No. E34701 and the Air Force Office of Scientific Research (AFOSR), Grant No. F49620-98-1-0291.

REFERENCES

[1] B. L. Gray and R. S. Fearing "A surface micromachined microtactile sensor array". *Proceedings of IEEE International Conference on Robotics and Automation*, Minneapolis, Minnesota, pp. 1-6, 1996.

[2] R. H. Liu, L. Wang, and D. J. Beebe "Progress towards a smart skin: fabrication and preliminary testing". *Proceedings of International Conference of the IEEE Engineering in Medicine and Biology Society*, Vol. 20, No. 4, pp. 1841-1844, 1998.

[3] B. J. Kane, M. R. Cutkosky and G. T. A. Kovacs "A traction stress sensor array for use in high-resolution robotic tactile imaging". *Journal of Microelectromechanical Systems*, Vol. 9, No. 4, pp. 425-434, 2000.

[4] G. T. A. Kovacs, *Micromachined Transducer Sourcebook*, WCB McGraw-Hill, New York, 1998.

[5] L. A. Liew, A. Tuantranont, and V. M. Bright "Modeling of thermal actuation in a bulk-micromachined CMOS micromirror". *Microelectronics Journal*, 31, pp. 791-801, 2000.

[6] D. A. Koester, R. Mahadevan, and K. W. Markus "Multi-user MEMS process (MUMPs): introduction and design rules, rev.4" MCNC MEMS Technical Application Center, Research Triangle Park, NC, 1996.

[7] A. Tuantranont, L. A. Liew, V. M. Bright, J. L. Zhang, W. Zhang, and Y. C. Lee "Bulk-etched surface micromachined and flip chip integrated micromirror array for infrared applications". *Proceedings of IEEE/LEOS International Conference on Optical MEMS*, Kauai, Hawaii, pp. 71-72, 2000.

Modelling of Microspring Thermal Actuator

J.K.Luo, A.J.Flewitt, S.M.Spearing*, N.A.Fleck and W.I.Milne
Eng. Dept. Cambridge University, Trumpington Street, Cambridge, CB2 1ZP U.K.
*Dept. of Aeronautics & Astronautics, MIT, Cambridge, MA, 02139, USA

Abstract

A new type of Microspring (μ-spring) electrothermal actuator consisting of multi-chevron structures was proposed. Finite element analysis was used to model the device performance, and the results compared with that of chevron thermal actuator. The simulation showed that the μ-spring actuator can deliver a much large deflection than that of chevron actuator. The deflection is proportional to the number of chevron structures linked together. By linking more chevron structures, this μ-spring can deliver a larger displacement than any other thermal actuators. The device has a much smaller footage than that of a single chevron device for a similar amount of deflection. Electroplated Ni thin film and SiO$_2$ were used as active and linkage materials to fabricate the μ-spring actuator.

Keyword: electrothermal actuator, microspring actuator, chevron, metal MEMS.

1. Introduction

Microelectrothermal actuators are one of the most attractive micro-moving actuators as they can deliver large forces and displacements compared to other types of actuators such as piezoelectric and electrostatic actuators. Electro-thermal actuators that utilises the difference in thermal expansion coefficient between two layers of a bilayer structure for vertical actuation (out of the plane of the layer) and others that utilises different structures made from the same material to induce varying expansions and motion in-plane have been reported previously [1,2]. The latter is preferential as it is a planar structure, suitable for the integration, especially on the CMOS. In order to obtain a large displacement and force for this type of flexure devices, a number of thermal actuators are typically connected in series [3]. Chevron type actuators [4], as shown in **Fig.1 a**, provide an even larger displacement, but such a single beam like device is difficult to integrate into a system. Here we report the development of a new type of planar microspring (μ-spring) thermal actuator. Finite element analysis (FEA) is used to simulate the device performance, and electroplated Ni thin film was used to fabricate the device.

2. Device concept and modelling

The deflection of a chevron actuator strongly depends on the length and the angle of the beams; a larger deflection is produced if a smaller angle is used. The drawback is that it is a single beam like device. To deliver a larger displacement, even longer beams are required for chevron actuators. This is problematical in term of fabrication and operation, as the long beams easily buckle or stick to the substrate. Therefore the beam length of a chevron device is limited, typically to a 200μm long, which in-turn limits the amount of the displacement and applications. In our new type of device, instead of using a single long beam, the arm is folded into a number of chevron sections to become a planar spring-like actuator as shown in **Fig.1.b**. (2 chevrons with 1 linkage bar are defined as 1 ring). An insulating beam with a very low thermal expansion coefficient is used to form a cross-linkage thereby constraining the displacement in the x-direction. As the actuator is heated up under bias, the thermal expansion of the active material leads to a displacement in the y-direction only.

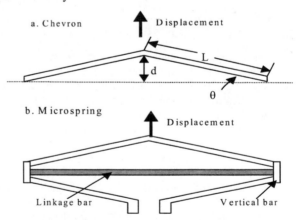

Figure.1 Schematic drawing of a).chevron and b). μ-spring actuators with 1 ring.

For a chevron device, the displacement in y direction under bias can be estimated as follows [4].

$$\Delta y \approx \frac{L\Delta L}{d} \approx \frac{\Delta L}{\sin \theta} \approx \frac{\alpha L \Delta T}{\theta} \qquad (1)$$

Here α is the thermal expansion coefficient of the beam material; d the height of a chevron device and L is the beam length. Assuming the convection and the radiation make little contribution to the heat loss, then the temperature distribution along the beams is governed by the classic heat transfer equation [1,4,5],

$$k\frac{d^2T}{dx^2} + J^2\rho = 0 \qquad (2)$$

$$J = \frac{V}{RA} = \frac{V}{\rho L} \qquad (3)$$

Where J is the current density, V the applied voltage, L the length of the resistor, R the resistance, A the cross section of the beam, ρ the resistivity and k the thermal conductivity. The maximum temperature is at the joint point of two beams, and the average temperature ΔT_{ave} of the beam is given by:

$$\Delta T_{ave} = \int \frac{T(x) - T_o}{2L}dx = \frac{V^2}{3k\rho} = \frac{PL}{3kA} \qquad (4)$$

where $P = IV$ the consumed power. The deflection can be expressed as follows,

$$\Delta y \approx \beta\frac{\alpha L\Delta T_{ave}}{\theta} = \beta\frac{\alpha LV^2}{3k\rho\theta} = \beta\frac{\alpha PL^2}{3kA\theta} \qquad (5)$$

where β is the coefficient, $\beta = 1$ for chevron device and $\beta < 1$ for spring actuator as will be discussed later. At a small angle, $\theta \approx d/L$, eq.(5) then becomes,

$$\Delta y \approx \beta\frac{\alpha L^2V^2}{3k\rho d} = \beta\frac{\alpha PL^3}{3kAd} \qquad (6)$$

At a fixed bias V and material properties, the deflection is proportional to the square of the beam length, while for a fixed power P, the deflection increases with the cubic of the beam length; therefore increasing the beam length is the most effective way to achieve a larger displacement. As mentioned above, there is a limitation to the beam length due to buckling and stiction problems, the maximum displacement of a chevron actuator is therefore limited. Reducing the height d of the structure has the similar effects to increase the risk of buckling and stiction as increasing the beam length. Therefore increasing the beam length and reducing the height is not the best way to improve the device performance On the other hand, by introducing a multi-chevron spring structure, the buckling and stiction issues can be solved readily, as the folded beams reduce the length in one direction.

The deflection of a μ-spring actuator is similar to that of a chevron device and the displacement in y-direction can also be expressed by eqs.(5) and (6) with a modified coefficient β, as the current does not pass through the insulator. The main contributions to a modified coefficient β are from the limited thermal expansion of the insulator

beam at an elevated temperature, which leads to a displacement in x-direction, and reduce the displacement in y-direction. Also introduction of vertical bars to link beams together causes a slight deflection in x-direction. When the heat loss through the surface is considered, then the linkage bars lose more heat owing to increased surface area. Therefore the coefficient β for a μ-spring actuator is always smaller than 1, as we will see later from FEA modelling.

From eq.(6), it is understood that active materials with a higher value of α/k are better for thermal actuators. Metals generally provide a larger displacement than Si based materials as most metals have a much larger thermal expansion coefficient [5]. Insulator such as SiNx and SiO$_2$ are among the best materials for cross linkage bars, as they have the lowest thermal expansion coefficients. Nickel and silicon nitride were chosen as the materials for beams and cross-linkage bars respectively for the simulation. Thermal radiation and convection are neglected here as they only contribute a few percent to the total heat loss in the temperature range of interest. Finite element analysis based on the FEMLAB software was used to simulate the performance of both chevron and μspring actuators. In the simulation, the temperature dependence of the resistivity was considered, and the voltage operation mode was used instead of the current operation mode. The constants used for this simulation are listed in Table.I, where the resistivity, its temperature coefficient and the Young's modulus for Ni thin films were obtained experimentally from our work [5].

Table.I

	$\alpha \ 10^{-6}$ (/C°)	ρ (Ω.cm)	κ (W/M.°C)	E (GPa)	$\xi \ 10^{-3}$ (/°C)
Ni	12.7	2×10^{-7}	83	210	~3.0
SiN$_x$	2.0	n/a	20	~150	n/a
SiO$_2$	0.4	n/a	2	~150	n/a

3. Microfabrication

Microspring actuators were fabricated using a two-mask process, and the process flow is shown in **Fig.2**. The first mask stage is the formation of the insulator cross-linkage bars. Stress free LPCVD Si$_3$N$_4$ or thermally grown SiO$_2$ were used as the linkage bars. Photoresist patterns were used as an etch mask to etch Si$_3$N$_4$ or SiO$_2$ with concentred HF or BHF. After the formation of linkage bars, a seed layer of Cr/Cu (10/50nm) was deposited on the wafer by sputtering. Positive photoresist AZ5214 was used to form the plating mould. A multilayer coating method was used to form a thick mould with a thickness of ~3μm [5]. Stress free Ni thin films were then electroplated using nickel sulphamate solution at an optimised conditions of 60°C and 2 ~ 4mA/cm^2 current density. The typical thickness of actuator devices is 1.5~2μm. After plating, the photoresist and the seed layer out side of device area were removed by acetone and acid respectively, and was followed by releasing process. For Si₃N₄ linkage bars, the actuators were released by a wet etching in a KOH (20% in wt) solution at 85°C for ~30mins, while for the SiO₂ bars, the devices were released by SF6 RIE dry etching as SF6 has a good etch selectivity between Si and SiO₂. More detailed description of the process can be found in Ref.5.

It was found that the adhesion between the Cr/Cu seed layer and the Si₃N₄ bar is very poor, and it detaches from each other at various process stages. Some special treatments can improve the adhesion, but the yield of successful devices was very low. On the other hand, the adhesion between the SiO₂ and the Cr/Cu seed layer is excellent; there is no delamination problem.

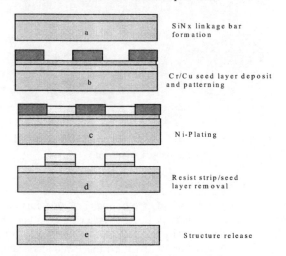

Fig.2 Process flow for fabricating actuators.

4. Results and discussions

Figure 3 shows the simulated displacement and the temperature distribution for a microspring actuator with 2-rings of chevron structures under a bias of 0.4V. The beam length and width are 200 and 4μm respectively. The angle is d/L=0.1 The maximum temperature is at the top of the middle point, T_{max}= 923K, the displacement is 33.7μm. The displacement of the lower chevron is small because of a lower temperature, while that for the upper chevron is large owing to a higher temperature. For a steady state, the linkage bar has a similar temperature to that of the joint points with the arms, and the temperature is uniform along the bar. The temperature distribution along the beams for a μ-spring was found to be similar to that of a chevron actuator, with a parabolic distribution. But there are temperature dips at the joint points of the beams and the linkage bars due to the heat loss through the linkage bars. For dynamic state, it was found that the temperature of a linkage bar is not uniform, and the temperature at the middle of the bar is lower than that on the edge owing to a finite time constant for the heat transfer.

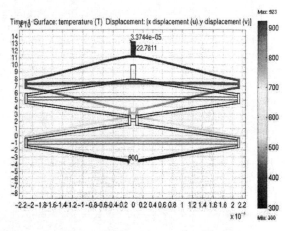

Figure 3 the temperature and displacement distribution of a 2-rings μspring at a bias of 0.4V. It delivers a deflection of ~34μm with T_{max} of 923K.

Simulation was also performed for chevron actuator devices with the same beam length and cross-section. The angle for chevrons is also set d/L=0.1. The deflection results are shown in **Fig.4** together with those of spring actuators with the total beam length as a variable, at a fixed bias of 0.4V. The displacement for a chevron device increases linearly with increasing the beam length as predicted by the analytical eq.(5). For a chevron actuator with L = 200μm, will deliver a displacement of 10.2μm. By introducing another chevron structure to form a 1-ring spring actuator with the same beam length 200μm (total length is 400μm), it delivers a displacement of 16.7μm, much larger than that of a chevron actuator. The displacement of a μ-spring increases with increasing the number of chevron structures connected in series. When 2-rings chevron structures are used to form a spring, it can produce a displacement up to 33.4μm. For this amount of displacement, a chevron actuator device with L=600μm is needed, which is not a desirable size for fabrication and operation. From the point of device area, this large chevron device occupies an area of 2x600x60μm², while the corresponding 2-rings spring actuator only takes an area of ~220x20x2x4μm², less than half the area of the chevron actuator. By connecting more chevron structures together, e.g. 6-rings, a spring actuator can produce a displacement as much as ~100μm, much larger than any planar actuators can produce. This has clearly showed the advantage of a μ-spring actuator over any other actuators.

Fig.4 also showed the dependence of the displacement of a spring actuator with L=100μm on the total beam length for comparison with that of L=200μm. It is clear that there exists a difference in displacement between the spring and the chevron actuators. The coefficient β decreases from chevron's β=1 to β=0.81 and β=0.71 for spring actuators with L=200 and L=100μm respectively. A smaller β for L=100μm springs is because more linkage and vertical bars are used. This result also indicates it is possible to maximize the device performance by optimizing the device structure with a β close to 1. From the FEA simulation it was found that the deflection and the maximum temperature increase with increasing the square of the bias, consistent with the analytical results shown by eqs.(4) and (6).

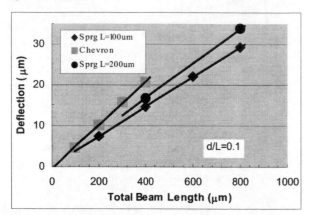

Fig.4 The deflection of chevron and spring actuators as a function of total beam length. The deflection of spring actuator increases with increasing the number of chevron connected.

Microsprings with varying beam lengths and the number of rings were fabricated. The beam lengths for chevron structures were 100, 200 and 400μm, while the ring numbers were 1, 2, 4 and 6. The length of vertical bars was 20μm. The width of the insulator bar was typically 4~6μm, while the beam width 3~4μm. At the tip of the actuator, a bar with multi-scales was also designed for measuring the deflection. It was found that with a Ni film thinner than 0.8μm, the beams easily buckle under the bias, a thicker Ni metal is therefore required for a proper operation of a μ-spring device.

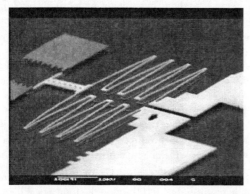

Fig.5 SEM picture of spring actuator without insulator linkage bars. Plating Ni was used, and released by SF6 dry etch.

Figure 5 shows a SEM micrograph of a fabricated μ-spring without insulator cross-linkages. The structure has 4-rings with the beam length of 200μm, and was released

by SF6 dry etch. The electroplated Ni structure is freestanding with no visible curvature, indicating the gradient stress is at its minimum. The total length of the spring arms on one side is 800μm.

Figure 6 shows a micrograph of a μ-spring with oxide constraint bars. The oxide thickness is ~1.0μm. These insulator bars hold the beams tightly without separating from the Ni-beams. Preliminary electrical test showed a motion of μ-spring along y-direction, and the displacement increases with increasing the ring number.

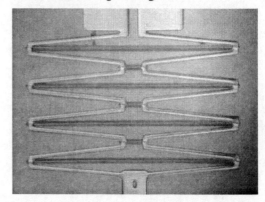

Fig.6 micrograph of a 4-rings spring actuator with SiO_2 (dark) linkage bars. L=200μm.

5. Conclusions

A new spring type of electrothermal actuator has been proposed and modelled by FEA. This spring type actuator delivers a much larger deflection than all other electrothermal actuators, but with a much-reduced device area. This type of device minimizes the buckling and stiction problem, which are associated with long beam devices, thus it is easy to integrate with the system. Spring actuator has been fabricated using electroplated Ni as active material and electrically tested.

Acknowledgement: This project was sponsored by the Cambridge-MIT Institute under grant number 059/P

References

1. Q.A.Huang and N.K.S.Lee, J. Micromech. MicroEng. **9**, 64 (1999).

2. L.Que, J.S.Park and Y.B.Gianchandani; J.Microelectro. mech.sys. **10**, 247 (01')

3. J.R.Reid, V.M.Bright and J.H.Comtois Proc. of SPIE 96 **2882**, 296 (1996).

4. R.Hichey, D.Sameoto, T.Hubbard amd M.Kujath; J. Micromech. MicroEng. 13, 40 (2003).

5. J.K.Luo, J.H.He, A.J.Flewitt, S.M.Spearing, N.A.Fleck and W.I.Milne; Proc. of SPIE 04 Conf. In press.

6. W.N.Jr.Sharpe, "Mechanical Properties of MEMS Materials" 2001, Johns Hopkins University: Baltimore, MD USA

The Application of Parametric Excitation to a Micro-Ring Gyroscope

B.J. Gallacher, J.S. Burdess, A.J. Harris and J. Hedley

Institute for Nanoscale Science and Technology, University of Newcastle upon Tyne

b.j.gallacher@ncl.ac.uk

ABSTRACT

A parametric excitation scheme that separates the drive and response frequencies of an electrostatically driven ring gyroscope in the approximate ratio of 10:1 is reported. This frequency separation permits reduction of troublesome electrical feedthrough common in many MEMS devices.

Keywords: parametric, excitation, gyroscope, feedthrough

1 INTRODUCTION

This paper reports on a parametric excitation scheme that separates the drive and response frequencies of a micro-ring gyroscope in the ratio 10:1 approximately, with the aim of minimising electrical "feedthrough". Electrical feedthrough of the drive signal due to parasitic capacitance is a common problem in many MEMS devices actuated electrostatically, inductively and piezoelectrically and in the case of the gyroscope, limits its sensitivity to applied angular velocities. To date research into the application of parametric excitation to MEMS/NEMS has been restricted to subharmonic and superharmonic parametric resonances [1,2,3,4]. In this case the ratio of the drive and response frequencies has a maximum value of two, for first-order parametric resonance. For a multi-dimensional system, combination resonances are excited when the frequency of modulation of a system parameter is a multiple of either the sum or difference of two of the natural frequencies of the system [5]. By ensuring the drive frequency is the sum of the natural frequency of a higher order mode and the natural frequency of the mode required to be excited, the ratio of the drive to response frequencies will be greater than two. In this paper, sum combination resonances between the 2nd and 5th order in-plane flexural modes are considered and are shown to be an alternative method of excitation suitable for the gyroscope.

2 EQUATION OF MOTION

Figure (1) illustrates the ring gyroscope consisting of planar ring of radius a, width b and thickness d. Actuation and sensing is performed electrostatically using a plurality of electrodes also of thickness d, displaced radially from the outer surface of the ring by a distance h_o. The parametrically excited ring gyroscope takes advantage of the dynamic instability possible when the condition for a sum combination resonance is met. Therefore, the dynamics

of the ring will be expressed in terms of two pairs of orthogonal flexural modes of order n and m, where m<n.

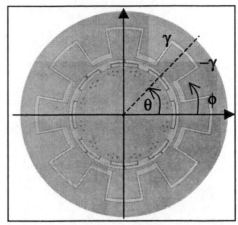

Figure (1) Photograph of ring gyroscope

Excitation of the ring structure into vibration is accomplished electrostatically through a plurality of p cyclically arranged electrodes. The electrical energy stored in the series of capacitors, formed between the ring (assumed to be held at Earth potential) and the electrodes, each biased with a voltage U_i is given by

$$E_E = \frac{\varepsilon_o a d}{2 h_o} \sum_{i=1}^{p} U_i^2 \int_{-\gamma+\theta_i}^{+\gamma+\vartheta_i} \left[1 + \frac{u}{h_o} + \left(\frac{u}{h_o}\right)^2 + \left(\frac{u}{h_o}\right)^3 + \cdots \right] d\phi \quad (1)$$

The radial displacement u, of a point on the centre line of the ring may be expressed in terms of the undamped, unforced flexural modes of an unsupported ring by:

$$u = q_1 \cos n\phi + q_2 \sin n\phi + q_3 \cos m\phi + q_4 \sin m\phi \quad (2)$$

where q_i and are generalised coordinates which represent the contribution made by each mode to the motion.

The radial displacement of the ring is assumed to be small compared to the nominal air gap separation thus terms of order greater than $(u/h_o)^3$ may be ignored. The generalised stiffness matrix and forcing vector may be obtained from substituting in the expression for the radial displacement given by equation (2) into equation (1) and by determining $\frac{\partial E_E}{\partial q_i}$ and $\frac{\partial^2 E_E}{\partial q_i^2}$ for i=1..4.

The equation of motion of the ring excited electrostatically by a voltage $U(t)$ is given by

$$[m]\underline{\ddot{q}}+[c]\underline{\dot{q}}+[k]\underline{q}=\frac{\varepsilon_o ad}{h_o^3}U(t)^2[K]\underline{q}(t)+\frac{\varepsilon_o ad}{2h_o^2}U(t)^2\underline{G}$$

where

$$[m]=\begin{bmatrix} m_n & & & \\ & m_n & & \\ & & m_m & \\ & & & m_m \end{bmatrix}, \quad [k]=\begin{bmatrix} k_n & & & \\ & k_n & & \\ & & k_m & \\ & & & k_m \end{bmatrix}$$

and

$$[c]=\begin{bmatrix} 2m_n\xi_1\omega_1 & & & \\ & 2m_n\xi_2\omega_2 & & \\ & & 2m_m\xi_3\omega_3 & \\ & & & 2m_m\xi_4\omega_4 \end{bmatrix}$$

and

$$m_n=m_r\left(1+\frac{1}{n^2}\right), \ m_m=m_r\left(1+\frac{1}{m^2}\right), \ m_r=\pi\rho abd$$

$$k_n=\pi\left(1-n^2\right)^2\frac{EI}{a^3}, \ k_m=\pi\left(1-m^2\right)^2\frac{EI}{a^3}$$

The matrix $[K]$ and vector $\underline{G}$ are the spatial parts of the generalised stiffness and forcing vector corresponding to the modal vector $\underline{q}=(q_1,\ q_2,\ q_3,\ q_4)^{\mathrm{T}}$.

It is convenient to introduce the non-dimensional parameter μ such that $\mu=U(t)^2/U_p^2$ where U_p is the pull-in voltage.

For typical drive voltages $\mu\ll 1$ and the equation of motion may be rewritten in the form

$$\underline{\ddot{q}}+[2\xi\omega]\underline{\dot{q}}+\left\{[\omega^2]-[D]\mu\right\}\underline{q}=\mu\underline{F}$$

where

$$[D]=\frac{\varepsilon_o ad}{h_o^3}U_p^2[m]^{-1}[K] \text{ and } \underline{F}=\frac{\varepsilon_o ad}{2h_o^2}U_p^2\underline{G}$$

The periodic voltage applied to ring may be conveniently represented by the Fourier series

$$\mu=\varepsilon\eta\left(u_o+\sum_{r=1}^{\infty}u_r e^{ir\omega t}+\sum_{r=1}^{\infty}u_r^* e^{-ir\omega t}\right)$$

where $\varepsilon\ll 1$ and η represents the voltage scaling factor. The equation of motion is therefore given by

$$q_i+2\varepsilon\upsilon_i\omega_i\dot{q}_i+\left\{\omega_i^2 q_i-\varepsilon\eta u_o\sum_{j=1}^{4}D_{ij}q_j\right\}-$$

$$\varepsilon\eta\left(\sum_{r=1}^{\infty}u_r e^{ir\omega t}+\sum_{r=1}^{\infty}u_r^* e^{-ir\omega t}\right)\sum_{j=1}^{4}D_{ij}q_j=$$

$$\varepsilon\eta\left(u_o+\sum_{r=1}^{\infty}u_r e^{ir\omega t}+\sum_{r=1}^{\infty}u_r^* e^{-ir\omega t}\right)F_i \tag{3}$$

where use was made of the relation $\zeta_i=\varepsilon\upsilon_i$ since operation is under vacuum. Equation (3) represents a set of coupled Hill's equations. As the flexural modes of vibration of the perfect ring occur in degenerate pairs, the natural frequencies of a slightly imperfect ring may be expressed in the form [6]

$$\omega_i=\omega_\alpha+\varepsilon\alpha_i \ \ i=1,2.$$
$$\omega_k=\omega_\beta+\varepsilon\beta_k \ \ k=3,4. \tag{4}$$

where α and β represent mis-tuning due to structural imperfections and $\alpha_1=\beta_3=0$.

3 PERTURBATION ANALYSIS

A multiple time scales perturbation analysis is performed [7] using ε as the perturbation parameter to investigate the dynamic stability and mode coupling. To determine the stability boundaries it is necessary to rewrite the condition for the sum resonance as

$$r\omega=\omega_\alpha+\omega_\beta+\varepsilon r\lambda$$

where λ represents slight mis-tuning from the sum resonance condition. Using the relations given by equation (4), this can be written as

$$r\omega=\omega_i+\omega_k+\varepsilon\left[r\lambda-(\alpha_i+\beta_k)\right]$$

In the multiple time scales method the solution is expressed as an asymptotic expansion

$$q_i(\bar{t},\hat{t})=q_i^{(0)}+\varepsilon q_i^{(1)}+\cdots+O(\varepsilon^2) \tag{5}$$

where $\hat{t}=t+\cdots O(\varepsilon^2)$ and $\bar{t}=\varepsilon t$. $\tag{6}$

As the generalised coordinates q_i are functions of both the slow and fast time scales $\bar{t}$ and $\hat{t}$, the time derivatives are given by

$$\dot{q}_i=\frac{\partial q_i^{(0)}}{\partial\hat{t}}+\varepsilon\frac{\partial q_i^{(1)}}{\partial\hat{t}}+\varepsilon\frac{\partial q_i^{(0)}}{\partial\bar{t}}+\cdots+O(\varepsilon^2) \tag{7}$$

$$\ddot{q}_i=\frac{\partial q_i^{(0)}}{\partial\hat{t}}+\varepsilon\frac{\partial^2 q_i^{(1)}}{\partial\hat{t}}+2\varepsilon\frac{\partial^2 q_i^{(0)}}{\partial\bar{t}\partial\hat{t}}+\cdots+O(\varepsilon^2) \tag{8}$$

Using of the equations (5,7 and 8) and equating powers of ε, equation (3) yields the recurrent equations

$$\frac{\partial^2 q_i^{(0)}}{\partial\hat{t}^2}+\omega_i^2 q_i^{(0)}=0 \tag{9}$$

$$\frac{\partial^2 q_i^{(1)}}{\partial\hat{t}^2}+\omega_i^2 q_i^{(1)}=-2\frac{\partial^2 q_i^{(o)}}{\partial\hat{t}\bar{t}}-2\upsilon_i\omega_i\frac{\partial q_i^{(0)}}{\partial\hat{t}}+$$

$$\eta\left(u_o+\sum_{r=1}^{\infty}u_r e^{ir\omega\hat{i}}+\sum_{r=1}^{\infty}u_r^* e^{-ir\omega\hat{i}}\right)\sum_{j=1}^{4}D_{ij}q_j^{(o)}+$$

$$\eta\left(u_o+\sum_{r=1}^{\infty}u_r e^{ir\omega\hat{i}}+\sum_{r=1}^{\infty}u_r^* e^{-ir\omega\hat{i}}\right)F_i \tag{10}$$

The general solution to equation (9) is

$$q_i^{(0)} = A_{oi}(\bar{t}) e^{i\omega_i \hat{t}} + B_{oi}(\bar{t}) e^{-i\omega_i \hat{t}}. \qquad (11)$$

A_{oi} and B_{oi} are determined by imposing the condition that the solution to equation (10) is uniform in $\hat{t}$. Equation (10) represents a system subjected to both external and parametric forcing through the F_i and D_{ij} terms, respectively. As any resonance producing terms in the equation of motion will lead to unbounded particular solutions with respect to $\hat{t}$, these must be removed to meet the condition for uniform solutions.

Parametric resonance occurs when $j=i$ or $j=k$.
Thus conditions for solutions uniform in $\hat{t}$ are given by

$$\frac{\partial A_{oi}}{\partial \bar{t}} + v_i \omega_i A_{oi} + i\eta u_o \frac{D_{ii}}{2\omega_i} A_{oi} + i\eta u_r \frac{D_{ik}}{2\omega_i} B_{ok} e^{i\bar{\lambda}\bar{t}} = 0$$

$$\frac{\partial B_{oi}}{\partial \bar{t}} + v_i \omega_i B_{oi} - i\eta u_o \frac{D_{ii}}{2\omega_i} B_{oi} - i\eta u_r^* \frac{D_{ik}}{2\omega_i} A_{ok} e^{-i\bar{\lambda}\bar{t}} = 0$$

$$\frac{\partial A_{ok}}{\partial \bar{t}} + v_k \omega_k A_{ok} + i\eta u_o \frac{D_{kk}}{2\omega_k} A_{ok} + i\eta u_r \frac{D_{ki}}{2\omega_k} B_{oi} e^{i\bar{\lambda}\bar{t}} = 0$$

$$\frac{\partial B_{ok}}{\partial \bar{t}} + v_k \omega_k B_{ok} - i\eta u_o \frac{D_{kk}}{2\omega_k} B_{ok} - i\eta u_r^* \frac{D_{ki}}{2\omega_k} A_{oi} e^{-i\bar{\lambda}\bar{t}} = 0$$

$$(12)$$

where $\bar{\lambda} = r\lambda - (\alpha_i + \beta_k)$ and use was made of the relation $\bar{t} = \varepsilon t$.
The solutions to the above set of equations are

$$A_{oi} = A_{oi}^{(o)} e^{\left(\sigma + i\frac{\bar{\lambda}}{2}\right)\bar{t}}, \quad A_{ok} = A_{ok}^{(o)} e^{\left(\sigma + i\frac{\bar{\lambda}}{2}\right)\bar{t}},$$

$$B_{oi} = B_{oi}^{(o)} e^{\left(\sigma - i\frac{\bar{\lambda}}{2}\right)\bar{t}} \text{ and } B_{ok} = B_{ok}^{(o)} e^{\left(\sigma - i\frac{\bar{\lambda}}{2}\right)\bar{t}} \qquad (13)$$

Substituting the solutions given by equation (13) into equation (12) yields the quadratic

$$\sigma^2 + (v_i \omega_i + v_k \omega_k)\sigma + v_i \omega_i v_k \omega_k +$$

$$\frac{\bar{\lambda}^2}{4} + \frac{\bar{\lambda}}{4} \eta u_o \varepsilon \left(\frac{D_{ii}}{\omega_i} + \frac{D_{kk}}{\omega_k}\right) + \frac{\eta^2 u_o^2 \varepsilon^2}{4} \frac{D_{ii}}{\omega_i} \frac{D_{kk}}{\omega_k} -$$

$$\frac{\eta^2 u_r u_r^* \varepsilon^2}{4} \frac{D_{ik}}{\omega_i} \frac{D_{ki}}{\omega_k} = 0$$

Clearly for stable solutions the real part of σ must be zero. This condition is met when λ is described by

$$r\lambda = -\frac{1}{2}\left[\begin{array}{c} \eta u_o \left(\frac{D_{ii}}{\omega_i} + \frac{D_{kk}}{\omega_k}\right) \pm \\ \sqrt{\eta^2 u_o^2 \left(\frac{D_{ii}}{\omega_i} - \frac{D_{kk}}{\omega_k}\right)^2 + } \\ \sqrt{4\eta^2 |u_r|^2 \frac{D_{ik}}{\omega_i} \frac{D_{ki}}{\omega_k} - 16 v_i \omega_i v_k \omega_k} \end{array}\right] + (\alpha_i + \beta_k)$$

$$(14)$$

4 EXCITATION FROM A SINGLE ELECTRODE

It can be shown that for any single drive electrode, the matrix D will be in the form

$$[D] = \begin{bmatrix} D_{11} & 0 & D_{13} & 0 \\ 0 & D_{22} & 0 & D_{24} \\ \frac{m_m}{m_n} D_{13} & 0 & D_{33} & 0 \\ 0 & \frac{m_m}{m_n} D_{24} & 0 & D_{44} \end{bmatrix} \qquad (15)$$

Therefore, the modes corresponding to coordinates q_1, q_3 and q_2, q_4 are parametrically coupled through the matrix elements D_{13} and D_{24}, respectively. For the perfect axisymmetric ring $\omega_i = \omega_\alpha$ and $\omega_k = \omega_\beta$ for $i=1,2$ and $k=3,4$. Therefore, when the condition for a sum resonance between the n and m-order modes is met, the relative magnitudes of D_{13}, D_{24} determine which mode pair is the first excited when the critical voltage is exceeded. It may be shown that the mode pair q_1, q_3 is excited first when $\gamma < \pi/(n+m)$. The ring gyroscope considered makes use of the $n=2$ modes to sense rate of rotation. Hence eight cyclically arranged electrodes with $\gamma = \pi/16$ radians provide the drive and sense functions. Thus the mode pair q_1, q_3 will be excited first for values of m up to a maximum value of 13.

5 STABILITY BOUNDARIES

Consider the case where excitation is provided by a single electrode assumed to be aligned with the nodal diameter of the $\sin(n\theta)$ mode. A square wave voltage waveform with Fourier components $u_o = 1/2$ and $|u_1| = 1/\pi$ provides the excitation. The modes utilised in the case have orders $n=2$ and $m=5$ and their mode shapes are shown in figure (2).

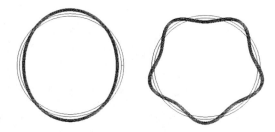

Figure (2) $\cos 2\phi$ and $\cos 5\phi$ mode shapes

The damping ratios of the n and m- order modes are assumed to be equal and have the value 2.5×10^{-5}. This is typical of the measured damping ratio of vacuum-packed devices for in-plane flexural modes. It is assumed that the ring is slightly imperfect with $\alpha_1 = \beta_3 = 0$, $\varepsilon\alpha_2 = 0.0001\omega_2$ and $\varepsilon\beta_4 = 0.0001\omega_4$. This constitutes a frequency split of 0.1%

between the otherwise degenerate modes and is similar to that measured in fabricated devices.

Using equation (14) the stability boundaries may be drawn. There are two stability boundaries for each order of mode. The two boundaries are displaced with respect to each other due to the mis-tuning between the otherwise degenerate modes and also to the asymmetric form of the electrostatic stiffness matrix when one drive electrode is used. As a result, the onset of instability in the q_1, q_3 mode pair occurs at a lower excitation voltage than the q_2, q_4 mode pair. Therefore, it will be possible to excite only the q_1, q_3 mode pair into oscillation. The minimum voltage amplitude required to induce instability in the q_1, q_3 mode pair is shown on Figure (3) and has the value $\eta_m = 162$. This value occurs at a excitation frequency described by $\omega = \omega_1^* + \omega_3^*$, where ω_1^* and ω_3^* are the damped natural frequencies of the $\cos(n\phi)$ and $\cos(m\phi)$ modes, respectively. When $n=2$ and $m=3$, $\omega \approx 10\omega_1^*$.

6 CONCLUSIONS

It has been shown that an electrostatically actuated ring gyroscope may be excited using combination parametric resonances between the in-plane flexural modes of order $n=2$ and $m=5$. This frequency separates the drive and response signals by approximately a decade and enables effective filtering of the electrical feedthrough of the drive signal which typically contaminates the sense signal produced in response to a rate of rotation and limits the device performance.

REFERENCES

[1] E.M. Abdel-Rahman and A.H. Nayfeh, "Superharmonic Resonance of an Electrically Actuated Resonant Microsensor", Technical Proceeding of the 2003 Nanotechnology Conference and Trade Show", vol.1, pp312-315.

[2] K.L. Turner, et-al., "Five Parametric Resonances in a Microelectromechanical System" Nature, vol. 396(6707), November 1998, pp149-152.

[3] Zalalutdinov, et-al., "Autoparametric Optical Drive for Micromechanical Oscillators". Applied Physics Letters 79, 2001, pp695-697.

[4] M-F Yu and G.J. Wagner "High Order Parametric Resonance and Nonlinear Mechanics of Nanowires": Technical Proceeding of the 2003 Nanotechnology Conference and Trade Show", vol.3, pp325 - 328.

[5] G. Schmidt and A. Tondl "Non-Linear Vibrations" Cambridge University Press, 1986.

[6] C.S. Hsu "Further Results on the Parametric Excitation of a Dynamic System" Journal of Applied Mechanics, Vol.32, 1965, p.373-377.

[7] E.J. Hinch "Perturbation Methods" Cambridge University Press, 1991.

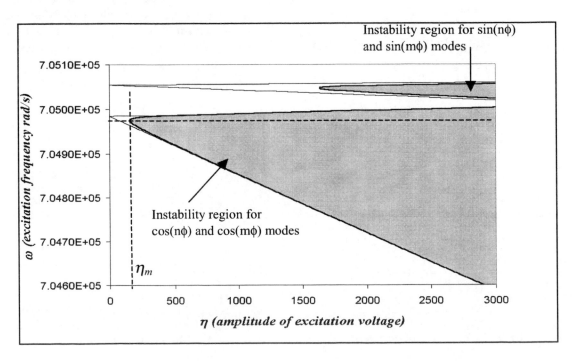

Figure (3) Stability map for $n=2$, $m=5$, one drive electrode

A Nanoscale Composite Material for Enhanced Damage Tolerance in MEMS Applications

A. Paranjpye, Noel C. MacDonald, Glenn E. Beltz

Department of Mechanical and Environmental Engineering,
University of California, Santa Barbara 93106

ABSTRACT

A two-dimensional laminar composite material has been shown [1] to exhibit enhanced damage tolerance due to zones of residual compressive stress. The damage tolerance appears as a threshold stress below which the structure does not fail catastrophically. We utilize this concept at the microscale, towards obtaining fracture resistant MEMS structures. Cantilever beams of this laminar composite material, manufactured by MEMS fabrication methods are deflected to fracture, and the measured fracture strength compared to single crystal silicon beams (SCS) of identical geometry. All structures are made on the same die, so uniform processing damage is likely for all of them. A wide range of fracture strength is seen in both the composite and SCS beams. However, the lowest value of fracture stress for all the composite beams is found to be above the predicted threshold strength, while some of the SCS beams fail at lower stresses. We conclude that this composite offers a useful design alternative as a MEMS material.

Keywords: MEMS, silicon, damage tolerance, composite

1 BACKGROUND

Brittle materials, like glasses and ceramics often undergo catastrophic failure with little or no warning even when stressed at levels below their elastic limit. These materials typically have no mechanisms that allow the rearrangement of atoms into more stable configurations to accommodate the applied load. Therefore, any flaw in the material structure that builds up a stress concentration around it can serve as the precursor to a crack. Such cracks can extend catastrophically through the entire structure to cause failure even though the applied stress may be well below the level to rupture atomic bonds. The failure strengths of these materials can be characterized probabilistically at best, since it is not possible to know the size, shape, position and orientation of every single flaw in the structure. Designing structures using these materials becomes a challenge, and proof testing of each component to the design load becomes necessary when the cost of failure is high.

Single Crystal Silicon, the material of choice for many bulk micromachined Micro-Electro-Mechanical Systems (MEMS), suffers from the bane of brittle materials, i.e. a low fracture toughness of 0.9 MPa.m$^{1/2}$ [2] and a sharp transition from elastic deformation to fracture. This effect is exacerbated by the presence of residual stresses and surface flaws from processing MEMS devices, which can lead to brittle behavior and a non-deterministic strength for finished structures. An earlier attempt to develop MEMS microsurgery tools was aborted when the finished devices failed unpredictably during testing. The low fracture toughness of silicon makes a silicon flexure susceptible to failure under almost any kind of shock or impact loading. As another example, in a micro aerial vehicle development project, [3] silicon was rejected as a material for the wing frames in favor of metal because the wings shattered every time the MAV crash-landed.

Regions of residual compressive stress in a material can be effective at improving material reliability by acting as "crack traps" because the driving force for extending a crack is lowered when a crack enters one of these zones. Specifically, thin compressive layers within a laminar bimaterial ceramic have been shown [1] to be effective at trapping large surface and internal cracks. The damage tolerance and reliability of the materials is manifested in terms of a threshold stress below which the probability of failure is zero. Mathematical modeling and finite element analysis of this geometry [4] shows that reducing the size scale of these compressive zones and their separation enhances the damage tolerance phenomenon.

Microfabrication technology employed for MEMS fabrication allows structures to be made with minimum feature sizes in the sub-micron to 100-nm size range. Consequently, a high threshold stress can be achieved, and this approach could present a useful means for improving upon the mechanical properties of bulk micromachined devices. The process flow envisaged allows a MEMS device, or critical stressed zones in a device e.g. flexures to be fabricated from this material using standard microfabrication techniques, without requiring a new substrate material or etching techniques and tools.

2 DAMAGE TOLERANCE MODEL

Rao [1] developed a model for estimating the stress intensity factor at the tip of a crack running across multiple layers of a laminate composite structure with alternating layers of residual biaxial compressive and tensile stress. The residual stresses may evolve due to a differential thermal expansion coefficient, a phase transformation, an increase in molar volume due to a chemical reaction or, in general, any phenomenon that leads to a strain mismatch between layers. Their analysis assumes a crack of length 2a extending across a tensile layer and into the two neighboring compressive layers. The stress intensity factor obtained is equated to the fracture toughness of the compressive layer material to obtain an expression for the threshold stress.

$$\sigma_{thr} = \frac{K_c}{\sqrt{\pi \frac{t_2}{2}(1+\frac{2t_1}{t_2})}} + \sigma_c \left[1 - \left(1 + \frac{t_1}{t_2} \right) \frac{2}{\pi} \sin^{-1} \left(\frac{t_2}{t_2 + 2t_1} \right) \right] \quad (1)$$

where:

$a \rightarrow$ length of crack

$\sigma_a \rightarrow$ applied tensile stress

$\sigma_c \rightarrow$ residual compressive stress

$t_1, t_2 \rightarrow$ compressive and tensile layer widths

$K_c \rightarrow$ critical stress intensity for compressive layer material

The material system under consideration consists of alternating thin layers of silicon dioxide with a residual biaxial compressive stress sandwiched between thicker layers of silicon with a residual biaxial tensile stress. The strain mismatch originates from the differential thermal expansion coefficient of the two materials, as the system is cooled from the process temperature to room temperature. The relevant material properties are shown in Table 1.

Material	CTE (x10⁻⁶/°C)	Young's Modulus (GPa)	Poisson's Ratio	Fracture Toughness K_c (MPa.m$^{1/2}$)
Silicon	2.60	130	0.25	0.9
SiO₂	0.50	50	0.16	1.0

Table 1: Material Properties for Si-SiO$_2$ system

For a Si-SiO$_2$ composite, the strain mismatch is assumed to start to build up below 960°C, the glass transition temperature. The maximum residual compressive stress that can be generated is ~208 MPa, which is relatively low. From (1), we can see that the threshold stress depends primarily on two factors, the magnitude of the residual compressive stress, and inversely with the size scale of the layers. The compressive stress generated depends upon the elastic coefficients of the two materials, and the strain mismatch. Since none of the other factors are under experimental control, reducing the size scale of the layer widths and optimizing the layer width ratio, becomes the primary route to obtaining a high threshold stress. The minimum size scale that can be fabricated is dictated by technological considerations. By using electron beam lithography and high plasma power etching tools, layer widths as small as 50-100 nm can be fabricated. Figure 1 shows predicted variance of the threshold stress with layer size scale.

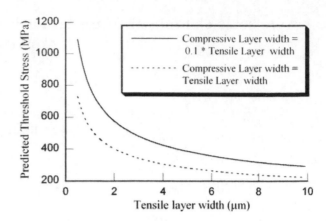

Figure 1: Predicted threshold stress for microcomposites

3 EXPERIMENT

Bulk silicon microfabrication techniques were used to fabricate cantilevers of this composite material. Monolithic silicon cantilevers are fabricated on the same die along with the composite cantilevers to compare the fracture strengths of the two materials. Each cantilever is 1.5mm long, 2mm wide and 40 μm deep and consists of a large number of alternating layers of silicon and thermally grown silicon oxide.

Tensile Layer		Compressive Layer		Predicted Threshold Strength (MPa)
Thickness (μm)	Residual Stress (T)(MPa)	Thickness (μm)	Residual Stress (C)(MPa)	
1.713	86.41	1.787	82.83	394
4.213	41.65	1.787	98.20	333
14.213	13.56	1.787	107.84	225
3.426	86.41	3.574	82.83	293
8.426	41.65	3.574	98.20	250
28.426	13.56	3.574	107.84	170

Table 2: Geometry and properties of fabricated cantilevers

The layer widths, as well as their ratio are varied to explore the effect of these parameters on the crack arrest properties. Each die consists of two sets of cantilevers, each set having a constant compressive layer thickness and with tensile layers of different thickness. A table of the various geometries fabricated, along with the fracture properties predicted from the model is shown in Table 2.

3.1 Microfabrication Process

After patterning the mask oxide, the silicon beams are etched by Deep Reactive Ion Etching (DRIE) using the Bosch process. The wafers are then thermally oxidized at 1100°C in steam. When silicon is oxidized, the resulting SiO$_2$ occupies a larger volume (2.27 times) than the silicon it consumes. This effect is utilized to fill the gaps between the silicon beams to form a material block of the laminar composite. A high temperature (1100°C) anneal allows the SiO$_2$ grown from both sides of the trench to flow and form a SiO$_2$ layer without an interface running midway through it. When cooled to room temperature, the differential thermal expansion between the Si and SiO$_2$ layers causes the residual stresses. The residual stresses, however, only build up below ~960°C, the glass transition temperature of SiO$_2$, below which it ceases to flow. A pattern of vias is then aligned with the cantilevers on the backside of the wafer. The vias are sized and aligned such that the composite structure extends beyond the fixed end of the cantilevers to ensure that the point of highest tensile stress is not at the boundary between the composite and monolithic silicon. The vias are then etched from the backside by DRIE to release the cantilevers from the wafer. The cantilevers are finally freed from the thin oxide film that covers the via floor by dipping them in hydrofluoric acid, which quickly dissolves the membrane-like film.

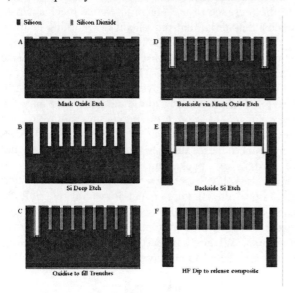

Figure 2: Schematic of fabrication process for composite

3.2 Fracture Testing

Fracture testing on the cantilevers is performed on a modified Tukon microhardness tester, which is well suited for performing fracture testing on these cantilevered samples because testing can be done under load control while keeping the displacement rate as low as desired to avoid any shock loading. The modified indenter consists of a 2mm wide tungsten carbide wedge that allows a uniform load to be applied across the width of the cantilevers. A laser extensometer attached to the instrument enables displacement measurement with up to 1μm accuracy. Loading is done under gravity and varying loads are applied simply by placing different small masses (1-100gm) on top of the indenter. Precise positioning of the samples under the indenter is done with the help of a stereoscopic microscope.

4   RESULTS

The first phase of testing involves verifying the accuracy of the experimental setup and the model to be used to transform the load-to-fracture data into fracture strength data. For this purpose, various loads are applied to the end of the cantilever beams, and the displacement at the point of load application measured to give a load-displacement curve for each sample. A simply supported cantilever beam model, assuming geometric and material linearity is used to obtain the Young's modulus for each sample. The cross section of the cantilevers is assumed to remain uniform over the length, and their height is assumed to correspond to the height as measured from scanning electron micrographs after fracturing one of the samples and imaging the cross section.

The measured values of Young's moduli are shown in Figure 3. The best fit to this data assuming a constant strain model as shown in (2) gives a Young's modulus of 132.8 GPa and 58 GPa for silicon and silicon oxide, respectively.

$$E_{composite} = V_{Si}E_{Si} + V_{Oxide}E_{Oxide} \qquad (2)$$

This matches well with the know values for silicon and silicon oxide, and validates our testing procedure.

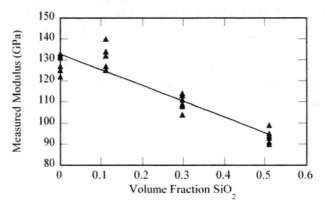

Figure 3: Young's Modulus results for composites with different layer width ratios

Load to fracture measurements are performed on a number of silicon and composite cantilevers to measure fracture strengths. The load is incrementally applied by increasing the mass on the load stage until the sample fractures. The deflection rate is kept at a relatively low value (approximately 7.5μm/sec) to ensure that no fracture occurs due to shock loading.

Initial indications of the efficacy of the laminate geometry with residual compressive stress layers for improved damage tolerance is seen in scanning electron micrographs of the crack path and crack surface on fractured samples as shown in Figure 4. The micrograph shows the crack deflecting each time it crosses the interface between a tensile and compressive layer. Crack deflection is an energy absorbing mechanism, because it creates more

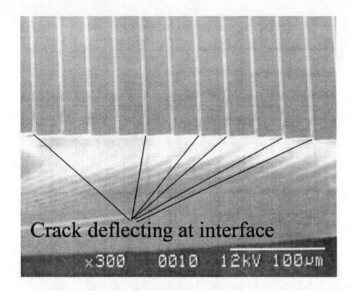

Figure 4: SEM image of fractured composite

surface area in the crack plane than would be formed for a crack growing in a straight line, perpendicular to the laminar structure. Cracks in single crystal silicon grow by cleaving apart adjacent atomic planes, leaving very smooth crack faces. From an energy viewpoint, this represents a higher toughness value G_c for the composite material in comparison to single crystal silicon.

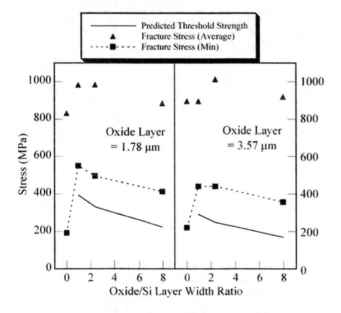

Figure 5: Fracture test data for composites

The data shown in Figure 5 indicates the active presence of a crack arrest phenomenon in the microscale composite structures. SCS cantilevers are shown as composites with the oxide/Si layer width ratio equal to zero. The solid line indicates the threshold stress predicted from the model for composites of different layer width ratios, and the average fracture stress indicated the mean of measured strength for a number of samples of identical composition. It is notable

that while the *average* fracture strengths for all the samples lie in a small range, the *minimum* strength has more variance. The minimum measured strength for the composite samples clearly lies above the predicted threshold strength, while the lowest failure stress in SCS is much lower.

A calculation of the flaw size distribution in the SCS samples from the fracture data gives a conservative estimate of 0.1-5 μm. These are reasonable values for flaw sizes in a millimeter scale structure. Two possible sources of these flaws are damage during various processing steps and surface damage due to handling. The measured values for monolithic silicon are also consistent with observations by Johansson [5,6] on the fracture strengths of monolithic silicon cantilevers.

5 CONCLUSIONS

The results obtained from fracture testing of the composite and single crystal silicon cantilevers clearly indicate the presence of a damage tolerance mechanism in the composite material that prevents it from failing below a certain threshold stress. As shown earlier in Figure 1, the threshold stress achievable varies inversely with the size scale of the layer widths. The capability exists for fabricating composites with layer widths between 50-100 nanometers which, according to the model, would only fail under stress levels of 1.8-2 GPa. Preventing failure to such a high level of stress would be invaluable to MEMS designers, since these devices or structures would not require extensive packaging. Another example of application is MEMS devices with comparatively short flexures, since the flexure material would be able to withstand a higher lever of stress. Shorter flexures would reduce the footprint of the device, allowing more devices per chip area. We are currently working on fabricating composites with sub-100 nm layer widths.

REFERENCES

[1]   M. P. Rao, A. J. Sanchez-Herencia, G. E. Beltz, R. M. McMeeking, and F. F. Lange, Science **286,** 102-105 (1999).
[2]   K. Yasutake, M. Iwata, K. Yoshii, M. Umeno, and H. Kawabe, Journal of Materials Science **21,** 2185-2192 (1986).
[3]   T. N. Pornsin-sirirak, Y. C. Tai, H. Nassef, and C. M. Ho, Sensors and Actuators a-Physical **89,** 95-103 (2001).
[4]   R. M. McMeeking and K. Hbaieb, Zeitschrift Fur Metallkunde **90,** 1031-1036 (1999).
[5]   S. Johansson, F. Ericson, and J. A. Schweitz, Journal of Applied Physics **65,** 122-128 (1989).
[6]   S. Johansson, J. A. Schweitz, L. Tenerz, and J. Tiren, Journal of Applied Physics **63,** 4799-4803 (1988).

Novel Monolithic Micro Droplet Generator

C.K. Chung*, C.J. Lin**, C.C. Chen***, S.W. Chuang***, Y.J. Fang*** and Y.Z. Hong*

*Dep't of Mechanical Engineering, National Cheng Kung University,
Tainan, Taiwan 701, ROC, ckchung@mail.ncku.edu.tw
**National Tsing Hua University, Taiwan 300, ROC, d897104@oz.nthu.edu.tw
***Industrial Technology Research Institute, Taiwan 310, ROC, chungchu@itri.org.tw

ABSTRACT

A novel monolithic off-shooter droplet generator has been demonstrated. The design, fabrication, bubble formation and droplet injection are described in this paper. It is different from the conventional droplet generators of top-shooter, side-shooter and back-shooter for injection. The droplet ejected from such off-shooter generator has the merits of straight trajectory with satellite improvement by two bubbles cutting and strong structure of stacked metal-polymer or polymer-polymer layers by integration of two-photoresist lithography and electroforming technologies. It is potentially used for inkjet printing, LCD color filter dispersion, microarray for bio application and so on.

Keywords: monolithic, droplet generator, SU8, MEMS

1 INTRODUCTION

The conventional inkjet chip needs complex alignment and bonding process to stick the nozzle plate and heating chip together [1]. It takes long time, high cost and much lower alignment resolution. Some monolithic inkjet chips [2-6] fabricated by the MEMS technology can improve the above drawbacks but suffer from the nozzle membrane broken during processes. Chen et al [2] used the method of anisotropic wet etching and chemical vapor deposition (CVD) sealing process to form the side-shooting nozzle plate. It requires the CVD high temperature process and the strength of sealed Si-based nozzle plate is a challenge. Westberg and Andersson [3] used CMOS process in combination with bulk etching and aluminum sacrificial etching to fabricate a monolithic CMOS compatible inkjet chip. However, the device was rather fragile and hard to handle during the process. Tseng et al [4] designed a monolithic microinjector with virtual chamber neck for satellite improvement but it still encountered the problem of membrane broken during processes.

In this paper, we report this monolithic droplet generator in combination of the advantages of conventional and monolithic inkjet technologies by changing the conventional inkjet heater shape from square into ring and integrating the electroforming and two photoresists (SU8 or JSR) technologies to form the strong metal or polymer nozzle plate. It avoids the membrane broken problem and simplifies process procedure with low cost and good droplet properties.

1. The 1st resist layer SU8 patterning

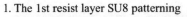

2. Absorption (/seed) layer deposition

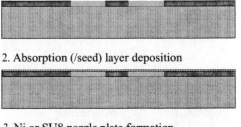

3. Ni or SU8 nozzle plate formation

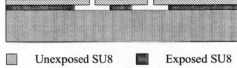

☐ Unexposed SU8   ■ Exposed SU8
☐ Si   ▨ Nozzle plate: Ni or SU8

Figure 1 Schematic process flow of the combination of thick resist and electroforming technologies for the formation of ink chamber and nozzle plate of the monolithic inkjet chip.

2 EXPERIMENTAL PROCEDURE

Figure. 1 shows the schematic process flow of combination of thick resist and electroforming technologies for the formation of ink chamber and nozzle plate of the monolithic inkjet chip. It integrates two kinds of thick photoresists (SU8 and JSR) lithography process and one-step nickel electroplating process to form the chambers and nozzles of monolithic inkjet chip. The first thick resist of SU8 of about 35 μm is used for the structure formation of ink channel and chamber. It is coated and patterned on the substrate but not developing. A light-absorbing CK-6020L resist layer is followed to coat on the SU8. The following process depends on the kind of nozzle plate material. The first kind of nickel metal nozzle plate is performed by second thick resist JSR mold patterning on the first SU8 chambers, followed by nickel electroforming and non-exposure resist stripping. The second kind of polymer nozzle plate utilized second thick resist of SU8 material to replace nickel metal. Second thick SU8 resist was patterned by photolithography and the overall non-exposure

SU8 was removed to form the monolithic SU8 nozzle plate. The quality of thick resist and electroformed Ni or SU8 nozzle plate are examined in optical microscopy (OM) and scanning electron microscopy (SEM). The thermal bubble formation in open pool test of heating chip was observed by a flow visualization experiment in this study. The droplet injection was also examined in this system. The function synthesizer produces two synchronized periodic pulses for the heating element and the LED stroboscope. The delay time is adjustable to freeze different stages of bubble formation to characterize its formation sequence.

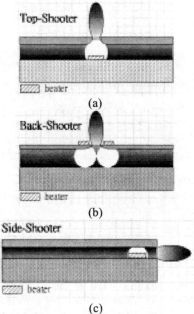

(a)

(b)

(c)

Figure 2 Schematic traditional three types of droplet generators: (a) top-shooter, (b) back-shooter, and (c) side-shooter

3   RESULTS AND DISCUSSION

Figures 2(a)- 2(c) show the schematic traditional three types of droplet generators: (a) top-shooter, (b) back-shooter, and (c) side-shooter, respectively. The top-shooter droplet generator means the droplet injected in the same direction as the thermal bubble formation while the back-shooter with the opposite direction of droplet and bubble. Our design and cross-section of off-shooter droplet generator are shown in Figures 3(a) and 3(b), respectively. The off-shooter means the droplet injecting direction is a little shift from the two thermal bubbles formed by the ring heater of inkjet chip. The cross-sectional structure of generator includes the TaAl heater on SiO_2/Si substrate and passivated with Si_3N_4/SiC as well as Al conductor, SU8 chamber and nozzle plate.

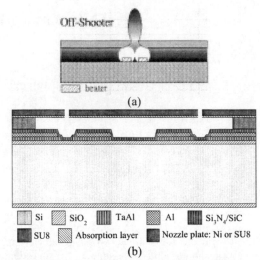

(a)

| Si | SiO_2 | TaAl | Al | Si_3N_4/SiC |
| SU8 | Absorption layer | | Nozzle plate: Ni or SU8 | |

(b)

Figure 3 Novel off-shooter droplet generator: (a) schematic diagram and (b) the detailed cross section of generator with TaAl heater on SiO_2/Si substrate and passivated with Si_3N_4/SiC as well as Al conductor, SU8 chamber and nozzle plate.

Figures 4(a)- 4(d) show the experimental result of SU8 structure, JSR molds, Ni and/or SU8 nozzle plate respectively by integrating the two thick resists and electroforming technologies for the formation of ink chamber and nozzle plate of the droplet generators. The structure layer is formed by first thick resist SU-8 due to its superior properties [7] while metal or polymer nozzle plate is formed by electroforming and second thick resist process. For stacked metal-polymer ie Ni-SU8 structure, the second JSR thick photoresist is used for forming the electroplating mold of nickel nozzle plate due to easily removal by acetone [8], followed by nickel electroforming and stripping overall non-exposure resist. For stacked polymer-polymer ie SU8-SU8 structure, the second thick SU8 resist is used for total SU8 polymer injecting microstructure, instead of both JSR resist and nickel electroforming process. The light-absorbing CK-6020L polymer layer is coated between two photoresist layers for protecting the SU-8 layer from overheating during the metal seed layer deposition and over-exposure during the second thick resist lithography process. Good quality of structure at each step during the fabrication of monolithic chip has been achieved. The visualization of the bubble formation by conventional square- and off-shooter ring-type heater are compared using open pool test as shown in Figures 5(a) and 5(b), respectively. The bubble starts to form around 2 µs after the beginning of the heating pulse. A single bubble is formed on the square heater at the off-center position while Two oval bubbles at the ring heater center, rather than a single 'donut' bubble, are generated by the ring heater. The two oval bubbles will generate the droplet with straighter trajectory than the single bubble at off-center position.

central areas to produce two thermal bubbles as shown in Figure 5(b). The two thermal bubbles will cut the droplet injection with straight trajectory and satellite improvement. Figure 7 shows the picture of droplet injected from off-shooter generator. Good quality of droplets have been generated in this novel droplet generator.

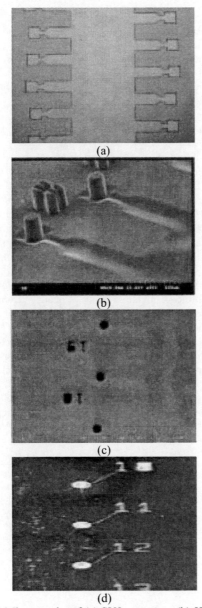

(a)

(b)

(c)

(d)

Figure 4 Micrographs of (a) SU8 structure, (b) JSR molds, (c) Ni nozzle plate, and (d) SU8 nozzle plate during the monolithic process.

Figures 6(a) and 6(b) show the temperature distribution of ANSYS simulation for square-type heater and ring-type heater after voltage applied, respectively. The highest temperature region is formed at one central area of square-type heater to produce one thermal bubble nucleation and growth. The real bubble position of square heater in Figure 5(a) is away from the center due to the fluidic flow direction and surface-dependent nucleation. The position away from the ink inlet neck is with lower cooling effect to get higher temperature than the inlet neck. But the highest temperature region at ring-type heater is formed at two

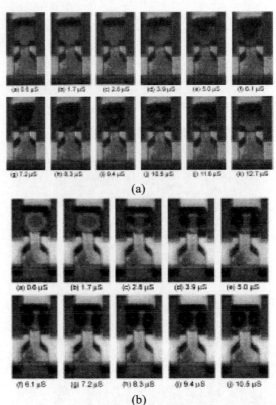

(a)

(b)

Figure 5 Time series of bubble formation with: (a) square-type heater and (b) off-shooter ring-type heater. A single bubble is formed on the square heater at the off-center position while Two oval bubbles is formed at the ring heater center.

4 SUMARRY

A novel monolithic off-shooter droplet generator has been designed and fabricated by changing the conventional square heater into ring one and integrating two thick resist photolithography and one-step electroforming technologies for the generator formation. This process integration based on silicon micromachining has the advantages of accurate alignment of nozzle plate to the substrate, simple process and low cost in comparison to the conventional hybrid technology of nozzle plate attached to the substrate. A light-absorbing polymer layer between two thick resist layers is good for protecting the first SU8 layer from overheating or over-exposure during the metal seed layer deposition and the second thick resist lithography process. The thermal bubble formation and droplet injection are also realized in this study. The off-shooter ring-type heater with two thermal bubbles will inject the droplet with straight trajectory and satellite improvement by bubble cutting mechanism.

ACKNOWLEDGEMENTS

This work is sponsored by National Science Council under contract No NSC 92-2212-E-006-125. We pay our sincere thanks to the Southern Regional MEMS Center in National Cheng Kung University and Common Laboratory of the Microsystems Technology Center of Electronic Research Service Organization in Industrial Technology Research Institute for the access of process equipments.

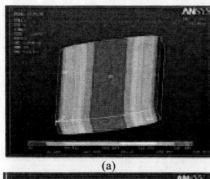
(a)

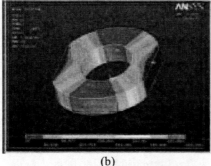
(b)

Figure 6 Temperature distribution of ANSYS simulation: (a) square-type heater and (b) ring-type heater after voltage applied.

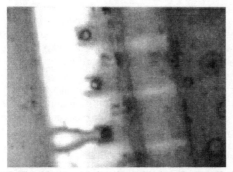

Figure 7 The picture of droplet injection from off-shooter generator.

REFERENCES

[1] C.A. Boeller, T.J. Carlin, P.M. Roeller and S.W. Steinfield, *J. Hewlett-Packard* 39 , pp. 6-15, 1988.

[2] J. Chen, and K.D. Wise, "A High-Resolution Silicon Monolithic Nozzle Array for Inkjet Printing", IEEE Transactions on Electron Devices. Vol. 44, No. 9, pp.1401-1409, 1997.

[3] D. Westberg, and G.I. Anderson, "A novel CMOS compatible inkjet head", Transducers '97, pp. 813-816, 1997.

[4] F.G. Tseng, C.J. Kim, and C.M. Ho, "A novel microinjector with virtual chamber neck", MEMS98, pp. 57–62, 1998.

[5] J.D. Lee, J.B. Yoon, J.K. Kim, H.J. Chung, C.S. Lee, H.D. Lee, H.J. Lee, C.K. Kim, C.H. Han, "A thermal inkjet printhead with a monolithically fabricated nozzle plate and self-aligned ink feed hole", *J. Microelectrmechanical Systems*, vol.8, pp. 229-236, 1999.

[6] S.W. Lee, H.C. Kim, K. Kuk and Y.S. Oh, "A monolithic inkjet print head Domejet", *Transducers'01*, pp.515-518, 2001.

[7] H. Lorenz, M. Despont, N. Fahrni, J. Brugger, P. Vettiger and P. Renaud, "High-aspect-ratio, ultrathick, negative-tone near-UV photoresist and its application for MEMS", Sensors and Actuators A64, pp.33-39, 1998.

[8] F.G. Tseng and C.S. Yu, "High aspect ratio ultrathick micro-stencil by JSR THB-430N negative photoresist", Sensors and Actuators A97-98, pp.764-770, 2001.

A Novel Fabrication Technique of Cylindrical Ion Traps using Low Temperature Co-fired Ceramic Tapes

A. Chaudhary*, **F.H.W. van Amerom#, S. Bhansali*** **and R. T. Short#**

*Dept. Of Electrical Engineering, University of South Florida, Tampa,
4202 East Fowler Av. ENB118, FL 33620, USA.
Center for Ocean Technology, University of South Florida, St. Petersburg,
140 7th Av. S, FL 33701, USA.
ashchau@eng.usf.edu

ABSTRACT

This work concentrates on a novel technique for the fabrication of miniaturized cylindrical ion trap structures (CIT) to be used as replacements of commercial hyperbolic structure ion trap mass spectrometers. Low temperature co-fired ceramic (LTCC) tapes (Dupont 951), commonly known as "green tape", were used to fabricate a CIT structure of 2.75 mm in diameter. The fabrication process included a stainless steel die, which was used to obtain the impressions, for automatic alignment of stainless steel end plates. The ring electrode structure made out of green tape and the stainless steel endplates, were plated with electro-less Ni and electro-less Au to obtain a conducting surface. The green tape CIT structures were tested for background contamination and out-gassing under vacuum in a Saturn 2000 MS Mass Spectrometer. Experiments showed that no considerable out-gassing or contamination is present at pressures of 10^{-4} to 10^{-6} Torr. Other techniques to fabricate CIT structure involving green tapes are being explored as an alternative.

Keywords: green tape, cylindrical ion trap, miniaturization, vacuum contamination.

INTRODUCTION

LTCC manufactured by Dupont, are soft and pliable ceramic tapes, often called green tapes in their unfired state. The soft and flexible nature of this material allows for the batch fabrication of a large variety of complex 3-D structures using lamination, punching, and drilling. 3-D structures fabricated out of laminated green tapes turn into ceramics when fired. The shrinkage during the firing process is repeatable and uniform in all directions making green tapes apposite for fabrication of high precision meso and micro-scale structures. Once turned into ceramics, green tape structures can be plated with electro-less Ni or Au to achieve uniform metallization required for electrical conduction.

FABRICATION PROCESS

The fabrication process consists of the following steps as shown in Figure 1.

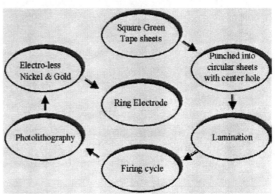

Figure 1: Schematic of the fabrication process.

First, circular green tapes, with a center hole, were punched with a hammer punch. 70 of such circular tapes were put in a stainless steel die (see Figure 2) and pressed in a hydraulic press at 8100 psi at 80°C for 15 minutes.

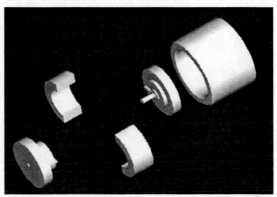

Figure 2: An exploded view of the Stainless Steel Die.

The thickness of the structure pressed in the green tape is defined by the design of the die and not the pressure. Beyond that thickness, any increase in pressure would be passed on to the stainless steel cage in which the die is placed. This gives a semi-hard 3-D structure with all the patterns from the die transferred on it. The green tape is still in its green state, ready to be fired. The sample is then fired in air according to the firing cycle provided by Dupont, see Figure 3.

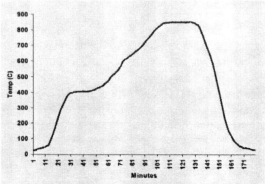

Figure 3: Recommended firing cycle [3].

The temperature is ramped up at 10°C/minute to 350°C and then kept constant for 30 minutes. At this temperature, green tape looses all its organic components. The temperature is again increased at the same ramp rate to 850°C where it is kept for 30 minutes. The sample is allowed to cool slowly to room temperature.

The green tape sample turns into hard ceramic during the firing cycle. Using an SR100 set up, photolithography is done on the sample to metallize specific areas on it using negative photo-resist PKP.

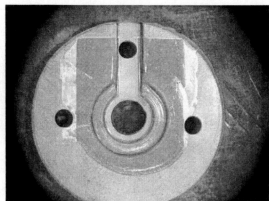

Figure 4: Green tape CIT structure with photolithography done.

The exposed sample is developed and then directly plated with electro-less Ni or Au to achieve metallization. The immersion solution plates the entire sample, which is then dipped in the photo-resist stripper to get rid of the unwanted metal as in the standard lift-off process. The photo-resist and metal present on the sample are removed. Ni was used as the first layer due to its better adhesion. The Au plating on top of the Ni ensured a highly conductive layer.

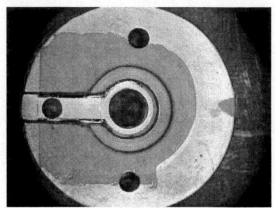

Figure 5: Green tape CIT structure plated with electro-less Ni.

The electro-less Ni and Au are plated for 2-3 minutes and 10 minutes, respectively to get sufficiently thick layers of metal required to operate at high voltage and frequency. Experiments were performed with test parameters up to 2.3 kV_{pp}, 0.01 A at 3MHz to confirm the quality of the plating and to investigate the breakdown voltage.

VACUUM FEASIBILITY TEST

An experiment was performed to confirm the feasibility of green tapes in vacuum applications. Out-gassing of the ceramics might interfere with or contaminate compound analysis instruments such as a mass spectrometer. The setup consisted of a Saturn 2000 Ion Trap Mass Spectrometer, which typically operates in the range of 10^{-6} to 10^{-4} Torr. A green tape sample was taken directly from the oven and placed inside a vacuum chamber extension, which replaced the inlet system of the mass spectrometer, to ensure a clean sample. The turbo pump current and startup time were consistent with the normal operation, indicating no significant out-gassing of the green tape structure.

Mass spectrometric analysis showed that the contaminants in the vacuum chamber extension, filled with the green tape structure, did not differ significantly from the background signal contaminants of the empty vacuum chamber extension. It may be concluded from above that there is no noticeable out-gassing from the fired green tape structure.

CONCLUSIONS

A fabrication process for a ring electrode of a CIT structure was designed and tested that uses a die to create the detailed impressions of an electrode channel, automatic alignment cut and includes a predefined distance from the endplate to the cylinder. Green tape laminated at 8100 Psi at 80°C seemed to be non-porous and no significant out-gassing was observed in vacuum at 10^{-6} to 10^{-4} Torr thus making them useful in vacuum applications.

REFERENCES

[1] Patricio Espinoza-Vallejos, Clara Dimas, and Jorge Santiago-Aviles, "Lithographic Processing of LTCC Tapes," IMAPS 99-Chicago

[2] Material safety data sheet, Dupont Microcircuits materials.

[3] Dupont, "951 Green Tape^{TM}" Dupont microcircuit materials.

[4] G. E. Patterson, A. J. Guymon, L. S. Riter, M. Everly, J Griep-Raming, B. C. Laughlin, Z. Ouyang and R. G. Cooks, "Miniature cylindrical ion trap mass spectrometer," Anal. Chem., 2002, 74, p6145-6153.

ACKNOWLEDGMENT

The U.S. Army, Space and Missile Defense Command, provided financial support to the University of South Florida through Grant DASG60-00-C-0089.

Micromachined III-V Multimorph Actuator Model and Design Using Simulated Annealing (SA)-Based Global Optimization

A. Ongkodjojo[a], F. E. H. Tay[a,b] and R. Akkipeddi[c]

[a]Micro- & Nano-Systems Cluster, Institute of Materials Research & Engineering (IMRE)
3 Research Link, Singapore, an-ong@imre.a-star.edu.sg
[b]Department of Mechanical Engineering, NUS, Singapore, mpetayeh@nus.edu.sg
[c]Opto- & Electronic Systems Cluster, IMRE, Singapore, ram-akki@imre.a-star.edu.sg

ABSTRACT

A new AlAs/GaAs-based multimorph microactuator is designed and modeled for micro-opto-electro-mechanical systems (MOEMS) applications with photonic devices. As the piezoelectric-based III-V materials such as AlAs/GaAs have small piezoelectric constants, we have developed the Simulated Annealing (SA)-based global optmization to obtain the optimum solution for the structure with a high sensitivity. This proposed optimization method can efficiently lead the design for achieving the highest sensitivity as the structure materials can be grown epitaxially with a minimum thickness of 10 nm for each layer. Therefore, the smallest thickness of the structure layers and its highest sensitivity can easily be achieved by using the optimization method. In addition, for more realistic design and model, we develop the constraint optimization method using the SA algorithm, which can satisfy the specified constraints.

Keywords: global optimization, Simulated Annealing (SA), III-V multimorph, MOEMS, AlAs/GaAs

1 INTRODUCTION

The micro-opto-electro-mechanical systems (MOEMS) are currently in a good progress in many fields [1] and using silicon material for their functionalities because of good mechanical properties, less processing difficulties, and low cost of the material. However, with no optoelectrical properties for the material, it is highly promising to consider III-V materials integrated with MEMS devices, which have already made a tremendous impact in optoelectronics and high frequency devices, and communication fields and industries. Detailed investigations of the materials have been dealt by various authors in the literature [2]. As we use the piezoelectric means for actuation, piezoelectric-based III-V materials such as AlAs/GaAs are used for our mechanical structures.

In this research work, a multimorph microactuator is proposed and developed, which provides high sensitivity, big generative force, good strength, and high natural frequency for MEMS applications with relatively small piezoelectric coefficients, especially for MOEMS applications. Besides, the multimorph models provide better validation to the experimental data as they consider more detailed models including all of the thin films [3]. According to our best knowledge, there is no literature on multimorph or bimorph using the same material such as AlAs/GaAs for MEMS applications.

As the piezoelectric-based III-V materials have small piezoelectric constants, the SA-based global optmization has been proposed to obtain the optimum solution without violating the specified constraints [4]. For more realistic designs, some constraints are imposed to satisfy mechanical and electrical requirements as well as micro-fabrication process capabilities and limitations.

2 MATHEMATICAL MODELING

In this section, we present a mathematical model of a novel piezoelectric multimorph structure, which is made of two piezoelectric layers, a middle layer, and some electrode layers with parallel electrical connections. Our proposed structure is shown in Fig. 1. The middle layer, which plays an elastic role in addition to its electrical function, has been inserted between the two piezoelectric layers for improving the strength of our structure.

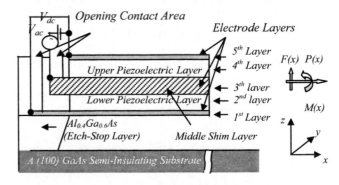

Figure 1: AlAs/GaAs-based multimorph microactuator

As large displacements are needed for our optical MEMS applications, the magnitude of deflection of the actuator will be investigated in the form of a sensitivity parameter. Therefore, we focus on the sensitivity as our single objective function in order to obtain high sensitivity for our actuator applications.

When a driving voltage is applied to a piezoelectric material, the material structure will be deflected. The electric field is non-zero only for piezoelectric layers. The structure will tend to contract in the x-y plane, and expand along the electric field, which is applied in parallel with the polarization of the material along the z-axis. For example, the upper piezoelectric material expands while the lower piezoelectric material contracts as a result of the electric field, resulting in a bending in the downward direction. However, the structure will deform in the opposite direction as an alternate driving voltage is applied to the multimorph. Thus, the microactuator will vibrate dynamically, when the AC driving voltage V_{ac} is applied with a driving frequency.

Generally, the piezoelectric effect of the structure is expressed by [5]:

$$S_{ij} = s_{ijkl}^E T_{kl} + d_{kij} E_k \tag{1}$$

$$D_i = d_{ikl} T_{kl} + \varepsilon_{ik}^T E_k \tag{2}$$

where S_{ij}, s_{ijkl}^E, T_{kl}, d_{kij}/d_{ikl}, E_k, D_i, and ε_{ik}^T are strain, elastic compliance constant at constant electric field, stress, the piezoelectric constant, electric field, electric displacement, and permittivity at constant stress, respectively.

By imposing force and moment equilibrium on the structure, the deflection of the proposed multimorph can be obtained as given by [6]:

$$z(x) = \frac{x^2}{2} \frac{\left(-\sum_{n=1}^{N} d_{31-n} E_n A_n Z_{n-g} E_{e-n} \right)}{\left(\sum_{n=1}^{N} E_n \left[I_n + A_n Z_{n-g}^2 \right] \right)} \tag{3}$$

where x is the length of the multimorph; d_{31-n} is the transverse piezoelectric coefficient for the nth piezoelectric layer ($d_{31-n} = d_{31-n}(1 + \nu_n)$); E_n is the effective Young's modulus for the nth layer ($E_n = E_n/(1 - \nu_n^2)$), where ν_n is Poisson's ratio for the nth layer; A_n is the area of nth layer cross section; $Z_{n-g} = (z_n - z_g)$ is the new coordinate, which considers the neutral or global axis; E_{e-n} is the electric field; and I_n is the moment inertia for the nth layer. For obtaining the deflection equation, we use a relationship between the curvature $1/R$ and the displacement $z(x)$ by $1/R(x) = d^2 z(x)/dx^2$ for small displacements, where R is the radius of the curvature.

Then, the natural frequency of the structure ω_n is derived using the equivalent stiffness k_{eq} and the mass of the multimorph m_b as expressed by:

$$\omega_n = (\beta l)_n^2 \left(\frac{k_{eq}}{m_b} \right)^{\frac{1}{2}} \tag{4}$$

where $(\beta l)_1^2$ is a constant of 3.5160 for the first mode of the natural frequency [7]. Finally, the quality factor due to the squeeze film damping of the structure considering gas rarefaction effects has been investigated [4, 8].

3 GLOBAL OPTIMIZATION

The optimal design of our proposed multimorph is carried out to find the maximum sensitivity while satisfying some constraints imposed by functional and geometrical constraints. A method of global optimal design with Simulated Annealing (SA) is used in this research work. The SA is an efficient and adaptive search method applicable to real-life constrained optimization problems. It is analogous to the physical annealing process where an alloy is cooled gradually so that a minimal energy state is achieved. The method avoids getting stuck in local optima and keeps track of the best overall objective function value overall, thereby, resulting in efficient global search strategies. In the design process, the optimization problem is formulated to find the design variables so that the objective function is maximized, while satisfying the multidisciplinary design constraints, including both mechanical and electrical requirements as well as microfabrication limitations/capabilities. More details on the optimization method have been reported [4].

3.1 Design Variables

To select the design variables, the influence of each parameter to the sensitivity should be analyzed. The design variables are the length, the width, and the height of each layer of the multimorph actuator as presented in Table 1. The influence of each design parameter can be decided through Eq. (3).

Variable	Description	Reasonable Range	
		Minimum	Maximum
l_b (μm)	Length	5×10^1	1×10^2
w_b (μm)	Width	5	1×10^1
h_{p1} (μm)	Height (AlAs)	1×10^{-2}	5
h_{p2} (μm)	Height (GaAs)	1×10^{-2}	5
h_m (μm)	Height (inner electrode)	1×10^{-2}	5
h_e (μm)	Height (outer electrode)	1×10^{-2}	3
V_{dc} (V)	DC bias voltage	0	5
V_{ac} (V)	AC voltage	0	1

Table 1: Design variables for the multimorph

3.2 Constraints

In the design process of the constrained nonlinear optimization, constraints need to be identified and defined correctly. Generally, constraint expressions involve both dependent and independent design variables as well as explicit bounds on the variables. Some constraints, which are used in this research work, have been derived from [4] such as the natural frequency, the quality factor, the output linearity, and the support beam buckling. Under certain fabrication conditions, the microstructure can collapse and permanently adhere to their underlying substrates. In order to avoid stiction, collapse, and contact with the substrate, real and practical design constraints of our proposed multimorph microactuator must be taken into account. Therefore, we present the additional constraints such as other geometrical constraints, stiction, contact during fabrication, and collapse in this paper.

Stiction may occur whenever flexible and smooth structures are brought in contact with the substrate. It is clear that stiction can easily cause malfunctioning in our actuator. In order for the actuator to snap back after contacting the substrate, the length of the beam must be shorter than the critical length (l_{cr-s}), which is expressed by [9]:

$$l_{cr-s} = \sqrt[4]{\frac{3}{8} \frac{E\left(h_{p1} + h_{p2} + h_m + 2h_e\right)^3 g^2}{\gamma_s}} \qquad (5)$$

where γ_s is adhesion energy per unit area or surface tension, which is assumed to be approximately 0.11 J/m^2 [10].

During the drying process after wet sacrificial layer etching, the structure is being pulled against the substrate surface by strong capillary forces. It is more practical to avoid contact during the drying process. To avoid contact, which is bridging a part of the beam area flat against the substrate surface, the critical length given by Eq. (6) must be considered.

$$l_{cr-c} = \sqrt[4]{\frac{3}{16} \frac{E\left(h_{p1} + h_{p2} + h_m + 2h_e\right)^3 g^2}{\gamma_{la} \cos\theta_c}} \qquad (6)$$

where $\gamma_{la} \cos\theta_c$ is the adhesion energy of the liquid bridge per unit area, which is assumed to be approximately 0.11 J/m^2. However, this constraint can be neglected, when a dry sacrificial etching process is chosen for the fabrication process.

As our actuator can be applied to a class of wavelength tunable electro-optic devices, a micro-mirror can be attached to the actuator for the real applications. Therefore, we need to consider our actuator to be loaded at its tip by an applied external force (*n* times its own weight). In this case, we can model that our actuator is loaded by a force of $(n\rho w_b(h_{p1} + h_{p2} + h_m + 2h_e)l_b a)$, when the substrate is accelerated by *a* in an upward direction. Thus, the critical length of the actuator is expressed by [9]:

$$l_{cr-load} = \sqrt[4]{\frac{Eg\left(h_{p1} + h_{p2} + h_m + 2h_e\right)^2}{4n\rho a}} \qquad (7)$$

4 RESULTS AND DISCUSSION

The computational results of the global optimization for the structure are shown in Fig. 2 and Table 2. These computational results are presented with a driving frequency of 530 kHz, an air pressure, and an applied force of 50 μN. Fig. 2 shows the results of the SA run, where the solution has converged to the global maximum. As the temperature falls, one can observe that the algorithm closes in on the global maximum. According to Table 2, the design variables give a maximum value of 3.4 nm/V for the sensitivity of the multimorph, while satisfying the imposed constraints.

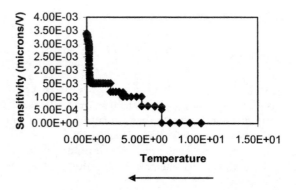

Figure 2: Sensitivity value converged using the SA

Variable	Description	Results
l_b (μm)	Length	7.175 x 10^1
w_b (μm)	Width	1.0 x 10^1
h_{p1} (μm)	Height (AlAs)	7.89 x 10^{-1}
h_{p2} (μm)	Height (GaAs)	4.98 x 10^{-1}
h_m (μm)	Height (inner electrode)	2.71
h_e (μm)	Height (outer electrode)	1.0 x 10^{-2}
V_{dc} (Volt)	DC bias voltage	4.3
V_{ac} (Volt)	AC voltage	7.136 x 10^2
Geometrical Constraints		
z (μm)	Deflection	1.705 x 10^{-2}
$h_{p1}+h_{p2}+h_m+2h_e$	Total height	4.013
R (μm)	Curvature (radius)	1.5096 x 10^5
Functional Constraints		
ω_z (rad/s)	Resonant frequency	3.3313 x 10^6
Q_z	Quality factor	2.525 x 10^2
$0.8*l_{cr-b}$ (μm)	Support beam buckling	1.324 x 10^2
$0.8*l_{cr-s}$ (μm)	Stiction	9.32 x 10^1
$0.8*l_{cr-c}$ (μm)	Contact (fabrication)	7.84 x 10^1
$0.8*l_{cr-l}$ (μm)	Collapse (loading)	7.175 x 10^1
Objective Function		
S (μm/V)	Maximum sensitivity	3.395 x 10^{-3}

Table 2: Global optimization results for the multimorph

To ensure that our models and designs are valid and accurate, the optimization results were verified with the FEA using ANSYS^{TM}. A coupled-field multiphysics analysis that accounts for the interaction between electric and structural fields was run. The 3-D model uses SOLID5 elements (brick 8-node element, 3-D coupled field element) for the piezoelectric layers and SOLID45 element (3-D structural element) for the middle shim layer as used by the modal analysis. Then, an input positive voltage is applied in parallel to the piezoelectric layers. Material property data for the dielectric constant matrix ($[\varepsilon/\varepsilon_0]$, where $\varepsilon_0 = 8.85 \times 10^{-12}$ F/m), the piezoelectric stress coefficient matrix $[e]$, and the anisotropic elastic coefficient matrix $[c]$ have been reported [2]. For obtaining the elastic coefficient matrix [e], $c_{11} = c_{22} = c_{33} = 1/s_{11}^E = E$, N/m²/Pa are inputs. While, the piezoelectric stress coefficient matrix $[e]$ has been calculated by using the relationship between the piezoelectric strain coefficient matrix $[d]$ and a compliance coefficient S_{44}, where $S_{44} = 1.7 \times 10^{-11}$ m²/N and 1.68×10^{-11} m²/N for Alas and GaAs respectively.

According to Table 3 and Fig. 3, there is an agreement among the mathematical modeling, the proposed models, and the finite element modeling. The discrepancies (as shown in Fig. 3) are caused by the fact that the derived mathematical modelings did not consider the piezoelectric effect of the structure width. Considering the piezoelectric effect across the model width, there is an excellent agreement between the SA and the FEA results. In addition, the validity of our designs have been confirmed by Fig. 5 (resonance frequency as a function of the cantilever microbeam length for GaAs) of the paper [11]. Therefore, very high resonance frequencies can be achieved by the small dimension (length) of the structure by using the growth control in III-V materials.

Mode	SA Result (Hz)	FEA Result (Hz)	Error (%)
1st	5.3162×10^5	5.4581×10^5	2.60
2nd	-	1.3472×10^6	-
3rd	-	3.3919×10^6	-

Table 3: Verification summary of the natural frequency

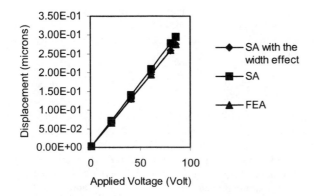

Figure 3: Comparison between our multimorph model and FEA simulation results using ANSYS™

5 CONCLUSION

Based on the global optimization, the best design has been globally the optimum design, while satisfying the multidisciplinary design constraints imposed by mechanical and electrical requirements as well as III-V microfabrication process capabilities and limitations. These constraints have made our proposed design to be visible and real for our applications.

Based on the design, the AlAs/GaAs-based microactuator has a sensitivity of 3.4×10^{-3} μm/V with a resonance frequency of 530 kHz and a quality factor of 253 in air without violating the multidisciplinary design constraints. In addition, the valid driving voltage range of the applied DC voltage (V_{dc}) is 0 – 87.5 Volt with a fixed applied AC voltage (V_{ac}) of 714 mV for deflecting the structure up to 10% of the gap. Considering the breakdown field of the piezoelectric materials, the valid driving DC bias voltage is up to about 20 Volts, which can be applied to the material structure.

The III-V materials-based multimorph shows potential applications for monolithic integration of optoelectronic components with MEMS devices enabling the realizations of MOEMS.

ACKNOWLEDGEMENTS

The authors are grateful to A*STAR (Agency for Science, Technology, and Research), Singapore for their funding. This work is supported by the A*STAR under the SERC Grant No.0221070006: III-V Optical MEMS Project.

REFERENCES

[1] H. Fujita and H. Toshiyoshi, IEICE Trans. Electron., E83-C, 9, 1427 – 1434, 2000.
[2] S. Adachi, J. Applied Physics, 58-3,1 – 29, 1985.
[3] D. L. DeVoe and A. P. Pisano, J. Microelectromech. Sys., 6-3, 266 – 270, 1997.
[4] A. Ongkodjojo and Francis E. H. Tay, J. Micromech. Microeng., 12, 878 – 897, 2002.
[5] IEEE Standard on Piezoelectricity – An American National Standard, ANSI/IEEE Std, 176-1987.
[6] E. B. Tadmor and G. Kosa, J. Microelectromech. Sys., in press, 2003.
[7] Thomson W and Dahleh M, "Theory of Vibrations with Applications," Englewood Cliifs, NJ: Prentice Hall, 1998.
[8] A. Ongkodjojo and Francis E. H. Tay, Int. J. Comp. Eng. Sci. (IJCES) – Symposium G:MEMS – ICMAT 2003, Singapore, in press.
[9] N. Tas, T. Sonnenberg, H. Jansen, R. Legtenberg, and M. Elwenspoek, J. Micromech. Microeng. 6, 385 – 397, 1996.
[10] F. Shi, S. MacLaren, C. Xu, K. Y. Cheng, and K. C. Hsieh, J. Applied Physics, 93-9, 5750 – 5756, 2003.
[11] H. Ukita, Y. Uenishi, and H. Tanaka, Science, 260-5109, 786 – 789, 1993.

Optimized distributive μ-mixing by ‚chaotic' multilamination

F. Schönfeld, K.S. Drese, S. Hardt, V. Hessel, C. Hofmann

Institut für Mikrotechnik Mainz –(IMM),
Carl-Zeiss-Str. 18-20, 55129 Mainz, Germany, schoenfeld@imm-mainz.de

ABSTRACT

A new micro mixer relying on the consequent utilization of the split-and-recombine (SAR) principle is presented. We show that the mixing can be characterized by a positive finite-time Lyapunov exponent although being highly regular and uniform. Furthermore, a spectral method based on the numerical solution of a pure diffusion equation is developed which allows a computation of concentration profiles and mixing residuals without artifacts from numerical diffusion. Eventually, we compare the performance of the SAR mixer to that of micro mixers relying on interdigital multilamination and hydrodynamic focusing and find it to be superior in terms of mixing efficiency, mixing time and pressure drop in the regime of small Reynolds numbers.

Keywords: micro mixing, split-and-recombine principle, spectral method

1 INTRODUCTION

Efficient mixing of liquid/liquid systems is one of the paramount tasks in the development of μ-TAS and Lab-on-a-chip systems as well as in micro chemical process engineering. We present a new type of micro mixer relying on a multi-step split-and-recombine (SAR) approach. In the last years, different SAR mixers have been fabricated (see e.g. [1-3]). However, all these mixers were mainly validated experimentally with regard to global process results such as yield, effective reaction times, or properties of produced particles. However, there is, to our best knowledge, neither an in-depth simulation of the mixing efficiency nor an experimental validation of the mixing process itself. Using computational fluid dynamics (CFD), a favorable mixer design was identified which allows for an almost ideal multilamination. The numerical results are corroborated by experimental visualization of multilamination and mixing. Furthermore, a semi-analytical approach is presented allowing for computation of the mixing efficiency. By comparison the mixing performance is found to be superior to that of micro mixers relying on the principle of interdigital multilamination and hydrodynamic focusing [4,5] in the regime of small Reynolds numbers.

2 SPLIT-AND-RECOMBINE MIXING

SAR mixing relies on a multi-step procedure. According to its name the basic operations are: splitting of a bi- or multi-layered stream perpendicular to the lamellae orientation into sub-streams and recombination of these. Usually, these basic steps are accomplished by one or more re-shaping steps. The mixing mechanism is schematically displayed in Fig.1.

Figure 1: Schematic of a SAR step: initial configuration, splitting, recombination, reshaping (from left to right).

In the framework of chaotic advection, the mixing performance is commonly characterized by interfacial stretching

$$\lambda(t) = \lim_{L_0 \to 0} L(t) \Big/ L_0 \, , \qquad (1)$$

where L_0 and $L(t)$ denote the characteristic dimensions of the interfacial area at $t=0$ and at a finite time t, respectively (see e.g. [6,7]). In the case of a 2D, incompressible flow the diffusive mass transport, which determines the mixing performance, depends quadratically on the interfacial stretching λ [6,8]. Chaotic flows ensure particular efficient mixing since they imply an exponential increase of stretching over time. Accordingly, the finite-time Lyapunov exponent σ may be defined via $\lambda(t) \sim e^{\sigma t}$. The repeated application of the SAR principle implies an exponential increase of interfacial area. The corresponding stretching factor is given by

$$\lambda(t) = (2^n - 1) \approx 2^{u \cdot t / l_{SAR}} \, , \qquad (2)$$

where n, u, and l_{SAR} denote the number of SAR steps, the mean velocity and the length of one SAR unit, respectively. According to Eq. (2) the SAR mixer has a finite-time Lyapunov exponent of $\sigma = \ln 2(u/l)$. Thus, even in case of a highly regular flow pattern a positive finite time Lyapunov exponent can be achieved, a characteristic feature of chaotic advection. Whereas chaotic advection generally induces regions of regular flows and poor mixing besides regions of high chaoticity where mixing predominantly occurs [6], here, in the case of an ideal SAR multi-lamination, a spatially homogeneous mixing is obtained.

Since the final lamella dimension does not solely depend on the channel width, but also on the number of SAR steps, thorough mixing can be achieved under moderate pressure drops.

2.1 CFD-aided layout and validation

The starting point for the layout was the numerical investigation of flow patterns of the so-called caterpillar mixer realized in 1999 [9] which has been successfully applied in various applications (e.g. see [10]). Details on CFD simulations relying on the simultaneous solution of the Navier-Stokes equations and a convection-diffusion equation for a tracer are given in [11]. The CFD model of the revised and optimized geometry is shown in Fig. 2a-c.

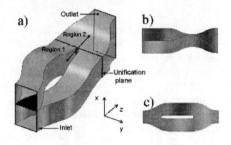

Figure 2: Geometry model of a single mixing step of the optimized SAR mixer - a) slanted view with the splitting layer shown in black; b) side- and c) top-view.

According to the simulation results the developed design leads to an almost prefect lamellae configuration within the SAR mixer for Reynolds numbers below about 30. Thus the mixer is particularly suited for mixing of highly viscous fluids. For higher Reynolds numbers the CFD simulations show more and more deviations from an ideal SAR multi-lamination pattern. Since inertial forces come into play a secondary flow pattern is superposed to the SAR velocity profile of the creeping flow regime.

The good agreement between simulation and experimental results is highlighted in Fig. 3, where the simulated streamline pattern for Re = 0.22 is superimposed to a micrograph of the first SAR step obtained from an iron-rhodanide reaction performed in water/glycerol solutions. Fig. 4 shows a series of micrographs obtained by feeding a transparent and a dyed glycerol/water solution to the mixer. Highly regular lamellae patterns are observed up to the 8th SAR step. The experiments were performed using a SAR mixer realised in PMMA: inlet dimension: 1 mm x 2 mm; length of one SAR unit: 12 mm; number of SAR steps: 8. Fabricational and experimental details are given in [11].

Figure 3: Comparison between CFD results (black lines) and experiment for 85% glycerol/water solutions at a flow rate of 0.2l/h. The black lines denote the numerically computed streamlines seeded at the lamellae interface at the inlet (taken from [11]).

Figure 4: Optical inspection of multi-lamination in the SAR mixer over 8 mixing steps [11]. The dyed (water-blue) and transparent lamellae of a 85% glycerol/water solution are shown in dark and light gray, respectively. The applied total volume flow rate of 0.2 l/h corresponds to Re = 0.22.

2.2 Analytical Approach

The CFD simulations allow to study the flow distribution and the arrangement of fluid lamellae. However, even if grids with more than 10^6 computational cells are used, a solution of the convection-diffusion equation for concentration fields in liquids can hardly be achieved with appropriate accuracy. The reason are discretization errors which induce an artificial diffusive transport [12-14]. A quantity characterizing their magnitude is the cell-based Peclet number, $Pe = ul_c/D$, where u denotes a typical flow velocity, l_c the length scale of a computational cell, and D the molecular diffusivity. As outlined above the SAR mixer is especially suited for mixing of highly viscous liquids. If a liquid of a viscosity of 0.1 Pa·s is assumed, characteristic diffusion constants of the order of 10^{-11} m^2/s are obtained according to the Stokes-Einstein relation. The length scale of the computational cells is of the order of 50 µm for typical grids. For a Reynolds number of 30 the average flow velocity is approximately 11 cm/s. These numbers imply a cell-based Peclet number of $5.5 \cdot 10^5$, indicating that even on grids with multi-million cells it is usually not possible to compute the mixing performance of the SAR mixer for highly viscous liquids using standard CFD techniques.

In order to approximately compute the time evolution of the concentration field inside the mixer, a method was developed which is based on the numerical solution of a pure diffusion equation using a spectral method. The method rests on the assumption that the mixing process can be regarded as a diffusive mass transfer in a frame-of-reference co-moving with the average flow velocity. That way the velocity profile inside the mixer is replaced by a plug-flow profile. Furthermore it is assumed that the fluid lamellae are aligned with the x-axis of Fig. 2 and are of equal thickness. Fig. 4 indicates that this assumption is reasonable for low Reynolds numbers although lamellae next to the channel walls are slightly thicker than the lamellae in the middle. In accordance with these approximations, the diffusion equation is solved in a plane perpendicular to the flow being swept through the mixer geometry with the average flow velocity. In order to study the two branches of the mixing step independently, each branch is assigned its own plane. For taking effects of hydrodynamic focusing into account, it is helpful to write down the diffusion equation using a non-dimensionalized spatial variable s

$$\frac{\partial c}{\partial t} = \frac{D}{L^2} \frac{\partial^2 c}{\partial s^2} \ , \tag{3}$$

where L is the channel width and $s = s'/L$, with s' being the spatial coordinate in width direction. At the inlet of a mixing step s' is identical to y. When the plane on which the diffusion equation is solved is swept through the mixer geometry, the length scale changes, i.e. $L = L(t)$.

In order to solve the diffusion equation on these planes perpendicular to the flow, the geometry of a mixing step is divided into two regions, as shown in Fig. 2. Region 1 is the region upstream of the unification plane, while region 2 comprises those parts downstream of the unification plane. In both of these regions, the concentration field is expanded in harmonics

$$c_n(s',t) = \frac{1}{2} + \sum_{i=0}^{\infty} a_i^{(n)}(t) \sin k_i s \tag{4}$$

where $n = 1, 2$ denotes the region where this expansion is applied and the wave numbers k_i are given by $(2i+1)\pi$. Assuming two inlet lamellae entering with identical flow rates with tracer concentrations of 0 and 1, respectively, no cosine functions appear in Eq. 4 due to antisymmetry with respect to inversion of the x-axis. For a given volume flow, the unification plane will be reached after a specific mean residence time t_u. The channel width before and after unification of the two streams are $L_- = L(t_u - \varepsilon)$ and $L_+ = L(t_u + \varepsilon)$, where ε is an infinitesimal number. Obviously, $L_+ = 2L_-$. When entering region 2, the concentration field is composed of two solutions for region 1 shifted by an amount ½ to the left and to the right, respectively. The resulting expression can be cast into the following form

$$c_2(s',t) = \frac{1}{2} + \sum_i a_i^{(1)}(t_u)(-1)^i [\theta(-s_-) - \theta(s_-)] \cos k_i s_- \tag{5}$$

where $\Theta(s)$ denotes the heaviside function and $s_- = s'/L_-$. The goal is now to write the solution in region 2 as an expansion in the Fourier modes of a domain with a width of L_+. Again, for reasons of symmetry only the sin-components contribute and the expansion coefficients are given as

$$a_i^{(2)}(t_u) = \frac{2}{L_+} \int_{-L_+/2}^{L_+/2} c_2(s',t) \sin k_i s_+ \, ds' \tag{6}$$

where $s_+ = s'/L_+$. When inserting the expression for $c_2(s',t)$ the integrals involving a product of a sin and a cos function can be computed analytically and straightforward algebra yields

$$a_i^{(2)}(t_u) = \sum_j (-1)^{j+1} a_j^{(1)}(t_u) \frac{k_i}{k_i^2/4 - k_j^2} \tag{7}$$

When inserting this expression into Eq. 4, the concentration field at the entrance to region 2 of the mixing step is obtained. Within each of the two regions, the time evolution of the concentration field is obtained when inserting the series expansion of Eq. 4 into Eq. 3. This way, an ordinary differential equation for the expansion coefficients is obtained which can be solved in a straightforward manner by integration to give

$$a_i^{(n)}(t) = a_i^{(n)}(0) \exp\left(-Dk_i^2 \int_0^t \frac{dt'}{L^2(t')} \right) \tag{8}$$

It should be noted that Eq. 8 can only be applied within either region 1 or 2. The transition between the two regions is described by Eq. 7. The translation between the time coordinate appearing in the previous equations and the spatial coordinate along the mixing channel is performed using the mean flow velocity u. For the investigation of mixing in the SAR mixer, it is necessary not only to compute the concentration profile in a single mixing step, but rather a sequence of steps has to be considered. The complete algorithm for determining the concentration profile in a multi-step mixer consists of the following steps which have to be performed iteratively:

- Computation of the time evolution of the Fourier coefficients describing the concentration distribution at the inlet of an SAR step via Eq. 8.
- Derivation of the corresponding coefficients in region 2 using Eq. 7.
- Computation of time evolution in region 2 (Eq. 8)
- Translation of the coefficients from the outlet of region 2 to the inlet of region 1.

Figure 5 shows the mixing residual, i.e. the average deviation from perfect mixing, deduced from the concentration profiles at the interface between region 1 and region 2 of Fig. 2. for a diffusion constant of $1.768 \cdot 10^{-11}$ m²/s and a total flow rate of 0.2 l/h.

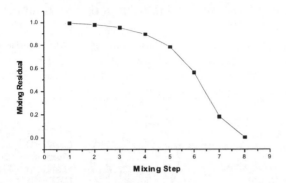

Figure 5: Mixing residual vs. number of SAR steps.

3 COMPARISON TO INTERDIGITAL MICRO MIXERS

The generic design of micro mixers relying on interdigital multilamination and hydrodynamic focusing as shown in Fig. 6 was analytically investigated in [5].

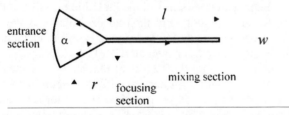

Figure 6: Mixer design analyzed in [5].

The two fluids enter through inlets equally distributed over the entrance section having a pitch dp and form an interdigitated lamellae pattern. With the assumptions outlined above concentration profiles are computed by solving Eq. 3 by means of a spectral method. Assuming a mixer geometry which reproduces a mixing residual of 0.007 and a mixing time of about 14 s as found in the case of the SAR mixer after the 8th step, one obtains a pressure drop of 0.2 bar in the case of multilamination and hydrodynamic focusing, whereas, a pressure drop of about 0.0038 bar is obtained for the SAR mixer for an highly viscous liquid having a kinematic viscosity $\eta = 80$ mPas.

4 SUMMARY ANS CONCLUSION

An improved mixer layout was derived by means of CFD simulations which assures an almost ideal realization of SAR mixing. Generally, a positive finite-time Lyapunov exponent which is a generic feature of chaotic advection is

associated with the SAR mixer. The corresponding exponential stretching of interfacial area together with the uniform lamellae distribution ensures efficient homogeneous mixing especially at Reynolds numbers below 30.

The mixing performance was approximately calculated by adapting a spectral method to the particular geometry. The semi-analytical approach allows to overcome problems due to discretization errors as well as the computation of concentration profiles and corresponding mixing residuals. A simulation run takes about 10 s on a standard PC compared to hours which are typically needed for CFD simulations.

Furthermore, in the low-Reynolds-number regime the multi-step nature of the SAR mixing concept leads to a superior mixing performance at equal pressure drop compared to the concept of interdigital multilamination and hydrodynamic focusing.

REFERENCES

[1] N. Schwesinger, T. Frank and H. Wurmus, J. of Micromech. and Microeng., 1996, 6, 99.
[2] J.P. Branebjerg, J., Gravesen, J.P. Krog and C.R. Nielsen, IEEE-MEMS '96, USA, 1996, 441.
[3] H. Mensinger, T. Richter, V. Hessel, J. Döpper and W. Ehrfeld, in Micro Total Analysis System, A. van den Berg, P. Bergfeld (Eds.), Kluwer Academic Publishers, Dordrecht 1995, 237.
[4] V. Hessel, S. Hardt, H. Löwe and F. Schönfeld, AIChE J., 2003, 49, 566.
[5] K.S. Drese, IMRET 7, Chem. Eng. J., accepted.
[6] J.M. Ottino, F.J. Muzzio, M. Tjahjadi, J.G. Franjione, S.C. Jana and H.A. Kusch, Science, 1992, 257, 754.
[7] D.R. Sawyers, M. Sen and H.-C. Chang, Chem. Eng. J., 1996, 64, 129.
[8] J.M. Ottino, "The Kinematics of Mixing: Stretching, Chaos, and Transport", Cambridge Univ. Press, Cambridge 1990.
[9] H. Löwe, W. Ehrfeld, V. Hessel, T. Richter and J. Schiewe, Proc. IMRET 4, Atlanta, USA, 2000, 31.
[10] V. Hessel, H. Löwe, C. Hofmann, F. Schönfeld, D. Wehle and B. Werner, Proc. IMRET 6, AIChE Pub. No. 164, New Orleans, USA, 2002, 39.
[11] F. Schönfeld, V. Hessel, and C. Hofmann, Lab on a Chip, in press.
[12] J.P. Boris, and D. L. Book, J. Comput. Phys., 11, 38 (1973).
[13] C.A. Fletcher, Computational Techniques for Fluid Dynamics; Springer-Verlag, Berlin (1991)
[14] S. Hardt and F. Schönfeld, AIChE J. 2003, 49, 578.

Analysis of Micromachined PZT Membranes for MEMS Power

J.H. Cho[*], M. J. Anderson[**], R.F. Richards[*], D.F. Bahr[*] and C.D. Richards[*]

[*]Washington State University Pullman, School of Mechanical and Materials Engineering
WA, USA, cill@wsu.edu
[**]University of Idaho, Department of Mechanical Engineering Moscow, ID, USA, anderson@uidaho.edu

ABSTRACT

The performance of a piezoelectric generator in terms of the quality factor Q, the electromechanical coupling coefficient k^2, and the efficiency were examined. The effect of design parameters such as membrane size, piezoelectric thickness, silicon thickness, and top electrode area are explored. The results show that both k^2 and Q are sensitive to PZT thickness and electrode size.

Keywords: pzt membrane, mems power, electromechanical coupling coefficient, quality factor, efficiency.

1 INTRODUCTION

The need for miniaturized power sources for MEMS and microelectronics devices is widely recognized. Micro-scale concepts to generate electrical power include devices which use the stored energy in fuels to those which harvest energy from the environment. Piezoelectric materials are increasingly employed in both types of devices to convert mechanical to electrical energy [1–5]. The use of piezoelectric materials yields significant advantages for micro power systems. Since piezoelectric materials convert mechanical energy into electrical energy via strain in the piezoelectric material, they lend themselves to devices that operate by bending or flexing which brings significant design advantages.

Recent work by our team at Washington State University has been directed at the design of a micro heat engine in which thermal power is converted to mechanical power through the use of a novel thermodynamic cycle. Mechanical power is then converted into electrical power through the use of a thin-film piezoelectric membrane generator [1].

In this paper we examine the performance of the piezoelectric generator in terms of the quality factor Q, the electromechanical coupling coefficient k^2, and the efficiency. The effect of design parameters such as membrane size, piezoelectric thickness, and silicon thickness are explored.

2 DEVICE FABRICATION

The structure of the thin-film piezoelectric membrane generator is a two-dimensional sandwich structure similar to that used for ultrasonic transducers [6]. Figure 1 shows a cross section of the membrane generator, which consists of a silicon membrane, a bottom platinum electrode, a thin-film of the piezoelectric ceramic PZT (Lead Zirconate Titanate) and a top gold electrode.

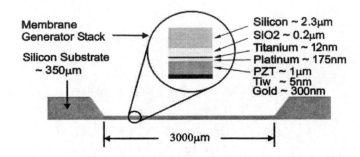

Figure 1: The structure of the PZT membrane generator.

The substrate for the PZT membrane generator is a (100) silicon wafer. Boron is doped into the one side of the silicon for an etch stop, and then an oxide layer is grown and patterned to provide a mask for anisotropic etching in ethylene diamine pyrochatechol (EDP). A 12nm layer of titanium and a 175nm thick layer of platinum are sputtered using DC magnetron sputtering. PZT is spun onto the platinum films in a sol-gel process [7]. A top electrode consisting of a 5nm TiW adhesion layer of gold and 300nm of gold is deposited by DC magnetron sputtering. Photolithography is then used to etch the top electrode and PZT around the top electrode. A completed membrane generator is shown in Fig. 2.

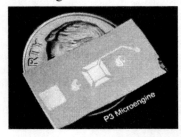

Figure 2: Piezoelectric membrane generator.

3 PIEZOELECTRIC MEMBRANE CHARACTERISTICS

A piezoelectric membrane generator produces power when it is strained in response to an applied pressure. In the micro heat engine that pressure is supplied by the periodic expansion and compression of a two-phase working fluid. The pressure pulse applied by the two-phase working fluid causes the piezoelectric membrane to flex out, straining the piezoelectric ceramic. This straining of the piezoelectric material causes the film to become thinner, due to Poisson's effect, and results in an electric potential across the electrodes. Since the pressure pulse is periodic, the membrane oscillates and produces an alternating current.

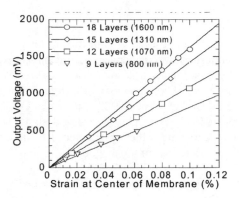

Figure 3: PZT output voltage vs. strain.

Important parameters to the output voltage produced by a piezoelectric membrane are the thickness of the PZT layer and the strain generated in the PZT layer during maximum membrane deflection. Fig. 3 illustrates these dependencies. In the figure, peak-to-peak open-circuit voltages are plotted against the tensile strain experienced by the PZT layer during maximum piezoelectric membrane deflection. Piezoelectric membrane output voltage increases linearly both with PZT layer thickness and with PZT maximum strain.

The chemistry of the PZT, specifically the ratio of Zr to Ti, can have a substantial impact on the specific output (V/%$\varepsilon\mu$m PZT). A higher specific voltage translates to a higher voltage output from a membrane at a given strain.

The effects of PZT thickness and chemistry on power

output from a 3mm membrane are shown in Fig. 4. All three piezoelectric membranes were driven with the same pressure and frequency. Power is greatest from the 30 layer, 40:60 PZT membrane at a load resistance of 15 kΩ.

3.1 Mechanical behavior

The mechanical resonance frequency of the membrane generator is very important for power generation and efficiency. The frequency response of an electrically excited piezoelectric membrane generator is shown in Fig. 5. Deflection measurements are acquired using an optical lever consisting of a laser beam and detector. The detector voltage is directly proportional to the generator deflection in both time and space. These measurements are acquired in air with the membrane suspended in a clamping jig.

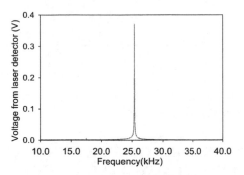

Figure 5: Membrane generator's frequency response.

The resonance frequency of membrane is function of membrane size, stiffness and mass [8]. Table 1 shows the variation of resonance frequency with changes in membrane size, piezoelectric thickness and silicon thickness. As membrane size is increased from 3 to 6 mm, the resonance frequency decreases from 28.4 kHz to 14.8 kHz.

3.2 Electrical behavior

The PZT oscillator can be modeled with an equivalent circuit as shown in Fig. 6. It consists of the resistances of top and bottom electrode, shunt capacitance C_o, and mechanical components, C_m, R_m and L_m. The shunt capacitance C_o is a function of the electrode size and separation, C_m is inversely proportional to the stiffness of the membrane, L_m is proportional to the effective mass, and

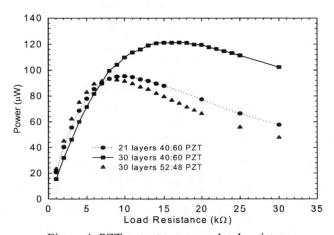

Figure 4: PZT output power vs. load resistance.

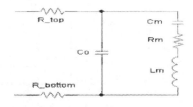

Figure 6: Equivalent electrical circuit of generator.

R_m is dependent upon the damping characteristics of the structure.

Impedance and phase data were acquired for each of the membranes using an impedance analyzer. The PZT membranes were poled by applying 120 kV/cm parallel to the direction of measurement for 10 minutes. The membranes were not tested until a minimum of 24 hours had elapsed after poling. With the impedance data, the equivalent circuit of each membrane was found by impedance data matching as shown in Fig 7.

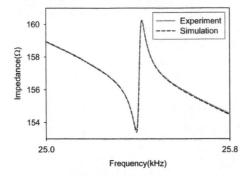

Figure 7: The impedance curves of experiment and simulation of membrane generator.

4 EXPRESSION FOR THE GENERATOR EFFICIENCY WITH K^2 AND Q

Although researchers conducting fundamental studies [9] have noted that high efficiency for piezoelectric conversion devices requires large quality and electromechanical coupling factors, they provide no discussion of the impact of Q and k^2 on efficiency η or guidelines for device development.

An exact expression for efficiency of a piezoelectric element may be derived from first principles [10]. The device is modeled as a rigid mass (m) coupled to a stationary surface through a spring (s), with a damper (b), and a piezoelectric element poled in the thickness direction. An applied force excites the system and energy is extracted by connecting an electric circuit to the electrodes on the piezoelectric element. Details of the derivation are provided in [10].

As shown in equation (1), the expression for the generator efficiency η depends only on the quality factor Q and the electromechanical coupling coefficient k^2. This relationship is shown graphically in Fig. 8.

$$\eta = \frac{\frac{1}{2}\frac{k^2}{1-k^2}}{\frac{1}{Q}+\frac{1}{2}\frac{k^2}{1-k^2}} \tag{1}$$

This expression can be used to gain further insight into the design of micro power generators using piezoelectric oscillators. For example, if a device efficiency of 85% is desired there are several combinations of k^2 and Q that will work: 1) $k^2 = 0.01$ and Q = 1000, 2) $k^2 = 0.05$ and Q = 200, and 3) $k^2 = 0.1$ and Q = 100. This gives the designer some flexibility in choosing the piezoelectric material, configuration and structure for the device.

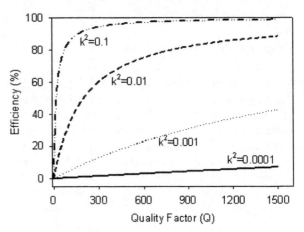

Figure 8: The relationship among η, Q and k^2.

The quality factor, $Q=(ms)^{1/2}/b$, depends on the effective mass, stiffness, and damping of the structure. In terms of the circuit analog $Q = 1/R_m(L_m/C_m)^{1/2}$. The electromechanically coupling coefficient, $k^2=C_m/(C_o+C_m)$, is a function of the material properties and electrode configuration through C_o. The parameter C_m is inversely

	PZT thickness (μm)	Membrane size (mm²)	Si thickness (μm)	Top electro size (mm²)	%change in C_o	%change in C_m	%change in L_m	%change in R_m	Resonance frequency [kHz]
Ref.	1	3x3	2	2x2	-	-	-	-	25.4
Case 1	2	3x3	1	2x2	- 50	- 44	47	214	28.4
Case 2	1	3x3	1	2x2	0	- 33	39	- 66	26.3
Case 3	1.8	3x3	2	2x2	- 45	- 8	- 8	- 69	27.7
Case 4	3	3x3	2	1.6x1.6	- 79	- 7	103	35	18.5
Case 5	1	6x6	2	2x2	0	- 69	826	100	14.8

Table 1: Membrane generator property according to design parameters.

proportional to the structural stiffness of the device. Thus the equivalent circuit model may be used to explore the effect of mechanical design changes to the electromechanical performance of the piezoelectric generator.

5 EFFECT OF PARAMETER CHANGE ON Q AND K^2

The design parameters of six piezoelectric membranes are provided in Table 1. PZT thickness, Si thickness, membrane size, and top electrode area were varied. The effect of these changes on the circuit parameters, C_o, C_m, L_m, and R_m are shown in the table. Fig. 9 shows the relative effect of these design changes on k^2 and Q. Decreasing the Si thickness while maintaining the PZT thickness results in the largest improvement in Q with a smaller decrease in k^2. Increasing the PZT thickness while maintaining the Si thickness resulted in increase in Q and k^2. Increasing the PZT thickness and decreasing the top electrode area resulted in an increase in Q and a remarkable increase in k^2. Increasing the PZT thickness and decreasing Si thickness resulted in a decrease in Q and a slight increase in k^2, and high resonance frequency because of high residual stress from the thin Si layer. Increasing the membrane size leads to a substantial increase in Q with a correspondingly substantial decrease in k^2.

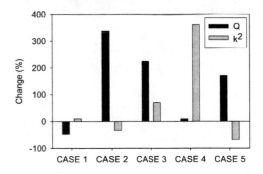

Figure 9: The effect of parameter change on Q and k^2.

6 SUMMARY

The effect of design parameters such as membrane size, piezoelectric thickness, silicon thickness, and top electrode area of a piezoelectric membrane generator were explored. The efficiency of a piezoelectric generator may be expressed in terms of the coupling coefficient, k^2 and the quality factor, Q. Thus the effect of design parameters on the electromechanical coupling coefficient and quality factor were evaluated. The results show that both k^2 and Q are sensitive to PZT thickness and electrode size.

ACKNOWLEDGEMENTS

Financial support was provided by DARPA MTO's MicroPower Generation Program and the US Army SMDC contract #DASG60-02-C0001.

REFERENCES

[1] S. Whalen, M. Thompson, D. Bahr, C. Richards and R. Richards, "Design, fabrication and testing of the P^3 micro heat engine, Sensors and Actuators, A: Physical, Vol. 104, No. 3, pp. 290-298, 2003.

[2] N.M. White, P.Glynne-Jones, and S.P. Beeby, "A novel thick-film piezoelectric micro-generator," Smart Materials and Structures, Vol 10, No. 4, pp 850-852, 2001.

[3] R. Sharaf, W. Badawy, W. Mausa, "Piezoelectrically m-generator for implantable bio-devices," Photonics West MEMS/MOEMS Conference, Proceedings of SPIE Vol. 5344, January 2004, San Jose, CA.

[4] H. Li, A. Lal, J. Blanchard, D. Henderson, "Self-reciprocating radioisotope-powered cantilever," Journal of Applied Physics, Vol. 92, No. 2, pp. 1122–1127, 2002.

[5] N.W. Hagood, et al. "Micro-hydraulic transducer technology for actuation and power generation", Proceedings of SPIE - The International Society for Optical Engineering, Vol. 3985, pp. 680-688, 2000.

[6] J.Baborowski, N. Ledermann, P. Muralt, "Piezoelectric Micromachined Ultrasonic Transducers based on PZT Films, in Materials," Research Society Symposium Proceedings, Nano- and Microelectromechanical Systems (NEMS and MEMS) and Molecular Machines eds. D. A. LaVan, A. A. Ayon, T. E. Buchheit, M. J. Madou, 741 J12.4.1-6, 2003.

[7] L.M.R. Eakins, B.W.Olson, C.D. Richards, R.F.Richards and D.F. Bahr, "Influence of structure and chemistry on piezoelectric properties of lead zirconate titanate in a microelectromechanical system power generation application," JMR, Vol.18, pp. 2079-2086, 2003

[8] Richard Haberman, "Applied partial differential equations," Prentice Hall, 4th, pp. 142-147, 2004

[9] M. Umeda, K. Nakamura, S. Ueha., "Analysis of the transformation of mechanical impact energy to electric energy using piezoelectric vibrator," Japanese Journal of Applied Physics, Part 1: Regular Paper & Short Notes & Review Papers, Vol. 35, No. 5B, pp. 3267-3273, 1996.

[10] C. D. Richards, M. J. Anderson, D. F. Bahr, and R.F. Richards, "Efficiency of Energy Conversion for Devices containing a Piezoelectric Component," submitted to J. Micromechanics and Microengineering.

Integrated High Frequency RF Inductors with Nano/micro Patterned Ferromagnetic Cores

Y. Zhuang, M. Vroubel, B. Rejaei, and J. N. Burghartz

Delft University of Technology, ECTM-DIMES, Mekelweg 4, 2600 GA Delft, The Netherlands,
y.zhuang@dimes.tudelft.nl

K. Attenborough[*]
OnStream MST, Lodewijkstraat 1, Eindhoven, The Netherlands

ABSTRACT

Integrated solenoid inductors with high operating frequency and low loss have been demonstrated by using nano-/micro- size granular $Ni_{80}Fe_{20}$ cores. The $Ni_{80}Fe_{20}$ films were deposited by electroplating on three types of seed layer, Cr, Ti, and Ti covered by TiN under a magnetic field ~ 80 mT to align the magnetization. The $Ni_{80}Fe_{20}$ film on Ti seed layer exhibits a large amount mostly disconnected islands with maximum diameter < 1.6 μm. By using the granular $Ni_{80}Fe_{20}$ layer as the magnetic core, the inductors show high operating frequency >6.5 GHz and high cut-off frequency >20 GHz. Systematically optimizing the device's geometrical parameters, a high inductance per area >0.20 μH/mm^2, and a high quality factor >4.5 have been reached.

Keywords: RF, ferromagnetic, micro-patterning, inductor, integration

1 INTRODUCTION

Developing high performance and small volume of on-chip inductive RF/microwave components like inductors is crucial for the cost-effective RF/BiCMOS and RF/CMOS technologies [1]. Considerable efforts are underway to develop on-chip inductors with ferromagnetic (*FM*) cores having a high inductance per area (*IPA*) with a sufficiently high maximum quality factor (Q_{max}), a high operating frequency $f(Q_{max})$ (where the quality factor Q reaches its maximum), and a high cut-off frequency ($f_{cut-off}$) that is related to the ferromagnetic resonance (*FMR*) frequency [2]-[4]. However, the *FM* core's high conductivity deteriorates the device performance at RF/Microwave frequencies manifested by the low Q_{max}, $f(Q_{max})$, and $f_{cut-off}$, even the principal superior solenoid-type inductors have been exploited [4]. Reduction of the effective *FM* film conductivity and thus of eddy currents, while maintaining a sufficiently high permeability and *FMR*, can be achieved by nano/micro-size patterning of the *FM* film. Recently, nano-granular *FM* films with low conductivity $\delta < 10^5$ S/m has been reported by using multiple-target sputtering

techniques [5]. In IC processing, however, a more cost-effective deposition method is more preferable.

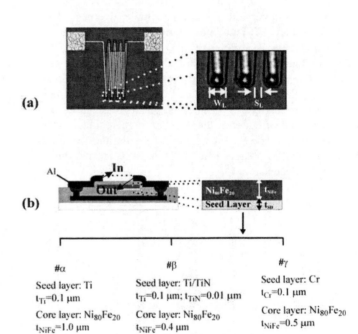

Fig.1 Plain view photograph (a), and cross-sectional sketch (b) of a 4-turn $Ni_{80}Fe_{20}$-core solenoid inductor. "In" and "Out" in (b) denote the direction towards inside and outside of the plane, respectively. $Ni_{80}Fe_{20}$ films (thickness indicated above) have been deposited by means of electroplating on three different kinds of seed layer, i.e. α−seed: 100nm Ti, β− seed: 100nm Ti covered by 10 nm TiN, γ− seed 100 nm Cr.

In this paper, we present a novel low-cost method to obtain nano/micro structured $Ni_{80}Fe_{20}$ film by electroplating in combination with an optimized seed layer. A series of on-chip solenoid inductors with $Ni_{80}Fe_{20}$ films on three types of seed layer were fabricated and compared. By optimizing the design of the devices, high $f(Q_{max})$ (>6.5 GHz), and high $f_{cut-off}$ (>20 GHz) have been obtained on inductors with the granular $Ni_{80}Fe_{20}$ core.

---
[*] Current address: Philips Research Leuven, Kapeldreef 75, B-3001 Leuven, Belgium

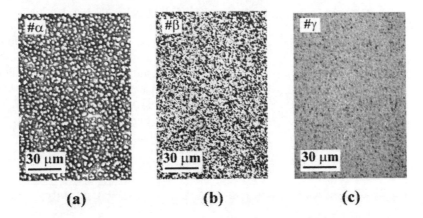

(a)   **(b)**   **(c)**

Fig.2 Micrograph of surface morphology of electroplated $Ni_{80}Fe_{20}$ films on the three kinds of seed layer described in Fig.1: (a) α– seed, (b) β– seed, (c) γ– seed. Mostly isolated grains with the maximum diameter less than 1.6 μm have been achieved with the α-seed by properly optimizing the seed layer and its deposition condition, together with the plating condition of $Ni_{80}Fe_{20}$ film

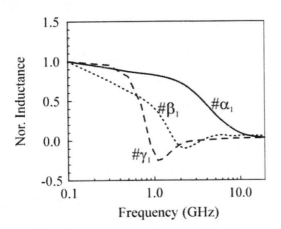

Fig.3 Comparison of normalized inductance *versus* frequency of 20-turn 1000×500μm^2 solenoid coils with the three magnetic cores. The decay of inductance results from the limits by the *FMR*, the eddy current loss, and the *LC*-resonance. The #α$_1$ exhibits the highest drop-off frequency.

2   FABRICATION

The process of integrated solenoid inductors with $Ni_{80}Fe_{20}$ core has been described in [4]. Fig. 1 shows a 4-turn solenoid inductor, as an example. Three types of seed layer 100nm *Ti* (α-seed), 100nm *Ti*/10 nm *TiN* (β-seed), and 100 nm *Cr* (γ-seed) have been deposited by magnetron DC sputtering. The core $Ni_{80}Fe_{20}$ films with thickness of 1.0 μm (α-core), 0.4 μm (β-core), and 0.5 μm (γ-core) have been deposited by electroplating on α-seed, β-seed and γ-seed, respectively. DC magnetic field ~ 80 mT has been applied during the electroplating to align the magnetization along the y-axis. The α–core with *Ti*-seed exhibits a

nano/micro pattern of mostly disconnected *NiFe* grains with a maximum diameter of $D \sim 1.6$ μm (Fig. 2(a); α– core). The β– core (Fig. 2(a); β– core) and γ– core (Fig. 2(a); γ–core) are homogeneously continuous films. The β– core, however, shows a much rough surface than the γ– core.

3   RESULTS AND DISCUSSIONS

Results for three 20-turn solenoid inductors #α$_1$, #β$_1$, #γ$_1$ with α-, β-, and γ- cores have been compared in Fig. 3. Their parameters are listed in Table I. The normalized inductance *versus* frequency of #α$_1$ clearly exhibits a higher drop-off frequency compared to #β$_1$ and #γ$_1$. The drop-off frequency is dependent on the effective permeability μ$_{effc}$ of the magnetic core [6] and the *LC* – resonance. The μ$_{effc}$

Table I: Geometrical structure parameters of solenoid inductors with $Ni_{80}Fe_{20}$ core. Here L_{FM} and W_{FM} denote the width of the $Ni_{80}Fe_{20}$ core (Fig. 1). W_L and S_L denote the line width and spacing of the coil. W_S and n are the spacing between the *FM*-core and the via connections, and the number of turns, respectively. Core denotes the type of core: α - 1.0 μm $Ni_{80}Fe_{20}$/100 nm Ti, β - 0.4 μm $Ni_{80}Fe_{20}$/100 nm Ti/10 nm TiN, γ- 0.5μm $Ni_{80}Fe_{20}$/100 nm Cr.

	W_{FM} (μm)	L_{FM} (μm)	W_L (μm)	S_L (μm)	W_S (μm)	n	Core
#α$_1$	500	1000	20	30	10	20	α
#α$_2$	30	60	6	10	10	4	α
#α$_3$	60	60	6	10	10	4	α
#α$_4$	120	60	6	10	10	4	α
#β$_1$	500	1000	20	30	10	20	β
#γ$_1$	500	1000	20	30	10	20	γ

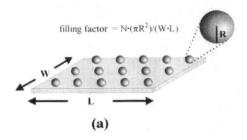

filling factor $= N \cdot (\pi R^2)/(W \cdot L)$

(a)

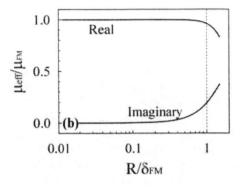

Fig.4 (a) Model of configuration of the granular film (α–core) shown in Fig.2 (a). The granular grains are simplified to a number (N) of spheres with radius R. (b) Calculation of the ratio of permeability with (μ_{effc}) and without (μ_{FM}) considering the eddy current loss. As R approach to the skin depth δ_{FM}, the eddy current loss become significant by manifest itself from the decrease and increase of the real and imaginary μ, respectively. The filling factor in the calculation is set to 1.

relates to the ferromagnetic resonance frequency (*FMR*), and the eddy current flowing in the magnetic core at radio frequency. Below the *FMR*, the real part of permeability is positive and the device with magnetic core shows inductor characteristic. Above the *FMR*, however, the real part of permeability becomes negative and the device behaves as a capacitor. Therefore, the inductor with magnetic core can only work below the *FMR*. The grains in α- core have arbitrary shape and size, which leads to a non-uniform randomly orientated magnetization (*NUROM*), owing to their large magnetic shape anisotropy. The *NUROM* causes an extraordinarily broad *FMR* peak, as a result, there is no clear *FMR* peak on both the real and imaginary part of permeability of the α-core, which are proved by extraction of permeability based on the method described in [7]. Due to the absence of *FMR*, the $\#\alpha_1$ has much less impact from the *FMR* compared to $\#\beta_1$ and $\#\gamma_1$. The *NUROM* also causes a generally low permeability of the α-core (around 10 from 1GHz to 10 GHz), which consequently weak the eddy current effects. The skin depth δ_{FM} of the α-core becomes larger than the β- and γ- core's. In addition, the calculation in Fig.4 points out that nano-/micro- patterning of the magnetic core into isolated islands can effectively

eliminate the eddy current effect, demonstrated by the constant ratio between μ_{effc} (permeability considering eddy current effect) and μ_{FM} (permeability without considering eddy current effect), when the skin depth is larger than the island size. Furthermore, because of the low permeability and smaller inductance, the *LC* resonance frequency of $\#\alpha_1$ is higher than that of $\#\beta_1$ and $\#\gamma_1$. Due to the combined effects of less impact from the *FMR*, weaker eddy current effect, and higher *LC* resonance frequency, the drop-off frequency of $\#\alpha_1$ is higher.

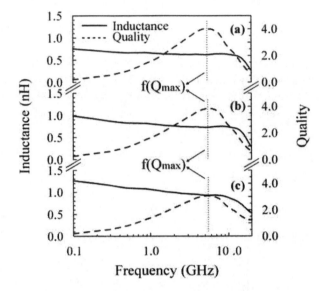

Fig.5 Inductances and quality factors versus frequency of 4-turn α-core inductors, (a) $\#\alpha_2$: 30×60μm², (b) $\#\alpha_3$: 60×60μm², (c) $\#\alpha_4$: 120×60μm². The device operating frequencies $f(Q_{max})$ have been shifted above 5 GHz by using the developed nano/micro granular $Ni_{80}Fe_{20}$ film, while the cut-off frequencies have been shifted above 20 GHz

Inductance and quality factor of three 4-turn α-core inductors $\#\alpha_2$, $\#\alpha_3$, and $\#\alpha_4$ as a function of frequency are shown in Fig.5. The $f(Q_{max})$ of $\#\alpha_2$, $\#\alpha_3$, and $\#\alpha_4$ stay almost constant, while the W_{FM} and the track capacitance has a 4-fold increase. This means the $f(Q_{max})$ is independent on the geometrical size of the devices and does not originate from the *LC* resonance, but mainly determined by the μ_{effc} of the $Ni_{80}Fe_{20}$ core. The *LC* resonance frequencies of $\#\alpha_2$, $\#\alpha_3$, and $\#\alpha_4$ have been estimated above 20 GHz. Compared to the control device with SiO_2 dummy core, $\#\alpha_2$ has a more than 40% increase of inductance from 0.1 GHz to 10 GHz. This indicates the high permeability of the α-core in spite of the granularity. As discussed above, the nano/micro granular $Ni_{80}Fe_{20}$ film α-core has advantages over the β- and γ- cores typically in high frequency applications. The device operating frequencies $f(Q_{max})$ have been reached above 5 GHz for $\#\alpha_2$, $\#\alpha_3$, and $\#\alpha_4$. By comparison to the control devices with SiO_2 dummy core

[4], the cut-off frequencies are well above 20 GHz (out off the measurement range of our equipment). The inductance of the devices increases when their size increases, for example from 0.65nH ($\#\alpha_2$) to 1.05nH ($\#\alpha_4$) at 1 GHz by ~60%, however, the density of inductance IPA and the maximum quality factor decrease. Apparently, there exists an optimized design of device to tailor the device's inductance, the quality factor, and the inductance per area.

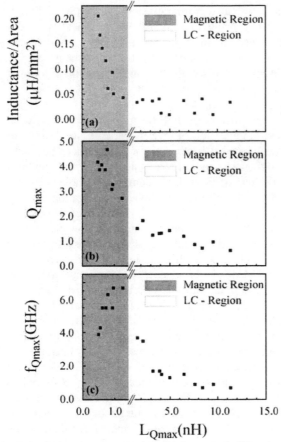

Fig.6 Inductance per area (IPA), Q_{max}, and $f(Q_{max})$ versus $L(Q_{max})$ (the value of inductance where Q shows the maximum). Device parameters are strongly dependent on the magnetic properties of the core film in the magnetic domain, and on the resonance in the LC-Region. The trade-off in inductance per area, Q_{max}, and $f(Q_{max})$ on the device size is demonstrated in the magnetic domain with high IPA, Q_{max} but low $f(Q_{max})$.

Systematic design optimization of inductors has been carried out by varying L_{FM}, W_{FM}, W_L, S_L, and n. The inductance per area (IPA), the maximum quality factor Q_{max}, and the operating frequency $f(Q_{max})$ are compared as a function of $L(Q_{max})$ (the value of inductance where Q shows the maximum), shown in Fig. 6. The *Magnetic Region* marks the domain, within which the devices' performance is mainly dependent on the magnetic properties of the core film, while in the LC-Region, the devices' performance depends on both the magnetic core and the structural LC-resonance. Generally, devices in the *Magnetic Region* have

higher IPA, Q_{max}, and $f(Q_{max})$ than those in the LC-Region. Additionally in the *Magnetic Region*, IPA and Q_{max} are favorite to the small size devices and are traded against the device dimension, while the $f(Q_{max})$ exhibits a maximum at $L(Q_{max}) \sim 1.0$ nH. The trade-off between IPA, Q_{max}, and $f(Q_{max})$ indicates how the full potential of FM films can be exploited by proper design of the solenoid coil and optimum micro/nano patterning of the FM core. After optimizing the design, the inductors with $IPA > 0.20$ $\mu H/mm^2$, $Q_{max} > 4.5$ and $f(Q_{max}) > 6.5$ GHz have been achieved (Fig.6).

4 SUMMARY

A novel nano/micro patterned $Ni_{80}Fe_{20}$ film, formed by using cost-effective electroplating on a Ti seed layer, has been demonstrated for RF applications. Optimum FM-core inductor characteristics can be achieved through a trade-off of IPA, Q_{max}, and $f(Q_{max})$, in combination with an appropriate micro/nano patterning of the FM film in order to reduce the effective conductivity and thus eddy currents in the FM film. Inductors with $IPA > 0.20$ $\mu H/mm^2$, $Q_{max} > 4.5$ and $f(Q_{max}) > 6.5$ GHz have been achieved.

REFERENCES

[1] B. Rejaei, M. Vroubel, Y. Zhuang, J. N. Burghartz, "Assessment of ferromagnetic integrated inductors for Si-technology", Proc. SIRF03, 2003, pp. 100-103.

[2] M. Yamaguchi, T. Kuribara, and K. I. Arai, "Two-port type ferromagnetic RF integrated inductor," *IEEE MTT-S Digest*, pp. 197-200, 2002.

[3] A. M. Crawford et al., IEEE Trans. Magn. 2002, pp. 3168-3170.

[4] Y. Zhuang, M. Vroubel, B. Rejaei, J. N. Burghartz, "Ferromagnetic RF inductors and transformers for standard CMOS/BiCMOS," *Techn. Dig. IEDM2002*, pp. 475-478..

[5] M. Munakata, M. Namikawa, M. Motoyama, M. Yagi, Y. Shimada, M. Yamaguchi, and K. I. Arai, "Magnetic properties and frequency characteristics of $(CoFeB)_x$-$(SiO_{1.9})_{1-x}$ and CoFeB films for RF application," *Transactions of Magnetic Society of Japan*, Vol. 2, pp. 388-393, 2002.

[6] J. Huijbregtse, F. Roozeboom, J. Sietsma, J. Donkers, T. Kuiper, and E. van de Riet, "High frequency permeability of soft magnetic Fe-Hf-O films with high resistivity", J. Appl. Phys., vol. 83, pp. 1569-1574, 1998.

[7] M. Vroubel, Y.Zhuang, B. Rejaei, J. N. Burghartz "Investigation of magnetic microstrips (2): modelling," *Transactions of Magnetic Society of Japan*, Vol. 2, pp. 371-376, 2002.

ACKNOWLEDGEMENTS
Y. Zhuang and M. Vroubel thank the Foundation for Fundamental Research on Matter (FOM), the Netherlands, for financial support.

Two-Regime Hollow PDMS Abbe Prism for μ-TAS Applications

A. Llobera*, R. Wilke* and S. Büttgenbach*

*Institut für Mikrotechnik, TU-Braunschweig,
Alte Salzdahlumer Str. 203, 38124 Braunschweig, Germany, a.llobera@tu-bs.de

ABSTRACT

A new robust and compact system, which comprises a PDMS-based hollow Abbe prism and microlenses, together with self-alignment channels for optical fibre positioning, is presented. This optical detection method has two different regimes for signal transduction, namely absorption and refractive index shift. μM concentrations of fluorescein are detected with this system using standard equipment (SLED and photodetector). The results clearly show the regions in which each physical effect is significant.

Keywords: micro total analysis systems, laser induced fluorescence, optical detection.

1 INTRODUCTION

The application of micromachining to microfluidic systems has been successfully proved in the development of several components, as could be mixers, valves and pumps. In the early 90's, the microfluidic concept was extended with the introduction of a new concept, called miniaturized chemical analysis systems, also known as micro total analysis systems. A scaling down of the analytical systems resulted not only in the reduction in size and price, but also in a significant improvement of the analytical performance [1]. Chip-based capillary electrophoresis (CE) systems have shown to be powerful tools for many promising applications [2],[3]. Very often, laser-induced fluorescence detection, a well-established detector in conventional CE-systems, is the method of choice due to its high sensitivity. For efficient measurements the excitation wavelength is focused into the channel. Pure emission fluorescence wavelength measurements are done by placing the collection system and the photodetector at 90° with respect to the direction of the excitation wavelength. Emission and excitation wavelengths are collected together if the photodetector is placed at 180°. In both cases, these set-ups require a considerable amount of optical components, such as lenses, collimators, mirrors, pinholes and filters that have to be aligned very carefully. In relation to the microfluidic device this assembly becomes fairly bulky. The tackling of this drawback goes together with the use of poly(dimethylsiloxane) (PDMS) finding its way into microfabrication. Recently, the integration of optical fibres and lenses for fluorescence detection by taking advantage of the elastomer's extraordinary properties has been reported, on the basis of forming complex three-dimensional structures from a given mold [4],[5]. Although the use of PDMS helps on solving some optical problems on LIF devices, the multiplexation of the excitation and emission wavelengths is a problem that still remains. So far, they are incorporated by either using conventional optical components (filters) or complex and expensive technological steps [6], [7].

Exploiting the merits of PDMS in combination with a prism-based optical set-up, we were able to design an optical transducer that is suitable for on-chip detection with a high degree of monolithic integration and without requiring the use of filtering. It comprises a hollow prism, that can be filled with the fluid under investigation via two fluidic ports, and a pair of 2D biconvex lenses that allows obtaining parallel beams both at the input and the output. All these components are fabricated in a single process step by replica molding with PDMS. To seal the microfluidic device, the PDMS is bonded to conventional soda-lime glass. Light is coupled into the system through a multimode optical fibre inserted into a channel. This channel has the same width as the optical fibre. Hence, the optical fibres are directly aligned with the biconvex lenses simply by placing them in the input/output channels. To demonstrate the utility of this detector we examined solutions with different concentrations of fluorescein, diluted in buffer solution, in a range between 5 and 1000 μM. Nevertheless, the use of such a hollow Abbe prism cannot be only restricted to fluorescence-based measurements, since it could also be used for absorption detection of chemical species that does not have appreciable fluorescence, or transparent fluids with different refractive index.

2 WORKING PRINCIPLE

One of the best known applications of the prisms is the measurement of the refractive index (RI). For a given prism, by measuring the incident light angle and the so-called minimum deviation [8], the RI as a function of the wavelength can be obtained. This technique is nowadays one of the most precise ways of measuring the refractive indices of substances (including transparent gases and liquids) in a wide range of wavelengths.

Among the different prisms already been designed, we have focused on the Abbe prism due to its simplicity. Its basic structure can be observed in Fig. 1. The Abbe prism is based upon the combination of three prisms (ADE, AEB and BEC), where δ stands for the total deviation of a ray propagating through the Abbe prism.

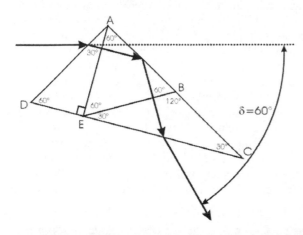

Figure 1: Configuration and ray-tracing in an Abbe prism.

Using the ray-tracing theory, it can be shown that if a monochromatic ray with wavelength λ_1 is injected, with given tilt, in one of the prism's facets, it crosses symmetrically - orthogonal to the AE facet – the ADE prism and is 30° reflected from the AB facet. A total deviation of 60° is obtained after crossing the BEC prism. Hence, the overall system can be thought as a 60° prism (ADE combined with BEC) with the ray propagating in the minimum deviation condition and a second prism that confers the Abbe prism the 60° deviation. With a small rotation of the Abbe prism, the former wavelength is no longer transmitted at 60°, but it is replaced by a second wavelength, λ_2, that matches the conditions for minimum deviation. This behavior causes this type of prisms to be called "constant deviation prisms".

3 DESIGN

After the Abbe prism principles have been described, a new compact detector, shown in Fig. 2, is presented. A hollow Abbe prism is connected to input/output reservoirs. Two multimode optical fibers with PDMS biconvex lenses at the end are defined. With the refractive indices of PDMS and air 1.41 and 1.00, respectively, the distances between the lenses and the optical fibers where chosen so as to have parallel beams at the surface of the prism. The tilt of the optical fibers in relation to the Abbe prism is chosen in order to have maximum intensity for the excitation wavelength (λ=460 nm) when the prism is filled with buffer solution (phosphate buffer, pH 7.4, 10 mM, n=1.334).

The expected behavior of the hollow Abbe prism as the fluorescein concentration, diluted in phosphate buffer, increases, is the following: for low concentration no appreciable RI change of the solution should occur. In this regime, only a decrease of the excitation wavelength peak, without shifting its wavelength, should be observed due to the absorption of light by fluorescein. At high fluorescein concentrations, together with this absorption effect, a change of the RI of the fluid inside the prism is produced, causing a change of the optical path. In turn, this causes the

modification of the collected wavelengths at the output optical fiber. Since the prism is designed for the excitation wavelength (460 nm), the high absorption, together with the modification of the optical path will cause a high increase of the sensitivity for high fluorescein concentrations.

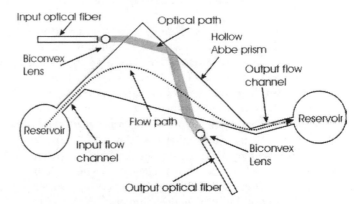

Figure 2: System for fluorescence detection with a hollow Abbe prism (not to scale).

For correct working of the lenses, the optical fiber has to be placed at a given distance from the lenses. Stoppers at the channels fix the distance at which the optical fiber should be placed to obtain parallel beams. Knowing the refractive index of PDMS and air, and designing the radii to be R_1=160 µm and R_2=-R_1, the distance between the fiber and the lens should be S_0=200 µm.

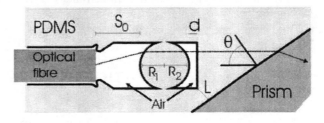

Figure 3: Detailed schematic of the optical fiber channel, the lenses and the prism. Fiber optics are stopped at a distance S_0 which, considering the RI of PDMS and air, together with the curvature of the lens (R_1 and R_2), allows having parallel beams at the biconvex lens output. Lens and channel are tilted an angle θ to have propagation with minimal deviation through the prism. An air gap, with a minimum distance d, separates the PDMS bulk region from the PDMS lens. L remains for the distance between the air gap and the prism in order to avoid leakage.

The propagation through the ADE and BEC prism with minimum deviation also requires the correct tilt of the optical fiber. Considering that the prism is filled with phosphate buffer, a tilt θ=54.8° with respect to the prism wall assures that light is propagating through the prism with minimum deviation. Biconvex PDMS lenses are differentiated from the bulk PDMS by a small air gap, with d=30µm. Since light emerging from the lenses has parallel beams, which have orthogonal incidence to this air gap, no ray deviation is produced at this gap. To avoid leakage, a security distance, fixed to be L=140 μm is left between the closest corner of the air gap and the Abbe prism.

4 FABRICATION

The microfluidic device was fabricated by casting of PDMS (Sylgard 184 elastomer kit, Dow Corning, Midland, MI, USA) against a master made of EPON SU8 (SU8-25, MicroChem Corporation, Newton, MA, USA). In order to obtain a sufficient height that allows a hassle-free insertion of the optical fibers we used a two-step spin-on process for SU8. After spin-coating and drying the negative photoresist was exposed to UV light through a mask. A post-exposure bake was followed by developing the SU8 in propylen glycol methyl ether acetat (PGMEA, MicroChem Corporation, Newton, MA, USA). The resulting master exhibited a height of 250 μm. A prepolymer of PDMS was prepared by mixing the curing agent and the elastomer base in a 1:10 ratio (v/v) and degassing in a vacuum chamber. After pouring the prepolymer over the master the PDMS was cured on a hotplate at 60°C for 1 hour. Prior to bonding the PDMS slab to glass the elastomer was peeled off from the mold and holes were punched out with a stencil for access to the fluidic channels. We used a bonding procedure based on a surface treatment in an oxygen plasma [9]. Both the PDMS slab and the glass were put in a barrel etcher (Surface Technology Systems, Newport, UK) and exposed to an oxygen plasma. Immediately after plasma oxidation the two surfaces were brought in contact and the fluidic system was irreversibly sealed. The total volume of the prism is 1.579 μL.

5 CHARACTERIZATION

The fabricated device can be seen in Fig. 4: light with the excitation wavelength, emitted from an SLED (λ=460 nm), is injected into a multimode waveguide with a core diameter of 200 μm. The optical fibre is placed at the input channel. Readout comprises an identical optical fibre that collects all the light emerging from the prism at 60° and that is connected to a spectrometer (P.117, STEAG MicroParts, Dortmund, Germany) with a spectral resolution of 12 nm and a time of integration of 2.5 sec.

Air entrapment at the unstable upper meniscus can affect the overall properties of the Abbe prism. Although it could be solved by liquid injection at low speed [10], for simplicity and measurement repeatability, the positioning of an auxiliary channel on this meniscus, directly connected to the output port, allows a faster and more homogeneous hollow Abbe prism filling.

Two different regimes can be observed in Fig. 5: For low concentrations (Fig. 5a), only a decrease of the collected intensity as the fluorescein concentration increases can be observed. Since the maximum peak remains fixed at the excitation wavelength, it can be

concluded that, in this region, there are no significant changes on the RI of the solution. Light emitted from the fluorescein is emitted with arbitrary directions and with a wavelength (510 nm) for which the Abbe prism has not been designed. Hence, fluorescence (or light at the emission wavelength) is mainly not collected by the output fibre optics. For high fluorescein concentrations (Fig. 5b), in addition to this absorption effect, a shift of the output intensity peak is obtained. This shift is due to a variation of the RI of the solution inside the hollow Abbe prism. As a result, the optical path is modified and the excitation wavelength no longer fulfils the minimum deviation condition, resulting in a sharp decrease of the measured intensity.

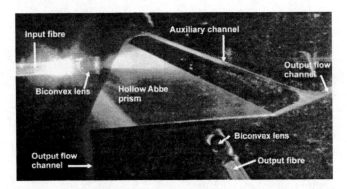

Figure 4: Picture of the system used, displaying the input/output PDMS biconvex lens, the optical fibres, the fluidic channels and the Abbe prism filled with buffer solution + 100 μM fluorescein.

From the previous data and considering the Lambert-Beer law, the absorbance as a function of the fluorescein concentration was calculated. As can be observed in Fig. 6, two regions can be clearly distinguished. For low concentrations, the behaviour of the prism fullfills the Lambert-Beer law, allowing to determine the limit of detection (LOD) to be 4.7±0.1μM. The observed abrupt change on the linear behaviour cannot be explained by only considering fluorescein absorption, since in this region, the RI shift also has to be considered. Due to its non-linear behaviour, no LOD can be calculated, and only the smallest fluorescein concentration (112.5 μm) that gives significant signal of the emission wavelength can be determined.

The confirmation of the RI changes for high fluorescein concentrations is confirmed with the help of Fig. 7. As can be observed, in the pure absorption regime, there is no shift of the excitation wavelength, in accordance to the non-variation of the designed optical path. Conversely, for fluorescein concentrations higher than 100 μM, a shift of the excitation and emission wavelengths collected at the output fibre optics is observed. This shift is caused by the modification of the optical path inside the prism, or, in other words, due to a shift of the RI.

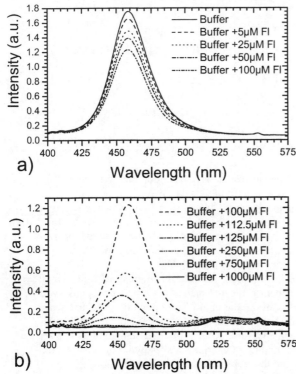

a)

b)

Figure 5: Output intensity as a function of the wavelength for different fluorescein concentrations in phosphate buffer a) at the pure absorption region, b) at the absorption-RI shift region.

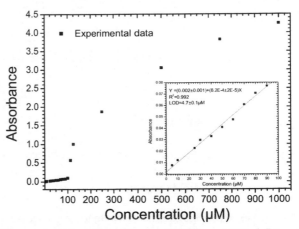

Figure 6: Absorbance as a function of the fluorescein concentration, Inset: Linear region, where the Lambert-Beer law applies.

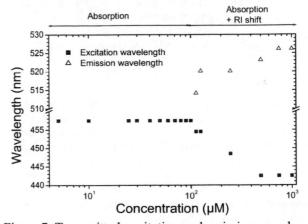

Figure 7: Transmitted excitation and emission wavelength as a function of the fluorescein concentration.

6 CONCLUSIONS

A PDMS-based hollow Abbe prism for detection based on absorption and RI shift has been presented. By the implementation of biconvex lenses and self aligned channels for optical fibers, the external parts of the setup have been reduced to a connected SLED and a spectrometer. Two different working regions can be distinguished: at the pure absorption region (fluorescein concentrations below 100 µM), the Abbe prism matches the Lambert-Beer law, with a calculated LOD of 4.7±0.1 µM. Conversely, for higher concentrations, the RI shift inside the prism cannot be neglected, since it causes a modification of the optical path. As a result, a sharp deviation of the linear behavior is observed. Due to the non-linear behavior, no LOD can be calculated for this region. However the minimum fluorescein concentration detectable, by ways of measuring the emission wavelength, is 112.5 µM.

7 ACKNOWLEDGEMENTS

The authors thank the German Research Foundation (DFG) for support of this work. A. Llobera would also like to thank the AGAUR (catalan council) for his grant Nanotec2002.

REFERENCES

[1] A.Manz, N.Graber, H.M.Widmer. Sens. & Act. B 1, 244-248. 1990.

[2] D.C.Duffy, J.C.McDonald, O.J.A.Schueller, G.M. Whitesides. Anal.Chem. 65, 2637-2642.

[3] J.Wang, M.P.Chatrathi, B.Tian, R.Polsky. Anal. Chem. 78, 2514-2518. 2000.

[4] M.L. Chabinyc, D.T.Chiu, J.C. McDonald, A.D. Stroock, J.F. Christian, A.M. Karger, G.M. Whitesides. Anal. Chem. 73, 4491-4498. 2001.

[5] S.Camou, H.Fujita, T.Fujii. Lab Chip 3, 40-45. 2003.

[6] J.R.Webster, M.A.Burns, D.T.Burke, C.H. Mastrangelo. Anal. Chem. 73, 1622-1626. 2001.

[7] N.A.Lacher, N.F.de Rooij, E.Verpoorte, S.M.Lunte. Jour. Chromat. A 1004, 225-235. 2003.

[8] E.Hecht, A.Zajac. *Optics* Addison-Wesley. 1986.

[9] B.H.Jo, L.M.Van Lerberghe, K.M.Motsegood, D.J. Beebe. Jour. Microelectromech. Syst. 9, 76-81. 2000.

[10] F.Goldschmidtböing, R.Schlosser, S.Schonhardt, P. Woias. Transducers '03, 1883-1886. 2003.

Device Characterization at the Wafer Level via Optical Actuation and Detection

J. Hedley[*], J. S. Burdess[*], A. J. Harris[*] and B. J. Gallacher[*]

[*]Institute for Nanoscale Science and Technology, Newcastle University,
Newcastle upon Tyne, NE1 7RU, England.
John.Hedley@ncl.ac.uk, J.S.Burdess@ncl.ac.uk, Alun.J.Harris@ncl.ac.uk, B.J.Gallacher@ncl.ac.uk

ABSTRACT

This paper investigates the technique of optical actuation of MEMS. It is shown that a device can be driven into all of its vibrational modes simply and non-destructively by a laser pulse, making this an ideal actuation methodology for device characterization at the wafer level. This actuation method is shown to be comparable to a mechanical impulse actuation.

Keywords: optical, actuation, MEMS, characterization

1 INTRODUCTION

It has become a routine operation to perform static and dynamic measurements on micromechanical structures [1] however the problem still exists of actuating these devices prior to the packaging stage. One solution is to mechanical shake the devices with a piezoelectric disk. Although effective, the mechanism by which the device is firmly attached to the piezo disk renders the device unusable after characterisation. An alternative method to actuate devices is optically.

The method of optical actuation has been investigated by several authors. Theoretical analysis has been performed by Fatah [2]. In this work, the heating effect of an intensity modulated laser spot on the centre of a micromechanical beam resonator was considered. It was assumed the laser produced a uniform temperature across the beam's cross-section with a temperature gradient along its length. Solving the one-dimensional heat-flow equation for this system led to a value for the beam's expansion and thus its resonating amplitude.

Experimental results have been obtained on a variety of structures [2-6]. In all these experiments, the actuating light beam was intensity modulated over the frequency range of interest. Although successful, this technique can be time consuming if a narrow bandwidth is being searched for over a large frequency range as is the case of locating a resonance condition with a high Q.

A quicker method is to use an impulse of light which drives all modes into resonance. An impulse response was investigated by Zhang et al [7] to optically drive and sense a micromechanical silicon bridge resonator. The bridge was excited and measured with a focused semiconductor laser which required that the bridge was coated with a metal film thus limiting this approach to general application.

In this work, a pulsed laser is used to excite an uncoated device and a laser vibrometer system used to measure the corresponding vibrations.

2 EXPERIMENTAL PROCEDURE

To actuate the devices, a frequency doubled (532 nm) Nd:YAG laser, pulse width of 4 ns with a 20 Hz repetition rate, was used. The resulting vibrations were detected with a Polytec OFV 501 laser vibrometer whose operational wavelength was 632 nm. The beams were merged with a Nd:YAG second harmonic mirror which reflected all the 532 nm light whilst allowing partial transmission of the vibrometer beam. A red filter was used to account for any stray light which may have damaged the vibrometer system. The signal from the vibrometer was digitized and collected on a computer, the trigger for this collection coming from a light dependent resistor incorporated into the system.

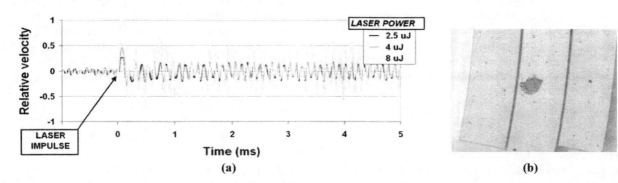

Figure 1: Focused optical impulses produce actuation (a) but resulted in device damage (b).

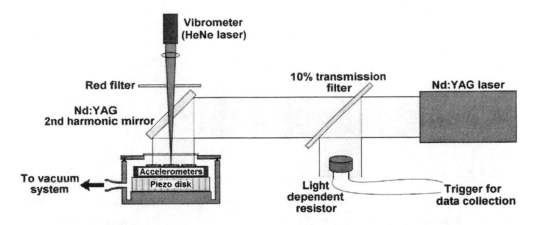

Figure 2: The setup used to optically actuate and measure a device.

In the initial experimental setup, both actuation and sensing beams were focused. The resulting traces for various actuation powers on a vibratory ring gyroscope (under a vacuum of <20 μbar) is given in figure 1(a). Although the device could clearly be seen to be resonating as a result of the impulse, the high power of the Nd:YAG laser resulted in device damage, as can be seen in figure 1(b). The beams were slightly defocused however this problem still persisted. It was concluded that the device was being driven through a conservation of momentum mechanism due to mass ejection rather than a heating effect.

The experimental arrangement was reconfigured so that only the vibrometer beam was focused; this configuration is shown in figure 2. Although this required a much larger working distance for the vibrometer beam, the actuation beam did not damage the device under test. This configuration allows for a much greater actuation power.

3 RESULTS AND DISCUSSION

The unfocused Nd:YAG beam was used to actuate a 300μm × 300μm × 3μm polysilicon accelerometer, see figure 3, the results of which are given in figure 4. The data is an average of 10 traces. As a comparison, the device was attached to a piezo disk and actuated mechanically with a 20 Hz saw-tooth waveform; this simulates the optical impulse drive. The results are also shown in figure 4. The

natural frequencies of the device are clearly visible with both actuation methods. A closer inspection of the peaks reveal that there is a 20 Hz modulation superimposed onto the resonances, this is a consequence of the 20 Hz drive frequency of the impulse.

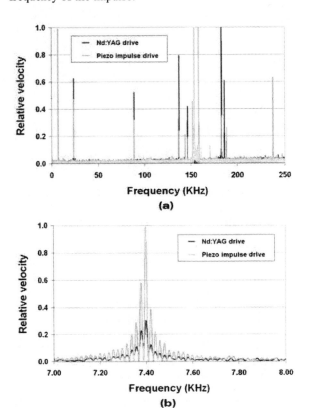

Figure 4: (a) The frequency response of a polysilicon accelerometer excited with mechanical and optical stimuli. The data is an average of 10 traces. (b) The 20 Hz repetition rate of the drive is superimposed onto the data.

Figure 3: The polysilicon accelerometer used in the experiment.

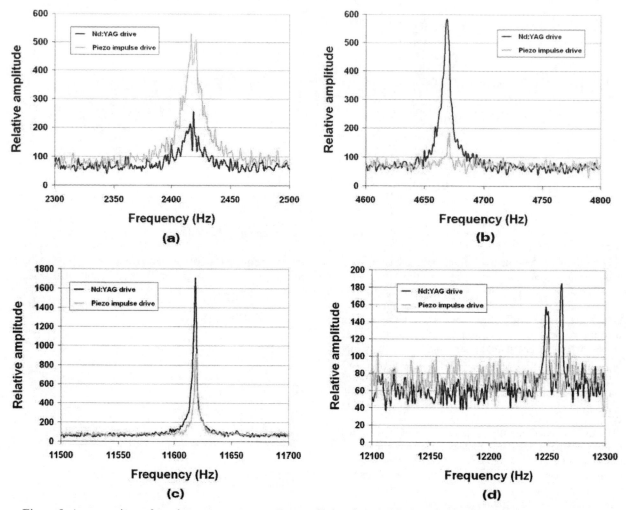

Figure 5: A comparison of accelerometer response to an optical and mechanical actuation (first four modes shown).

The measurements were repeated with another accelerometer at a repetition rate of 0.9 Hz, data collection beginning immediately after the impulse. This has resolved the modulation issue; the data for the first four modes is shown in figure 5. It is noteworthy that the optical and mechanical actuations produce different relative amplitudes between the various modes of vibration. Indeed both of the degenerate modes in figure 6(d) are visible with the optical actuation but not with the piezo drive. This difference in effective actuation is believed to be due to the non-uniform intensity of the light falling onto the device, this will tend to drive degenerate modes (i.e. rocking modes) more efficiently than the piezo whereas the uniform nature of the piezo drive should drive non-degenerate modes (i.e. the bounce mode) more effectively. This work is on-going.

4 CONCLUSIONS

Actuation of devices by a high energy nanosecond pulse has been demonstrated as a practical methodology in the dynamic characterization of MEMS at the wafer level. Being a non-contact technique, the method is both simplistic and non-destructive allowing for rapid testing. This technique has been shown to give comparable response data to currently used actuation techniques.

5 ACKNOWLEDGEMENTS

The authors wish to acknowledge EPSRC for the research grants GR/M31026 and GR/M13756 and the industrial sponsors DERA for supply of the test devices.

REFERENCES

[1] J. Hedley, A. J. Harris and J. S. Burdess, "The development of a 'workstation' for optical testing and modifications of IMEMS on a wafer", Design, Test, Integration and Packaging of MEMS/MOEMS 2001 : Proceedings of SPIE 4408, 402-408, 2001.

[2] R. M. A. Fatah, "Mechanisms of optical activation of micromechanical resonators", Sensors and Actuators A 33, 229-236, 1992.

[3] H. Yu, Y. Wang, C. Ding, Y. Wang and Y. Xu, "The characteristics of point-heating excitation in silicon micro-mechanical resonators", Sensors and Actuators 77, 187-190, 1999.

[4] S. Venkatesh and B. Culshaw, "Optically activated vibrations in a micromachined silica structure", Electronics Letters 21, 315-317, 1985.

[5] M. V. Andres, K. W. H. Foulds and M. J. Tudor. "Optical activation of a silicon vibrating sensor", Electronics Letters 22, 1097-1099, 1986.

[6] D. Uttamchandani, K. E. B. Thornton, J. Nixon and B. Culshaw, "Optically excited resonant diaphragm pressure sensor", Electronics Letters 23, 152-153, 1987.

[7] L. M. Zhang, D. Uttamchandani and B. Culshaw, "Transient excitation of silicon microresonator", Electronics Letters 25, 149-150, 1989.

A MOEMS Electrothermal Grating

E. P. Furlani

Integrated Materials and Microstructures Laboratory, Eastman Kodak Company
Rochester, New York 14650-2011, edward.furlani@kodak.com

ABSTRACT

We present a thermoelastically driven MOEMS light modulator in the form of an ElectroThermal Grating (ETG). The ETG consists of an array of equally spaced bilayer microbeams suspended at both ends above a substrate (Fig. 1). The bilayer microbeams have a conductive/reflective top layer that has highly resistive end sections, and a central portion with low resistance, and a nonconductive bottom layer. In an unactivated state, the microbeams are flat and the ETG reflects incident light like a mirror (Fig. 2a). Light modulation occurs when a potential difference is applied across the conductive layers of alternate microbeams. This causes current to flow through the resistive end sections which, in turn, expand due to joule heating. The top sections expand more than the corresponding bottom sections, which have a lower coefficient of thermal expansion. This causes a thermoelastically induced deformation of the heated microbeams downward, toward the substrate. A diffraction pattern is produced when the heated microbeams deform a distance of $\lambda/4$ from their rest position (λ is the wavelength of the incident light) (Fig. 2b). Since the deformation depends on the level of current, an ETG can operate in an analog mode wherein it selectively diffracts a range of wavelengths. In this presentation, we demonstrate the performance of an ETG. We present an analytical design formula for rapid parametric optimization, and demonstrate ETG viability at operating voltages below one volt.

Keywords: MOEMS grating, electrothermal diffraction grating, thermoelastic grating, electrothermal light modulator

1 INTRODUCTION

The ETG is functionally similar to the Grating Light Valve (GLV) which is an electrostatically driven reflective phase grating [1]. Electrostatic actuation has drawbacks in that the activated microbeams have a limited range of motion due to pull-in (1/3 of the gap between the microbeam and the substrate or ground). This limits analog operation where the microbeams need to be deformed to range of depths to modulate a range of wavelengths. On the other hand, if the microbeams are operated in a contact mode (pulled down to the substrate), then only a single wavelength is efficiently diffracted and contact stiction is an issue. Electrostatic actuation has additional disadvantages: it requires relatively high operating voltages (tens of volts) and narrow operating gaps, which can give rise to deleterious squeeze-film effects. Moreover, performance degradation can occur over time due to charge build up in dielectric layers.

The ETG overcomes these limitations. Specifically, thermoelastically actuated microbeams have a full range of motion (no pull-in), which is ideal for analog operation, and their motion is independent of the gap size. Moreover, there is no need for the microbeams to contact the substrate, and therefore, stiction is not an issue. Additionally, the ETG activation voltage is relatively low (less than one volt).

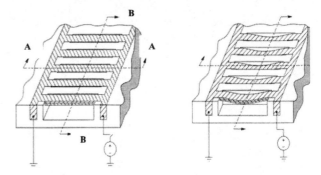

Figure 1: ETG device: (a) unactivated, (b) activated.

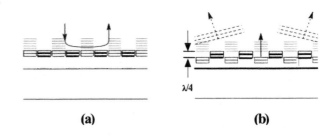

(a)   **(b)**

Figure 2: ETG device: (a) unactivated, (b) activated.

The behavior of the ETG is governed by a coupled system of thermal, thermoelastic, and fluidic (gas flow) equations that cannot be solved in closed-form. Nevertheless, analytical models can be developed for evaluating concept viability.

In the following sections we present a one-dimensional quasistatic thermoelastic analysis for predicting ETG performance, and for performing rapid parametric design studies that take into account material properties, dimensions, and operating conditions. We apply the theory to various ETG designs.

2   QUASISTATIC ANALYSIS

The quasistatic thermoelastic deformation of an N-layer microbeam with heated end sections can be predicted to first-order by assuming that a uniform thermal moment M(T) exists along the end sections, labeled by x_1 and x_2 in Fig. 3b. For a given temperature distribution the deformation is governed by the following thermoelastic equation [2]:

$$D_{eq}\frac{d^4 y}{dx^4} = \frac{d^2 M(T,x)}{dx^2} , \qquad (1)$$

where y(x) is the vertical deformation along the length of the microbeam (Fig. 3b). T is the temperature distribution along the microbeam, which is assumed to be uniform throughout the thickness of the beam, and M(T,x) is the thermally induced moment. The solution to Eq. (1) for a microbeam, which is rigidly supported ($y = \frac{dy}{dx} = 0$) at either end, is given by

$$y(x) = \sum_{j=1}^{2}(-1)^j\left[-\frac{M_j x^2}{2EI} + \frac{R_j y^3}{6EI} - \frac{M\langle x-x_j\rangle^2}{2EI}\right], \qquad (2)$$

where x_j (j=1,2) denotes the edges x_1 and x_2, respectively,

$$M_j = \frac{M(T)\left[2x_j\left(L-x_j\right)-\left(L-x_j\right)^2\right]}{L^2} , \qquad (3)$$

$$R_j = \frac{6M(T)\left[2x_j\left(L-x_j\right)\right]}{L^3} , \qquad (4)$$

and

$$\langle x-x_j\rangle^2 = \begin{cases} 0 & x < x_j \\ \\ (x-x_j)^2 & x > x_j \end{cases} \qquad (5)$$

In these equations

$$M(T) = \frac{C}{B}\sum_{k=1}^{N}\frac{E_k\alpha_k T_k}{1-\nu_k}(h_k - h_{k-1}) - \sum_{k=1}^{N}\frac{E_k\alpha_k T_k}{1-\nu_k}\left(\frac{h_k^2 - h_{k-1}^2}{2}\right), \qquad (6)$$

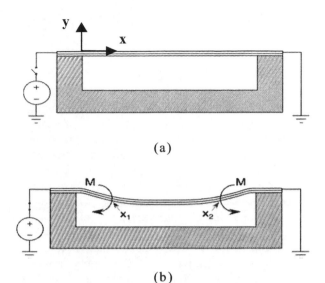

(a)

(b)

Figure 3: Cross-section of microbeam (A-A in Fig. 1).

and

$$D_{eq} = \frac{BD - C^2}{B} , \qquad (7)$$

where

$$B = \sum_{k=1}^{N}\frac{E_k}{1-\nu_k^2}\left(\frac{h_k - h_{k-1}}{2}\right), \quad C = \sum_{k=1}^{N}\frac{E_k}{1-\nu_k^2}\left(\frac{h_k^2 - h_{k-1}^2}{2}\right) \qquad (8)$$

$$D = \sum_{k=1}^{N}\frac{E_k}{1-\nu_k^2}\left(\frac{h_k^2 - h_{k-1}^2}{2}\right),$$

and $E_i, \nu_i,$ and α_i are the Young's modulus, poison ratio, and coefficient of thermal expansion of the i'th layer of the microbeam, h_i is the distance from the surface of the top layer (y=0) to the bottom of the i'th layer, and T_i is the temperature rise in the end sections of the i'th layer ($T_i = \Delta T$ in this analysis).

The solution (2) applies to an N-Layer microbeam with constant cross sectional dimensions, uniform material properties along its length, and negligible heating across its midsection.

The optically active region of an ETG microbeam is the unheated central portion ($x_1 \leq x \leq x_2$). The average deformation of this section is

$$y_{ave} = \frac{1}{(x_2 - x_1)}\int_{x_1}^{x_2} y(x)\, dx . \qquad (9)$$

Therefore, to diffract incident light of wavelength λ, the ETG is designed so that $y_{ave} = \lambda / 4$.

3 DESIGN

There are myriad issues to consider in determining a robust ETG design. Many of these center around the microbeams themselves, their size, spacing, and material properties (structural and thermal). In an ideal microbeam, the resistive end sections should quickly, uniformly, and efficiently heat when activated, and rapidly dissipate their heat to the substrate between activations. Moreover, the end sections should have a high resistance relative to the midsection to maximize thermal deformation. In addition, since the midsection is the optically active region, it should be highly reflective over the entire range of operating wavelengths. Some different microbeam configurations are shown in Fig. 4.

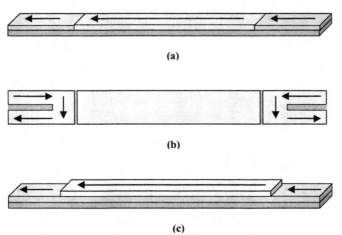

(a)

(b)

(c)

Figure 4: Microbeam configurations showing layer structure and current flow: (a) bilayer structure with low resistance midsection, (b) top view of bilayer structure with isolated midsection (current through end sections only), (c) trilayer structure with low resistance shunt layer.

In Fig. 4 (a) the microbeam has two layers: a nonconductive bottom layer, and a top layer with a conductive/reflective midsection and resistive end sections. For example, the bottom layer could be low stress silicon nitride, and the top layer could have an aluminum midsection with TiAl or TiN end sections. If the structural properties of the top layer vary along its length, then Eq. (2) does not rigorously predict the deformation. Nevertheless, it still can be used to estimate upper and lower bounds for the beam deflection for a given thermal load. It is instructive to determine the temperature rise (ΔT) required to deform this structure to diffract light of wavelength $\lambda =$

520 nm. Consider a microbeam that is 3 µm wide and 80 µm long with a 0.25 µm thick TiAl/Al/TiAl top layer and an equally thick, low stress, SiN bottom layer. The end sections are 15 µm long resulting in a 50 µm long optically active midsection. The CFDRC ACE+ multiphysics program was used to compute the deformation of this structure (www.CFDRC.com). The analysis shows that a temperature rise of $\Delta T = 14°C$ along (and confined to) the end sections of the beam is sufficient to produce an average deformation of $y_{ave} = 130$ nm in the optically active region (Fig. 5). The voltage and current required to achieve this ΔT can be crudely estimated using $V = \sqrt{mc\Delta T R / \Delta t}$, where V is the voltage across the microbeam, R is the series resistance of both end sections (midsection ignored), Δt is the duration of the heat pulse (taken to be 1 µs), m is the total mass of both end sections (top and bottom layers), and c is the weighted average of the heat capacities of both end section layers. The voltage, current, and energy required to obtain $\Delta T = 14°C$ are estimated to be 294 mV, 5.2 mA, and 1.53 nJ, respectively.

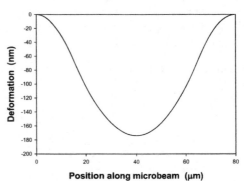

Figure 5: Deformation profile for the microbeam structure of Fig. 4 (a) ($\Delta T = 314°C$).

In Fig. 4 (b) the microbeam has uniform material properties except for the gap regions that separate (electrically isolate) the midsection of the top layer from its end sections. During activation, the end sections experience joule heating and the gaps slow the diffusion of heat into the center of the beam. The midsection is coated with a thin (40 nm) layer of aluminum (not shown) so that it will have a high reflectance. The deformation of this structure can be estimated using Eq. (2). Consider a 5 µm wide, 80 µm long microbeam with 0.25 µm thick layers: TiAl on top and low stress SiN on the bottom. Each end section is 15 µm long with a 1 µm wide gap centered width-wise that extends 13 µm along its length starting at the support edge. This gives the end section a U-shaped conductive path. There are 1 µm wide gaps separating the end sections from the midsection, which results in a 48 µm long optically active region. The deformation of this microbeam was predicted using both the analytical model with no gaps, and the

CFDRC numerical program with the gaps included. A comparison of these results is shown in Fig. 6. In this case, $\Delta T = 19^{\circ}C$ yielded $y_{ave} = 130$ nm for diffracting light at

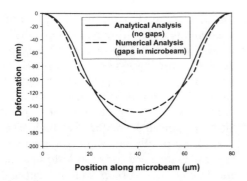

Figure 6: Deformation profile for the microbeam structure of Fig. 4. (b) with 1 μm gaps ($\Delta T=19^{\circ}C$).

$\lambda = 520$ nm. Notice that the gaps produce a flattening of the midsection, which should improve diffraction efficiency relative to the no gap case. This same flattening occurs in electrostatic light modulators when there is only partial electrode coverage [3]. The ΔT of 19° C seems very modest but could have been even lower if the optically active midsection were aluminum (E = 70 GPa) rather than TiAl (E=187 GPa). The voltage required to obtain $\Delta T = 19^{\circ}C$ can be estimated using $V = \sqrt{mc\Delta TR / \Delta t}$ where V is the voltage across one of the U-shaped heater elements, R is its resistance, Δt is the duration of the heat pulse (taken to be 1 μs), m is the mass of the end section (both layers), and c is the weighted average of the heat capacities of both end section layers. For the case studied, the voltage and current required for $\Delta T = 19^{\circ}C$ are 419 mV and 4.13 mA, respectively. The energy required to activate the microbeam is 3.36 nJ. The operating voltage, current, energy and ΔT depend on the length, thickness, and material properties of the microbeam. A parametric analysis of this structure with the TiAl and SiN materials is shown in Fig. 7. The thickness of both layers was set to H, the end sections were fixed at a length of 15 μm, and $y_{ave} = 130$ nm. The analysis shows that ΔT (and hence energy) decrease with length and thickness as expected.

In Fig. 4 (c) the microbeam has three layers: a bottom support layer, a highly resistive middle layer, and a partial top layer that is reflective and has low resistance. The top layer shunts the current that would otherwise pass through the central portion of the middle layer. Thus, it greatly reduces joule heating in this section. A numerical simulation of this structure was performed in order to determine a viable operating temperature. A 3 μm wide, 80 μm long microbeam was analyzed. The microbeam has

three 0.25 μm thick layers: a 50 μm long Al top shunt layer, a TiAl middle layer, and a low stress SiN bottom layer. The end sections are 15 μm long. The analysis showed that $\Delta T = 25^{\circ}$ C was required to produce $y_{ave} = 130$ nm. This ΔT is higher than in the previous two cases because the microbeam has three layers as compared to two, which makes it less compliant.

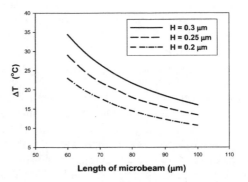

Figure 7: ΔT vs L for the microbeam structure of Fig. 4 (b) ($y_{ave} = 130$ nm, H = layer thickness).

4 CONCLUSIONS

The ETG shows promise for use in low frequency light modulator applications. Preliminary quasistatic calculations suggest that light modulation can be achieved with modest temperature increases in the active elements ($\sim 20^{\circ}C$). The ETG should operate at low voltages (< 1V), and small currents (< 10 mA). Further research is needed to characterize the ETG's frequency response, damping, optical efficiency, and thermal management.

ACKNOWLEDGEMENTS

The author would like to thank J. A. Lebens of Eastman Kodak for valuable discussions on material properties and device fabrication, and Amit Saxena of CFDRC Corp. for his suggestions on structural analysis using the CFDRC program.

REFERENCES

[1] R. B. Apte, "Grating Light Valves for High Resolution Displays," Ph.D. Dissertation, Standard University, 1994..

[2] K. S. Pister and S. B. Dong, "Elastic Bending of Layered Plates," J. Eng. Mech. Div. Proc. Amer. Soc. Civil Eng., 2194, 1959.

[3] E. P. Furlani, E. H. Lee and H Luo, "Analysis of Grating Light Valves with Partial Surface Electrodes," J. Appl. Phys., **83**, 629, 1999.

Practical Techniques for Measuring MEMS Properties

J. V. Clark, D. Garmire, M. Last, J. Demmel, S. Govindjee

University of California at Berkeley, Berkeley, CA, USA, jvclark@eecs.berkeley.edu

ABSTRACT

We propose practical analysis techniques to accurately measure geometric, dynamic, and material properties of MEMS. Analytical methods and test structures are presented to extract over a dozen properties by electric probing in a minimal chip area. Geometric properties include fabrication error with respect to layout geometry such as beam widths, gap spacings, etch holes, and beam lengths. Dynamic properties include mass, damping, stiffness, bulk compliance, quality factor, exponential damping factor, displacement amplitude, velocity amplitude, comb drive force, and fringing field factor. Material properties include Young's modulus, residual stress, shear modulus, Poisson's ratio, and material density. These techniques differ from those currently available in practicality and that no geometric, dynamic, or material property values are presumed. Only capacitively-based measurements are required.

Keywords: property extraction, characterization, metrology

1 INTRODUCTION

Although microsystems technology is becoming a well-established industry, testing standards for MEMS properties are still under development. S. Brown [1] states that the industry has suffered due to an absence of standard testing methods, making it difficult for manufacturers to specify their materials and for customers to specify their needs. J. Marshall [2] states that industrially accepted standards are necessary to facilitate international commerce. Just recently, the first three MEMS testing standards were offered by the National Institute of Standards and Technology (NIST) and the American Society for Testing and Materials (ASTM) for beam length and residual stress [2]. A difficultly in developing MEMS testing standards is due to the numerous uncertain properties of MEMS. These uncertainties may cause testing methods to yield only a couple of significant digits of accuracy, and different methods can yield different results [7]. Additionally, microscale properties are extremely sensitive to process variations. Accuracy and precision of standard processes are difficult to control – properties vary by recipe, fabrication run, wafer-to-wafer, and across the wafer.

Although there are few testing standards, there are plenty of nonstandard testing methods available in the MEMS literature. Most methods are expensive with respect to cost and time, which make them nonideal for a competitive industry. Most methods inevitably rely on unconfirmed properties, which reduce confidence in the results. Examples of previous efforts include full chips of test arrays [3-4], vacuum chambers to reduce damping [5], scanning electron microscopy for geometry [7], large structures or highly specialized equipment for stress [2, 6], and FEA-based geometrical extraction [11]. W. Sharpe Jr. provides an extensive overview [7]. Due to the impracticality of most methods, properties are often coarsely measured, or presumed from the literature. The properties of a device are typically left as roughly-known, and the device is electronically tuned for good performance. Although electronic tuning has worked well for getting simple systems to market quickly, progress toward practical and accurate measurement techniques has been slow. Measurements of fundamental properties are critical to the study of complex phenomena, and to the improved modeling and prediction of more complex microsystems.

To address the above issues, we propose practical analysis techniques which do not rely on presumed properties. Our techniques differ from previous work as follows. Instead of obtaining one to three properties from large arrays of structures, we obtain over a dozen properties from small test sets. We refine the numerous uncertainties by deducing fundamental properties through measurement. For precision, measurements are based on capacitance [10], which is the most sensitive type of sensing to date. As an exceptional example, capacitive changes on the order of zeptofarads from the deflection of beams that have displaced a few femtometers have been reported [12]. We also identify the first use of velocity resonance in MEMS as a substitute for resonance in vacuum. The following sections provide the theoretical basis and formulas for property measurements.

2 FABRICATION VS LAYOUT

The entire analysis is centered on a basic test structure that was fabricated in the standardized poly-MUMPs process. Properties are extracted by analyzing the effect of simple modifications to the design. It is assumed that the only run data provided is layer thickness h which is typically the only well-controlled process parameter. An optical image of the test structure is shown in Figure 1. The simplicity of the design allows properties to be more clearly targeted and increases the workability. Two SEM insets show magnified views of geometrical nonidealities such as "webbing" instead of sharp $90°$ corners at all beam-to-base interfaces, coarse sidewalls instead of smooth surfaces, beam widths which have recessed ~20%, and gaps which have increased ~20%. Base compliance in the plate and anchor, labeled as κ_{bc} in the figure, permits a planar rotation of $\Delta\phi$ at the beam-to-base interface such that support beams do not meet the base at right angles upon bending. This effect is magnified by beam length. The webbing effect appears at all acute corners. Its presence at the interface between the support beams and the plate and anchor increases the apparent stiffness of the support beams by κ_{web}. Furthermore, enlarged gaps reduce capacitance and comb drive force; thinned beams reduce system stiffness; and residual stress offsets comb finger alignment. These effects are considered below.

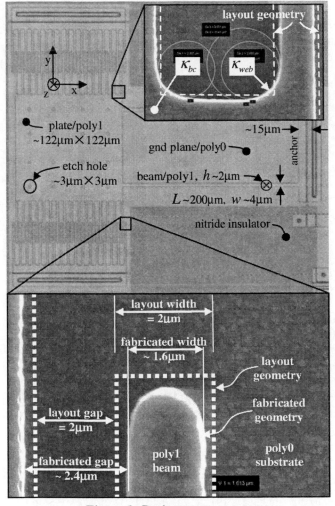

Figure 1: Basic test structure.

3 THEORY

If device components are fabricated within relative close proximity to each other and have the same orientation such that deposition, lithographic, and etchant reactions among corresponding components are alike, then it is safe to assume that material properties and geometric inaccuracies are locally consistent. We assume that small deflection theory applies such that the system can be effectively modeled as a 1D second order nonhomogeneous ODE with constant coefficients. However, we do not assume a parallel plate approximation for the comb drive.

3.1 Measuring planar geometry

We begin by extracting the fabrication error in beam width Δw, which we define as the difference between the layout width and the fabricated width. Let there be two test structures, a and b, of the type shown in Figure 1. Assume that all layout geometries are identical except for the widths of the parallel support beams. Let the layout beam widths of a and b differ by a multiplicative layout parameter n_b, such that $w_{b,layout} \equiv n_b w_{a,layout}$. We express the fabricated widths in terms of layout and error as,

$$w_a = w_{a,layout} + \Delta w, \quad \text{and} \quad w_b = n_b w_{a,layout} + \Delta w. \quad (1)$$

The case of a rigid plate and anchor is considered first. If a potential ΔV_0 is applied to a and b, the balance of static equilibrium $F_{spring} = F_{electrostatic}$ is,

$$k_{eff,a}\Delta y_a = \frac{1}{2}\Delta V_0^2 \frac{\Delta C_a}{\Delta y_a}, \quad \text{and} \quad k_{eff,b}\Delta y_b = \frac{1}{2}\Delta V_0^2 \frac{\Delta C_b}{\Delta y_b} \quad (2)$$

where k_{eff} is the effective stiffness, and ΔC is the linear change in capacitance of the comb drive due to a small lateral displacement Δy. Since a and b have identical comb drives, then their electrostatic forces are identical as well. Equating the expressions in (2) and taking their ratio to be unity, we find

$$\frac{\Delta C_a}{\Delta C_b} = \frac{w_b^3}{w_a^3} = \frac{\left(n_b w_{a,layout} + \Delta w_{eff}\right)^3}{\left(w_{a,layout} + \Delta w_{eff}\right)^3}, \quad (3)$$

where $k_{eff} \propto w^3$. Solving for the effective error, we have

$$\Delta w_{eff} = \left(\frac{n_b \left(\Delta C_b/\Delta C_a\right)^{1/3} - 1}{\left(\Delta C_b/\Delta C_a\right)^{1/3} - 1}\right) w_{a,layout}. \quad (4)$$

If compliancy in the plate and anchor is considered, then the effective stiffness can be dissected as

$$k_{eff} = k + 4\kappa \quad (5)$$

where k is the stiffness of the parallel beam supports (18), and κ is defined as $\kappa \equiv \kappa_{bc} w^2 + \kappa_{web}$ for each beam-to-base interface [13]. The additional unknowns κ_{bc} and κ_{web} require two additional test structures that differ from a and b in layout beam width only. Then by taking ratio combinations and equating forces, the quantities κ_{bc}, κ_{web}, and Δw, may be expressed solely in terms of changes in capacitance and layout geometry [13].

Although the effective error Δw_{eff} and the actual error Δw are fitted to match experiment, Δw_{eff} absorbs base compliancy, while Δw filters it out to capture the average width of the beam. Similar arguments apply to k_{eff} and k.

From the introductory assumptions, it can be inferred that the error in beam width identically applies to all planar geometries as well. Such as the lengths of support beams, gaps, and cantilevers are respectively,

$$L = L_{layout} - \Delta w, \quad g = g_{layout} - \Delta w, \text{ and } L_f = L_{f,layout} \quad (6)$$

where cantilevers are simply translated toward their base. The areas of rectangular plates and comb fingers are

$$A_{plate} = \left(L_{1,layout} + \Delta w\right)\left(L_{2,layout} + \Delta w\right) - N_H \left(L_H - \Delta w\right)^2,$$

$$A_{comb} = N_f L_{f,layout} \left(w_{layout} + \Delta w\right), \quad A_{beam} = \left(L_{layout} - \Delta w\right)\left(w_{layout} + \Delta w\right) \quad (7)$$

where N_H and N_f are the numbers of holes and fingers. The fabricated geometry (6)-(7) with respect to error Δw is exemplified in Figure 2.

3.2 Measuring dynamic properties

Modifying the basic test structure to include displacement stops (labeled as ds in Figure 2) allows a precisely defined displacement to be obtained. Upon closure of this gap, the comb drives will be displaced by

$$\Delta y_{ds} = g_{layout} - \Delta w, \quad (8)$$

which corresponds to a change in capacitance of ΔC_{ds}.

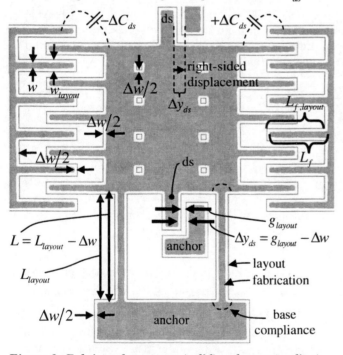

Figure 2: Fabricated geometry (solid) vs layout (outline)

The potential that is applied to close the gap will most likely be greater than what is necessary. The precise voltage required to displace Δy_{ds} may be obtained by reverse-biasing the system such that the structure deflects in the opposite direction, which is not limited by the displacement stop. Upon reaching an identical change in capacitance ΔC_{ds}, the potential that produces this change ΔV_{ds} corresponds to the prescribed deflection in (8). Hence, a measurement of the comb drive force is revealed as

$$F_0 = \frac{1}{2}\Delta V_{ds}^2 \frac{\Delta C_{ds}}{\left(g_{layout} - \Delta w\right)} \quad (9)$$

which includes comb finger misalignments and fringing field effects. Designers often estimate the comb drive force using the parallel plate approximation, $\alpha\frac{1}{2}V^2 N_f h / g_{layout}$, where $\alpha \approx 1.12$ is the fringing field correction factor. Using (9), α may be precisely measured as $\alpha = \Delta C_{ds} g_{layout} / N_f h \varepsilon_0 \left(g_{layout} - \Delta w\right)$.

Now let ΔV_{ds} be applied sinusoidally such that (9) is the amplitude of the driving force. Then the lumped-model equation of motion for the system is represented by

$$m_{eff}\ddot{y} + d_{eff}\dot{y} + k_{eff}y = F_0 e^{i\omega t}, \quad (10)$$

where d_{eff} includes damping of the plate, comb drives, and support beams, and ω is the frequency of the voltage source. It can be shown [13, 8] that the displacement and velocity amplitudes are maximized when the driving frequency equals the displacement and velocity resonance frequencies, ω_r and $\omega_0 = \sqrt{k_{eff}/m_{eff}}$ respectively. These resonances are related to the natural frequency ω_d by

$$\omega_r^2 = 2\omega_d^2 - \omega_0^2 \quad (11)$$

where $\omega_d^2 \equiv \omega_0^2 - \left(d_{eff}/2m_{eff}\right)^2$, $\omega_r^2 \equiv \omega_0^2 - 2\left(d_{eff}/2m_{eff}\right)^2$. Displacement resonance may be measured by maximizing the amplitude of the output signal of the capacitive divider as a function of frequency. Velocity resonance may be measured by maximizing the amplitude of the output signal of a capacitive divider that is sent through a differentiator as a function of frequency [13]. It is important to note that only the velocity resonance in (11) is independent of damping – even for critically damped systems. It is also shown that damping is related to the displacement amplitude y_0 and natural frequency by

$$d_{eff} = F_0 / y_0 \omega_d \quad (12)$$

where the displacement and velocity amplitudes are

$$y_0 = F_0 \Big/ 2\gamma m_{eff}\sqrt{\omega_0^2 - \gamma^2}, \text{ and } \dot{y}_0 = F_0 / 2\gamma m_{eff} \quad (13)$$

Realizing that (2), (9), (12), and (13) are expressions relating to the same force amplitude in (10), we measure the effective damping, mass, and stiffness as

$$d_{eff} = \frac{\Delta V_{ds}^2 \Delta C_0}{2 y_0^2 \omega_d}, \quad m_{eff} = \frac{d_{eff}}{2\gamma}, \text{ and } k_{eff} = m_{eff}\omega_0^2 \quad (14)$$

where the exponential damping factor is defined as $\gamma^2 \equiv \left(\omega_0^2 - \omega_r^2\right)/2$, and we have assumed that the ratio of the change in capacitance to displacement for over the linear range of the comb drive is the constant $\Delta C_0 / y_0 = \Delta C_{ds} / \Delta y_{ds}$. γ^{-1} is the time for which the amplitude of motion decays by a factor of e^{-1} once the driving force is removed.

3.3 Measuring material properties

Now that the geometric and dynamic properties have been determined, material density can be found by taking the ratio of the effective mass to the effective volume,

$$\rho = \frac{m_{eff}}{v_{eff}} = \frac{m_{eff}}{h\left(A_{plate} + A_{comb} + 2 \times \frac{13}{35} A_{beam}\right)} \quad (15)$$

where geometry (7) and a Rayleigh-Ritz method [13] is used to find the effective volume of the system.

We define the quality factor of the system as

$$Q \equiv \frac{\omega_d}{\sqrt{\left(\omega_0^2 - \omega_r^2\right)/2}} = \frac{m_{eff}\omega_d}{d_{eff}} . \quad (16)$$

The Young's modulus of the material can be found by substituting geometry, k_{eff}, and κ into equation (5), [13]. By absorbing the effects of base compliance, an effective Young's modulus is more simply expressed as

$$E_{eff} = \frac{k_{eff}\left(L/2\right)^3}{3I} = \frac{k_{eff}\left(L_{layout} - \Delta w\right)^3}{2h\left(w_{a,layout} + \Delta w\right)^3} . \quad (17)$$

where I is the second moment of area of the fitted model.

Once Young's modulus is uncovered, we find the stiffness of the parallel beam supports to be

$$k = 2Eh\left[\left(w_{layout} + \Delta w\right)/\left(L_{layout} - \Delta w\right)\right]^3 . \quad (18)$$

The shear modulus G is related to Poisson's ratio ν by $G \equiv E/2(1+\nu)$. Assuming small deflection theory applies, shear rotation at the beam-to-base interface is [13]

$$\Delta\phi = 4LF_0/\pi hGw_a^2 = \Delta y_{ds}/L \quad (19)$$

which leads us to a shear modulus and Poisson's ratio of,

$$G = \frac{4L^2F_0}{\pi h\Delta y_{ds}\left(w_{a,layout} + \Delta w\right)^2} , \text{ and } \nu = \frac{E}{2G} - 1 . \quad (20)$$

Lastly, planar residual stress may be obtained by capacitively measuring the stress induced elongation of a beam. If the structure in Figure 1 is modified such that one comb drive is replaced with a beam that laterally extends for some length L_σ before it is anchored, then its change in length Δy_σ may be detected by a change in capacitance ΔC_σ in the remaining comb drive, where the unmodified test structure is the reference. Since the Δy_σ is related to capacitance and stress by $\Delta C_\sigma/\Delta y_\sigma = \Delta C_{ds}/\Delta y_{ds}$ and $\sigma = E\Delta y_\sigma/L_\sigma$, we write the stress and elongation as

$$\sigma = \frac{E\Delta y_{ds}\Delta C_\sigma}{\left(L_{\sigma,layout} - \Delta w\right)\Delta C_{ds}} , \text{ and } \Delta y_\sigma = \frac{\Delta C_\sigma}{\Delta C_{ds}}\Delta y_{ds} . \quad (21)$$

If planar stress is significant, then large geometric quantities in (6)-(7) must be adjusted accordingly.

4 CONCLUSION / FUTURE

Practical analysis techniques for measuring over a dozen MEMS properties have been proposed. Accuracy was addressed by formulating all properties in terms of precise electrical measurements and exactly known layout geometry. Confidence was addressed by uncovering the underlying dependencies. Possible benefits of these methods are as follows. Small test sets reduces chip real-estate costs and allows structures to be positioned alongside primary MEMS devices for localized testing. Electronic probing allows packaged testing to be performed in-the-field, such as monitoring the changes in properties due to the environment, fatigue, and recalibration after long-term dormancy. These methods may be applied to other designs.

Future work includes experimental verification, and the analysis of sensitivities and possible errors due to small deflection theory. Length and stress measurements will be verified by ASTM standards. The experimentally-accurate lumped models can be used in Sugar [13] and verified by comparing simulations to the performance of more complex designs on the chip. Measurements of other properties including nonlinear effects are being considered as well.

REFERENCES

[1] Small Times, "MEMS Standards, While Small, May Mean Much for the Industry," Aug 26, 2003

[2] J. Marshall, NIST, Gaithersburg, Md. www.astm.org/ SNEWS/OCTOBER_2003/micro_oct03.html

[3] P. M. Osterberg, S. D. Senturia, "M-Test: A Test Chip for MEMS Material Property Measurement Using Electrostatically Actuated Test Structures," JMEMS, June 1997, v.6, no.2:107-118

[4] H. Guckel, et. al., "Diagnostic Microstructures for the Measurement of Intrinsic Strain in Thin Films," J. Micromech. and Microeng., 6 1992, v.2, no.2:86-95

[5] K. E. Peterson, et al., "Young's Modulus Measurements of Thin Films Using Micromechanics," J. Appl. Physics, 50, 1979, pp6761-6766.

[6] L. Lin, et. al., "A Micro Strain Gauge with Mechanical Amplifier" JMEMS, 6, 1997, pp313-321.

[7] W. Sharpe Jr., "Mechanical Properties of MEMS Materials" The MEMS Handbook, 2002, Ch3 pp1-33.

[8] G. R. Fowles, "Analytical Mechanics," Saunders, 1990

[9] S. D. Senturia, "Microsystem Design," Kluwer, 2001.

[10] M. Elwenspoek, R. J. Wiergerink, Mechanical Microsensors, Berlin, New York, Springer, 2001.

[11] R. K. Gupta, "Electronically Probed Measurements of MEMS Geometries," JMEMS v.9, no.3, 2000, p380.

[12] J. Green, et. al., "New iMEMS Angular Rate-Sensing Gyroscope," Analog Devices, Dialogue, 37-03, 2003.

[13] J. V. Clark, dissertation in progress, "Modeling, Simulation, and Design of Complex Microelectromechanical Systems," U.C. Berkeley, CA.

Pumping Capacity and Reliability Study on Silicon-Based Cryogenic Micro Pump

Yi Zhao, Biao Li, Daryl Ludlow, and Xin Zhang

Department of Manufacturing Engineering and Fraunhofer USA Center for Manufacturing
Innovation, Boston University; 15 Saint Mary's Street, Boston, MA 02215
E-mail: achelous@bu.edu; Tel: 617-358-1631; Fax: 617-353-5548

ABSTRACT

As part of a program to develop a cooling system for satellite instrumentation capable of working at cryogenic temperature, we present herein pumping capacity and material reliability study on a silicon-based cryogenic micro pump which will be applied into a future cooling system to satisfy both active and remote cooling requirements.

A test rig for actuating silicon diaphragm, the main functional component of the micro pump, was built by using compressive gas actuation. A Dewar was utilized to cool the diaphragm down to cryogenic temperature. The deflection of silicon diaphragms was measured using both WYKO and ZYGO interferometer. As a result, pumping capacity was derived. The maximum deflection of the silicon diaphragm was found to vary linearly with differential pressure, and the pumping capacity decreased at the cryogenic temperature. Additionally, micro-Raman spectroscopy was employed for stress mapping. As expected, the diaphragm edge centers are most vulnerable to fracture. Finally, a cryogenic fatigue test was conducted. The diaphragm suffered no damage during 10^6 cycles for ~10 days, thereby demonstrating the viability of the silicon-based system for space applications.

Keywords: micro pump, pumping capacity, silicon reliability

1. INTRODUCTION

Satellite instrumentation suffers from lack of heat dissipation [1-3]. Heat is basically generated by the electronic systems, and if exposed, by radiation absorption as well. Therefore, the heat sensitive instrumentation needs an active cooling system, other than natural gas flow over devices or passive atmospheric heat exchanger, in order to maintain at an acceptable operating temperature [4, 5].

Several solutions have been employed. NASA has applied a capillary pumped loop (CPL), utilizing a turbo pump for cooling infrared cameras on the Hubble telescope [6]. Another current option is the heat pipe technology [7], which has been used extensively on space-based missions.

Fulfilling the active cooling requirement, however, these current options are not compatible with remote cooling without some sort of additional pump loop. Some satellite instrumentations require as many degrees of freedom as possible, in order to move and rotate smoothly. A silicon-based micro pump array is desirable due to the cost associated

with weight and volume in space based applications. The concept shown in Figure 1a will serve in our proposed cooling system, connecting with a long coiled tubing to provide the flexibility so as to transfer the cryogen from a cryocooler to the instrumentation.

The cryogenic micro pump consists of multiple silicon wafers, making up the diaphragm (Figure 1b). It is expected that the pumping capacity depends on the deformation of the diaphragm, and the system viability associates with diaphragm fracture and fatigue.

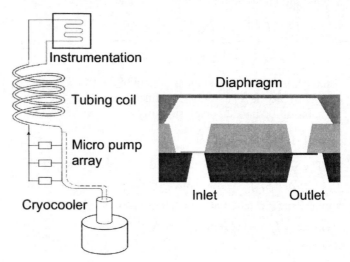

Figure 1: Schematics of cooling transport loop using micro pump array (a) and silicon-based micro pump design (b).

2. EXPRIMENTAL SETUP

The realization of the pumping capacity requires some sort of actuation mechanism to deform the silicon diaphragm. These actuation approaches include piezo disc [8], differential pressure [9], electrostatic [10], thermopneumatic [11], and etc. Unfortunately, most of them are either incompatible with the cryogenic operation requirements or too complicated to be integrated with the system. In this work, differential pressure actuation was utilized due to its simplicity and cost efficiency.

2.1 Pumping rig

The compressive gas was first desiccated to avoid gas pipe blockage due to ice formation. Then the gas was led through a solenoid valve to adjust the differential pressure. An optional three-way valve was used for fast switching in the dynamic deflection measurement. The silicon diaphragm was clamped in a two-part sample holder using bolts. The upper part has an opening in the center to allow the interferometer measurement. The compressive gas reaches the backside of diaphragm through the chamber of the lower part for actuation.

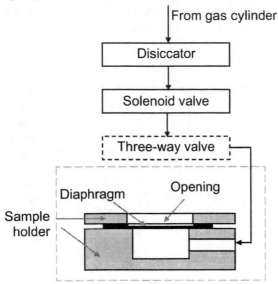

Figure 2: *Pumping rig for diaphragm actuation.*

2.2 Cryogenic Dewar

In order to achieve a space-based operation environment, a Dewar test rig was designed and further utilized for cooling the silicon-based diaphragm to cryogenic temperature (Figure 3). As shown in the figure, the sample holder was first mounted on the cold finger and then aligned to the quartz window. Prior to each measurement, cryogen (liquid nitrogen) was fed into the inner layer, and the dewar was pumped to vacuum in order to reduce the heat transfer and to keep the temperature in a narrow range.

In this work, long coil tubing was winded onto the inner layer to cool the compressive gas to cryogenic temperature before it reached the sample holder. With such configuration the gas volume change with temperature was minimized.

3. EXPERIMENT AND RESULTS

3.1 Diaphragm preparation

Square-shape diaphragms were fabricated on 4", single crystalline silicon wafers using standard microfabrication process (Figure 4). An array of indicators was designed along with the diaphragm to indicate the etch depth. The thickness of then diaphragm was measured using Fourier Transform Infrared spectroscopy; the lateral dimension of the diaphragm was examined using a microscope with accurate grids.

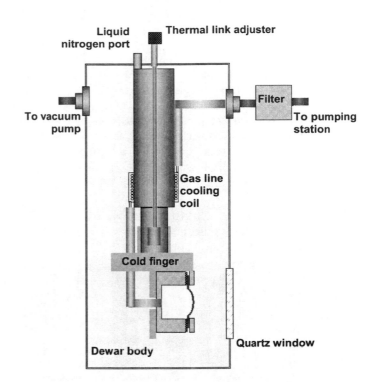

Figure 3: *Schematic of Cryogenic Dewar Setup.*

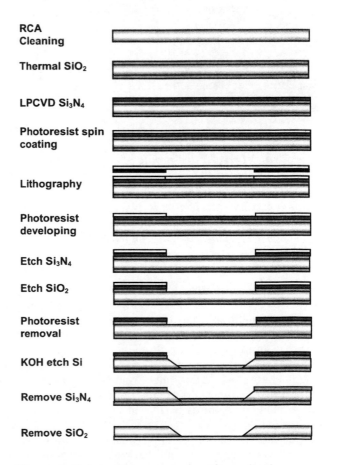

Figure 4: *Fabrication process for silicon diaphragm.*

3.2 Deflection mapping

It is expected that the pumping capacity depends on the actuation frequency and volume stroke of silicon diaphragm during single cycle. The volume stroke can be obtained by integrating the deflection over the diaphragm region.

In order to measure deflection distribution, WYKO interferometer was applied. Matching with theoretical analysis, this result indicates a linear relationship with the maximum deflection and differential pressure. Figure 5 shows deflection mapping of a 6.3 mm x 6.3 mm diaphragm achieved at room temperature. The thickness is about 110 μm. The maximum deflection was found to be 30.1μm under 40 psi differential pressure. The measured deflection profile has a fairly good agreement with our FEM simulation.

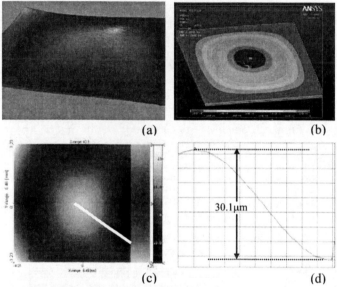

(a)   (b)

(c)   (d)

Figure 5: (a) 3-D deflection profile obtained from WYKO interferometer under 40 psi differential pressure; (b) 3-D deflection profile obtained from finite element simulation (The calculated maximum deflection is 29.5 μm at 40 psi); (c)-(d) Deflection mapping and profile obtained from the WYKO interferometer (The measured maximum deflection is 30.1 μm).

Smaller deflection was observed at decreased temperature. For example, the maximum deflection of a diaphragm was 25 μm at room temperature, and 18 μm at 77 K. Given the lateral dimension of the diaphragm (6.3 mm x 6.3 mm), it can be derived that during a single cycle the diaphragm can transfer ~0.36 mm³ liquid cryogen under 40 psi differential pressure at room temperature, while this value would drop down to ~0.24 mm³ at 77 K. In other words, the volume change at 77 K would be about two thirds as that at 293 K, with the same actuation condition.

Moreover, in order to investigate the influence of actuation frequency, ZYGO interferometer measurement was subsequently conducted for dynamic deflection observation.

The diaphragm was actuated at different frequency, and the deflection was recorded. The results showed that decreasing deflection associates with increasing frequency, which is probably due to the gas resistance in the tubing pipes.

3.3 Stress distribution

To transfer cryogen for cooling, a large volume stroke with high stresses is desired. Therefore, a closer examination of the stress distribution in the diaphragm was conducted to predict the possible diaphragm failure and longevity working viability.

Micro-Raman was employed for stress mapping; stress distribution was also estimated using FEM simulation. Since the thickness of the silicon diaphragm is much smaller than its lateral dimension, the X-axis (Y-axis) stress was found much higher than the out-of-plane stress (Z-axis) and in-plane shear stress. As such, stress distribution within a diaphragm region can be regarded as a biaxial case, and the in-plane stress can be linearly interpreted from the Raman shift, if measured.

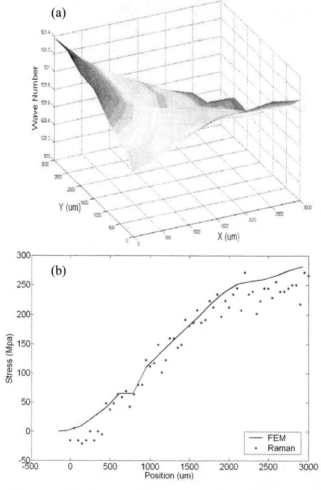

Figure 6: (a) Raman shift mapping within a quarter region of the diaphragm. (b) Comparison of stress achieved from Micro Raman spectroscopy and simulation analysis.

Raman shift distribution within a quarter region of the diaphragm was achieved (Figure 6a). The negative shift occurs in the diaphragm center, indicating tensile stress, while the positive shift occurs in the edge centers of the diaphragm, indicating compressive stress. Moreover, a line scanning was carried out from diaphragm edge to the edge centers, and Figure 6b shows a good match with the FEM simulation. As expected, the edge centers (red region) suffered the maximum stress, and will be most vulnerable for diaphragm failure.

3.4 Fatigue testing

There is no fatigue observed in single crystalline silicon diaphragm at room temperature, while little information was found at cryogenic temperature. Therefore, to investigate the long-term reliability of micro pump at cryogenic temperature, fatigue testing was carried out. Figure 7 shows the setup for longevity tests. The silicon diaphragm was clamped in the sample holder, and then immersed into the liquid nitrogen. Compressive gas was induced to actuate the diaphragm. Preliminary results show that there is no crack observed in the diaphragm after 1.8×10^6 cycles (about 10 days). Efforts are currently underway to evaluate intensively the viability of silicon-based micro pumps for space missions.

Figure 7: *Preliminary fatigue test setup for cryogenic temperature operations.*

4. CONCLUSIONS

This paper presented pumping capacity and reliability studies of a silicon-based cryogenic micro pump for space applications. Fabricated using standard processes, a square-shape diaphragm served as the major functional component of the micro pump. A linear relationship between the differential pressure and maximum deflection of the diaphragm was found. The volume stroke of the diaphragm during single cycle at cryogenic temperature (77 K) is about two thirds as that at room temperature (about 293 K). Moreover, the pumping capacity during single cycle was found to decrease with increasing actuation frequency.

The stress distribution was further examined using micro-Raman interferometer. It was observed that the tensile stress concentrates in the diaphragm center while the compressive stress concentrates in the edge centers of the diaphragm. Accordingly, the edge center regions are most susceptible to the future system failure. Finally, silicon-based diaphragms survived a cryogenic operation for more than 10 days (after 1.8×10^6 cycles), thereby preliminarily demonstrating the viability of silicon-based micro pumps for space missions.

ACKNOWLEDGEMENT

This work has been supported in part by Air Force Research Laboratory (AFRL) under Contract No. F29601-02-C-0163 and Air Force Office of Scientific Research (AFOSR) under Contract No. F49620-03-1-0078. The authors would like to thank our partners, Mr. David Walker and Mr. Andrew Harvey at the Foster-Miller Inc., for the help of fatigue testing. We would also like to thank Professor Thomas Bifano and his team members at Boston University for the help with the WYKO and ZYGO measurements, and Dr. Anna Swan for the help with the micro-Raman characterization. Finally, the authors wish to thank all the other members of the Boston University Laboratory for Microsystems Technology (LMST) for their help and support, especially Matt Freddo, David Mabius, and Jeremy Moyers.

REFERENCES

[1] T. Nakagawa, HIDES, 2003.
[2] S. Arnon and N.S. Kopeika, Proceeding of the IEEE, vol. 85, no. 10, Oct. 1997, pp. 1646-1661.
[3] NHK Science & Technical Research Laboratories, Annual Report 2001.
[4] M.A. Greenhouse, P. Bely, R. Bitzel, et al, NASA NGST, 1999.
[5] Sidney W.K. Yuan, Pacific Rim/ASME International, Electronic Packaging Technical/Business Conference, 1999.
[6] D. Zimbelman, Goddard Space Flight Center: NASA, 2002.
[7] Thermacore Inc, NASA SBIR, 2000.
[8] N.-T. Nguyen and X. Huang, Technical Proceedings of the 2000 International Conference on Modeling and Simulation of Microsystems, 2000, pp. 636-639.
[9] G.T.A Kovacs, N.I. Maluf, and K.E. Petersen, Proceedings of the IEEE, vol. 86, no. 8, Aug. 1998, pp. 1536 - 1551.
[10] T.Y. Jiang, T.Y. Ng and K.Y. Lam, Technical Proceedings of the 2000 International Conference on Modeling and Simulation of Microsystems, 2000, pp. 632-635.
[11] O.C. Jeong and S.S. Yang, Journal of the Korean Physical Society, vol. 37, no.6, Dec. 2000, pp. 873-877.
[12] K. Brunner, G. Abstreiter, B. O. Kolbesen, and H. W. Meul, Appl. Surf. Sci., 39, 1989, pp. 116.

System issues in arrays of autonomous intelligent sensors

Robert M. Newman, Elena I. Gaura,

School of Mathematical and Information Sciences
Coventry University
Coventry CV1 5FB, United Kingdom
bobn@coventry.ac.uk

ABSTRACT

This paper considers some of the systems level issues of concerned when building of large scale sensing systems composed of a large number of autonomous intelligent sensors. The work is motivated by the potential offered by monolithic, integrated, intelligent MEMS sensors. They provide very low cost sensing and computational power in a form which may be integrated into 'smart structures', in which structural diagnostic capability is deeply embedded into the structure itself. However, in order to realise them, some intrinsic systems issues must be addressed. Ultimately, the goal of this work is to develop a systems architecture which enables the designer of application systems to deploy intelligent networked sensors as a generic systems component, concentrating on applications issues, rather than the detailed system design problems associated with decentralised, distributed sensor systems.

Keywords: sensor, smart, intelligent, cogent, MEMS, artificial intelligence

1. INTRODUCTION

Developments in intelligent MEMS sensors may be expected to revolutionise the design of all systems based on sensors. Sometimes the claims made for them can be sweeping. Kumar et. al. [1] state:

> Networked microsensors technology is a key technology for the 21st century. Cheap, smart devices with multiple, on board sensors, networked through wireless links and the Internet and deployable in large numbers, provide unprecedented opportunities for instrumenting, and controlling to our advantage, homes, cities, the environment, and battlefields.

While some of these claims may appear at first sight to be overstated, there certainly are some application areas where the effects of the deployment of very large numbers of intelligent MEMS sensors will be profound. One domain where there appears to be very great potential is that of 'smart materials', an application area in which there is particular interest being airframe design. This application will be used as an exemplar throughout the paper, in order to illustrate some of the real systems design issues involved in the deployment of very large arrays of intelligent sensors.

Smart materials in airframes are being proposed to address a number of design goals for future advanced aircraft, two major goals being adaptive aerodynamics and self diagnosis, and possibly repair of the structural integrity of an airframe. The latter is aimed at extending the life of airframes, possibly ultimately leading to the 'eternal aircraft' as proposed in the NASA CTIP [2] project. Both applications depend on the location of sensors throughout the structure of the airframe, with possible need for actuators as well, dependent on the precise nature of the application.

Structural diagnosis systems are derived originally from structural health monitoring systems. If the sensors are not smart, the installation of such systems is also problematic. Each sensor needs to be individually wired to the data processing unit, and the connections must usually be of very high quality to cope with low level analogue signals. The complexity of this wiring, and its sensitivity often means that such systems cannot be permanently fitted.

By providing some redundancy in provision of sensors, reliability may be enhanced at a systems level, even if the reliability of an individual sensor is not better (although a monolithic intelligent sensor might be expected to be more reliable than a multiple component one).

The systems design implications of intelligent MEMS based systems, using redundant provision of sensors, and distributed data processing within the intelligent MEMS sensor, are very different from those of conventional systems. These design problems are also very subtle and complex, particularly for those not familiar with the field. In the more general domain of information systems, where the default system is now a distributed one, system designers have developed generic distributed operating systems which free the applications designer from the need to consider each and every consequence of distribution. Our eventual goal is to develop such an 'operating system' for distributed sensor systems, but the and components of that system will be different from those for more general systems. Several authors have surveyed these system level requirements [3,4], and have considered the importance of the issues outlined above. The major considerations that must be taken into account, and will contribute to the specification for 'operating systems' for intelligent networked sensors are briefly presented below.

2. SENSOR CALIBRATION AND COMPENSATION

Existing systems are dependent on a largely manual process of sensor identification, calibration and compensation. There is a considerable manufacturing spread over the sensors used. Individual sensors need to be calibrated, but if only sensors which calibrate within a small range can be used, the cost of the system rises considerably, thus it is often necessary to provide compensation for an individual sensor's characteristics. The characterisation and calibration of individual sensors is undertaken by providing a controlled stimulus (in an accelerometer, via a calibrated shaker) and measuring the output of the sensor under test. Compensation requires the same process under a range of stimuli and the production of correction tables, which need to be provided individually to the data processing system with which the sensor is to be used. Gaura et. al. [5,6] have described the use of artificial neural networks for the identification and compensation of some commonly found sources of error in micro-machined accelerometers. Sachenko et al. describe the use of an artificial neural network to predict the development of a particular type of fault, sensor parameter drift [7]. Using historical data from sensors the ANN is trained to predict the drift of a sensor over time. The prediction is used to correct the output from the sensors.

Application designers will certainly not wish to be burdened with the detail of sensor compensation, so this is a facility that will need to be included in the sensor operating system.

3. SENSOR PLACEMENT AND STRUCTURE FABRICATION COSTS

. There is a complex, four way trade-off between number of sensors used, degree of redundancy, cost of installation and locality of sensors to the data points of interest. In a 'traditional' pre-intelligent MEMS system the trade off was very much weighted against using a large number of sensors and they would be strategically placed on the structure to be monitored, at sites known to show symptoms of structural stress. The sites for the sensors have to be predetermined by laboratory stress experiments or computer simulation. Staszewski, et al, discuss the use of AI techniques for determining an optimum sensor placement for acoustic impact sensors distributed on the surface of a composite material [8].

Using intelligent MEMS sensors, the cost and processing constraints on the number of sensors used may be relaxed. Instead of placing a few sensors at strategic points of interest, sensors are distributed all over the structure. Although there may not be a sensor exactly at a point of interest, the data value for that sensor can be synthesised using a process of data fusion. Jungert discusses the issues of distributed data fusion, as would be required in an array of intelligent sensors where the only processing power for the system is that contained in the sensors themselves [9]. Koushanfar, Potknojak and Sangiovanni-Vincentelli describe the use of multimodal data fusion [10]. In their example, they obtain a system of fifteen equations to yield the values of twelve variables, giving a redundancy of three equations. Several methods for data fusion have been reported including Kallman filters and artificial neural networks [11,12].

It is not necessary for sensors to be distributed in an orderly way. If sensors are so inexpensive and small that installation is a relatively high cost comparatively, it may be advantageous to use a very low cost installation method, such as embedding them in the matrix of a composite, and allowing them to be randomly distributed through the matrix as the composite is built up. In this case, the se sensors form an 'ad-hoc' network, and one of the first tasks on initialisation is establishment of a virtual topology for the network, as discussed in section 6.

The design of a sensor operating system will need to be sufficiently flexible to account for different types of sensor array, from orderly, placed arrays to random, ad-hoc networks. The application designer is more likely to be interested in the desired data points, than the actual sensor locations, so the operating system will need to have inbuilt the facilities for data fusion, in order to map the actual network topology to the one required by the application.

4. POWER AND HEAT

Passive or conventional smart sensors consume little or no power, but active sensors, particularly those endowed with a large amount of processing capability, may consume significant amounts of power and dissipate substantial amounts of heat. In many applications (see, for instance, Schweibert et al [13]) power, and the associated heat production may be a limiting factor in the design of a system. The design considerations for intelligent sensor networks are therefore often dominated by power considerations. This can be apparent in the detailed design of the sensor electronics, favouring selection of low power processing methods. It can also affect network design and topology establishment. The sensor operating system will require the ability to configure its own internal workings according to the power and heat constraints of a particular application.

5. DATA TRANSMISSION AND COMMUNICATIONS

The wiring requirements are simplified to provision of a simple passive multi-drop network, or, in many cases, wireless connection. The bulk of the computation occurs within the array itself, and all that a structural diagnostic system has to do is to extract the information from a convenient location within the array. The passive network has advantages in terms of simplicity, reliability and flexibility with respect to adjustment of the connectivity of the network to handle sensor failures or other dynamic reconfiguration needs. On the other hand point to point switched networks can have advantages in terms of power use, and this is always an important consideration in sensor networks. Formal descriptions of re-routing algorithms in such networks have been presented previously [14]. Although at first sight wired sensors would seem to be appropriate to smart structure applications (because wires are needed to distribute power to intelligent sensors) wireless networks would be much simpler to connect, particularly if a matrix-embedded random sensor deployment has been adopted.

The specification of the sensor operating system must deal with all of these types of connectivity, in a way that is transparent to the applications designer.

6. NETWORK INITIALISATION AND TOPOLOGY

In placed, wired networks the initialisation of the network, and its topology is established by the wired connections. However a wireless, ad-hoc network is unusable until the 'real' network has been established. There is a great deal of interest in this topic, particularly with reference to 'pervasive computing'. Most of this work, in the sensor arena, has been to do with ad hoc radio networked sensors for applications such as environmental and battlefield sensing [15,16,17], however it can also be used in structural sensing, for the reasons discussed above. In a wireless network, every sensor within the range of another sensor's transmitter is connected, but selecting particular favored routings can have a substantial effect on overall transmitter power required. Other factors such as bandwidth and reliability can also be optimised. The literature above reports several topology establishment methods based on AI and other optimisation techniques.

In ad-hoc networks the sensor operating system will be responsible for initialisation and maintenance of the network, and the discovery of the topological an location information on which higher level functions such as data fusion and fault management depend.

7. FAULT MANAGEMENT

In conventional systems sensors are mechanically attached to the structure under test using screws or other removable attachment. Failed sensors can be removed and replaced, and since they have been pre calibrated, or, in the case of a smart sensor, internally compensated, the new sensor fits into the rest of the system just as the old one did. This is not the case for a sensor embedded in the fabric of the structure under test. Instead, the array needs to be provided with some degree of redundancy, and a means for detecting and swapping out failed sensors. Data fusion techniques [18,19] are used to extract the required data from this array and sensor failure determined by deviation from expected values based on models of the specific application. The authors have proposed an application independent fault management strategy based on self-diagnosis of sensors, in which sensors are removed from the array if faulty. False alarms are also possible, and sensors must be re-introduced to the array if found to be healthy. It is possible to calculate the lifetime of an array in this case [20]. If, in an array consisting of n sensors, the mean time between failures of a sensor is t_{mbf}, the false error detection rate is f, there is a proportion r of redundant sensors and the time to correct a false fault is some multiple C of the fault detection time, then the lifetime L_{rff} of the whole array is given by

$$L_{rff} = t_{mbf}(r - fC)$$

This indicates some of the parameters that are important in the selection of a fault management strategy. The fault detection time of any strategy is an important figure of merit, since it directly affects the error rate from the array as a whole. Any strategy must either have a zero false fault detection rate or must include the re-introduction of a sensor into the array should a fault diagnosis subsequently prove to be faulty. The time to re-introduce a sensor must be a small multiple of the fault detection time, preferably unity or less.

An effective, application independent and transparent fault management system is an important component of the sensor operating system.

8. APPLICATION LEVEL FUNCTIONS

We have put forward the concept of a 'cogent sensor' [21], one that participates in the process of transformation from data to information. Of course, what is 'information' depends on the requirements of the application, so each application will require some different processing within a cogent sensor. This requirement conflicts with our goal of a generic systems component unless the sensor part of the system can take responsibility for the downloading of applications level data processing to the sensor web and its distribution within it in a way that is optimal for the overall operating constraints, be they power consumption, computational speed or other constraints. There are several ideas from distributed systems technology that can be used to achieve this goal, agent technology being a currently favoured option.

9. CONCLUSIONS

Several functions, namely, sensor calibration and compensation, data fusion, data transmission and communications, network initialisation and topology, fault management and distribution of application level functions are common requirements for very nearly all intelligent sensor networks, and therefore form the basic components for the system support for a generic intelligent sensor system. Presently we are undertaking a programme to address all of them, and ultimately integrate them into an 'operating system' for distributed sensor networks, providing the engineer using these systems with a straight forward and standardised way of controlling and gathering the information from sensor mechanisms. In our work to date, mainly in the area of fault management, artificial intelligence techniques, particularly artificial neural networks, have provided elegant solutions to some of the design problems for our system. Other researchers have found them of use in the topology establishment and data fusion. Along with intelligent agent technologies, we expect them to play a central part in the workings of our system.

REFERENCES

[1]   Kumar, S. , Shepherd, D., Zhao, F., Collaborative Signal and Information Processing in Micro-Sensor Networks, *IEEE Signal Processing Magazine,* March 2002.

[2]   Miller, W. M. Peterson, K. A. (1998) MEMS reliability: The challenge and the promise, NASA Technical Reports.

[3]   Peti, P, Obermaisser, R, Elmenreich, W, Losert, T (2002) An Architecture Supporting Monitoring and Configuration in Real-Time Smart Networks.

[4]   Martinez, D, Gruber, M, (2002) Next Generation Technologies to Enable Sensor Networks.

[5]   Gaura, E. Rider, R.J. Steele, N. (2000) Closed-loop neural network controlled accelerometer, Proceedings of the I. Mech. E, Part I, Journal of Systems and Control Engineering, vol. 214, no.I2, pp.129-138.

[6]   Gaura, E. Rider, R.J. Steele, N. (2000). Developing smart micromachined transducers using feed-forward neural networks: a system identification and control perspective. The IEEE International Joint Conference on Neural Networks, IJCNN'2000, Proceedings, ISBN 0-7695-0619-4, Vol. IV, pp. 353-358, Como, Italy.

[7]   A.Sachenko et al. (2002) Metrological Automatic Support in Intelligent Measurement Systems Computer Standards & Interfaces vol. 24, issue 2, pp. 123-131.

[8]   Staszewski, W. J., Worden, K., Wardle, R., Tomlinson, G. R. (2000), Fail-safe sensor distributions for impact detection in composite materials. Smart Material Structures, 9 (2000), pp. 298-303. IOP Publishing, London.

[9]   Jungert, E (2000) A data fusion concept for a query language for multiple data sources, Proc 3rd International Conference on Information Fusion (Fusion 2000), Paris, Jul 10-13

[10] Koushanfar, F, Potknojak, M, Sangiovanni-Vincentelli, A, (2002) Fault Tolerance Techniques for Wireless Ad-hoc Sensor Networks.

[11] Clarke, J, Yuille, A, (1990) Data Fusion for Sensory Information Processing Systems, Kluwer.

[12] Brookes R, Iyengar, S, (1997) Multi-Sensor Fusion: Fundamentals and Application with Software, Prentice-Hall.

[13] Schweibert, L, Gupta, S, Auner, P, Abrams, G, Iezzi, R, McAllister, P (2002) A Biomedical Smart Sensor for the Visually Impaired.

[14] Gaura, E.I, Newman, R.M., (2003) Intelligent sensing: Neural Network based health diagnosis for sensor arrays, Proc ASME'03 , Kobe, Japan.

[15] Pister, K.S.J., Boser, B.E., (1999)Smart Dust: Wireless Networks of Millimeter-Scale Sensor Nodes. University of California, Berkeley Electronics Research laboratory Research Summary.

[16] Catterall E, Van Laerhoven, K., and Strohbach, M.. (2002) SelfOrganization in Ad Hoc Sensor Networks: An Empirical Stud. In Proc. Alife VIII: the 8 th International Conference on the Simulation and Synthesis of Living Systems, Sydney, Australia. MIT Press, 2002.

[17] NASA, Advanced Architectures and Automation Branch, Sensor Web Application Prototype (SWAP). NASA GSFC Code 588.

[18] Clark, J.L. , Yuille. A.L.,(1990). Data Fusion for Sensory Information Processing Systems. Kluwer Academic Publishers.

[19] Brooks, R. R. , Iyengar, S. S.(1996). Robust distributed computing and sensing algorithms. IEEE Computer, pages 53--60, June 1996.

[20] Newman, R.M., Gaura, E.I., (2004) A systems level perspective of fault management in large sensor networks. To be published.

[21] Gaura, E. I., Newman, R.M., Smart, Intelligent and Cogent Microsensors – Intelligence for Sensors and Sensors for Intelligence, Proc. MSM'04, Boston, MA, March 2004.

Simulation of Micromachined Inertial Sensors with Higher-Order Single Loop Sigma-Delta Modulators

Yufeng Dong, and Michael Kraft

University of Southampton, School of electronics and computer science,
Highfield, Southampton, SO17 1BJ, UK, yd02r@ecs.soton.ac.uk

ABSTRACT

Micromachined capacitive inertial sensors incorporated in sigma-delta force-feedback loops have been proven to improve linearity, dynamic range and bandwidth, and also provide a direct digital output. Previous work mainly focused on using only the sensing element to form a 2nd-order single loop sigma-delta modulator ($\Sigma\Delta M$). Therefore, the advantages of higher-order (4th-order and 5th-order) single loop electro-mechanical $\Sigma\Delta M$ have not been explored, especially for inertial sensors that require higher Signal to Quantization Noise Ratio (SQNR), wide-band signal and low power dissipation. This paper presents architecture for higher-order single loop electro-mechanical $\Sigma\Delta M$ with optimal stable coefficients that lead to better SQNR. Simulations show the maximum SQNR of 3rd-order, 4th-order and 5th-order $\Sigma\Delta M$ is 88dB, 105dB and 122dB, respectively, using an Oversampling Ratio (OSR) of 256.

Keywords: higher-order, micromachined inertial sensors, noise transfer function, sigma-delta modulation.

1 INTRODUCTION

There is increasing demand for high performance (micro-g accuracy) micromachined inertial sensors that comes from inertial navigation/guidance systems, space micro-gravity, unmanned aerial vehicles, and seismometery for oil exploration and earthquake prediction. High performance inertial sensors usually exploit the advantages of closed loop control strategy to increase the dynamic range, linearity and bandwidth of the sensors. To avoid the electrostatic pull-in problems in purely analogue force feedback closed loop scheme, a $\Sigma\Delta M$ closed loop force feedback control scheme has become very attractive for capacitive inertial sensors. Its output is digital in the format of a pulse density modulated bitstream and can directly interface to a digital signal processor.

Previous work mainly focused on using a sensing element to form a 2nd-order, single loop, electro-mechanical $\Sigma\Delta M$ [1]-[2]. A sensing element can be approximated by a second order mass-damping-spring system and can be regarded as analogous to the two-cascaded electronic integrators commonly used in 2nd order electronic $\Sigma\Delta M$ A/D converters. However, the equivalent d.c. gain of the

mechanical integrator functions is considerably lower than their electronic counterparts. This leads to a much lower SQNR (for example, most below 66dB at OSR=256) for electro-mechanical $\Sigma\Delta M$ compared with a purely 2nd order electronic implementation. The most obvious method to increase the SQNR is to increase the sampling frequency of the system, but increasing the sampling frequency will also increase the electronic thermal noise and power dissipation. Recently, some researchers used an additional electronic integrator to form a 3rd-order $\Sigma\Delta M$ [3]-[4]. Kajita et al [3] demonstrated the higher noise shaping at low frequencies by adding an electronic integrator to a 2nd-order single loop electro-mechanical $\Sigma\Delta M$, but the SQNR is around 90dB at OSR=2500, which may be still low for high performance applications. Furthermore, Kraft et al [4] presented a 3rd-order multi-stage noise shaping (MASH) electro-mechanical $\Sigma\Delta M$, but it is known that MASH loop is more sensitive to the coefficient variations than single loop [5]. The microfabrication usually exhibits large manufacturing variations, the characteristics of a micromachined inertial sensor are not precisely fixed without calibration, therefore, the cascaded MASH structures, which rely on accurate quantization noise cancellation, may be problematic in implementation. Therefore, single loop electro-mechanical $\Sigma\Delta M$ architecture seems the most practical configuration for high performance inertial sensors. However, the advantages of higher-order (4th-order and 5th-order) single loop electro-mechanical $\Sigma\Delta M$ have not been explored, especially for inertial sensors that require higher SQNR, wide-band signal and low power dissipation. This work is to design a higher-order single loop electro-mechanical $\Sigma\Delta M$ with optimal stable coefficients that lead to better SQNR. To achieve the same performance as 2nd-order $\Sigma\Delta M$, a higher-order $\Sigma\Delta M$ can use lower OSR and this leads to lower electronic thermal noise and power dissipation.

2 DESIGN METHODOLOGY

2.1 Topology of a Higher Order Single Loop Electro-Mechanical $\Sigma\Delta M$

Although interpolative topologies with multiple feedback and forward paths are very successful approaches to implement single loop high-order delta-sigma A/D converters, these topologies can not be directly applied to an electromechanical $\Sigma\Delta M$ because the sensing nodes can not be connected to feedback & feedforward paths. Here, we propose a practical topology to maximize the SQNR using the fewest feedback paths that give the simplest implementation and thus the lowest circuit complexity. Such a topology in a 5th order loop is shown in Figure 1. K_1 to K_3 are coefficients of the electronic integrators, K_4 is the variable gain of the quantizer, K_{dv} is the gain of displacement to voltage of the sensor, K_{po} is the pick-off gain of capacitance position sensing circuitry, and K_{fb} the gain through the voltage to force conversion in the feedback path. $K_m = K_{po}K_{fb}$ is an equivalent coefficient. In fact, K_{po} and K_1 are effective one coefficient, splitting them is meaningful only on convenient comparisons among different order electro- mechanical $\Sigma\Delta M$.

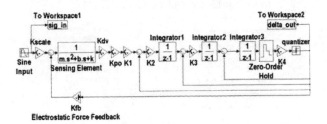

Fig. 1: Topology of a 5th order electromechanical $\Sigma\Delta M$.

2.2 Noise Transfer Function of a nth Order Electro-Mechanical $\Sigma\Delta M$

The linearized quantizer, modeled by a white noise source and variable gain, is still good approximation for a higher order $\Sigma\Delta M$ with great OSR to determine its properties [5]. The noise transfer function (NTF) $H_e(z)$ of nth-order electromechanical $\Sigma\Delta M$ in a topology of Figure 1 is given by:

$$H_e(z) = \frac{1}{1 + \prod_{i=1}^{n-1} K_i K_m M(z)[H(z)]^{n-1} + \sum_{i=2}^{n-1} \prod_{j=i}^{n-1} K_j [H(z)]^{n-i}} \quad (1)$$

where M(z) is the mechanical transfer function of the sensing element by transforming to discrete-time z-domain:

$$M(z) = K_f \cdot \frac{(z - a_f)}{(z - b_f) \cdot (z - c_f)} \quad (2)$$

where K_f, a_f, b_f and c_f are gain, zero and poles of a sensing element, which are a function of the sampling frequency. According to the z-transformation $z = e^{j\omega}$ and assuming $f \le f_b \ll f_s$:

$$z^{-1} = e^{-j2\pi f/f_s} = \cos(\frac{2\pi f}{f_s}) - j\sin(\frac{2\pi f}{f_s}) \approx 1 - j\frac{2\pi f}{f_s} \quad (3)$$

$$\left|1 - z^{-1}\right| \approx \left|j\frac{2\pi f}{f_s}\right| = \left|j\pi\frac{f}{f_b}\frac{2f_b}{f_s}\right| = \left|j\pi\frac{f}{f_b}\right| \cdot \left|\frac{1}{OSR}\right| \ll 1 \quad (4)$$

where f_b is the signal bandwidth, f_s the sampling frequency, and OSR=$f_b / f_s /2$ the oversampling ratio.

After omitting the items containing $|1-z^{-1}|$ in the denominator of (1), the NTF $H_e(z)$ can thus be approximated at low frequencies:

$$\left|H_e(z)\right| = \frac{\left|1-z^{-1}\right|^n + \left|2 - b_f - c_f\right|\left|1-z^{-1}\right|^{n-1} + \left|(1-c_f)(1-b_f)\right|\left|1-z^{-1}\right|^{n-2}}{K_m K_f (1-a_f)\prod_{i=1}^{n-1} K_i + \left|(1-c_f)(1-b_f)\right|\prod_{i=2}^{n-1} K_i} \quad (5)$$

It can be seen from (5) that the ability of nth order noise shaping will degrade by the denominator of (5) in an nth order electromechanical $\Sigma\Delta M$. It is also worth to be noted from (5) that for a given sensing element, the in-band noise power degrades with increasing the product of the loop coefficients $\prod_{i=1}^{n-1} K_i$, but there is an upper limit on the product due to the stability of the $\Sigma\Delta M$.

2.3 Stability and Optimal Coefficients

Similar to the design of higher-order sigma-delta A/D converters [6], there is no precise analytic approach to ascertain the stability of a modulator without resorting to simulation due to a highly nonlinear element, the quantizer. For the time being, the most reliable method for verifying stability of high-order loop is simulation [7]. Any gain introduced before the quantizer does not affect the output, the coefficient of the last integrator is also irrelevant for simulation purpose. In modern VLSI fabrication, the mismatches in switch-capacitor amplifiers may be between 0.1% and 1%, however, the tolerance in fabrication of MEMS will be large (3%) depending on technology, device size and circuit topology. In order to guarantee the optimal SQNR with robust stability, an empirical approach is suggested to find an optimal product of loop coefficients after exhaustive simulations. We use a normalized sinusoidal input with amplitude of -10dB to avoid overloading, and the SQNR and gain of NTF are calculated by Hann window 128*1024 bin FFT. The constraints set for simulations are: ①SQNR loss less than 3dB due to 3% variation of the coefficients, ②maximization of SQNR, ③ stability criteria (|NTF|'s gain <1.5) and ④ coefficients should be between 0.001 and 1 according to normalized input-output signal amplitude of 1 volt, which are practical in circuit implementation. The approach to find stable and optimal coefficients can be illustrated in three steps:
1. Simulations start from a 2nd order electro-mechanical $\Sigma\Delta M$ to identify the stability of the loop. If the loop is unstable, as for an underdamped sensing element, add a lead filter to stabilize it.

2. Next is to find the stable and optimal coefficient $K_{po}*K_1$ in a 3rd-order electro-mechanical $\Sigma\Delta M$: sweeping the coefficient across a range of values 0.001 to 1 in steps of 0.05, meanwhile the data in simulation process are saved to a file according to the $K_{po}*K_1$, SQNR and gain of NTF.
3. Use a similar strategy for 4th-order and 5th-order loops to get $K_{po}*K_1$, K_2 and K_3. After applying the constraints to these files, post-processing can find a relatively maximum of the SQNR.

To illustrate this procedure, Figure 2 shows the SQNR distribution of a 4th order electro-mechanical $\Sigma\Delta M$ as function of the coefficients K_1, K_2. It was found that when $K_1 \approx 0.2$, $K_2 \approx 0.5$, the SQNR=82dB is optimized using an OSR=64.

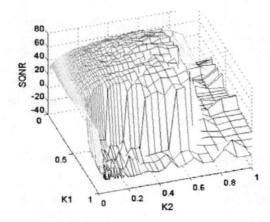

Fig. 2: SQNR distribution of a 4th order electro-mech. $\Sigma\Delta M$.

A stable and optimal 3rd-order electro-mechanical $\Sigma\Delta M$ is fundamental for designing a higher-order loop. The constraint that fabrication uncertainty 3% leading to SQNR variation less than 3dB is the most important stability criterion in practical implementation. Based on extensive simulations of 3rd-order to 5th-order loops and careful post processing, it was found that the product of the electronic pick-off gain and integrator coefficients is found to be nearly a constant according to order level. The SQNR of the high-order electro-mechanical $\Sigma\Delta M$ is maximized with stable constraints when the integrator coefficients are approximately:

$$K_{po} \cdot \prod_{i=1}^{n-2} K_i \approx K_{po} \cdot (\frac{1}{2}, \frac{1}{10}, \frac{1}{50}), \ (n = 3, 4, 5) \quad (7)$$

The coefficients of the electronic integrators are chosen by borrowing the expertise from a mature high-order $\Sigma\Delta M$ A/D converter [6]:

$$K_1 \approx 0.5, \qquad\qquad (n = 3)$$
$$K_1 \approx 0.2, \ K_2 \approx 0.5, \qquad (n = 4) \qquad (8)$$
$$K_1 \approx 0.2, K_2 \approx 0.2, K_3 \approx 0.5, \quad (n = 5)$$

3 SIMULATION RESULTS

The sensor used to demonstrate the approach is a bulk micromachined overdamped accelerometer (mass m = 1.5x 10^{-6} kg, damping coefficient b=0.0867 Nm/s, and spring stiffness k=98.1 N/m) provided by QinetiQ. The input is sinusoidal and signal bandwidth 1kHz.

3.1 3rd Order Electro-Mechanical $\Sigma\Delta M$

The simulation model of a 3rd-order electro-mechanical $\Sigma\Delta M$ is shown in Figure 3. The maximum SQNR obtained from this architecture is 88dB at OSR=256. Figure 4 shows the power spectrum density (PSD) of noise shaping in this 3rd order electro-mechanical $\Sigma\Delta M$.

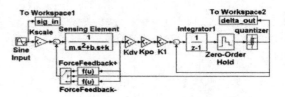

Fig. 3: Model of a 3rd-order electro-mechanical $\Sigma\Delta M$.

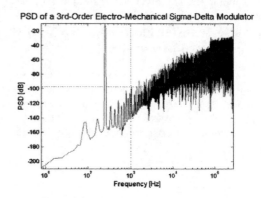

Fig. 4: Noise shaping: a 3rd-order electro-mechanical $\Sigma\Delta M$.

3.2 4th Order Electro-Mechanical $\Sigma\Delta M$

The simulation model of a 4th-order electro-mechanical $\Sigma\Delta M$ is shown in Figure 5. The maximum SQNR obtained from this architecture is 105dB at OSR=256. Figure 6 shows the PSD of noise shaping in this 4th-order electro-mechanical $\Sigma\Delta M$.

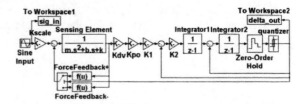

Fig. 5: Model of a 4th-order electro-mechanical $\Sigma\Delta M$.

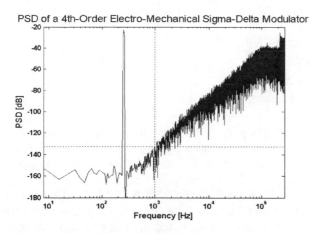

PSD of a 4th-Order Electro-Mechanical Sigma-Delta Modulator

Fig. 6: Noise shaping: a 4th-order electro-mechanical $\Sigma\Delta M$.

3.3 5th Order Electro-Mechanical $\Sigma\Delta M$

The simulation model of a 5th-order electro-mechanical $\Sigma\Delta M$ is shown in Figure 7. The maximum SQNR obtained from this architecture is 122dB at OSR=256. Figure 8 is the comparison of the PSD of noise shaping between a 5th-order and a 2nd-order electro-mechanical $\Sigma\Delta M$ at OSR=256. The SQNR difference between the two loops is about 60dB.

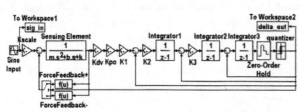

Fig. 7: Model of a 5th-order electro-mechanical $\Sigma\Delta M$.

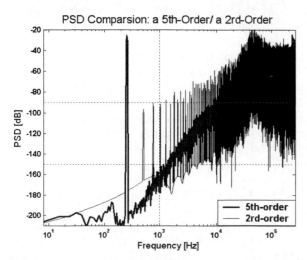

Fig. 8: Noise shaping comparison: a 5th-order /a 2nd-order electro-mechanical $\Sigma\Delta M$.

The coefficients and performance for higher-order electro-mechanical $\Sigma\Delta M$ are summarized in Table 1. DR (dynamic range) refers to the maximum SQNR achievable for sinusoidal input signal, and OL the maximum input signal amplitude for which the SNR degrades less than 6dB from the maximum SQNR.

Order	3		4		5	
Integrator Coefficients	K1=0.5		K1=0.2 K2=0.5		K1=0.2 K2=0.2 K3=0.5	
OSR	DR	OL	DR	OL	DR	OL
64	69dB	0.95	79dB	0.85	88dB	0.75
128	79dB	0.9	91dB	0.80	105dB	0.65
256	88dB	0.85	105dB	0.75	122dB	0.55

Table 1: Coefficients and performance in a higher-order electro-mechanical $\Sigma\Delta M$.

4 CONCLUSIONS

Using the same OSR, a 5th-order electro-mechanical $\Sigma\Delta M$ can improve the SQNR by 60dB compared with a 2nd-order electro-mechanical $\Sigma\Delta M$ at OSR=256. To achieve the same performance as 2nd-order electro-mechanical $\Sigma\Delta M$, higher-order $\Sigma\Delta M$ can use lower oversampling frequency to reduce electronic thermal noise and power dissipation. The present work involves simulation of the optimal higher-order single loop electro-mechanical $\Sigma\Delta M$. To validate these architecture, design and fabrication of a high performance inertial sensor and associated interface ASIC are currently underway. Other important issues in a higher-order single loop electro-mechanical $\Sigma\Delta M$ such as force feedback linearization, various noises analysis and fabrication mismatch on performance degradation by Monte-Caro analysis will be published in the near future.

REFERENCES

[1] W. Henrion et. al. IEEE Solid-State Sensor and Actuator Workshop, pp. 153-157, Hilton Head Island, June 1990.
[2] M.A. Lemkin, et. al. IEEE Journal of Solid-State Circuits, Vol. 34, No. 4, pp. 456-68, 1999
[3] T. Kajita, et. al. IEEE ISCAS 2000, pp. IV.337-IV.340, Geneva, Switzerland, May 2000.
[4] M. Kraft, et al. Proc. 4th Conf. On Modeling and Simulation of Microsystems, pp.104-107, Hilton Head, March 2001.
[5] S.N. Norsworth et. Al. Delta-Sigma Data Converter: Theory, Design and Simulation, IEEE Press, 1997.
[6] Augusto Marques, et. al. IEEE Trans. on Circuits and Systems II, Vol. 45, No. 9, pp.1232-1241, 1998.
[7] Richard Schreier, IEEE Trans. on Circuits and Systems II, Vol. 40, No. 8, pp. 461-466, 1993.

Design and Development of an Integrated MEMS Sensor for Real Time Control of Plasma Etching

Bryan G Morris* and Gary S. May
Georgia Institute of Technology
Atlanta, Georgia, USA
*morrisbg@ece.gatech.edu

ABSTRACT

This paper explores a novel technique for monitoring film thickness in reactive ion etching that incorporates a micromachined sensor. The prototype sensor correlates film thickness with the change in resonant frequency that occurs in the micromachined platform during etching. The prototype sensor consists of a platform that is suspended over a drive electrode on the surface of the substrate and electrically excited into resonance. As material is etched from the platform, its resonant vibrational frequency shifts by an amount proportional to the amount of material etched, allowing etch rate to be inferred. The micromachined sensor is simulated using ANSYS 7.0. Simulation shows a direct correlation between platform film thickness and resonant frequency, as well as between the platform thickness and its capacitance. Modeling the sensor as a variable capacitor in an auto-zeroing floating gate amplifier (AFGA) circuit using HSPICE reveals that the deflections of the platform are amplified as expected.

Keywords: ANSYS, micromachined sensor, reactive ion etching, real-time control, resonant frequency

I. INTRODUCTION

State-of-the-art integrated circuits currently employ upwards of 60 million transistors, six layers of metal, and operate at clock frequencies over 1 GHz. As device dimensions continue to shrink and the speed of computing and communications systems increases, the effect of fluctuations in the manufacturing process becomes critical. Achieving tight specifications, given the continuing trend toward even further miniaturization, represents a major challenge in process control, an issue of critical importance to the semiconductor industry.

Reactive ion etching (RIE) has emerged as a crucial process in IC manufacturing because of the need to process devices with extremely small geometries and its ability to provide etches necessary to define submicron features. Ideally one would like to monitor the RIE process in-situ and control practical manufacturing parameters such as etch rate, film thickness, uniformity and anisotropy in real time. However, the current lack of robust on-line RIE monitoring and control methodologies can lead to unacceptable large volumes of defective material to be processed, resulting in significant wasted material and lower yield.

In this paper, a MEMS device for real-time RIE monitoring is evaluated using ANSYS and HSPICE simulation software. Ultimately, this device is expected to facilitate closed-loop feedback control using hardware and algorithms designed to integrate the sensor output signals with a neural network based control scheme. As an application vehicle, process control will be demonstrated in the Plasma Therm SLR series RIE system located in the Georgia Tech Microelectronics Research Center.

II. DESIGN CONSIDERATIONS

There are several key elements to consider in the design of a resonant sensor. The sensor typically has an element vibrating at resonance, which changes its output frequency as a function of a physical or chemical parameter. The vibrating element must be sensitive enough to adequately respond to the impact of what is being measured. The conversion from the measured quantity to the resonant frequency is based on the changes in internal forces and stresses in the resonator, which will change the resonant frequency. Another method of conversion is by mass loading on the vibrating element, where changes in the equivalent mass of the resonator change the resonant frequency. It is also possible to let the quantity to be measured change the shape of the resonator and thereby the stiffness of the vibrating structure, where an increased stiffness structure causes the resonant frequency to increase [1].

The mechanical Q-factor is an important parameter in the design and operation of resonators. The Q-factor depends on different damping mechanisms such as imbalances, viscous and acoustic radiation losses. The sensor should be designed with a very high Q, since the sensor stability and performance is then almost entirely dependent on the mechanical properties of the resonator elements [2].

III. PRINCIPLE OF OPERATION

The RIE sensor relies on a principle of electrostatic excitation and capacitive detection. This technique requires two electrodes that are in close proximity to a vibrating structure. A voltage applied across the electrodes creates an electrostatic pulling effect on the vibrating element and a displacement current result from the air gap capacitance between the electrodes and the vibrating structure.

Figure 1 shows the prototype micromachined sensor designed to assess RIE etch rate. The sensor is composed of a micro-machined platform 700um long, 140 um wide and extends 4 um above drive and sense electrodes [3]. During operation, the MEMS platform is coated with the same material to be etched from the substrate and positioned above drive and sense electrodes. As etching takes place, the sensor is excited into resonance by the drive electrodes

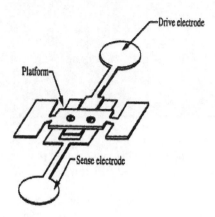

Figure 1: Prototype micromachined sensor showing the platform suspended over drive and sense electrodes [3].

As the mass loading of the platform decreases, its resonant vibration frequency shifts by an amount proportional to the amount of material etched, allowing etch rate to be inferred. The sensor provides direct real-time feedback on the wafer state during the etching by correlating film thickness with resonant frequency. The RIE sensor operability has been previously evaluated and the resonant frequency was determined experimentally to be approximately 33 kHz [3].

IV. MODELING

The design of the RIE sensor was analyzed and optimized by finite element analysis (FEA). This computer-based numerical technique for calculating the strength and behavior of engineering structures was conducted using ANSYS 7.0. With the ANSYS model, we were able to simulate deflection, stress, vibration and other phenomena. The MEMS structure was deformed and examined to establish the relationship between applied loads and deflections. A static ANSYS simulation of the RIE sensor yielded a fundamental frequency of 50.4 kHz, which is within the desired range. We were also able to verify the expected correlation between the platform height and its resonant frequency.

Steady state, transient and random vibration behavior was also analyzed to study the effects caused by the applied load to the sensor. Boundary conditions were employed to indicate where the structure was constrained and restricted against movement or in the case of our symmetric RIE

structure when only a portion of the sensor needed to be modeled.

The RIE sensor in [3] was composed of a polyimide platform 700 μm long and 140 μm wide that extends 4 μm above drive and sense electrodes. The platform was modeled as a uniform beam with both ends clamped and suspended above a ground plane. The ANSYS parameters are given in Table 1 below.

Parameter	μMKSV units
Platform Length	700 μm
Platform Width	140 μm
Platform Height	9 μm
Young's Modulus	8.3e3 GPa
Poisson Ratio	0.34
Density	1400e-18 kg/μm

Table 1: ANSYS parameters for RIE sensor model

The sensor was first modeled and meshed using a 2-D element type, and then extruded into the 3-D model shown in Figure 2.

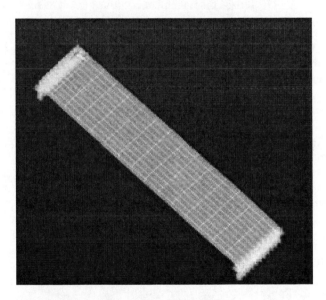

Figure 2: 3-D meshed model of RIE sensor

We obtained the profile of the first three mode shapes and resonant frequencies and found that the fundament and harmonics were within the expected range. Simulations compare closely with the expected value of 33.3 kHz. The results are summarized in Table II.

Harmonics	Frequency
f1	50.4 kHz
f2	137.3 kHz
f3	137.3 kHz

Table 2: RIE sensor resonant frequencies

To confirm the expected response of the sensor, an electrostatic force was applied. The application of a range of voltages between 10-100 volts caused the platform to deflect. The resonant frequency and displacements varied with the application of the load. Figure 3 show that the platform had its maximum deflection at the center, as expected, with a maximum displacement of 0.03216 μm.

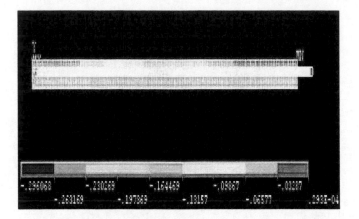

Figure 3: RIE sensor displacement contour plot showing maximum displacement of 0.03216 μm

Changes in the mass loading of the platform are simulated by varying the height of the platform which produces a change in the capacitance between the platform and electrodes, resulting in the change in the resonant frequency of the structure. The platform height was varied from 4μm to 16μm to simulate the changes in film thickness that occurs during etching. The height increases as material is etched since there is reduced bending in the middle of the beam due to its own weight. When the height of the platform increased, the resonant frequency shifts to the right on the frequency spectrum and vice versa.

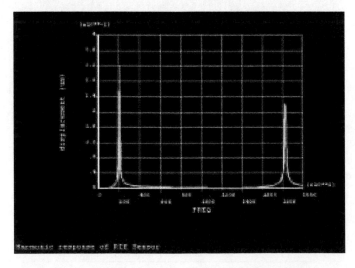

Figure 4: First harmonics f_1 = 20.45 kHz. Beam height is 9 μm. Maximum displacement at platform center is 0.032166 μm.

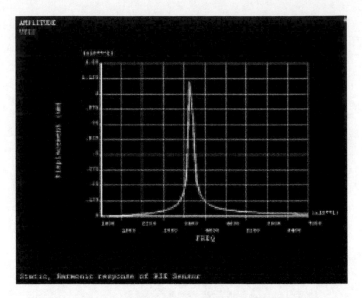

Figure 5: First harmonics f_1 = 35.65 kHz. Beam height is 16 μm. Maximum displacement at platform center is 0.011 μm.

The frequency spectra in Figures 4 and 5 illustrate how changing the height of the platform from 9μm to 16μm affects resonant frequency. The 9 μm platform has a resonant frequency of 20.45 kHz and a maximum displacement at the center of 0.03216 μm. The resonant frequency occurred at 35.65 kHz for a platform of height of 16μm. This value is closer to the expected resonance frequency of 33.1 kHz. The maximum displacement at the platform center was reduced to 0.011 μm. The direct correlation of platform thickness and resonant frequency is evident, potentially providing the desired mechanism for real-time feedback on the wafer state during etching.

The platform is modeled as a 3-D beam that is positioned above the drive and sense electrodes, represented by the ground plane. An applied voltage coupled to the beam causes the beam to deflect and change the air gap between the beam and the ground plane.

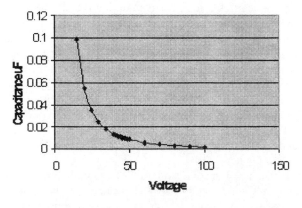

Figure 6: RIE sensor voltage vs. air gap capacitance

A plot of the air gap capacitance versus applied voltage provides insight into the stability of the sensor. Figure 6 shows that the device appears to be well behaved, since it exhibits a nearly linear dependence on the applied voltage for voltages below approximately 40 V.

V. AFGA IMPLEMENTATION

Figure 7 shows the proposed implementation of the sensor in an auto-zeroing floating gate amplifier (AFGA) circuit [4]. The variable capacitor represents the micromachined RIE structure. The change in capacitance of the sensor is directly related to the output voltage of the amplifier. The capacitance values were obtained from the ANSYS simulation and represent the effects of the mass loading of the top plate of the capacitive structure that occurs during etching.

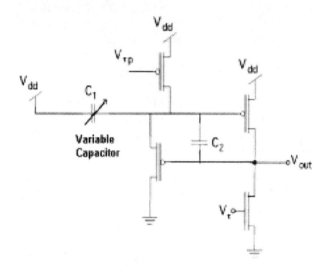

Figure 7: RIE sensor as variable capacitor in the AFGA circuit

This modified AFGA circuit with the RIE sensor as the variable capacitor is modeled using HSPICE to determine whether the small scale deflections of the platform and the small current signals that are used to locate the resonant condition will be properly sensed and amplified. From the voltage waveform shown in Figure 8, it appears that the unique sensor implementation results are consistent with what is expected from this amplifier circuit, since the small scale deflections of the platform are indeed amplified.

VI. DISCUSSION

Preliminary design of a MEMS sensor for RIE monitoring has been conducted and verified using ANSYS. ANSYS simulation is an effective tool for verification of the RIE sensor design and for illustrating potential areas for improvement. Ansys Parametric Design Language (APDL) input files were useful for building and solving the RIE

sensor models. APDL is an extension of the ANSYS command mode which allows the user to create an input text file and execute the commands in batch mode rather than line by line. The model is created in terms of parameters (variables), which makes it easy to change the design later.

Overall, the sensor design appears to be feasible. The relationship between the height of the platform and resonant frequency has been established. There may also be potential advantages in minimizing the platform width or varying the platform height. Optimization of the design will result in the best properties for the minimum cost and impact future fabrication processes. The successful development of this sensor could lead to significant productivity enhancements in RIE.

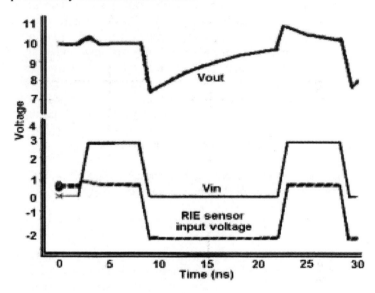

Figure 8: Voltage waveforms from RIE sensor as variable capacitor in the AFGA circuit

REFERENCES

[1] G. Stemme, "Resonant Silicon Sensors," J. Micromech. Microeng., vol. 1, pp. 113-125, 1991.

[2] R. Buser and N.F. DeRooij, "Capacitively Activated Torsional High-Q Resonator," J. Microelectro. Mech. Sys., pp. 132-135, February 1990.

[3] M. Baker and G. May, "A Novel In Situ Monitoring Technique for Reactive Ion Etching Using a Surface Micromachined Sensor," IEEE Trans. Semi. Manufac., vol. 11, no. 2, May 1998.

[4] P. Hasler, B. Minch, C. Diorio, and C. Mead, "An Autozeroing Floating-Gate Amplifier," IEEE Trans. Circuits and Systems, vol. 48, no.1, January 2001.

GEMS: A Revolutionary System for Environmental Monitoring

J. Manobianco[*], R. J. Evans[*], K. S. J. Pister[**], and D. M. Manobianco[***]

[*]ENSCO, Inc., Cocoa Beach, FL, USA, manobianco.john@ensco.com
[**]Dust, Inc., Berkeley, CA, USA, info@dust-inc.com
[***]Mano Nanotechnologies, Research, and Consulting, Melbourne, FL, USA, mano-nano@cfl.rr.com

ABSTRACT

This paper describes a revolutionary, new observing system designed for environmental monitoring that will integrate MicroElectroMechanicalSystems (MEMS) and nanotechnologies. The concept, known as Global Environmental MEMS Sensors (GEMS), features an integrated system of airborne probes that will remain suspended in the atmosphere for hours to days and take measurements as they are carried by wind currents. Previous efforts focused on defining the major feasibility issues associated with system design and development in the decadal time frame. The current 2-year project focuses on studying the major feasibility issues including costs/benefits and developing a technology roadmap.

Keywords: environmental monitoring, airborne probes, microelectromechanical systems, nanotechnology

1 INTRODUCTION

Technological advancements in MicroElectro MechanicalSystems (MEMS) and nanotechnology have inspired a revolutionary, multi-purpose observing system known as Global Environmental MEMS Sensors (GEMS). The GEMS concept features in situ, micron-scale airborne probes that can monitor all regions of the Earth with unprecedented spatial and temporal resolution.

The probes will be designed to remain suspended in the atmosphere for hours to days and take measurements of pressure, temperature, humidity, and wind velocity as they are carried by atmospheric currents. Once the probes settle out of the atmosphere, they could continue taking surface measurements over land or water. In addition to gathering meteorological data, the probes could be used for monitoring and predicting the dispersion of particulate emissions, organic and inorganic pollutants, ozone, carbon dioxide, and chemical, biological, or nuclear contaminants. GEMS could also be modified for use in a variety of defense applications including intelligence gathering and space situational awareness as well as aid in space and planetary exploration [1].

A phase I proof-of-concept effort funded by the NASA Institute for Advanced Concepts (NIAC) in May 2002 focused on validating the viability of GEMS and defining the major feasibility issues related to system design and development [2]. The goal of the 2-year, phase II project, which began in September 2003, is to study the major

feasibility issues in detail, examine the potential performance and cost benefits of the system, and develop a technology roadmap that NASA could use to integrate GEMS and its enabling micro and nanotechnologies into future missions and programs in the decadal time scale. An experiment is also planned in late 2004 to demonstrate the deployment of a sample network of airborne probes, built from commercial-off-the-shelf hardware and software. This paper highlights results from the phase I project and describes the details for studying the meteorological and engineering issues during the phase II effort.

2 DESCRIPTION

GEMS consist of an integrated system of MEMS-based probes that are envisioned to be mass-produced at very low per unit cost. Based on specific applications, the probes will be as small as 50–100 microns in one or more dimensions and lightweight enough to pose virtually no danger upon contact with persons or property. The blending of materials science and biology could produce new materials that significantly limit probe mass and potentially make them biodegradable or bioinert. Continued advancements in materials science and nanotechnology could pave the way for the design and development of morphing probes that could change shape using smart materials and structures. Such morphing capabilities may enable larger probes to remain airborne for longer periods of time.

The evolutionary successes found throughout nature suggest that nanobiotechnology be explored as a possible means to design and create micro and nanoscale components for the system. Nanobiotechnology merges biological systems with nanofabrication through the creation of tools that enable researchers to better understand how organic cells function. This knowledge can open a door for mimicking and integrating proven biological cellular devices and components into probe designs [3].

The size, mass, aspect ratio, component geometry, buoyancy control, and aerodynamic design will all determine how long probes remain airborne. Buoyancy control could be the most effective way to reduce the terminal velocity of probes and keep them suspended for much longer periods of time. Recent work by [4] and [5] demonstrated that hydrogen storage on nanocrystalline materials may be used to regulate buoyancy. Depending on the size and shape of the probes, aerodynamic design could also reduce their ballistic coefficient and terminal velocity.

Many examples of such design exist in the natural world, including simple dandelion spokes and threads of balloon spiders, as well as sophisticated evolved forms like auto-rotating samaras [6].

The probes will be capable of sensing a number of parameters including temperature, moisture, pressure, velocity, chemical/biological compounds, and radiation. Microfabricated devices for sensing temperature, humidity, and pressure are well developed [7]. Atmospheric wind velocity will be measured based on changes in probe position that will be determined using global positioning system-aided inertial navigation systems and network localization. With network localization, only several probes have knowledge of their locations and the remaining ones estimate their relative positions dynamically using on-board algorithms [8]. There has been a recent surge in microcantilever research for high-sensitivity chemical, biological, and radiation sensing [9]. In most cases, these existing technologies can be modified for integrated, low-power applications such as GEMS.

In a step towards their ultimate goal of 1-cubic millimeter wireless devices [10], University of California Berkeley researchers integrated a radio transmitter, sensor interface electronics, microprocessor, and memory into a single silicon chip measuring 5 mm^2 (Fig. 1). Single-chip integration of communication, sensing, and computation reduces the cost per device since post fabrication steps can be eliminated. A continued decrease in size and cost through mass production provides ample opportunities for low-cost deployment of large numbers of probes.

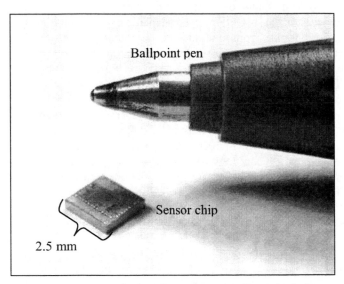

Figure 1. Sensor chip for the latest generation of wireless devices from University of California Berkeley.

The probes will communicate with other probes, intermediate nodes, data collectors (i.e. "mother ships"), and remote receiving platforms using radio and optical frequency transmissions to form a wireless, mobile, in situ network (Fig. 2). As part of a wireless network, the probes will not require recovery to collect data and therefore will be disposable. The data collectors are envisioned to be larger than individual probes and distributed throughout the atmosphere. There will likely be intermediate nodes in the hierarchy as well, capable of longer-range communication and/or more data processing than the individual probes, but still autonomous. The largest data collectors are envisioned to be a series of ground- and space-based receiving platforms that relay data from the intermediate nodes and provide a means to acquire and process data within wired regional and/or global terrestrial networks.

Figure 2. GEMS conceptualization illustrating possible probe design, deployment, dispersion, communication, and networking.

Deployment of GEMS is envisioned from both ground sites and airborne platforms such as aircraft, unmanned aerial vehicles, conventional weather balloons, and high altitude balloons [11]. There are numerous scenarios where high spatial and temporal resolution data from GEMS would be valuable to assess the potential for and development of high impact weather such as tornadoes, severe thunderstorms, and hurricanes, or to support special operations in data limited or data denied regions.

3 PHASE I: KEY ISSUES AND ENABLING TECHNOLOGIES

The phase I project focused on defining the major feasibility issues and determining the primary enabling technologies that must be clearly addressed for the future design and development of GEMS. A framework was also established for an interdisciplinary simulation-design-test (SDT) cycle that will be used extensively in phase II as part of detailed cost-benefit analyses to study the major feasibility issues. A summary of the key issues and enabling technologies is presented in Table 1 along with selected results from the phase I study. More details are available in [2].

Major Feasibility Issues	Primary Enabling Technologies
Probe design	Materials science, nanotechnology, biomimetics
Power	Batteries, micro fuel cells, solar energy
Communication	MEMS-based radio frequency and/or free-space optical systems
Networking	Artificial intelligence (autonomous self-healing networks)
Measurement	MEMS-based sensors
Data collection/management	Artificial intelligence, data mining
Cost	MEMS mass production/packaging, deployment strategies, data collection infrastructure

Table 2. Key issues and enabling technologies affecting GEMS design and development.

Several critical issues were identified in phase I relating to probe design. Power for the probes will likely come from a combination of batteries, fuel cells, and solar energy. Fuel cells are a very promising long-term solution for autonomous system power generation because their energy densities are dramatically higher than that of batteries. Both battery and fuel cell operation present a challenge at the colder temperatures in the upper atmosphere.

The probes will have to be robust enough to ensure that the electronics are isolated from the effects of liquid and frozen water, dust, chemical pollutants, radiation, and static electricity so MEMS packaging is an issue [12]. It will also be necessary to shield the sensors from the electronics so that they are not affected by thermal noise in the circuits.

At a minimum, the probes will need to communicate with some form of data collection nodes. If probe communications are handled using active Radio Frequency (RF) transmissions, the power requirements are projected to be 100 times greater than free-space optical communication. Communication from airborne probes using optical frequencies can have some advantages over RF due to the high antenna gains available when the wavelength is small; however, optical signals cannot penetrate dense cloud cover.

The issues related to probe deployment and dispersion were explored using a numerical weather prediction (NWP) model coupled with a Lagrangian particle model (LPM). To study probe deployment and dispersion over larger spatial and temporal scales, the models were configured to run over most of the northern hemisphere for a period of 24 days, and simulate probe release from a hypothetical network of high altitude balloons. This simulation demonstrated that atmospheric circulation patterns can disperse probes throughout the northern hemisphere in just 10 days with such a deployment scenario.

Observing system simulation experiments (OSSEs) were used to assess the impact of simulated probe data on forecasts of a selected weather event within a regional domain. The assimilation of simulated probe data resulted in a substantial reduction of forecast errors and improvements in predicted precipitation patterns. Excluding temperature, moisture, or wind data in the OSSE framework significantly degraded the model forecast not only for the parameter excluded, but for the precipitation forecast as well. This sensitivity test indicated that, at least for the case studied, the maximum data impact resulted when GEMS provided a full suite of temperature, moisture, pressure, and wind measurements.

4 PHASE II DEVELOPMENT PLAN

Pursuit of the GEMS concept is necessarily interdisciplinary, guided by realistic projections of progress in the miniaturization of probe components and demonstrable improvements in weather forecast accuracy by simulated meteorological observations from a mobile network of airborne probes. Interdisciplinary collaboration is the key element of the SDT cycle, quantifying trade-offs between weather forecast impacts and probe characteristics. The large number of possible trade-offs based on design characteristics such as sensor accuracy and sampling frequency, weather scenarios, and other factors constitute a multi-dimensional parameter space. Because of the high dimensionality and complexity of the parameter matrix, it is not reasonable to attempt a mechanical examination of every possible combination and permutation. Instead, extensive use will be made of modeling in the SDT cycle as a cost-effective and controlled way to explore the trade-offs, mapping out logical and self-consistent pathways for further detailed exploration. An overall system model of GEMS will be formulated in phase II following guidelines for systems engineering that will link together results from the component SDT cycles and provide a pathway for developing a technology roadmap. The system model will facilitate understanding how changes in one subsystem will affect other subsystems and the cost-benefit of the system.

4.1 Meteorological Issues

The meteorological portion of the development plan will include two parts. The first part will study probe dispersion and resulting distributions for several potential deployment strategies on a hemispheric scale. The second part of the meteorological plan will use the OSSE framework to study the data impact issues. A number of potential deployment strategies will be explored including probe release from high-altitude balloons, surface stations assuming positive buoyancy, manned/unmanned aircraft for targeted observations and vertical profiles similar to weather balloon measurements. Each of these deployment strategies will be simulated using the NWP/LPM models on a hemispheric scale. A comprehensive analysis of the resulting dispersion patterns will be conducted for variations of each deployment scenario. This analysis is critical to understand the scientific feasibility and cost practicality of each scenario.

The second part of the meteorological development plan will consist of regional data-impact analyses of specific deployment scenarios using OSSEs. OSSEs have been conducted for decades in meteorology to evaluate the potential impact of proposed remote and in situ observing systems, determine trade-offs in instrument design, and evaluate the most effective data assimilation methodologies to incorporate the new observations into regional and global NWP models [13].

4.2 MEMS and Engineering Issues

The major MEMS and engineering issues identified in Table 1 will be studied by focusing on physical limitations for measurement and signal detection based on fundamental and technical noise sources, scaling as the individual probes become smaller and the number of probes in the network becomes larger, and disciplined engineering optimization of the trade-offs which affect the cost, practicality, and feasibility of the overall system.

A detailed cost-benefit analysis will be performed to explore the costs associated with building and implementing GEMS, weighed against the benefits derived from the improved weather forecasts that will result from the data. Included in this cost analysis will be a quantitative comparison of GEMS with future observing systems.

When assessing the economic value of weather forecasts, [14] advocated a multidisciplinary approach that includes meteorology, economics, psychology, statistics, management science, and operations research. Likewise, the cost of the GEMS will vary based on the established requirements of achieving certain levels of data collection. These levels include, but are not limited to, probe density in space and time and measurement accuracy. Probe costs will also be driven by the engineering and manufacturing challenges discussed previously.

5 SUMMARY

This paper describes a new, in situ observing system known as Global Environmental MEMS Sensors (GEMS) that features airborne probes suspended in the air for hours to days. Results from the first phase of the project focused on identifying the major feasibility issues and key enabling technologies necessary for system development. The goals of the phase II project are to study these issues in detail including cost/benefits and development time and construct a technology roadmap. The presentation at the conference will highlight progress during the first 6 months of the phase II effort highlighting simulated probe dispersion and

deployment scenarios, power budgets, and limitations of micromachined circuits and sensors for this application.

6 ACKNOWLEDGEMENTS

This work was supported by the Universities Space Research Association's NASA Institute for Advanced Concepts.

REFERENCES

[1] J. Manobianco, J. Bickford, S. George, K. S. J. Pister, and D. Manobianco, "GEMS: A revolutionary concept for planetary and space exploration," Space Technology and Applications International Forum, 8-12 February 2004, Institute for Space and Nuclear Power Studies, Albuquerque, NM, 2004.

[2] J. Manobianco, "Global Environmental MEMS Sensors (GEMS): A revolutionary observing system for the 21st century," NIAC phase I final report, 59 pp., available online at http://www.niac.usra.edu /studies/, 2002.

[3] R. K. Soong, and coauthors, Science, 290, 1555-1558, 2000.

[4] A. Chambers, C. Park, R. T. K. Baker, and N. M. Rodriguez, Journal of Physical Chemistry, B 102, 4253-4256, 1998.

[5] P. Chen, X. Wu, J. Lin, and K. L. Tan, Science, 285, 91-93, 1999.

[6] J. Walker, Sci. American, 226-237, 1981.

[7] G. T. Kovacs, "Micromachined Transducers Sourcebook," McGraw-Hill, New York, 1-944, 1998.

[8] A. Savvides, C. Han, and M. Srivastava, "Dynamic fine-grained localization in ad-hoc networks of sensors," Proceedings, ACM MobiCom, Rome, Italy, 166-179, 2001.

[9] T. Thundat, Scanning, 23(2), 129, 2001.

[10] B. Warneke, M. Last, B. Liebowitz, and K. S. J. Pister, Computer, 34, 44-51, 2001.

[11] A. Pankine, E. Weinstock, M. K. Heun, and K. T. Nock, "In-situ science from global networks of stratospheric satellites," Preprints, Sixth Symp. On Integrated Observing Systems, Amer. Meteor. Soc., Orlando, FL, 260-266, 2002.

[12] K. James, "For MEMS, the key to lowering cost is all in the packaging," Small Times, available on line at http://www.smalltimes.com/document_ display.cfm?document_id=5079, 2002.

[13] M. S. F. V. De Pondeca, and X. Zou, Tellus, 33A, 192-214, 2001.

[14] R. Katz, and A. Murphy, "Economic Value of Weather and Climate Forecasts," Cambridge University Press, 222 pp., 1997.

Natural Frequency Shift of Rectangular Torsion Actuators due to Electrostatic Force

Zhixiong Xiao*, Yan Sun, Baoqing Li, Sanghui Lee, Karmjit S Sidhu, Ken K. Chin, and K. R. Farmer

New Jersey Institute of Technology
456 Tiernan Hall, Dept. of Physics, 161 Warren Street, Newark, NJ 07102, USA
*E-mail: xiao@adm.njit.edu

ABSTRACT

We have investigated the effect of natural frequency shift in single-crystal silicon, rectangular, electrostatic torsion actuators due to the electrostatic force. The electrostatic actuator was fabricated by using ultra-thin silicon wafer, SU-8 thermal compression bonding, and deep reactive ion etching. Natural frequency and Q factor of the actuator were tested in a high vacuum ambient. An analytical model was proposed to simulate the effect of natural frequency shift due to the electrostatic force. It showed that the natural frequency decreased with the increase in applied voltage, which could be used to tune the natural frequency of the torsion actuator. At a voltage near pull-in, the natural frequency was close to zero due to the negative electrostatic spring constant. The pull-in curve of the actuator was tested in air. Good agreement was found by comparing the results of analytical model with those of Coventor 2001 finite element analysis software and experiments.

Keywords: natural frequency shift, electrostatic torsion actuator, pull-in, SU-8 bonding, ultra-thin silicon wafer

1. INTRODUCTION

Electrostatic torsion actuators are important components in microelectromechanical systems (MEMS), especially for telecommunications applications such as cross-connects, switches and variable attenuators [1], and for use in optical scanning technology for imaging and display applications [2]. In [2-18], the rectangular torsion actuators and the related pull-in characteristics have been well addressed. Beyond the pull-in voltage, the applied electrostatic torque overcomes the spring restoring torque, and the movable structure will be snapped down to the bottom electrode plate.

The electrostatic force is widely used for the tuning of spring constant and natural frequency in comb-drive resonators and other parallel plate actuators [20-23]. By controlling the natural frequency of a MEMS device, e.g., accelerometer or gyroscope, the sensitivity of this device can be actively controlled to a broad band of excitation frequencies [21]. For electrostatic torsion actuators, the

natural frequency can also be shifted due to the electrostatic torque related to the applied bias. No systematic study on this issue was found in available literatures. In this research, a type of full electrode micro electrostatic actuators were designed and fabricated by using ultra-thin silicon wafer, SU-8 bonding and deep reactive ion etching to examine this natural frequency shift at different voltages.

2. DESIGN AND FABRICATION

Figure 1 shows a schematic view of a typical rectangular torsion actuator, where a_3 and b are the actuator length and width, d_0 is the air gap separation, a_2 is the length from the rotation axis to the far end of the fixed electrode, and a_1 is the length from the rotation axis to the near end of the electrode. The electrode was fabricated on oxidized silicon or other dielectric materials. In this study, actuators with full electrode ($\beta=a_2/a_3=1$, $\theta_{pin}=\alpha_{pin}/\alpha_{max}=0.44$) were designed and fabricated [11]. Table 1 shows the designed and fabricated actuator parameters. The parameter differences were mainly due to the photolithography and deep reactive ion etching.

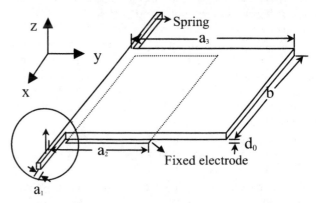

Figure 1 Schematic view of electrostatic rectangular torsion actuators

When a signal with a combination of DC and AC components is applied on the actuator, i.e., on the bottom electrode and the movable plate, and if the torsion mode is dominant, the electrostatic torque can be written as [10, 11]

$$T_e(\alpha) = \frac{\varepsilon_0 (V_{DC} + v_{AC} \sin wt)^2 b}{2\alpha^2} \left[\frac{1}{1-\beta\theta} - 1 + \ln(1-\beta\theta) \right] \quad (1)$$

$$\propto (V_{DC}^2 + 0.5 v_{AC}^2 + 2V_{DC}v_{AC} \sin wt - 0.5 v_{AC}^2 \cos 2wt)$$

where ω is the angular frequency of the AC component, and α is the tilt angle at a voltage of $V_{DC}+v_{AC}\sin\omega t$. In the above equation, $\gamma = a_1/a_3$ is omitted because a_1 is usually much smaller than a_2 and a_3.

According to equation (1), the vibration amplitude of the actuator will reach a peak value when the signal frequency f or 2f equals the natural frequency f_0.

The natural frequency will decrease with the increase in applied voltage because of the induced negative spring constant, K_e, due to the external electrostatic force. K_e can be written as

$$K_e(\alpha) = \frac{\partial T_e(\alpha)}{\partial \alpha} = -\frac{\varepsilon_0 V^2 b}{2\theta^3} \frac{a_3^3}{d^3} \times \quad (2)$$

$$\left[\frac{1}{1-\beta\theta} - 1 + \ln(1-\beta\theta) - \frac{1}{2}\frac{\beta\theta}{(1-\beta\theta)^2} + \frac{1}{2}\frac{\beta\theta}{(1-\beta\theta)} \right]$$

The natural frequency, f_e, at any voltage before pull-in can be written as

$$f_e = \frac{1}{2\pi} \sqrt{\frac{K_m - K_e}{I}} \quad (3)$$

where I and K_m are the mass moment of the inertia about the rotation axis and the mechanical spring constant of torsion mode. I can be written as [11]

$$I_\alpha = \iiint \rho(y^2 + z^2) dx dy dz = \frac{1}{3}\rho a_3 bw(a_3^2 + w^2)$$

$$+ \frac{2}{3}\rho twl(t^2 + w^2) \approx \frac{1}{3}\rho a_3 bw(a_3^2 + w^2) \quad (4)$$

where ρ is the density of silicon.

When t<w, the torsion mode mechanical spring constant K_m can be written as [19]

$$K_m = \frac{2}{3}G\frac{wt^3}{l}\left[1 - \frac{192t}{\pi^5 w} \sum_{n=1,3,5...} \frac{1}{n^5} \tanh\left(\frac{n\pi w}{2t}\right) \right] \quad (5)$$

where G is the shear modulus and equals 0.73×10^{11}Pa for silicon[10], and w, l and t are the spring width, length, and thickness, respectively.

Table 1 Designed and fabricated actuator parameters. l, w, and t are respectively the length, width, and thickness of the two folded springs. Folded spring was used in the design of actuators.

	l (μm)	w	t	a_3	b	d_0	a_2
Design	200+490+530 =1220μm	25	5.0	700	700	20	700
Test	200+490+530 =1220μm	23.4	4.26	700	700	21.2	700

Figure 2 shows the fabrication process [15] for the test devices by using ultrathin silicon wafers [24], SU-8 bonding [25] and deep reactive ion etching (DRIE) technology. The main process steps are illustrated. The oxidized silicon wafer was sputtered with 5000Å thick aluminum, followed by photolithography and aluminum

etching in phosphoric acid at 40°C. The wafer was spin-coated with SU-8 negative photoresist at 5000rpm followed by photolithography. An ultra-thin silicon wafer with approximate thickness of 25μm was bonded to the patterned Su-8/SiO$_2$/Si substrate through SU-8 thermal compressive bonding at 105°C. 5000Å of aluminum was then sputtered and patterned on the bonded ultra-thin silicon surface using photolithography and phosphoric wet etching at 40°C. Deep reactive ion etching (DRIE) was used to finally release the structure, by using photoresist as a mask material. Finally the photoresist was removed in oxygen plasma.

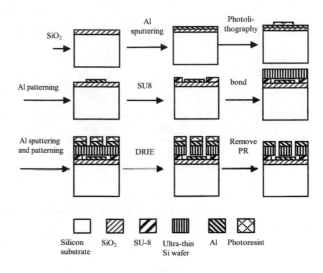

Figure 2 Main fabrication process steps for the electrostatic actuators.

3. DEVICE TEST

Figure 3 shows an optical profilometry image of the electrostatic actuator with VEECO NT 3300 system at an applied voltage of 15.8V, which is just 0.1V smaller than the pull-in voltage V_{pin} of 15.9V. As illustrated in Table 1, a folded spring was used to save actuator area.

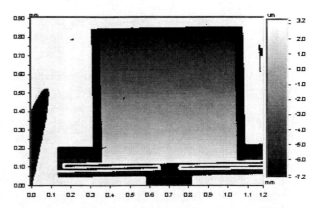

Figure 3 An optical profilometry image of the electrostatic actuator used for pull-in and natural frequency shift test with an applied voltage of 15.8V.

Figure 4 shows the tilt angle versus applied voltage, simulated with Coventor 2001, measured in air, and calculated by using the analytical model in [11]. Good agreement was found for these results. The fractional angle ratio at pull-in, θ_{pin}, was 43.2%, which was close to the theoretical prediction of 44% for full electrode rectangular electrostatic actuator [11].

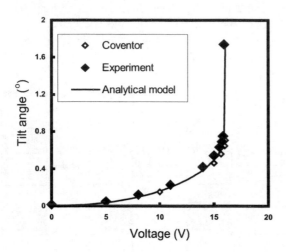

Figure 4 Tilt angle versus applied voltage for the electrostatic actuator.

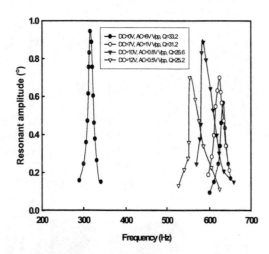

Figure 5 Vibration amplitude versus frequency of the AC component in vacuum.

Figure 5 shows the vibration amplitude of the device versus the frequency of AC signal. The actuator was tested in a high vacuum of 10^{-3}mBar. The Q factor was found to be around 20 to 30. We attribute this low Q factor to the SU-8 bonding. SU-8 is a kind of negative photoresist. The bonding quality is worse than anodic bonding and silicon fusion bonding. For next generation devices, we may use anodic bonding to replace the SU-8 thermal compression bonding to achieve a better Q factor. Figure 5 shows two curves at V_{DC}=0V and V_{AC}=6Vpp at 316Hz and 632Hz. At other voltages, it only shows one

curve at high frequencies, the reason being that the device is very sensitive at high DC voltages. Because when the actuator operated near the pull-in, the amplitude of AC component must be small to avoid pull-in and then the vibration amplitude was small to get a continuous curve. But the resonant peaks were observed at all half frequency points.

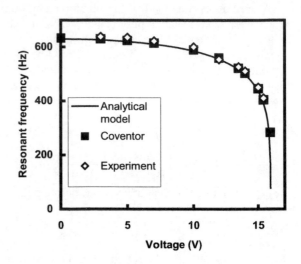

Figure 6 Natural frequency of the actuator versus the applied voltage.

Figure 6 shows the natural frequency versus the applied DC voltage obtained by testing in vacuum, simulating with Coventor 2001 and using analytical model of equation (3). During the test, the amplitude of AC component was kept as small as possible to suppress the influence of AC signal on natural frequency. Good agreement was found among these results. The natural frequency decreased with the increase in the applied voltage. At a voltage near pull-in, the natural frequency was close to zero. The natural frequency at pull-in was zero in theory. DC voltage could be used to tune the natural frequency of the electrostatic torsion actuators, which might be useful for mirror, switch and other applications.

4. CONCLUSIONS

This paper has examined the natural frequency shift of rectangular electrostatic torsion actuators due to the applied electrostatic force. We found that the natural frequency decreased with the increase in applied voltage based on a proposed equation, measurement of fabricated device, and simulation with finite element analysis software, Coventor 2001. At a voltage near pull-in, the natural frequency was observed to be close to zero. Good agreement was found among all the results. The electrostatic actuator was fabricated by using ultra-thin silicon wafer, SU-8 thermal compression bonding and deep reactive ion etching. The Q factor was found to be quite low even at very high vacuum due to the loose SU-8 bonding.

REFERENCES

[1] J. A. Walker, "The future of MEMS in telecommunications networks," *J. Micromech. Microeng.*, 10, R1-R7(2000).

[2] J. Buhler, J. Funk, J. G. Korvink, F. P. Steiner, P. M. Sarro, and H. Baltes, "Electrostatic aluminum micromirrors using double-pass metallization," *IEEE J. Microelectromech. Syst.*, 6, 126-135(1997).

[3] H. Toshiyoshi and H. Fujita, "Electrostatic micro torsion mirrors for an optical switch matrix," *IEEE J. Microelectromech. Syst.*, 5, 231-237(1996).

[4] S. W. Chung, J. W. Shin, Y. K. Kim, and B. S. Han, "Design and fabrication of micromirror supported by electroplated nickel posts," *Sens. Actuators A*, 54, 464-467(1996).

[5] Y. Uenishi, M. Tsugai, and M. Mehregany, "Micro-opto-mechanical devices fabricated by anisotropic etching of (110) silicon", *J. Micromech. Microeng.*, 5, 305-312(1995).

[6] M. Fischer, M. Giousouf, J. Schaepperle, D. Eichner, M. Weinmann, W. von Munch, and F. Assmus, "Electrostatically deflectable polysilicon micromirrors-Dynamic behavior and comparison with results from FEM modeling with ANSYS," *Sens. Actuators A*, 67, 89-95(1998).

[7] L. J. Hornbeck, "Deformable-mirror spatial light modulators", *SPIE Critical Review Series, vol.1150, Spatial Light Modulators and applications III*, pp. 86-102,1989.

[8] P. M. Osterberg, "Electrostatically actuated micromechanical test structures for material property measurements", Ph.D dissertation, Dept. Elec. Eng. Comput. Sci., Massachusetts Institute of Technology, September, 1995.

[9] R. K. Gupta, "Electrostatic pull-in test design for in-situ mechanical property measurements of microelectromechanical systems(MEMS)", Ph.D dissertation, Dept . Elec. Eng. Comput. Sci., Massachusetts Institute of Technology, June, 1997.

[10] O.Degani, E. Socher, A. Lipson, T.Leitner, D. J. Setter, S. Kaldor, and Y. Nemirovsky, "Pull-in study of an electrostatic torsion microactuator," *IEEE J. Microelectromech. Syst.*, 7, 373-379(1998).

[11] Z. Xiao, X. Wu, W. Peng and K. R. Farmer, "An angle-based design approach for rectangular electrostatic torsion actuators", *IEEE J. Microelectromech. Syst.*, 10, 561-568(2001).

[12] Z. Xiao, W. Peng, X. T. Wu, and K. R. Farmer, "Pull-in study for round double-gimbaled electrostatic torsion actuators", *J. Micromech. Microeng.*, 12, 77-81(2002).

[13] Y. Nemirovsky and O. Bochobza-Degani, "A Methodology and model for the pull-in parameters of electrostatic actuators", *IEEE J. Microelectromech. Syst.*, 10, 601-615(2001).

[14] O.Degani, and Y. Nemirovsky, "Design considerations of rectangular electrostatic torsion actuators based on new analytical pull-in expressions", *IEEE J. Microelectromech. Syst.*, 11, 20-26(2002).

[15] Z. Xiao and K. R. Farmer, "Instability in micromachined electrostatic torsion actuators with full travel range", Transducers'2003, Boston, pp. 1431-1434, June, 2003.

[16] Jin-Ho Lee, Y.-C. Ko, H.-M. Jeong, et al., "SOI-based fabrication processes of the scanning mirror having vertical comb fingers", Sens. Actuators A, 102, 11-18(2002).

[17] G. -D. J. Su, S.-S. Lee, and M. C. Wu, "Optical scanners realized by surface-micromachined vertial toraion mirror", IEEE Photonics Technology Letters, 11, 587-589(1999).

[18] H. Toshiyoshi, D. Miyauchi, and H. Fujita, "Electromagnetic torsion mirrors for self-aligned fiber-optic crossconnectors by silicon micromachining", IEEE J. of Selected Topics in Quantum Electronics, 5, 10-17(1999).

[19] S. P. Timoshenko and I. N. Goodier, *Thoery of Elasticity*. New York: McGraw-Hill, 1970, ch. 3, pp.40-46

[20] Francis Tay E. H., Kumaran R. B. L. Chau and Logeeswaran VJ, "Electrostatic spring effect on the dynamic performance of microresonators", Technical Proceedings of the 2000 International Conference on Modeling and Simulation of Microsystems.

[21] Yie He, J. Marchetti, C. Gallegos, and F. Maseeh, "Accurate fully-coupled natural frequency shift of MEMS actuators due to voltage bias and other external forces", The twelfth IEEE International Conference on Micro Electro Mechanical Systems, pp.321-325, Jan. 1999.

[22] K. B. Lee and Y. -H. Cho, "Frequency tuning of a laterally driven microresonator using electrostatic comb array of linearly varied length", 1997 International Conference on Solid-State Sensors and Actuators, pp.113-116, 1999.

[23] K. Y. Park, C. W. Lee, Y. S. Oh and Y. H. Cho, "Laterally oscillated and force-balanced micro vibratory rate gyroscope supported by fish hook shape springs", Proceedings of IEEE Tenth Annual International Workshop on Micro Electro Mechanical Systems, pp.26-30, Jan. 1997.

[24] Ultrathin silicon wafers are available from Virginia Semiconductor, Inc., Fredericksburg, VA.

[25] S. K. Sampath, L. St.Clair, X. Wu, D. V. Ivanov, Q. Wang, C. Ghosh and K. R. Farmer, "Rapid MEMS Prototyping using SU-8, Wafer Bonding and Deep Reactive Ion Etching," Proceedings of the 14th Biennial University/Government/Industry Microelectronics Symposium, Richmond, VA, 2001, p. 158.

Electrostatic Tuning of a Micro-Ring Gyroscope

B.J.Gallacher[*], J. Hedley[*], J. S. Burdess[*], A. J. Harris[*], A. Rickard[**] and D. King[**]

[*1]Institute for Nanoscale Science & Technology, Newcastle University,U.K., b.j.gallacher@ncl.ac.uk
[**]QinetiQ (Malvern), St Andrews Rd, Malvern, Worcs, U.K.

ABSTRACT

This paper describes a procedure for identifying and correcting anisoinertia and anisoelasticity present in a micro-ring gyroscope. The major disadvantages of popular tuning methods such as mass trimming using laser ablation or focussed ion beam is that it can only be employed prior to packaging. Therefore tuning to correct for environmental or structural changes during operation is not possible. However, electrostatic tuning may be implemented on-chip during operation thus enabling active tuning. Furthermore, mass trimming procedures introduce a degree of damage to the structure, which may reduce the intrinsic high quality factor desirable in the ring gyroscope.

Keywords: tuning, electrostatic, gyroscope, ring

1 INTRODUCTION

The isotropic inertial and elastic properties of a perfectly formed axisymmetric ring ensure that the flexural modes of vibration exist in degenerate pairs. Thus the natural frequencies of the modes within each pair are identical and their orientation is indeterminate. This "tuned" phenomenon may be utilised in order to improve the performance of ring gyroscopes. Any structural imperfection introduced in the ring or its suspension during fabrication or operation will result in anisotropic inertial and elastic properties [1,2]. This is manifested as a frequency split between the otherwise "tuned" natural frequencies, which reduces the sensitivity of the gyroscope to rates of rotation. In addition, the modes of vibration are coupled both statically and inertially. As the ring gyroscope utilises Coriolis coupling between the modes of vibration in order to sense rate of rotation, the anisoinertia and anisoelasticity must be minimised in order to prevent spurious measurements. Further, as the quality factor of typical ring gyroscope may be in excess of 10000, precise tuning of the modes is essential in order to take full advantage of the amplification possible through matching the mode frequencies. With such tuning control the overall performance of the ring gyroscope will be greatly improved and may lead to inertial grade MEMS gyroscopes.

2 DYNAMICS OF THE IMPERFECT RING

Figure (1) shows a schematic of the QinetiQ ring gyroscope of ring of radius a, width b and thickness d and the electrode displaced radially from the ring by a distance h_o. The dynamics of the ring are expressed with respect to an axis X passing through the ring centre O and the midpoint of an electrode.

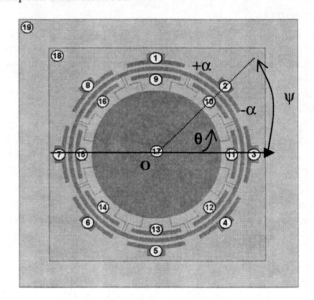

Figure (1) Schematic of Gyroscope

The radial displacement u of a point on the centre line of the ring may be expressed in terms of the undamped, unforced flexural modes of an unsupported ring by

$$u = q_1 \cos n\theta + q_2 \sin n\theta. \qquad (1)$$

The electrical energy stored in the capacitor, formed between the ring (assumed to be held at earth potential) and the drive electrode, biased with a voltage V is given by

$$E_E = \frac{\varepsilon_o a d}{2 h_o} V_i^2 \int_{\psi-\alpha}^{\psi+\alpha} \left[1 + \frac{u}{h_o} + \left(\frac{u}{h_o}\right)^2 + \left(\frac{u}{h_o}\right)^3 + \cdots \right] d\theta. \qquad (2)$$

It is assumed that the air-gap h_o, is small with respect to the ring radius and so the ring and the electrode form a parallel plate capacitor. As the radial displacement of the ring is assumed to be small compared to the nominal air gap separation, terms of order greater than $(u/h_o)^3$ may be ignored. Fringing effects between adjacent electrodes are neglected. The generalised stiffness matrix and forcing vector for the arbitrary drive electrode may be obtained from substituting in the expression for the radial displacement given by equation (1) into equation (2) and by determining $dE_E\big/dq_i$ and $d^2E_E\big/dq_i^2$ for i=1..2.

Consider the general case where driving is performed by activating the $\overline{Z}$ electrodes positioned at the antinodal postions of the $\sin n\theta$ mode and tuning is performed using a set consisting of t electrodes. The equation of motion for the in-plane flexural modes of order n may be written in the form

$$\begin{bmatrix} 1+\varepsilon_1 & \varepsilon_2 \\ \varepsilon_2 & 1-\varepsilon_1 \end{bmatrix}\begin{bmatrix} \ddot{q}_1 \\ \ddot{q}_2 \end{bmatrix} +$$

$$\frac{1}{m_o}\left\{ k_o\begin{bmatrix} 1+\mu_1 & \mu_2 \\ \mu_2 & 1-\mu_1 \end{bmatrix} - \sum_{j=1}^{t}[K]_j \right\}\begin{bmatrix} q_1 \\ q_2 \end{bmatrix} = \frac{1}{m_o}\overline{Z}\beta V_z^2\begin{bmatrix} 0 \\ 1 \end{bmatrix} \quad (3)$$

where $\beta = \dfrac{\varepsilon_o ad}{h_o^2}\dfrac{1}{n}\sin(n\alpha)$

and $\varepsilon_1, \varepsilon_2, \mu_1$ and μ_2 represent perturbations in the generalised mass and stiffness values, m_o and k_o of the ideal ring. In the perfectly axisymmetric case the natural frequencies of both modes corresponding to generalised coordintes q_1 and q_2 are equal thus $\omega_1=\omega_2$. Damping is assumed small and has been neglected. The electrostatic stiffness matrix from the tuning electrodes may be expressed in the form

$$\sum_{j=1}^{t}[K]_j = \sum_{l}^{L}[K_c]_l + \sum_{m}^{M}[K_s]_m$$

$$= Lk_o\lambda_c\begin{bmatrix} 1 & 0 \\ 0 & \sigma \end{bmatrix} + Mk_o\lambda_s\begin{bmatrix} \sigma & 0 \\ 0 & 1 \end{bmatrix} \quad (4)$$

where

$$\sigma = \frac{\left(2\alpha - \dfrac{1}{n}\sin 2n\alpha\right)}{\left(2\alpha + \dfrac{1}{n}\sin 2n\alpha\right)}, \quad \sigma \ll 1,$$

and

$$k_o\lambda_{c,s} = \left(2\alpha + \frac{1}{n}.\sin 2n\alpha\right)\frac{\varepsilon_o ad}{2h_o^3}U_{c,s}^2$$

The cyclic arrangement of tuning electrodes has been separated into two distinct sets with each set consisting of a maximum of $2n$ electrodes. The parameters L and M are

the number of tuning electrodes employed. The two sets of electrodes corresponding to stiffness matrix K_c and K_s may be employed to reduce the natural frequencies of the $\cos(n\theta)$ and $\sin(n\theta)$ modes respectively. Which set of electrodes is used depends on which mode has the higher natural frequency. For example, suppose $\omega_1>\omega_2$, then reduction of the natural frequencies of the $\cos(n\theta)$ mode is achieved by employing the set corresponding to stiffness matrix K_c. Similarly, if $\omega_2>\omega_1$ then reduction of the natural frequencies of the $\sin(n\theta)$ mode is achieved by employing the set corresponding to stiffness matrix K_s. However, it will be shown later that if any coupling between the $\cos(n\theta)$ and $\sin(n\theta)$ modes is present then it will be impossible to completely eliminate the frequency split such that $\omega_1=\omega_2$. Removal of the coupling between the modes is not possible with this electrode configuration for the $n=2$ case.

The external force and thus the response are assumed to be harmonic with frequency ω. Therefore,

$$V_1^2 = V_{01}^2 e^{i\omega t} \text{ and } q_1 = q_{01}e^{i(\omega t-\delta)}, \quad q_2 = q_{02}e^{i\omega t} \quad (5)$$

where $\delta=0,\pi$ and represents the phase angle between the responses. Assuming excitation occurs near resonance then

$$\omega^2 = p^2(1+\eta) \text{ where } \eta \ll 1 \quad (6)$$

and p is the natural frequency of the undamped perfect ring. By making use of equations (4), (5) and (6), equation (3) may be written in the form

$$[Y]\begin{bmatrix} q_{01}e^{-i\delta} \\ q_{02} \end{bmatrix} = \frac{1}{m_o}\overline{Z}\beta V_z^2\begin{bmatrix} 0 \\ 1 \end{bmatrix}$$

where $[Y] = p^2\begin{bmatrix} A+B-\eta & C \\ C & A-B-\eta \end{bmatrix}$. $\quad (7)$

The terms A, B and C defined as
$$A = -(1+\sigma)\left[L\lambda_c + M\lambda_s\right]$$
$$B = f - \frac{L\lambda_c}{2}(1-\sigma) - \frac{M\lambda_s}{2}(1-\sigma)$$
$$C = g$$
where
$$f = \varepsilon_1 - \mu_1 \text{ and } g = \varepsilon_2 - \mu_2.$$

The natural frequencies of the imperfect ring may be obtained from the requirement that $|Y = 0|$, which yields the expression

$$\eta = A \pm \sqrt{B^2 + C^2}. \quad (8)$$

The frequency split between the modes is therefore given by
$$|\omega_1 - \omega_2| = \Delta\omega \approx p\sqrt{B^2 + C^2}. \quad (9)$$

The conditions $B=0$ and $C=0$ are necessary for identical natural frequencies and yield

$$f = \frac{1}{2} L \lambda_c (1-\sigma) - \frac{1}{2} M \lambda_s (1-\sigma) \qquad (10)$$

Clearly for this electrode configuration the coupling between the modes determined by g cannot be altered and thus $C \neq 0$. Appropriate values for λ_c, λ_s, must to chosen in order to remove the direct imperfection parameters f.

3 DETERMINING THE MIS-TUNING AND IMPERFECTION PARAMETERS

When $\lambda_c = \lambda_s = 0$, equation (8) reduces to $\eta = \pm \sqrt{f^2 + g^2}$. Therefore the damped natural frequencies of the imperfect ring are

$$\omega_1 \approx p \left(1 \pm \frac{1}{2} \eta_c \right) \qquad (11)$$

$$\omega_2 \approx p \left(1 \mp \frac{1}{2} \eta_c \right) \qquad (12)$$

where $\eta_c = \sqrt{f^2 + g^2}$. $\qquad (13)$

In addition, an expression for η_c in terms of the undamped natural frequencies may be obtained using equations (11) and (12). Thus

$$\eta_c = 2 \left| \frac{(\omega_1 - \omega_2)}{\omega_1 + \omega_2} \right|. \qquad (14)$$

To determine f and g, recall the equation of motion described by equation (7). Multiplying both sides of equation (7) by the inverse of the matrix Y, when $\lambda_c = \lambda_s = 0$ yields

$$\begin{bmatrix} q_{01} e^{-i\delta} \\ q_{02} \end{bmatrix} = \frac{\beta \bar{Z} V_z^2 p^2}{m_o |Y|} \begin{bmatrix} -(f+\eta) & -g \\ -g & (f-\eta) \end{bmatrix} \begin{bmatrix} 0 \\ \pm 1 \end{bmatrix} \qquad (15)$$

where

$$|Y| = \eta^2 - f^2 - g^2. \qquad (16)$$

From equation (16), it is clear that identical values of $|Y|$ are obtained for $\eta = \pm \eta_c$. The ratio of the amplitudes of the modes when driven at their respective natural frequencies corresponding to $\eta = \pm \eta_c$ is given by

$$r = \frac{q_{01}(\pm \eta_c)}{q_{02}(\mp \eta_c)} = \frac{-g}{f \pm \eta_c} e^{i\delta}. \qquad (17)$$

The phase angle δ between the modes may be determined from Nyquist plots of displacement or velocity frequency response. The substitution of g from equation (17), into the expression for η_c given by equation (13) yields a quadratic equation in f of which the permissible roots are

$$f = \pm \eta_c \left[\frac{1 - r^2}{1 + r^2} \right]. \qquad (18)$$

Equations (17) and (18) show that the imperfection parameters f and g are now expressed in terms of measurable quantities r, η_c and δ. The signs of f and g are determined from r and δ.

For $r = \frac{q_{01}(\eta_c)}{q_{02}(-\eta_c)}$, $f > 0 \Leftrightarrow r < 1$ and $g > 0 \Leftrightarrow \delta = \pi$. $\quad (19)$

For $r = \frac{q_{01}(-\eta_c)}{q_{02}(+\eta_c)}$, $f > 0 \Leftrightarrow r > 1$ and $g > 0 \Leftrightarrow \delta = 0$. $\quad (20)$

Clearly from equation (18), $f=0$ when $r=1$.

4 CORRECTION OF ANISOINERTIA AND ANISOELASTICITY

The frequency split due to the imperfection parameter f may be removed using the set of electrodes corresponding to stiffness matrix K_c or K_s. The decision as to which set of electrodes are employed is determined by the sign of f. The sign of f may be determined from equations (19) and (20). From equation (10) the condition $B=0$ is met when

$$\lambda_c = \frac{2f}{L(1-\sigma)} \quad \text{if } f > 0 \qquad (21)$$

$$\lambda_s = \frac{-2f}{M(1-\sigma)} \quad \text{if } f < 0. \qquad (22)$$

Thus, the voltages required to be applied to the two tuning electrode sets to remove the imperfection due to f are given by

$$U_c^2 = \frac{1}{\left(2\alpha + \frac{1}{n}.\sin 2n\alpha \right)} X \omega_{1,2}^2 m_o \lambda_c \text{ and } U_s^2 = 0 \text{ for } f > 0 \qquad (23)$$

$$U_s^2 = \frac{1}{\left(2\alpha + \frac{1}{n}.\sin 2n\alpha \right)} X \omega_{1,2}^2 m_o \lambda_s \text{ and } U_c^2 = 0 \text{ for } f < 0. \qquad (24)$$

where $\omega_{1,2}$ corresponds to either of the undamped natural frequencies ω_1 and ω_2 of the imperfect ring.

5 PRACTICAL DEMONSTRATION

The drive and sense circuitry is shown in figure (2). Mode (1) and mode (2) correspond to the $\sin(n\theta)$ and $\cos(n\theta)$ modes, respectively. A bias of 30 V was placed on the ring; the silicon substrate, chip and the two unused sense electrodes were also set to this voltage. In order to investigate the electrostatic tuning process over a larger range, artificial mis-tuning was introduced by applying a -60 V bias (with respect to the ring) to the set of tuning electrodes corresponding to the $\sin(n\theta)$ mode. For this device $\omega_1 > \omega_2$. Thus a variable voltage supply was connected to the set of tuning electrodes corresponding to the $\cos(n\theta)$ mode. In this case four tuning electrode were used to adjust the direct terms so $L=4$. Table (1) shows describes the device dimensions.

a (mm)	2.05
b (μm)	100
d (μm)	100
α (deg)	14.25
h_o (μm)	4.7+/-0.25

Table 1: Device dimensions

The $\sin(n\theta)$ mode of the gyroscope was driven with 0.25 V a.c with a 25 V d.c. component. Two drive electrodes were employed thus $\overline{Z} = 2$. The current detected by the sense electrodes of each mode was converted to a voltage and fed into a HP-3562 dynamic signal analyser together with the a.c. drive signal. The remaining drive and sense electrodes were held at the same potential as the ring. All dynamic measurements were obtained at a pressure of 0.02 mbar.

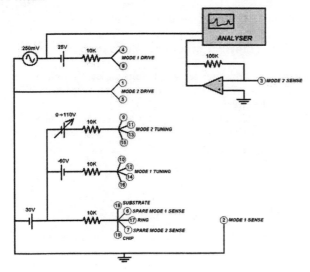

Figure (2) Experimental set-up

5.1 Tuning voltage

Table (2) shows the measured values of r, η_c, δ and. also the imperfection parameters f and g determined from r, η_c and δ by equations (17) and (18). The electrode configuration of the device does not permit adjustment of the coupling between the $\cos(n\theta)$ and $\sin(n\theta)$ modes for the case where $n=2$. Therefore the imperfection parameter g will remain constant throughout the tuning procedure and will ultimately determine the minimum frequency split. For this device $f > 0$ and the value of the voltage U_c required to eliminate f is described by equation (23). The tuning voltage U_c corresponding to the electrode gap size of 4.7+/- 0.25 μm has the value of 59.4 +/- 4.9 V.

parameter	value
η_c	0.0076
r	0.1025
δ	π
f	0.0074
g	0.0015

Table 2:Imperfection parameters

5.2 Tuning cuves

Figure (3) illustrates the theoretical and experimental tuning curves obtained using equations (9) and (23) with M=0. Two theoretical curves corresponding to the +/- 0.25 μm tolerance on the nominal value of the electrode gap have been drawn to create the band shown in the figure.

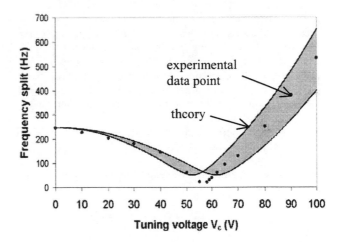

Figure (3) Frequency split versus voltage with coupling between modes present

6 CONCLUSION

The predicted tuning voltage to remove the direct imperfections was found theoretically to be 59.4 +/- 4.9 V. Experimentally the voltage at which the frequency split was a minimum occured at a voltage of 58 V. Therefore there is agreement between the theoretical model and experiment. The cross-coupling between the modes is not altered for the electrode arrangement considered when $n=2$. Additional electrodes would have to be used in order to permit adjustment of the cross-coupling and hence completely eliminate the frequency split between the modes.

REFERENCES
1. B.J. Gallacher et al. "Multimodal tuning of a vibrating ring using laser ablation" *Journal of Mechanical Engineering Science*, Part C, vol.217, p557-576.
2. C. Painter and A. Shkel. "Identification of anisoelasticity for electrostatic trimming of rate integrating gyroscopes", *2002 SPIE Annual International Symposium on Smart Structures and Materials*, San Diego, CA, March 2002.

Macromodel of Intelligent Sensor Structure with Accelerometer

M. Husak, J. Jakovenko, P. Kulha, J. Novak, V. Janicek,

Department of Microelectronics
Faculty of Electrical Engineering, Czech Technical University in Prague
Technicka 2, CZ – 166 27 Prague 6, CZECH REPUBLIC
husak@feld.cvut.cz, tel. +420 224 352 267, fax: +420 224 310 792

ABSTRACT

The paper describes conceptual approach to the design of a tilt structure determined for integration on a chip. An accelerometer for measurement of inclination in x and y axes is used as a sensor. There is described approach to the development of functional and system model of the structure. A hardware model has been developed for verification of properties. Model properties have been simulated using simulators of electronic circuits. Basic parts of the hardware model are: accelerometer, microprocessor, interface.
The system is temperature-compensated. The values of referential position are set automatically.

Keywords: accelerometer, tiltmeter, microsystem, interface, microprocessor

1 INTRODUCTION

During the process of sensor system or microsystem design determined for integration, it is necessary to use modern design approaches. Models on different levels represent an inseparable part of this process. The models enable to develop the system starting from logical functions up to layout design. It is necessary to use a complex approach based on system engineering [1].

Development of the integrated tiltmeter using accelerometer as a sensitive sensor element has been divided into several parts. The topic of this paper is description of the model development of the whole system and verification of its properties using electronic hardware model. For simulation of the properties, there have been used tools for simulation of electronic circuits. Simulation and modeling have been focused on optimization of circuit layout of the system. There has been performed, for example, sensitivity analysis for identification of properties of transmission chain. At the system input there is an analogue quantity – tilt. At the system output there is an electric quantity in digital form.

2 CONCEPTUAL MODEL OF THE SENSOR SYSTEM

When designing the model we start from basic blocks. These blocks ensure basic functions of signal processing.

The blocks are characterized by input and output quantities. The models of individual blocks and the whole system comprise basic algorithms for signal processing according to designed functions. The aim is to optimize mutual interconnection of signals between individual blocks and guarantee of mutual compatibility of signals between these blocks – see figure 1.

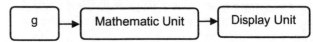

Figure 1: Complex model of tiltmeter system.

2.1 Mathematical model of the system

The mathematical model starts from the description of vector division of forces and acceleration of deflected mathematical weight.
From the model of mathematical pendulum it is possible to derive the relation between tilt and static acceleration.

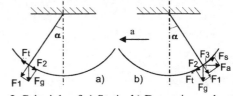

Figure 2: Principle of a) Static, b) Dynamic acceleration

If the hinge attachment of the mathematical pendulum moves with a certain acceleration a, the hinge deflects to the opposite side with certain angle α, see figure. 2a). Value of this angle depends on weight of the bob and value of the acceleration. In tilted state three forces influence the bob: force F_2 caused by gravitational force F_g that returns the bob to original equilibrium state, force F_s caused by inertial force F_a, deflecting the bob in the opposite direction and friction force F_t acting against force F_s. Values of these forces are given by following equations:

$$F_2 = F_g \cdot \sin(\alpha), \quad F_S = F_a \cdot \cos(\alpha), \quad F_t = k \cdot F_S \qquad (1)$$

where k is coefficient of friction that corresponds to beam stiffness in the accelerometer construction, $F_g = mg$ and $F_a = ma$. At each point equality of forces must hold, according to the equation (2).

$$F_S = F_2 + F_t \tag{2}$$

Equivalent expression is

$$ma \cdot \cos(\alpha) = mg \cdot \sin(\alpha) + k \cdot ma \cdot \cos(\alpha) \tag{3}$$

After modifications - dividing the equation by mass m and merging members containing expression cos(α) - we get expression for the value of deviation in dependence on dynamic acceleration:

$$\alpha = arctg\left[\frac{a}{g}(1-k)\right] \tag{4}$$

We do not consider moving of hinge attachment but only its tilt by angle α. Thanks to gravitational force, the hinge returns to vertical position. This situation is illustrated in figure 2b) where the whole system is only turned back by angle α.

This time only two forces influence the tilted ball: force F_2 caused by gravitational force F_g, returning the ball to its equilibrium state and friction force F_t in the opposite direction, but having the same value. This equality follows from the equality of forces in each point of the trajectory of the ball. If we compare figure 2a) and figure 2b) we arrive at conclusion that the tilt angle of the hinge at tilt of the hinge plane is caused by imaginary (fictitious) acceleration. Acceleration value is proportional to the value of force F_t. Considering the fact that we have turned the whole system at the beginning by angle α, we find out that this acceleration is proportional directly to the value of force F_t and not to its product with the function cos(α). Thus we can write the equation:

$$F_t = F_2 = mg \cdot \sin(\alpha) = ma \tag{5}$$

From (4) the relation between static acceleration and tilt value can be acquired:

$$\alpha = \arcsin\left(\frac{a}{g}\right) \tag{6}$$

Using accelerometer as a sensor of static acceleration or tilt, it is necessary to consider this non-linear dependence of acceleration on tilt angle and correct it in an appropriate way.

2.2 Model of the accelerometer for tilt measurement in both axes x and y

Design of the system with accelerometer comes from the equation (6). Mathematical expression calculates the tilt angle from the measured acceleration in axis x and axis y.

The calculations of tilt angle α_x in axis x and α_y in axis y are realized in mathematical unit. According to requirements laid on accuracy and complexity of the

system, the mathematical unit can be realized either in analog or digital version. Analog version is simpler but it offers fewer possibilities for signal processing. Digital version of the mathematical unit is significantly more complex for the design but it offers more possibilities for signal processing and realization of functions. In our model we have proposed communication between individual blocks using digital signals. Digital signals are not prone to interference.

2.3 Model of mathematical unit

In mathematical unit the calculation of tilt is realized. The input data are signals from accelerometer in a suitable form. The output signal from mathematical unit gives the information on tilt in x axis and y axis. These signals are led to information unit for final display, recording or further processing.

3 REALIZATION OF THE HARDWARE MODEL

For verification of basic functions of the designed model there has been chosen a hardware model of the system. For realization of the model commercially available parts have been used. Individual blocks realize designed functions. The signal shape between individual system parts has been modified according to properties of used parts.

For mathematical processing of information from the accelerometer, there has been used a microprocessor and PC as an information unit. The hardware model has been completed with communication circuits that ensure correct data transfer between individual parts of the system. Another block ensures thermal compensation. Correct operation of the system is controlled by control software. Hardware model concept is shown in figure. 3.

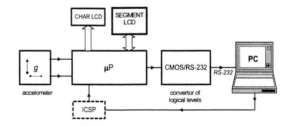

Figure 3: Hardware model of sensor system for measurement of tilt in X and Y axis

3.1 Accelerometer

There has been used a sensitive two-axes accelerometer with analog and digital output with PWM modulation with adjustable period length. It enables direct connection to evaluating microprocessor without A/D convertor. The sensor with range of ±2g can be used for sensing static acceleration – gravitational effect.

Digital outputs have nominal sensitivity value of 12.5 (±2.5) %/g for both axes. Internal convertor realizing both outputs has discrimination of 14 bits, which exceeds discrimination of accelerometer itself. Connected microprocessor must be able to measure mark-space ratio of modulated signal in the range 35-65% with sufficient accuracy for acquisition of information about tilt. It is ideal to use a microprocessor with integrated 16-bit counter/timer and double capture system as its peripheral accessory for this purpose. It is not necessary to measure with high accuracy because discrimination is limited especially by the noise of sensor itself (up to 1 mg/$\sqrt{Hz}$). Not used analog outputs have given sensitivity value of 312 mV/g and with respect to interference they are blocked with capacitors to ground (GND) [2], [3].

3.2 Microprocessor as mathematical unit

Since it is necessary to measure mark-space ratio of output signals of used accelerometers and send measured data via serial link, a microprocessor with integrated 16-bit counter/timer and double capture system, circuit for synchronous and asynchronous transmission and sufficient number of INPUT/OUTPUT (I/O) outlets have been chosen. There has been chosen a microcontroller. The processor is of hardware type, i.e. it has separate memory for data and program (8 kB, word length 14 bits). Timing can be done by external signal of frequency up to 20 MHz. It can reach 5 MIPS (5.10^6 instructions/sec).

On the chip there are peripheral circuits [4], [5]: Two 8bit and one 16bit counter/timer with programmable predivider, two capture, comparing and PWM output modules, 10bit A/D convertor with fivefold multiplex input, synchronous serial port (SSP) supporting SPI™ and I²C™ transfer protocols in both Master and Slave modes, universal synchronous/asynchronous receiver-transmitter (USART/SCI) with 9-bit address discrimination, Brown-out Reset for monitoring decrease of supply voltage.

3.2.1 ICSP™ - option to program in application

PIC microcontrollers of medium performance class, are nowadays equipped with ICSP™ function (In-Circuit Serial Programming) as standard. This function enables microcontroller programming in finished application in the following way: at first the application with microcontroller (hardware) is designed and then it is programmed (software); in case of FLASH program memory it can be programmed repeatedly. This property brings substantial advantage in design and compilation of application execution because it is not necessary, for example, to take the microcontroller out of precise socket and put into

programmer socket for re-programming and back. Instead of continuous strain of its outlets, the adjusted application can be directly connected with the programmer by five-wire cable. If needed it is possible to update the program saved in FLASH memory, add new functions and properties.

3.2.2 Hardware model of temperature compensation

For precise setting of temperature compensation connection with a concrete circuit, there has been chosen a way of precise measurement of temperature dependence of output signal, i.e. mark-space ratio of the given accelerometer sample in the temperature range –15 °C to +80 °C. The mark-space ratio has been measured for X and Y axes in 4 boundary positions. Measured data have been placed to EEPROM memory of the microprocessor. The temperature compensation has been realized in software in the microprocessor using data acquired from measurement of temperature dependence of tilt of X and Y axes. Calibration is performed automatically using data on measured surrounding temperature.

3.3 Software control of the system

The model includes system autocalibration. The autocalibration enables to define initial setup of the inclination of the tiltmeter as referential position before measurement. The calibration cycle saves the information into memory and further measurements are related to these values. The algorithm for control of measurements with autocalibration functions is illustrated in figure 5.

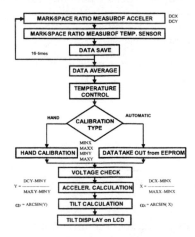

Figure 4: Algorithm for control of measurement, tilt calculation and calibration

4 RESULTS

Hardware model of the tiltmeter system has been realized. The circuit functions have been simulated and modeled using SPICE program. Possible errors in tilt are caused by numerical errors of calculation and storing of measured data, and by temperature variations that represent parasitic parameter of the whole system. Using temperature model, sensitivity of feedback and the whole system to temperature has been simulated.

Hardware model of the system has been tested for tilt in both axes and for tilt in the range of ±90° for each axis.

The measurements have been performed for "manual" and "automatic" calibration of initial positional setup of the system. The measured data have been compared with simulated ones. The deviations between measured and real values are presented in figure 5.

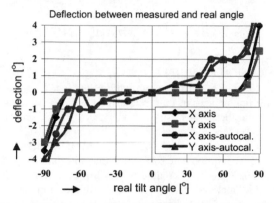

Figure 5: Measurement error in dependence on tilt angle (manual or autocalibration)

The dependence is shown for environment temperature of 24 °C. Greater error of automatic calibration is caused both by error of temperature measurement and by error done by inaccuracy calibration value stored in EEPROM.

It is obvious from the curve of conversion function arcsin that in the areas around the values ±90° there are so-called dead zones, namely zones with zero sensitivity. This fact is reflected in measurement accuracy in the highest ranges. There have been measured temperature dependences as well. Since the sensor itself is temperature dependent, then the whole sensor system is temperature dependent. When testing stability of the measuring system at constant measuring temperature there have not been detected practically any time changes in measured data. Due to intrinsic sensor noise the mark-space ratio of output signals varies in certain interval that can be decreased by averaging, however choosing discrimination value of 0.5° the measured value is sufficiently stable.

There has been measured deviation between measured and real adjusted tilt value In the range of temperature -10 °C to +80 °C at 5 tilt values (0°, ±45° and ±80°). Deviations for tilt values of 0°, 45° and 80° for X and Y axes are shown in figure 6. Absolute accuracy value of the devices is practically temperature independent, it is given by precision of measurement of tilt angles.

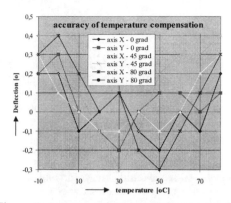

Figure 6: Accuracy of temperature compensation

5 CONCLUSION

There has been developed a system model of tiltmeter system using accelerometer as a tilt sensor. In the model there have been defined signals and their processing in individual blocks of the system. There have been performed functional and logical simulations. For verification of simulated results, there has been designed and realized a hardware model using commercial parts. Circuit and functional simulations using SPICE program have been performed. The model includes temperature compensation and automatic calibration of the referential position of the system as well.

Local display or PC connected via RS 232 can be used for evaluation of information about measured tilt. PC serves for measured data storage and processing. If the system is well calibrated, then it can be used for tilt measurement in both axes in the rage of ±80°, tilt angle 90° represents "dead" point of measurement. Measurement accuracy value in the given range has been ±0.5° at the laboratory sample; hysteresis has been negligible. Setting of the sensor system for measurement at given location is enabled thanks to fast and simple calibration (the system itself evaluates data).

At present design, modeling and simulation for the system realization in integrated version are being performed using ConvetorWare and CADENCE tools.

Acknowledgements

This research has been supported by the Grant Agency of the Czech Republic project No. 102/03/0619 "Smart Microsensors and Microsystems for Measurement, Control and Environment".

REFERENCES

[1] Doscher, J. - Innovations in Acceleration Sensing Using Surface Micro-Machining
[2] http://www.analog.com (Analog Devices)
[3] cataloque information ADXL202/ADXL210 fy Analog Devices
[4] http://www.microchip.com (Microchip)
[5] http://www.hw.cz (Hardware Server)

Modeling and Practical Verification of the Ionophore Based Chemically Modified Field Effect Transistor

Marcin Daniel, Michal Szermer, Marcin Janicki, Andrzej Napieralski

Department of Microelectronics & Computer Science, Technical University of Lodz
Al. Politechniki 11, 93-590 Lodz, POLAND, daniel@dmcs.p.lodz.pl

ABSTRACT

Proper CAD design of electronic microsystems oriented for environment monitoring requires accurate models of various sensors, which would be compatible with the existing behavioral simulators. In this paper the analysis of the ion sensitive transistor was performed and two models of the CHEmically Modified Field Effect Transistors (CHEMFETs) were applied to the sensor model. Two models of the membrane potential were also compared. The models were implemented in Hardware Description Language (VHDL-AMS) in order to provide the future possibility for the mixed domain, mixed signal simulation of the microsystem incorporating the sensors, operating circuit and analog to digital conversion circuits. Both models were validated by the measurements of the real structures. The most important parameters were identified as a result of parameter extraction procedure.

Keywords: CHEMICAL SENSORS, VHDL–AMS, MICROSYSTEMS, MIXED SIGNAL SIMULATION

1 INTRODUCTION

Pollution of the natural environment is one of the most serious problems of industrialized countries. Consciousness of importance of environment protection gave rise to the program of sustainable development, which sets up the standards of economical growth and technological progress with consideration given to environmental care. This idea is being realized by the 5th Framework Programme of the European Union, which supports the SEWING project. The main aim of the project is to provide a cheap and efficient system dedicated to water pollution monitoring. This project is based on a sensing element - the ion sensitive transistor. The base structure is Ion Sensitive Field Effect Transistor (ISFET) dedicated to detection of the hydrogen ion. In order to obtain a sensor dedicated to ion other than hydrogen, the sensor has to be covered with special ion selective membrane. The membrane sensors are called CHEMFETs. The set of such sensors will be incorporated with data processing and acquisition unit, as well as a data transmission system. Data from the sets of sensors will be further transmitted to the acquisition center. Then, according to the sensor indication appropriate alarms will be generated. This approach to the subject of distributed system is especially efficient due to low cost and fast analysis of the measured samples. It allows constant supervision and fast counteraction in case of alarms. In the next section the operation of the sensor as well as existing models are described. After that the measurements of the manufactured structures are presented. On the basis on presented models the parameter estimation is performed.

2 CHEMFET MODELING

OPERATION PRINCIPLE
CHEMFET- Chemically Modified Field Effect Transistor is the miniature sensor that bridges the silicon technology with analytical chemistry. The base of the sensor is represented by the gateless Field Effect Transistor. This kind of sensor can be efficiently applied as the ISFET sensor [2,7,10] dedicated to hydrogen ion monitoring. The CHEMFET sensor is created by covering the ion sensitive inorganic material of ISFET with polymer ion selective membrane and the reference electrolyte (polyHEMA) layer preventing the membrane from mechanical replacement and ensuring thermodynamical stability of the electrolyte-membrane-semiconductor system.

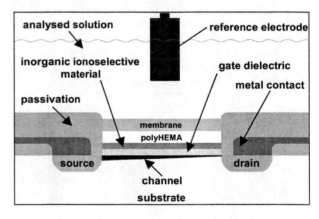

Figure 1. Cross section of the CHEMFET sensor

The crucial part of the sensor is the ion selective membrane. Among various kinds of ion selective membranes the most promising are the polymeric ones with ionophore. The ionophore is responsible for reversible compellation of the main ion and its transport through the membrane. The chemical constants characterizing the ionophore determine the measurement range and selectivity of the whole sensor. The other component of significant importance is lipophylic salt, which prevents the anti-ions from into the membrane phase. All the mentioned components are suspended in a polymer matrix. The modeling of the CHEMFET sensor requires a common approach to the modeling of the semiconductor part of the sensor as well as the ion selective membrane. In this paper an analysis of available models of the membrane is performed and a simple but accurate model of semiconductor part is proposed.

The sensor system can be described as a multi phase system consisting of the phase of reference electrode, measured solution, and sensor system. The schematic representation of such system with potential distribution on the phase boundaries is presented in Figure 2.

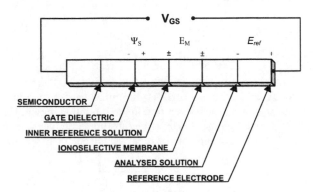

Figure 2.

The operation of the CHEMFET sensor can be explained by considering the system of potential drops on the phase boundaries. The only potential, which depends on the activity of the certain ion in the measured solution, is the membrane potential. It can be formulated regarding the two potential drops on the boundaries membrane-reference electrolyte, membrane-measured electrolyte and diffusion potential. From the CHEMFET point of view, the membrane potential is the most important component, since it is the only one, which is directly dependent on the composition of the analyzed solution.

$$E_M = E_{B1} - E_{B2} + E_D \qquad (1)$$

where:

E_M – membrane potential
$E_{B1,2}$ – boundary (Donnan) potential drops
E_D – diffusion potential

Bearing in mind that the membrane potential is a part of the gate circuit of the transistor, this modifies the threshold voltage formula as follows:

$$V_T^{CHEMFET} = E_{OFFSET} - E_M + V_T^{MOSFET} \qquad (2)$$

Where $V_T^{CHEMFET}$, V_T^{ISFET}, E_{OFFSET} – threshold voltage of CHEMFET sensor, MOSFET transistor and offset voltage respectively.

Usually, the FET based chemical sensor operates in a compensation circuit in the constant current mode. It means that changes of the drain current of the CHEMFET sensor are compensated by additional potential applied to the reference electrode. The output signal of the chemical sensor is represented by the changes of the reference electrode potential keeping the drain current constant. However, in order to provide a complete model of the sensor as a part of the bigger system it is necessary to characterize not only the sensorical properties of the sensor but the electrical ones as well.

ION SELECTIVE MEMBRANE MODELING

The authors have considered two description of the membrane potential. The first one is based on the well-known Nikolski-Eisenman formula [3] and can be applied to describe all kinds of membrane (even inorganic ones). Despite its simplicity the Nikolski-Eiseman model offers good accuracy in a wide range of operation as well as numerical efficiency.

$$E_{Ms} = 2.303 \frac{RT}{z_i F} \log[a_i + \sum_{j \neq i} k_{ij} a_j^{\frac{z^i}{z^j}}] \qquad (3)$$

As presented above, the change of the membrane potential (directly influencing the threshold voltage) can be calculated using Equation 3. It depends on the activity of the ion to be detected a_i (called the main ion), but it also depends on the concentration of some other ions a_j, called interfering ions. The influence of an interfering ion is multiplied in the formula by the relative selectivity coefficient k, which models the membrane preference to disturbing ion in reference to the main one.

The other model describing the membrane potential is based on a physical representation of the electrochemical processes, which take place both at the boundaries of the membrane phase and inside the membrane. It was first developed by van der Bergh and continued by other authors [2,4,5]. The detailed description of the model can be found elsewhere and will not be discussed in this paper.

The most important problem refers to the selectivity of the membrane. In Figure 3 the simulation results of the membrane potential were shown for two different models of the membrane for certain activity of the interfering ion concentration. The NH_4^+ sensor with presence of exemplary interfering ion in concentration of $0.01M/dm^3$ was simulated with application of Nikolski-Eisenman formula (NI) and van der Berg model (VDB). As can be seen in Figure 4, for the range of small to medium concentration, both models offer comparable accuracy. For the very high concentration range, which is rarely encountered in real conditions, the difference can be easily noticeable due to the phenomenon of anionic error, which can be adequately modeled only by the van der Bergh model. It can be also seen that, due to the limited selectivity, the linear part of the curve (called the Nerstian range) flattens out. As a result the sensor becomes insensitive to the main ion at low activities. The location of the bending point and the measurement range are determined by the selectivity of the membrane to the main ion. In case of physical model the Nerstian range of the membrane potential is limited also from the high concentration side.

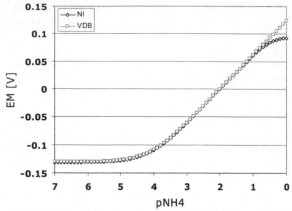

Figure 3. Simulation results of two membrane potential models: physical (VDB) and empirical (NI). Simulated sensor for main NH_4^+ ion, interfering ion concentration $0.01M/dm^3$

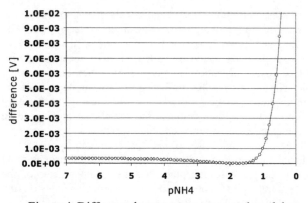

Figure 4. Difference between two presented models

SEMICONDUCTOR PART MODELING

Providing accurate description of the semiconductor part of the sensor is an equally important issue from the CAD point of view. For the electrical simulations, the authors adapted the existing SPICE Level 3 compatible MOSFET model, in which the threshold voltage is chemically modified through the surface potential. The fact that the transistor under consideration was wide and long allowed the simplification of the model by neglecting most of the terms describing short channel effect. This approach is not entirely new and was proposed already in [7].

The transistor output curves can be well described by the following equation:

$$I_D = \frac{C_{ox}\,\mu_{eff}}{2} \frac{W_{eff}}{L_{eff}} (2\,(V_{GS} - V_T) - V_{DS})V_{DS} \quad (4)$$

where:

I_D – transistor drain current, C_{ox} – gate dielectric capacitance per unit area, μ_{eff} – effective carrier mobility, W_{eff}, L_{eff} –

effective channel width and length, V_{DS}, V_{GS} – drain-source and gate-source voltages, V_T – transistor threshold voltage

The carrier mobility depends on the electric fields generated both by the gate and the drain potentials. Then, the effective carrier mobility μ_{eff} can be expressed by the following equation:

$$\mu_{eff} = \frac{\mu_s}{1 + \mu_s \dfrac{V_{DS}}{v_{max}\,L}} \quad (5)$$

where μ_s is equal to:

$$\mu_s = \frac{\mu_0}{1 + \theta\left(U_{GS} - V_T\right)} \quad (6)$$

Θ is the mobility modulation coefficient, which has to be determined in the simulations. v_{max} is the maximal carrier velocity in the channel. μ_0 is the initial carrier mobility. When the transistor is operating in the saturation region, the drain-source voltage V_{DS} must be substituted by the voltage V_{DSsat}. The saturation voltage of an ISFET can be expressed as:

$$V_{DSsat} = V_{GS} - V_T + v_{max}\frac{L_{eff}}{\mu_{eff}} + \sqrt{\left(V_{GS} - V_T\right)^2 + \left(v_{max}\frac{L_{eff}}{\mu_{eff}}\right)^2} \quad (7)$$

The threshold voltage of an ISFET modified by the membrane potential as it shows Equation 2. More detailed implementation of this model as well as its experimental verification can be found in [10].

3 SENSOR MEASUREMENTS

The object of the measurements was the CHEMFETs developed by the Institute of Electron Technology from Warsaw (Poland). The NH_4^+ selective membranes were deposited by the Department of Analytical Chemistry of WUT (Warsaw University of Technology). The detailed technological description of the semiconductor part as well as the chemical membrane can be found in [1,2,10]. The measurements were performed using the specially designed measurement stand described in [8]. The stand allows fully automated measurements of several transistor characteristics with various ion concentrations at different temperatures. The presented measurements were taken simultaneously for 10 CHEMFETs with various main ion concentration and single disturbing ion concentration. Two interfering ion were taken into account: Na^+ and Ca^{2+}.

The aim of the CHEMFET measurements was to capture the output electrical characteristics of the transistor so as to verify the model of the electrical part of the sensor. On the basis of these curves the chemo-sensorical curves were possible to obtain. Obtained results are shown in the next section.

4 ELECTROCHEMICAL MODEL SIMULATION AND MODEL PARAMETER EXTRACTION

The presented measurements provided a base for the CHEMFET model verification as well as model parameter extraction. The parameters were extracted using the Newton-Gauss method based automatic procedure already implemented in the MATLAB environment. The procedure uses the least mean squares criterion so as to find the best possible set of model parameters. The detailed description of the model parameter procedure can be found in [10]. The extracted parameters identifying the CHEMFET transistor are presented in Table 1. The value of threshold voltage was extracted separately from the measurements of the CHEMFET structure before the ion selective membrane was deposited. The measurements of such structure were performed for the buffer solution of pH=7. For this value the potential drop on the border the solid-liquid interface is close to 0 (point of zero charge for the silicon nitride) thus V_T can be efficiently determined.

Parameter	Value	Unit
Channel width W	638	μm
Channel length L	14	μm
Gate capacitance	4.42e-4	F/m^2
Threshold voltage V_T	-0.368	[V]
Low field carrier mobility μ_0	0.0970	$[m^2/Vs]$
Maximal carrier velocity v_{max}	$1.00*10^5$	[m/s]
Mobility modulation coefficient Θ	0.176	$[V^{-1}]$

Table 1. The most important parameters of the semiconductor part of the sensor

In Figures 5 and 6 the simulation results are compared to the measured values. As can be seen, the simulation results match the measurement results well. It is worth noticing that such accuracy is obtained with the application of relatively simple transistor model (SPICE Level 3).

In Figures 7 and 8 the changes of the reference electrode voltage were showed. The changes of this voltage compensate the influence of the measured solution on the drain current. Initially the measured reference electrode changes were fitted with application of the Nikolski-Eisenman model and afterwards the van der Bergh model was implemented so as to fit the measured curves. As it was mentioned before two interfering ions were taken into account: calcium and sodium. Their activity maintained the same level of $0.01M/dm^3$ during the measurements.

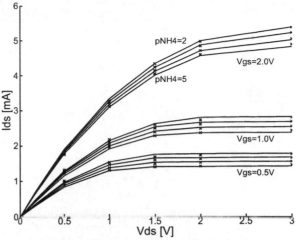

Figure 5. CHEMFET output characteristics with calcium interfering ions (dots) fitted with model (solid)

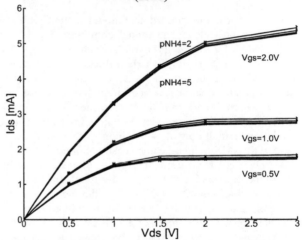

Figure 6. CHEMFET output characteristics with sodium interfering ions (dots) fitted with model (solid)

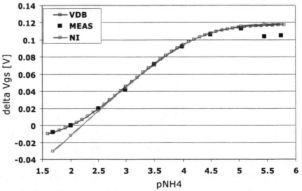

Figure 7. Measurements of NH_4^+ selective CHEMFET interfered by Na^+ ions, and simulation results for both Nikolski (NI) and van der Bergh model (VDB)

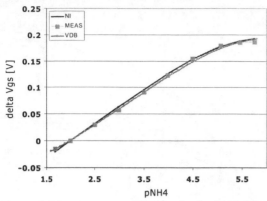

Figure 8. Measurements of NH_4^+ selective CHEMFET interfered by Ca^{2+} ions and simulation results for both Nikolski (NI) and van der Bergh model (VDB)

5 CONCLUSSIONS

In this paper a simple but efficient model of the CHEMFET sensor was presented. The model combines the standard SPICE MOSFET model and two electrochemical models of the membrane potential, which influences the threshold voltage of the semiconductor model. In the model at least two different domains are considered - electrical and chemical ones. Apart from these, the thermal one can be easily introduced to the model. For the purpose of efficient simulation a multidomain simulation with application of the VHDL-AMS standard can be easily applied.

The presented measurement results indicated that the membrane potential for higher activity of the main ion cannot be accurately modeled by the semi empirical Nikolski Eiseman model. In order to obtain satisfying accuracy for the wide range of activities, it is advisable to use rather the more complex, physical van der Bergh model. On the other hand, such high values of the activities normally do not occur in the natural environment so they can probably be neglected in the case of river or lake water monitoring. What seems much more problematic is the poor selectivity of the NH_4^+ membrane (for example to K^+ ions). Nevertheless, if this sensor is to be used for analysis of highly polluted waters, then it is necessary to apply the physical model.

6 THE AUTHORS

Dr. Marcin Daniel, Mr. Michal Szermer, Dr. Marcin Janicki and Prof. Andrzej Napieralski, are with the Department of Microelectronics and Computer Science, Technical University of Łódź, Politechniki 11, 93-590 Łódź, Poland. E-mail: daniel, szermer, janicki, napier@dmcs.p.lodz.pl

7 ACKNOWLEDGEMENTS

The presented research is supported by the 5th European Framework Program SEWING, IST–2000–28084 and the grant of the Polish State Committee for Scientific Research No. 4 T11B 012 24. The authors would like to extend their gratitude to all the SEWING partners, especially to prof. R. Jachowicz and Mr. J. Sochoń from the Institute of Electronic Systems at the Warsaw University of Technology where the measurements were performed. Special thanks are also directed towards Prof. Z. Brzózka, Prof. W. Wróblewski and Prof. A. Dybko form the Institute of Chemistry at the Warsaw University of Technology, for the preparation of the electrolytes and membranes.

8 REFERENCES

[1] Bergh A. van der (1988). *Ion Sensors Based on ISFETs with Synthetic Ionophores.* PhD thesis, University of Twente, The Netherlands.

[2] Daniel M., Szermer M., Napieralski A., Wróblewski W. and Dybko A. (2002). Ion-selective sensors modeling for CAD. In: *9th International Conference MIXDES*, Wroclaw, Poland, pp. 147-150.

[3] Eisenman G., Rudin D. and Casby J. (1957). Glass electrode for measuring sodium ion. *Science*, **61**, 831-834.

[4] Morf W., Badertscher M., Zwickl T., Rooij N. de and Pretsch E. (1999). Effects of Ion Transport on the Potential Response of Ionophore-Based Membrane Electrodes: A Theoretical Approach. *J. Phys. Chem. B*, **103**(51), 11346 – 11356.

[5] Ogrodzki J. (2002). CHEMFET Sensors Modeling for Water Monitoring System Design. In: *9th International Conference MIXDES*, Wroclaw, Poland, pp. 151-156.

[6] Szermer M., Daniel M. and Napieralski A. (2003). Design and Modelling of Smart Sensor Dedicated for Water Pollution Monitoring. In: *NANOTECH Conf.*, San Francisco, CA, USA, Vol. 1, pp. 110-114.

[7] M. Grattarola, G. Massobrio, S. Martinoia, Modelling H^+ Sensitive FET's with SPICE, IEEE Trans. on Elec. Devices, Vol.39, No.4, April 1992, pp. 813-819

[8] R. Jachowicz, J. Weremczuk, J. Sochoń, Automatic stand for IS-FET sensors parameters identification, Proceeding of the 9th International Conference Mixed Design of Integrated Circuits and Systems – MIXDES 2002, 20-22 June 2002, Wrocław, Poland, pp. 199-202

[9] W. Torbicz, Theory and Properties of FETs as Biochemical Sensors (in Polish), Institute of Biocybernetics and Biomedical Engineering Polish Academy of Science, 1988

[10] M. Daniel, M. Janicki, A. Napieralski, "Simulation of Ion Sensitive Transistors using a SPICE Compatible Model", 2nd IEEE international conference on sensors IEEE Sensors, 22-24 October 2003, Toronto, Canada

Smart, Intelligent and Cogent Microsensors – Intelligence for Sensors and Sensors for Intelligence

Elena Gaura, Robert M. Newman

Coventry University, Priory St, CV1 5FB, UK
e.gaura@coventry.ac.uk

ABSTRACT

The paper develops a terminology to describe sensors which have been enhanced in some way by the integration of some additional processing circuitry. Several terms, current in the literature, including 'smart sensors' and 'intelligent sensors' are discussed and the 'cogent' sensor is introduced. This is followed by a brief review of existing and potential applications of Artificial Intelligence (AI) to microsystems, in terms of technology integration, device level performance enhancement and system level added functionality.

Keywords: sensor, smart, intelligent, cogent, MEMS, artificial intelligence

1 INTRODUCTION

The fast maturing MEMS field is presently seen as a core enabling technology both in the civil and military application domains.

A particularly buoyant sector in the MEMS industry is that of sensors. Traditionally, the main sensor requirements were in terms of metrological performance, i.e. the (most often) electrical signal produced by the sensor needed to match relatively accurately the measurand. If such basic sensor functionality was adequate several years back, this is no more the case. The use of sensors by industry has seen a gradual shift, away from large systems incorporating relatively few and expensive transducers towards the utilization of more and more sensors as components, or in subsystems. A new set of requirements was therefore generated which can potentially be met by using micromachining as a sensor fabrication technology. MEMS devices can be designed and produced to achieve: low cost, low mass and low power consumption, plug and play, digital output, enhanced reliability, sensitivity and selectivity and high accuracy, to name a few, general, desired sensor features.

Whilst some of the above requirements can be met by advances in the manufacturing techniques of the sensing element itself (to produce linear, accurate, highly sensitive and reliable sensing devices), others have to rely on efficient and clever processing of data generated by the sensing device, before such data reaches the outer world.

It is here, in the area of data processing and extraction of information, that the authors propose to clarify some commonly used terminology and introduce new terms. The definitions are supported by examples.

2 SMART, INTELLIGENT AND COGENT SENSORS

The MEMS/micro sensors field has become an interdisciplinary one, pulling together researchers from microelectronics, mechanics, physics and computer sciences. As a consequence, "borrowed" technologies, methods and terminology from the macrosystems domain began to be used in conjunction with the development of microsystems. Several terms are current in the literature, including 'smart sensors' and 'intelligent sensors'. 'Adaptive', 'distributed', 'autonomous' and other adjectives are also routinely applied to pick out the function of a particular sensor enhanced by some processing capability from the common herd of 'dumb' sensors. (Such terminology, a few years ago, was solely dedicated to macrosystems, with a much stronger meaning.)

By comparison with the usage of these terms in other fields, it would appear that the sensor community is over-selling the 'intelligence' of their products. The phrase 'intelligent sensor' often merely indicates that the sensor is integrated with a digital processor, it may say nothing about the intuitive abilities of the functionality programmed into the sensor. Given that the terms 'smart' and 'intelligent' have become somewhat confusing and meaningless with respect to sensors, we introduce a new term, the 'cogent sensor'. In contrast to the above terminology, we define the meaning of cogent sensor by what it does, rather than what came before it or how it is constructed.

Smart sensors -The original 'smart sensors' [1] came about with the opportunity to enhance the capabilities of sensors by the integration of some signal conditioning circuitry within the sensor. According to the IEEE 1451 smart transducer interface standards (which describe a set of open, common, and network-independent communication interfaces for smart transducers), "smartness" means on-board data storage/processing capability, interfaced/integrated with the and/or digital sensor [2]. Variations of the term referred to the hardware implementation of the sensor, as follows: when a microsensor is integrated with signal processing circuits in a single package, it is referred to as an integrated sensor. A monolithic integrated sensor has the signal processing circuitry fabricated on the same chip as the sensor, while a hybrid integrated sensor has the signal processing circuit on the same hybrid substrate as the sensor chip.

'Smart' is however a term which, although used in conjunction with nearly every newly designed/produced sensor, means different things to different people. To some, it just means sensors which can communicate digitally, while to others it means sensors that have serious computing power integrated within them, to include calibration, non-linearity correction, offset elimination, failure detection, communication and decision making ability. It is important to mention that, such designs have been developed strictly for a sensor type and manufacturing technique and are, mostly, highly application oriented, so much so that, it is often difficult to assess whether it is the sensor or the application that is 'smart'.

Intelligent sensors - As the capacity of VLSI techniques increased, it became possible to integrate substantial digital or analogue processing capability onto a sensor. The term 'intelligent sensors' was often used incrementally from 'smart sensor'. Initially "intelligence" meant and served the same purposes as the electronics in a smart sensor: enhancement of the function of the sensor itself. In macrosensors, the term intelligent sensor was coined in 1992 [3]. Trade journal and popular news articles still use smart and intelligent interchangeably [4].

There are however, several examples of work here, where by using the term intelligent the authors implied the use of macro-scale intelligent techniques (i.e. artificial intelligence and most often neural networks (NN)). Such works are discussed in Section 3. It should be noted however that the nonlinear signal processing abilities of NNs were merely exploited rather than more evolved 'thinking/decision making in new situations' aspects of the technique.

Cogent sensors - What is common to the previous classes of sensors is that they provide raw data. Essentially, the readings of the original sensor are passed on, albeit linearized, temperature corrected, hysteresis corrected, packetised, network routed or re-packaged in one of many ways specific to the 'intelligent sensor' in question. What these sensors do not do is reduce the data to information, neither do they do processing to remove unneeded data and convert it to the particular form that the application requires. We term a sensor that performs the above two functions a 'cogent sensor'. Below we give some examples of existing and proposed cogent sensors and discuss some ways in which artificial intelligence techniques may be used to implement them.

3 AI FOR MEMS INTELLIGENCE

Amongst the "borrowed" macrosystems design techniques and tools, AI appears to be receiving increasing attention [5, 6] in the microsensors world. Despite the promise of application of 'intelligence', in many cases the AI techniques contribute only to a 'smart' sensor, let alone a cogent one. One exception to this is the case of virtual/software sensors, where AI techniques act on a system/array of microsensors and infer new information from multiple sensor data. Examples for each sensor category defined in Section 2 follow.

3.1 AI in smart sensors

We classify the applications below as 'smart' because no additional functionality has been added to the sensor, rather the quality of the data is enhanced by the AI techniques.

Sensor metrological performance enhancement

Optimizing analogue and digital methods for a transducer's characteristic interpolation and/or linearization is a field where constant research is being done. Different calibration methods have been applied for a variety of macro and microsensors (for example Newton, Lagrange, LMS regression and ANN). It has been found that ANN interpolation is more accurate than polynomial interpolation, especially when multivariable extrapolation or nonlinearity characteristics are under analysis. The extrapolation errors with ANNs are lower both inside and outside the calibration range. The computational load with ANN interpolation is of the same order as polynomial interpolation; however, ANN training requires more computational resources. This is not important in NN applications where training can be performed in a central host or implemented locally based on previous training weights and biases [7,8].

Sensor Data Validation

In this application of AI the 'intelligence' is applied to validate [10] or restore missing or corrupted data [11], by exploiting either physical sensor redundancy or other observable states within the sensor or the application itself. The output is still data, as opposed to information.

3.2 AI in intelligent sensor systems

In this category the system as a whole uses sensors and AI but the sensor itself is not endowed with the data to information transformation.

Multiple sensor systems – actuator control

The adventurous Silicon Active Skin project [12,13], lead by the Center for Neuromorphic Systems Engineering at Caltech, aims to integrate MEMS sensors and actuators, neural network sensory processing, and control circuits all on the same silicon substrate to form a "smart skin", capable of reducing drag on an aircraft wing. This is one of the best examples of intelligent microsystems. The system senses the shear stress, while a neural controller with feedback mechanism efficiently actuates robust micro actuators for surface stress reduction. The neural network controller is trained off-line to predict actuation using data from near-wall controlled experiments. The controller is allowed to adapt on-line, as it is included in an on-line adaptive inverse model scheme. In simulation, a 20% shear stress reduction was obtained.

3.3 AI in cogent sensors

Our category of 'cogent sensors' includes sensors integrated with the means to reduce raw data to 'information', of the type required by a specific application. Up to date, most examples of cogent sensors in the literature are of a multidimensional nature, so they are, strictly speaking cogent sensor systems. Two categories here are: the mono-type sensor systems (all sensors measure the same physical quantity) and the multi-type ones. Examples follow for each category.

At individual sensor level, cogent functions have been reported as work in progress and two examples are offered here.

Multi-type sensor systems

In some applications, the information has to be inferred from available measurements of observable quantities using a statistical model, which is referred to as a "software" or "virtual" sensor [14]. The actual measurement devices within such virtual sensors are often MEMS components. Software sensors were the first and major application of ANN to the instrumentation field and today they continue the lead in terms of number of research contributions. The innovative aspects of such works are primarily application specific and reside in the integration of various techniques in a global system allowing for prediction of the quantity of interest. Some contributions extend the role of ANN (or add on specific NN modules) to enhance the designed system with functions such as data validation, reconstruction, and analysis of uncertainties. It is not the scope here to survey the multitude of such applications. The work by Roppel et. al. is however mentioned as it is a good example of design for hardware integration of MEMS and AI. They successfully designed a low-power, portable sensor system using mixed analog and digital VLSI circuitry for on-board data pre-processing together with pulse coupled neural networks for feature extraction and also for pattern recognition [15]. The design is aimed at minimizing cost, size, weight, power and post-sensing computational burden. The sensor testbed is a 30 nodes, MEMS sensor array consisting of tin-oxide gas sensors and the target is to discriminate among 7 odors (acetone, ammonia, beer, etc.). Spatio-temporal encoding is used for pre-processing of the dynamic sensor outputs. The system provides correct identification rates of 96% for the odor data sets considered. Other good example of a gas sensing array with NN processing and a VLSI-MEMS was presented in [16].

Mono-type sensor arrays, smart sensor webs

Several researchers have noted the potential benefits of organizing/designing and exploiting large sensor systems based on the same principles as those governing both natural and artificial neural processing. This has led to the proposal that functions be distributed across a network of 'intelligent' sensors. The network itself then, collaboratively extracts information from the data field provided by the sensor network. Sensors are seen as ANN nodes and ANNs are designed to have architectures approximating those of the biological NN that performs analogous functions [17]. Such systems are part of a "third wave of computing" that could use NN to build sensors and other machines capable of the unsupervised learning exhibited by the human brain [18].

Fault detection and classification.

Over the past 10 years, neural networks have found wide application in systems that are designed to recognize fault conditions from sensor data, the derived information being whether the data is trustworthy or not [19,20]. If information of this kind is derived by a microsensor, about its own data and independent of the application of which the sensor is part of, such a microsensor would be subscribing to our "cogent" property.

The authors' work in this respect includes the development of a microsensor with such ability [21]. The fault detection function is intrinsic to the sensor functionality and provides the cogent sensor not only with the means of detecting its healthy/faulty state but also with the power of decision about its appropriateness of contributing its data to the specific application. Such a sensor can be used when two or more sensors of the same type are linked in an intelligent array. The method is based on the use of Neural Networks (NN) both for detecting faults and for classifying them. When implementing this function, use is made of the existing hardware resources in an array of intelligent sensors. The outmost merits of the method are:

-the low communication level necessary for performing the diagnosis: at any one time, only two neighboring sensors need to exchange data in order for the diagnosis decision on a particular sensor to be made;

-the real-time element – the diagnosis is performed based on a 3 steps back only history of the sensor readings;

-no a priory information is needed on the devices themselves as the sensor signatures are to be analyzed and learnt by the NN;

-the method is application independent and easily scalable.

Work is in progress to add new cogent features to the proposed sensor, apart form the diagnosis ability.

Real time monitoring

As part of the UMCP Small Smart Systems Program, a team of researchers aims at developing MEMS-VLSI single chip sensors where the smart portion of the system is a NN [22]. In the proposed designs, a fluid or gas acts within a portion etched out of the VLSI chip to activate the sensor. Upon NN processing of the sensor data, decisions are taken by the cogent sensor and feedback is produced for other components of the application (biomedical or environmental determinations for ex.)

4 CONCLUSIONS

As we have seen, the term 'intelligent' is applied to a wide variety of MEMS systems, ranging from those that have processing capability, through those that use artificial intelligence to those that, in a sense, think, sense, act and communicate. We reserve the term 'cogent' for the latter.

Cogent sensors are of great interest in a variety of industries. MEMS technology offers new ways of realizing cogent sensors by combining sensing, signal processing and actuation on a microscopic scale.

MEMS will open up a broad new array of cost-effective solutions only if they prove sufficiently reliable and their use does not pose insuperable systems design problems. The basic purpose of sensors is measurement. Achieving high metrology performance is the primary design aim, which has been achieved/solved in two ways: by technological perfection (which is inherently expensive and difficult to achieve) or by the application of structural or structural-algorithmic methods.

Relaxing the requirement for technological perfection allows designers to achieve the same performance with lower cost, design effort and on shorter time-scales. The second approach not only allows for increased measurement accuracy but also allows the extension of functional capabilities of such systems. The concept of the cogent sensor extends this principle further. What is important is not so much the quality of measurement itself, but the quality of the information derived from it. With cogent sensors the information required by the application may be available, in the form required by the application directly from the sensor or network of sensors. The information may directly reflect the sensed data or it may have been deduced and sifted by the application of degrees of 'intelligence'.

REFERENCES

[1] Varadan, Vijal K. (2001) MEMS- and NEMS based complex adaptive smart devices and systems, Proc. SPIE Vol. 4512, p. 25-45, Complex Adaptive Structures, William B. Spilman Ed.

[2] P.G. Ranky, Smart sensors, Sensor review, Vol. 22, Issue 4

[3] W. H. Ko and Fung, VLSI and intelligent transducers, Sensors and Actuators 2-pp. 239-250

[4] http://mems.cwru.edu/shortcourse/partI_3.html – Physical microsensors

[5] Brukner, S. Rudolf, S. (2000) Neural Networks Applied to Smart Structure Control, Proc. SPIE Aerosense 2000 Conf. On Applications and science of Computational Intelligence III, Orlando, Florida, pp. 24-28, vol.4.

[6] Miller, W. M. Peterson, K. A. (1998) MEMS reliability: The challenge and the promise, NASA Technical Reports

[7] Dias Pereira, J.M. Silva Girao, Postolache O. (2001) Fitting transducer characteristics to measured data, IEEE Instrumentation and Measurement Mag., Dec., pp. 26-39.

[8] Ref 2: Patra, J.C. Van den Bos, A. Kot, A.C. (2000) An ANN-based smart capacitive pressure sensor in dynamic environment, Sensors and Actuators 86, pp. 26-38

[9] Almodarresi, S. White, N. M. (1999) Application of artificial neural networks to intelligent weighing systems, IEE Proc.-Sci. Meas. Technol., Vol. 146, No. 6, pp.265-269.

[10] Narayanan, S. Marks, R. J. &all, (2002) Set constraint discovery: Missing sensor data restoration using auto-associative regression machines, Proc of the IEEE World Congress on Computational Intelligence, Hawai, pp. 2872-2877.

[11] Bohme, T. Cox, C. et al. (1999) Comparison of autoassociative neural networks and Kohonen Maps for signal failure detection and reconstruction http://www.hds.utc.fr/~tdenoeux/congres/annie99.pdf

[12] Gupta, B., Goodman, R., Jiang, F., Tai, Y.C., Tung, S., Ho, C.M., (1996) Analog VLSI For Active Drag Reduction, IEEE Micro, Vol. 16, No. 5, pp 53-59.

[13] Koosh, V. Babcock, D. et. al. (1996) Active drag reduction using neural Networks, Proc. of International Workshop on Neural Networks for Identification, Control, Robotics, and Signal/Image Processing, http://www.rodgoodman.ws/Active%20Skin.htm.

[14] Valentin, N. Denoeux, T. (2001) A Neural network-based software sensor for coagulation control in water treatment plant, Intelligent Data Analysis, 5:23-39.

[15] Roppel, T. Wilson, D. et al. (1999) Design of a low-power sensor system using embedded neural networks and hardware preprocessing, Proc. of IJCNN99, Paper no. 357, USA.

[16] Dae_sik Lee et al. (2002) SnO2 gas sensing Array for Combustible and Explosive gas Leakage Recognition, IEEE Sensors Journal, Vol. 2, No. 3, pp. 140-149.

[17] Hand C. (2002) ANNs for organizing sensor webs, http://www.nasatech.com/Briefs/july02/NPO30317.html

[18] Leopold, G., US Military deploys neural network technology, EETimes, March 6th, 2003

[19] NASA (1998) Neural network based sensor validation,http://technology.larc.nasa.gov/scripts/nls_ax.dll/twDispTOPSItem(111;TOP3-3;0;1)

[20] Hines, J.W. Gribock, A.V. et al (2000) Improved Methods for on-line sensor calibration verification Proc. of 8th Int. Conf. on Nuclear Engineering, USA, http://web.utk.edu/~hines/publications.html

[21] Gaura, E.I, Newman, R.M., (2003) Intelligent sensing: Neural Network based health diagnosis for sensor arrays, Proc Advanced Intelligent Mechatronic, Japan, pp.36—365.

[22] Newcomb, R.W. VLSI-MEMS smart pulse-coded biosensors,http://www.ece.umd.edu/~newcomb/memsensr.html

Soft and Probe Lithography Without Ink Transfer

Jurriaan Huskens, Xuemei Li, Mária Péter, and David N. Reinhoudt

Laboratory of Supramolecular Chemistry and Technology, MESA+ Institute for Nanotechnology,
University of Twente, Enschede, PO Box 217, 7500 AE Enschede, The Netherlands.
Phone: +31-53-4892995; fax: +31-53-4894645; E-mail: j.huskens@utwente.nl

ABSTRACT

The present study shows examples of novel strategies for eliminating diffusion as an underlying principle of μCP and DPN. The preparation and use of catalytically active stamps was shown with a catalyst attached directly to the stamp surface. With these stamps it is possible to form patterned SAMs without any transfer of ink. These stamps were more active than the earlier developed oxidized stamps and provide an ink-free μCP scheme allowing more versatile catalytic reactions, and also allow physical and chemical pattern functionalization. Furthermore, an extension to catalytic probe lithography was given. Here line patterns with sub-100 nm widths were created by scanning a catalyst-functionalized tip across a reactive SAM substrate.

Keywords: nanotechnology, soft lithography, microcontact printing, probe lithography, self-assembled monolayers

1 INTRODUCTION

Microcontact printing [1] (μCP) is a versatile technique for the creation of patterned surfaces. It commonly employs the transfer of an ink (e.g. a thiol) onto a surface (e.g. a gold-coated substrate) thus forming a self-assembled monolayer (SAM). Patterns of different thiols can be created using self-assembly of a second thiol from solution, or the printed patterns can be used directly as etch resists allowing the structuring of the underlying substrate itself. This technique has been extended to probe lithography in the recently developed dip-pen nanolithography (DPN) [2].

Resolution of these techniques is typically around 200 nm, diffusion of the ink during the printing or writing stage being the main limiting factor [3]. Diffusion can be restricted when heavy inks are used, such as high-molecular-weight thioethers [4], proteins [5], or nanoparticles [6]. The present study will show examples of novel strategies for eliminating diffusion as an underlying principle of μCP and DPN.

2 RESULTS AND DISCUSSION

2.1 Microcontact Printing with Catalytically Active Nanoparticles

Microcontact printing with nanoparticles leads to well-defined patterns because of the lack of diffusion of the printed particles [6]. For the same reason, however, the order of the printed particle pattern is usually poor and does not provide etch resistance. In order to obtain well-ordered, patterned SAMs but at the same time keep the advantage of diffusion-less printing of nanoparticles, we have coated gold nanoparticles with catalytically active acid groups (Scheme 1) to hydrolyze protecting silyl ether (either trimethylsilyl, TMS, or tert-butyl-dimethylsilyl, TBDMS) groups present at a homogeneous SAM when transferred onto this SAM by μCP (Scheme 2) [7]. This has led to patterned SAMs with an edge resolution below 100 nm (Figure 1).

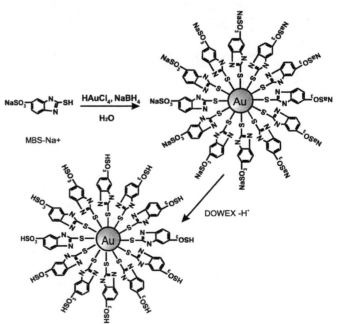

Scheme 1. Preparation of catalytically active gold nanoparticles.

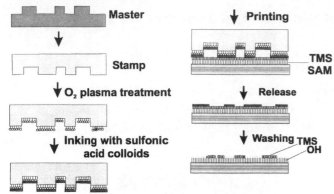

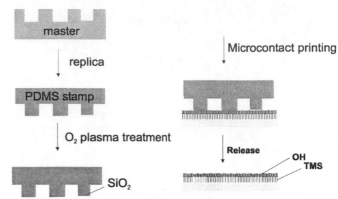

Scheme 2. General µCP scheme for printing catalytically active gold nanoparticles onto a reactive SAM.

Scheme 3. General µCP scheme for printing with catalytically active, oxidized PDMS stamps onto a reactive SAM.

Figure 1. An AFM phase image (30 × 30 µm²) of a TMS SAM patterned by µCP with catalyst-functionalized gold nanoparticles (bright areas: TMS, dark areas: hydrolyzed TMS).

Figure 2. An AFM friction image (25 × 25 µm²) of a TBDMS SAM patterned by µCP with oxidized PDMS stamps (dark areas: TBDMS, bright areas: hydrolyzed TBDMS).

2.2 Microcontact Printing with Oxidized Stamps

Ideally, in order to completely eliminate diffusion as the underlying principle of µCP, one could envisage pattern formation without the transfer of ink but rather by contact of the stamp with a substrate only. This idea was pursued by using oxidized PDMS stamps that can catalytically cleave silyl ether groups of reactive SAMs in an ink-free µCP process (Scheme 3) [8]. Pattern formation on both TMS and TBDMS (Figure 2) SAMs was shown to occur with sub-100 nm edge resolution. On the former, more reactive SAMs, pattern formation was achieved in only a few min of contact, whereas for the latter SAMs contact times needed to be above 10 min to be visible by AFM.

2.3 Microcontact Printing with Catalyst-Functionalized Stamps

In order to achieve more active stamps, a catalyst was attached directly to the surface of a gold-coated PDMS stamp (Scheme 4). With these stamps it was possible to form patterned SAMs without any transfer of ink (Figure 3). These stamps were more active than the earlier developed oxidized PDMS stamps. Such a general approach for attaching catalysts to a stamp in principle provides an ink-free µCP scheme allowing more versatile catalytic reactions as well.

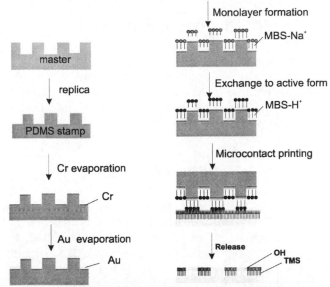

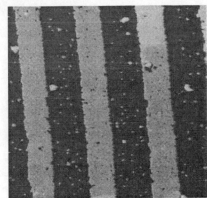

Scheme 4. General µCP scheme for printing with catalyst-functionalized PDMS stamps onto a reactive SAM.

Figure 4. An AFM height image (25×25 µm^2) after adsorption of amino-functionalized polystyrene nanoparticles (height 120 nm) onto a SAM patterned by µCP with an OH thiol (narrow areas) and filling with a TMS adsorbate (wide areas).

Figure 3. An AFM friction image (25×25 µm^2) of a TMS SAM patterned by a catalyst-functionalized PDMS stamp (bright areas: TMS, dark areas: hydrolyzed TMS).

2.4 Pattern Enhancement

The versatility of the current schemes employing TMS and TBDMS SAMs was shown by using such patterned SAMs in guided deposition of nanoparticles by physisorption and in pattern enhancement by chemical functionalization. Amino-functionalized polystyrene nanoparticles (120 nm diameter) were deposited from solution onto a patterned SAM with TMS- and OH-functionalized areas created by µCP. The particles showed a large preference for adsorption onto the OH areas forming a layer with a single-particle thickness (Figure 4), probably owing to the formation of multiple hydrogen bonds. Versatile chemical functionalization of patterned TBDMS SAMs was possible using reaction with a diisocyanate, followed by an amine (Scheme 5). Also this approach showed preferential reaction of the OH areas (Figure 5), in this case because the TBDMS groups are protecting the other SAM areas.

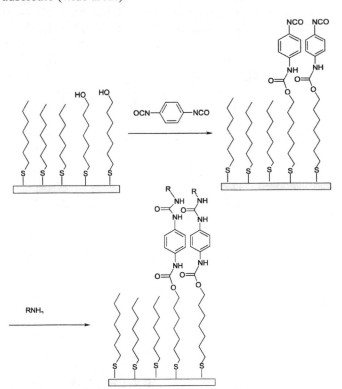

Scheme 5. Chemical amplification of patterned SAMs by reaction of exposed OH functionalities.

Figure 5. AFM height (left) and friction (right) images (25 × 25 μm²) after reaction of the OH-functionalized areas of a SAM patterned by μCP with an OH thiol (wide areas) and filling with a TBDMS adsorbate (narrow areas).

2.5 Catalytic Probe Lithography

The catalytic μCP process shown in Scheme 4 was extended to catalytic probe lithography by using catalytically active AFM tips instead. Hereto, gold-coated AFM tips were functionalized with an acid catalyst (Scheme 6). Scanning such a catalytically active tip across a reactive TMS SAM substrate led to hydrolysis of the TMS groups in the contacted areas as shown by higher friction (Figure 6, left). Sub-100 nm lines were written as well (Figure 6, right).

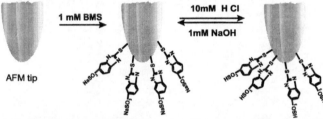

Scheme 6. Functionalization of gold-coated AFM tips with an acid catalyst.

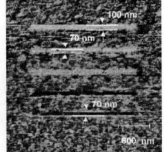

Figure 6. AFM friction images of box (left, 6 × 6 μm²) and line patterns with sub-100 nm widths (right, 3 × 3 μm²) created on TMS SAMs by catalytic probe lithography using an acid-functionalized AFM tip (bright box and lines: with an active tip, dark lines: with an inactive tip).

3 CONCLUSIONS

Catalytic schemes have been developed for μCP and probe lithography. Not only can such schemes lead to improved resolution because of the complete elimination of diffusion, but they may also lead to more versatile nanofabrication schemes in general.

REFERENCES

[1] Xia, Y. N.; Whitesides, G. M. *Angew. Chem. Int. Ed.* **1998**, *37*, 551-575; Michel, B.; Bernard, A.; Bietsch, A.; Delamarche, E.; Geissler, M.; Juncker, D.; Kind, H.; Renault, J.-P.; Rothuizen, H.; Schmid, H.; Schmidt-Winkel, P.; Stutz, R.; Wolf, H. *IBM J. Res. Dev.* **2001**, *45*, 697-719.

[2] Piner, R. D.; Zhu, J.; Xu, F.; Hong, S.; Mirkin, C. A. *Science* **1999**, *283*, 661-663; Mirkin, C. A.; Hong, S.; Demers, L. *ChemPhysChem* **2001**, *2*, 37-39.

[3] Libioulle, L.; Bietsch, A.; Schmid, H.; Michel, B.; Delamarche, E. *Langmuir* **1999**, *15*, 300-304.

[4] Liebau, M.; Huskens, J.; Reinhoudt, D. N. *Adv. Funct. Mater.* **2001**, *11*, 147-150.

[5] Bernard, A.; Renault, J. P.; Michel, B.; Bosshard, H. R.; Delamarche, E. *Adv. Mater.* **2000**, *12*, 1067-1070.

[6] Hidber, P. C.; Helbig, W.; Kim, E.; Whitesides, G. M. *Langmuir* **1996**, *12*, 1375-80; Hidber, P. C.; Nealey, P. F.; Helbig, W.; Whitesides, G. M. *Langmuir* **1996**, *12*, 5209-5215.

[7] Li, X.-M.; Paraschiv, V.; Huskens, J.; Reinhoudt, D. N. *J. Am. Chem. Soc.* **2003**, *125*, 4279-4284.

[8] Li, X.-M.; Péter, M.; Huskens, J.; Reinhoudt, D. N. *Nanolett.* **2003**, *3*, 1449-1453.

Ferrofluid Masking for Lithographic Applications

Benjamin Yellen, Gary Friedman

Drexel University, ECE Department, Philadelphia, PA 19104

ABSTRACT

A novel self-aligned "soft masking" method that is compatible with traditional photolithographic processes is demonstrated. This method uses a suspension of ultra-fine iron oxide grains (ferrofluid) to protect or de-protect selected areas of a magnetically patterned substrate according to a programmable sequence. Automatic mask formation and registration is controlled by ferromagnetic alignment marks patterned on a substrate. External magnetic field bias applied to the system causes ferrofluid to aggregate only over designated areas on the surface, thereby masking those areas from UV or chemical exposure.

Keywords: ferrofluid, self-assembly, self-aligning, masking

INTRODUCTION

Heterogeneous substrates of the kind used for combinatorial chemistry, drug discovery and genomics, where characteristic features are below a few microns, are difficult to manufacture economically. Most approaches have utilized photolithography to produce these substrates[1], however photolithography is an expensive, chemically intensive, and laborious process typically requiring multiple steps. For the substrate to be functional, many different patterns must be aligned on top of one another with high precision. Ensuring proper geometric relation between the patterns on the surface requires repeating the alignment and registration steps for each new pattern, which becomes increasingly difficult to control when the number of patterns increases and the critical feature size becomes microns or even sub-micron in resolution.

A novel "self-aligned" soft masking method is presented here that overcomes many of the problems of traditional photolithographic masking. The masking methodology uses an array of programmable alignment marks built into the substrate to drive the assembly of sequences of ferrofluid masking patterns. The ferrofluid used in these experiments is composed of ultra-fine iron oxide grains suspended in fluid (ferrofluid). For decades, accumulation of ferrofluid on has been used to detect domain structures in thin films[2]. In this paper, by contrast, the accumulation of ferrofluid is used to mask selected areas of the surface according to a programmable sequence. Controlled photoresist patterns are produced by this methodology for the first time, and the unique advantage of ferrofluid masking is demonstrated in the creation of self-aligned electrical interconnects.

However, the lithographic applications of ferrofluid are not limited to just optical masking. Ferrofluid masking can also be applied as a diffusion mask directly to the surface during solution-based synthesis. This approach is attractive because it does not employ high temperatures, chemical solvents, or UV radiation, while combining mask formation and material deposition into a single step. Potentially, this gentle synthesis approach can be used to pattern multiple layers of delicate biological materials like proteins and cells.

This work represents a departure from previous research, in which micron-sized colloidal particles were programmed to carry various molecules to precise locations on a surface[3]. It was experimentally demonstrated that the number of particles deposited (or not deposited) at each lattice site could be reliably controlled through a combination of magnetic and morphological template features. Regular heterogeneous colloidal patterns were assembled by this technique using only physical forces (i.e. magnetic, hydrodynamic and surface forces). The theoretical process of particle assembly onto magnetic surfaces has also been analyzed previously in order to guide experimental investigations[4-7]. Models indicate that magnetically driven assembly is a plausible patterning approach for a wide range of conditions.

In this paper, by contrast, nanometer sized ferrofluid particles are directed not to deliver molecules but rather to mask particular locations of the surface from the delivery of molecules. A single island by itself attracts ferrofluid to both of its poles. Useful ferrofluid masking, however, occurs only in the presence of a bias field, which causes ferrofluid to prefer specific locations on the alignment marks. Depending on the external field orientation and hence particle magnetization, ferrofluid particles will be preferentially deposited at various locations with respect to the alignment marks, as illustrated in Figure 1.

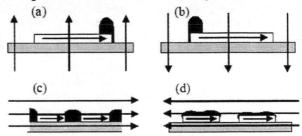

FIG 1. Illustration (a) depicts ferrofluid (black) attracted to the right-side of a ferromagnetic island (white rectangle with arrow denoting island magnetization) by external magnetic field bias applied perpendicularly to the island magnetization. When the external field bias is reversed, the ferrofluid is repelled from the right-hand pole and attracted to the left-hand pole as illustrated in (b). Ferrofluid will deposit in between the islands when external magnetic fields are applied parallel to the island magnetization, as illustrated in (c), and on top of the island when applied field is reversed, as illustrated in (d).

For example, when the external field is applied normal to the plane, ferrofluid particles accumulate over one or the other pole as demonstrated in Figure 1(a,b). Reversing the magnetization of the island rather than the external field bias can also achieve the effect of ferrofluid switching from one end of the island to the other in Figure 1a and 1b. When the external field is applied in the plane with orientation parallel or anti-parallel to the film magnetization, the ferrofluid assembles in between or on top of the film as demonstrated in Figure 1(c,d). Preferential deposition is used to selectively mask certain locations on the surface. After masking and exposure to UV radiation or chemicals, the ferrofluid can be re-suspended by simply reversing or removing the applied magnetic field bias. The ferrofluid can also be easily washed away. Spatial programming of the mask in each step can be accomplished by recording the magnetization direction of individual alignment marks, and thus the desired locations to be masked.

METHODS AND MATERIALS

Experiments were performed using Silicon or Pyrex substrates patterned with 100 nm thick Cobalt alignment marks by conventional photolithographic lift-off process. In the lift-off process, an image is defined by exposing photoresist to ultraviolet radiation through a patterned optical mask. The photoresist pattern is developed, and then Cobalt is evaporated onto the patterned wafer. Finally, the remaining photoresist is stripped to produce a pattern of Cobalt film. While negative or multi-layer resists are frequently utilized to obtain an undercut resist profile, positive Shipley 1813 photoresist yielded satisfactory results.

Individual islands acting as alignment marks in these experiments have planar dimension of 4 μm by 20 μm. Aqueous based ferrofluid domain detection fluid (EMG 807 purchased from Ferrotec) was applied to the substrate in concentrations ranging from 10% to 100% concentration in order to determine optimal masking conditions. When uniform external fields were applied to the system in the range of 10 - 200 G, one of the masking patterns shown in Figure 1 resulted. The assembled ferrofluid masking patterns were used on three different photoresist varieties of both positive and negative tone. The positive tone photoresist was Shipley 1813, and the two negative tone photoresists were NR-7 and SU-8. Ferrofluid masking films were observed as a thin coating of the substrate formed by placing a cover glass slide on top of a drop of ferrofluid.

RESULTS

The masking behavior of ferrofluid in the presence of external magnetic fields applied perpendicularly to the substrate is confirmed in Figure 2. In situations where all islands are magnetized in the same direction, the ferrofluid masks should uniformly assemble on the same side of the

island as shown in Figure 1a. In Figure 2, however, the ferrofluid is not accumulating on the same side for all the islands, because four islands in the left column are magnetized in the opposite direction of the rest of the islands. Hence, it is logical why ferrofluid should accumulate on one end for islands magnetized in one direction, and the other end for islands magnetized in the opposite direction. Effectively, this image demonstrates that the location of ferrofluid accumulation on a surface can be programmed with magnetic alignment marks.

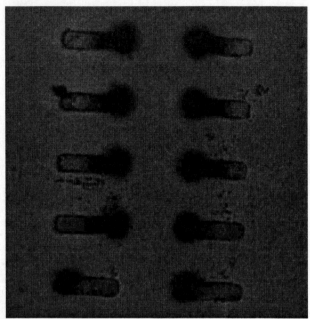

FIG 2. Optical image of ferrofluid accumulation on Cobalt islands. The accumulation is clearly dependent on the island's magnetization, since the 4 islands on the left have ferrofluid accumulating on the opposite side of the island as the 4 islands on the right.

Possible lithographic applications of ferrofluids include optical masking and diffusion masking. To demonstrate the ability of ferrofluid to act as an optical mask, Shipley 1813 photoresist was spun onto a patterned wafer of Cobalt islands, and ferrofluid was dispensed directly on top of the photoresist film. A transparent glass slide was placed on top of the ferrofluid in order to promote the formation of a uniformly thin ferrofluid masking film. When external magnetic fields were applied as shown in Figure 1a, ferrofluid accumulated over one pole of the island while avoiding the other pole. The optical contrast between the two poles was used as a photolithographic mask against flood exposure by UV radiation. A diagram of this process is illustrated in Figure 3. The wafer with ferrofluid mask was exposed to I-line UV radiation ($\lambda \approx 365$nm).

Ferrofluid masking of positive photoresist materials proved to be interesting because of the very nature of ferrofluid masking. For example, the total exposure time was found to depend strongly on the ferrofluid masking concentration. Highly concentrated ferrofluid masking films required longer exposure times, whereas dilute ferrofluid masking films required shorter exposure times. This discrepancy is due to the discovery that higher ferrofluid concentrations cause the fluid outside the areas of particle aggregation to be less transparent to UV radiation. Typically, 90-120 mJ/cm^2 is sufficient to fully expose Shipley 1813 photoresist film. In these experiments, UV exposure in the range of 150-1500 mJ/cm^2 was needed to expose the photoresist through the various ferrofluid masking film concentrations. Example photoresist patterns produced by this masking methodology on positive photoresist materials are shown in Figure 4. In all experiments, the bake times, spin speeds, and other photoresist processing parameters were kept constant.

The images in Figure 4(a,b) are examples of photoresist patterns produced by concentrated ferrofluid masks. The gray-scale photoresist patterns are the result of highly non-uniform masking of the surface. Photoresist patterns produced by dilute ferrofluid masks, on the other hand like that shown in Figure 4(c,d), were found to create more sharply defined features. The photoresist spot sizes over the poles produced by dilute ferrofluid masks averaged 8-μm in diameter with great uniformity among spots across the wafer. The slope of the photoresist sidewalls was approximately 45° as measured by AFM.

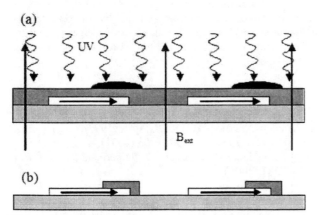

(a)

UV

B_{ext}

(b)

FIG 3. This diagram depicts ferrofluid acting as a photomask. External magnetic fields (B_{ext}) are applied perpendicularly in (a) to the micro-magnet's magnetization, causing the ferrofluid (black) to be attracted to one of the micro-magnet's poles. The accumulation of ferrofluid at the one pole is used to mask exposure of the underlying positive photoresist from ultraviolet (UV) radiation. The resulting positive photoresist structure after development is shown in (b).

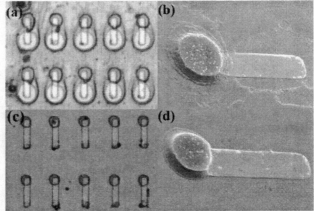

(a) (b)
(c) (d)

FIG 4. The four images (a-d) depict the results of ferrofluid optical masking of Shipley photoresist. Optical and SEM images of photoresist patterns formed by concentrated ferrofluid masking suspensions are shown in (a) and (b), and by dilute ferrofluid masking suspensions are shown in (c) and (d).

Useful structures, such as arrays of metallic spots and electrical interconnects were fabricated by masking negative tone NR-7 photoresist, which is especially formulated for lift-off. A two-step process was used to produce a self-aligned array of Au spots. First, ferrofluid masks were assembled on the substrate in a configuration like that shown in Figure 1a, and the underlying photoresist was exposed to 400 mJ/cm^2 of radiation through a ferrofluid masking film at 33% concentration. The resulting photoresist patterns were highly regular as shown in Figure 5(a,b). Next, Au metal was evaporated on top of the photoresist pattern and the remaining photoresist stripped, thus producing an array of Au spots. Such an array of may find use in the assembly and fabrication of heterogeneous substrates. Electrical interconnects can be created similarly by applying the masking configuration like that shown in Figure 1c to lift-off photoresist formulation. The photoresist structures produced from this masking configuration are shown in Figure 5(d,e). After Au evaporation and lift-off, the electrically interconnected Co islands are shown in Figure 5f. One of the main problems with self-assembly of useful devices is the question of how to interface with the assembled components. Self-aligned photolithographic patterning may provide one route to electrically connecting self-assembled device.

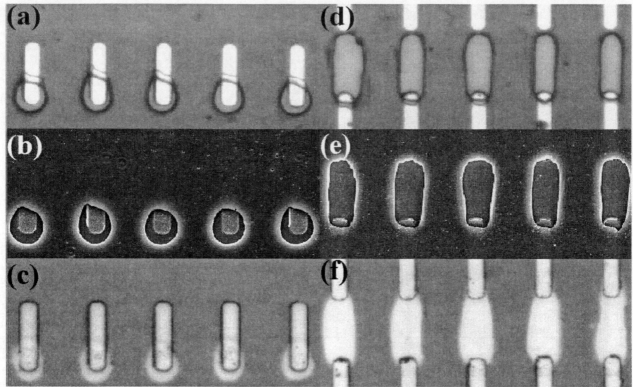

FIG. 6 (a-f) demonstrates some lithographic capabilities of ferrofluid optical masking with negative tone NR-7 photoresist. Photographs (a), (b), (d) and (e) shows the optical and SEM images of photoresist patterns produced by different applied magnetic field bias. Images (c) and (f) are the results after Gold evaporation and lift-off. The sequence of images (a-c) is the resulting pattern due to magnetic fields applied normal to the plane, while the sequence of images (d-f) is the resulting pattern due to magnetic fields applied parallel to the major axis of the islands.

CONCLUSION

In summary, patterning by ferrofluid masking enjoys significant advantages over other masking techniques. It makes resist free lithography possible. Another advantage is automatic alignment of the ferrofluid to the pre-patterned programmable ferromagnetic alignment marks. Single domain magnetic islands have been fabricated as small as 20 nm in width[8]. Such ferromagnetic alignment marks can potentially be used to build heterogeneous patterns with 50 nm resolution or better. Ferrofluid has been shown in this paper to perform consistent optical masking of microscopic features through photoresist micrometers in thickness. The formation of reliable ferrofluid masks at a relatively long range from the alignment features suggests that masking can also be accomplished with alignment marks that are embedded in the substrate or located some distance on top of the substrate. This would allow patterning to be performed on surfaces with homogeneous material characteristics or uniform surface chemistry.

ACKNOWLEDGEMENTS
The authors are thankful to NSF for support under the ECS-0304453 grant.

REFERENCES

[1] G. McGall, A. Barone, M. Diggelmann, S. Fodor, E. Gentalen, N. Ngo. *JACS*, **119**, 5081, (1997).

[2] F. Bitter, *Phys. Rev.* **38**, 1903 (1931).

[3] B. Yellen, G. Friedman, A. Feinerman, *J. Appl. Phys.*, **93**, 7331 (2003).

[4] B. Yellen, G. Friedman, A. Feinerman, *J. Appl. Phys.*, **91**, 855 (2002).

[5] B. Yellen, G. Friedman. *J. Appl. Phys.* **93**, 8447 (2003).

[6] A. Plaks, I. Tsukerman, G. Friedman, B. Yellen, *IEEE Trans Mag.*, **39**, 1436 (2003).

[7] O. Hovorka, B. Yellen, G. Friedman, *IEEE Trans. Mag.*, **39**, 2549 (2003).

[8] M. Park, C. Harrison, P. Chaikin, R. Register, D. Adamson. *Science* **276**, 1401 (1997).

Micro actuator array based on ionic polymer with patterned self-assembled gold electrode

Marie Le Guilly[*], Chunye Xu[*], Minoru Taya[*] and Yasuo Kuga[**]

[*]Center for Intelligent Materials and Systems, Department of Mechanical Engineering,
[**]Department of Electrical Engineering,
University of Washington, Seattle, WA, USA, tayam@u.washington.edu

ABSTRACT

A new fabrication technique is proposed to create electrodes on thin ionic polymer membranes for micro actuator applications. The technique consists in plasma deposition of a thin (a few nm) allylamine coating on the membrane followed by stepwise assembly of Au colloids on the amine surface into a conductive electrode. Patterned gold electrodes can be created by using a stencil mask during allylamine plasma deposition. The influence of the allylamine plasma conditions on the electrode and the subsequent actuator is studied. A Flemion membrane (14μm) with a low power plasma allylamine coating and 10 sequential assemblies of Au colloids showed a fast, repeatable actuation at +/-1.5V.

Keywords: ionic polymer, self assembly, plasma polymerization, array, patterning, MEMS.

1 INTRODUCTION

Ionic polymer metal composite actuators (ion exchange membrane sandwiched by highly conductive compliant electrodes, see Figure 1) offer large deflection, fast, yet soft actuation, and a very good durability at low voltage in wet conditions. Therefore they can be used in pumps [1], valves and for manipulation in solution.

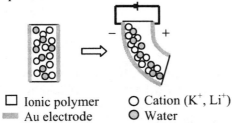

Ionic polymer · ○ Cation (K[+], Li[+])
▨ Au electrode · ◉ Water

Figure 1: Ionic polymer metal composite actuator

Previously, we reported [2] the fabrication of an actuator array based on a square of ionic polymer with a patterned electrode independently activating nine disk (diaphragm) actuators. To increase the spatial density of actuators we wish to scale down the actuator array. This requires a high resolution patterned electrode which cannot be achieved with chemical plating. In the more general aim of integrating ionic polymer actuators into MEMS, chemical plating is not suitable as an electrode fabrication technique because it requires lengthy immersions of the polymer in different ionic solutions with heating [3], which is not applicable if the polymer is already cast onto a MEMS. Zhou et al. used gold sputtering to create the electrodes of a Nafion microactuator anchored on a MEMS [4]. However, the large strain mismatch between the sputtered electrodes and the ionic polymer when the latter swells in solution is bound to create major durability issues. Therefore, a new fabrication technique yielding durable electrodes on ionic polymer with high resolution is very desirable. It would pave the way for the integration of ionic polymer actuators in MEMS, micro fluidic devices, and biochips.

Here, a new electrode fabrication technique is developed based on plasma treatment and self assembly (SA) of gold colloids. The fabrication scheme is presented in figure 2. Au-binding functional groups are created on the surface of an ionic polymer membrane by briefly exposing the membrane to an allylamine (AA) plasma which yields polymerization of AA (NH$_2$ functionality) on the surface.

By applying a mask on the membrane during the treatment one can pattern the allylamine, resulting in

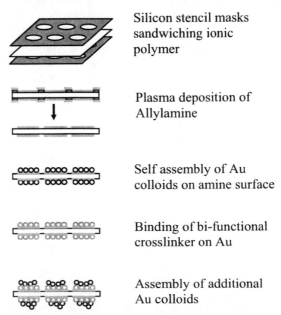

Silicon stencil masks sandwiching ionic polymer

Plasma deposition of Allylamine

Self assembly of Au colloids on amine surface

Binding of bi-functional crosslinker on Au

Assembly of additional Au colloids

Figure 2: micro-patterned self assembled electrodes fabrication scheme.

patterned binding of gold in the following step. Immediately after this treatment, the membrane is soaked in colloidal Au followed by 2-mercaptoethanol or 2-mercaptoethylamine (MEA) and this process is repeated about 15 times (a technique developed by Natan's group [5]). After the surface water has evaporated, the Au colloid multilayer film changes from dark blue/black to shiny gold. With this technique, metal deposition is done in solution which results in no strain mismatch between the polymer and metal and thus, better durability. It is easily adaptable to microfabrication, and it allows high resolution patterning of the metal. Using this process, thin gold electrodes were created on an ionic polymer and actuation is demonstrated at +1.5V/-1.5V.

2 EXPERIMENTAL

2.1 Allylamine plasma polymerization

The plasma deposition is done on the surface of the dry ionic polymer membrane and so it will be affected once the membrane swells in solution (possible cracking and loss in coverage). In order to limit the expansion of the membrane after the plasma treatment, the ionic polymer membrane is pre-stretched before the plasma treatment. The membrane (Flemion, Asahi Glass Corporation, 14 μm thick, 1.8 meq/g, K^+ counter ion) is soaked in deionized water to reach a stable swelling state at room temperature and then the edges are clamped and the membrane is allowed to dry. By this process the polymer chains elongated by osmotic pressure in the wet state are forced to remain in this stretched condition in the dry state, and when the membrane is swollen again, it will not expand further.

NH_2 functional groups (which will bind to the Au colloids) are created on both sides of the pre-stretched membrane by plasma polymerization of allylamine. The plasma polymerization apparatus is similar to the one described in Ref 10. The allylamine monomer flows into the chamber via a needle valve. Allylamine is degassed by freezing-thawing 3 times before the polymerization experiment. The Flemion membrane is placed inside the chamber on a glass holder that allows coating of both sides of the membrane. The membrane is first exposed to an argon plasma for one minute at 15W, 150 mtorr. Then the membrane is exposed to the allylamine plasma with varying conditions (table 1) and an allylamine pressure of 150mtorr. Allylamine is allowed to flow in the chamber for 5 minutes after the plasma experiment for quenching. The allylamine coating is characterized by XPS. In order to create a gold pattern on the ionic polymer, the membrane may be sandwiched by stencil masks during the allylamine coating (figure 2). Silicon masks with micron size features can be made by photolithography and etching. The feasibility of patterning is proved here by the fact that the areas of the membrane that are covered by the glass holder during the plasma treatment do not show any gold assembly afterwards.

2.2 Stepwise assembly of gold colloids

The gold colloid solution is prepared according to Natan et al. [6]. After allylamine plasma treatment, the membrane is immediately soaked in 80 ml of the Au colloid overnight to let the colloid assemble on the amine surface. The gold monolayer on Flemion looks transparent blue. The membrane is thoroughly rinsed and soaked in 80ml 10mM 2-mercaptoethanol or 2-mercaptoethylamine for 10 minutes. The membrane is thoroughly rinsed again and put back in the colloid for 40 minutes. This sequence is repeated about 10 times or until enough gold has assembled. Finally the water on the surface of the membrane is allowed to evaporate and the coating turns shiny gold. The surface and cross section of the fabricated electrodes are observed by SEM-EDS.

2.3 Actuation

A strip (actuating length 3mm, total size 1x6mm^2) of Flemion with SA gold electrodes is tested as an actuator. It is previously soaked in 0.1N LiCl so that the cation inside is lithium. The electrodes are connected to a potentiostat (CH Instruments) and a 0.2Hz +1.5/-1.5V square voltage is applied. The actuation is recorded by a CCD camera.

3 RESULTS AND DISCUSSION

3.1 Allylamine coating

The allylamine polymer coating should be strongly bonded to the Flemion, it should be thin, and it should retain the most amine groups possible. The strong bonding was realized by cleaning and etching the surface of Flemion by argon plasma. A long exposure (5 minutes) to a 15 W Argon plasma was damaging the thin Flemion membrane so the duration was reduced to 1 minute. Different plasma conditions were used for allylamine polymerization. Low power continuous wave (CW) plasma and low equivalent power pulse plasma have been shown to yield polymers with well preserved monomer structure [7, 8]. Here, the amine group should be preserved in order for the film to bind with the Au colloids. Besides, it has been shown that polymer films deposited at low power tend to be less cross linked than the high power plasma polymers, and therefore they are more flexible [9] and they swell significantly in aqueous solution [10]. The swelling behavior of low power plasma polymer films may turn out to be very beneficial in the fabrication of the electrode as explained later. The conditions used here are summarized in table 1.

Table 1: Plasma polymerization conditions

#	Duty cycle	Peak power	Eq. power	duration
1	CW	22W	22W	2min
2	3ms/10ms	66W	20W	1min
3	CW	26W/7.5W	7.5W	10s/1min50s

The first sample (sample 1) was analyzed by XPS. Figure 3a) is the spectrum of the Flemion after Argon etching. The present elements are C, F, K, O and Si. C, F and O are the elements in Flemion, K is the counter ion in the Flemion and Si likely comes from surface contamination during pre-stretching where silicone rubber is used to clamp the thin membrane. Figure 3 b) is the spectrum of the membrane after allylamine polymerization. The only detected elements are C, N and O, suggesting that the membrane is completely covered by allylamine. This means that a brief (2 min) allylamine treatment is enough to

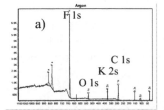 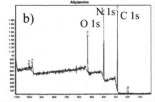

Figure 3: XPS Spectra of Flemion after argon plasma a), and after allylamine plasma b).

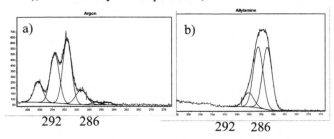

Figure 4: C1s spectra of Flemion after argon plasma a), and after allylamine plasma b) (energy references in eV).

completely cover the membrane. The presence of a significant amount of oxygen in the allylamine film may suggest some air leaking in the plasma chamber, resulting in incorporation of oxygen in the polymer coating. Figure 4 shows the C1s spectra of Flemion after argon plasma, and after allylamine plasma. In the first case the C1s peak is centered around 292eV which is the binding energy of Carbon in Teflon (structure similar to that of Flemion). There are distinct peaks at higher energy, the peak at 294eV may be partly due to CF₃ groups in Flemion and there is probably also a contribution of potassium (K2p3/2 at 293.9eV and K2p1/2 at 296.7eV). Figure 4-b) is the C1s spectrum after allylamine polymerization, the broad peak is now centered at 286eV which is the binding energy associated with amine groups. This peak is composed of several peaks likely due to C-C (285eV), C-N (286eV), C-O (286.5eV) and to a lower extent amide (288.8eV). By limiting the leaks in the system, it should be possible to significantly decrease the amount of oxygen in the coating. Besides, by using a lower power CW or a pulse plasma, it should be possible to preserve better the monomer structure, resulting in a C1s peak rid of the C-O (286.5eV) component and higher energy components.

Based on all the above, two more samples where prepared. For these experiments, the allylamine monomer was contained in a small tube with a high vacuum connector instead of the large spherical flask used in the first experiment, in order to significantly limit air leaking. One sample was prepared by pulse plasma (sample 2) and the other was prepared by low power (7.5W) continuous plasma. These samples are characterized in the following part in view of the SA Au films created on them.

3.2 Self assembly of Au colloids

After the first soaking in the Au colloid solution, all 3 samples showed strong assembly of Au, they turned dark blue. This suggests that the Allylamine plasma polymer films are very effective to bind Au colloids, probably more so than silanized glass substrates which appear pale red after the first Au layer and turn dark blue after several additional layers [11].

Figure 5 shows the Au colloids assembled on sample 2 after 4 soakings in the colloid solution and MEA. The spherical colloids are very distinct, they seem to assemble but not coalesce (figure 5-a)). Two issues arose with the use of MEA. First MEA appears to penetrate very easily in Flemion (maybe by an ion exchange process, MEA being positively charged in aqueous solution) and then to be released in the Au colloid solution in the following step, leading to precipitation of the colloid. Musick et al. reported this issue even for their glass samples but they were able to overcome the problem by an automated thorough rinsing [12]. The second issue was the very little binding of Au on MEA as opposed to mercaptoethanol. Wong et al. studied the orientation of mercaptoethylamine on Au and found that the molecule tends to adopt a gauche conformation to bring the positively charged amine closer to the gold surface [13]. This might indeed limit further Au binding. For samples 1 and 3, mercaptoethanol was used as cross linker between Au colloids. Au appeared to bind massively on mercaptoethanol and a resistance of about 50Ω (measurement in 2-point fashion with the probes 5mm apart) was obtained after 10 Au colloid layers for both samples 1 and 3. More work is underway to better understand the mechanisms of binding of Au and the different cross-linkers.

In sample 2, pulse plasma with an equivalent power of 20W was used to polymerize allylamine and no Au colloid seems to penetrate in the Flemion (figure 5). The same property was observed in sample 1 (not shown). On the other hand, in sample 3 (figure 6), a 7.5W continuous wave plasma was used and it appears that the gold colloids are able to diffuse through the coating in the Flemion. Zhang et al. reported that allylamine films prepared by low power continuous plasma were able to swell and showed DNA binding all throughout the film [9]. The allylamine coating on sample 3 is probably less cross linked than sample 2 and it is able to swell, therefore Au colloids are able to penetrate through the coating into the substrate. This characteristic is very beneficial towards making electrodes because this increases the gold contact area with Flemion.

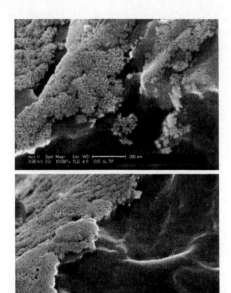

Figure 5: SEM micrographs of Au colloids assembled on Flemion sample 2 (cross section view), the cross linker is mercaptoethylamine (MEA).

Figure 6: Au colloids assembled on Flemion sample 3 (cross section view), the cross linker is mercaptoethanol.

3.3 Actuation

Sample 3 showed a fast, repeatable actuation at +/-1.5V (figure 7), the bending occurring mostly near the clamp probably because of the voltage drop along the sample. This actuation is remarkable given the resistivity of the electrode (still high relative to chemically plated electrodes [3]). Note that a voltage of +/-3V was necessary to get actuation of sample 1. The gel-like nature of the low power plasma (easily swollen) likely allows Au binding all throughout the coating and into the subsurface of the Flemion, insuring a better contact between the electrode and the membrane. The nature of the fabricated electrode (actually a collection of nanoelectrodes) makes it very effective at activating the polymer.

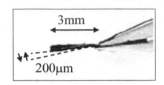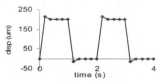

Figure 7: Flemion with SA gold electrodes, tip displacement (μm) vs time.

4 CONCLUSION

Low power plasma polymerization of allylamine and stepwise assembly of Au/mercaptoethanol appears effective to create electrodes on ionic polymer. This new technique is readily applicable for microfabrication and allows patterning of the electrode. The actuation obtained is very promising and suggests that the prepared electrodes, despite their relatively high resistivity, are very effective at activating the ionic polymer, behaving like a collection of spherical nanoelectrodes with very high capacitance.

REFERENCES

[1] Guo, Hata, Sugumoto, Fukuda and Oguro, Proc. IEEE International Conference on Robotics and Automation, 2171-2176, 1999.
[2] Le Guilly, Uchida and Taya, Proc. SPIE, Vol. 4695, 78-84, 2002
[3] Fujiwara, Asaka, Nishimura, Oguro and Torikai, Chem. Mater., 12, 1750-1754, 2000.
[4] Zhou, Li and Mai, Proc. SPIE, Vol. 5051, 332-337, 2003.
[5] Musick, Keating, Keefe and Natan, Chem. Mater., 9, 1499-1501, 1997.
[6] Grabar, Griffith Freeman, Hommer and Natan, Anal. Chem., 67, 735-743, 1995.
[7] Alexander and Duc, J. Mater. Chem., 8, 937-943, 1998.
[8] Mackie, Castner and Fisher, Langmuir, 14, 1227, 1998.
[9] Butoi, Mackie, gamble, Castner, Barnd, Miller and Fisher, Chem. Mater., 12, 2014-2024, 2000.
[10] Zhang, Chen, Knoll, Foerch, Holcomb and Roitman, Macromolecules, 36, 7689-7694, 2003.
[11] Musick, Keating, Lyon, Botsko, Pena, Holliway, McEvoy, Richardson and Natan, Chem. Mater., 12, 2869-2881, 2000.
[12] Musick, Pena, Botsko, McEvoy, Richardson and Natan, Langmuir, 15, 844-850, 1999.
[13] Wong and Vilker, Langmuir, 11, 4818-4822, 1995.

The authors gratefully acknowledge the support of the National Science Foundation Grant No ECS-0218805 and they thank Asahi Glass Corporation especially Mr Hideyuki Kurata and Mr Yoshiaki Higuchi for providing Flemion.

Batch fabrication of microsensor arrays by TiO₂ nanoparticle beam deposition

P. Milani[1*], E. Barborini[1], P. Piseri[1], S. Iannotta[2], P. Siciliano[3]

[1]CIMAINA-Dipartimento di Fisica, Università di Milano, Via Celoria 16, 20133 Milano, Italy
[2]CNR – Institute of Photonics and Nanotechnology – Trento Division – 38050 Povo di Trento, Italy
[3]CNR – Institute of Microelectronics and Microsystems – Lecce Division – Via Arnesano, Lecce, Italy

ABSTRACT

Here we present the production and characterization of chemical microsensors based on nanostructured TiO₂ deposited on microfabricated platforms. Aerodinamically filtered SCBD allows to control the nanoparticle size and to deposit in batch on microplatform arrays by using stencil masks. Since the material retains the memory of the precursor clusters, it is possible, by thermal annealing, to prepare arrays with elements characterized by different crystalline phase, grain size and porosity. This allows a novel combinatorial approach to the large-scale fabrication of multi-element micro- and nanosensing devices.

keywords: nanostructure, oxides, film, patterning, sensors

Among nanostructured transition metal oxides, TiO₂ is attracting a growing interest due to its widespread technological applications, ranging from photocatalysis to solar energy conversion, to gas sensing. The key point is the coexistence of suitable functional properties of TiO₂ with the huge surface available in nanostructured systems.

For the fabrication of gas- and bio-sensors the TiO₂ surface area, the micro- and nanostructure play a key role since the performances of these devices depend on reactions occurring at the surface and at the grain boundaries [1,2,3]. Engineering the surface structure and morphology, therefore, is of paramount importance for the integration of the nanostructured material into an effective device.

Up to now a deep comprehension of the role of film characteristics (stoichiometry, crystallographic phase, grain size distribution, porosity, roughness) in determining gas sensing properties is still lacking. The reason lays on the absence of independent control on each variable, due to their reciprocal influence. Many efforts have been directed to the control of TiO₂ surface nanostructure, however several problems still remain to be solved. For example, annealing significantly changes at the same time the microstructure, the crystalline phase and the nanoparticle dimensions of nanostructured titania [4,5]. This means that these parameters cannot be controlled independently. Another major issue is that conventional production and processing methods based on solid state or wet chemistry are not suitable for integration with batch microfabrication technology typical of microelectromechanical systems (MEMS) [6]. In particular the fabrication of arrays of microsensors for environmental monitoring, hazardous substances detection and food monitoring require a very precise control of the sensor material structure coupled to a combinatorial approach to explore and produce a larger number of different structures [6].

Alternatively to conventional synthesis methods, deposition of clusters from the gas phase is becoming an enabling technology for the production of nanostructured materials. Among different experimental approaches, supersonic clusters beam deposition (SCBD) has been shown as a viable route for the production of nanostructured systems ranging from organized nanoislands to nanostructured thin films [7,8]. Moreover, SCBD offers the possibility to synthesise films controlling the grain size distribution, the phase, the micro and nano porosity, and the width of the optical gap. Films with engineered properties can thus be synthesized.

Here we present the production and characterization of chemical microsensors based on nanostructured TiO₂ deposited on microfabricated platforms. Aerodinamically filtered SCBD allows to control the nanoparticle size and to deposit in batch on microplatform arrays by using stencil masks [9,10]. Since the material retains the memory of the precursor clusters, it is possible, by thermal annealing, to prepare arrays with elements characterized by different crystalline phase, grain size and porosity [10,11]. This allows a novel combinatorial approach to the large-scale fabrication of multi-element micro- and nanosensing devices [11]

Clusters are generated by a pulsed microplasma cluster source (PMCS). The principle of operation of a PMCS consists in the sputtering of a titanium rod by inert gas plasma ignited with a pulsed electric discharge [12]. After each pulse the ablated material inside the source condenses to form clusters. The clusters-inert gas mixture exits the source through a nozzle producing a seeded supersonic beam that can be intercepted by a substrate kept at room temperature. Deposition occurs under high vacuum conditions. The as-deposited films, after exposition to air, are stoichiometric [10,13].

During the expansion of the cluster beam an aerodynamic mass separation takes place causing the divergence of smaller clusters away from the beam axis, while larger clusters concentrate in the central part of the beam [14,15]. The exploitation of aerodynamic mass separation enabled us to produce cluster-assembled TiO₂ films characterized by a gradual change in size of deposited clusters [10].

Surface morphology can be also controlled acting on size distribution of cluster precursors by means of inertia-based filtering devices (aerodynamic lenses) fitted to cluster beam source. Strong removal of large mass clusters leads to flat films with grain size of ~15 nm and surface roughness of ~20 nm. At the opposite, the unfiltered beam leads to the synthesis of films with very high roughness extending up to micrometric scale, due to the presence in the beam of re-suspended large size agglomerates. Atomic force microscopy and scanning electron microscopy have been used to characterize the film morphology.

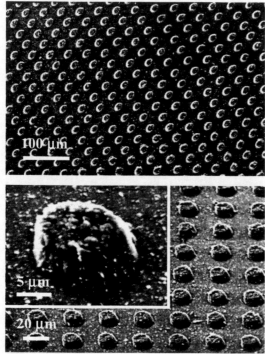

Figure 1: Microarrays produced by SCBD

Aerodynamic mass separation also affects the phase composition of the films. Films grown with the particulate close to the beam axis are composed of nanometric rutile crystals embedded into an amorphous matrix, while collecting the extreme periphery of the beam the films simply have an amorphous structure. Phase composition can be completely controlled by annealing processes: films containing rutile nanocrystals undergo grain growth by adding material from amorphous fraction; amorphous films follow the known phase transition route, switching to anatase at 400 °C and subsequently to rutile at 600 °C. We demonstrated that annealing processes up to 1000 °C do not destroy the nanostructure. Raman spectroscopy and X-rays diffraction have been used to characterize the phase composition and the size of nanocrystals.

One of the key feature of supersonic expansions is the high collimation of the particles beam. Exploiting this feature we were able to deposit regular patterns with a submicrometric lateral resolution by the use of stencil masks [9,10] (Fig. 1).

Aerodynamic separation effects can also be exploited to deposit in different regions different cluster mass distributions. It will be thus possible to batch fabricate arrays of sensors on micromachined platforms by depositing the clusters in selected areas. The arrays can then undergo to a single annealing step to create a large number of devices based on the same architecture but with the sensing material with different nanostructure and hence different performances.

REFERENCES

[1] S. Mo, W. Y. Ching, Phys. Rev. B 51, 13023 (1995).

[2] P. K. Nair, F. Mizukami, J. Nair, M. Salou, Y. Oosawa, H. Izutsu, K. Maeda, T. Okubo, Mater. Res. Bull. 33, p.1495 (1998).

[3] S. C. Liao, W. E. Mayo, K. D. Pae, Acta Mater., 45, 4027 (1997).

[4] B. Panchapakesan, D.L. DeVoe, M.R. Widmaier, R. Cavicchi, S. Semancik, Nanotechnology 12, 336 (2001).

[5] S. C. Liao, W. E. Mayo, K. D. Pae, Acta Mater., 45, 4027 (1997).

[6] B. Panchapakesan, D.L. DeVoe, M.R. Widmaier, R. Cavicchi, S. Semancik, Nanotechnology 12, 336 (2001).

[7] P. Milani, S. Iannotta, *Cluster beam synthesis of nanostructured materials,* Springer Series in Cluster Physics, Springer Verlag, Heidelberg-Berlin, (1999).

[8] L. Ravagnan, F. Siviero, C. Lenardi, P. Piseri, E. Barborini, P. Milani, C. Casari, A. Li Bassi, C.E. Bottani, Phys. Rev. Lett. 89, 285506-1 (2002)

[9] E. Barborini, P. Piseri, A. Podesta', P. Milani, Appl. Phys. Lett. 77, 1059 (2000).

[10] E. Barborini, I. N. Kholmanov, P. Piseri, C. Ducati, C.E. Bottani, P. Milani, Appl. Phys. Lett. 81, 3052 (2002).

[11] I. N. Kholmanov, E. Barborini, S. Vinati, P. Piseri, A. Podestà, C. Ducati, C. Lenardi, P. Milani, Nanotechnology 14, 1168 (2003).

[12] E. Barborini, P. Piseri, P. Milani, J. Phys. D: Appl. Phys. 32. L105 (1999).

[13] E. Barborini, I. N. Kholmanov, A. M. Conti, P. Piseri, S. Vinati, P. Milani, C. Ducati, Eur. Phys. J. D 24, 277 (2003).

[14] E. Barborini, P. Piseri, A. Li Bassi, A. C. Ferrari, C. E. Bottani, P. Milani, Chem. Phys. Lett. 300, 633 (1999).

[15] H. Vahedi Tafreshi, P. Piseri, S. Vinati, E. Barborini, G. Benedek, P. Milani, Aerosol Sci. Technol., 36, 593 (2002).

*email: pmilani@mi.infn.it
Topic: Micro and Nano Structuring and Assembly

DIRECT SCANNING LASER WRITING OF 3D MULTULAYER MICROSTRUCTURES

Hui Yu, Biao Li, and Xin Zhang

Department of Manufacturing Engineering and Fraunhofer USA Center for Manufacturing
Innovation, Boston University; 15 Saint Mary's Street, Boston, MA 02215
E-mail: ghy@bu.edu; Tel: 617-358-1631; Fax: 617-353-5548

ABSTRACT

Material processing with lasers is an expanding field since it not only makes manufacturing cheaper, faster, cleaner, and more accurate but also opens up entirely new technologies and manufacturing methods that are simply not available using standard techniques. In this paper, we demonstrate the use of a simple beam-scanning laser system for the rapid prototyping of 3D microstructures with different materials as well as the reentrant geometries favorable for photoresist lift-off technique. With the aid of an auxiliary laser alignment setup, this technology could be readily adopted for flexible fabrication of 3D multilayer microstructures.

1. INTRODUCTION

The interest in microstructures with true 3D geometries has dramatically increased in microsystems applications in recent years [1]. To promote the wide applications of such 3D microstructures, the materials and patterning technologies for their fabrication should be inexpensive. A good example would be disposable chips in biomedical applications. Advanced lithographic techniques have been developed to create multidimensional 3D structures with SU-8 negative photoresist such as embedded micro-fluidic channels [2,3], structural parts of micro motors [4] for (bio-) chemical and medical applications. However, those 3D microstructures are fabricated with tedious fabrication processes. Alternative technique, proton beam, has been adopted as direct writing micro-machining method to form embedded micro-channels; but it requires costly facility [5].

Material processing with lasers is an expanding field [6] since it not only makes manufacturing cheaper, faster, cleaner, and more accurate but also opens up entirely new technologies and manufacturing methods that are simply not available using standard techniques. This paper introduces a new, revolutionary 3D manufacturing approach for rapid processing of freeform multi-layer microstructures using a scanning laser system. This technique combines the best features of photolithography techniques in multi-layer processing with the versatility of existing 3D prototyping technologies. The fabrication process is extremely versatile. An apparatus has been built for precise multilevel microstructure alignment, enabling manufacturing of a large number of multi-layered, complex microstructures. In addition, the new fabrication process provides the capability to generate flexible freeform structures using CAD which greatly reduces the development cycle, potentially reaching hours or less design-to-fabrication turnaround time.

2. EXPERIMENT SETUP

2.1 Scanning Laser Direct Writing System

Figure 1a describes experimental apparatus. A diode-pumped, high repetition rate (~100KHz), nanosecond pulse duration 3rd harmonic Nd:YAG laser BL8-355Q operating at a wavelength of 355nm was used as the light source. The laser produces a maximum average power of larger than 400mW at 20kHz. The output energy can be adjusted by varying the pump diode current or by altering the repetition rate. An x-y-z-θ stage was built to translate a sample in four degrees-of-freedom. The laser light is directed through a 2-axis Scan Head before it reaches the sample. The principle of the 2-axis Scan Head is as follows: the laser beam passes through and is steered by a set of x and y mirrors that are coupled to galvanometers. The orthogonal arrangement of the x and y mirrors directs the beam down toward the work piece and over the length and width of the scan field. Field distortion is compensated with an *F-Theta* lens after the two-mirror system. This enables both a large scan field (100×100 mm^2) and a small spot size (~5 µm). ScanWare laser scanning control software developed by Nutfield Technology was used to generate the desired micro-patterns.

In previous work, we have demonstrated various 3D microstructures such as micro-peg array, T-plug, embedded channels fabricated using multi-step inclined scanning laser writing with various incident angles [7].

2.2 Apparatus for Multi-layer Alignment

Since the pulsed UV laser is not easily visible, an auxiliary red alignment laser was used as a target for multilevel microstructure alignment. Figure 1b shows the schematic diagram of the experiment setup. The alignment procedure between different layers is as follows: An alignment mark was permanently carved on a target sample using pulsed laser writing. A cross pattern was then generated with the alignment laser and aligned to the mark.

By maintaining the position of the alignment laser cross pattern in place, the sample can be easily unloaded for multilayer processing and then reloaded on the stage for further laser processing. This method works for both front-to-front and front-to-backside alignment.

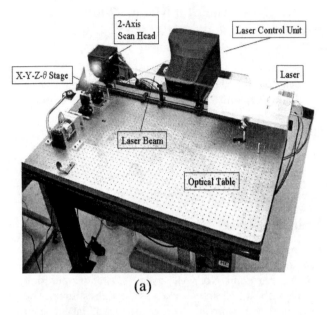

(a)

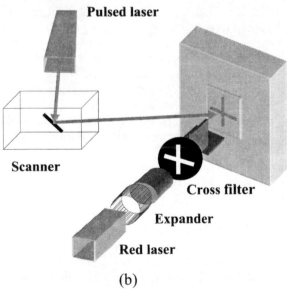

(b)

Figure 1: *(a) Photograph of the scanning laser direct writing system. (b) Schematic diagram of the experiment setup for multi-layer alignment.*

The best-established application of laser processing is abrasive laser machining such as drilling, cutting and trimming etc. With the auxiliary laser, we can align the pulsed laser to micro-machined structures for micro-surgery such as trimming of resistors, which can significantly increase the yield in the processing of resistive elements. In electronics industry, resistors are fabricated with intentionally low values of resistance and then trimmed by

removing material from the conducting path due to the difficulty of controlling the value of the resistance to within the tolerances required by the circuit. Laser trimming offers the advantages of better cleanliness and better control over the final resistance. Figure 2a illustrates resistance modification of poly-Si resistors realized by using a second source of laser (auxiliary red laser) for alignment, and the UV pulsed laser for trimming. Much more functionalities were demonstrated, for example metal sputtered etching on Au film shown in Figure 2b and micro-hole drilling illustrated in Figure 2c. The different geometries of ~4µm diameters holes created on poly-Si and Si substrates may be caused by thermal conductivity difference between the two materials.

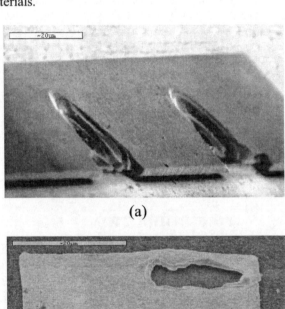

(a)

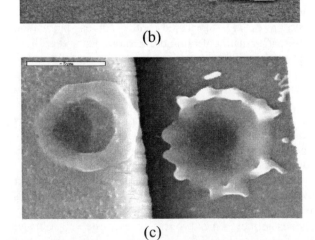

(b)

(c)

Figure 2: *(a) Resistance modification for a poly-Si resistor using laser trimming. (b) Metal sputter etching. (c) Etching holes in Si and poly-Si using one laser pulse.*

3. CHARACTERISTICS AND APPLICATIONS

3.1 Microfabrication Using Gaussian Laser Beam

Since the Gaussian-distributed light intensity of UV laser pulse can generate a Gaussian geometry in photoresist, negative wall angle will be formed which is favorable for metal lift-off technique (Figure 3). As can be seen, positive photoresist is subjected to laser pulsing from the backside of a glass substrate. After photoresist developing, an inverse funnel (reentrant) structure is formed. Figure 4a shows the SEM photography of the Gaussian geometry in photoresist generated by Gaussian intensity laser pulsed from backside. Figure 4b demonstrates a micro-heater on glass substrate fabricated using lift-off technique. The fabrication process is as follows: a glass substrate was prepared by cleaning with Piranha (3 sulfuric acid: 1 hydrogen peroxide) for 10 minutes. HMDS was applied by spin-coated at 900 rpm for 5 s, followed by 4000 rpm for 60 s. Shipley 1813 resist was then spin-coated at 900 rpm for 5 s; ramp to 2000 rpm and hold for 60s. The resist was soft-baked and allowed to cool back to room temperature. After the exposure of resist to UV light from backside and developing in MF319 for 1 minute, the sample was hard-baked for 3 minutes at 120°C. A thin film of Au (~230 nm) was sputtered onto the patterned resist. Finally, Acton was applied to strip the resist. When the resist is stripped, the metal on top of the resist is "lifted-off".

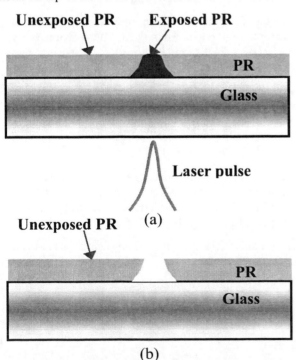

Figure 3: *The inverse funnel (reentrant) structure favorable for metal lift-off technique. (a) Positive photoresist subjected to backside laser pulsing. (b) Reentrant geometry formed after photoresist developing.*

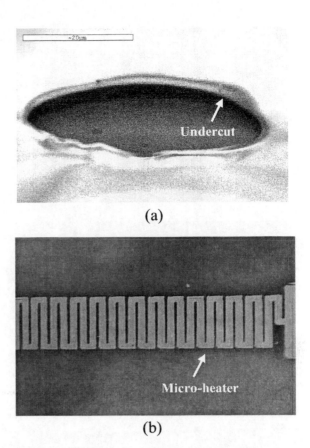

(a)

(b)

Figure 4: *(a) SEM photography of the Gaussian geometry in photoresist generated by Gaussian intensity laser pulse from backside. (b) Picture of micro-heater fabricated using our lift-off technique illustrated in Figure 3.*

3.2 Three-dimensional Multi-layer Manufacturing

In general, complex 3D structures fabricated from single layer deposition are limited. Multi-layer alignment and fabrication are required with the increase of geometry complexity. The alignment function in a conventional photolithography is realized by using alignment marks to align structures in different layers. Similar to those align marks, with the auxiliary red alignment laser system, a variety of multi-layer 3D structures, which otherwise remain challenges for other scanning laser 3D micro-fabrication techniques, could be readily realized especially for those need the alignment from both the front-side and backside. A schematic diagram of double-side laser writing on two SU-8 layers with a thin metal interfacial layer is illustrated in Figure 5a. A glass sample was first prepared by cleaning in a piranha solution (sulphuric acid:hydrogen peroxide, 3:1) for 10 minutes and dehydrating on a hot plate at 95°C for 5 minutes. SU-8 2025 was then spin-coated to produce a film approximately 60 μm in thickness. The resist was soft-baked and a thin film of Au (~230 nm) was coated by sputtering. Following the sputtering, the spin-coating and pre-bake steps were repeated to coat a second layer of

SU-8 onto the Au film. Then the first layer of SU-8 was exposed from the back-side, while the second layer of SU-8 was exposed from the front side. After the post-exposure baking, both resist layers were finally developed in propylene glycol methyl ether acetate (PGMEA). Figures 5b and 5c illustrate double-side-aligned microstructures using laser pulsing from both sides and a magnified embedded channel, respectively.

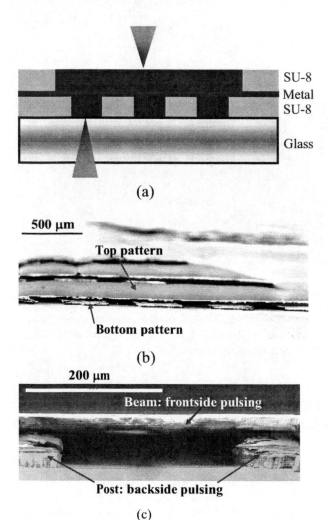

Figure 5: *(a) Schematic diagram of double-side laser writing on two SU-8 layers with a thin metal interfacial layer. (b) Proof of concept double-side-aligned structures fabricated using laser pulsing from both sides. (c) The magnified embedded channel.*

4. CONCLUSIONS

In summary, we have developed a novel beam-scanning laser system for 3D microstructures fabrication with different materials. By applying the laser-scanning direct writing technique, it was demonstrated that the high resolution, high speed and low cost of the scanning laser direct writing system opened up a variety of applications for example embedded channels, metal lift-off favorable structures etc. Moreover, it is readily to be extended to other materials and microstructures manufacturing which are challenging to achieve by conventional lithography and etching techniques.

ACKNOWLEDGEMENT

This work has been supported in part by National Science Foundation CAREER Award under Contract No. DMI 0239163 and Air Force Office of Scientific Research (AFOSR) under Contract No. F49620-03-1-0078. The authors would like to thank all the other members at the Laboratory of Microsystems Technology (LMST), Boston University for their help and support, especially Alexander Gruentzig, Robert Baranowski, David Mabius, and Jeremy Moyers.

REFERENCES

[1] M. Han, W. Lee, S. Lee, and S.S. Lee, "Microfabrication of 3D Oblique Structures by Inclined UV Lithography," Proceedings of μTAS 2002, 106, 2002.

[2] R.J. Jackman, T.M. Floyd, R. Ghodssi, M.A. Schmidt, and K.F. Jensen, "Microfluidic Systems with On-line UV Detection Fabrication in Photodefinable Epoxy," J. Micromech. Microeng. 11, 1, 2001.

[3] Y. Chuang, F. Tseng, J. Cheng, and W. Lin, "A Novel Fabrication Method of Embedded Micro-channels by Using SU-8 Thick-film Photoresists," Sensors and Actuators A103, 64, 2003.

[4] V. Seidemann, S. Butefisch, and S. Buttgenbach, "Fabrication and Investigation of In-plane Compliant SU8 Structures for MEMS and Their Application to Micro Valves and Micro Grippers," Sensors and Actuators A97-98, 457, 2002.

[5] F.E.H. Tay, J.A.V. Kan, F. Watt, and W.O. Choong, "A Noval Micro-machining Method for the Fabrication of Thick-film SU-8 Embedded Micro-channels," J. Micromech. Microeng. 11, 27, 2000.

[6] Dieter Bäuerle, Laser Processing and Chemistry, Springer, 2000.

[7] Hui Yu, Alexander Grüntzig, Yi Zhao, Andre Sharon, Biao Li, and Xin Zhang, "Rapid Prototyping 3D Microstructures on Soft and Rigid Templates Using A Scanning Laser System", Proceedings of μTAS 2003, 347, 2003.

Bottom-Up meets Top-Down – A Novel Approach towards 3-D Optical Devices

R. Houbertz, S. Cochet, G. Domann, M. Popall, J. Serbin*, A. Ovsianikov*, and B.N. Chichkov*

Fraunhofer ISC, Neunerplatz 2, 97082 Würzburg, Germany

*Laser Center Hannover LZH, Hollerithallee 8, 30419 Hannover, Germany

ABSTRACT

Building materials from the bottom-up is complementary to conventional top-down materials processing. Inorganic-organic hybrid polymers designed for optical applications, were synthesized by hydrolysis/ polycondensation reactions, resulting in storage-stable resins built-up by nm-size organically functionalized primary particles (oligomers). The organically functional groups are typically oligo-(meth)acryl or styryl which allows one to pattern the resin by two-photon polymerization (2PP) processes using femtosecond lasers. With this method, any arbitrary structure can be directly written in three dimensions into the hybrid polymer with a resolution down to approx. 100 nm. The materials and the patterning process will be discussed with respect to applications in photonic devices.

Keywords: inorganic-organic hybrid polymers, two-photon polymeriaztion, photonic structures

1 Introduction

During the last decade, inorganic-organic hybrid polymers (ORMOCER®s) have attracted considerable attention for optical (microoptics, waveguides) or microelectronical (dielectrics, passivation) applications [1-3]. The properties of these multifunctional materials can be tuned towards the respective application which thus makes them promising also for photonics. The processing of ORMOCER®s consists of two steps: first, an inorganic-organic storage-stable sol is established by hydrolysispolycondensati on (sol-gel) reactions, followed by an organic cross-linking of the polycondensed alkoxysilanes due to their organic functionalities (e.g., methacryl or styryl moieties). The latter allows one to pattern the material by conventional photolithography [4] as well as using femtosecond laser pulses [5]. Both methods result in polymerized (solid) structures where the non-polymerized parts are removed by standard developers, whereas arbitrary 3-D structures can be written with the latter method.

2 Inorganic-organic hybrid polymers

Figure 1 shows a schematic sketch of the multifunctional precursors for the hybrid polymer synthesis as well as a brief overview about the variation of material properties. Generally, first an Si-O-Si network is established via polycondensation reactions of alkoxysilanes which yields organically modified nanoscaled inorganic-oxidic units.

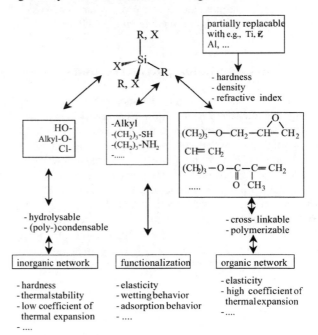

Figure 1: Schematic sketch of multifunctional precursors for hybrid polymers and the variation of properties.

The size of these inorganic units (typically 1 to 5 nm) is dependent on the alkoxysilanes and the reaction conditions (catalyst, concentration, solvent, temperature). Bi-substituted silanes yield chain or ring polymers, while tri-substituted silanes might result in three-dimensional networks. The Young modulus as well as the mechanical and thermal stability can be increased by increasing the amount of inorganics in the network. In addition, the coefficient of thermal expansion (CTE) and the optical losses in the near-infrared (NIR) regime are reduced as well.

A modification of the individual units influences the Young modulus and simultaneously the CTE of the

ORMOCER®s. In order to achieve highest transparency in the NIR regime, the material can be partially fluorinated or specific groups are replaced by perfluoroaryl groups, respectively. The organic polymerizable units are chosen depending on the requirements of the application. If UV lithography is applied for patterning, e.g., methacryl or styryl functionalities are chosen. Moieties such as alkyl or aryl groups connected to the Si also influence the material properties. It is likely that an increase in their amount within the hybrid polymer will reduce the degree of polymerization due to sterical reasons, thus resulting in a reduced density within the coated layers. In addition, these groups influence also the refractive index and the dielectrical properties.

Upon sol-gel processing, solvents are usually present in the resin. Typically, these are removed from the material in order to achieve highest processing flexibility. For thin-film applications, other solvents can be added to adjust the viscosity of the material. The resins storage stability without initiators at room temperature is about one year; for longer storage the resins should be cooled down (- 18 ℃).

2   Two-photon polymerization

In this section, two-photon polymerization technique allowing the fabrication of arbitrarily complicated 3D nanostructures is described. It has recently been demonstrated by several groups [5-9] that two-photon-polymerization (2PP) of photosensitive materials allows the fabrication of three-dimensional microstructures. When tightly focussed into the volume of a liquid resin (which is transparent in the infrared), femtosecond laser pulses can initiate two-photon polymerization and produce structures with a resolution better than 0.1 μm. Compared to conventional photo-lithography which is a planar process, 2PP is a real three-dimensional volume micro- and nano-fabrication technique.

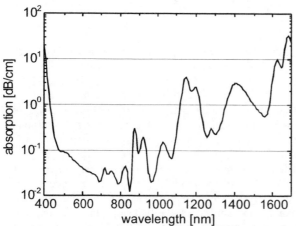

Figure 2: Absorption spectrum of an inorganic-organic hybrid polymer (ORMOCER®I) resin.

There is a strong demand for cheap and reliable materials such as, for example hybrid polymers, which can be photochemically patterned in two and three dimensions. Important advantages of these materials are: adjustable refractive index (in the range of 1.46 to 1.6), high optical transparency (Fig. 2) with low losses at data and telecommunication wavelengths (<0.06 dB/cm at 830 nm, <0.2 dB/cm at 1310 nm, and <0.6 dB/cm at 1550 nm), good thermal and mechanical properties, high chemical resistance, and quite low cost. These properties allow one to use ORMOCER®s for fabrication and rapid prototyping of different photonic structures and integrated optical devices.

These hybrid polymers are designed for ultraviolet (UV) lithography and contain initiators (e.g., Irgacure 369) sensitive to UV radiation with a single-photon absorption maximum at about 330 nm wavelength. In the near infrared, especially at 780 nm, these materials are transparent (Fig. 2) which allows one to focus femtosecond Ti:Sapphire laser pulses into the volume of the liquid resin. The experimental setup for real 3-D lithography is shown in Fig. 3.

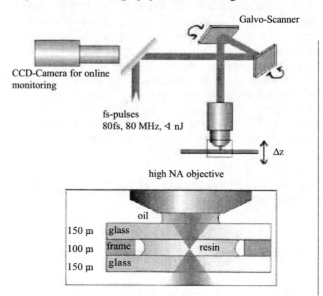

Figure 3: Experimental setup for 3D nanostructuring by means of 2PP.

The number of laser pulses and the irradiation time are controlled by an acousto-optical shutter with a minimum switching time of a few nanoseconds. Femtosecond laser pulses are focussed by a 100x immersion lens microscope objective having a numerical aperture of NA=1.4 and filled with a refractive index-matching oil (n_{oil}=1.515).

The high photon density in the focal volume triggers two-photon absorption by initiator molecules and results in the generation of radicals which, in turn, initiate the polymerization of the resin. All these processes are confined to a highly localized area at the focal point due to the quadratic dependence of the rate of radical formation on the laser intensity

$$\frac{\partial \rho}{\partial t} = (\rho_0 - \rho)\sigma_2 N^2 \quad , \tag{1}$$

where σ_2 is the effective two-photon cross-section for the generation of radicals, $N(r,z,t)$ is the photon flux, ρ_0 is the radical density at total conversion of initiator into radicals and $\rho\ (r,z,t)$ is the density of radicals [5].

When the laser focus is moved through the resin in three dimensions, the polymerization occurs along the trace of the focus, i.e., any computer-generated 3-D structure can be fabricated by direct laser "writing" inside the volume of the hybrid polymer. The non-irradiated liquid resin can be, analogously to planar processing dissolved in a solvent (e.g., acetone), leaving the polymerized copy of the computer model.

In order to realize real 3-D microstructuring with femtosecond lasers, the laser beam was scanned with an x-y galvo-scanner and the sample was moved in the z direction with a translational stage. As a computer model, a Venus statue as shown in Fig. 4 (a) was used. To fabricate a 3-D microcopy of this statue, the laser beam is scanned in the horizontal plane along the contours shown in the central figure of Fig. 4. On the right side [Fig. 4 (c)], a scanning-electron microscope (SEM) image of a μ-scale Venus statue fabricated by 2PP is shown. In order to speed up the fabrication process which took approximately 5 minutes, only the outer shell was irradiated with femtosecond laser pulses. After that, the liquid resin was washed out and the statue was subsequently irradiated with UV light for final polymerization of the inner body.

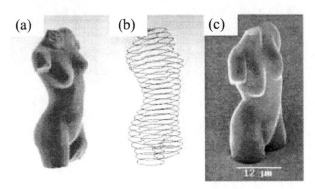

Figure 4: (a) Venus model used for the fabrication of 3-D micro-statue, (b) trace of the laser focus, and (c) SEM image of a μ-scale Venus fabricated by 2PP in ORMOCER®I.

Because of the threshold behavior and nonlinear nature of the 2PP process, a resolution beyond the diffraction limit can be realized by control of the laser pulse energy and the number of applied pulses. This is demonstrated in Fig. 5, where the measured data for the polymerized voxel (volume pixel) diameter are shown for different laser powers and irradiation times [5].

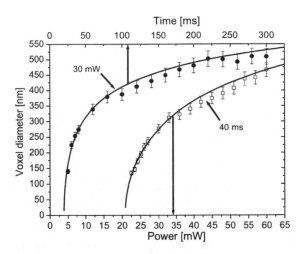

Figure 5. Measured data for diameter of the polymerized volume (voxel) as a function of the average laser power (for constant irradiation time $t = 40$ ms) and as function of the irradiation time (for constant laser power $P = 30$ mW).

Since the optical losses of ORMOCER®s are lower than 0.5 dB⁄cm at data and telecommunication wavelengths [1], the 2PP microstructuring technique is very attractive for the fabrication of micro-optical components and devices. In Fig. 6, an example of a 'photonic crystal' structure fabricated by 2PP in an ORMOCER® is shown. The structure was build up from individual rods with a diameter of about 200 nm and a spacing of 450 nm, respectively. The total fabrication time was 4 minutes.

Figure 6: SEM images of 'photonic crystal structures' fabricated by 2PP in ORMOCER®I.

A major problem one always has to cope with when working with photosensitive resins in general is that the material shrinks upon cross-linking. For the 2PP process, shrinkage does not affect the 3-D geometry since each voxel shrinks to the same extent and remains exactly at the position of the controlled laser focus. During the final UV exposure process, however, further cross-linking takes place, leading to a pronounced shrinkage [c.f., Fig. 6 (a)]. This manifests itself in a distortion of the final 3-D structure compared to the computer model. This is due to a very good adhesion of the material to the substrate which can be seen as boundary condition, whereas this boundary condition is less present above the interface glassORMOCER®. Thus, the material can easily shrink upon cross-linking and distort the model, while at the interface the material is strongly bound to the substrates surface.

The more laser power is used, i.e. the higher the amount of polymerized units after laser irradiation, the less shrinkage occurs during the UV irradiation. Different pre- and post-treatments such as annealing the sample up to 80 ℃ have been tested in order to suppress the shrinkage which was not successful. For a compensation of the distortion caused by the shrinkage, the original model structure which is used to control the laser scanner was adapted such that the final structure shows no distortion after UV irradiation (Fig. 6 a and b). In Fig. 6 (c), a zoom into the structure depicted in Fig. 6 (b) is given which clearly shows the woodpile structure of the model.

Compared to other polymer materials, ORMOCER®s do not exhibit parasitic polymerization due to their microstructure (inorganic backbone), and the surface quality (particularly the rms roughness) is much better than for commercial acrylate systems [9]. It has to be mentioned, however, that the refractive index of the material has to be tuned towards higher refractive indices in order to account for a 3-D photonic bandgap. This is currently under investigation.

3   Conclusions

Inorganic-organic hybrid polymers and real 3-D patterning by two-photon absorption using femtosecond lasers seem to complement one another perfectly. The use of a bottom-up material with a top-down method promise not only to be a low-cost alternative for fabrication of components for applications in optics, medicine, and biology, but also opens a wide range for producing arbitrary three-dimensional micro- and nanostructures for the development of novel devices. By tuning the materials refractive index and by choosing a more suitable initiator with a high two-photon absorption cross-section, the fast fabrication of 3-D photonic crystal structures at any place on an optical chip will be feasible at fairly low cost.

4   Acknowledgements

We would like to acknowledge financial support from the Deutsche Forschungsgemeinschaft (Schwerpunktpro-gramm Photonic Crystals). We would like to thank C. Cronauer for technical assistance with the material.

REFERENCES

[1]   U. Streppel, P. Dannberg, C. Wähter, A. Bräer, L. Frölich, R. Houbertz, and M. Popall, Opt. Mat. **21**, 475 (2002).

[2]   L. Frölich, R. Houbertz, S. Jacob, M. Popall, R. Mueller-Fiedler, J. Graf, M. Munk, and H. von Zchlinski, Mat. Res. Soc. Symp. **726**, 349 (2002).

[3]   R. Houbertz, L. Frölich, J. Schulz, and M. Popall, Mat. Res. Soc. Symp. **665**, 321 (2001).

[4]   U. Streppel, P. Dannberg, Ch. Wähter, A. Bräer, and R. Kowarschik, Appl. Opt. **42**, 3570 (2003).

[5]   J. Serbin, A. Egbert, A. Ostendorf, B.N. Chichkov, R. Houbertz, G. Domann, J. Schulz, C. Cronauer, L. Frölich, and M. Popall, Opt. Lett. **28**, 301 (2003).

[6]   S. Maruo, O. Nakamura, and S. Kawata, *Opt. Lett.*, vol. 22, pp. 132-134, 1997; S. Kawata, H.-B. Sun, T. Tanaka, and K. Takada, Nature **412**, pp. 697 698, 2001.

[7]   B. H. Cumpston, S. P. Ananthavel, S. Barlow, D. L. Dyer, J. E. Ehrlich, L. L. Erskine, A. A. Heikal, S. M. Kuebler, I.-Y. S. Lee, D. McCord-Maughon, J. Qu, H. Rökel, M. Rumi, XL. Wu, S. R. Marder, and J. W. Perry, *Nature* **398**, pp. 51-54, 1999.

[8]   H.-B. Sun, S. Matsuo, and H. Misawa, *Appl. Phys. Lett.*, Vol. 74, pp. 786- 789, 1999.

[9]   P. Galajda and P. Ormos, *Appl. Phys. Lett.*, vol. 78, pp. 249-251, 2001.

A Novel Method for determining Optimum Etch Times for the One step Dry Release Process

P.T Docker P. K. Kinnell, M.C. Ward

University of Birmingham Edgbaston Birmingham B15 2TT England

ABSTRACT

This paper details a novel method for determining the optimum etch times for releasing structures when using the authors' one step dry release process. By using self releasing structures known as 'waffles' the exact point when a device is released can be determined without having to etch numerous chips for different times and sectioning them. This solution allows a worker to determine when a device is released by a simple visual check.

Keyword: One step DRIE dry process, STS, SOI Waffles

1. INTRODUCTION

This paper details an optimisation technique for the one step dry release process developed by the authors[1] from the work of Ranglow[2] and Ayon[3]. The authors' process facilitates the manufacture of released structures using STS Deep reactive ion etching[4,5,6] and silicon on insulator (SOI) wafer technology[7,8] without the requirement for a second wet chemistry process step requiring the use of Hydrofluoric acid. It utilises the notching phenomena[9,10] experienced when DRIE reaches the buried oxide when etching SOI wafers. The authors developed design rules, which harness this notching phenomenon and if followed carefully can be used to release structures[1]. Initially there was a severe limitation to this process whereby any structure to be released by this process had to be bounded by a narrow trench typically 5 microns wide which was a limiting factor if a device was to achieve large actuation.

Further work led to the development of 'waffle structures'[11] which allow for the one step process to be utilised to obtain released structures now bound by large cleared areas. These waffle structures radically increase the applications for the one step process technology, which now can be used to manufacture artefacts such as large throw actuators or even large reservoirs for micro fluidics applications. It has now come to light that theses waffles can have a further application. As the waffles are removed by the same mechanism that releases structures, their removal occurs at the same time as a device adjacent to it is released. Thus they can be used to indicate when the device is released. The merits of this new application for waffles are detailed in this paper.

2. WAFFLES

A 'Waffle' is an area of silicon populated with a matrix of square holes and isolated from other design features by a perimeter trench. Figure 1 shows a micrograph of one of these sacrificial structures, 'waffles' that has been partly etched.

Figure 1 micrograph of a partly etched waffle (holes are 10 microns square)

Once etching is completed, the 'waffle' structures are released by the same undercutting effect[1,2,3] used by the one step process to release devices. This instigates their removal. The design rules for these sacrificial waffles are the same as those for any other feature integral to the device that needs to be released. Figure 2 Shows a waffle after it has been released from the oxide layer and its removal is nearly complete.

Figure 2 Waffle after it has released and dropped to the trench floor

To date the mechanism for the removal of the waffles is not fully understood. The proposed hypothesis for their removal is as follows. While a waffle is still attached to the oxide layer its orientation is assured to be perpendicular to the plasma (see Figure 3). Ion bombardment is collimated and only hits the trench floor and the top surface of the waffle, which is protected by photoresist. This orientation ensures the side walls and their passivation layers are not subject to ion bombardment and therefore they remain unetched either mechanically or chemically.

Once the waffle is released it loses this perpendicular orientation when it falls to the oxide layer. This is believed to be due to irregularities in the underside of the waffle occurring as a result of the release process. Subsequent etch steps expose the sidewalls of the waffles to ion bombardment removing the protective polymer layer (see Figure 3), mechanically etching them and then leaving them exposed to chemical etching. In this work the waffle walls are only 5 microns thick and the etch rate for the DRIE process is 3 microns per minute. In these conditions a disorientated waffle would only last approximately 1.5 minutes. This would explain why once a waffle is released its failure is radical and catastrophic.

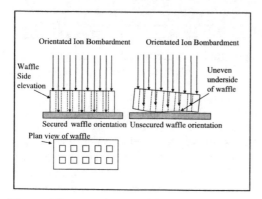

Figure 3 Proposed method for waffle removal

As previously stated these waffles have only been used to clear larger areas in close proximity to a released structure. Figure 4 shows a patterned wafer prior to etching containing a structure surrounded by waffles and then in figure 5 a micrograph of the device after its release and the waffles have been removed.

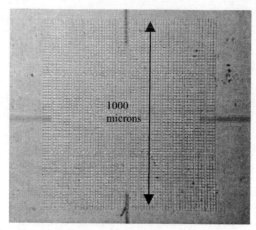

Figure 4 Optical Micrograph of the design with waffles prior to etching

Figure 5 A micrograph of a device released and its surrounding area cleared using waffles.

3. NEW PROPOSED USE FOR WAFFLES

This paper details work that successfully illustrates how these waffles can be used for another purpose, to determine when a structure is released. When the one step process was being developed, determining when a structure had been released required sectioning of the sample and checking visually using an SEM. This method was time consuming and wasteful of chips.

It was then decided that the waffle structures could be utilised for an additional purpose. By placing a waffle in close proximity to a structure its removal could be used as an indicator as to when the released structure was released. As their removal is so radical and catastrophic it also allows the workers to accurately judge when structures are released. This allows for the production of released structures with a minimum notch size, and hence an optimum released structure.

The following micrographs show an example of a structure as it releases which is indicated as the waffle releases, fails and is finally cleared. Figure 6 shows a micrograph of a dual beam resonator after it has been etched for 8 minutes. To the side of it can clearly be seen the release indicating waffle still intact and unreleased there fore indicating the device is still unreleased (see also figure 7 for close up). Figure 8 shows micrograph of the same device after etching for a further minute and it can be seen that waffle removal has begun indicating that the device is releasing. Finally figure 9 shows the device and the complete elimination of the waffle indicating the device is fully released after a further 30 seconds of etching.

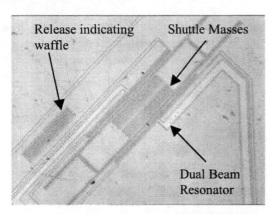

Figure 6 Micrograph of complete device after 8minutes of etching (device is 800 microns long)

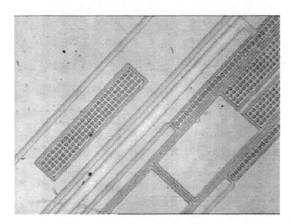

Figure 7 Close up of release indicating waffle (after 8 minutes etching)

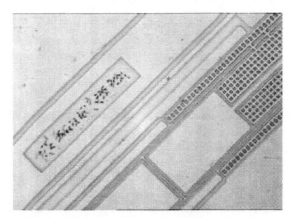

Figure 8 Release indicating waffle after catastrophic failure has commenced (after 9 minutes of etching)

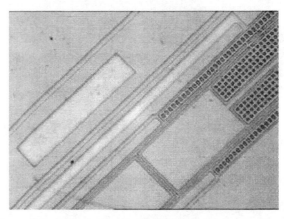

Figure 9 Release indicating waffle has been completely eradicated indicating that the device is completely released. (after 9.30 minutes)

To date the resonator has been excited using a piezoelectric device to prove it was in fact released.

From this work further information can be drawn to reinforce the hypothesis offered by the authors for the removal of the waffles. The time taken for them to be removed directly correlates to the time predicted by the hypothesis.

For production environments larger areas of these waffles that can be seen with a simple unmagnified visual check could be placed at various locations on a wafer. This would allow global releasing of devices across a wafer to be checked visually from the observation window of the STS DRIE. This could be a valid control check for a production environment, which would not involve costly downtime if removal of the wafer from the etching chamber were required. It can also be used by anyone already running an STS DRIE facility without the need for new and potential costly additional equipment.

4. CONCLUSION

This paper presents a technique that further improves the case for using the one step process over the traditional two step process.

The dry one step process has now a number of advantages over the more traditional two-step process. The first is the elimination for any wet bench chemistry, often a costly facility to install and often hazardous. It is able to clear large areas on a wafer, in fact its only limitation is the wafer area itself! Not only is it able to do this it does it without encountering grass formation[12] in these larger features that is encountered when using the two step process. It now has the facility to indicate when a structure is released. In addition to the previously known advantages, this new application of the waffle technique is a novel and valid asset to anyone wanting to take advantage of the new and vibrant one step dry process.

[1] P.T.Docker, P. Kinnell, M.C.L.Ward, 'A dry single step process for the manufacture of released MEMS structures', Journal of Micro-mechanics and Micro-engineering, 13,2003,p790-794

[2] B.E. Volland, H. Heerlein, I. Kostic, I. W. Rangelow, ' The Application of secondry effects in high aspect ration dry etching for the fabrication of MEMS', Microelectronic Engineering, 57-58, 2001, p641-650

[3] A.A. Ayon, K.Ishira, R. A. Braff, H. H. Sawin, M. A. Schmidt,' Microfabrication and testing of suspended structures compatible with silicon on insulator technology', J. Vac. Sci. technol. A 17(A), ul/Aug 1999, p2280-2285

[4] A.A. Ayon, X. Zhang, R.Khanna, 'Anisotropic silicon trenches 300-500um deep employing time multiplexing deep etching (TMDE)', Sensors and Actuators A, 91, 2001, p381-385

[5] A.A. Ayon, R. Braiff, C.C.Lin, H.H.Sawin, and M.A. Scmidt, 'Characterisation of a time multiplexed inductively coupled plasma etcher', Journal of the electromechanical Society, 146, 1, 1999, p339-349

[6] J. Bhardwaj, H. Ashraf, A. McQuarrie, 'Dry silicon etching for MEMS', Surface Technology Systems Ltd, Prince of Wales Industrial Park, Abercam, gwent, NP1 5AR

[7] S.A. McAuley, H. Asraf, L.Atabo, A. Chambers, S.Hall, J. Hopkins, G.Nicolls, 'Silicon micromachining using a high density plasma source', J. Physics D: Applied Physics. 34 (2001) p2769-2774

[8] C. Gormley, K.Yallop, W.A.Nevin, J. Bhardwaj, H. Asraf, P.Huggegett, S. Blackstone, 'State of the art deep silicon anisotrophic etching on SOI bonded substrates for dielectric isolation and MEMS applications', Presented at the fifth symposium on semiconductor wafer bonding science, technology and application. The fall meeting of Electrochemical Society, Hawaii. USA. October 17-22, 1999

[9] A. Rickard, M. McNie, ' Characterisation and optimization of deep dry etching for MEMS applications' Microsystems and Microengineering, Defence evaluation and Research Agency, Malvern, UK

[10] J. Karttunen, J. Kiiharmaki, S. Franssila, 'Loading effects in deep silicon etching', 'Helsinki University of Technology, Microelectronics Centre, P.O.B. 3000, FIN-02015 TKK Finland

[11] P.T.Docker, P. K. Kinnell, M.C.L. Ward,'Development of the one step DRIE dry process for unconstrained fabrication of released MEMS devices', Submitted to Journal of Micro-mechanics and Micro-engineering November 2003

[12] Verbal communication with Dr J. Hopkins at Surface Technology Systems, may 2001

Planarize the sidewall ripples of silicon deep reactive ion etching

Kuo-Yao Weng*, Mei-Ya Wang, Po-Hao Tsai

Electronics Research & Service Organization (ERSO)
Industrial Technology Research Institute, Hsinchu, Taiwan
195 Sec. 4, Chung Hsing Rd., Chutung, Hsinchu, Taiwan 310, ROC
*E-mail: ALLAN_WENG@ITRI.ORG.TW

ABSTRACT

In order to diminish the sidewall defects in silicon deep reactive ion etching process, depositing doped silicon dioxide and post-annealing processes are applied. Compared with conventional approaches, the filling-reflow surface shows nearly optical quality. This novel scheme is not restricted to design layout without silicon crystal orientation dependence in the wet chemical etching methods. The unique integrated processes are expected to implement in the micro devices or replica master with optical surfaces.

Keywords: silicon deep RIE, sidewall roughness

1 INTRODCUTION

High aspect ratio technology is used to realize structures with feature sizes up to hundreds of microns in MEMS fabrication. Bosch silicon deep reactive ion etching (DRIE) based technologies are generally satisfying the demands of the MEMS fabrication. This process is based on the technique invented by Laermer and Schilp[1] and uses a variant of the sidewall passivating technique, which is deliberately segregated by using sequentially alternating etching and deposition steps. First a sidewall passivating polymer is deposited and subsequently the polymer and silicon are etched from the base of the trench, to allow the etching to proceed directionally. In this cyclic way the etching and deposition can be balanced to provide accurate control of the anisotropy. However, the process leaves sidewalls with ripples, or scallops, because of the nature of the process sequence that is cycled between etch and passivation phases. The defect limits the process virtues in optical device such as micro cavity lasers, and some micro fluidic component applications.

Reactive ion etching (RIE) is the combination of mechanical ion bombardment (as in sputter etching) and the chemical process of plasma etching. DRIE is also based on reactive ion etching, but an important consideration in the decision of using reactive ion etching vs. deep reactive ion etching is the aspect ratio of the trench. The aspect ratio is defined as the maximum depth versus the maximum width. When aspect ratios are too high (>3:1), etching with RIE may lead to lag, bowing, bottling, micrograss, and tilting. These issues are a major concern because trenches will not result in perfectly straight walls. Many microfluidic and micromechanical devices require very deep trenches or very tall beams, with straight sidewalls. In order to achieve high aspect ratios (up to 30:1) using reactive ion etching, the sidewalls must be prevented somehow from being etched.

In the time-multiplexing scheme of DRIE, the etching and passivating gas monomers are flowed independently one at a time, and the machine alternates between an etching cycle and a passivating cycle. The etch step forms a shallow trench in the silicon substrate, with an isotropic profile characteristic of fluorine-rich glow discharges. This process, succinctly described as time-multiplexed deep etching, TMDE, offers the advantage of exploiting the high silicon etching rate. Surface roughness issues are particularly important in TMDE tools. Because of the alternating etching and passivating cycles and the spontaneous nature of the etch in fluorinated chemistries, structures fabricated by means of TMDE exhibit a characteristic scalloped sidewall roughness that can be unacceptable in some applications, see in the figure 1.

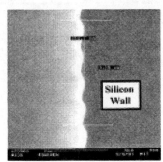

Figure 1, the sidewall ripples of DRIE structure

Lin etc. [2] proposed the technique in fabricating a solid polymer dye microcavity laser, which is molded with the etched Si cavity. By a combination of deep-RIE and EPW (the ethylenediamine pyrocatechol and water etchant) wet etching of silicon, optically smooth surfaces have been obtained. Alternately, Nillson etc. [3] removed the surface roughness of the sidewalls with adding a short etch in KOH+IPA after the deep etching processes in optimized parameters. Early, Volland etc. [4] reported on the development of a novel gas chopping etching technique (GCET)process in order to achieve a smooth (rippled free) sidewall surface. However, the modified GCET has lower selectivity (in range of 30) than the conventional approaches.

The sidewall ripples are decreased in the above techniques, but some compromises are made. The microstructure features created by wet etching methods, for example, depend on silicon crystal orientation. The ripple-free sidewall can be obtained by anisotropic etching instead of isotropic mode during etching- passivating switch cycle, but the modification must loss the benefit of selectivity in Volland's process. We have established a novel scheme to diminish the sidewall ripples and kept the virtues of the DRIE process.

2 EXPERIMENTAL

Figure 2 shows the conventional process sequence for the fabrication of silicon deep reactive ion etching. The substrate of the mold is a Si (100) wafer with a thickness of about 500μm. Before photolithography, a layer of low pressure chemical vapor deposition, LPCVD, silicon oxide as an etch mask for etching is put on the wafer. The design layout is patterned by photolithography and the resist and oxide masks are stripped in sequential processes.

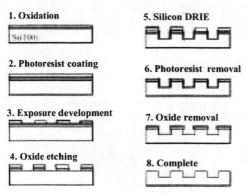

Figure 2, DRIE fabrication processes

Borophosphosilicate glass, BPSG, films have been used as fusible deposited dielectrics in silicon-gate MOS integrated circuits, with fusion tapering achieved at temperatures 100 °C to 200 °C lower than is normally possible with phosphosilicate glass, PSG, films. Both glass films can be deposited by LPCVD or plasma enhanced chemical vapor deposition, PECVD processes. The former produce conformable film profile over the microstructures, and overhang rise by the latter way, see in the figure 3 (a) and (b).

BPSG film of 3μm thickness is deposited by PECVD process. Soften the layer 90 minutes with the high temperature of 800°C. To improve the step coverage on the etching structure, deposit PSG of to fill in the scallop structure. Annealing PSG film is processed at 1000 °C for

two hours; see in the figure 3(c). Compare the sidewall roughness of different schemes by cross-section views of scanning electron microscope, SEM, and by ripple profiles of atomic force microscopy, AFM.

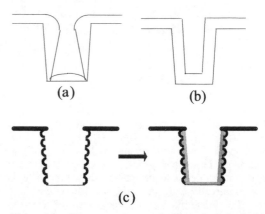

Figure 3, Step coverage of film deposited by PECVD (a) and LPCVD (b) processes; doped silicate glass depositing and annealing after DRIE etching(c).

3 RESULTS

3.1 BPSG film with PECVD process

The microstructures of 5μm wide and 15μm depth are etched and remove the passivated polymer layer of the etching process by oxygen plasma. The sidewall surfaces are shown in the figure 4(a) and ripples of 0.5μm size is obvious in the SEM images. The structures covered by BPSG show the smooth profile and the overhang produced by PECVD process exits on the top corner in figure 4(b).

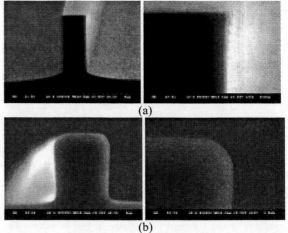

Figure 4, the sidewall ripples of DRIE structure: prior to BPSG film deposition (a); afterwards film deposition and annealing (b).

3.2 PSG film with LPCVD process

The same etched structures are deposited with 2μm PSG film. The good step coverage profile is observed by SEM and the sidewall performance shows in the Figure 5(b). No overhangs appear on the corner of the etched structures. The surface roughness monitored by AFM indicates 77 and 22 nm in the etched structure and 37 and 8 nm for the PSG-deposited structures; respectively peak-to-valley difference and root mean square, which is satisfied for the optical device application, see in the figure 6.

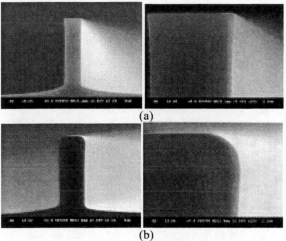

(a)

(b)

Figure 5, the sidewall ripples of DRIE structure: prior to PSG film deposition (a); afterwards film deposition and annealing (b).

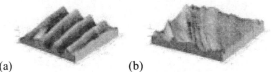

(a)   (b)

Figure 6, Surface roughness measurements by AFM; Prior to PSG film deposition (a):77nm (peak-valley), 22 nm (mean square). Afterwards film deposition and annealing (b): 37nm (peak-valley); 8nm (mean square)

4   DISCUSSIONS

The proposed strategy of doped silicon dioxide deposited and annealing films processes after etching procedures can refill the sidewall defects, competed with conventional approaches including replacing of the isotropic etching step by an anisotropic etching step, shortening etching and passivating cycle time in recipes and using a combination of DRIE and anisotropic wet etching with KOH. Our approach avoids the limit of optimizing the

etching parameters in Nillson [3], who reduced the etching-passivating switch time. To improve the roughness by reducing the switch time, the time is too short to be completely evacuated the atmosphere of the etching or passivating respectively, before entering the next state, see in the Table 1. The post-DRIE wet etching strategies suffer the etching selectivity of silicon crystal orientation and restrict the freedom of the layout designer. Our scheme keeps all benefits of the Bosch's DRIE technique and improves the sidewall to the optical surface quality. Furthermore, the processes can also be used to smooth the vertical mirror roughness [5] or replica the nickel embossing mold by tuning the de-mold angle.

	Recipe I	Recipe II
Etch time	7.5 sec	3.75 sec
Passivation time	5 sec	2.5 sec
Cycles	20	30
Process time	3 min 9 sec	3 min 9 sec
Resulting etch depth	20 μm	20 μm
Resulting scallops vertical size	1 μm	700 nm
Resulting scallops lateral size	300 nm	130 nm

Table 1, selection of optimized parameters for silicon deep RIE and the resulting features. [3]

5   CONCLUSIONS

The reflow process of doped silicate glass produces liquid free surfaces to get the optical quality and is not limited to the crystal orientation in design layout. We successfully diminish the sidewall ripples of silicon etched structures. The unique integrated processes are expected to implement in the micro devices or replica master with optical surfaces.

REFERENCES

[1] F. Laermer and A. Schilp, "Method of Anisotropical- ly Etching Silicon", German Patent: DE4241045. and US Patent: US5498312 and US5501893

[2] Y. Lin, M. Sasaki, and K. Hane ,"Fabrication and testing of solid polymerdye microcavity lasers based on PMMA micromolding", J. Micromechanics and Microengineering, 11,234-238, 2001

[3] D. Nilsson, S. Jensen, A. Kristensen and A. Menon, " Fabrication of Silicon Molds for Polymer Optics", Proceedings of the Micromechanics Europe conference, MME 2002, 111-114, Sinaia, Romania, October, (2002) and J. Micromechanics and Microengineering, 13, S57-S61, 2003

[4] B. Volland, F. Shi, P. Hudek, H. Heerlein, and Ivo. W. Rangelow, "Dry etching with gas chopping without rippled sidewalls", J. Vac. Sci. Technol. B 17(6), 2768, 1999

[5] C. Marxer etc.,"Vertical Mirrors Fabricated by Deep Reactive Ion Etching for Fiber-Optic Switching Applications" J. Microelectromechanical systems, 6(3), 277-284, 1997

Process Factors in the Reduction of Output Conductance in Sub-micron CMOS

N. Che May*, H. S. Tan** and A. V. Kordesch**

Silterra Malaysia Sdn Bhd, Kulim Hi-Tech Park, Kulim, Kedah 09000 Malaysia,
norhafizah@silterra.com*, hooisin_tan@silterra.com**, al_kordesch@silterra.com***

ABSTRACT

CMOS Analog circuits require transistors with low output conductance (g_{ds}) in order to achieve high gain. Submicron MOSFETs with halo implants and retrograde wells are designed to have high transconductance (g_m) but often suffer from poor output conductance. In this paper we investigated the process factors affecting g_{ds} and we show how to optimize g_{ds}. Our experimental results from 180nm CMOS are compared with 2D simulations in order to understand the mechanisms involved. Output conductance is the derivative of the I_D-V_D curve, $g_{ds} = dI_D/dV_D$. In saturation, several effects contribute to the increase of I_D with V_D, namely channel length modulation (CLM), drain-induced barrier lowering (DIBL), and substrate current body effect (SCBE). We have mainly focused on PMOS transistors at voltages where the substrate current is not significant.

Keywords: channel length modulation, drain-induced barrier lowering, early voltage, output conductance, output resistance

1 INTRODUCTION

Successful integration of mixed signal circuits incorporating high performance and analog circuits are vital for system on chip applications (SoC). However, CMOS device design scale-down is particularly challenging as the requirement for analog and digital devices are often conflicting and thus usually result in trade-offs. High performance logic devices are optimized for good drive current, low leakage and SCE control which incorporate super halo and double-pocket structures. These structures however, often result in low output resistance, device gain, transconductance-to-drive current ratio and matching properties [1].

Low output resistance is the result of increase I_D with V_D in saturation regime. Three components are associated with this increase, namely channel length modulation (CLM), drain-induced-barrier-lowering (DIBL) and substrate current body effect (SCBE). Novel structures such as single pocket implant [1,2,3] and shadow-mask implant [4] were proposed to increase analog performance while maintaining good high performance characteristics. Both structures refer to asymmetric channel doping profile that has good SCE control for high drive current. It also has good control of CLM and DIBL for low output conductance. However, asymmetric device also means

higher cost of production and increase in device complexity. Therefore, the paper is aimed to study the existing process factors that can be tuned in for the reduction of g_{ds}. We have mainly focused on PMOS transistors at voltages where the substrate current is not significant.

2 EXPERIMENTAL/SIMULATION

PMOS and NMOS transistors were fabricated with gate length of 180nm. We systematically varied the process factors LDD implant, halo implant, well anneal, spacer and anti punch-though (APT) implant to study their effects on output conductance. Experiments in silicon were run as well as 2D simulations. Process simulator, TSUPREM4 and MEDICI were used for simulations. Each factor was evaluated by comparing the Early Voltage (V_A) as a figure of merit rather than g_{ds} directly because g_{ds} varies as a function of gate voltage and drain current. Early Voltage is defined as $V_A = I_D/g_{ds} - V_D$. It is always desired to get high V_A, which corresponds to low output conductance. At the same time, it is also desirable to maintain high drain current.

3 RESULTS AND DISCUSSION

Table 1 shows the parameters and the variables that were varied to study the various process factors. The experimental and simulation results shown are mainly from PMOS with gate length of 180nm. Even though NMOS results are not shown, their results are consistent with PMOS.

Parameter	Variation
LDD implant	dose
Halo implant	dose and tilt
Nwell implant	dose
Spacer	thickness
Well Anneal	anneal temperature
APT implant	dose and energy

Table 1: Parameters and their variations used in this study.

3.1 LDD Implant

Figure 1 shows the sensitivity of V_A to LDD dose for a PMOS transistor with gate length of 180nm. Increasing the LDD dose decreases V_A because the effective channel becomes shorter, which is shown in Figure 2. Shorter channel length results in larger residual DIBL thus causing output resistance to decrease [5]. The doping profiles of the different LDD dose in the channel are illustrated in Figure 3.

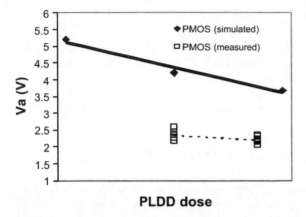

Figure 1: V_A dependence on PLDD dose. Note as dose increases, V_A decreases.

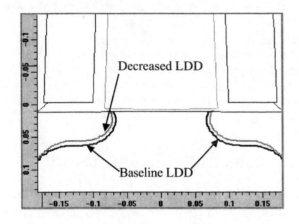

Figure 2: Junction profile for different LDD dose (MEDICI). Higher LDD dose shortens the channel length

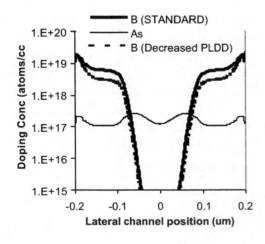

Figure 3: PLDD and channel doping profiles (TSUPREM4)

3.2 Halo Implant

Figure 4 shows that increasing the halo dose increases V_A. Increasing the halo tilt angle also increases V_A as shown in the inset. The larger tilt angle placed the halo implant almost at the center of the channel as in Figure 5. The additional arsenic in the channel lessens the effect of DIBL as shown in Figure 6. However, increasing halo dose also causes Idsat to decrease. As can be seen in Figure 7, increasing halo dose has opposite effect on V_A and Idsat. Experimental studies have also shown that pocket implant has tradeoff effects on V_A and Idsat [5].

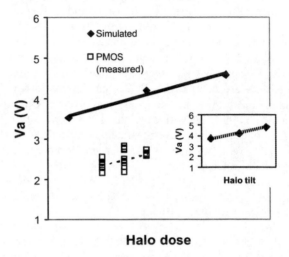

Figure 4: V_A dependence on halo dose. Higher halo dose produces higher V_A

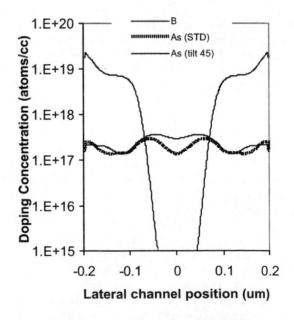

Figure 5: Halo doping profiles in the channel (TSUPREM4). Higher halo dose in the channel at higher tilt

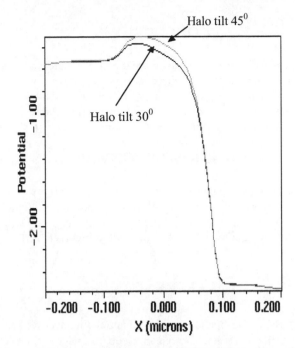

Figure 6: Potential barrier shifts at different halo tilt angle (MEDICI). Higher tilt has higher potential barrier, thus decreases DIBL.

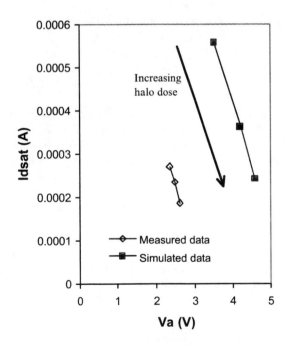

Figure 7: Idsat vs. Va plot. Effect of increasing halo dose on Idsat and Va. Va increases but Idsat decreases as halo dose is increased.

3.3 Nwell Implant

Figure 8 shows that increasing well dose has no effect on V_A. As shown in the inset, the well is far deeper than the channel, and the surface doping is dominated by channel implants.

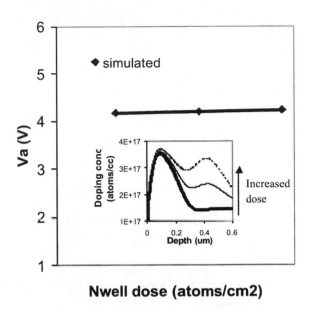

Figure 8: V_A dependence on Nwell dose. No effect of Nwell dose on V_A.

3.4 Spacer Thickness

Figure 9 shows that spacer width also has very little effect on V_A. Increasing the spacer width makes the LDD longer towards the source-drain (Fig. 8 inset), thus increasing the resistance, but does not affect DIBL or CLM.

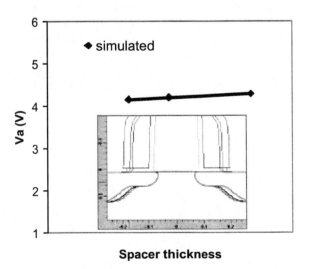

Figure 9: V_A dependence on spacer thickness. Note insignificant effect of spacer thickness on V_A

3.5 Well Anneal

Figure 10 shows the effect of well anneal temperature against V_A. An optimum well anneal temperature is observed for each device. It was found that output resistance is very sensitive to well anneal temperature due to degradation of retrograde implant shape at very high and very low anneal temperature [6]. At very high temperature, large thermal diffusion and pile-up at SiO/Si interface causes the degradation. On the other hand, transient enhanced diffusion (TED) occurs at gate oxidation process at low temperature [6].

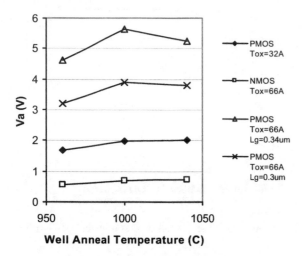

Figure 10: Measured V_A dependence on well anneal temperature. Optimum temperatures are observed.

3.6 Anti Punch-through Implant

Figure 12 shows the plot of Idsat versus V_A at different anti punch-through (APT) implant conditions. As APT energy implant is increased, V_A shows a contradict trend depending on the implant dose used. At high dose and high energy, the plot is shifted to the up-left when energy is increased. However, the plot shifted to the up-right as energy is increased at low APT implant dose and energy. At low dose and energy, increasing the APT implant energy forms super steep retrograde (SSR) channel, which has positive effects for analog applications [1,6].

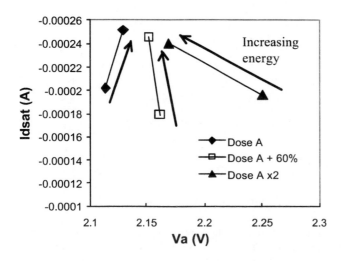

Figure 12 : Effects of APT implant on V_A and Idsat.

4 CONCLUSIONS

In this work we have shown what process factors affect g_{ds} and we have optimized these factors to achieve low g_{ds}. We found that g_{ds} is most sensitive to LDD dose, halo dose, halo tilt and APT dose and energy. Our experimental results confirmed the trends of the 2D simulations.

ACKNOWLEDGEMENT

Authors would like to express gratitude to Silterra for the support given to the project. Also would like to thank the people in Manufacturing, Dept of Integration and Dept of Device Modeling who were indirectly involved in this project.

REFERENCES

[1] H. V. Deshpande et. al, "Channel Engineering for Analog Device Design in Deep Submicron CMOS Technology for System on Chips Applications," *IEEE Trans. On Electron Devices*, vol.49, no.9, 1558-1565, 2002.

[2] H. V. Deshpande et. al, "Analog Device Design for Low Power Mixed Mode Applications in Deep Submicron CMOS Technology," *IEEE Electron Device Letters*, vol.22, no.12, 588-590, 2001.

[3] H.V. Deshpande et. al, "Deep Sub-Micron CMOS Device Design for Low Power Analog Applications," *Symp. on VLSI Tech Digest of Tech Papers*, 87-8, 2001.

[4] T. B. Hook et. al, "High-Performance Logic and High-Gain Analog CMOS Transistors Formed by a Shadow-Mask Technique with Single Implant Step," *IEEE Trans. on Electron Devices*, vol.49, no.9, 1623-27, 2002.

[5] A. Chatterjee et. al, "Transistor Design Issues in Integrating Analog Functions with High Performance Digital CMOS," *Symp. On VLSI Tech. Digest of Tech. Papers*, 147-8,1999.

[6] E. Morifuji et.al, "An 1.5V High Performance Mixed Signal Integration with Indium Channel for 130nm Technology Node," *IEDM*, 459-462, 2000.

Inverse RIE Lag of Silicon Deep Etching

C.K. Chung and H.N. Chiang

Dep't of Mechanical Engineering, National Cheng Kung University,
Tainan, Taiwan 701, ROC, ckchung@mail.ncku.edu.tw

ABSTRACT

This paper reports for the first time that the phenomena and reaction mechanism of inverse reactive ion etching (RIE) lag occurs in the silicon deep RIE process without feature coalescence. This phenomenon is related to two important parameters of feature area and process pressure. Increasing the process pressure in different patterns with equivalent feature area will increase the etching rates as well the smaller width has higher increasing etching rate ratio. So, the smaller feature pattern has the chance to get higher etching rate than the larger one as the pressure increased is high enough. This will lead to three stages of RIE lag transition: reduced RIE lag, lag elimination and inverse RIE lag. A possible new reaction mechanism exists in the inverse RIE lag phenomena. The increasing etching rate ratio of different feature sizes is affected more obvious by C_xF_y radicals dissociated from C_4F_8 passivation gas than F radicals or SF_x ions dissociated from SF_6 etching gas.

Keywords: inverse, RIE lag, deep etching, ICP, MEMS

1 INTRODUCTION

Silicon deep reactive ion etching (DRIE) for deep and high aspect ratio microstructure is a very important technology in microfabrication and microelectromechanical system (MEMS) industry [1-3]. RIE lag in ICP etching is a frequently seen defect in semiconductor or microfabrication processes and appears in MEMS feature sizes up to hundreds of micrometers. It will affect the etching micro-uniformity and is much dependent on the pattern geometry. The effect is more severe as the feature width becomes smaller. Many publications reported the related study of silicon DRIE including effect of process parameters [4, 5] and pattern shape [6, 7] on RIE lag, 2D simulator [8], and curved surface using RIE lag [9]. For example, Jansen et al. [4] reported that RIE lag could be improved by lower pressure. Ayon et al. [5] proposed that high SF_6 flow rate, the dominant variable, can minimize RIE lag. Kiihamaki et al. [6] reported that the RIE lag is related to the feature length-to-width ratio (L/W ratio), width and pattern shape. Jiao et al. [7] proposed that the etching rate under silicon deep etching process is much affected by the mask layout related area including the other features around it. All reports are related to the normal RIE lag, no inverse RIE lag and less discussion on area effect. In this paper, we report the special inverse RIE lag phenomenon in DRIE process correlated to the effect of area and pressure parameters. It will be much helpful for the process control of etching uniformity, 2D-3D simulator establishment, and the application for different level or curved surface device.

2 EXPERIMENTAL PROCEDURE

Boron-doped p-Si(100) wafers of 4- 10 Ω-cm were initially cleaned in the H_2SO_4:H_2O_2 = 3:1 solution. Standard lithography was used to transfer the photo mask pattern to silicon surface and the etching mask was photoresist. The different feature geometry of rectangles, squares, circles and donuts is designed to realize how the pattern geometry dimension affects the ICP etching properties. Three major kinds of patterns of constant length, constant width and constant area with different feature sizes of 2 to 100 μm are divided to understand the dominating factor of geometry for RIE lag in ICP etching. Silicon DRIE was performed by ASE™ process in STS Multiplex ICP system [10-11]. The etching and passivation gases were SF_6 and C_4F_8 respectively and switched during the process. Little O_2 was added to SF_6 during etching process. Process pressure was adjusted by auto pressure control (APC) from 30% to 75% to understand the process pressure effect on RIE lag. The plasma source was generated by 600 W 13.56 MHz RF generator and biased by 10- 14 W 13.56 MHz RF generator at platen during the etching step. The wafer was electrostatically clamped and cooled by backside helium flow. Scanning electron microscope (SEM) was used to examine the etching result after ASE™ process.

3 RESULTS AND DISCUSSION

Figures. 1(a) and 1(b) show the SEM micrographs of the etched donut trench with constant area of of $\pi \times 52.5^2$ μm^2 at pressures of APC (a) 30% and (b) 75%, respectively. The RIE lag exists at low pressure of APC 30% while an inverse RIE lag appears at much high pressure of APC 75%. Normal RIE lag of trenches at constant length in Figure 2 shows that smaller feature size leads to lower etching rate. Increasing process pressure to the feature in Figure. 2, we still get the RIE lag, not inverse RIE lag. Figures 3(a) and 3(b) show the etching depth of trenches with feature length from 100 to 500 μm at constant width of 5-10 and 25 μm, respectively. The etching depths of trenches are nearly the same at the identical feature width and vary with width dimension. For example, the etching depth is 57.91 μm for the fixed feature width of 5 μm while the etching depth is 64.35 μm for the fixed feature width of 10 μm as shown in

Figure 3(a). And the etching depth is 72.76 μm for the fixed feature width of 25 μm, Figure 3(b). The etching depth of rectangular trench increases with the increasing feature width and is insensitive to the feature length. It indicates that the feature width of pattern geometry is a dominating factor affecting the RIE lag in ICP etching.

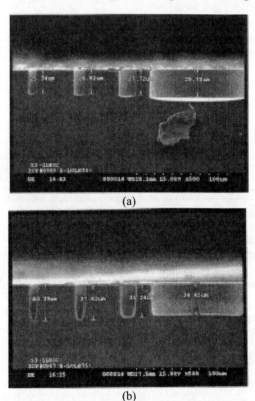

(a)

(b)

Figure 1 SEM micrographs of the etched donut trench with constant area of π x 52.5² μm² at pressures: (a) APC 30% and (b) APC 75%, respectively. The RIE lag exists at low pressure of APC 30% while an inverse RIE lag appears at much high pressure of APC 75%.

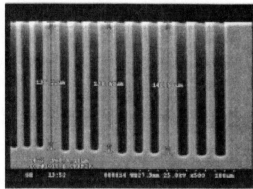

Figure 2 SEM micrograph of trenches with feature width of 8, 9 and 10 μm at constant length. The smaller feature size, the slower etching rate.

Besides the feature width, the continuous feature area or pattern shape also affects the etching rate. As the shape of feature with constant width is changed from rectangles to donuts, the etching depth increases with the continuous feature area as shown in Figure 4. Etching rates of rectangular trenches are sensitive to width while the donut trenches are sensitive to both width and area. The larger continuous feature area, the deeper etching depth. It implies that the feature area of pattern geometry is a secondly factor affecting the RIE lag.

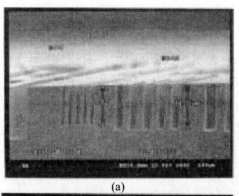

(a)

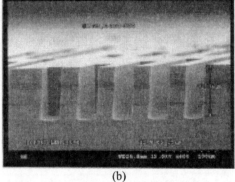

(b)

Figure 3 SEM micrographs of trenches with feature width of 100-500 μm at constant width of: (a) 5, 10 μm, and (b) 25 μm. Etching depths of trenches are sensitive to width, not length or area.

Figure 4 SEM micrographs of rectangular and circular donut trenches with constant width or radius of 5 μm.

The ICP etching behavior is relevant to the pattern geometry and process condition. To study the process parameter effect on RIE lag in ICP etching, we fixed the secondary factor of feature area to decrease the extra geometry contribution to RIE lag. The process pressure is a crucial factor on the RIE lag control. Figures 5(a)-5(c) show SEM micrographs of etched donut trench with constant area of $\pi \times 105^2$ μm^2 at pressures of APC (a) 50%, (b) 70%, and (c) 75%, respectively. The etching time is for 15 min. Obvious RIE lag is present at APC 50% and elimated at 70% while inverse RIE lag trenches are obtained at APC 75%. The inverse RIE lag means the higher etching rate occurs at smaller feature width. It implies that other mechanism exists in Si DRIE with respect to the previous reports [4-7]. A possible new reaction mechanism is C_xF_y radical dissociated from C_4F_8 passivation gas playing an important role to affect the etching rate ratio of different feature sizes. The total etching rate is related to the formation of C_4F_8 passivation film on the Si trench bottom by C_xF_y radicals and its removal by S_xF_y ions from SF_6 etching gas, in addition to the arrival of F radicals dissociated from SF_6 etching gas to etch Si surface at the bottom. The formation and removal of C_4F_8 passivation film impedes the increasing of etching depth while the arrival of SF_6 etching gas enhances the etching processing. These two contrary reaction mechanisms compete by density and mass transport of radicals and ions dissociated from C_4F_8 and SF_6 reaction gases and affect the total etching rate. The increasing pressure will proportionally increase the C_xF_y and F radical density and decrease the S_xF_y ion density [12]. The aspect ratio will affect the formation and removal of C_4F_8 passivation film and Si etching at different feature width due to the mass transport. The smaller width with larger aspect ratio will lead to thinner passivation film than larger width. The removal of passivation film by ions under bias will be similar for different features. So Si etching dominated by F radical density will start earlier in the small width than large one. As the pressure increases, the increasing C_xF_y radical density and reducing S_xF_y ion density will delay longer to start etching Si surface at bottom in large feature than small one. The etching rate is enhanced more at small width than large one. This will lead to the reduction and elimination of RIE lag. If the pressure is much high, the inverse RIE lag occurs. It indicates that the effect of C_4F_8 passivation gas on RIE lag is more important than SF_6 etching gas.

4 SUMARRY

A RIE lag phenomenon occurs in many patterns and becomes more severe at the smaller feature width and higher etching depth. In general, the smaller feature size, the lower etching rate and the more obvious RIE lag. A RIE lag elimination can be achieved for different feature size at constant area by higher process pressure of APC 70 % and the inverse RIE lag appears at much higher pressure

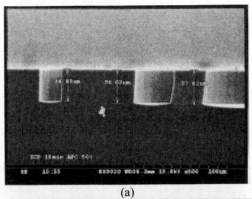

(a)

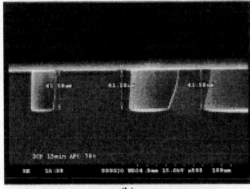

(b)

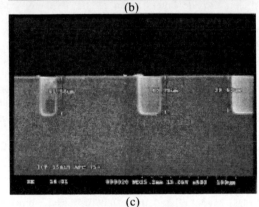

(c)

Figure 5 SEM micrographs of etched rectangular trench structure with constant area at pressures of APC: (a) 50%, (b) 70%, and (c) 75%, respectively. Obvious RIE lag is present at APC 50% and elimated at 70%, as inverse RIE lag trenches are obtained at APC 75%.

of APC 75 %. This will lead to three stages of RIE lag transition: reduced RIE lag, lag elimination and inverse RIE lag. As the pressure increases, the increasing C_xF_y radical density and reducing S_xF_y ion density will delay longer to start etching Si surface at bottom in large feature than small one. So, the etching rate is enhanced more at small width than large one to result in the reduction, elimination and inverse of RIE lag. It indicates that the effect of reaction gas of C_4F_8 passivation on RIE lag in Si deep etching is more important than SF_6 etching gas. It is an interesting phenomenon. The etching rate could be adjusted by the pattern geometry and process parameter. Different structures with RIE lag, RIE lag-elimination or inverse RIE lag would be fabricated by proper pattern and process design in future.

ACKNOWLEDGEMENTS

I would like to express my sincere thanks to Miss H.C Lu and Mr. K.S. Yen for their technical support. I also pay my sincere thanks to Common Laboratory of the Microsystems Technology Center of Electronic Research Service Organization in Industrial Technology Research Institute and the Southern Regional MEMS Center in National Cheng Kung University for the process equipment support.

REFERENCES

[1] R. Kassing and I.W. Rangelow, "Etching process for High Aspect ratio Micro Systems Technology (HARMST)", *Microsystems Technology* **3** pp. 20-27, 1996.

[2] W.H. Juan and S.W. Pang, "High-aspect-ratio Si etching for microsensor fabrication". *J. Vac. Sci. Technol. A* **13,** pp. 834-838, 1995.

[3] C. Zhang and K. Najafi, "Fabrication of thick silicon dioxide layers using DRIE, oxidation and trench refill", *MEMS2002*, pp. 160-163, 2002.

[4] H. Jansen, M. de Boer, R. Wiegerink, N. Tas, E. Smulders, C. Neagu and M. Elwenspoek, "RIE lag in high aspect ratio trench etching of silicon", *Microelectronic Engineering*, **35**, pp. 45-50, 1997.

[5] A.A. Ayon, R. Braff, C.C. Lin, H.H. Sawin and M.A. Schmidt, "Characterization of a time multiplexed inductively coupled plasma etcher", *J. Electrochem. Soc.*,**146**(1), pp.339-349, 1999.

[6] J. Kiihamaki and S. Franssila, "Pattern shape effects and artifacts in deep silicon etching", *J. Vac. Sci. Technol.*, **A** 17(4), pp.2280-2285, 1999.

[7] J. Jiao, M. Chabloz, T. Matsuura and K. Tsutsumi, "Mask layout related area effect on HARSE process in MEMS application", *Transducers'99*, pp.546-549, 1999.

[8] R. Zhou, H. Zhang, Y. Hao, D. Zhang and Y. Wang, "Simulation of profile evolution in etching-polymerization alternation in DRIE of silicon with SF_6/C_4F_8", *MEMS2003*, pp.161-164, 2003.

[9] T-K A. Chou and K. Najafi, "Fabrication of out-of-plane curved surface in Si by utilizing RIE lag", *MEMS2002*, pp.145-148, 2002.

[10] J.K. Bhardwaj and H. Ashraf, "Advanced silicon etching using high density plasma", Proc. *SPIE Micromachining and Microfabrication Process Technology*, **2639, pp.** 224- 233, 1995.

[11] A.M. Hynes, H. Ashraf, J.K. Bhardwaj, J. Hopkins, I. Johnston, and J.N. Shepherd, "Recent advances in silicon etching for MEMS using the ASE™ process", *Sensor and Actuators* **74,** pp.13-17, 1999.

[12] M.A. Blauw, T. Zijlstra, and E. van der Drifta, "Balancing the etching and passivation in time-multiplexed deep dry etching of silicon", *J. Vac. Sci. Technol. B* **19(6)**, pp. 2930-2934, 2001.

Microscope Slide Electrode Chamber for Nanosecond, Megavolt-Per-Meter Biological Investigations

Y. Sun[1], P. T. Vernier[2], M. Behrend[3], L. Marcu[4], and M. A. Gundersen[5]

[1]Department of Materials Science, University of Southern California,
Los Angeles, CA, USA, yinghuas@usc.edu
[2]MOSIS, Information Sciences Institute, USC, Marina del Rey, CA, USA, vernier@mosis.org
[3]Department of Electrical Engineering-Electrophysics, USC, Los Angeles, CA, USA, behrend@usc.edu
[4]Cedars-Sinai Medical Center, Los Angeles, CA, USA, marculaura@cshs.org
[5]Department of Electrical Engineering-Electrophysics, USC, Los Angeles, CA, USA, mag@usc.edu

ABSTRACT

Nanosecond pulsed electric fields pass through the external membranes of biological cells (which have typical response times much greater than nanoseconds) and perturb fast-responding intracellular structures and processes. To enable real-time imaging and investigation of these phenomena, a microchamber with integral electrode walls and optical path for observing individual cells exposed to electric pulses was designed and fabricated utilizing photolithographic and microelectronic fabrication methods. SU-8 photoresist was patterned to form straight sidewalls 10 to 30 μm in height, and gold was deposited on the opposing walls and the top for a conductive, non-reactive electrode surface. Results from observations with the microchamber in real-time electroperturbation imaging experiments include intracellular calcium bursts and membrane phospholipid translocation. Real-time nanoelectropulse investigations with microfabricated imaging tools open new pathways for the study and engineering of biological systems.

Keywords: SU-8 photoresist, liftoff, megavolt electric field, microchamber, electroperturbation, microelectronic fabrication technology, microscopy

1 INTRODUCTION

Responses of biological systems to pulsed electric fields are diverse and highly dependent on the amplitude and duration of the field. Electropermeabilization (electroporation) technology utilizes pulses with durations in the millisecond and microsecond range for facilitated drug delivery and the introduction of genetic material into cells [1, 2]. Recently it has been demonstrated that nanosecond pulsed electric fields with amplitudes greater than 1 MV/m can cause apoptosis and other intracellular physiological changes not associated with traditional electroporation [3,4,5]. In the ultra-short high-field regime, the cytoplasmic membrane may be considered to be a capacitor with a charging time constant of 100 ns (approximate value for a typical mammalian cell). Pulses that are shorter than this time are felt inside the cell —

across intracellular membranes (nucleus, mitochondria, storage vacuoles and other compartments). Under appropriate conditions nanosecond pulsed electric fields can thus penetrate the plasma membrane and perturb intracellular structures and processes with little or no effect on the external membrane. Ultra-short electric pulses, remotely controlled and delivered in precisely tailored amplitudes, durations, and patterns, provide a tool for manipulating the intracellular environment, inducing apoptosis or modulating gene expression noninvasively, without a requirement for direct contact with cells or tissues.

Electroperturbation has specific requirements for the pulse duration in nanosecond level and electric field over 1 MV/m. Real-time observations of electroperturbation phenomena, combining optical microscopy with the delivery of ultra-short, high-field electric pulses presents a challenge for pulse generator and exposure chamber design. In this paper we describe the design and fabrication of a microchamber –100 μm wide, 10 μm deep and 12,000 μm long – on a glass microscope slide using standard microelectronic technology, and the integration of the microchamber with an advanced technology, compact pulse

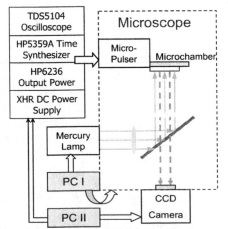

Figure 1: Real-time electroperturbation microscopy system.

generator.

2 MICROCHAMBER DESIGN AND FABRICATION

The microchamber is an integral component of a real-time electroperturbation imaging system consisting of a Zeiss Axiovert 200 microscope, a compact pulse generator designed and constructed at USC [6] and a LaVision Imager QE camera. The system diagram is shown in Figure 1.

2.1 Structure Design

Five specific material and structural requirements were identified. They are depicted in the diagram of a cross-section of an ideal channel in Figure 2.

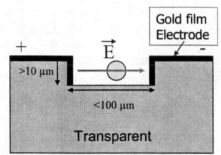

Figure 2: Ideal microchamber cross section.

1) Since the microchamber is designed for the optical study of living cells with fluorescence microscopy, the channel must be transparent to visible and ultraviolet light and may not fluoresce under excitation wavelengths from ultraviolet to infrared. To meet this condition a standard glass microscope slide was used as the substrate of the chamber. 2) The physical dimensions must accommodate the cell types under investigation, with an interelectrode distance appropriate for generating megavolt-per-meter electric fields with commercially available MOSFETs. The lymphocytes that were used in the experiments reported here have diameters up to 10 μm, so the depth of the channel must be at least 10 μm. To achieve fields up to 4 MV/m with the USC MicroPulser (400 V maximum), which is mounted along with the microchamber on the stage of an inverted microscope in a standard slide holder, the interelectrode distance must be 100 μm or less. 3) In order to provide a non-corroding conductive surface that is unreactive with biological buffers and growth media and harmless to cells, gold was chosen as the electrode material. 4) To form the smooth channel sidewalls that are perpendicular to the chamber floor, required for delivering a uniform electric field to all cells in the channel, SU-8 photoresist from MicroChem was used to build up the walls of the microchamber. 5) The length of channel is longer than 1 cm to permit the observation of reasonable numbers of living, metabolizing cells. In order to improve operational efficiency and provide more cells for experiments, four separated channels were patterned on a glass slide. The pattern is shown in Figure 3. Each channel is 12000 μm x 100 μm x 10 to 30 μm. Electrical signals may be switched independently to each channel.

Figure 3: Design schematic of the microchamber.

2.2 Fabrication

The microchamber is constructed with microelectronic fabrication technology. The key requirement of the straight sidewall was achieved by using a special material – SU-8 photoresist [7]. SU-8 is a chemically amplified, epoxy-based, negative resist. Standard formulations are offered to cover a wide range of film thicknesses from 1 to 200 μm. SU-8 has a high optical transparency and is sensitive to near UV radiation. High-aspect-ratio structures with straight sidewalls are readily formed by contact-proximity or projection printing. Cured SU-8 is highly resistant to solvents, acids, and bases and has excellent thermal stability, making it well suited for applications in which cured structures are a permanent part of the device. A double-layer photoresist liftoff step was developed to achieve satisfactory gold coverage of the channel tops and sidewalls (Figure 4).

All processing was carried out on the microscope slide substrate, which required pre-treatment to provide the smooth, clean surface needed for the foundation of the microchamber. Following a rinse with acetone and methanol, and drying with nitrogen, the slides were etched in buffered oxide etchant (10:1) for 10 minutes at room temperature, and then rinsed in DI water and acetone again. Before the lithography, samples were baked at 120 °C for at least 20 minutes. The special chemical and mechanical properties of SU-8 thick films necessitate longer exposure and development times and lower bake temperatures than those for normal photoresists. For 12 μm high channels SU-8 5 was spun on the substrate at 1000 rpm for 30 s and exposed for 35 s at 3 mW/cm², 365 nm. For 30 μm channels the SU-8 50 formulation was used as follows: 4000 rpm spin for 30 s, 54 s exposure at 3 mW/cm², 365 nm. Sharply defined edges and nearly vertical sidewalls were produced in this manner.

Because the sharp, vertical and high sidewalls were not sufficiently protected by regular photoresist in later wet etching steps, a special liftoff process was used to remove the deposited Au/Ti film in the areas outside the electrodes seen in figure 4. Before the metal film deposition, the positive photoresist AZ5214 (1 μm thickness) was developed for liftoff with the same mask that was used for the SU-8 (negative).

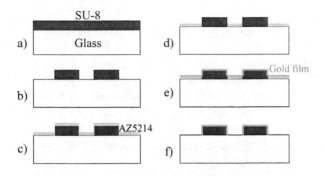

Figure 4: Double-layer photoresist liftoff processes.

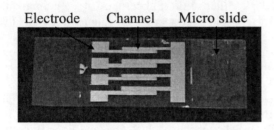

Figure 5: Microchamber on a standard microscope glass slide.

Before gold film deposition, a 50 nm Ti film was deposited on the surface of the glass substrate and photoresist channel in order to improve gold adhesion. To cover the channel sidewalls two-layer Ti/Au films were deposited with +45° and -45° angle with respect to the target source instead of the parallel position. The glass slides were rotated 90 degrees between the two steps. Different thicknesses of gold electrode from 100 nm to 250 nm were tested in experiments. The thicker film made the liftoff more difficult but improved the mechanical reliability of the microchamber in the biological experiments.

Samples were immersed in acetone for 15 to 30 minutes to lift off the metal film in the areas outside the electrodes. Ultrasonic treatment was used to accelerate the pattern forming and clean the edge in the last 5 minutes. The final microchamber pattern is shown in Figure 5.

3 EXPERIMENTAL

3.1 Cell Culture

Human Jurkat T lymphocytes (ATCC TIB-152) were grown in RPMI 1640 (Irvine Scientific) containing 10% heat-inactivated fetal bovine serum (Irvine Scientific), 2 mM Lglutamine (Gibco-BRL), 50 units/mL penicillin (Gibco-BRL), and 50 μg/mL streptomycin. Live cells stained with calcium indicator Calcium Green (Molecular Probes), which exhibits a strong increase in fluorescence

emission (531 nm) intensity upon binding to calcium, were monitored to observe the response of cells to the nanosecond electric pulses.

3.2 Test System

For real-time microscopic investigation, the microchamber was loaded with cells, mounted on the microscope stage in ambient atmosphere at room temperature, and electrically mated with the USC MicroPulser. The MicroPulser is a compact, solid-state pulse generator designed to provide nanosecond pulses with amplitudes variable up to 400 V into the microchamber load (200 to 400 ohms). With an interelectrode distance of 100 μm the electric field can reach 4.0 MV/m. The MicroPulser-microchamber assembly is mounted on a standard microscope stage frame so that the cells in the exposure chamber can be accessed optically and mechanically like any field on a standard slide. Images were recorded by the LaVision Imager QE CCD camera with high sensitivity working at -12 °C, controlled by image acquisition and analysis software DaVis by LaVision.

4 RESULTS AND DISCUSSION

An example of real-time observation of electroperturbation is shown in Figure 6 for Jurkat T cells in the 100 μm x 30 μm microchamber, a representative image from a dose-response experiment with 30 ns pulses at fields from 0.5 MV/m to 3.5 MV/m. A series of electroperturbation experiments have been conducted with the micropulser microscopy system. Among the most

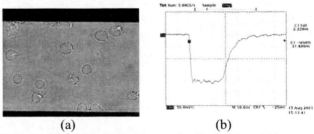

(a)   (b)

Figure 6: (a) White light image of cells inside a 100 μm x 30 μm channel, (b) 30 ns pulse applied to a cell suspension in the microchamber.

interesting observations have been the intracellular calcium bursts induced by nanosecond pulse exposure, visualized in cells loaded with the fluorescent calcium indicator Calcium Green. Calcium Green-stained cells exhibit a uniform and rapid release of calcium throughout the cell within milliseconds of the leading pulse edge. The plot in Figure 7 shows the fluorescence intensity changes in several cells after exposure to 30 ns 2.5 MV/m pulses. This set of curves represents 5 cells with Calcium Green fluorescence intensity increases from 8.5% to 40% after a single pulse. EGTA, a calcium chelating agent, in the external medium has no effect on calcium bursts, indicating that the source of the calcium causing the fluorescence intensity increase is

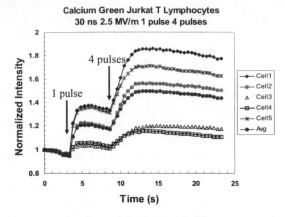

Figure 7: Fluorescence intensity change of Calcium Green after 30 ns 2.5 MV/m electric pulses.

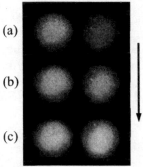

Figure 8: Fluorescent images of Jurkat T Lymphocytes stained with Calcium Green during 30 ns 2.5 MV/m electric pulses (a) cell before the shock, (b) 2 seconds after 1 pulse, (c) 4 seconds after 4 pulses.

inside the cell. The intensity change is strongly inhibited after incubation with thapsigargin, which blocks the normal filling of intracellular calcium compartments. These two sets of observations support the hypothesis that nanosecond pulses affect intracellular structures and processes (thapsigargin-inhibited calcium bursts) inside cell and not the plasma membrane (no influence from extracellular calcium).

During hundreds of hours of service some properties of the microchamber have been observed. 1) SU-8 5 photoresist has better long-term adhesion than SU-8 50 in aqueous environments. Hard-bake processing can effectively improve adhesion property, but it may cause the deformation of a large photoresist pattern due to thermal expansion, especially at high bake temperatures. 2) Weak fluorescence from SU-8 was recorded in a broad range from 450 to 600 nm. Since most areas of the channel are covered by gold film, this does not significantly affect imaging.

Further improvements may include the addition of microfluidic connections for better control of the entry and exit of cells from the channel, regulation of temperature and amosphere, and development of increased integration of the instrumentation, with full software control of pulse generation and microscope and imaging operations. The microfabricated imaging tools are opening new pathways for the investigation of biological systems and biological engineering.

5 ACKNOWLEDGEMENTS

This work was supported by grants from the Air Force Office of Scientific Research and the Army Research Office. The authors gratefully acknowledge Song Han and Xiaolei Liu for stimulating discussions and technical assistance in microchamber processing, and Sarah Salemi, Mya Thu, Jingjing Wang and Yushun Zhang for cell manipulations and culture.

REFERENCES

[1] K. J. Miller, V.L. Sukhorukov, U. Zimmermann, "Reversible eletropermeabilization of mammalian cells by high-intensity, ultra-short pulse of submicrosecond duration," J. Membrane Biology, vol. 184, pp. 161-170, 2001

[2] J. C. Weaver, Y. A. Chizmadzhev, "Theory of electroporation: review," Bioelecrochemistry and Bioenergetics, vol. 41, pp. 135-160, 1996

[3] K. H. Schoenbach, S. J. Beebe, and E. S. Buescher, "Intracellular effect of ultrashort electrical pulses," Bioelectromagnetics, vol. 22, pp. 440-448, 2001.

[4] P. T. Vernier, A. Li, L. Marcu, C. M. Craft, and M. A. Gundersen, "Ultrashort pulsed electric fields induce membrane phospholipid translocation and caspase activation: differential sensitivities of Jurkat T lymphoblasts and rat glioma C6 cells," IEEE Transactions on Dielectrics and Electrical Insulation, vol. 10, pp. 795-809, 2003.

[5] P. T. Vernier, Y. Sun, L. Marcu, S. Salemi, C. M. Craft, and M. A. Gundersen, "Calcium bursts induced by nanosecond electric pulses," Biochemical and Biophysical Research Communications, vol. 310, pp. 286-295, 2003.

[6] M. Behrend, A. Kuthi, X. Gu, P. T. Vernier, L. Marcu, C. M. Craft, and M. A. Gundersen, "Pulse generators for pulsed electric field exposure of biological cells and tissues," IEEE Transactions on Dielectrics and Electrical Insulation, vol. 10, pp. 820-825, 2003.

[7] W. W. Flack, W. P. Fan, and S. White, "The optimization and characterization of ultra-thick photoresist films," SPIE, #3333-67, 1998.

Numerical Analysis of Nano-imprinting Process Based on Continuum Hypothesis

H. C. Kim[*], Y. S. Woo[*], W. I. Lee[*], S. I. Oh[*] and B. S. Kim[**]

[*]School of Mechanical and Aerospace Engineering, Seoul National University, Korea
sesese72@snu.ac.kr, yswoo72@snu.ac.kr, wilee@snu.ac.kr, sioh@snu.ac.kr
[**]Korea Institute of Machinery and Materials, Changwon, Kyungnam, Korea
kbs@kimm.re.kr

ABSTRACT

Nano-imprint lithography (NIL) is a processing technique capable of transferring nano-scale patterns onto a thin film of thermoplastics such as polymethyl methacrylate (PMMA). Feature sizes down to 10 nm have been demonstrated to be made possible using this process. In NIL, it is imperative to thoroughly understand the flow of resin to ensure complete transfer of the pattern. Due to the size of the pattern, experimental observation may be very difficult, if not totally impossible. In this study, a numerical simulation tool to calculate the two dimensional resin flows and heat transfer during NIL has been developed. The code is based on the continuum hypothesis with some modification to accommodate the size effect, such as slip along the boundary. In order to fully account for the surface tension effect, which may play a dominant role, a moving grid system is adopted. Numerical simulations have been performed for different sets of process parameters. Parametric study shows that the surface tension effect indeed is a dominant factor. With change in the values of the surface tension as well as the contact angle, filling patterns and the resulting free surface become different.

Keywords: Nanoimprint, Surface tension effect, Slip, Numerical simulation

1 INTRODUCTION

There are different fabrication methods available for the production of nanostructures. Optical lithography technique, which is widely used in semiconductor industry, is limited by the light diffraction at nano length scale. Electron beam and scanning probe microscope (SPM) lithographies suffer from low productivity while the X-ray lithography requires a major capital investment. The nanoimprint lithography (NIL) is known as a low cost alternative to these methods of fabricating nano-scale patterns as small as 10nm [1].

The NIL process consists of four stages, namely spin coating, embossing, de-molding and dry etching (Fig. 1). As the structures become smaller and more complex, it is very important to understand the behavior of thin polymer film during embossing and de-molding process. Due to the size of the pattern, experimental observation may be very difficult. Therefore, numerical simulation may serve as a viable and useful tool to design and optimize the processing conditions [2].

There are two ways to simulate nano scale flows, namely top-down and bottom-up approaches. In the top-down approach, it is assumed that the polymeric melt flow is a continuum. On the contrary, the bottom-up approach employs the molecular dynamics (MD) technique.

Even though the bottom-up approach may provide more accurate details of the polymer melt behavior, comprehensive modeling is not yet available. Moreover, calculation overhead is still prohibitively large. Therefore, as a practical consideration, the top down approach was used in this study. For the numerical analysis, two dimensional moving grid system was chosen to obtain the accurate prediction of the flow front. Numerical mesh was regenerate at every time step. Effects of surface tension as well as the slip at the boundary were also considered.

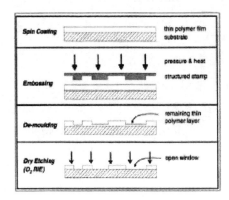

Figure. 1 NIL Process

2 CONTINUUM MODEL

Behavior of polymer melt during NIL process can be obtained by solving the following the continuity, momentum and energy conservation equations [3].

Mass conservation equation:

$$\frac{D\rho}{Dt} + \rho \nabla \cdot u = 0 \tag{1}$$

Momentum conservation equation:

$$\rho \frac{Du}{Dt} = \nabla \cdot \sigma + \rho f \tag{2}$$

where $\sigma = -PI + 2\mu D$

Energy conservation equation:

$$\rho C_v \frac{DT}{Dt} = -\nabla \cdot q + Q \qquad (3)$$

In the above equations, t is time, u is the velocity, ρ is density, μ is the dynamic viscosity, f is the body force and σ denote the stress.

The penalty function method was employed to solve the above equations. Using this method, Eq.(1) and (2) can be reformulated into a single equation by expressing pressure in terms of velocity as follows [4]

$$P = -\gamma_e \nabla \cdot u \qquad (4)$$

where γ_e is a penalty parameter.

2.1 Surface Tension Modeling

Surface tension is usually neglected in macro-scale analysis. However, in NIL, surface tension may play a dominant role due to the size effect.

The surface stress boundary condition at an interface between two fluids is given as [5]

$$(p_1 - p_2 + \sigma\kappa)\vec{n}_i = (\tau_{1ik} - \tau_{2ik})\vec{n}_k + \frac{\partial\sigma}{\partial x} \qquad (5)$$

where σ is the fluid surface tension coefficient, p_α is the pressure in fluid α for $\alpha = 1, 2$, is the viscous stress tensor, $\tau_{\alpha ik}$ is the unit normal at the interface and κ is the local surface curvature($= -(\nabla \cdot \vec{n}) = 1/R$).

Assuming that the viscous effect neglected on a free surface and the fluid is incompressible, we can rearrange the surface boundary condition constant σ as follows [5]

$$p_s \equiv p_2 - p_1 = \sigma\kappa \qquad (6)$$

2.2 Wall Adhesion

Effect of wall adhesion at fluid interface can be estimated in terms of the contact angle between cavity wall and free surface. The static contact angle is not simply a material property of the fluid. It depends on the smoothness and geometry of the wall as well.

For still fluids, the normal vector to the interface for calculating the curvature at boundary points can be expressed using the unit normal to the interface at points on the wall as follows [6]

$$\vec{n} = \vec{n}_w \cos\theta_{eq} + \vec{n}_t \sin\theta_{eq} \qquad (7)$$

where θ_{eq} is the contact angle between the interface and the wall, $\vec{n}_t$ lies in the wall and is the unit normal to the contact line and $\vec{n}_w$ is the unit wall normal directed into the wall.

Wall adhesion boundary condition becomes more complicated when the contact line is in motion. In NIL, because the stamp wall moves downward, the dynamic contact angle was used instead of the static contact angle to estimate the velocity effect on the contact angle [7].

$$v = \frac{\sigma}{\pi} \frac{-\dfrac{\bar{\mu}}{\mu}(\sin^2\theta_d - \theta_d^2) + \sin^2\theta_d - \theta_d(\theta_d + \pi)}{(\mu - \bar{\mu})\sin^2\theta_d}(\cos\theta_d - \cos\theta_s) \qquad (8)$$

where θ_s is the static contact angle, θ_d is the dynamic contact angle, v is velocity near the contact point, μ is the fluid viscosity and $\bar{\mu}$ is the air viscosity.

2.3 Slip Boundary Condition

In macroscopic flow, the no-slip condition at a fluid-solid interface is enforced. However, in nano-scale, it is known that slip takes place between the fluid and the contact wall. For both liquids and gases, the following Navier boundary condition empirically relates the tangential velocity slip at the wall to the local shear [8].

$$\Delta u\big|_w = u_{fluid} - u_{wall} = L_s \frac{\partial u}{\partial y}\bigg|_w = L_s \chi \qquad (9)$$

where L_s is the constant slip length and χ is the strain rate estimated at the wall. The slip length for nano-scale fluid can be defined using Lennard-Jones potential model as follows [9]

$$L_s = L_s^0 \left[1 - \frac{\chi}{\chi_c}\right]^{-\frac{1}{2}}, \quad \chi_c^{-1} \approx \tau^{wf} = \left(\frac{m\sigma^{wf}}{\varepsilon^{wf}}\right)^{\frac{1}{2}} \qquad (10)$$

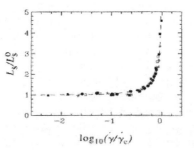

Figure. 2 Slip length as a function of shear rate

where τ^{wf} is the characteristic time, σ^{wf} is length of wall-fluid coupling, ε^{wf} is the strength of wall-fluid coupling and m is mass of a molecule. L_s^0 ranges from 0 to $17\sigma^{wf}$. Relationship between the slip length and the shear rate is shown in Fig. 2.

2.4 Rheological behavior of polymeric melts

The viscosity of polymeric melts changes according to the shear rate. To describe this shear rate dependency of viscosity, the Cross Williams-Lendal-Ferry (WLF) model was used [10].

3 VALIDATION AND RESULTS

3.1 Validation

In order to validate the numerical code, mesh size sensitivity was evaluated. Identical flow front shapes were obtained independent of mesh size for the same process conditions.

3.2 Result

To simplify numerical studies, we assumed that polymeric melt is incompressible. The stamp moves down with a constant velocity, and the density of PMMA is 1.17×10^3 kg/m^3. The geometry of the stamp and PMMA layer is shown in Fig. 3.

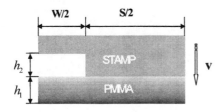

Figure. 3 Geometry of stamp and PMMA layer

First, calculation was done for the no-slip boundary condition without surface tension. These conditions are the ones encountered in the most macroscopic flow simulations. Calculated flow front shape is shown in Fig.4. The geometry considered was as follows: S=500nm, W=100nm, h$_1$=70nm, h$_2$=50nm and v=0.2nm/s. The temperature of the stamp as well as the PMMA layer was taken to be 150℃.

Figure. 4 Flow front shape at t=20 sec. Simulated result for no slip boundary condition without surface tension.

In order to evaluate the effect of slip length, calculations were done for different values of slip length ranging from 0.01nm to 0.1nm. Geometry and process conditions were same as the previous example. Surface tension was also considered (θ_s =30°, and σ = 30mN/m). Numerical simulations were performed for two different cases, L_s^0=0.01nm and L_s^0=0.1nm. Even though the results were not identical as shown Fig. 5, the effect of slip length on the change in the flow front shape was found to be minimal.

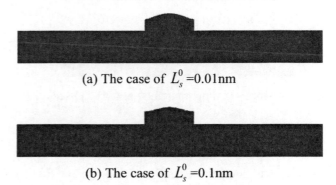

(a) The case of L_s^0=0.01nm

(b) The case of L_s^0=0.1nm

Figure. 5 Effect of slip length on the flow front shape at t=20 sec.

To study the effect of temperature, three different cases were considered for 140℃, 150℃ and 160℃. In these cases, different geometry was used as follows: S=100nm, W=300nm, h$_1$=150nm, h$_2$=50nm, v=3nm/s. Parameters for the surface tension and the slip at the boundary were given as θ_s =30°, σ =30mM/m and L_s^0=0.1nm. As can be seen from Fig.6, increase in the temperature resulted in completely different filling patterns. This change in the filling pattern is attributable to the decrease of viscosity with temperature increase. At lower viscosity of PMMA, the effect of surface tension and slip length became relatively larger than the shear stress caused by the viscosity. Therefore, the flow tends to rise faster along the boundary resulting in a depression in the center (Fig.6c).

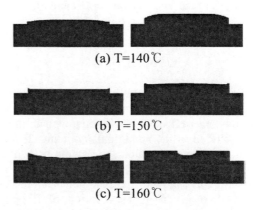

(a) T=140℃

(b) T=150℃

(c) T=160℃

Figure. 6 Effect of temperature on the flow front shape at t=6 and 12 sec.

For the same geometry, effect of the stamp velocity was also investigated. The process temperature was set to be T=140℃. Two different stamp velocities were considered, v =3nm/s and 1nm/s. As the stamp velocity is decreased, contribution of surface tension to drive the flow becomes more important. Therefore, as can be seen in Fig. 7, the flow front near the stamp wall rose faster.

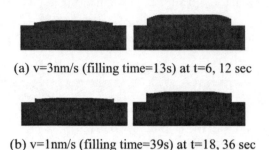

(a) v=3nm/s (filling time=13s) at t=6, 12 sec

(b) v=1nm/s (filling time=39s) at t=18, 36 sec

Figure. 7 Effect of stamp velocity on the flow front shape

Finally, we compared the flow front shape for the change of the surface tension coefficient. Same geometry was used as the last two examples. Temperature was 140℃ and the stamp velocity was 3nm/s. As the surface tension coefficient becomes larger, the surface tension and capillary force increase. Thus with increased surface tension, flow front near the border ascended faster as shown Fig. 8.

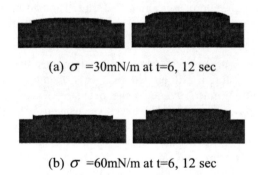

(a) σ =30mN/m at t=6, 12 sec

(b) σ =60mN/m at t=6, 12 sec

Figure. 8 Effect of surface tension coefficient σ on the flow front shape at t=6 and 12 sec.

4 CONCLUSION

Nano-scale flow arising in NIL process was simulated using a code based on continuum hypothesis. In order to account for the size effect, surface tension and the slip length were included in the numerical modeling. Effect of parameters such as the temperature, the stamp velocity, the slip length and the surface tension coefficient were investigated. It was shown that the change in these parameters resulted in different evolution of flow front pattern. The code developed in this study can be used in optimizing the process variables in NIL.

REFERENCES

[1]   S. Chou et al., 1996, "Imprint lithography with 25 nanometer resolution", *Science*, Vol. 272, pp.85-87.

[2]   L.J. Heyderman et al., 2000, "Flow behaviour of thin polymer films used for hot embossing lithography", *Micro electronic Engineering*, Vol. 54, pp.229-245.

[3]   L. D. Landau and E.M. Lifshitz, 1959, Fluid Mechanics, Pergamon, New York.

[4]   J.N. Reddy and D.K. Gartling, 1994, The Finite Element Method in Heat Transfer and Fluid Dynamics, CRC Press.

[5]   J. U. Brackbill et al., 1992, "A continuum method for modeling surface tension", *Journal of Computational Physics*, Vol. 100, pp.334-354.

[6]   Jun-Ho Jeong et al., 2002, "Flow Behavior at the Embossing Stage of Nanoimprint Lithography", *Fiber and Polymer 2002*, Vol. 3, No3, pp.113-119.

[7]   Arian Novruzi et al., 2003, "An analytical study on the dynamical contact angle of a drop in steady state motion", Department of Mathematics and Statistics, University of Ottawa, Canada.

[8]   Mohamed Gad-el-Hak, 2001, "Flow physics in MEMS", *Mec. Ind.* , Vol. 2, pp.313-341.

[9]   Thompson P.A., Troian S.M., 1997, "A general boundary condition for liquid flow at solid surfaces", *Phys. Rev. Lett,*, Vol. 63, pp. 766-769.

[10] Ferry J.D., 1989, Viscoelastic Properties of Polymers, John Wiley& Sons

Index of Authors

Index of Keywords

swiss_nanotech_™

welcomes you

to the **Nanotech 2004**
Trade Show & Conference

Switzerland has been at the forefront of the nanotechnology development curve, contributing enormously to the field through inventions such as the Atomic Force Microscope, the Scanning Tunneling Microscope and many other groundbreaking discoveries.

Through the strong leadership, excellent research, and innovative business environment, Switzerland has become a destination for the global nanotechnology market so come visit the Swissnanotech Booth 200.

Exhibitors include:

Nanoworld AG	*Swissnanotech*
Nanosensors	*Swiss Business Hub USA*
Nanofair	*Consulate of Switzerland, SHARE*
Greater Zurich Area	*Location:Switzerland*
Reinhardt Microtech AG	*Basel Area Business Development*
	Economic Development Western Switzerland

Experts from Switzerland will also be participating as keynote speakers as well as panelists, including representatives from the Swiss Federal Institute of Technology, Zuerich, M.E. Mueller Institute of Basel, EPFL, University of Neuchatel, and others. More information is available in the official program.

The Swiss delegation at NANOTECH 2004 also includes representatives from Nestlé, UBS Private Banking, NanoDimension, Euresearch, Swiss Re, the Swiss American Business Council, PriceWaterhouseCoopers, CSM Instruments, and Semasopht.

You are also cordially invited to join us for a **panel discussion and reception** hosted by the **Consulate of Switzerland, SHARE** on **Tuesday, March 9th at 6:15 p.m.**

The event will start with short presentations from Prof. Peter Seeberger (ETHZ) and Prof. Andreas Engel (Biozentrum at Uni Basel) and will be followed by a reception gathering a large group of leading nanotech researchers and industry experts from Switzerland as well as many members of the local nanotech community.

SHARE is located at 420 Broadway in Cambridge. For details on the program and directions, please visit our web site at **www.shareboston.org**.

The **swiss**_nanotech_™ initiative at NANOTECH 2004 is organized, coordinated and sponsored by:

swiss business hub
usa

member of business network
switzerland

Swiss House
for Advanced Research and Education

Consulate of Switzerland

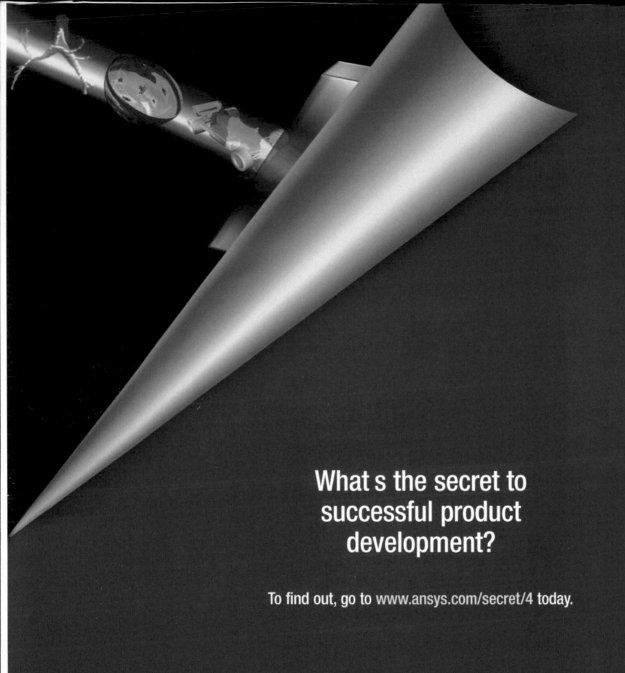

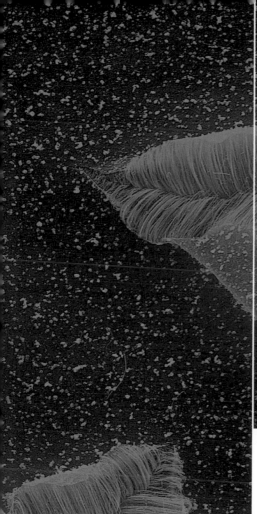

Nanotechnology Investment Opportunities

Australian Government

Invest Australia

To find out more about Australian nanotechnology opportunities or set up a meeting with an Investment Advisor, visit the Nanotechnology Australia stand at Nanotech 2004, or contact:

Hill Helen
Invest Australia - New York Office
Phone: **+1 212 351 6575**
Cell Phone: +1 917 608 3430
helen.hill@investaustralia.gov.au

Michael Claessens
Invest Australia - Canberra Office
Phone: **+61 2 6213 6704**
Cell Phone: +61 412 177 526
michael.claessens@investaustralia.gov.au

Australia has an active and rapidly growing nanotechnology capability, with a growing number of specialist nanotechnology firms, MNCs, and foreign government agencies investing in the development and commercialisation of breakthrough Australian technologies.

World-class scientific expertise and R&D facilities, combined with strong government support and a robust economy, have provided investors in Australian nanotechnology with a wide range of commercialisation and partnership opportunities. Practical outcomes are being delivered in manufacturing, materials processing, energy generation, and environmental management.

Products which exemplify this solutions-based philosophy include supercapacitors as light weight power sources in portable electronic devices, low energy filtration membranes, and clear ultraviolet screens for cosmetics, paints, and glass.

Assistance for investors

Invest Australia is the national inward investment agency, set up by the Australian Government to help foreign investors establish or expand a business in Australia. It leads a national approach to investment promotion and facilitation, providing businesses with an initial point of contact for investment enquiries.

Invest Australia understands the challenges of investing in nanotechnology and is committed to attracting new investment which will enable sustainable industry growth and development.

Working in partnership with business, Invest Australia provides organisations with the information and contacts needed to establish or expand your nanotechnology capability into Australia, including:

- expert advice on Australia's nanotechnology capabilities and opportunities for potential and current investors;

- the arrangement of site visits and links to potential joint venture partners;

- information on business costs, skills availability, taxation, R&D and other business related infrastructure;

- advice on investment regulations, streamlining major project approval processes and contacts with key Australian government agencies; and

- the identification of any relevant government industry assistance schemes.

Image courtesy CSIRO Health Sciences and Nutrition: **A Nanostar**